Birkhäuser Advanced Texts

John J. Benedetto
Wojciech Czaja

Integration and Modern Analysis

Birkhäuser
Boston • Basel • Berlin

John J. Benedetto
Department of Mathematics
University of Maryland
College Park, MD 20742
jjb@math.umd.edu

Wojciech Czaja
Department of Mathematics
University of Maryland
College Park, MD 20742
wojtek@math.umd.edu

ISBN 978-0-8176-4306-5 e-ISBN 978-0-8176–4656-1
DOI 10.1007/978-0-8176-4656-1

Library of Congress Control Number: 2009934518

Mathematics Subject Classification (2010): 03Exx, 22Bxx, 26-xx, 28-xx, 42-xx, 43-xx, 46-xx, 49Q15, 58B10, 58Cxx, 60Axx, 60Gxx

Printed on acid-free paper

Birkhäuser Boston is part of Springer Science+Business Media (www.birkhauser.com)

To my wife, Catherine

Moim Rodzicom

Table of Contents

List of Figures

Preface

This book is both a text and a paean to twentieth-century real variables, measure theory, and integration theory. As a text, the book is aimed at graduate students. As an exposition, extolling this area of analysis, the book is necessarily limited in scope and perhaps unnecessarily unlimited in idiosyncrasy.

More than half of this book is a fundamental graduate real variables course as we now teach it. Since there are excellent textbooks that generally cover the course material herein, part of this Preface renders an apologia for our content, presentation, and existence. The following section presents our syllabus properly liberated from too many demands. Subsequent sections deal with outline, theme, features, and the roles of Fourier analysis and Vitali, respectively.

Mathematics is a creative adventure driven by beauty, structure, intrinsic mathematical problems, extrinsic problems from engineering and the sciences, and serendipity. This book treats integration theory and its fascinating creation through the past century.

What about the rest of our title?

"Analysis" is many subjects to many mathematicians. "Modern analysis" is hardly a constraint for a single volume such as ours; one can argue the opposite. Differentiation and integration are still the essence of analysis, and, along with "integration", the title could very well have included the word "differentiation" because of our emphasis on it. Guided by the creativity of mathematics, our title is meant to assert that the technology we have recorded is a basis for many of the analytic adventures of our time.

Syllabus

We shall outline the material we have used in teaching a first-year graduate course in real analysis. Sometimes a student will take only the first semester of this two-semester sequence.

- Chapter 1. Sections 1.1, 1.2, 1.3.1, 1.3.2, 1.3.3.
- Chapter 2. Sections 2.1, 2.2, 2.4, 2.5, with emphasis on Sections 2.2 and 2.5.

- Chapter 3. Sections 3.1, 3.2, 3.3, 3.7.
- Chapter 4. Sections 4.1, 4.2, 4.3, 4.4, 4.5.
- Chapter 5. Sections 5.1, 5.2, 5.3, 5.4, 5.5.

Exercises are listed in the problem sections of the first five chapters. The problems range in difficulty from the routine to the challenging; and they are computational, theoretical, and often provide perspective. There are suggested problems listed at the beginning of each of these five chapters. These problems are appropriate for the course. There are no problem sections in the remaining chapters, which can be considered as providing special topics material in a full-year course.

Typically, finer points will come later in a section than basic, essential theory. For example, in recommending Section 3.3 in the course syllabus, we note that the items are numbered 3.3.1–3.3.15 in Section 3.3, whereas items 3.3.1–3.3.6 are sufficient for the course. This caveat to the recommended sections listed above is crucial for presenting the essential topics for a graduate course in real analysis.

Chapters 1, 2, 3 and parts of Chapter 4 and Appendix A are the usual content of the first-semester graduate course in real analysis. Chapter 5 is essential for the second-semester graduate course in real analysis along with other parts of Chapter 4 and Appendix A. There will usually be some time remaining in the second semester, and the authors have presented Fourier analysis (Appendix B) or some of the material in Chapters 6–9. Naturally, instructors should present their favorite applications of the theory in this remaining time.

Students are encouraged to read the Potpourri and Titillation sections at the ends of chapters. These sections are meant to be educational and fun, providing some breadth without depth.

The book may be used not only for a graduate course in real analysis, but also for advanced independent study, or as a source for student projects.

Outline

The basic text, Chapters 1 through 5, begins with the fundamental classical problems that led to Lebesgue's definition of the integral, and develops the theory of integration and the structure of measures in a measure-theoretical format. Chapter 1 is meant to be a pointed account of classical real variables theory. We have inserted details for some of the elementary material; but there is a bloc of more advanced matter, with details omitted, which should be read for perspective. Of course, the instructor may choose to develop this material more fully at any time.

Chapters 2 through 5 form the core of the course, and they are presented in full detail. Chapters 2 and 3 are Lebesgue's theory of measure and integration. Chapter 4 presents the fundamental theorem of calculus

(FTC), and Chapter 5 deals with spaces of measures and the Radon–Nikodym theorem (R–N). The motivation for Lebesgue's theory viewed from the FTC is based on Volterra's example in Chapter 1 of a differentiable function whose derivative is not Riemann integrable.

The remaining chapters, Chapters 6 through 9, are also systematically presented, but could be viewed as eclectic topics by some readers. We view them as essential to our intellectual vision of twentieth-century real analysis.

Chapter 6 is devoted to deep results by Vitali, Nikodym, Hahn–Saks, Dieudonné, Dunford–Pettis, and Grothendieck as they relate to interchanging limits and integration, and to the characterization of weak sequential convergence of measures. Our goal has been to highlight the importance of this material and to show how it fits centrally into integration theory.

We prove the Riesz representation theorem (RRT) in Chapter 7 in the setting of locally compact Hausdorff spaces. We begin historically with Riesz' original proof and are led to Radon measures and Laurent Schwartz' theory of distributions. The RRT establishes the equivalence of the set-theoretic measure theory of the previous chapters with the theory of Radon measures considered functional-analytically as continuous linear functionals. It is striking that on $\mathbb{R}$, the RRT asserts that a Schwartz distribution is a bounded Radon measure if and only if it is the first distributional derivative of a function of bounded variation. We view this material as a quantitative approach to apply measure theory in harmonic analysis, partial differential equations, and distribution theory. In Schwartz' obituary (Notices Amer. Math. Soc. 50 (2003), 1072–1084), it is noted that Schwartz became disenchanted with Bourbaki's presentation of measure theory in the setting of Radon measures on locally compact Hausdorff spaces, even though he was on the Bourbaki writing group for this material. His main objection was concerned with its inadequacy in dealing with the research of probabilists, e.g., Paul Lévy and Joseph L. Doob, establishing measure theory on infinite-dimensional spaces such as $C([0,1])$. Notwithstanding our awe of Schwartz, we have presented the Bourbakist point of view. We think that it is spectacularly beautiful and unifying as far as it goes. Further, as illustrated in Chapter 7, its capacity to transmute the Riesz representation theorem from theorem to definition is stunning. It is also a natural setting for developing Schwartz' own theory of distributions.

Chapter 8 develops differentiation theory on Euclidean space $\mathbb{R}^d$, and proves the d-dimensional version of the Lebesgue differentiation theorem on $\mathbb{R}$, which itself was proved in Chapter 4. Substantial technology is required, which includes the notion of bounded variation on $\mathbb{R}^d$, Vitali and Besicovich covering lemmas, maximal functions, rearrangement inequalities, and a semimartingale maximal theorem. This material has also had significant impact on other areas of mathematics. We close in Chapter 9 by analyzing self-similar sets and fractals. In a sense, this material comes full circle from Chapter 1 by presenting a modern treatment and generalization of our analysis of Cantor sets, which were vital in classical real variables and the development of measure theory.

We have included two appendices: functional analysis (Appendix A) and Fourier analysis (Appendix B). Functional analysis is a significant area of twentieth-century analysis, and it is an integral part of the structure and language of modern analysis. Appendix A provides those topics that we use throughout Chapters 1–9, and it is also a standalone outline of a broadly useful part of functional analysis. Appendix B is more of a luxury, meant to provide the background for some of the Fourier-analytic examples we give, as well as a prelude for harmonic analysis, establishing its dependence on the material in the book.

Theme

One of our themes is the notion of *absolute continuity* and its role as the unifying concept for the major results of the theory, viz., the fundamental theorem of calculus (FTC) (Chapter 4), the Lebesgue dominated convergence theorem (LDC) (Chapter 3), and the Radon–Nikodym theorem (R–N) (Chapter 5). The main mathematical reason that we have written this book is that none of the other texts in the area stresses this issue to the extent that we think it should be stressed.

Let us be more specific.

The problem of taking limits under the integral sign, that is, "switching limits", is in a very real sense the fundamental problem in real analysis. Lebesgue's axiomatization that formulates and proves LDC in an optimal way yields the most important general technique for examining such problems. This material is developed in Chapter 3. Shortly after Lebesgue's initial work, Vitali gave necessary and sufficient conditions for interchanging limits in terms of uniform absolute continuity. Vitali's result led to research that has culminated in the Vitali–Hahn–Saks theorem and in Grothendieck's study of weak convergence of measures. This material, found in Chapter 6, is usually not included in most texts; in particular, its relationship to LDC is not emphasized.

Knowledge of the structure of measures provides an important tool in potential theory, harmonic analysis, probability theory, and nonlinear dynamics, and it plays a role in a host of other subjects from number theory to representation theory. Its scope of application ranges from establishing a mathematical model for the continuous spectrum of white light to formulating the action of the stock market as Brownian motion in terms of the Wiener measure.

A key theorem in this milieu is R–N, and the major results involve decompositions of a given measure into various parts with specific properties. R–N can be considered as a generalization of the FTC for the case of functions defined on the real line or of the Green and Stokes theorems in Euclidean space. Of course, the FTC, which is the basic and amazing relationship by which integration and differentiation are formulated as inverse operations, is characterized in terms of absolute continuity. We have dwelled on these issues in Chapters 4 and 5. We give the classical point function results, study the abstract setting, examine their relation, and spend a good deal of time with examples. The Fubini–Tonelli theorem (Section 3.7), one of the most important theorems in analysis and a classical case of interchanging operations, can be related to R–N from the point of view of conditional probabilities; and LDC is used in its proof.

Features

Besides the theme of absolute continuity there are features of a more secular nature. We have included some extensive historical and motivational passages. Integration theory did not develop in a vacuum, and we have presented information on the development of Fourier series because of its close relation with many of the notions from real analysis; see the following section. Our problem sets include certain types of problems that abound in the folklore (e.g., the Amer. Math. Monthy), but which are generally omitted from a full-year real analysis text whose purpose is to present systematically n topics. As mentioned in describing the syllabus, some of the problems are quite difficult and will probably challenge even the most mathochistic student.

We hope that the historical remarks, the sections on Potpourri and Titillation, and the problems (especially the harder ones) are *read*, since we believe that they provide relevant perspective. A list of these and other features follows.

- Historical approach and technical perspective on the development of real variables
- Role of classical topics in modern analysis
- Extensive array of problems, ranging from the routine to the challenging, with background, related issues, and hints
- Brief, illuminating biographies
- Many examples, from straightforward to profound
- Treatment of symmetric perfect sets, used as a foundation for the introduction of fractal analysis
- Important role of Vitali for original results and modern proofs given 100 years ago
- Applications to Fourier analysis and fractal geometry
- Comprehensive appendix on elementary functional analysis
- Introductory appendix on Fourier analysis
- Complete indexes of terms, names, and notation
- Substantial selection of references, including original works by the founders of real analysis as well as relevant modern publications

With regard to references, there are a disproportionate number of references to the first-named author, mostly because of perspectives he has written on various topics. These references should generally be viewed as literature sources as opposed to research results.

Topics

- Distributional formulation of the Riesz representation theorem
- A unified theory of measure and integral
- Haar measure
- Functional-analytic equivalence of spaces of measures
- Vitali–Nikodym–Hahn–Saks characterization for interchanging limits
- Weak convergence of measures
- Hausdorff measure
- Maximal functions and Lebesgue differentiation theorem on $\mathbb{R}^d$

Fourier Analysis

In discussing Grothendieck's idea of bringing certain cohomological concepts into algebraic geometry, Fields' medalist David Mumford wrote, "It completely turned the field upside down. It's like analysis before and after Fourier. Once you get Fourier techniques, suddenly you have this whole deep insight into a way of looking at a function" (Notices Amer. Math. Soc. 51 (2001), 1052). It is for this reason, and going beyond the notion of a function to differentiation and integration, where Fourier analysis was also a driving force, that we have included as much Fourier analysis in the text as we have. Of course, this intellectual rationale has to be coupled with the authors' ignorance of so many other aspects of modern analysis, such as partial differential equations, vector measures, stochastic integration, ergodic theory, geometric measure theory, and probability theory. (For measure-theoretic motivation, we do, however, outline some elementary probability theory in the Potpourri and Titillation sections of several chapters.)

Historically, Fourier series were developed in the nineteenth-century for the analysis of some of the classical partial differential equations (PDEs) of mathematical physics; and these series were used to solve such equations. In order to understand Fourier series and what sorts of solutions they could represent, some of the most basic notions of analysis were defined, e.g., the concepts of "function" and "bounded variation". The simplest linear PDEs were eventually understood to be convolutions, which could be Fourier transformed into algebraic equations. When these latter equations could be solved, an inverse Fourier transform was required to write the solution of the original PDE. Uniqueness questions naturally arose. Further, since the coefficients of Fourier series are integrals, it is no surprise that Riemann integrals were conceived to deal with uniqueness properties of trigonometric series. Cantor's set theory was developed because of such uniqueness questions.

Nineteenth-century real variables and integration theory went hand in hand with Fourier analysis into the twentieth-century, along with new stimulation from probability theory and statistical mechanics. Lebesgue created his magnificent theory, which is the centerpiece of this book, and which has become the setting for harmonic analysis, as well as the inspiration and language for so much analysis.

Vitali

The catalyst for writing this book occurred many years ago, during the academic year 1970–1971, when one of the authors was a guest at the Scuola Normale Superiore in Pisa. At that time he discovered Vitali's work at a different level from what he had previously known. Vitali is responsible for the notion of absolute continuity, the first nonmeasurable set, the first necessary and sufficient conditions for LDC, the first statement and proof of Lusin's theorem, "modern" proofs given 100 years ago, and more, as you will read in the text.

We have decided to engage in some advertising for this most important figure in integration theory.

Acknowledgments

A number of colleagues and students have generously helped us. From a long-ago version, we thank A. Beardon, R. Burckel, A. Bruckner, S. Espelie, R. Johnson, E. Korn, D. Madan, D. Schmidt, and R. Warner. For the present book, we appreciatively thank R. Benedetto, S. Dineen, L. C. Evans, C. Flake, I. Konstantinidis, W. Luxemburg, A. Powell, F. Ricci, D. Speegle, S. Sumetkijakan, and J. Tanis for their mathematical insights. We are especially appreciative of Ann Kostant's restraint in the face of extended delays due to the older author, and of the superbly professional work and support of Ms. Kostant, Ms. Jessica Belanger, Mr. Thomas Grasso, and the staff at Birkhäuser Boston.

John J. Benedetto and Wojciech Czaja

The Norbert Wiener Center, Department of Mathematics,
University of Maryland, College Park

1 Classical Real Variables

1.1 Set theory: a framework

There are mathematicians who claim that there is no difference between mathematics and set theory, but I believe this claim can be dismissed. No mathematician of my acquaintance would abandon his field if an apparently insurmountable contradiction were discovered in the general concept of subset ANDREW M. GLEASON.

We make this appeal to authority (taken from the preface to [198]) because of the tendency to build mathematics from basic notions such as set theory, as opposed to the point of view herein. In fact, notwithstanding such a tendency, this book treats set theory superficially as a necessary language, and not as the intellectual foundation of integration theory. On the other hand, we do assume a working relationship with naive set theory and the axioms for the real numbers, as treated in an advanced calculus course; see, e.g., [7], [213], [407].

Of course, set theory is fundamentally important, and the following three articles give readable expositions, including a survey of results, in the area of "foundations":

- L. Henkin, *Mathematical foundations for mathematics*, Amer. Math. Monthly 77 (1970), 463–486.
- J. D. Monk, *On the foundations of set theory*, Amer. Math. Monthly 77 (1970), 703–710.
- W. Marek and J. Mycielski, *Foundations of mathematics in the twentieth century*, Amer. Math. Monthly 108 (2001), 449–468.

The last reference has a suggested reading list on different topics in the foundations of mathematics.

The letters $\mathbb{C}$, $\mathbb{R}$, and $\mathbb{Q}$ denote the fields of complex, real, and rational numbers, respectively; $\mathbb{Z}$ is the ring of integers, and the quotient group $\mathbb{T} = \mathbb{R}/\mathbb{Z}$ is identified with $[0,1) \subseteq \mathbb{R}$. The sets $\mathbb{R}^+$ and $\mathbb{Z}^+$ are the nonnegative elements of $\mathbb{R}$ and $\mathbb{Z}$, respectively, and $\mathbb{N} = \{1, 2, \ldots\}$. The symbol $\emptyset$ denotes the *empty set*. For a set $A \subseteq X$, $A^{\sim}$ is the *complement* of the set A in X, that is, the collection of all elements of X that are not in A, and

J.J. Benedetto and W. Czaja, *Integration and Modern Analysis*, Birkhäuser Advanced Texts, DOI 10.1007/978-0-8176-4656-1_1,

$A\backslash B = \{x \in A\} \cap \{x \notin B\}$. The cardinality of A is denoted by card A and we write card $\mathbb{Z} = \aleph_0$. "$S_1 \Longrightarrow S_2$" is read as "S_1 implies S_2", and "$S_1 \iff S_2$" is read as "S_1 if and only if S_2".

1.2 The topology of $\mathbb{R}$

The notion of *topology* is defined in Appendix A.1; for now, when you see the word, it is meaningful to translate it as "convergence criterion".

Definition 1.2.1. Limits
a. $x \in \mathbb{R}$ is the *limit* of the sequence $\{x_n : n = 1, \ldots\} \subseteq \mathbb{R}$ if

$$\forall\, \varepsilon > 0,\ \exists\, N \text{ such that } \forall\, n \geq N,\ |x_n - x| < \varepsilon\,.$$

In this case we write $\lim_{n\to\infty} x_n = x$.

b. $x \in \mathbb{R}$ (respectively, "$x = \infty$") is a *limit point* of the sequence $\{x_n : n = 1, \ldots\} \subseteq \mathbb{R}$ if

$$\forall\, \varepsilon > 0,\ \forall\, N,\ \exists\, n \geq N \text{ such that } |x_n - x| < \varepsilon \text{ (respectively, } x_n \geq \varepsilon).$$

Definition 1.2.2. $\overline{\lim}$ and $\underline{\lim}$
a. Given $\{x_n : n = 1, \ldots\} \subseteq \mathbb{R}$, we define

$$\overline{\lim_{n\to\infty}}\ x_n = \inf_n \sup_{k\geq n} x_k, \tag{1.1}$$

$$\underline{\lim_{n\to\infty}}\ x_n = \sup_n \inf_{k\geq n} x_k. \tag{1.2}$$

Thus, $x \in \mathbb{R}$ (respectively, "$x = \infty$") is $\overline{\lim}\ x_n$ if and only if

$$\forall\, \varepsilon > 0,\ \exists\, N \text{ such that } \forall\, n \geq N,\ x_n < x + \varepsilon$$

and

$$\forall\, \varepsilon > 0,\ \forall\, N,\ \exists\, n \geq N \text{ such that } x_n > x - \varepsilon$$

(respectively, $\forall\ \varepsilon > 0$, $\forall\ N$, $\exists\ n \geq N$ such that $x_n > N$). Naturally, the above notions are also defined for "$x = -\infty$".

b. Intuitively, $\overline{\lim}\ x_n$ is the *largest limit point* of $\{x_n : n = 1, \ldots\}$ and $\underline{\lim}\ x_n$ is the *smallest limit point* of $\{x_n : n = 1, \ldots\}$. Note, for example, that

$$\overline{\lim_{n\to\infty}}\ x_n = -\infty \text{ if and only if } \lim_{n\to\infty} x_n = -\infty.$$

c. The following facts are proved by routine calculations:

$$\begin{aligned}\underline{\lim_{n\to\infty}}\ x_n + \underline{\lim_{n\to\infty}}\ y_n &\leq \underline{\lim_{n\to\infty}}\ (x_n + y_n) \leq \overline{\lim_{n\to\infty}}\ x_n + \underline{\lim_{n\to\infty}}\ y_n \\ &\leq \overline{\lim_{n\to\infty}}\ (x_n + y_n) \leq \overline{\lim_{n\to\infty}}\ x_n + \overline{\lim_{n\to\infty}}\ y_n,\end{aligned}$$

provided "$\infty - \infty$" does not occur; and

$$\overline{\lim_{n\to\infty}}\ (x_n + y_n) = \overline{\lim_{n\to\infty}}\ x_n + \lim_{n\to\infty}\ y_n, \tag{1.3}$$

provided $\lim\ y_n$ exists.

We shall use the following result in the construction of Cantor sets, and Cantor sets will be an intriguing source of examples for us.

Proposition 1.2.3. *Let $p > 1$ be an integer.*

a. *If $x \in (0,1)$, then there is a sequence $\{a_n : n = 1, \dots$ and $0 \le a_n < p\} \subseteq \mathbb{Z}$ such that*

$$x = \sum_{n=1}^{\infty} a_n/p^n, \tag{1.4}$$

and the representation (1.4) is unique except when $x = q/p^n$, $q = 1, \dots, p^n - 1$, in which case there are exactly two such representations.

b. *If $\{a_n : n = 1, \dots$ and $0 \le a_n < p\} \subseteq \mathbb{Z}$, then $\sum_{n=1}^{\infty} a_n/p^n$ converges to some $x \in [0,1]$.*

Proof. We shall not give a careful proof. On the other hand, a formal proof can be constructed based on the following calculation for the case $p = 3$ and $x = 1/2$ in (1.4):

$$\frac{1}{2} = \frac{1}{3} + r_1 \quad \text{and} \quad \frac{1}{3^2} < r_1 = \frac{1}{6} = \frac{1}{2} - \frac{1}{3} < \frac{1}{3},$$

$$\frac{1}{2} = \frac{1}{3} + \frac{1}{3^2} + r_2 \quad \text{and} \quad \frac{1}{3^3} < r_2 = \frac{1}{2} - \frac{1}{3} - \frac{1}{3^2} < \frac{1}{3^2}.$$

In this way,

$$\frac{1}{2} = \sum_{j=1}^{n} \frac{1}{3^j} + r_n, \quad \frac{1}{3^{n+1}} < r_n < \frac{1}{3^n},$$

and so

$$\frac{1}{2} = \sum_{j=1}^{\infty} \frac{1}{3^j}.$$

Consequently, we write

$$\frac{1}{2} = .111\dots \quad (3). \tag{1.5}$$

The symbol "(3)" designates that we have expanded a given number, "1/2" in this case, in the base 3. In the same way we have

$$\frac{1}{3} = \begin{cases} .100\dots & (3), \\ .022\dots & (3). \end{cases}$$

□

We shall use notation analogous to (1.5) for binary expansions. However, when we write decimal expansions $a = .a_1a_2a_3 \ldots$ (10) we shall suppress the "(10)" and write $a = .a_1a_2a_3 \ldots$.

A set $U \subseteq \mathbb{R}$ is *open* if for each $x \in U$ there is an open interval I such that $x \in I \subseteq U$. The *interior* of $X \subseteq \mathbb{R}$, denoted by $\operatorname{int} X$, is the union of all open intervals contained in X. It is easy to prove that *the finite intersection of open sets is open* and that *the arbitrary union of open sets is open.*

Theorem 1.2.4. Characterization of open sets in $\mathbb{R}$
If $U \subseteq \mathbb{R}$ is open then $U = \bigcup_{j=1}^{\infty} I_j$, where I_j is an open interval and $I_j \cap I_k = \emptyset$ for $j \neq k$.

Proof. The set U being open implies that for each $x \in U$ there is $y > x$ such that $(x, y) \subseteq U$. Set

$$a = \inf \{z : (z, x) \subseteq U\} \quad \text{and} \quad b = \sup \{y : (x, y) \subseteq U\}.$$

Note that a and b depend on x. Then, $x \in (a, b)$ and $I_x = (a, b)$ is an open interval. We shall prove:

i. $U = \bigcup\{I_x : x \in U\}$;

ii. $\{I_x : x \in U\}$ is disjoint unless $I_x = I_y$;

iii. $\operatorname{card}\{I_x : x \in U\} \leq \aleph_0$, again excluding the case $I_x = I_y$.

(With regard to claim *ii*, we say that a family of sets is *disjoint* if the elements of the family are pairwise disjoint.)

It is necessary to make the following preliminary observation: *if $x \in U$ and $I_x = (a, b)$ then $I_x \subseteq U$ and $a, b \notin U$.* To see this let $w \in (a, b)$, so that by the definition of b there is $y > w$ for which $(x, y) \subseteq U$ and hence $w \in U$. Now if $b \in U$ there would exist $\varepsilon > 0$ such that $(b-\varepsilon, b+\varepsilon) \subseteq U$. Take ε small enough that $(x, b - (\varepsilon/2)) \subseteq U$. Consequently, $(x, b + \varepsilon) \subseteq U$, contradicting the definition of b.

i. Claim *i* is now obvious since $x \in I_x$ for each $x \in U$.

ii. Define $I_x = (a, b)$ and $I_y = (c, d)$, and assume $I_x \cap I_y \neq \emptyset$. Then, $a < d$ and $c < b$. Since $c \notin U$ we have $c \notin (a, b)$, and so $c \leq a$ because $c < b$. Since $a \notin U$ we have $a \notin (c, d)$, and so $a \leq c$ because $a < d$. Hence $a = c$, and, similarly, $b = d$. Claim *ii* follows.

iii. For each I_x there is $q_x \in I_x \cap \mathbb{Q}$. If $I_x \neq I_y$ then $q_x \neq q_y$ from part *ii*, and we obtain claim *iii* since $\operatorname{card} \mathbb{Q} = \aleph_0$. □

The characterization of open sets in $\mathbb{R}$ in terms of open intervals does not extend so elegantly to higher dimensions. For $\mathbb{R}^2$ we have the following result.

Proposition 1.2.5. *Every open set in the plane $\mathbb{R}^2$ can be represented as a disjoint union of closed straight line segments, see Definition 1.2.6b.*

A false proof of Proposition 1.2.5 is given in Problem 1.10; a correct proof is given as Solution III of Problem E 1434 in Amer. Math. Monthly 68 (1961), 381.

Definition 1.2.6. Limit points

a. Let $E \subseteq \mathbb{R}$. A point $x \in \mathbb{R}$ is a *limit point* of E if

$$\forall\, \varepsilon > 0,\ \exists\, y \in E \text{ such that } |x - y| < \varepsilon .$$

Recall that we defined the notion of a limit point for sequences $\{x_n : n = 1, \ldots\}$ in Definition 1.2.1*b*. These two notions are not identical.

b. We denote by $\overline{E}$ the set of limit points of E, and E is *closed* if $E = \overline{E}$. If $E \subseteq \mathbb{R}$ is closed then $X \subseteq E$ is *dense* in E if $\overline{X} = E$.

c. A point $x \in \mathbb{R}$ is an *accumulation point* of a set $E \subseteq \mathbb{R}$ if $x \in \overline{E \setminus \{x\}}$, that is, if x is a limit point of the set $E \setminus \{x\}$. E' is the set of accumulation points of E. We say that $x \in E$ is an *isolated point* of E if $x \notin E'$, or, in other words, if x is not a limit point of the set $E \setminus \{x\}$.

The following facts are easy to check, and this is the content of Problem 1.11:

- $A \subseteq B$ *implies* $\overline{A} \subseteq \overline{B}$;
- $\overline{A \cup B} = \overline{A} \cup \overline{B}$;
- $\overline{\overline{A}} = \overline{A}$;
- *the finite union of closed sets is closed*;
- *the arbitrary intersection of closed sets is closed*;
- *the complement of an open set is closed*;
- *the complement of a closed set is open.*

Example 1.2.7. The ternary Cantor set

a. We now define the *ternary Cantor set* C. This set and its generalizations were crucial motivation for the development of integration theory. GEORG CANTOR's work in set theory began with his work in trigonometric series; and in this latter research, on the problem of unique representation in trigonometric series, CANTOR had discussions with his colleague HEINRICH E. HEINE at Halle, who worked on the same problem; cf. Section 3.8. After describing C and a generalization of it we shall prove the *Heine–Borel theorem*, and observe later, see Section 1.3.2, that C is an intriguing example related to this result.

b. C is the set of all $x \in [0, 1]$ with ternary expansion

$$x = .a_1 a_2 \ldots \quad (3), \quad a_j = 0 \text{ or } 2.$$

Thus,

$$\frac{1}{3} = .0222\ldots \quad (3), \quad \frac{2}{3} = .2000\ldots \quad (3) \in C.$$

We observe that $x \notin C$ if $x \in (1/3, 2/3)$. In fact,

$$x = \frac{1}{3} + r_1, \quad 0 < r_1 < \frac{1}{3},$$

and so

$$x = .1000\ldots \quad (3) + .0a_2a_3\ldots \quad (3) = .1a_2a_3\ldots \quad (3),$$

where some a_j is nonzero. Consequently, $x \notin C$ because of the uniqueness assertion in Proposition 1.2.3. In the same way,

$$\text{if } x \in \left(\frac{3k+1}{3^n}, \frac{3k+2}{3^n}\right), \ k \geq 0, \ \text{and} \ n \geq 1, \ \text{then } x \notin C. \tag{1.6}$$

c. The observation (1.6) leads to the following geometrical description of C. Define

$$\begin{cases} C_1 = C_1^1 \cup C_1^2 = [0,1/3] \cup [2/3,1] \\ C_2 = C_2^1 \cup \ldots \cup C_2^4 = [0,1/9] \cup [2/9,1/3] \cup [2/3,7/9] \cup [8/9,1] \\ \quad \vdots \\ C_n = C_n^1 \cup \ldots \cup C_n^{2^n} \\ \quad \vdots \end{cases} \tag{1.7}$$

Then

$$C = \bigcap_{n=1}^{\infty} C_n, \tag{1.8}$$

and, in particular, C is *closed.*

d. From the characterization of C in part *c* it is clear that C contains a countably infinite set with elements of the form $\{k/3^n : k < 3^n, n = 1, \ldots\}$, although there are integers k such that $k/3^n \notin C$, e.g., $4/9 \in (1/3, 2/3)$. We now prove that $\text{card}\, C > \aleph_0$. Define $f : C \to [0,1]$ such that if $x = .a_1a_2\ldots$ (3), where a_j is 0 or 2, then $f(x) = .b_1b_2\ldots$ (2), where $b_j = a_j/2$. For example, $f(1/3) = f(.022\ldots (3)) = .0111\ldots (2) = 1/2$. Clearly, f maps C *onto* $[0,1]$, so that since $\text{card}\, [0,1] > \aleph_0$, C is uncountable; see Example 1.3.17 for an alternative description of f. Thus, C is an *uncountable set.*

e. What is the "length" of C? We denote the length of an interval I by $m(I)$, so that $m([0,1]) = 1$. To construct C we have discarded the following lengths:

$$\text{first step} \quad \frac{1}{3}$$

$$\text{second step} \quad 2\left(\frac{1}{9}\right)$$

$$\vdots$$

$$n\text{th step} \quad 2^{n-1}\left(\frac{1}{3}\right)^n.$$

Note that

$$1 = \sum_{n=1}^{\infty} \frac{2^{n-1}}{3^n}, \tag{1.9}$$

since the series in (1.9) is

$$\sum_{n=0}^{\infty} \frac{1}{3}\left(\frac{2}{3}\right)^n = \frac{1}{3}\frac{1}{1-\frac{2}{3}} = 1.$$

Consequently, the natural "length" or "measure" of C is 0. We shall make this assertion more precise in (1.10) and in Chapter 2.

Example 1.2.8. Perfect symmetric sets
a. We generalize the construction of C and define the *perfect symmetric set* $E \subseteq [0,1]$ determined by $\{\xi_k : k = 1, \ldots\} \subseteq (0, 1/2)$ as follows. Let

$$E_1 = E_1^1 \cup E_1^2,$$

where

$$E_1^1 = [0, \xi_1], \quad E_1^2 = [1 - \xi_1, 1];$$
$$E_2 = E_2^1 \cup \cdots \cup E_2^4,$$

where

$$E_2^1, E_2^2 \subseteq E_1^1, \quad E_2^3, E_2^4 \subseteq E_1^2,$$

and

$$E_2^1 = [0, \xi_1\xi_2], \quad E_2^2 = [\xi_1 - \xi_1\xi_2, \xi_1],$$
$$E_2^3 = [1 - \xi_1, 1 - \xi_1 + \xi_1\xi_2], \quad E_2^4 = [1 - \xi_1\xi_2, 1];$$

continuing this procedure we set

$$E = \bigcap_{k=1}^{\infty} E_k, \quad \text{where} \quad E_k = \bigcup_{j=1}^{2^k} E_k^j.$$

It is not difficult to prove that E is a closed uncountable set. The proof can be modeled on the method used in Example 1.2.7.

b. Observe that the length of each E_k^j is

$$m\left(E_k^j\right) = \xi_1\xi_2\cdots\xi_k,$$

and so it makes sense to define the *measure* of E as

$$m(E) = \lim_{k\to\infty} 2^k \xi_1 \cdots \xi_k. \tag{1.10}$$

c. Note that if $\xi_k = 1/3$ for each k then $E = C$, and we sometimes write $C = E_{1/3}$. Consequently, $m(C) = \lim (2/3)^k = 0$, which, of course, is the same result that we previously computed for the "length" of C.

d. We now observe that *for each* $\xi \in [0,1)$ *there is* $\{\xi_k : k = 1,\ldots\} \subseteq (0,1/2)$ *such that* $m(E) = \xi$ *for the corresponding perfect symmetric set* E. In fact, begin by choosing $\xi_1 \in (0,1/2)$ such that $0 < 2\xi_1 - \xi < 1$. Then, take $\xi_2 \in (0,1/2)$ such that $0 < 2^2\xi_1\xi_2 - \xi < 1/2$, and ξ_k such that $0 < 2^k\xi_1 \cdots \xi_k - \xi < 1/k$; this does it. To see that the second step (for example) is possible, we know that there is $\alpha \in (0,1)$ such that $0 < 2\xi_1\alpha - \xi < 1/2$, and therefore we write $\alpha = 2\xi_2$.

e. In order that $m(E) > 0$ for a perfect symmetric set E we need $\{\xi_k : k = 1,\ldots\}$ to approach $1/2$ very quickly. For example, take $\xi_k = 1/2 - 1/2^{k+1}$.

f. Observe that C has no isolated points. To see this, assume that $x \in C$ is isolated and let $(x-\delta, x+\delta) \cap C = \{x\}$ for some $\delta > 0$; since $x \in C$ there is $\{C_n^{j_n} : n = 1,\ldots\}$ for which $x \in \bigcap_n C_n^{j_n}$. Because $m(C_n^{j_n}) \to 0$ as $n \to \infty$, $C_n^{j_n} \subseteq (x-\delta, x+\delta)$ for all but finitely many n, and this is the desired contradiction.

Also, C is *totally disconnected*, that is, it contains no open intervals; for if an open interval I were contained in C we would be able to find an n and j for which

$$(j/3^n, (j+1)/3^n) \subseteq I,$$

a contradiction.

The same arguments work for any perfect symmetric set E.

Summarizing the observations in Example 1.2.8*b–f*, we have the following result.

Theorem 1.2.9. Properties of perfect symmetric sets
Let E *be a perfect symmetric set determined by* $\{\xi_k : k = 1\ldots\} \subseteq (0,1/2)$. *Then* E *is a closed, uncountable, totally disconnected set without isolated points. Also, for each* $\xi \in [0,1)$ *there is a perfect symmetric set* $E \subseteq [0,1]$ *for which* $m(E) = \xi$.

Our next aim is to prove the *Heine–Borel theorem.* This theorem is important for several reasons not the least of which is that it serves as the proper motivation to define a compact set.

Theorem 1.2.10. Heine–Borel theorem
a. *Let* $F \subseteq \mathbb{R}$ *be a closed and bounded set. Every open covering* $\mathcal{O}$ *of* F *has a finite subcovering. (*$X \subseteq \mathbb{R}$ *is bounded if it is contained in an interval of finite length; and an open covering* $\mathcal{O}$ *of a set* $X \subseteq \mathbb{R}$ *is a collection of open sets whose union contains* X.*)*

b. *Let* $\mathcal{C}$ *be a collection of closed sets* $F \subseteq \mathbb{R}$ *such that every finite subcollection of* $\mathcal{C}$ *has a nonempty intersection. Assume that at least one element of* $\mathcal{C}$ *is bounded. Then,*

$$\bigcap_{F \in \mathcal{C}} F \neq \emptyset.$$

Proof. **a.** Let $F = [a, b]$ and define $E = \{x \le b : [a, x]$ can be covered by a finite number of elements of $\mathcal{O}\}$. Then $E \ne \emptyset$ since $a \in E$. Thus, since $E \ne \emptyset$ and is bounded above by b, we can apply the completeness axiom for $\mathbb{R}$ and assert that E has a least upper bound $c \in [a, b]$. (See the Remark after Definition 1.2.11.) We shall verify that $c = b$.

Assume $c < b$. There is $U \in \mathcal{O}$ such that $c \in (c - \varepsilon, c + \varepsilon) \subseteq U$ for some $\varepsilon > 0$ satisfying $c + \varepsilon \le b$. Since $c - \varepsilon$ is not an upper bound of E, we take $x \in E$ for which $x > c - \varepsilon$. By the definition of E, there exist $U_1, \ldots, U_n \in \mathcal{O}$ such that

$$[a, x] \subseteq \bigcup_{j=1}^{n} U_j,$$

and, thus,

$$[a, c + \varepsilon] \subseteq \left(\bigcup_{j=1}^{n} U_j \right) \cup U.$$

Consequently, $c + (\varepsilon/2) \in E$, a contradiction; hence, $c = b$.

Next let $F \subseteq \mathbb{R}$ be an arbitrary closed bounded set. That F is bounded implies $F \subseteq [a, b]$, for some $a, b \in \mathbb{R}$. Take the open cover $\mathcal{O} \cup \{F^{\sim}\}$ of $[a, b]$. From the first part there are $U_1, \ldots, U_n \in \mathcal{O}$ such that $F \subseteq [a, b] \subseteq U_1 \cup \cdots \cup U_n \cup F^{\sim}$. Since $F \cap F^{\sim} = \emptyset$, $F \subseteq U_1 \cup \cdots \cup U_n$.

b. Part *b* is immediate from part *a*. □

Definition 1.2.11. Compact sets

a. A set $X \subseteq \mathbb{R}$ is *compact* if every family of open sets that covers X has a finite subfamily that is a covering of X; see Appendix A.1. From Theorem 1.2.10 it follows that every closed and bounded subset of $\mathbb{R}$ is compact.

b. It is a basic result from advanced calculus, e.g., [7], [178], [407], that the following assertions are equivalent for $X \subseteq \mathbb{R}^d$:

i. X is compact;

ii. X is closed and bounded;

iii. Every infinite subset of X has a convergent subsequence whose limit is in X.

This last assertion, *iii*, is the *Bolzano–Weierstrass theorem* for the case $d = 1$ and $X = [a, b]$. Also, properties *i* and *iii* are equivalent in any metric space; see Definition A.1.4. The spaces $L^1_m([a, b])$, which are the subject of this book, are metric spaces that do not have the property that conditions *i* and *ii* are equivalent. Can you find a counterexample $X \subseteq \mathbb{Q}$?

c. We say that a collection $\mathcal{C}$ of closed sets has the *finite intersection property* if every finite subcollection of $\mathcal{C}$ has a nonempty intersection. Thus, a set $X \subseteq \mathbb{R}$ is *compact* if and only if every collection of closed subsets of X with the finite intersection property has a nonempty intersection.

Remark. We have not defined $\mathbb{R}$ rigorously, and there are several ways to do so.

One approach is to define $\mathbb{R}$ abstractly, through a set of axioms that it must satisfy. We say that an ordered field $\mathbb{K}$ is *Archimedean* if, *for any* $0 < a < b \in \mathbb{K}$, *there exists* $n \in \mathbb{N}$ *such that* $na \geq b$. An ordered field $\mathbb{K}$ is *complete* and is said to satisfy the *completeness axiom* if *every nonempty subset of* $\mathbb{K}$ *that is bounded above has a least upper bound.* The set $\mathbb{R}$ of real numbers is a *complete Archimedean ordered field.* All such fields are isomorphic in the sense that any two of them are in a one-to-one correspondence that preserves the group actions and the order.

Another approach is to construct a model of a field with the above properties. Such constructions start with the field of rational numbers $\mathbb{Q}$, which is ordered and Archimedean but not complete. Thus, the goal is to "add" enough elements to make it complete. This can be done, for example, with the use of *Dedekind cuts* or by defining an equivalence relation for sequences of rational numbers as KARL WEIERSTRASS did. A readable exposition of the structure and construction of the real number system is [110]; cf. [463].

1.3 Classical real variables

1.3.1 Motivation for the Lebesgue theory

In 1881, VITO VOLTERRA was a student of ULISSE DINI at the Scuola Normale Superiore in Pisa. At this time he published an example of a function on $(0, 1)$ whose derivative exists everywhere, is a bounded function, and is not Riemann integrable [489]. In his thesis [312] (1902), HENRI LEBESGUE noted that because of the Volterra example one cannot consider differentiation and integration as inverse operations for a large enough class of functions; and thus he was motivated to try to find a notion of integration such that integration and differentiation are inverse operations for a much larger class of functions than is usually included by BERNHARD RIEMANN's theory. (HENRY J. SMITH had published a Volterra example in 1875, but LEBESGUE was apparently unaware of SMITH's work in 1902.) There were several other important reasons for extending the notion of integration; generally, one could lump them all together and say that there was a need for an integration theory that included many "natural" results, e.g., the validity of term-by-term integration of a series of functions, for many functions. We shall see that LEBESGUE's theory fills this criterion better than any other that has come along so far.

Now for the above-mentioned example due to VOLTERRA.

Example 1.3.1. Volterra example
Let $E \subseteq [0, 1]$ be a perfect symmetric set with $m(E) > 0$. On each contiguous interval (a, b) we define

$$\phi(x, a) = (x - a)^2 \sin\left(\frac{1}{x - a}\right), \quad x \in [a, b],$$

noting that $\phi(a, a) = 0$. (The interval (a, b) is *contiguous* to the closed set $F \subseteq \mathbb{R}$ if $a, b \in F$ and $(a, b) \subseteq F^{\sim}$.) For $x \in (a, b)$,

$$\phi'(x, a) = 2(x - a) \sin\left(\frac{1}{x - a}\right) - \cos\left(\frac{1}{x - a}\right).$$

Consequently, for all large k and $x = a + (1/k\pi)$,

$$\phi'(x, a) = -\cos(k\pi);$$

and so ϕ' achieves ± 1, and hence 0, infinitely often. Let $a + y$ be the largest zero of ϕ' less than or equal to $(a + b)/2$ and define

$$f(x) = \begin{cases} 0, & \text{if } \ x \in E, \\ \phi(x, a), & \text{if } \ a \leq x \leq a + y, \\ -\phi(x, b), & \text{if } \ b - y \leq x \leq b, \\ \phi(a + y, a), & \text{if } \ a + y \leq x \leq b - y \end{cases}$$

on $[0, 1]$, where (a, b) is a generic contiguous interval of E. It is now straightforward to verify the following:

i. f' exists on $(0, 1)$ and is 0 on E;
ii. f' is discontinuous on E.

To check *ii*, for example, if $x \in E$ then there are contiguous intervals (a_j, b_j) such that $\lim a_j = x = \lim b_j$; thus, since ϕ' takes on both ± 1 values, we have the required discontinuity.

The state of the art was such at the time of the Volterra example that it was realized that he had constructed a function f with a bounded non-Riemann-integrable derivative, and, in particular, that there was an everywhere differentiable function f for which the (Riemann) fundamental theorem of calculus did not work. The key thing is that f' is discontinuous on a set with strictly positive "length". The technicalities of the Volterra example are based on a simple idea. It was well known that $g(x) = x^2 \sin(1/x)$, $x \neq 0$, and $g(0) = 0$ has a discontinuous derivative at $x = 0$; VOLTERRA constructed f in terms of such functions g, and so spread the discontinuities of f' to all of E.

Because of Example 1.3.1 it became important to study set-theoretic properties of continuous and differentiable functions, and to take more care in defining notions of small sets. As such we shall recall some well-known notions from the calculus.

1.3.2 Continuous functions

Definition 1.3.2. Limits and continuous functions

a. Let $f : X \to \mathbb{R}$ be a function, where $X \subseteq \mathbb{R}$, and let $x \in \overline{X}$. We define the *limit*, $\lim_{y \to x, y \in X} f(y) = L$, to mean that

$$\forall\, \varepsilon > 0,\ \exists\, \delta > 0 \text{ such that } \forall\, y \in X,$$
$$0 < |x - y| < \delta \implies |L - f(y)| < \varepsilon.$$

b. A function $f : X \to \mathbb{R}$, where $X \subseteq \mathbb{R}$, is *continuous* at $x \in X$ if

$$\begin{aligned} &\forall\, \varepsilon > 0,\ \exists\, \delta > 0 \text{ such that } \forall\, y \in X, \\ &|x - y| < \delta \implies |f(x) - f(y)| < \varepsilon. \end{aligned} \tag{1.11}$$

Thus, f is continuous at x if and only if $\lim_{y \to x} f(y) = f(x)$

c. A function $f : X \to \mathbb{R}$, where $X \subseteq \mathbb{R}$, is *continuous* on X if it is continuous at each $x \in X$.

d. A function $f : X \to \mathbb{R}$, where $X \subseteq \mathbb{R}$, is *right (left) continuous* at $x \in X$ if the inequality $|f(x) - f(y)| < \varepsilon$ in (1.11) holds for $y > x$ $(y < x)$.

Proposition 1.3.3. *Let $F \subseteq \mathbb{R}$ be compact (closed and bounded), and assume that the function $f : F \to \mathbb{R}$ is continuous. Then f is bounded on F and it achieves its maximum and minimum there.*

Proof. i. To show that f is bounded.

Since f is continuous on F, then for each $x \in F$ there is an open interval I_x such that $x \in I_x$ and

$$\forall\, y \in I_x \cap F, \quad |f(x) - f(y)| < 1.$$

Since $F \subseteq \bigcup\{I_x : x \in F\}$, we can apply the Heine–Borel theorem, and so there are intervals $I_{x_1}, \ldots, I_{x_n}$ that cover F. Consequently, $|f(y)| \leq 1 + \max\{|f(x_1)|, \ldots, |f(x_n)|\}$ and f is bounded.

ii. To show that f achieves its maximum, respectively, minimum, on F.

Since f is bounded, $-\infty < \sup\, \{f(x) : x \in F\} = M < \infty$. We must find $x_1 \in F$ such that $f(x_1) = M$. If this is not the case then $f(x) < M$ for each $x \in F$; we obtain a contradiction to this possibility.

From the continuity of f,

$$\forall\, x \in F,\ \exists\, I_x, \text{ an open interval, such that } x \in I_x$$

and

$$\forall\, y \in I_x \cap F, \quad |f(x) - f(y)| \leq \frac{1}{2}(M - f(x)).$$

Consequently,

$$\forall\, x \in F \text{ and } \forall\, y \in I_x \cap F, \quad f(y) < \frac{1}{2}(M + f(x)).$$

Because of the Heine–Borel theorem there is a finite subcover $I_{x_1}, \ldots, I_{x_n}$ from $\{I_x : x \in F\}$ of F with the property that

$$\forall\, y \in F, \quad f(y) < \frac{1}{2}(M + A),$$

where $A = \max\{f(x_1), \ldots, f(x_n)\}$. Therefore, $(1/2)(M + A)$ is a bound for f on F; and this is a contradiction, since $(1/2)(M + A) < M$. □

The following result is not difficult to prove.

Proposition 1.3.4. *Let $f : X \to \mathbb{R}$ be a function on $X \subseteq \mathbb{R}$. Then f is continuous on $X \subseteq \mathbb{R}$ if and only if for each open set $U \subseteq \mathbb{R}$, $f^{-1}(U)$ is open in X.*

With regard to the characterization of continuous functions in terms of open sets we now define a function $f : X \to \mathbb{R}$, where $X \subseteq \mathbb{R}$, to be *open* if for every open set $U \subseteq X$, $f(U)$ is open. We say that a function $f : X \to \mathbb{R}$ is *closed* if for each closed set $F \subseteq X$, $f(F)$ is closed.

It is easy to find open discontinuous functions $f : \mathbb{R} \to \mathbb{R}$ if we make the topology on the domain strictly weaker than on the range. The following is a more interesting example.

Example 1.3.5. Open discontinuous function
With the usual topology on $\mathbb{R}$ we give an example of an open discontinuous function $f : \mathbb{R} \to \mathbb{R}$. Let f be strictly increasing, continuous, and with range $\mathbb{R}$ on each contiguous interval of the Cantor set C. Set $f = 0$ on C. Thus, $f : [0,1] \to \mathbb{R}$ is certainly not continuous. Without loss of generality we show that $f(I)$ is open for each open interval $I \subseteq [0,1]$. The result is clear if I is contained in a contiguous interval. If I covers some endpoint c of a "middle-third interval" then there is a "middle-third interval" $J \subseteq I$ such that $\mathbb{R} = f(J) \subseteq f(I)$, and thus f is open. The fact that there is such a J follows since I is open and the endpoints of smaller and smaller "middle-third intervals" will converge to c.

A countable union of closed sets in $\mathbb{R}$ is an $\mathcal{F}_\sigma$, and a countable intersection of open sets is a $\mathcal{G}_\delta$. Thus, the complement of an $\mathcal{F}_\sigma$ is a $\mathcal{G}_\delta$ and vice versa. We can define much more complicated sets. For example, an $\mathcal{F}_{\sigma\delta}$ is the intersection of a countable family of $\mathcal{F}_\sigma$s. All of these sets are special cases of *Borel sets*, of which we shall have more to say in Section 2.1.

For a given function f, $C(f)$ is its *set of points of continuity* and $D(f)$ is its *set of discontinuities*. The *oscillation* of a function $f : \mathbb{R} \to \mathbb{R}$ over a closed bounded interval I is

$$\omega(f, I) = \sup_{x\in I} f(x) - \inf_{x\in I} f(x);$$

and the *oscillation* of $f : \mathbb{R} \to \mathbb{R}$ at x is

$$\omega(f, x) = \inf \{\omega(f, I) : I \text{ a closed bounded interval}, x \in \operatorname{int} I\}.$$

Clearly, $\omega(f, x) \geq 0$.

Proposition 1.3.6. *Let $f : \mathbb{R} \to \mathbb{R}$ be a function.*

a. $x \in C(f) \iff \omega(f, x) = 0$.

b. *$C(f)$ is a $\mathcal{G}_\delta$ and $D(f)$ is an $\mathcal{F}_\sigma$.*

Proof. ***a.*** This equivalence follows from the definition of continuity.

b. For $n = 1, \dots,$ define

$$E_n = \{x : \omega(f, x) \geq 1/n\}.$$

It is easy to check that E_n is closed.

Now if $x \in D(f)$, then $\omega(f, x) > 0$, and so $x \in E_n$ for some n. Thus,

$$D(f) \subseteq \bigcup_{n=1}^{\infty} E_n.$$

Conversely, if $x \in \bigcup E_n$, then x is in some E_n, which means that f is discontinuous at x by part *a*. Consequently,

$$\bigcup_{n=1}^{\infty} E_n \subseteq D(f).$$

□

RIEMANN defined the *ruler function* $r : [0, 1] \to [0, 1]$ as

$$r(x) = \begin{cases} 0, & \text{if} \quad x \in [0,1] \quad \text{is irrational,} \\ 1, & \text{if} \quad x = 0, \\ 1/q, & \text{if} \quad x = p/q \in (0,1],\ (p,q) = 1. \end{cases}$$

(The notation "$(p, q) = 1$" indicates that p and q are relatively prime.) He proved the following result.

Proposition 1.3.7. *The function r is continuous at each irrational number and discontinuous at each rational number.*

Proof. Take any rational p/q satisfying $(p, q) = 1$, and let $0 < \varepsilon < 1/q$. Since there is an irrational number in any interval about p/q, there is no interval I about p/q with the property that

$$\forall\, x \in I, \quad |r(x) - r(p/q)| < \varepsilon;$$

in fact, $|r(x) - r(p/q)| = 1/q > \varepsilon$ for irrational $x \in I$. Let $x \in [0, 1]$ be irrational and let $\varepsilon > 0$. We must find $\delta > 0$ such that

$$\forall\, y \in (x - \delta, x + \delta), \quad |r(x) - r(y)| < \varepsilon.$$

Clearly there are only finitely many $p/q \in [0, 1]$, $(p, q) = 1$, for which $1/q \geq \varepsilon$. Call these $x_1, \dots, x_n$. Choose δ such that $x_j \notin (x - \delta, x + \delta) = I$ for any j. Therefore, for all $y \in I$, $|r(x) - r(y)| = |r(y)| < \varepsilon$. □

We generalize r as follows.

Example 1.3.8. General ruler function
Let $\gamma = \{\gamma_q : q = 1, \ldots\}$ be a sequence of nonnegative real numbers. Define $r_\gamma : [0,1] \to \mathbb{R}$ as

$$r_\gamma(x) = \begin{cases} 0, & \text{if } \; x \in [0,1] \; \text{ is irrational,} \\ 1, & \text{if } \; x = 0, \\ \gamma_q, & \text{if } \; x = p/q \in (0,1],\ (p,q) = 1. \end{cases}$$

As in the proof of Proposition 1.3.7 we see that r_γ is continuous at irrational x if $\gamma_q \to 0$ as $q \to \infty$, and discontinuous at each rational x where $\gamma_q \neq 0$. We shall have more to say about r_γ; see Proposition 3.4.7.

Example 1.3.9. A Hardy function
GODFREY H. HARDY [215], page 190, considered the function $h : \mathbb{R} \to \mathbb{R}$ defined as

$$h(x) = \begin{cases} x, & \text{if } \; x \text{ is irrational,} \\ \left(\frac{1+p^2}{1+q^2}\right)^{1/2}, & \text{if } \; x = p/q,\ (p,q) = 1. \end{cases}$$

Then h is discontinuous on $(-\infty, 0) \cup (\mathbb{Q} \cap (0,\infty))$ and continuous on $(\mathbb{R} \setminus \mathbb{Q}) \cap (0, \infty)$.

It is now natural to ask whether there is a function $f : \mathbb{R} \to \mathbb{R}$ such that f is continuous on the rationals and discontinuous on the irrationals. We shall show that the answer is no. In light of Proposition 1.3.6, this reduces to showing that the rationals are not only an $\mathcal{F}_\sigma$, which is obvious, but are not a $\mathcal{G}_\delta$. To prove this latter property it is convenient to introduce the notion of a nowhere dense set. A set $X \subseteq \mathbb{R}$ is *nowhere dense* if

$$\operatorname{int} \overline{X} = \emptyset.$$

Nowhere dense sets are totally disconnected, and the two notions are equivalent on $\mathbb{R}$ for closed sets. In particular, any perfect symmetric set is nowhere dense. A set $X \subseteq \mathbb{R}$ is a set of *first category* if it can be written as a countable union of nowhere dense sets. The major result about category is the *Baire category theorem*, which we shall discuss in Appendix A.6. For now we assume its statement for $\mathbb{R}$:

$$\mathbb{R} \text{ is not a set of first category.} \tag{1.12}$$

Because of this, $\mathbb{R} \setminus \mathbb{Q}$ is not a set of first category, for if it were we would obtain a contradiction to (1.12), since $\mathbb{Q}$ clearly is a set of first category.

Proposition 1.3.10. ***a.*** *$\mathbb{Q}$ is not a $\mathcal{G}_\delta$.*
b. *There is no function $f : \mathbb{R} \to \mathbb{R}$ continuous on the rationals and discontinuous on the irrationals.*

Proof. ***a.*** If $\mathbb{Q}$ were a $\mathcal{G}_\delta$ then $\mathbb{R} \setminus \mathbb{Q} = \bigcup F_n$, F_n closed. Since $F_n \subseteq \mathbb{R} \setminus \mathbb{Q}$, it contains no intervals and so is nowhere dense. Thus, $\mathbb{R} \setminus \mathbb{Q}$ is a set of first category, the required contradiction.

b. This part is clear from Proposition 1.3.6, the above discussion, and part *a.* □

Example 1.3.11. Let $f : \mathbb{R} \to \mathbb{R}$ be a nonconstant function with the property that every point is a local minimum, i.e.,

$$\forall\, x \in \mathbb{R},\ \exists\, I,\ \text{ an open interval about } x,\ \text{ such that } \forall\, y \in I,\ \ f(x) \le f(y).$$

Here $\mathbb{1}_A$ denotes the *characteristic function* of a set A, and it is defined by

$$\forall\, x \in \mathbb{R}, \quad \mathbb{1}_A(x) = \begin{cases} 1 & x \in A, \\ 0 & x \notin A. \end{cases}$$

It can be shown (Problem 1.19) that $D(f)$ is nowhere dense. The argument proceeds by assuming that the result is false and generating a nested sequence of closed intervals in the proper way whose intersection $\{x\}$ is not a local minimum; the details are neither obvious nor impossible. As a sort of converse, take any closed nowhere dense set E and define

$$f = \mathbb{1}_{\mathbb{R} \setminus E}.$$

Then, $D(f) = E$ and f has a local minimum at each $x \in \mathbb{R}$.

Definition 1.3.12. Uniform continuity
Let $f : X \to \mathbb{R}$ be a function on $X \subseteq \mathbb{R}$. We say that $f : X \to \mathbb{R}$ is *uniformly continuous on* $X \subseteq \mathbb{R}$ if

$$\begin{gathered} \forall\, \varepsilon > 0,\ \exists\, \delta > 0 \text{ such that } \forall\, x, y \in X, \\ |x - y| < \delta \implies |f(x) - f(y)| < \varepsilon. \end{gathered}$$

The notion of uniform continuity arose in the following context to define properly the Riemann integral of a continuous function on $[a, b]$.

Proposition 1.3.13. *Let f be continuous on a closed and bounded set $F \subseteq \mathbb{R}$. Then f is uniformly continuous.*

Proof. For each $\varepsilon > 0$ and $x \in F$ there is $\delta_x > 0$ such that

$$|x - y| < \delta_x \Longrightarrow |f(x) - f(y)| < \varepsilon/2.$$

Let $U_x = (x - (1/2)\delta_x, x + (1/2)\delta_x)$, so that $\{U_x : x \in F\}$ is an open cover of F. From the Heine–Borel theorem there are sets $U_{x_1}, \ldots, U_{x_n}$ whose union covers F. Take $\delta = (1/2)\min\{\delta_{x_1}, \ldots, \delta_{x_n}\}$ and let $|z - y| < \delta$.
If $y \in U_{x_i}$ we have

$$|y - x_i| < \frac{1}{2}\delta_{x_i} < \delta_{x_i}$$

and thus

$$|z - x_i| \leq |z - y| + |y - x_i| < \delta + \frac{1}{2}\delta_{x_i} < \delta_{x_i}.$$

Consequently, $|f(z) - f(x_i)|$, $|f(y) - f(x_i)| < \varepsilon/2$, and so $|f(z) - f(y)| < \varepsilon$. □

1.3.3 Uniform convergence

Definition 1.3.14. Convergence
Let $f, f_n : X \to \mathbb{R}$ be functions on $X \subseteq \mathbb{R}$.

a. $\{f_n : n = 1, \ldots\} : X \to \mathbb{R}$ *converges pointwise* to f if

$$\forall\, x \in X,\ \forall\, \varepsilon > 0,\ \exists\, N \text{ such that } \forall\, n \geq N, \quad |f_n(x) - f(x)| < \varepsilon.$$

b. $\{f_n : n = 1, \ldots\} : X \to \mathbb{R}$ *converges uniformly on* X to f if

$$\forall\, \varepsilon > 0,\ \exists\, N \text{ such that } \forall\, x \in X \text{ and } \forall\, n \geq N, \quad |f_n(x) - f(x)| < \varepsilon.$$

The following result is not difficult to prove. Part *b* is due to DINI.

Proposition 1.3.15. *Let $f_n : X \to \mathbb{R}$, $n = 1, \ldots$, be continuous on $X \subseteq \mathbb{R}$.*

a. *If $\{f_n\}$ converges uniformly to a function $f : X \to \mathbb{R}$, then f is continuous on X.*

b. *Let X be compact. If $\{f_n\}$ is a monotone sequence converging pointwise to a continuous function $f : X \to \mathbb{R}$, then $\{f_n\}$ converges uniformly to f on X.*

The following result, called the *Cauchy criterion*, provides an equivalent condition for uniform convergence.

Theorem 1.3.16. Cauchy criterion
Let $\{f_n : n = 1, \ldots\}$ be a sequence of real-valued functions on $X \subseteq \mathbb{R}$. Then $\{f_n\}$ converges uniformly on X (to some function f) if and only if

$$\forall\, \varepsilon > 0,\ \exists\, N > 0 \text{ such that } \forall\, m, n > N \text{ and } \forall\, x \in X, \quad |f_m(x) - f_n(x)| < \varepsilon.$$

Example 1.3.17. Cantor function
Let $E \subseteq [0, 1]$ be a perfect symmetric set determined by $\{\xi_k : k = 1, \ldots\} \subseteq (0, 1/2)$. We shall define the *Cantor function* C_E for E. Set $f_1 = 1/2$ on $[\xi_1, 1 - \xi_1]$, and extend to $[0, 1]$ so that it is continuous, increases from 0 to 1 in range values, and is linear on E_1^1 and E_1^2, e.g., Figure 1.1.

We define f_n as the continuous, increasing function with $f_n(0) = 0$ and $f_n(1) = 1$ that is linear on each E_n^j, $j = 1, \ldots, 2^n$, and takes the value $j/2^n$

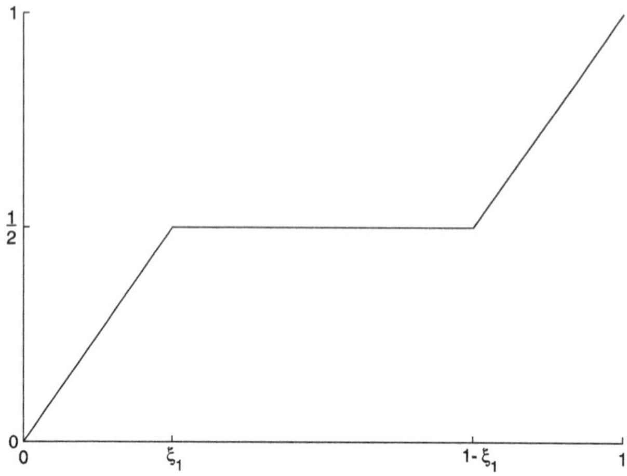

Fig. 1.1. Cantor function approximant.

on the contiguous interval immediately to the right of E_n^j. The sequence $\{f_n : n = 1, \ldots\}$ converges uniformly to a function f, so f is continuous. It is also easy to check that f increases from 0 to 1 in range values and is constant on each interval contiguous to E. The existence of f follows from the Cauchy criterion, Theorem 1.3.16. We denote f by C_E.

Note that the Cantor function C_E, for the case of $E = C$, is the restriction to C of the function f defined in Example 1.2.7*d*. The details of the proof of this fact are left as an exercise (Problem 1.23).

A classical problem in calculus is the problem of existence and interchangeability of iterated limits of multi-indexed sequences. One may consider problems involving sequences of functions as a special case. For example, the statement of Proposition 1.3.15*a* may be rephrased as follows:

Let $\{f_n : n = 1, \ldots\}$ be a sequence of continuous functions on $X \subseteq \mathbb{R}$. If $\{f_n\}$ converges uniformly to a function $f : X \to \mathbb{R}$, then

$$\forall\, x_0 \in X, \quad \lim_{x \to x_0} \lim_{n \to \infty} f_n(x) = \lim_{n \to \infty} \lim_{x \to x_0} f_n(x). \tag{1.13}$$

This result should be compared with the *Moore–Smith theorem* in Appendix A.4.

To see that the existence of single limits $\lim_{n\to\infty} f_n(x)$ and $\lim_{x\to x_0} f_n(x)$ is not enough to imply (1.13), let $f_n(x) = x^n$ and $f(x) = 0$ for $x \in [0,1]$. Clearly,

$$\lim_{n\to\infty} f_n(x) = \begin{cases} 0, & \text{if } x < 1, \\ 1, & \text{if } x = 1, \end{cases}$$

and $\lim_{x\to x_0} f_n(x) = f_n(x_0)$ for all $x_0 \in [0,1]$, whereas $\lim_{x\to 1} \lim_{n\to\infty} f_n(x) = 0 \neq 1 = \lim_{n\to\infty} \lim_{x\to 1} f_n(x)$. It is not difficult to see that in this example the sequence $\{f_n\}$ does not converge uniformly to f.

Proposition 1.3.18. *Let $\{f_n : n = 1, \ldots\}$ be a sequence of real-valued Riemann integrable functions on $[a,b] \subseteq \mathbb{R}$. If $\{f_n\}$ converges uniformly to a function f on $[a,b]$, then f is Riemann integrable and*

$$\lim_{n\to\infty} R\int_a^b f_n(x)\,dx = R\int_a^b f(x)\,dx.$$

For the definition of the Riemann integral $R\int_a^b f(x)\,dx$ see Section 3.1.

In light of Proposition 1.3.15*a*, or equivalently (1.13), and Proposition 1.3.18, it is not unexpected that if $\{f_n\}$ converges uniformly to f on $[a,b]$ and f_n' exists then f' exists and $\{f_n'\}$ converges, at least pointwise, to f'. Such is not the case. For example, let $f_n(x) = xe^{-nx^2}$ on $[-1,1]$.

Proposition 1.3.19. *Let $\{f_n : n = 1, \ldots\}$ be a sequence of real-valued differentiable functions on $(a,b) \subseteq \mathbb{R}$. If $f_n(x_0)$ converges for some $x_0 \in (a,b)$ and if $\{f_n'\}$ converges uniformly on (a,b) to some function g, then there exists a function f such that $\{f_n\}$ converges uniformly to f on (a,b) and*

$$\forall\, x \in (a,b), \quad f'(x) = \frac{d}{dx}\left(\lim_{n\to\infty} f_n(x)\right) = \lim_{n\to\infty} f_n'(x).$$

In particular, $f'(x) = g(x)$ on (a,b); cf. Problem 1.31.

A special example of the limit of a sequence of functions is an infinite series of the form $\sum_{n=1}^{\infty} f_n(x)$.

The following result is an analogue of Proposition 1.3.15*b*.

Proposition 1.3.20. *Let $X \subseteq \mathbb{R}$ be compact. If $\sum_{n=1}^{\infty} f_n$ is a series of continuous, nonnegative, real-valued functions, converging pointwise on X to a continuous function f, then $\sum_{n=1}^{\infty} f_n$ converges uniformly on X.*

The results we have just stated in this subsection are standard fare in an advanced calculus course, e.g., [7], [407], [504].

1.3.4 Sets of differentiability

Example 1.3.21. Nondifferentiability of the ruler function

We show that the *ruler function r is nowhere differentiable.* Recall that r is continuous only on $[0,1] \setminus \mathbb{Q}$, not on all $[0,1]$. Let $0 < x < 1$ be irrational, and note that

$$\frac{r(x+h) - r(x)}{h} = \frac{r(x+h)}{h},$$

where $h \neq 0$. Let $h_i \in \mathbb{R}$ have the property that $x + h_i$ is irrational and $h_i \to 0$ as $i \to \infty$. Thus,

$$\forall\, i = 1, \ldots, \quad \frac{r(x+h_i)}{h} = 0.$$

Write the decimal expansion $x = .a_1a_2\dots$ and set $h_i = .a_1\dots a_i - x$. Since x is nonzero and irrational, $a_i \neq 0$ for infinitely many i. Let N be the smallest integer such that $a_N \neq 0$. Now,

$$\forall\, i \geq N, \quad r(x+h_i) = r(.a_1\dots a_i) \geq 10^{-i},$$

by definition of r. Also, $|h_i| = .0\dots 0a_{i+1}\dots \leq 10^{-i}$. Hence,

$$\forall\, i \geq N, \quad \left|\frac{r(x+h_i)}{h_i}\right| \geq 1.$$

Consequently, $\lim_{h\to 0} r(x+h)/h$ does not exist.

With this example it is natural to inquire whether some r_γ, for $\gamma_q \to 0$, is differentiable anywhere or even possibly at all irrationals. We shall see in Example 1.3.22 that there is r_γ, for $\gamma_q \to 0$, such that $r'_\gamma(x)$ exists for some (irrational) x, but that there is no $\gamma_q \to 0$ such that $r'_\gamma(x)$ exists for all irrational x. To prove this last assertion we need the following result of MARION K. FORT, JR., see [64], pages 126–127, where the second inequality on line 6, page 127 should be reversed, or [64], 2nd edition:

$$\begin{gathered}\text{Let } f:[0,1]\to\mathbb{R} \text{ be discontinuous on } S,\ \overline{S}=[0,1];\\ \text{then } \{x : \exists\, f'(x)\} \text{ is a set of first category.}\end{gathered} \tag{1.14}$$

An interesting elaboration of FORT's theorem is found in [31].

Example 1.3.22. Differentiable points of general ruler functions
a. Let $\gamma_q = 1/q^4$ and define r_γ. A *quadratic irrational* is an irrational number that is a root of a quadratic equation with integer coefficients. It is well known [221], Theorem 188, that if x is a quadratic irrational, then for all large q,

$$\left|x - \frac{p}{q}\right| > \frac{1}{q^3}.$$

Consequently, for such an $x \in (0,1)$,

$$\left|\frac{r_\gamma(x) - r_\gamma(p/q)}{x-(p/q)}\right| < \frac{1}{q},$$

and so $r'_\gamma(x) = 0$.

b. If $\gamma_q \to 0$ then r_γ is discontinuous on a dense set (the rationals), and so r'_γ exists on a set of first category by (1.14); but the irrationals are not of first category by (1.12). Thus, there is no r_γ, where $\gamma_q \to 0$, for which r'_γ exists on all of the irrationals.

Example 1.3.23. Differentiability at single points
We give examples of functions $f : \mathbb{R} \to \mathbb{R}$ that are differentiable precisely at a single point.

a. Let $f(x) = x^2 \mathbb{1}_{\mathbb{Q}}(x)$. Clearly f is discontinuous everywhere except at the origin, and $f'(0) = 0$.

b. Assume for the moment that there are everywhere continuous nowhere differentiable functions $g : \mathbb{R} \to \mathbb{R}$; in fact, such functions exist, as we shall see in Examples 1.3.25 and 1.3.26. Define $f(x) = xg(x)$. Now

$$\frac{f(0+h) - f(0)}{h} = \frac{hg(h)}{h} = g(h) \to g(0), \quad \text{as} \quad h \to 0,$$

and so $f'(0) = g(0)$. If $x \neq 0$ then

$$\frac{f(y) - f(x)}{y - x} = x\frac{g(y) - g(x)}{y - x} + g(y),$$

and, consequently, $f'(x)$ does not exist, since $g(y) \to g(x)$ and $g'(x)$ does not exist.

Example 1.3.24. Existence of $|f|'$
Let $f : [0, 1] \to \mathbb{R}$ and suppose $f'(x)$ exists and is not zero for each x in an uncountable set S. We observe that $|f|'$ exists on an uncountable set X. In fact, define

$$X = \{x \in S : f(x) \neq 0\},$$

and first note, arguing by contradiction, that $\operatorname{card} X > \aleph_0$. Also, since f has the same sign in some neighborhood of $x \in X$ and since $f'(x)$ exists, we have the existence of $|f|'(x)$. For a bit more of a challenge consider the corresponding question for countably infinite sets.

For the sake of clarity, when we say that f is not differentiable at x we shall mean that the corresponding difference quotient does not approach a *finite* limit.

Example 1.3.25. Nowhere differentiability and Hadamard sets
We discuss the fact that

$$H(x) = \sum_{k=1}^{\infty} \frac{1}{2^k} \cos(2\pi 2^k x)$$

is an everywhere continuous nowhere differentiable function on $\mathbb{T}$. Let $\{n_k : k = 1, \ldots\}$ be a sequence of positive integers with the property that

$$\inf_{k \geq 1} \frac{n_{k+1}}{n_k} > 1.$$

Such sequences were first used by JACQUES HADAMARD to prove that every point on the circle of convergence of a power series with such exponents is a singularity of the function represented; and we refer to such a sequence $\{n_k : k = 1, \ldots\}$ as a *Hadamard set*. Such sequences and certain generalizations have been studied extensively in modern Fourier analysis, e.g., [404].

The fact that H is everywhere continuous nowhere differentiable follows immediately from the following result.

$$\begin{aligned}
&\text{Let } f(x) = \sum_{k=1}^{\infty} r_k \cos(2\pi n_k x), \text{ where } \sum_{k=1}^{\infty} r_k < \infty, \\
&r_k \geq 0, \text{ and } \{n_k : k = 1, \ldots\} \text{ is a Hadamard set;} \\
&\text{if } f \text{ is differentiable at a point then } \lim_{k\to\infty} n_k r_k = 0.
\end{aligned} \tag{1.15}$$

The proof of (1.15) is found in [267] and is generally simpler than the classical methods used to prove that a function is everywhere continuous nowhere differentiable.

One of the most popular and elementary examples of an everywhere continuous nowhere differentiable function is due to BARTEL L. VAN DER WAERDEN (1930) [471], Section 11.23; we also refer ahead to Problem 1.32 and Problem 4.9. We shall designate the van der Waerden function by W, which is defined as follows:

$$W(x) = \sum_{j=0}^{\infty} 10^{-j} f(10^j x), \quad x \in [0, 1],$$

where $f(x) = |x - k|$, $x \in [k - (1/2), k + (1/2)]$, $k \in \mathbb{Z}$. When "10" is replaced by "2" the corresponding function is referred to as the *Takagi function*; see Figure 1.2.

Example 1.3.26. Nondifferentiability and nowhere dense sets
We now indicate how, for every closed nowhere dense set $E \subseteq [0, 1]$, there is an everywhere continuous nowhere differentiable function f with the properties that

$$f = 0 \quad \text{on} \quad E$$

and

$$f > 0 \quad \text{on} \quad [0, 1] \setminus E;$$

cf. Problem 1.18. We first note that there are problems if we proceed "naturally", i.e., if we set $f = 0$ on E and define it to be a contraction of the van der Waerden function W on each contiguous interval, noting that $W(0) = W(1) = 0$. In this case, f is continuous but has points of differentiability.

Let $\{(a_j, b_j) : j = 1, \ldots\}$ be the set of contiguous intervals for E and enumerate all of the intervals $(2^{-j}(k-1), 2^{-j}k)$, $k = 1, 2, \ldots, 2^j$, $j = 1, 2, \ldots$, as $\{A_n : n = 1, \ldots\}$. Next define F_1 to be any (a_j, b_j) in A_1 in case such intervals exist, and $\emptyset$ otherwise. Generally, let F_n be any (a_j, b_j) in A_n, where $F_n \neq F_j$ for $j = 1, \ldots, n-1$, in case such intervals exist, and $\emptyset$ otherwise. If $x \in F_j = (a_{n_j}, b_{n_j})$ we define

$$f(x) = m(A_j) W\left(\frac{x - a_{n_j}}{b_{n_j} - a_{n_j}}\right);$$

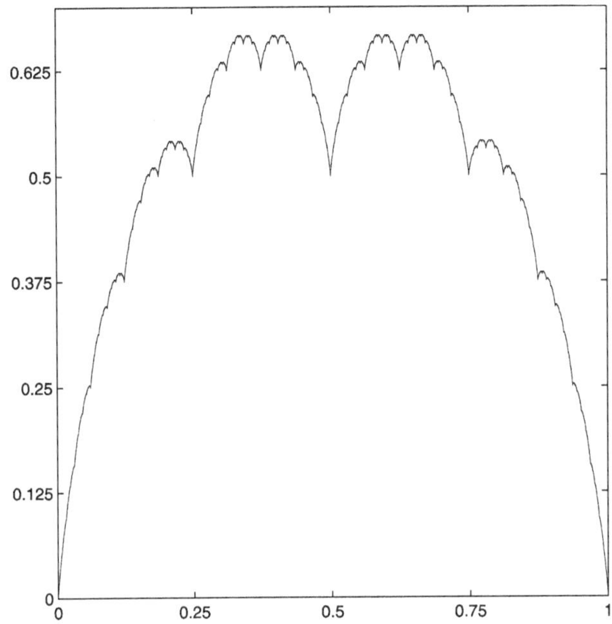

Fig. 1.2. Takagi function.

and if $x \in (a_n, b_n)$, a contiguous interval that is not equal to any F_j, $j = 1, \dots,$ we set

$$f(x) = (b_n - a_n) W\left(\frac{x - a_n}{b_n - a_n}\right).$$

Finally, set $f = 0$ on E. This function is due to JAN S. LIPIŃSKI [324]. The verification that f is everywhere continuous and nowhere differentiable is Problem 1.33.

Remark. STEFAN BANACH asked to show that the collection of all functions on $(0,1)$ that are continuous and that have a derivative everywhere does not form a Borel set in the space $C((0,1))$ of all continuous functions on $(0,1)$; see [344] Problem 52. The proof of this fact was given by STEFAN MAZURKIEWICZ [346]. It can also be shown that this set forms a set of first category [362]. In the "opposite" direction, R. DANIEL MAULDIN [343] proved that the set of all everywhere continuous nowhere differentiable functions forms a subset of $C((0,1))$ that is not a Borel set.

We now shift our emphasis and consider the situation that f' exists for each $x \in (0,1)$. We want to find the set of discontinuities of f'.

Example 1.3.27. Sets of discontinuities of f'
We shall construct a function $f : (0,1) \to \mathbb{R}$ such that f' exists on $(0,1)$ and $D(f') = \mathbb{Q} \cap (0,1)$. Let $g(x) = x^2 \sin(1/x)$, $x \neq 0$, and $g(0) = 0$, so that g' is

discontinuous at $x = 0$. If $\{r_n : n = 1, \ldots\} \subseteq (0,1)$ is an enumeration of the rationals, then set

$$f(x) = \sum_{n=1}^{\infty} \frac{1}{2^n} g(x - r_n).$$

Clearly, $\sum(1/2^n)g'(x - r_n)$ converges uniformly, and so

$$f'(x) = \sum_{n=1}^{\infty} \frac{1}{2^n} g'(x - r_n), \quad x \in (0,1).$$

Thus, f' is discontinuous at each rational by the definition of g; cf. the Volterra example.

Proposition 1.3.28. ***a.*** *$X = D(f')$ for some function $f : [0,1] \to \mathbb{R}$ whose derivative exists at each $x \in [0,1]$ if and only if X is an $\mathcal{F}_\sigma$ set of first category.*

b. *There is no function $f : [0,1] \to \mathbb{R}$ whose derivative exists on $[0,1]$ but that is discontinuous everywhere.*

Proof. ***a.*** ($\Longrightarrow$) f' is discontinuous on an $\mathcal{F}_\sigma$ by Proposition 1.3.6, and it is easy to check that this $\mathcal{F}_\sigma$ is of first category.
($\Longleftarrow$) We can write $X = \bigcup F_n$, $F_n \subseteq F_{n+1}$, F_n closed and nowhere dense. As in the Volterra example (Example 1.3.1) we take f_n differentiable on $[0,1]$ and with the property that $f_n' = \pm 1$ infinitely often in any neighborhood of any $x \in F_n$ and $|f_n'| \leq C < \infty$ for all n. Consequently, e.g., Example 1.3.27, $f = \sum f_n/3^n$ is differentiable and f' is discontinuous precisely on X.

b. Part *b* is a consequence of part *a*. □

There is no function $f : [0,1] \to \mathbb{R}$ whose derivative exists everywhere on $[0,1]$ but that is discontinuous on the complement of a countable set; see Appendix A.6.

For surveys of the remarks made in Section 1.3.2 and Section 1.3.4 we list [64], [80], [444].

Basically we have seen that some of the problems in nineteenth-century analysis led to various methods of characterizing the size of a given set, e.g., countability, category, and "length". As we shall see, an in-depth study of "length", i.e., measure, was crucial in getting past the problems of RIEMANN's integration theory that were focused upon by the likes of the Volterra example. ÉMILE BOREL was the mathematician who heralded measure theory in this regard; and LEBESGUE was the only one who discovered the associated integration theory, which, in a very real way, obtains more important theorems than possible by any other method. Of course, LEBESGUE came to his integral after considerably extending the Borel measure; cf. LEON W. COHEN's analysis [108]; and this issue was the major cause for the ensuing disagreement between BOREL and LEBESGUE; cf. ANTONIE MONNA's study (see Section 1.4) for another aspect of this polemic.

1.4 References for the history of integration theory

The following contains several listings on Fourier series because of the intimate relation in the development of integration theory and Fourier series.

- G. A. Bliss, *Integrals of Lebesgue*, Bull. Amer. Math. Soc. 24 (1917), 1–47.
- N. Bourbaki, *Intégration (Introduction to Chapters I–IV, and the historical remarks for succeeding chapters)*, Paris 1965.
- J. Dauben, *The trigonometric background of G. Cantor's theory of sets*, Arch. Hist. Exact Sci. 7 (1971), 181–216.
- M. Fréchet, *La vie et l'œuvre d'Émile Borel*, Enseignement (2) 11 (1965), (in particular, pages 53–63).
- H. Freudenthal, *Did Cauchy plagiarize Bolzano?* Arch. Hist. Exact Sci. 7 (1971), 375–392.
- G. A. Gibson, *On the history of the Fourier series*, Proc. Edinburgh Math. Soc. 11 (1893), 137–166.
- I. Grattan-Guinness, *The Development of the Foundations of Mathematical Analysis from Euler to Riemann*, MIT 1970 (if you read pages 76–78 then you have a moral obligation to read Freudenthal's article, above, and Waterhouse's review, below).
- I. Grattan-Guinness, *From the Calculus to Set Theory, 1630–1910. An Introductory History*, Princeton University Press, Princeton, NJ, 2000.
- R. Henstock, *A short history of integration theory*, Southeast Asian Bull. Math. 12 (2) (1988), 75–95.
- T. Hawkins, *Lebesgue's Theory of Integration. Its Origins and Development*, AMS Chelsea Publishing, Providence, RI, 2001.
- H. Lebesgue, *Intégrale, longeur, aire*, Annali di Matematica 7 (1902), 231–359.
- H. Lebesgue, *Leçons sur l'Intégration et la Recherche des Fonctions Primitives* (1st and 2nd ed.), Paris 1904 and 1928.
- E. R. Lorch, *Continuity and Baire functions*, Amer. Math. Monthly 78 (1971), 748–762.
- F. A. Medvedev, *Henri Lebesgue's works on the theory of functions (on the hundredth anniversary of his birth)* (Russian), Russian Math. Surveys 30 (1975), 179–191 (Uspehi Mat. Nauk 30 (1975), 227–238).
- A. F. Monna, *The concept of function in the 19th and 20th centuries, in particular with regard to the discussion between Baire, Borel, and Lebesgue*, Arch. Hist. Exact Sci. 9 (1972), 57–84.
- A. B. Paplauskas, *L'Influence de la théorie des séries trigonométriques sur le développement du calcul intégral*, Arch. Internat. Hist. Sci. 84–85 (1968), 249–260.
- I. Pesin, *Classical and Modern Integration Theories*, New York 1970.
- J.-P. Pier, *Intégration et mesure*, 517–564, in *Development of Mathematics 1900–1950*, Birkhäuser Verlag, Basel, 1994.

- M. Plancherel, *Contribution à l'étude de la représentation d'une fonction arbitraire par des intégrales définies*, Rend. Circ. Mat. Palermo, 30 (1910), 289–335.
- F. Riesz, *L'Évolution de la notion d'intégrale depuis Lebesgue*, Ann. Inst. Fourier, Grenoble 1 (1949–50), 29–42.
- S. Saks, *Zarys teorii całki*, Kasa Mianowskiego, Warszawa 1930, *Théorie de l'intégrale. Avec une note de Stefan Banach*, Monografie Matematyczne 2, Warszawa 1933, *Theory of the Integral*, 2. ed., English translation by L. C. Young, Monografie Matematyczne 7, Warszawa, Lwów 1937; Dover Publications, New York, 1964.
- C.-J. de la Vallée Poussin, *Intégrales de Lebesgue* (reprint of the second (1934) edition), Éditions Jacques Gabay, Sceaux, 1995.
- W. C. Waterhouse, *Book Reviews*, Bull. Amer. Math. Soc. 78 (1972), 385–391.

1.5 Potpourri and titillation

1. In Examples 1.3.23*b*, 1.3.25, and 1.3.26 we mentioned everywhere continuous nowhere differentiable functions.

In a lecture in 1861, RIEMANN is supposed to have asserted that

$$R(x) = \sum_{k=1}^{\infty} \frac{\sin(k^2 \pi x)}{k^2}$$

is everywhere continuous nowhere differentiable. The continuity follows from the uniform convergence of the series. In his attempt to prove RIEMANN's claim for R, WEIERSTRASS proved (although PAUL DU BOIS-REYMOND first published it in 1874) that

$$K(x) = \sum_{k=1}^{\infty} a^k \cos(b^k \pi x)$$

is everywhere continuous nowhere differentiable, where $b \in \mathbb{Z}$ is odd, $a \in (0,1)$, and $ab > 1 + 3\pi/2$; cf. [82]. In 1916, HARDY proved that R is not differentiable at any irrational and some rationals; and he also showed that if $b > 1$, $0 < a < 1$, and $ab \geq 1$, then K is everywhere continuous nowhere differentiable.

There have been two rather exciting chapters in the business of finding everywhere continuous nowhere differentiable functions.

In 1916, GRACE C. YOUNG brought attention to CHARLES CELLERIER's example of an everywhere continuous nowhere differentiable function,

$$\sum_{k=1}^{\infty} \frac{1}{a^k} \sin(a^k \pi x), \quad a > 1,$$

discovered before 1850; and BERNHARD BOLZANO had constructed an everywhere continuous nowhere differentiable function about 1830, a fact not discovered until 1921. It should be noted that BOLZANO did not realize the full potential of his example.

The second incident concerning everywhere continuous nowhere differentiable functions is more recent and involves R. In 1969, JOSEPH L. GERVER, at the time an undergraduate at Columbia University, proved the differentiability of R at certain rational points [192], [193], [194]; and there have been elaborations and generalizations by others. In particular, STÉPHANE JAFFARD and YVES MEYER, e.g., [253], [254], [259], have studied the exact regularity of R using methods based on wavelet transforms; see Sections 9.6.2 and 9.6.3. GERVER proved that *R is differentiable at points of the form* $(2p+1)/(2q+1)$, $p, q \in \mathbb{Z}$. He accomplished this in terms of the Hölder exponent. By definition the *Hölder exponent* of R is a number α such that there exists $C > 0$ for which

$$|R(x) - R(x_0)| \leq C|x - x_0|^\alpha,$$

for all x in some neighborhood of x_0. GERVER proved that at the points $(2p+1)/(2q+1)$ the Hölder exponent of R equals $3/2$, and at all other points it varies between $1/2$ and $3/4$.

Finally, we would like to mention that several elementary examples of everywhere continuous nowhere differentiable functions have been published in the Amer. Math. Monthly, e.g., JOHN MCCARTHY, 60 (1953), 709, M. MARK LYNCH, 99 (1992), 8–9, and LIU WEN, 107 (2000), 450–453.

2. An interesting problem was posed around 1928 by STANISŁAW MAZUR. It is now called the *Banach–Mazur game*. However, although this problem is at the foundation of game theory, we are interested in it because of its relation to Cantor sets. We cite its statement from *The Scottish Book*, Problem 43, as it is found in [344].

Given a set $E \subseteq \mathbb{R}$, a game between two players A and B is defined as follows: A selects an arbitrary interval d_1; B then selects an arbitrary interval d_2 contained in d_1; then A selects an arbitrary segment d_3 contained in d_2, and so on. A wins if the intersection $\bigcap_n d_n$ contains a point of the set E; otherwise, A loses. If E is a complement of a set of first category, there exists a method by which A can win; if E is a set of first category, there exists a method by which B will win.

Problem. *Is it true that there exists a method of winning for the player A if and only if the set E has its complement of first category in some interval; similarly, does a method of winning exist for B if and only if E is a set of first category?*

This problem was solved positively by BANACH in 1935, and his proof was published by JOHN C. OXTOBY; see [362]. In particular, if the set E has the *Baire property*, i.e., if it can be represented as a symmetric difference of an open set and a set of first category, then either player A or player

B has a winning strategy. This does not mean that the game is completely deterministic, and this is what makes it interesting.

The Banach–Mazur game has many generalizations. One of the most striking examples of such a generalization is due to STANISŁAW ULAM; see [344], Problem 43:

Given a set $E \subseteq \mathbb{R}$, players A and B provide in turns the digits 0 *and* 1. *A wins if the number formed by these digits in a given order (in the binary system) belongs to E. For which sets E does there exist a winning method for player A (player B)?*

JAN MYCIELSKI has shown that the Banach–Mazur game has an equivalent game in the ULAM scheme.

ULAM's generalization of the Banach–Mazur game plays a role in modern set theory; see, e.g., [350]. It is a basis for the definition of the axiom of determinacy, i.e., that for every subset of $[0, 1]$ the Ulam game is determined. DONALD A. MARTIN and JOHN R. STEEL (1989) proved that the axiom of determinacy follows from a large-cardinal hypothesis.

1.6 Problems

Some of the more elementary problems in this set include Problems 1.1, 1.2, 1.3, 1.4, 1.5, 1.11, 1.14, 1.22, 1.25, 1.27, 1.28, 1.30, 1.35.

1.1. If $\{\{a\}, \{a, b\}\} = \{\{c\}, \{c, d\}\}$ prove that $a = c$ and $b = d$.

1.2. Given $f : X \to Y$, where f is a function and X and Y are sets.

a. Prove

$$f\left(\bigcup A_\alpha\right) = \bigcup f(A_\alpha),$$

$$f^{-1}\left(\bigcup B_\alpha\right) = \bigcup f^{-1}(B_\alpha),$$

$$f^{-1}\left(\bigcap B_\alpha\right) = \bigcap f^{-1}(B_\alpha),$$

$$f^{-1}(B^\sim) = [f^{-1}(B)]^\sim.$$

b. Prove that the following are generally proper inclusions:

$$f\left(\bigcap A_\alpha\right) \subseteq \bigcap f(A_\alpha),$$

$$A \subseteq f^{-1}(f(A)), \quad \text{and} \quad f\left(f^{-1}(B)\right) \subseteq B.$$

c. Show that if f is surjective then $f\left(f^{-1}(B)\right) = B$.

1.3. ***a.*** Let X be a set and let $\mathcal{P}(X)$ be the set of all subsets of X. Prove that there is no surjection $f : X \to \mathcal{P}(X)$.

[*Hint.* Set $B = \{x \in X : x \notin f(x)\}$.]

b. Let $S = \{\{a_n : n = 1, \ldots\} : a_n = 0 \text{ or } 1\}$. Prove $\operatorname{card} S > \aleph_0$.

[*Hint.* Use the Cantor diagonal process.]

1.4. ***a.*** Prove that there is an uncountable closed set of irrationals in $[0, 1]$; also see Example 2.2.14.

b. Prove that there are uncountable sets of real numbers that do not contain uncountable closed subsets.
[*Hint.* One can show the existence of a set $X \subseteq \mathbb{R}$ that does not contain any subset Y having the properties that Y is closed and contains no isolated points; see [224], pages 201–202.]

1.5. Find a discrete subset of $\mathbb{R}$ with uncountable closure. Recall that X is *discrete* if for each $x \in X$ there is $I \subseteq \mathbb{R}$, an open interval, such that $I \cap X = \{x\}$.

1.6. ***a.*** (SCHRÖDER–BERNSTEIN) Prove that if $f : A \to B$ and $g : B \to A$ are injections, then there is a *bijection*, i.e., an injection that is also a surjection, $h : A \to B$.
[*Hint.* The basic decomposition required here is perhaps not the first idea that will come to mind; it might be more efficient to read the proof in [279].]

b. Using part a show that a countably infinite set has an uncountable family $\mathcal{F}$ of subsets, the intersection of any two of which is finite.
[*Hint.* Besides using part a there is also the following more geometric solution. Let $P = \{(n, m) : n, m \in \mathbb{Z}\} \subseteq \mathbb{R} \times \mathbb{R}$ and take S_ϕ to be the "strip" in P with angle of inclination ϕ and width greater than 1. There are uncountably many S_ϕ by the uncountability of the ϕ, and so, if we put our given countable set in one-to-one correspondence with P, we obtain the uncountability of $\mathcal{F}$. Clearly $S_\phi \cap S_\psi$ is finite, since this intersection is a bounded parallelogram in $\mathbb{R}^2$.]

1.7. Let $\mathcal{F}$ be the field of all functions $f = p/q$, where p and q are polynomials with real coefficients. Define

$$P_a = \{f \in \mathcal{F} : \exists\ \varepsilon > 0 \text{ for which } f > 0 \text{ on } (a, a + \varepsilon)\}$$

and

$$P'_a = \{f \in \mathcal{F} : \exists\ \varepsilon > 0 \text{ for which } f > 0 \text{ on } (a - \varepsilon, a)\}$$

for each $a \in \mathbb{R}$, and

$$P_\infty = \{f \in \mathcal{F} : \exists\ N > 0 \text{ for which } f > 0 \text{ on } (N, \infty)\}$$

and

$$P'_\infty = \{f \in \mathcal{F} : \exists\ N > 0 \text{ for which } f > 0 \text{ on } (-\infty, -N)\}.$$

Clearly, with each such P, $\mathcal{F}$ is an *ordered field* with pointwise addition and multiplication. This means that P is closed under addition and multiplication, and that if $f \in \mathcal{F}$ then exactly one of the following is true: $f \in P$, $-f \in P$, or $f = 0$. Prove that the above Ps give rise to distinct non-Archimedean ordered fields, and that these are the only subsets of $\mathcal{F}$ for which $\mathcal{F}$ is an ordered field (Archimedean fields were defined in Section 1.2). Also, prove that if $f \in \mathcal{F}$ is in every P, then it can be written as a sum of squares.

1.8. Let $\mathcal{F}$ be an ordered field with "positive elements" P; see Problem 1.7. The canonical linear ordering on $\mathcal{F}$ is given by defining $f < g$ to mean that $g - f \in P$. Define

$$[f, g] = \{h \in \mathcal{F} : f \leq h \leq g\}$$

and

$$(f, g) = \{h \in \mathcal{F} : f < h < g\}.$$

Then $[f, g]$ is a *closed interval* and (f, g) is an *open interval*. A set $\mathcal{S}$ of open intervals *covers* $[f, g]$ if each $h \in [f, g]$ is contained in at least one member of $\mathcal{S}$. Now assume that $\mathcal{F}$ is *order-complete*, i.e., each nonempty subset of $\mathcal{F}$ that has an upper bound has a supremum. Prove the *Heine–Borel theorem*: given $[f, g]$ and $\mathcal{S}$ a cover of $[f, g]$, there is a finite subcover.

1.9. Let $S \subseteq \mathbb{R}^d$ have the property that $\rho(s_1, s_2)$ is rational for every $s_1, s_2 \in S$, where ρ is the usual Euclidean distance defined as $\rho(s_1, s_2) = \sqrt{\sum_{j=1}^{d} |s_1^j - s_2^j|^2}$. Prove that S is countable.

1.10. Find the error(s) in the following proof of Proposition 1.2.5.

Every open set $G \subseteq \mathbb{R}^2$ is the disjoint union of open linear sets, e.g., the intersection of G with the set of all horizontal lines in $\mathbb{R}^2$; but it is known that every open linear set is the disjoint union of open straight line segments (Theorem 1.2.4). Thus, the problem is reduced to showing that every open straight line segment can be represented as a disjoint union of closed straight line segments; and this is clear, e.g.,

$$(0, 1) = [1/3, 2/3] \cup [1/9, 2/9] \cup [7/9, 8/9] \cup \cdots .$$

1.11. Prove the following set-theoretic assertions:

- $A \subseteq B$ implies $\overline{A} \subseteq \overline{B}$;
- $\overline{A \cup B} = \overline{A} \cup \overline{B}$;
- $\overline{\overline{A}} = \overline{A}$;
- the finite union of closed sets is closed;
- the arbitrary intersection of closed sets is closed;
- the complement of an open set is closed;
- the complement of a closed set is open.

1.12. Let C be the Cantor set. Can you find $\{x_n : n = 1, \ldots\} \subseteq \mathbb{R}$, bounded and infinite, such that

$$\forall\, x \in \mathbb{R}, \quad \text{card}\,(C \cap \{x_n + x : n = 1, \ldots\}) < \aleph_0\,?$$

1.13. Let E_ξ be a perfect symmetric set with $\xi_k = \xi$ for each k.

a. Prove that $E_\xi - E_\xi = \{x - y : x, y \in E_\xi\} \supseteq [0, 1]$ if $1/3 \leq \xi < 1/2$.

b. Prove that $m(E_\xi - E_\xi) = 0$ if $0 < \xi < 1/3$.

[*Hint.* First, try the case $0 < \xi < 1/4$. HUGO STEINHAUS proved part *a* for $C = E_{1/3}$ by considering $C \times C \subseteq [0, 1] \times [0, 1]$, and STEINHAUS' trick can be used to prove part *b* for the $E_{1/4}$ case; cf. Problem 3.6. T. ŠALAT settled the general case for $1/4 < \xi < 1/3$. In 1987, JACOB PALIS conjectured that *the difference of two perfect symmetric sets is either of Lebesgue measure zero, or it contains an interval* [364]. This conjecture was answered negatively by ATSURO SANNAMI [417]. However, in case of perfect symmetric sets E_ξ, the conjecture is true; see [84].]

In this problem it is first necessary to make a reasonable guess at the definition of "measure 0" or else wait until Chapter 2.

1.14. Prove (1.3).

1.15. ***a.*** Prove that $\mathbb{1}_{\mathbb{R}\setminus\mathbb{Q}}$ is not the pointwise limit of continuous functions. [*Hint.* Assume that a sequence $\{f_n : n = 1, \ldots\}$ of continuous functions converges pointwise to $\mathbb{1}_{\mathbb{R}\setminus\mathbb{Q}}$, and define

$$E_n = \{x : f_n(x) \geq 1/2\} \quad \text{and} \quad F_N = \bigcap_{j=N}^{\infty} E_j.$$

Prove that $\mathbb{R} \setminus \mathbb{Q} = \bigcup_{N=1}^{\infty} F_N$. This yields the desired contradiction since $\mathbb{R} \setminus \mathbb{Q}$ is not an $\mathcal{F}_\sigma$.]

b. Let $\{f_n : n = 1, \ldots\}$ be a sequence of continuous functions $[0,1] \to \mathbb{R}$ that converges pointwise to a function f. Prove that $\overline{C(f)} = [0,1]$. The solution to this is nontrivial; the presentation in [64], pages 99–101, uses the Baire category theorem three times.

c. Let $\{f_n : n = 1, \ldots\}$ be a sequence of continuous functions $[0,1] \to \mathbb{R}$. Prove that $\{x : \exists \lim_{n\to\infty} f_n(x)\}$ is an $\mathcal{F}_{\sigma\delta}$.

1.16. Prove that a monotone (increasing or decreasing) function $f : \mathbb{R} \to \mathbb{R}$ has at most countably many points of discontinuities and all of these discontinuities are jump discontinuities.

1.17. Let $f : \mathbb{R} \to \mathbb{R}$ be differentiable. Assume that

$$\forall\, r \in \mathbb{R}, \quad \{x : f'(x) = r\} \ \text{ is closed.}$$

Prove that f' is continuous on $\mathbb{R}$.

1.18. Construct a C^∞*-function*, i.e., infinitely differentiable function, $f : (0,1) \to \mathbb{R}$ such that $\{x : f(x) = 0\}$ is nowhere dense and uncountable.

1.19. Prove that the set of discontinuities of the function described in Example 1.3.11 is nowhere dense.
[*Hint.* As suggested in Example 1.3.11, assume that the result is false. Then, show that such a function must have a minimum on each closed interval. Start with an interval where the set of discontinuities is dense and use this property to construct a sequence of closed intervals, with increasing values of their minima, that converge to a single point that cannot be a local minimum.]

1.20. In light of Example 1.3.11, can the following be proved: for every nowhere dense $\mathcal{F}_\sigma$ set $E \subseteq \mathbb{R}$, there is $f : \mathbb{R} \to \mathbb{R}$ such that $D(f) = E$ and f has a local minimum at each $x \in \mathbb{R}$?

1.21. Let $f : [0,1] \to \mathbb{R}$ be a function with the property that $\overline{C(f)} = [0,1]$. Prove that f is continuous except on a (possibly empty) set of first category; cf. Problem 1.15*b* and Proposition 1.3.6.

1.22. Verify the claim of Example 1.3.1 (Volterra example) that f' exists on $(0,1)$ and is 0 on E.

1.23. Verify the claim at the end of Example 1.3.17 that the two Cantor functions we have defined are equal.

1.24. Find a dense set $S \subseteq \mathbb{C}$ such that for any line segment $L \subseteq \mathbb{C}$, $\overline{S \cap L} \neq L$.
[*Hint.* An *algebraic number* is a complex root of a polynomial with integer coefficients; cf. Example 1.3.22. Let $A \subseteq \mathbb{C}$ be the set of algebraic numbers, and note that A is a countable set $\{\alpha_n : n = 1, \ldots\}$; cf. Section 3.8.2. Define $s_1 = \alpha_1$, $s_2 = \alpha_2$, and

$$\forall\, n \geq 3, \quad s_n = \alpha_n + \frac{1}{2^n} e^{2\pi i / f(n)},$$

where $f(n)$ is the smallest positive integer for which s_n is not collinear with any two of $s_1, \ldots, s_{n-1}$. Set $S = \{s_n : n = 1, \ldots\}$.]

1.25. Order the rationals $\{r_n : n = 1, \ldots\} \subseteq (0,1)$ à la Cantor, i.e.,

$$\frac{1}{2}, \frac{1}{3}, \frac{2}{3}, \frac{1}{4}, \frac{1}{5}, \frac{3}{4}, \frac{2}{5}, \frac{1}{6}, \ldots,$$

and cover each r_n by an interval of length $1/2^{n+1}$ and center r_n. The total "length" covered is less than or equal to 1/2, and so there are real numbers $r \in [0,1]$ that are not covered by this process. Write down one such r explicitly.
[*Hint.* Let $r = 1/\sqrt{2}$ and note that r is a root of $f(x) = 2x^2 - 1$. For any rational $p/q \in (0,1)$, the mean value theorem allows us to write

$$|f(p/q)| = |f(p/q) - f(r)| = 4\xi|r - (p/q)|,$$

where $\xi \in (0,1)$. Since $\sqrt{2}$ is irrational, $|2p^2 - q^2|$ is a positive integer, and so

$$|f(p/q)| = \frac{|2p^2 - q^2|}{q^2} \geq \frac{1}{q^2}.$$

Consequently,

$$|r - (p/q)| \geq \frac{1}{4\xi q^2} \geq \frac{1}{4q^2}. \tag{1.16}$$

From properties of the Farey series in terms of the Euler function $\phi(n)$ (see the proof of Proposition 3.4.7 for the definition of ϕ as well as [221], Theorems 330 and 331), it can be seen that if p/q is the nth rational then $n+2 > q^2/4$. This fact combined with (1.16) gives the result.]

Remark. This technique was first used by JOSEPH LIOUVILLE to construct *transcendental*, i.e., nonalgebraic, numbers; e.g., [362], pages 7–8. We say that $r \in \mathbb{R} \setminus \mathbb{Q}$ is a *Liouville number* if for every positive integer n there are integers p and q such that $|r - (p/q)| < 1/q^n$, $q > 1$. *Every Liouville number is transcendental.*

1.26. Define the function $f : (0,1) \to \mathbb{R}$ as

$$f(x) = \sum_{0 \leq \frac{p}{q} \leq x} \frac{1}{q^3}, \quad (p,q) = 1.$$

a. Prove that f is increasing on $(0,1)$, continuous on the irrationals, and discontinuous on the rationals.

b. Since f is increasing, we shall see in Chapter 4 that f' exists for many points. Show that there are points x for which $f'(x)$ does not exist.
[*Hint.* Take x to be a Liouville transcendental number, mentioned in Problem 1.25, and verify that

$$\overline{\lim_{u \to x}} \frac{f(u) - f(x)}{u - x} = \infty.]$$

1.27. Let $f : \mathbb{R} \to \mathbb{R}$ be a function, $r > 0$, and let $E_r = \{x : \omega(f,x) \geq r\}$, where $\omega(f,x)$ is the oscillation of f at x. Show that E_r is closed.

1.28. ***a.*** Let $f : [0,1] \to \mathbb{R}$ be a continuous function, and for each n choose some $x_{k,n} \in [(k-1)/n, k/n]$, $k = 1, \ldots, n$. Prove that

$$R\int_0^1 f = \lim_{n \to \infty} \frac{1}{n} \sum_{k=1}^{n} f(x_{k,n})$$

exists, where the integral $R\int_0^1$ denotes the usual Riemann integral.

b. Calculate

$$\lim_{n \to \infty} \frac{1}{n} \left(\sin\left(\frac{\pi}{n}\right) + \sin\left(\frac{2\pi}{n}\right) + \cdots + \sin\left(\frac{n\pi}{n}\right) \right).$$

Remark. The partition of $[0,1]$ in part *a* is "uniformly distributed". Such evaluations of Riemann integrals play an important role in number theory. This relationship was first developed by HERMANN WEYL in 1914 and 1916, and we refer to [304], [288] for developments in this area. Also, see Proposition 3.4.7 and Problem 3.29.

1.29. Construct a continuous function $f : \mathbb{R} \to \mathbb{R}$ that has a *proper* local minimum (or maximum) at each point in $\mathbb{Q}$. (We say that f has a *proper* local minimum at some point q if $f(q) < f(x)$ for all x in some neighborhood of q.)
[*Hint.* Start with the function $g(x) = \min(|x|, 1) - 1$, and let

$$f(x) = \sum_{n=1}^{\infty} g_n(x),$$

where $g_n(x) = A_n g((x - r_n)/w_n)$, $\{r_n\}$ denotes an enumeration of the rationals, and A_n, w_n are to be specified [377].]

1.30. Prove the following assertions for a given continuous function $f : [a, b] \to \mathbb{R}$.
 a. There is $F : [a, b] \to \mathbb{R}$ such that $F' = f$ on $[a, b]$, and
 b.

$$R\int_a^b f = F(b) - F(a).$$

Remark. This result implies that integration and differentiation can be considered to be inverse processes, i.e., $D\int = \int D = Id$, where D symbolizes differentiation and Id is the identity operator.

This is the *fundamental theorem of calculus* for the first 1900 years A.D. We shall generalize it into a deep result, the *Vitali–Lebesgue–Radon–Nikodym theorem*, which forms a major part of the Lebesgue theory.

1.31. Let $\{f_n : n = 1, \ldots\}$ be a sequence of real-valued, twice differentiable functions on $[a, b] \subseteq \mathbb{R}$. Assume that $f_n \to f$ uniformly, and that $\{f_n''\}$ is uniformly bounded in n and x. Prove that $f_n' \to f'$ uniformly, and that there exists $C > 0$ such that $|f'(x) - f'(y)| \leq C|x - y|$ for all $x, y \in [a, b]$. This property is called a *Lipschitz condition*, and functions that satisfy it are called *Lipschitz functions*.

1.32. VAN DER WAERDEN's everywhere continuous nowhere differentiable function W, mentioned in Section 1.3.4, is defined as

$$W(x) = \sum_{j=0}^{\infty} 10^{-j} f(10^j x), \quad x \in (0, 1),$$

where $f(x) = |x - k|$, $x \in [k - (1/2), k + (1/2)]$, $k \in \mathbb{Z}$. Prove that this function is everywhere continuous nowhere differentiable.
[*Hint.* Use results from Section 1.3.3 to prove the continuity of W. To show that W is not differentiable at any $a \in [0, 1]$, it is enough to find a sequence $\{x_n\}$ converging to a, such that $\lim_{n\to\infty}(W(x_n) - W(a))/(x_n - a)$ does not exist. Write $a = 0.a_1 \ldots a_n \ldots$, and define $x_n = 0.a_1 \ldots a_{n-1} b_n a_{n+1} \ldots$,

where $b_n = a_n + 1$ if $a_n \neq 4, 9$, and $b_n = a_n - 1$ if $a_n = 4$ or 9. Observe that for any n we have

$$f(10^j x_n) - f(10^j a) = \begin{cases} \pm 10^{j-n}, & \text{if } j = 0, \ldots, n-1, \\ 0, & \text{if } j \geq n. \end{cases}$$

Use this to calculate $(W(x_n) - W(a))/(x_n - a)$.]

1.33. Prove that the function defined in Example 1.3.26 is everywhere continuous nowhere differentiable.

1.34. Verify that the function $f : \mathbb{R} \to \mathbb{R}$ defined by

$$f(x) = \sum_{k=1}^{\infty} \frac{1}{2^k} g\left(2^{2^k} x\right)$$

is everywhere continuous nowhere differentiable, where $g : \mathbb{R} \to \mathbb{R}$ has period 4 and is defined by

$$g(x) = \begin{cases} 1 + x, & \text{if } \quad -2 \leq x \leq 0, \\ 1 - x, & \text{if } \quad 0 \leq x \leq 2. \end{cases}$$

1.35. Prove that an open interval cannot be the countable disjoint union of closed sets; cf. Problem 1.10.

1.36. Let $C \subseteq [0, 1]$ be the Cantor set. Is there an irrational number x such that

$$\{c + x : c \in C\} \subseteq \mathbb{R} \setminus \mathbb{Q}\ ?$$

1.37. Consider

$$\sum_{n=1}^{\infty} \prod_{j=1}^{n} r_j, \quad r_j \in \mathbb{Q} \cap (0, 1). \tag{1.17}$$

a. If $\{r_j : j = 1, \ldots\} = \mathbb{Q} \cap (0, 1)$ is enumerated à la CANTOR (see Problem 1.25), can you prove that (1.17) converges?

b. Find an enumeration of $\mathbb{Q} \cap (0, 1)$ such that (1.17) diverges.

2 Lebesgue Measure and General Measure Theory

2.1 The theory of measure prior to Lebesgue, and preliminaries

For the historical development of measure theory prior to LEBESGUE we refer to ARTHUR SCHOENFLIES' work on set theory (1900) and [226].

One of the first people to recognize the connection between an integration theory and a precise theory for measuring the "length" of sets was GIUSEPPE PEANO, and his work on these matters is contained in his book *Applicazione geometriche del calcolo infinitesimale*, published in 1887 in Torino. Taking a nonnegative, bounded function $f : [a,b] \to \mathbb{R}$, he let E be the point set bounded by the graph of f and the lines $x = a$, $x = b$, $y = 0$, and he then defined

$$c_i(E) = \sup \{a(A) : A \subseteq E, A \text{ a polygonal region}\},$$
$$c_e(E) = \inf \{a(A) : A \supseteq E, A \text{ a polygonal region}\},$$

where $a(A)$ is the area of A. In these terms he observed that *f is Riemann integrable if and only if E is "measurable", i.e., $c_e(E) = c_i(E)$.*

When c_e and c_i are equal we write c for their common value, and PEANO actually developed the theory of finitely additive set functions for such c.

Mind you, there was a great deal of activity in the general problem of measuring sets prior to PEANO. For example, using the fact that a closed set $E \subseteq \mathbb{R}$ can be written as

$$E = \left(\bigcup_{j=1}^{\infty} I_j \right)^{\sim},$$

where $\{I_j : j = 1, \ldots\}$ is a disjoint family of open intervals and $m(I_j)$ denotes the length of the interval I_j, CANTOR and IVAR BENDIXSON were led to define

$$\text{“}m(C)\text{''} = 1 - \sum_{j=1}^{\infty} m(I_j),$$

whose value, as we showed in Example 1.2.7, is 0.

Five years after PEANO's work, CAMILLE JORDAN also developed the theory of finitely additive measures, and, although there are no references to

J.J. Benedetto and W. Czaja, *Integration and Modern Analysis*, Birkhäuser Advanced Texts, DOI 10.1007/978-0-8176-4656-1_2,

PEANO, it seems likely that JORDAN knew of his work. In any case, mathematically, JORDAN carried finitely additive measure theory very far, and the relation between measure and integrability was quite clearly explicated in JORDAN's theory.

The next major step in the evolution of the fundamental ideas leading to the present notions of integral and measure was taken by ÉMILE BOREL from the time of his doctorate in 1894. To discuss his work we need to define the notion of a *σ-algebra.*

Definition 2.1.1. Rings and algebras
Let X be any set. The power set $\mathcal{P}(X)$ is the set of all subsets of X and it was initially defined in Problem 1.3.

a. We say that a collection of sets $\mathcal{R} \subseteq \mathcal{P}(X)$ is a *ring* if, for all $A, B \in \mathcal{R}$, the following properties are satisfied:

$$A \cap B \in \mathcal{R},$$
$$A \cup B \in \mathcal{R},$$
$$A \setminus B \in \mathcal{R}.$$

A ring $\mathcal{R}$ is a *σ-ring* if, for any sequence $\{A_n : n = 1, \ldots\} \subseteq \mathcal{R}$,

$$\bigcap_{n=1}^{\infty} A_n \in \mathcal{R}$$

and

$$\bigcup_{n=1}^{\infty} A_n \in \mathcal{R}.$$

b. We say that a ring (respectively, σ-ring) $\mathcal{A}$ is an *algebra* (respectively, *σ-algebra*) if

$$X \in \mathcal{A}.$$

In this case, $\emptyset \in \mathcal{A}$, and $A \in \mathcal{A}$ implies $A^{\sim} \in \mathcal{A}$.

Proposition 2.1.2. *For each collection $\mathcal{C} \subseteq \mathcal{P}(X)$ there exists the smallest ring (respectively, σ-ring, algebra, and σ-algebra) that contains $\mathcal{C}$.*

Proof. We shall only show the existence of the smallest ring containing $\mathcal{C}$. Let $\mathcal{F}$ be the family of all rings in $\mathcal{P}(X)$ that contain $\mathcal{C}$, and let $\mathcal{R} = \bigcap\{\mathcal{D} : \mathcal{D} \in \mathcal{F}\}$. Clearly $\mathcal{C} \subseteq \mathcal{R}$. Moreover, $\mathcal{R}$ is a ring. In fact, let $A, B \in \mathcal{R}$. Then, for each $\mathcal{D} \in \mathcal{F}$, $A, B \in \mathcal{D}$. Since $\mathcal{D}$ is a ring, $A \cup B, A \cap B, A \setminus B \in \mathcal{D}$. This is true for each $\mathcal{D} \in \mathcal{F}$, and so $A \cup B, A \cap B, A \setminus B \in \mathcal{R}$. By definition, $\mathcal{R}$ is the smallest ring containing $\mathcal{C}$. □

We say that the smallest ring (σ-ring, algebra, σ-algebra) that contains a given family $\mathcal{C}$ is *generated* by $\mathcal{C}$.

In light of the importance of open and closed sets we make the following definition.

Definition 2.1.3. Borel sets
The collection $\mathcal{B} = \mathcal{B}(X)$ of *Borel sets* in $X \subseteq \mathbb{R}$ is the smallest σ-algebra in $\mathcal{P}(X)$ containing the open sets in X.

If X is any topological space, defined in Appendix A.1, we define the collection $\mathcal{B} = \mathcal{B}(X)$ of *Borel sets* in X in the same way; $\mathcal{B}$ is the *Borel algebra*.

Example 2.1.4. Borel sets
a. We have defined $\mathcal{F}_\sigma$, G_δ, and $F_{\sigma\delta}$ sets in Section 1.3.2. These sets are all Borel sets, as are

$$\mathcal{G}_{\delta\sigma}, \mathcal{G}_{\delta\sigma\delta}, \ldots, \mathcal{F}_{\sigma\delta\sigma}, \mathcal{F}_{\sigma\delta\sigma\delta}, \ldots.$$

b. There are Borel sets $B \subseteq \mathbb{R}$ that are not of the form $\mathcal{F}_n$ or $\mathcal{G}_m$, where, for example,

$$\mathcal{F}_n = \mathcal{F}_{\sigma\delta\sigma\ldots}.$$

In fact, set

$$B = \bigcup_{n=1}^{\infty} A_n,$$

where A_n is an $\mathcal{F}_n$ but not an $\mathcal{F}_{n-1}$, e.g., [224], page 182.

c. There are Borel sets $B_1, B_2 \in \mathcal{B}(\mathbb{R})$ such that $B_1 - B_2 = \{x - y : x \in B_1, y \in B_2\} \notin \mathcal{B}(\mathbb{R})$ [401].

d. There is precisely a continuum of Borel sets in $\mathbb{R}$; see, e.g., [307].

Besides first proving the Heine–Borel theorem in the form of Theorem 1.2.10, BOREL gave a reasonable definition of measure 0 and stressed countable additivity for his measures. In fact, by dividing $[0, 1]$ into equal parts JORDAN had reached the conclusion that "$m(\mathbb{Q} \cap [0, 1])$"$= 1$. BOREL, on the other hand, attached to each $r_n \in \mathbb{Q} \cap [0, 1]$ the segment of length ε/n^2. Consequently, he concluded that "$m(\mathbb{Q} \cap [0, 1])$"$< \varepsilon \sum 1/n^2$ for each ε; and so "$m(\mathbb{Q} \cap [0, 1])$"$= 0$. This example led BOREL into his study of measure.

We shall see how countable additivity and integration are related very soon. The countable additivity of BOREL, as opposed to the finite additivity of PEANO–JORDAN, was crucial in LEBESGUE's theory for attaining many fundamental results.

Definition 2.1.5. Finitely additive and σ-additive set functions and measures
a. Let $\mathcal{A} \subseteq \mathcal{P}(X)$ and let μ be a set function

$$\mu : \mathcal{A} \to \mathbb{R}^+ \cup \{\infty\}.$$

Then μ is *finitely additive* on $\mathcal{A}$ if for each finite sequence $\{A_n : n = 1, \ldots, N\} \subseteq \mathcal{A}$ of mutually disjoint sets, we have

$$\mu\left(\bigcup_{n=1}^{N} A_n\right) = \sum_{n=1}^{N} \mu(A_n).$$

b. Let $\mathcal{A} \subseteq \mathcal{P}(X)$ and let μ be a set function

$$\mu : \mathcal{A} \to \mathbb{R}^+ \cup \{\infty\}.$$

Then μ is *σ-additive* or *countably additive* on $\mathcal{A}$ if, for each sequence $\{A_n : n = 1, \ldots\} \subseteq \mathcal{A}$ of mutually disjoint sets, we have

$$\mu\left(\bigcup_{n=1}^{\infty} A_n\right) = \sum_{n=1}^{\infty} \mu(A_n).$$

c. Let $\mathcal{A}$ be a σ-algebra on X. A nonnegative, σ-additive set function $\mu : \mathcal{A} \to \mathbb{R}^+ \cup \{\infty\}$ is called a *countably additive measure* or *σ-additive measure* or, just simply, a *measure*.

It is not a priori clear that a nontrivial measure exists. LEBESGUE solved this existence problem in a very strong way by constructing a σ-algebra $\mathcal{M}(\mathbb{R}) \supseteq \mathcal{B}(\mathbb{R})$ and a measure m on $\mathcal{M}(\mathbb{R})$ with the reasonable properties that

$$\forall\, x \in \mathbb{R} \text{ and } \forall\, A \in \mathcal{M}(\mathbb{R}), \quad m(A + x) = m(A), \tag{2.1}$$

and

$$\forall\, I \subseteq \mathbb{R},\ I \text{ an interval, } m(I) \text{ is the length of } I; \tag{2.2}$$

see Section 2.2 for a proof. We call m *Lebesgue measure*, and conditions (2.1) and (2.2) are specific to m.

2.2 The construction of Lebesgue measure on $\mathbb{R}$

We start our venture into the business of constructing Lebesgue measure by first considering the special case of the real line $\mathbb{R}$. The difference between this case and the general case of $\mathbb{R}^d$ is that the ordering of $\mathbb{R}$ enables us to make simple arguments for comparison between Lebesgue measure on $\mathbb{R}$, which we shall construct, and the notion of length of intervals. This, in turn, allows us to use the family of open sets of $\mathbb{R}$ as our starting point.

As we have indicated, if I is an interval then $m(I)$ will designate its length. For any $A \subseteq \mathbb{R}$ the *Lebesgue outer measure* of A is

$$m^*(A) = \inf \Big\{ \sum_{n=1}^{\infty} m(I_n) : A \subseteq \bigcup_{n=1}^{\infty} I_n \text{ and } \{I_n : n = 1, \ldots\} \text{ is a countable family of open intervals} \Big\}. \tag{2.3}$$

Clearly,

i. if $A \subseteq B$ then $m^*(A) \leq m^*(B)$,
ii. $m^*(\emptyset) = 0$.

Proposition 2.2.1. *Let $I \subseteq \mathbb{R}$ be an interval. Then $m^*(I) = m(I)$.*

Proof. Let $I = [a, b]$, $-\infty < a \le b < \infty$. For each $\varepsilon > 0$, $I \subseteq (a - \varepsilon, b + \varepsilon)$, and so

$$m^*(I) \le m((a - \varepsilon, b + \varepsilon)) = b - a + 2\varepsilon. \tag{2.4}$$

Since (2.4) is true for each $\varepsilon > 0$,

$$m^*(I) \le b - a.$$

To prove $m^*(I) \ge b - a$, we shall show that, for any sequence $\{I_n : n = 1, \ldots\}$ of open intervals,

$$I \subseteq \bigcup_{n=1}^{\infty} I_n \Longrightarrow \sum_{n=1}^{\infty} m(I_n) \ge b - a. \tag{2.5}$$

By the Heine–Borel theorem each such sequence $\{I_n : n = 1, \ldots\}$, for which $I \subseteq \bigcup I_n$, has a finite subcollection $I_{n_1}, \ldots, I_{n_k}$ covering I, and, in particular,

$$\sum_{n=1}^{\infty} m(I_n) \ge \sum_{j=1}^{k} m(I_{n_j}).$$

Consequently, we need only prove (2.5) for finite covers $\{J_1, \ldots, J_k\}$. Let $a \in \bigcup_{j=1}^{k} J_j$. Without loss of generality, we suppose $a \in J_1 = (a_1, b_1)$. Thus, if $b_1 \ge b$,

$$\sum_{j=1}^{k} m(J_j) \ge b_1 - a_1 > b - a,$$

in which case we are finished. If $a < b_1 < b$, then since $b_1 \in I$ there is $J_2 = (a_2, b_2)$ in our finite subcollection such that $a_2 < b_1 < b_2$. Again, if $b_2 \ge b$ we are finished. If $b_2 < b$ we choose $J_3 = (a_3, b_3)$ from our finite collection, etc. Since the collection is finite, this process ends with $J_n = (a_n, b_n)$, $n \le k$; and by hypothesis and construction we have $a_n < b < b_n$. Hence,

$$\sum_{j=1}^{k} m(J_j) \ge \sum_{j=1}^{n} (b_j - a_j) \tag{2.6}$$
$$= b_n - (a_n - b_{n-1}) - (a_{n-1} - b_{n-2}) - \cdots - (a_2 - b_1) - a_1.$$

Now $a_j < b_{j-1}$ (recall $a_2 < b_1 < b_2$, etc.), and so, from (2.6),

$$\sum_{j=1}^{k} m(J_j) > b_n - a_1 > b - a;$$

this gives (2.5).

For the case of a bounded interval I, take any $\varepsilon > 0$ and choose a closed interval $J \subseteq I$ for which

$$m(J) > m(I) - \varepsilon.$$

Hence,

$$m(I) - \varepsilon < m(J) = m^*(J) \leq m^*(I) \leq m^*(\bar{I}) = m(\bar{I}) = m(I);$$

that is, for each $\varepsilon > 0$,

$$m(I) - \varepsilon < m^*(I) \leq m(I),$$

and so $m(I) = m^*(I)$.

If I is an infinite interval then for each $r > 0$ there is a closed interval $J \subseteq I$ such that $m(J) = r$. Therefore, $m^*(I) \geq m^*(J) = m(J) = r$, and so, "letting $r \to \infty$", we have $m^*(I) = +\infty$. □

Proposition 2.2.2. *Let* $\{A_n : n = 1, \ldots\} \subseteq \mathcal{P}(\mathbb{R})$. *Then*

$$m^*\left(\bigcup_{n=1}^{\infty} A_n\right) \leq \sum_{n=1}^{\infty} m^*(A_n). \tag{2.7}$$

Proof. If $m^*(A_n) = \infty$ for some n we are finished. Therefore, assume $m^*(A_n) < \infty$ for each $n \in \mathbb{N}$.

For a given $\varepsilon > 0$ and n, there is a sequence $\{I_{n,i} : n = 1, \ldots, i = 1, \ldots\}$ of open intervals such that

$$A_n \subseteq \bigcup_{i=1}^{\infty} I_{n,i} \quad \text{and} \quad \sum_{i=1}^{\infty} m(I_{n,i}) < m^*(A_n) + 2^{-n}\varepsilon,$$

by the definition of m^*. Now, card $\{I_{n,i} : i, n\} \leq \aleph_0$ and $\bigcup_n A_n \subseteq \bigcup_{n,i} I_{n,i}$. Therefore,

$$\begin{aligned} m^*\left(\bigcup_{n=1}^{\infty} A_n\right) &\leq \sum_{n,i=1}^{\infty} m(I_{n,i}) = \sum_{n=1}^{\infty}\sum_{i=1}^{\infty} m(I_{n,i}) \\ &\leq \sum_{n=1}^{\infty} m^*(A_n) + \varepsilon \sum_{n=1}^{\infty} \frac{1}{2^n} = \sum_{n=1}^{\infty} m^*(A_n) + \varepsilon. \end{aligned}$$

Since this is true for each ε, we have (2.7). □

Taking A_n to be a one-point set in Proposition 2.2.2 we have that

i. if A is countable then $m^*(A) = 0$.

This fact, combined with Proposition 2.2.1, gives an ingenious though complicated proof that

ii. card $[0, 1] > \aleph_0$.

The following result is straightforward to prove and the proof is left as an exercise (Problem 2.1).

Proposition 2.2.3. *Let $A \subseteq \mathbb{R}$.*

a. *For each $\varepsilon > 0$ there is an open set $U \subseteq \mathbb{R}$ such that*

$$A \subseteq U \quad \text{and} \quad m^*(U) \leq m^*(A) + \varepsilon.$$

b. *There is a $\mathcal{G}_\delta$ set G such that*

$$A \subseteq G \quad \text{and} \quad m^*(A) = m^*(G).$$

We use the Constantin Carathéodory approach (1914) to define Lebesgue measure. A set $A \subseteq \mathbb{R}$ is *Lebesgue measurable* if

$$\forall\, E \subseteq \mathbb{R}, \;\; m^*(E) = m^*(E \cap A) + m^*(E \cap A^\sim).$$

Thus, A is Lebesgue measurable if "no matter how you cut it" (with any E), m^* is nicely additive just as we hoped it would be. It is clear that

i. if $m^*(A) = 0$ then A is Lebesgue measurable;

ii. A is Lebesgue measurable if and only if $A^\sim$ is Lebesgue measurable;

and

iii. if A_1, A_2 are Lebesgue measurable then $A_1 \cup A_2$ is Lebesgue measurable.

We shall use $\mathcal{M}(\mathbb{R})$ to denote the collection of all Lebesgue measurable subsets of $\mathbb{R}$. We have the following result by part *iii.*

Proposition 2.2.4. *$\mathcal{M}(\mathbb{R})$ is an algebra.*

More important is the following theorem.

Theorem 2.2.5. Lebesgue and Borel σ-algebras on $\mathbb{R}$

a. *$\mathcal{M}(\mathbb{R})$ is a σ-algebra.*

b. *$\mathcal{B}(\mathbb{R}) \subseteq \mathcal{M}(\mathbb{R})$.*

Proof. **a.***i.* We first show that if $E \subseteq \mathbb{R}$ and $\{A_1, \ldots, A_n\} \subseteq \mathcal{M}(\mathbb{R})$ is a disjoint family then

$$m^*\left(E \cap \left(\bigcup_{j=1}^{n} A_j\right)\right) = \sum_{j=1}^{n} m^*(E \cap A_j). \tag{2.8}$$

We shall prove (2.8) by induction, noting that it is obvious for $n = 1$. Assume that (2.8) is true for $A_1, \ldots, A_{n-1}$ and take $A_1, \ldots, A_n$. Then

$$E \cap \left(\bigcup_{j=1}^{n} A_j\right) \cap A_n = E \cap A_n \tag{2.9}$$

and

$$E \cap \left(\bigcup_{j=1}^{n} A_j\right) \cap A_n^\sim = E \cap \left(\bigcup_{j=1}^{n-1} A_j\right). \tag{2.10}$$

Equation (2.9) is clear, and (2.10) follows since $\left(\bigcup_{j=1}^n A_j\right) \cap A_n^{\sim} = \bigcup_{j=1}^n (A_j \cap A_j^{\sim}) = \left(\bigcup_{j=1}^{n-1} (A_j \cap A_n^{\sim})\right) \cup (A_n \cap A_n^{\sim}) = \bigcup_{j=1}^{n-1} A_j$, where we have used the fact that $\{A_1, \ldots, A_n\}$ is disjoint in the last step. Thus, $A_n \in \mathcal{M}(\mathbb{R})$ implies, from (2.9) and (2.10), that

$$\begin{aligned} m^*\left(E \cap \left(\bigcup_{j=1}^n A_j\right)\right) &= m^*(E \cap A_n) + m^*\left(E \cap \left(\bigcup_{j=1}^{n-1} A_j\right)\right) \\ &= m^*(E \cap A_n) + \sum_{j=1}^{n-1} m^*(E \cap A_j) \end{aligned}$$

by the induction hypothesis. Consequently, (2.8) holds.

a.ii. We now prove that $\mathcal{M}(\mathbb{R})$ is a σ-algebra. Hence, given $\{B_j : j = 1, \ldots\} \subseteq \mathcal{M}(\mathbb{R})$ we must show that $A = \bigcup_{j=1}^\infty B_j \in \mathcal{M}(\mathbb{R})$. Since $\mathcal{M}(\mathbb{R})$ is an algebra, there is a disjoint family $\{A_j : j = 1, \ldots\} \subseteq \mathcal{M}(\mathbb{R})$ such that

$$A = \bigcup_{j=1}^\infty B_j = \bigcup_{j=1}^\infty A_j.$$

In fact, let $A_1 = B_1$, $A_2 = B_2 \setminus B_1$, $A_3 = B_3 \setminus (B_1 \cup B_2), \ldots$.

Let $C_n = \bigcup_{j=1}^n A_j$, so that $C_n \in \mathcal{M}(\mathbb{R})$, again using the fact that $\mathcal{M}(\mathbb{R})$ is an algebra. Next note that $A^{\sim} \subseteq C_n^{\sim}$ because

$$C_n = \bigcup_{j=1}^n A_j \subseteq A.$$

Taking any $E \subseteq \mathbb{R}$ we calculate

$$m^*(E) = m^*(E \cap C_n) + m^*(E \cap C_n^{\sim}) \geq m^*(E \cap C_n) + m^*(E \cap A^{\sim}). \quad (2.11)$$

From part $a.i$ and the disjointness of $\{A_j : j = 1, \ldots\}$ we obtain

$$m^*(E \cap C_n) = m^*\left(E \cap \left(\bigcup_{j=1}^n A_j\right)\right) = \sum_{j=1}^n m^*(E \cap A_j). \quad (2.12)$$

Combining (2.11) and (2.12) gives

$$m^*(E) \geq \sum_{j=1}^n m^*(E \cap A_j) + m^*(E \cap A^{\sim}),$$

and, since the left-hand side is independent of n,

$$m^*(E) \geq \sum_{j=1}^\infty m^*(E \cap A_j) + m^*(E \cap A^{\sim}).$$

Thus, by the subadditivity of m^* (Proposition 2.2.2),

$$m^*(E) \geq m^*\left(E \cap \left(\bigcup_{j=1}^{\infty} A_j\right)\right) + m^*(E \cap A^{\sim}). \tag{2.13}$$

The opposite inequality to (2.13) is always true, and so

$$A \in \mathcal{M}(\mathbb{R}).$$

b.i. We first show that $(a, \infty) \in \mathcal{M}(\mathbb{R})$. Take $E \subseteq \mathbb{R}$ and set

$$E_1 = E \cap (a, \infty) \quad \text{and} \quad E_2 = E \cap (-\infty, a].$$

It is sufficient to prove that $\varepsilon + m^*(E) \geq m^*(E_1) + m^*(E_2)$ for each $\varepsilon > 0$, and without loss of generality (obviously) we take $m^*(E) < \infty$. Fix $\varepsilon > 0$. Since $m^*(E) < \infty$ there is a sequence $\{I_n : n = 1, \ldots\}$ of open intervals covering E such that

$$\sum_{n=1}^{\infty} m(I_n) \leq \varepsilon + m^*(E).$$

We observe that $I'_n = I_n \cap (a, \infty)$ and $I''_n = I_n \cap (-\infty, a]$ are intervals (possibly empty), and

$$m(I_n) = m(I'_n) + m(I''_n) = m^*(I'_n) + m^*(I''_n)$$

by Proposition 2.2.1. Clearly, $E_1 \subseteq \bigcup I'_n$ and $E_2 \subseteq \bigcup I''_n$. Consequently,

$$m^*(E_1) \leq m^*\left(\bigcup_{n=1}^{\infty} I'_n\right) \leq \sum_{n=1}^{\infty} m^*(I'_n)$$

and

$$m^*(E_2) \leq m^*\left(\bigcup_{n=1}^{\infty} I''_n\right) \leq \sum_{n=1}^{\infty} m^*(I''_n)$$

from Proposition 2.2.2. Therefore,

$$m^*(E_1) + m^*(E_2) \leq \sum_{n=1}^{\infty} (m^*(I'_n) + m^*(I''_n)) = \sum_{n=1}^{\infty} m^*(I_n) \leq \varepsilon + m^*(E).$$

b.ii. $\mathcal{B}(\mathbb{R})$ is the smallest σ-algebra containing each (a, ∞); but we have just proved that $\mathcal{M}(\mathbb{R})$ is a σ-algebra containing each (a, ∞). □

The following theorem is elementary to prove, and is the content of Problem 2.3.

Theorem 2.2.6. Properties of sequences of Lebesgue measurable sets
Let $\{A_n : n = 1, \ldots\} \subseteq \mathcal{M}(\mathbb{R})$.

a. $m\left(\bigcup_{n=1}^{\infty} A_n\right) \leq \sum_{n=1}^{\infty} m(A_n)$.

b. *If* $\{A_n : n = 1, \ldots\}$ *is a disjoint family, then* $m\left(\bigcup_{n=1}^{\infty} A_n\right) = \sum_{n=1}^{\infty} m(A_n)$.

c. *If* $A_1 \supseteq A_2 \supseteq A_3 \supseteq \cdots \supseteq A_n \supseteq \cdots$ *and* $m(A_1) < \infty$, *then* $m\left(\bigcap_{n=1}^{\infty} A_n\right) = \lim_{n\to\infty} m(A_n)$.

d. *If* $A_1 \subseteq A_2 \subseteq A_3 \subseteq \cdots \subseteq A_n \subseteq \cdots$, *then* $m\left(\bigcup_{n=1}^{\infty} A_n\right) = \lim_{n\to\infty} m(A_n)$.

We shall prove Theorem 2.2.6*c,d* in a more general setting in Theorem 2.4.3.

Definition 2.2.7. Lebesgue measure
The set function $m : \mathcal{M}(\mathbb{R}) \to \mathbb{R}^+ \cup \{\infty\}$, defined by

$$\forall\, A \in \mathcal{M}(\mathbb{R}), \quad m(A) = m^*(A),$$

is *Lebesgue measure* on $\mathbb{R}$.

The following result is a consequence of the definition of Lebesgue measure. However, its usefulness in problem-solving cannot be overemphasized; and so we also state it as a theorem.

Theorem 2.2.8. Measure zero
A set $A \in \mathcal{M}(\mathbb{R})$ *has Lebesgue measure zero if and only if*

$$\forall\, \varepsilon > 0,\ \exists\ \{I_n : I_n \text{ an open interval and } n = 1, \ldots\} \text{ such that}$$

$$A \subseteq \bigcup_{n=1}^{\infty} I_n \quad \text{and} \quad \sum_{n=1}^{\infty} m(I_n) < \varepsilon.$$

NORBERT WIENER (AMS semicentennial) has made an historical case for the definition of measure 0 based on considerations from statistical mechanics; cf. [93] by LENNART CARLESON.

The following result is interesting in light of the ruler function discussed in Section 1.3. We have only outlined its proof; cf. Problem 2.9.

Proposition 2.2.9. *There are a function* $f : [0,1] \to \mathbb{R}$ *and a set* $D \subseteq [0,1]$ *that is dense in* $[0,1]$ *such that* $D \in \mathcal{M}(\mathbb{R})$, f *is continuous on* D, f *is discontinuous on* $[0,1] \setminus D$, *and* $m(D) = 0$.

Proof. Let $\{r_n : n = 1, \ldots\}$ be an enumeration of $\mathbb{Q} \cap [0,1]$, denote the open interval about r_j of radius $1/(k2^j)$ by I_{jk}, and set

$$D = \bigcap_{k=3}^{\infty} U_k, \quad \text{where } U_k = \bigcup_{j=1}^{\infty} I_{jk}.$$

Then D is not of first category by an application of the Baire category theorem, and $m(D) = 0$, e.g., Theorem 2.2.6*c*. Define $f = \sum_{k\geq 3} f_k$, where

$$f_k(x) = \begin{cases} 0, & \text{if } x \in U_k, \\ 1/2^k, & \text{if } x \in [0,1] \setminus U_k. \end{cases}$$

□

Example 2.2.10. Measure of perfect symmetric sets
Because of Theorem 2.2.6*c*, if E is a perfect symmetric set then the definition of $m(E)$ that we gave in Chapter 1 is precisely the Lebesgue measure of E.

Proposition 2.2.11. *Let $A \subseteq \mathbb{R}$ have the property that there is $q \in (0,1)$ such that for all $(a,b) \subseteq \mathbb{R}$ there exist intervals $I_n = (a_n, b_n)$, $n = 1, \ldots,$ such that*

$$A \cap (a,b) \subseteq \bigcup_{n=1}^{\infty} I_n \quad \text{and} \quad \sum_{n=1}^{\infty} m(I_n) \leq q(b-a).$$

Then $m(A) = 0$.

Proof. It is sufficient to prove that $m(A \cap (a,b)) = 0$, since we have $m(A) = m(A \cap (\bigcup J_i)) \leq \sum m(A \cap J_i)$, where $\{J_i : i = 1, \ldots\}$ is a cover of $\mathbb{R}$ by open intervals. Now cover $A \cap (a,b)$ by intervals (a_n, b_n) having the property that $\sum (b_n - a_n) \leq q(b-a)$. Then cover each $A \cap (a_n, b_n)$ by a countable collection of intervals having total length $q(b_n - a_n)$. Thus, we have covered A by open intervals of total length L, where

$$L \leq q(b_1 - a_1) + q(b_2 - a_2) + \cdots = q((b_1 - a_1) + \cdots) \leq q(b-a).$$

Repeating this process we obtain the inequality $m(A) \leq q^n(b-a)$, and so $m(A) = 0$, since $0 < q < 1$. □

With regard to Proposition 2.2.3 we have the following theorem.

Theorem 2.2.12. Properties of Lebesgue outer measure
The following are equivalent.

a. $A \in \mathcal{M}(\mathbb{R})$.
b. $\forall\, \varepsilon > 0$, $\exists\, U \supseteq A$, *open, such that* $m^*(U \setminus A) < \varepsilon$.
c. $\forall\, \varepsilon > 0$, $\exists\, F \subseteq A$, *closed, such that* $m^*(A \setminus F) < \varepsilon$.
d. $\exists\, G$, *a* $\mathcal{G}_\delta$ *set, such that* $A \subseteq G$ *and* $m^*(G \setminus A) = 0$.
e. $\exists\, F$, *an* $\mathcal{F}_\sigma$ *set, such that* $F \subseteq A$ *and* $m^*(A \setminus F) = 0$.

The straightforward proof of Theorem 2.2.12 is left as an exercise (Problem 2.11).

Let $A \in \mathcal{M}(\mathbb{R})$. Suppose $G \subseteq A$ is Lebesgue measurable, so that $A \setminus G \in \mathcal{M}(\mathbb{R})$. Then

$$A = G \cup (A \setminus G).$$

Because of this and Theorem 2.2.12 we obtain the following important fact.

Theorem 2.2.13. Lebesgue measurable sets and Borel sets
Let $A \in \mathcal{M}(\mathbb{R})$. Then,

$$\exists\, B \in \mathcal{B}(\mathbb{R}) \text{ and } \exists\, E \in \mathcal{M}(\mathbb{R}),\ m(E) = 0, \text{ such that}$$
$$A = B \cup E,\ B \cap E = \emptyset, \text{ and } m(A) = m(B),$$

i.e., every Lebesgue measurable set is a Borel set up to a set of measure zero.

Example 2.2.14. Closed uncountable sets of irrationals
In Problem 1.4a we wanted to find a closed uncountable subset of the irrationals. This is most easily done by taking an open interval of radius $\varepsilon/2^n$ about the nth rational and then looking at the complement of the union of these intervals. Theorem 2.2.12c gives another proof, since $m^*([0,1] \cap (\mathbb{R} \setminus \mathbb{Q})) = 1$, i.e., take $A = [0,1] \cap (\mathbb{R} \setminus \mathbb{Q})$ and use the facts that its outer measure is positive and that it is measurable.

With regard to Problem 1.4b, if $X \in \mathcal{M}(\mathbb{R})$ and $m(X) > 0$, then we can find a closed uncountable set $F \subseteq X$ by the same reasoning, i.e., using Theorem 2.2.12c, since $m(X \setminus F) = m(X) - m(F)$.

Example 2.2.15. A set of measure 0
Let A be the set of those $x \in [0,1]$ whose decimal expansion consists of not more than 9 distinct digits. Then, card $A > \aleph_0$ and $m(A) = 0$. To prove this, let $S_j \subseteq A$ consist of those elements whose decimal expansions do not contain j. Thus, $\bigcup_0^9 S_j = A$. Now S_j is equivalent, bijectively, to $[0,1]$ when we view $x \in [0,1]$ as having an expansion to the base 9. Thus, card $S_j > \aleph_0$ and so card $A > \aleph_0$.

The proof that $m(A) = 0$ reduces to showing that for each j, $m(S_j) = 0$. Fix j and consider the ten intervals $[k/10, (k+1)/10)$, $k = 0, \ldots, 9$. Then throw away the $(j+1)$st interval and so dispense with all decimals whose first term is j; note that the length of this interval is $1/10$. Next we divide each of the remaining 9 intervals into 10 intervals, each of length $1/10^2$. From each of these divisions we throw away the $(j+1)$st interval and thus we have removed all decimals whose second term is j; at this stage, then, we dispense with lengths totaling $9/10^2$. Continuing this process we end up with precisely S_j; on the other hand, we have thrown away, altogether, lengths totaling

$$\frac{1}{10} + \frac{9}{10^2} + \frac{9^2}{10^3} + \cdots = \frac{1}{10}\left(\frac{1}{1-(9/10)}\right) = 1,$$

and so $m(S_j) = 0$.

We have indicated that there is a continuum of Borel subsets of $\mathbb{R}$ (Example 2.1.4). Assuming this for the moment, we can assert that $\mathcal{M}(\mathbb{R}) \setminus \mathcal{B}(\mathbb{R}) \neq \emptyset$. In fact, all subsets of the ternary Cantor set C are measurable. Since the Cantor set is uncountable, the number of subsets of C (i.e., the cardinality of the power set $\mathcal{P}(C)$) is strictly greater than a continuum, and so there exist measurable subsets of C that are not Borel. See Example 2.4.14 for another proof that $\mathcal{M}(\mathbb{R}) \setminus \mathcal{B}(\mathbb{R}) \neq \emptyset$.

In 1904, LEBESGUE [313] posed the *problem of measure*: *Does there exist a nontrivial σ-additive set function on $\mathcal{P}(\mathbb{R})$ that is translation-invariant on $\mathcal{P}(\mathbb{R})$ and satisfies* (2.2)? GIUSEPPE VITALI settled this question in the negative in 1905 [483]. In particular, this means that $\mathcal{P}(\mathbb{R}) \setminus \mathcal{M}(\mathbb{R}) \neq \emptyset$.

Example 2.2.16. Nonmeasurable sets

a. The following is our translation of VITALI's ingenious original argument of 1905 [483]. We include it, since the publication is difficult to obtain. Subsequently, there have been many other published proofs, e.g., [438], frequently similar, and sometimes more complicated.

On the problem of the measure of subsets of a line by G. VITALI.

The problem of [measuring all bounded] subsets $A \subseteq \mathbb{R}$ is that of determining, for every such A, a positive number $\mu(A)$ that could be called the *measure* of A. This measure should satisfy the following properties (see [313], page 103):

i. $\forall\, x \in \mathbb{R}$ and $\forall\, A \subseteq \mathbb{R}$, bounded,

$$\mu(A) = \mu(A + x);$$

ii. $\forall\, \{A_n \subseteq \mathbb{R} : A_n \text{ bounded}\}$, a disjoint sequence,

$$\sum_{n=1}^{\infty} \mu(A_n) = \mu\left(\bigcup_{n=1}^{\infty} A_n\right);$$

iii. $\mu((0,1)) = 1$.

Let $x \in \mathbb{R}$. The points of $\mathbb{R}$ that differ from x by some element of $\mathbb{Q}$ form a denumerable set $A_x \subseteq \mathbb{R}$. If A_{x_1} and A_{x_2} are two such sets then

$$A_{x_1} \cap A_{x_2} = \emptyset \quad \text{or} \quad A_{x_1} = A_{x_2}. \tag{2.14}$$

Let us consider the various sets A_x, well defined by (2.14), and let $\mathcal{G}$ be the family of these sets. If $r \in \mathbb{R}$, then there is a unique element $A \in \mathcal{G}$, i.e., A is some A_x, for which $r \in A$. For $A \in \mathcal{G}$, choose a point $r_A \in (0, 1/2)$ for which $r_A \in A$; and let us denote by V_0 the set of such points r_A. If $q \in \mathbb{Q}$, we denote by V_q the set of points $r_A + q$, i.e.,

$$V_q = \{r_A + q \in (0, 1/2) + q : A \in \mathcal{G}\}.$$

The sets V_q corresponding to various $q \in \mathbb{Q}$ are pairwise disjoint, there are countably infinitely many of them, and by the translation-invariance [property *i*] they all have the same measure.

The sets

$$V_0, V_{1/2}, V_{1/3}, V_{1/4}, \ldots \subseteq (0, 1);$$

and therefore their union has measure $m \leq 1$ [by properties *ii* and *iii*, if, in fact, the V_q are "measurable"].

However,

$$m = \mu(V_0) + \sum_{n=2}^{\infty} \mu(V_{1/n}) = \lim_{n\to\infty} n\mu(V_0)$$

must also hold, and therefore $\mu(V_0) = 0$. But then the union of all the V_q, $q \in \mathbb{Q}$, must likewise have measure 0. However, this union is $\mathbb{R}$ and therefore must have infinite measure. Thus, we may conclude that *it is impossible to define* $\mu : \mathcal{P}(\mathbb{R}) \to \mathbb{R}^+$ *satisfying the properties i, ii, iii.*

b. Next we include the VITALI example written in more modern language.

For $x, y \in \mathbb{R}$ we write $x \sim y$ if $x - y \in \mathbb{Q}$. Clearly, "$\sim$" is an *equivalence relation*, i.e., it is

i. Reflexive: $\forall\, x \in \mathbb{R},\ x \sim x$,

ii. Symmetric: $\forall\, x, y \in \mathbb{R},\ x \sim y \Longrightarrow y \sim x$,

iii. Transitive: $\forall\, x, y, z \in \mathbb{R},\ x \sim y$ and $y \sim z \Longrightarrow x \sim z$.

The *equivalence class* corresponding to $x \in \mathbb{R}$ is $\{y \in \mathbb{R} : y \sim x\}$.

We now make explicit use of the axiom of choice and define A to be a subset of $(0, 1)$ that contains exactly one point from each equivalence class, noting, for each $x \in \mathbb{R}$, that an equivalence class is of the form $A_x = x + \mathbb{Q}$. We shall prove that $A \notin \mathcal{M}(\mathbb{R})$. Clearly,

$$\text{if } x \in (0,1) \text{ then } \exists\, r \in \mathbb{Q} \cap (-1,1) \text{ such that } x \in A + r;$$

also, by the definition of A we argue by contradiction to prove that

$$\text{if } r, s \in \mathbb{Q},\ r \neq s, \text{ then } (A + r) \cap (A + s) = \emptyset.$$

Assume $A \in \mathcal{M}(\mathbb{R})$ and define

$$S = \bigcup \{r + A : r \in \mathbb{Q} \cap (-1,1)\}.$$

From the translation invariance and the fact that $(A + r) \cap (A + s) = \emptyset$ for $r \neq s$, where $r, s \in \mathbb{Q}$, we have that $S \in \mathcal{M}(\mathbb{R})$, and

$$m(S) = \infty \quad \text{or} \quad m(S) = 0.$$

Since $A \subseteq (0, 1)$, $S \subseteq (-1, 2)$, and so $m(S) \leq 3$. Hence, $m(S) = 0$. On the other hand, $(0, 1) \subseteq S$; and so $m(S) \geq 1$, the desired contradiction. Therefore, $A \notin \mathcal{M}(\mathbb{R})$, i.e., $\mathcal{M}(\mathbb{R})$ cannot be all of $\mathcal{P}(\mathbb{R})$ if m is to satisfy (2.2) as well as being translation invariant.

The following result is true in any measure space (see Section 2.4) having nonmeasurable sets, and indicates how difficult it is to approximate nonmeasurable sets with measurable ones.

Proposition 2.2.17. *Let* $E \subseteq \mathbb{R}$ *be nonmeasurable. There is* $\varepsilon > 0$ *such that if* $E \subseteq A$ *and* $E^{\sim} \subseteq B$, *where* A *and* B *are measurable, then* $m(A \cap B) \geq \varepsilon$.

Proof. Assume that the result is false. Then for each n there are sets G_n, $D_n \in \mathcal{M}(\mathbb{R})$ such that

$$E \subseteq G_n, \ E^\sim \subseteq D_n, \text{ and } m(G_n \cap D_n) < 1/n.$$

The sets $G = \bigcap G_n$ and $D = \bigcap D_n$ are Lebesgue measurable, $E \subseteq G, E^\sim \subseteq D$, and $m(G \cap D) = 0$. Since $G \in \mathcal{M}(\mathbb{R})$, $m^*(S) = m^*(S \cap G) + m^*(S \cap G^\sim)$ for each $S \subseteq \mathbb{R}$. Now, $D \subseteq (G \cap D) \cup G^\sim$, and so

$$S \cap D \subseteq (S \cap G \cap D) \cup (S \cap G^\sim).$$

This implies that

$$\begin{aligned} m^*(S \cap D) &\leq m^*(S \cap G^\sim) + m^*((S \cap G) \cap D) \\ &\leq m^*(S \cap G^\sim) + m^*(G \cap D) = m^*(S \cap G^\sim). \end{aligned}$$

Consequently, $m^*(S \cap D) + m^*(S \cap G) \leq m^*(S \cap G^\sim) + m^*(S \cap G) = m^*(S)$. Observe that

$$E \subseteq G \Longrightarrow m^*(S \cap E) \leq m^*(S \cap G)$$

and

$$E^\sim \subseteq D \Longrightarrow m^*(S \cap E^\sim) \leq m^*(S \cap D).$$

Thus, for all $S \subseteq \mathbb{R}$,

$$m^*(S \cap E) + m^*(S \cap E^\sim) \leq m^*(S),$$

and so $E \in \mathcal{M}(\mathbb{R})$, a contradiction. □

2.3 The existence of Lebesgue measure on $\mathbb{R}^d$

In the previous section we have established the existence of Lebesgue measure m on $\mathbb{R}$. This measure is translation-invariant, it measures all Borel sets, and the Lebesgue measure of an interval is equal to its length.

In this section we shall construct an analogue of Lebesgue measure on $\mathbb{R}$ for $\mathbb{R}^d$ for any $d \geq 1$. The procedure is general in the sense that we shall actually give a method of construction of a family of measures on $\mathbb{R}^d$. Then, a choice of a specific ring and a specific set function on that ring will yield the desired analogue of Lebesgue measure. In this context, note that the family of open sets in $\mathbb{R}^d$ does not form a ring of subsets of $\mathbb{R}^d$.

Remark. We shall show that it is possible to extend an arbitrary σ-additive function on a ring $\mathcal{R}$ to a σ-additive function on a σ-ring generated by $\mathcal{R}$. Thus, formally, our construction is not a construction of an arbitrary general measure, which by definition is a σ-additive function on a σ-algebra. However, this approach combined with the properties of the standard volume function on $\mathbb{R}^d$ will yield Lebesgue measure on $\mathbb{R}^d$.

Let $\mathcal{R} \subseteq \mathcal{P}(\mathbb{R}^d)$ be a ring and let $\mu : \mathcal{R} \to \mathbb{R}^+ \cup \{\infty\}$ be a σ-additive set function. Our goal is to extend the function μ from the ring $\mathcal{R}$ to a σ-ring $\mathcal{R}_\sigma$ that contains $\mathcal{R}$. As before, for $A \subseteq \mathbb{R}^d$ we define the *outer measure* of A as

$$\mu^*(A) = \inf_{\{I_n\}} \sum_{n=1}^{\infty} \mu(I_n), \tag{2.15}$$

where the infimum is taken over all countable families $\{I_n\}$ of elements of the ring $\mathcal{R}$, which form a cover of the set A, i.e., $A \subseteq \bigcup_n I_n$ and $I_n \in \mathcal{R}$, $n = 1, \ldots$; cf. (2.3).

Example 2.3.1. Note that depending on the choice of the set X and the ring $\mathcal{R}$, it may happen that it will not be possible to define the outer measure on all of $\mathcal{P}(X)$. Consider X to be any uncountable set and let

$$\mathcal{R} = \{A \subseteq X : \operatorname{card} A < \infty\},$$

the ring of all finite sets, and let $\mu : \mathcal{R} \to \mathbb{R}^+$ be defined as

$$\mu(A) = \operatorname{card} A.$$

Then the outer measure μ^* can be extended only to the collection of countable subsets of X.

However, it is not difficult to check that the collection

$$\mathcal{H}(\mathcal{R}) = \left\{ A \subseteq \mathbb{R}^d : \; A \subseteq \bigcup_{n=1}^{\infty} I_n, \; I_n \in \mathcal{R} \right\}$$

is a σ-ring. Thus, μ^* is defined on the σ-ring $\mathcal{H}(\mathcal{R}) \subseteq \mathcal{P}(\mathbb{R}^d)$.

The σ-ring $\mathcal{H}(\mathcal{R})$ has the following property: if $A \subseteq B$ and $B \in \mathcal{H}(\mathcal{R})$ then $A \in \mathcal{H}(\mathcal{R})$. This explains the use of letter "h", as in "heritage". If $\mathcal{R} \subseteq \mathcal{P}(\mathbb{R}^d)$ is a ring and $\mathcal{R}_\sigma$ is the σ-ring generated by $\mathcal{R}$, then $\mathcal{R}_\sigma \subseteq \mathcal{H}(\mathcal{R})$.

Remark. We note that if the ring $\mathcal{R}$ contains all open subsets of $\mathbb{R}^d$, then $\mathcal{H}(\mathcal{R}) = \mathcal{P}(\mathbb{R}^d)$. In this case, the simpler arguments of Section 2.2 can be used to prove the existence of Lebesgue measure on $\mathbb{R}^d$. However, in the case of extensions of a ring to a σ-algebra for arbitrary measures in $\mathbb{R}^d$, as well as for the general measure theory developed in Section 2.4, we need the notion of the ring $\mathcal{H}(\mathcal{R})$.

Proposition 2.3.2. *Let $\mu : \mathcal{R} \to \mathbb{R}^+ \cup \{+\infty\}$ be a σ-additive set function on a ring $\mathcal{R} \subseteq \mathcal{P}(\mathbb{R}^d)$ and let $\{A_n : n = 1, \ldots\} \subseteq \mathcal{H}(\mathcal{R})$. Then*

$$\mu^*\left(\bigcup_{n=1}^{\infty} A_n\right) \leq \sum_{n=1}^{\infty} \mu^*(A_n). \tag{2.16}$$

The proof of Proposition 2.3.2 follows the proof of Proposition 2.2.2.

Proposition 2.3.3. *Let $\mu : \mathcal{R} \to \mathbb{R}^+ \cup \{+\infty\}$ be a σ-additive set function on a ring $\mathcal{R} \subseteq \mathcal{P}(\mathbb{R}^d)$. If $A \in \mathcal{R}$ then $\mu^*(A) = \mu(A)$.*

Proof. Clearly $A \subseteq A$ and thus $\mu^*(A) \leq \mu(A)$. To prove the opposite inequality, consider a collection $\{I_n : n = 1, \dots\} \subseteq \mathcal{R}$ such that $A \subseteq \bigcup_{n=1}^{\infty} I_n$. From the σ-additivity of the set function μ on $\mathcal{R}$ it follows that, for any such collection $\{I_n\}$,

$$\mu(A) \leq \sum_{n=1}^{\infty} \mu(I_n);$$

see Problem 2.14. Therefore, by the definition of outer measure we have $\mu(A) \leq \mu^*(A)$. □

We again use CARATHÉODORY's approach to define measurable sets. Let $\mu : \mathcal{R} \to \mathbb{R}^+ \cup \{+\infty\}$ be a σ-additive set function on a ring $\mathcal{R} \subseteq \mathcal{P}(\mathbb{R}^d)$ and let $A \in \mathcal{H}(\mathcal{R})$. Then A is *measurable* with respect to μ and $\mathcal{R}$ if

$$\forall\, E \in \mathcal{H}(\mathcal{R}),\ \mu^*(E) = \mu^*(E \cap A) + \mu^*(E \cap A^{\sim}).$$

Let $\mathcal{A}$ denote the family of all sets measurable with respect to μ and $\mathcal{R}$.

By the definition of the outer measure μ^*,

$$A,\, B \in \mathcal{H}(\mathcal{R}) \text{ and } A \subseteq B \implies \mu^*(A) \leq \mu^*(B).$$

Proposition 2.3.4. *Let $\mu : \mathcal{R} \to \mathbb{R}^+ \cup \{+\infty\}$ be a σ-additive set function on a ring $\mathcal{R} \subseteq \mathcal{P}(\mathbb{R}^d)$. The elements of $\mathcal{R}$ are measurable with respect to μ and $\mathcal{R}$, i.e., $\mathcal{R} \subseteq \mathcal{A}$.*

Proof. From Proposition 2.3.2 it is clear that $\mu^*(E) \leq \mu^*(E \cap A) + \mu^*(E \cap A^{\sim})$. To prove the opposite inequality, fix $\varepsilon > 0$ and take a collection $\{I_n : n = 1, \dots\} \subseteq \mathcal{R}$ such that $E \subseteq \bigcup_{n=1}^{\infty} I_n$ and

$$\sum_{n=1}^{\infty} \mu(I_n) < \mu^*(E) + \varepsilon.$$

Since $E \cap A \subseteq \bigcup (I_n \cap A)$ and $E \cap A^{\sim} \subseteq \bigcup (I_n \setminus A)$, we obtain

$$\begin{aligned}\mu^*(E) &> \sum_{n=1}^{\infty} \mu(I_n) - \varepsilon = \sum_{n=1}^{\infty} (\mu(I_n \cap A) + \mu(I_n \setminus A)) - \varepsilon \\ &\geq \mu^*(E \cap A) + \mu^*(E \cap A^{\sim}) - \varepsilon.\end{aligned}$$

This is true for each ε, and so $\mu^*(E) \geq \mu^*(E \cap A) + \mu^*(E \cap A^{\sim})$. □

Theorem 2.3.5. $\mathcal{A}$ is a ring

Let $\mu : \mathcal{R} \to \mathbb{R}^+ \cup \{+\infty\}$ be a σ-additive set function on a ring $\mathcal{R} \subseteq \mathcal{P}(\mathbb{R}^d)$. Then $\mathcal{A}$ is a ring.

Proof. We need to show only that, for any two measurable sets A, $B \in \mathcal{A}$, we have $A \cup B, A \setminus B \in \mathcal{A}$.

For any $E \in \mathcal{H}(\mathcal{R})$,

$$\begin{aligned}\mu^*(E) &= \mu^*(E \cap A) + \mu^*(E \cap A^\sim)\\ &= \mu^*(E \cap A \cap B) + \mu^*(E \cap A \cap B^\sim) + \mu^*(E \cap A^\sim)\\ &\geq \mu^*(E \cap A \cap B^\sim) + \mu^*(E \cap (A \cap B) \cup (E \cap A^\sim))\\ &= \mu^*(E \cap (A \setminus B)) + \mu^*(E \cap (A \setminus B)^\sim).\end{aligned}$$

Since the reverse inequality is always true, it follows that $A \setminus B \in \mathcal{A}$.

In a similar manner we obtain

$$\mu^*(E) = \mu^*(E \cap A \cap B) + \mu^*(E \cap A \cap B^\sim) + \mu^*(E \cap A^\sim \cap B) + \mu^*(E \cap A^\sim \cap B^\sim),$$

and, replacing E with $E \cap (A \cup B)$, we have that

$$\mu^*(E \cap (A \cup B)) = \mu^*(E \cap A \cap B) + \mu^*(E \cap A \cap B^\sim) + \mu^*(E \cap A^\sim \cap B).$$

Thus,

$$\mu^*(E) = \mu^*(E \cap (A \cup B)) + \mu^*(E \cap (A \cup B)^\sim),$$

and so $A \cup B \in \mathcal{A}$. □

To finish the construction of a measure on $\mathbb{R}^d$, we need the following observation: if $A_1, \ldots, A_n \in \mathcal{A}$ are pairwise disjoint and if $E \in \mathcal{H}(\mathcal{R})$, then

$$\mu^*\left(E \cap \left(\bigcup_{j=1}^{n} A_j\right)\right) = \sum_{j=1}^{n} \mu^*(E \cap A_j). \tag{2.17}$$

We leave the proof as an exercise (Problem 2.15).

Theorem 2.3.6. $\mathcal{A}$ is a σ-ring

Let $\mu : \mathcal{R} \to \mathbb{R}^+ \cup \{+\infty\}$ be a σ-additive set function on a ring $\mathcal{R} \subseteq \mathcal{P}(\mathbb{R}^d)$. Then $\mathcal{A}$ is a σ-ring.

Proof. In view of Theorem 2.3.5 we need to show only that $\mathcal{A}$ is closed under countable disjoint unions; see Problem 2.16.

Let $E \in \mathcal{H}(\mathcal{R})$, let $\{A_n : n = 1, \ldots\} \subseteq \mathcal{A}$ be a sequence of pairwise disjoint measurable sets, and let $A = \bigcup_{n=1}^{\infty} A_n$. We may assume without loss of generality that $\mu^*(E \cap A) < \infty$. Indeed, if $\mu^*(E \cap A) = \infty$ then $\mu^*(E) = \infty$, and A is measurable.

From Proposition 2.3.2, we have

$$\mu^*(E \cap A) \leq \sum_{n=1}^{\infty} \mu^*(E \cap A_n).$$

Now it follows from (2.17) that for every $\varepsilon > 0$ there exists N such that

$$\mu^*(E \cap A) \leq \sum_{n=1}^{N} \mu^*(E \cap A_n) + \varepsilon = \mu^*\left(E \cap \left(\bigcup_{n=1}^{N} A_n\right)\right) + \varepsilon.$$

Because of this and the fact that $\mathcal{A}$ is a ring, in particular $\bigcup_{n=1}^{N} A_n \in \mathcal{A}$, we have

$$\begin{aligned}
&\mu^*(E \cap A) + \mu^*(E \cap A^{\sim}) \\
&\leq \mu^*\left(E \cap \left(\bigcup_{n=1}^{N} A_n\right)\right) + \varepsilon + \mu^*\left(E \cap \left(\bigcup_{n=1}^{N} A_n\right)^{\sim}\right) \\
&= \mu^*(E) + \varepsilon.
\end{aligned}$$

Since this is true for each ε, $\mu^*(E) \geq \mu^*(E \cap A) + \mu^*(E \cap A^{\sim})$. The opposite inequality is easily verified, and so $A = \bigcup_{n=1}^{\infty} A_n \in \mathcal{A}$. □

Theorem 2.3.7. The σ-additive function μ^* on the σ-ring $\mathcal{A}$
Let $\mu : \mathcal{R} \to \mathbb{R}^+ \cup \{+\infty\}$ be a σ-additive set function on a ring $\mathcal{R} \subseteq \mathcal{P}(\mathbb{R}^d)$. Then μ^* is σ-additive on the σ-ring $\mathcal{A}$.

Proof. We shall prove that μ^* is σ-additive on $\mathcal{A}$. Let $A = \bigcup_{n=1}^{\infty} A_n$, where each $A_n \in \mathcal{A}$ and the A_ns are pairwise disjoint. We shall show that for any $E \in \mathcal{H}(\mathcal{R})$ we have

$$\mu^*(E \cap A) = \sum_{n=1}^{\infty} \mu^*(E \cap A_n). \tag{2.18}$$

Then, substituting A for E, we shall have the desired σ-additivity,

$$\mu^*(A) = \sum_{n=1}^{\infty} \mu^*(A_n).$$

Since $E \cap A = E \cap (\bigcup_{n=1}^{\infty} A_n)$, Proposition 2.3.2 implies

$$\mu^*(E \cap A) \leq \sum_{n=1}^{\infty} \mu^*(E \cap A_n),$$

and we have one direction of (2.18). For the opposite inequality, note that, for every N,

$$\sum_{n=1}^{N} \mu^*(E \cap A_n) = \mu^*\left(E \cap \left(\bigcup_{n=1}^{N} A_n\right)\right) \leq \mu^*(E \cap A).$$

Thus,

$$\sum_{n=1}^{\infty} \mu^*(E \cap A_n) \leq \mu^*(E \cap A).$$ □

Therefore, we have extended a nonnegative, σ-additive set function μ from a ring $\mathcal{R}$ to a σ-ring $\mathcal{A}$. This process is in fact parallel to our construction of Lebesgue measure on $\mathbb{R}$, where the set function was defined as the length of an interval.

Example 2.3.8. Volumes of parallelepipeds
Let $\mathcal{Q}$ be the collection of parallelepipeds in $\mathbb{R}^d$, i.e., sets of the form $Q = (a_1, b_1] \times \cdots \times (a_d, b_d]$, for any real $a_1 < b_1, \ldots, a_d < b_d$. Let $\mathcal{A}_{\mathcal{Q}}$ be the family of all finite unions of pairwise disjoint collections of such parallelepipeds. Then $\mathcal{A}_{\mathcal{Q}}$ is a ring.

Let $m^d : \mathcal{A}_{\mathcal{Q}} \to \mathbb{R}^+$ be defined as

$$m^d\left(\bigcup_{j=1}^{n} Q_j\right) = \sum_{j=1}^{n} \text{volume}\,(Q_j) = \sum_{j=1}^{n}\prod_{k=1}^{d}(b_k^{(j)} - a_k^{(j)}),$$

where $Q_j = (a_1^{(j)}, b_1^{(j)}] \times \cdots \times (a_d^{(j)}, b_d^{(j)}]$, $j = 1, \ldots, n$, are disjoint and where $\bigcup_{j=1}^{n} Q_j \in \mathcal{A}_{\mathcal{Q}}$. It is not difficult to see that m^d is a σ-additive set function on $\mathcal{A}_{\mathcal{Q}}$. Thus, we define the *Lebesgue outer measure* m^{d*}, according to (2.15). Then m^{d*} is defined on the corresponding σ-ring $\mathcal{H}(\mathcal{A}_{\mathcal{Q}})$, which in this case is equal to $\mathcal{P}(\mathbb{R}^d)$. Therefore, $\mathcal{H}(\mathcal{A}_{\mathcal{Q}})$ is a σ-algebra. In particular, when $d = 1$, the definition of the outer measure m^{1*} coincides with the definition of the Lebesgue outer measure on $\mathbb{R}$ in Section 2.1.

We now apply Theorem 2.3.7 to the setting of Example 2.3.8. In particular, Lebesgue outer measure m^{d*} is σ-additive on a σ-ring $\mathcal{M}(\mathbb{R}^d)$. Since $\mathbb{R}^d \in \mathcal{H}(\mathcal{A}_{\mathcal{Q}}) = \mathcal{P}(\mathbb{R}^d)$, we can write

$$\forall\, E \in \mathcal{H}(\mathcal{A}_{\mathcal{Q}}), \quad m^{d*}(E \cap \mathbb{R}^d) + m^{d*}(E \setminus \mathbb{R}^d) = m^{d*}(E) + m^{d*}(\emptyset).$$

Clearly $m^{d*}(\emptyset) = 0$ and so $\mathbb{R}^d \in \mathcal{M}(\mathbb{R}^d)$, which implies that $\mathcal{M}(\mathbb{R}^d)$ *is in fact a σ-algebra*. Therefore, we have the following result.

Theorem 2.3.9. Lebesgue measure on $\mathbb{R}^d$
There exist a σ-algebra $\mathcal{M}(\mathbb{R}^d) \supseteq \mathcal{B}(\mathbb{R}^d)$ and a measure m^d on $\mathcal{M}(\mathbb{R}^d)$ such that

$$\forall\, x \in \mathbb{R}^d \text{ and } \forall\, A \in \mathcal{M}(\mathbb{R}^d), \quad m^d(A + x) = m^d(A),$$

and

$$\forall\, Q \subseteq \mathbb{R}^d,\ Q \text{ a cube},\ m^d(Q) \text{ is the volume of } Q.$$

A measure that extends the set function m^d from Example 2.3.8, defined on $\mathcal{A}_{\mathcal{Q}}$, to the σ-algebra $\mathcal{M}(\mathbb{R}^d)$ of sets measurable with respect to $\mathcal{A}_{\mathcal{Q}}$ is unique and is called *Lebesgue measure* on $\mathbb{R}^d$. It is obviously translation invariant, and the σ-algebra $\mathcal{M}(\mathbb{R}^d)$ contains all Borel subsets of $\mathbb{R}^d$. The uniqueness follows from Problem 2.20; cf. the more general uniqueness result asserted in Problem 2.19.

Naturally, for $d = 1$, $m^1 = m$, where m is Lebesgue measure as defined in Section 2.2. In fact, our construction of the Lebesgue measure m^d is analogous to our treatment of measure for $\mathbb{R}$, differing only in the fact that we were using d-dimensional "cubes" instead of open intervals (which do not form a ring of subsets of $\mathbb{R}$).

Example 2.3.10. Lebesgue–Stieltjes measure associated with *f*
Let $f : \mathbb{R} \to \mathbb{R}$ be an increasing and right-continuous function. Let $Q = (a_1, b_1] \times \cdots \times (a_d, b_d] \subseteq \mathbb{R}^d$, and define the set function

$$\mu_f(Q) = (f(b_1) - f(a_1)) \cdots (f(b_d) - f(a_d)).$$

It is not difficult to see that μ_f is a σ-additive set function on the ring $\mathcal{Q}$ of disjoint unions of parallelepipeds in $\mathbb{R}^d$ (Problem 2.21). The procedure culminating in Theorem 2.3.7 allows us to extend μ_f to a measure on a σ-algebra of measurable sets generated by $\mathcal{A}_{\mathcal{Q}}$. This measure, also denoted by μ_f, is the *Lebesgue–Stieltjes measure* associated with f (see Section 3.5), and *Lebesgue measure is the Lebesgue–Stieltjes measure associated with* $f = 1$. Lebesgue–Stieltjes measures will be identified with Radon measures through the Riesz representation theorem in Section 7.5.

Systematic treatments of measure constructions on $\mathbb{R}^d$, from various points of view, are given in the books on Lebesgue integration by EDGAR ASPLUND and LUTZ BUNGART [15], JACQUES DIXMIER [142] (course at the Sorbonne), WENDELL H. FLEMING [179], pages 136–205, B. L. GUREVICH and GEORGII E. SHILOV [208], FRANK JONES [264], ELLIOT H. LIEB and MICHAEL LOSS [321], WALTER RUDIN [405], pages 49–52, KENNAN T. SMITH [445], RICHARD L. WHEEDEN and ANTONI ZYGMUND [502], and JOHN H. WILLIAMSON [512]. Also, see [135], [180], [287], [302].

Example 2.3.11. Lebesgue measure of the unit ball
The measure of the unit ball

$$B_d = \left\{ x \in \mathbb{R}^d : \|x\| = \left(\sum_{j=1}^d x_j^2 \right)^{1/2} \leq 1 \right\}$$

is

$$m^d(B_d) = \frac{\pi^{d/2}}{(d/2)\Gamma(d/2)}, \quad d = 1, 2, \ldots,$$

where the *Euler gamma function* is defined as

$$\Gamma(x) = \int_0^\infty t^{x-1} e^{-t} dt, \ x > 0.$$

The proof is by induction. First we observe that $m(B_1) = 2$. Next, we calculate

$$\frac{m^d(B_d)}{m^{d-1}(B_{d-1})} = 2\int_0^1 (1-x^2)^{(d-1)/2}dx.$$

Using a change of variables we obtain

$$\frac{m^d(B_d)}{m^{d-1}(B_{d-1})} = \int_0^1 y^{-1/2}(1-y)^{(d-1)/2}dy = \beta\left(\frac{1}{2}, \frac{d+1}{2}\right),$$

where the *Euler beta function* is

$$\beta(x,y) = 2\int_0^{\pi/2} \cos^{2x-1}(\theta)\sin^{2y-1}(\theta)\, d\theta = \frac{\Gamma(x)\Gamma(y)}{\Gamma(x+y)}.$$

Thus,

$$m^d(B_d) = m^{d-1}(B_{d-1})\frac{\Gamma(\frac{1}{2})\Gamma(\frac{d+1}{2})}{\Gamma(\frac{d}{2}+1)}.$$

In order to finish the proof we need to observe that

$$\Gamma\left(\frac{1}{2}\right) = \sqrt{\pi}$$

and

$$\Gamma(x+1) = x\Gamma(x);$$

see, e.g., [179].

2.4 General measure theory

In the previous two sections we presented methods of constructing σ-additive set functions on σ-rings of subsets of $\mathbb{R}^d$. A similar approach may be used to extend finitely additive functions on more general spaces X. As we shall later see, σ-additivity is crucial for proving general and good theorems about taking limits under integral signs and switching iterated limits. Thus, as an efficiency move, we now proceed directly to this setting.

Let X be a set and let $\mathcal{A}$ be a *σ-algebra* in $\mathcal{P}(X)$, i.e., $\emptyset \in \mathcal{A}$ and $\mathcal{A}$ is closed with respect to taking complements and countable unions. Then $(X, \mathcal{A})$ is *measurable space*. A *measure*, respectively, *finitely additive measure*, μ on $(X, \mathcal{A})$ is a function $\mu : \mathcal{A} \to \mathbb{R}^+ \cup \{+\infty\}$ such that

$$\mu(\emptyset) = 0$$

and

$$\mu\left(\bigcup_{n=1}^{\infty} A_n\right) = \sum_{n=1}^{\infty} \mu(A_n),$$

where $\{A_n : n = 1, \ldots\} \subseteq \mathcal{A}$ is a countable, respectively, finite, disjoint family. A *measure space* $(X, \mathcal{A}, \mu)$ is a measurable space $(X, \mathcal{A})$ and a measure

μ. In this case the elements of $\mathcal{A}$ are *μ-measurable sets* or, more simply, *measurable sets*.

The techniques and results of Section 2.3 generalize to this more abstract setting, and we can state the following theorem, which allows us to construct measures on many spaces.

Theorem 2.4.1. Carathéodory theorem
Let X be a set, let $\mathcal{R} \subseteq \mathcal{P}(X)$ be an algebra, and let $\mu : \mathcal{R} \to \mathbb{R}^+ \cup \{+\infty\}$ be a σ-additive set function. Then μ extends to a measure on the σ-algebra of measurable sets $\mathcal{A}$, and so also to a measure on the σ-algebra generated by $\mathcal{R}$.

We leave the details of the proof of Theorem 2.4.1 to Problem 2.20, where the issue of uniqueness of such extensions is also discussed.

If X is a topological space (see Appendix A.1) and $\mathcal{A}$ contains the Borel sets then a measure μ on $(X, \mathcal{A})$ is a *Borel measure*.

JOHANN RADON was the first to define a general measure space (in 1913). The idea was certainly in the air because of VITALI's and LEBESGUE's work to extend the fundamental theorem of calculus to $\mathbb{R}^d$; in fact, in such a setting it became important to consider set functions; cf. Section 5.6.3. The key result of RADON's paper was the first form of the now crucial Radon–Nikodym theorem, which is a natural generalization of the fundamental theorem of calculus.

Remark. Let X be a metric space (Appendix A.1), which is also a measure space. Two Borel subsets of a measure space X are *congruent* if there exists an isometry of one set onto the other. One of the simplest examples of such isometry and congruence is translation and translation invariance, which is required in defining Lebesgue measure. BANACH and ULAM asked whether, in compact metric spaces, one can always define a finitely additive measure for which all congruent Borel sets have the same measure; see [344], Problem 2. The answer is positive if one makes the additional assumption that X is countable; see [124]. On the other hand, it follows from works of MYCIELSKI [352], [353] that, for each compact metric space X, there exists a Borel measure with the property that all open congruent sets have the same measure.

Example 2.4.2. Measure spaces
We now give some examples of measure spaces.

a. The Lebesgue measure space $(\mathbb{R}^d, \mathcal{M}(\mathbb{R}^d), m^d)$.

b. For any set X and any σ-algebra $\mathcal{A} \subseteq \mathcal{P}(X)$, fix x and define

$$\forall\, A \in \mathcal{A}, \;\; \delta_x(A) = \begin{cases} 0, & \text{if } x \notin A, \\ 1, & \text{if } x \in A. \end{cases}$$

Then $(X, \mathcal{A}, \delta_x)$ is a measure space. Here δ_x is the *Dirac measure* at x associated with $\mathcal{A}$.

c. $(\mathbb{R}^d, \mathcal{B}(\mathbb{R}^d), m^d)$ is a measure space.

d. For any set X and any σ-algebra $\mathcal{A} \subseteq \mathcal{P}(X)$, define

$$\forall\, A \in \mathcal{A}, \quad c(A) = \begin{cases} \operatorname{card} A, & \text{if } \operatorname{card} A < \aleph_0, \\ \infty, & \text{if } \operatorname{card} A \geq \aleph_0. \end{cases}$$

Then $(X, \mathcal{A}, c)$ is a measure space and c is called *counting* measure. For $\mathcal{A} = \mathcal{P}(\mathbb{R})$, c is translation-invariant, i.e., c satisfies (2.1) on $\mathcal{P}(\mathbb{R})$; compare this with Example 2.2.16.

Theorem 2.4.3*c,d* are the results we mentioned after Theorem 2.2.6.

Theorem 2.4.3. The measure of unions and intersections
Let $(X, \mathcal{A}, \mu)$ be a measure space.

a. *If $A, B \in \mathcal{A}$, $A \subseteq B$, then $\mu(A) \leq \mu(B)$.*

b. *For each sequence $\{A_n : n = 1, \ldots\} \subseteq \mathcal{A}$,*

$$\mu\left(\bigcup_{n=1}^{\infty} A_n\right) \leq \sum_{n=1}^{\infty} \mu(A_n).$$

c. *If $\{A_n : n = 1, \ldots\} \subseteq \mathcal{A}$ satisfies the conditions that $\mu(A_1) < \infty$ and $A_n \subseteq A_{n-1}$ for each $n \geq 2$, then*

$$\mu\left(\bigcap_{n=1}^{\infty} A_n\right) = \lim_{n\to\infty} \mu(A_n).$$

d. *If $\{A_n : n = 1, \ldots\} \subseteq \mathcal{A}$ is a sequence with the property that $A_n \subseteq A_{n+1}$ for each ≥ 1, then*

$$\mu\left(\bigcup_{n=1}^{\infty} A_n\right) = \lim_{n\to\infty} \mu(A_n).$$

Proof. **a.** $B = A \cup (B \setminus A)$, and so $\mu(B) = \mu(A) + \mu(B \setminus A)$. Since $\mu(B \setminus A) \geq 0$ we have $\mu(A) + \mu(B \setminus A) \geq \mu(A)$.

b. Set $G_n = A_n \setminus \left(\bigcup_{j=1}^{n-1} A_j\right)$. Thus, $\{G_n : n = 1, \ldots\}$ is a disjoint family and $\bigcup_{n=1}^{\infty} G_n = \bigcup_{n=1}^{\infty} A_n$. Also, $G_n \subseteq A_n$, and so

$$\mu\left(\bigcup_{n=1}^{\infty} A_n\right) = \mu\left(\bigcup_{n=1}^{\infty} G_n\right) = \sum_{n=1}^{\infty} \mu(G_n) \leq \sum_{n=1}^{\infty} \mu(A_n).$$

c. Let $A = \bigcap_{n=1}^{\infty} A_n$, so that A_1 is the disjoint union

$$A_1 = A \cup \left(\bigcup_{n=1}^{\infty} (A_n \setminus A_{n+1})\right);$$

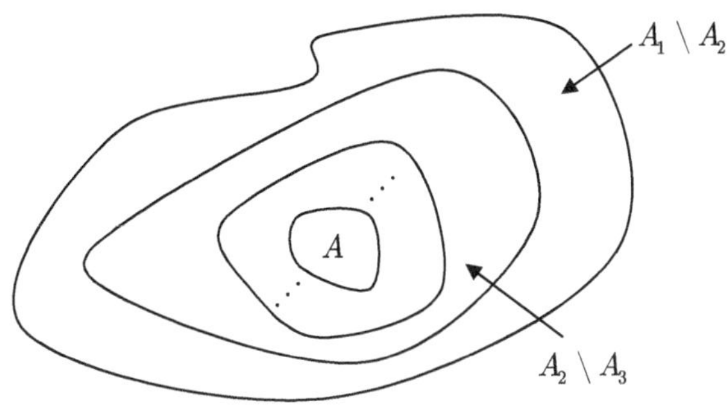

Fig. 2.1. Measure of intersections.

see Figure 2.1. Consequently,

$$\mu(A_1) = \mu(A) + \sum_{n=1}^{\infty} \mu(A_n \setminus A_{n+1}).$$

Also, $A_n = A_{n+1} \cup (A_n \setminus A_{n+1})$ is a disjoint union, from which we conclude that

$$\mu(A_n) - \mu(A_{n+1}) = \mu(A_n \setminus A_{n+1}),$$

since each of the terms $\mu(A_n)$ is finite. By our hypothesis on A_1, $\sum \mu(A_n \setminus A_{n+1})$ converges and $\mu(A) < \infty$; thus

$$\begin{aligned}\mu(A_1) &= \mu(A) + \sum_{n=1}^{\infty}(\mu(A_n) - \mu(A_{n+1}))\\ &= \mu(A) + \lim_{m\to\infty} \sum_{n=1}^{m}(\mu(A_n) - \mu(A_{n+1}))\\ &= \mu(A) + \mu(A_1) - \lim_{m\to\infty} \mu(A_{m+1}).\end{aligned}$$

d. We write

$$\bigcup_{n=1}^{\infty} A_n = A_1 \cup \left(\bigcup_{n=1}^{\infty}(A_{n+1} \setminus A_n)\right),$$

where all sets on the right-hand side of the above equation are mutually disjoint. Thus,

$$\begin{aligned}\mu\left(\bigcup_{n=1}^{\infty} A_n\right) &= \sum_{n=1}^{\infty} \mu(A_{n+1} \setminus A_n) + \mu(A_1)\\ &= \lim_{m\to\infty}\left(\sum_{n=1}^{m} \mu(A_{n+1} \setminus A_n) + \mu(A_1)\right) = \lim_{m\to\infty} \mu(A_m). \qquad \square\end{aligned}$$

Theorem 2.4.4. First Borel–Cantelli lemma
Let $(X, \mathcal{A}, \mu)$ be a measure space. If $\{A_n : n = 1, \ldots\} \subseteq \mathcal{A}$ satisfies the condition $\sum_{n=1}^{\infty} \mu(A_n) < \infty$, then

$$\mu\left(\bigcap_{m=1}^{\infty} \bigcup_{n=m}^{\infty} A_n\right) = 0.$$

Proof. From Theorem 2.4.3*b* it follows that for each m the set $\bigcup_{n=m}^{\infty} A_n$ is μ-measurable, and $\mu\left(\bigcup_{n=1}^{\infty} A_n\right) < \infty$. Thus, $\{\bigcup_{n=m}^{\infty} A_n\}$ satisfies the assumptions of Theorem 2.4.3*c* and we may conclude that

$$\mu\left(\bigcap_{m=1}^{\infty} \bigcup_{n=m}^{\infty} A_n\right) = \lim_{m\to\infty} \mu\left(\bigcup_{n=m}^{\infty} A_n\right) \le \lim_{m\to\infty} \sum_{n=m}^{\infty} \mu(A_n) = 0,$$

since the series is convergent. □

The conclusion of the First Borel–Cantelli lemma may also be understood as follows: *The collection of all those $x \in X$ that belong to infinitely many sets A_n has measure* 0.

Remark. For a given sequence $\mathcal{A} = \{A_n : n = 1, \ldots\} \subseteq [0, 1]$ of sets, ULAM asked for necessary and sufficient conditions in order to define a σ-additive measure on the Borel algebra generated by $\mathcal{A}$, with the property that the measure of the union $\bigcup A_n$ is 1 and single points are of measure 0; see [344], Problem 145.

This question was answered by BANACH [20], who introduced the notion of *atoms* associated with the *characteristic function of the sequence* $\mathcal{A} = \{A_n : n = 1, \ldots\}$:

$$\mathbb{1}_{\mathcal{A}}(x) = 2\sum_{n=1}^{\infty} \frac{1}{3^n} \mathbb{1}_{A_n}(x).$$

BANACH's solution is in terms of a condition for the set of values of $\mathbb{1}_{\mathcal{A}}$.

Let $(X, \mathcal{A}, \mu)$ be a measure space. We say that $\{A_n : n = 1, \ldots, N\}$ is a finite collection of *independent sets* if

$$\mu(A_{n_1} \cap \ldots \cap A_{n_k}) = \mu(A_{n_1}) \cdots \mu(A_{n_k}),$$

for all $k \le N$ and $n_1 < \ldots < n_k \le N$. An infinite collection of measurable sets is *independent* if each of its finite subcollections is independent.

Theorem 2.4.5. Second Borel–Cantelli lemma
Let $(X, \mathcal{A}, \mu)$ be a measure space such that $\mu(X) = 1$. If $\{A_n : n = 1, \ldots\} \subseteq \mathcal{A}$ is a sequence of independent sets and $\sum_{n=1}^{\infty} \mu(A_n) = \infty$, then

$$\mu\left(\bigcap_{m=1}^{\infty} \bigcup_{n=m}^{\infty} A_n\right) = 1.$$

Proof. The statement of the theorem is equivalent to

$$\mu\left(\bigcup_{m=1}^{\infty}\bigcap_{n=m}^{\infty} A_n^{\sim}\right) = 0.$$

Thus, it is enough to show that $\mu\left(\bigcap_{n=m}^{\infty} A_n^{\sim}\right) = 0$ for all m. Because of the independence and since $1 - x \leq e^{-x}$,

$$\mu\left(\bigcap_{n=m}^{m+k} A_n^{\sim}\right) = \prod_{n=m}^{m+k}(1-\mu(A_n)) \leq \prod_{n=m}^{m+k} e^{-\mu(A_n)};$$

and this last expression converges to 0 as $k \to \infty$ because of our assumption that $\sum_{n=1}^{\infty}\mu(A_n) = \infty$. □

Measure spaces satisfying the property $\mu(X) = 1$ are often called *probability spaces* and μ is called a *probability measure*. The Borel–Cantelli lemmas are important and useful results in probability theory, as is the following result.

Theorem 2.4.6. Kolmogorov zero–one law
Let $(X, \mathcal{A}, \mu)$ be a probability space and let $\{A_n : n = 1, \ldots\} \subseteq \mathcal{A}$ be a sequence of independent sets. Let $\mathcal{A}_m$ denote the σ-algebra generated by $\{A_n : n = m, \ldots\}$. For each $A \in \bigcap_{m=1}^{\infty}\mathcal{A}_m$, either $\mu(A) = 0$ or $\mu(A) = 1$.

For the proof of this result and many other interesting applications of measure theory in probability we refer the interested reader to [61], [266].

Definition 2.4.7. Finite, σ-finite, and complete measure spaces
a. Let $(X, \mathcal{A}, \mu)$ be a measure space. We say that $(X, \mathcal{A}, \mu)$ or X is a *finite measure space* and μ is a *bounded measure* if $\mu(X) < \infty$. We say that X or μ is *σ-finite* if

$$\exists\, \{A_n : n = 1, \ldots\} \subseteq \mathcal{A} \text{ such that } \forall\, n, \quad \mu(A_n) < \infty \quad \text{and} \quad \bigcup_{n=1}^{\infty} A_n = X.$$

Clearly, we can always take $\{A_n : n = 1, \ldots\}$ to be a disjoint family.

b. A measure space $(X, \mathcal{A}, \mu)$ is *complete* if

$$\forall\, A \in \mathcal{A}, \text{ for which } \mu(A) = 0, \text{ and } \forall\, B \subseteq A, \text{ we have } B \in \mathcal{A},$$

and thus $\mu(B) = 0$.

It follows from the definition of Lebesgue measure that $(\mathbb{R}, \mathcal{M}(\mathbb{R}), m)$ is complete. In Example 2.4.14 we shall "construct" a set $L \in \mathcal{M}(\mathbb{R}) \setminus \mathcal{B}(\mathbb{R})$ such that $L \subseteq C$, where C is the ternary Cantor set (Example 1.2.7). Thus, $(\mathbb{R}, \mathcal{B}(\mathbb{R}), m)$ is not a complete measure space.

Note that a complete measure space $(\mathbb{R}, \mathcal{M}(\mathbb{R}), m)$ is formed by "adding" all of the sets A having Lebesgue measure $m(A) = 0$ to $\mathcal{B}(\mathbb{R})$. This phenomenon is general in the following sense.

Theorem 2.4.8. Completeness theorem
Let $(X, \mathcal{A}, \mu)$ be a measure space. There is a measure space $(X, \mathcal{A}_0, \mu_0)$ such that

i. $\mathcal{A} \subseteq \mathcal{A}_0$,

ii. $\mu = \mu_0$ *on* $\mathcal{A}$,

iii. $A \in \mathcal{A}_0 \iff A = B \cup E$, *where* $B \in \mathcal{A}$ *and* $E \subseteq D$, *for some* $D \in \mathcal{A}$ *that satisfies* $\mu(D) = 0$,

iv. if $A \in \mathcal{A}_0$, $\mu_0(A) = 0$, *and* $S \subseteq A$, *then* $S \in \mathcal{A}_0$ *and* $\mu_0(S) = 0$.

Proof. ***a.*** We first show that $\mathcal{A}_0$, defined by *iii*, is a σ-algebra. First note that $\mathcal{A} \subseteq \mathcal{A}_0$ by *iii*, and so we have *i*. Let $A_n = B_n \cup E_n$, where $B_n \in \mathcal{A}$ and $E_n \subseteq D_n$, for some $D_n \in \mathcal{A}$ that satisfies $\mu(D_n) = 0$.

We have $\bigcup A_n = \bigcup (B_n \cup E_n) = (\bigcup B_n) \cup (\bigcup E_n)$ and $\bigcup B_n \in \mathcal{A}$, since $\mathcal{A}$ is a σ-algebra. Clearly, $\bigcup E_n \subseteq \bigcup D_n$, and $\bigcup D_n \in \mathcal{A}$, since $D_n \in \mathcal{A}$; also $\mu(\bigcup D_n) \leq \sum \mu(D_n) = 0$ by Theorem 2.4.3*b*. Consequently, $\mathcal{A}_0$ is closed under countable unions. Obviously, $\emptyset \in \mathcal{A}_0$.

Finally, we must show that $A^\sim \in \mathcal{A}_0$ if $A \in \mathcal{A}_0$. Note that $A^\sim = B^\sim \cap E^\sim = (B^\sim \cap D^\sim) \cup (D \cap E^\sim \cap B^\sim)$, $B^\sim \cap D^\sim \in \mathcal{A}$, $D \cap E^\sim \cap B^\sim \subseteq D \in \mathcal{A}$, and $\mu(D) = 0$. Thus, $A^\sim \in \mathcal{A}_0$ by the definition of $\mathcal{A}_0$. Therefore, $\mathcal{A}_0$ is a σ-algebra.

b. If $A = B \cup E$, with notation as in *iii*, we define $\mu_0(A) = \mu(B)$. In particular, we have *ii*. We must check that μ_0 is well defined. Letting $A = B_1 \cup E_1 = B_2 \cup E_2$, it is sufficient to prove that $\mu(B_1) = \mu(B_2)$. Clearly, $B_1 \subseteq B_2 \cup E_2 \subseteq B_2 \cup D_2$, and so $\mu(B_1) \leq \mu(B_2) + \mu(D) = \mu(B_2)$. Similarly, we compute $\mu(B_2) \leq \mu(B_1)$, and hence μ_0 is well defined.

c. Next we show that μ_0 is a measure. From part *b*, $\mu_0 \geq 0$. Now consider $\bigcup A_n$, where $A_n = B_n \cup E_n$ is decomposed as in *iii* and $\{A_n : n = 1, \ldots\}$ is a disjoint collection. Then

$$\mu_0\left(\bigcup_{n=1}^{\infty} A_n\right) = \mu_0\left(\left(\bigcup_{n=1}^{\infty} B_n\right) \cup \left(\bigcup_{n=1}^{\infty} E_n\right)\right) = \mu\left(\bigcup_{n=1}^{\infty} B_n\right)$$
$$= \sum_{n=1}^{\infty} \mu(B_n) = \sum_{n=1}^{\infty} \mu_0(A_n).$$

d. Finally, we verify *iv*. Let $\mu_0(A) = 0$ for a given $A \in \mathcal{A}$ and take $S \subseteq A$. Assume that $A = B \cup E$ is decomposed as in *iii*. Then $\mu(B) = 0$, since $\mu_0(A) = 0$. We write $A = \emptyset \cup (B \cup E)$, noting that

$$B \cup E \subseteq B \cup D \in \mathcal{A} \quad \text{and} \quad \mu(B \cup D) \leq \mu(B) + \mu(D) = 0.$$

Then, $S = \emptyset \cup S$ and $S \subseteq A = B \cup E \subseteq B \cup D \in \mathcal{A}$, where $\mu(B \cup D) = 0$. Thus, $S \in \mathcal{A}_0$ and $\mu_0(S) = \mu(\emptyset) = 0$. □

We call $(X, \mathcal{A}_0, \mu_0)$ the *complete measure space corresponding to* $(X, \mathcal{A}, \mu)$.

Theorem 2.4.9. Induced measure spaces
a. *Let* $\{(X_\alpha, \mathcal{A}_\alpha, \mu_\alpha)\}$ *be a collection of measure spaces. Define the triple* $(X, \mathcal{A}, \mu)$ *as* $X = \bigcup X_\alpha$,

$$\mathcal{A} = \{A \subseteq X : \forall\, \alpha,\ A \cap X_\alpha \in \mathcal{A}_\alpha\} \quad \text{and} \quad \forall\, A \in \mathcal{A},\ \mu(A) = \sum \mu_\alpha(A \cap X_\alpha).$$

Then $(X, \mathcal{A}, \mu)$ *is a measure space; and it is* σ*-finite if and only if all but a countable number of the* μ_α *are zero and the remainder are* σ*-finite.*

b. *Let* $(X, \mathcal{A}, \mu)$ *be a measure space and let* $Y \in \mathcal{A}$. *Set*

$$\mathcal{A}_Y = \{A \in \mathcal{A} : A \subseteq Y\}.$$

Define $\mu_Y(A) = \mu(A)$ *for* $A \in \mathcal{A}_Y$. *Then* $(Y, \mathcal{A}_Y, \mu_Y)$ *is a measure space.*

The proof of Theorem 2.4.9 is left as an exercise (Problem 2.30). In light of Theorem 2.4.9*b* we shall make the following notational convention. If we wish to consider Lebesgue measure m^d restricted to the Lebesgue measurable sets contained in a fixed set $X \subseteq \mathbb{R}^d$, we shall write $(X, \mathcal{M}(X), m^d)$ for the corresponding measure space. A similar remark applies to a measure μ defined on the Borel sets of a topological space X, in which case we shall write $(X, \mathcal{B}(X), \mu)$.

Let $\mathbb{R}^* = \mathbb{R} \cup \{\pm\infty\}$. By convention, $\infty + \infty = \infty$, $-\infty - \infty = -\infty$, $\infty \cdot (\pm\infty) = \pm\infty$, $-\infty \cdot (\pm\infty) = \mp\infty$, and $\mp\infty \pm \infty$ is undefined.

Proposition 2.4.10. *Let* $(X, \mathcal{A}, \mu)$ *be a measure space and let* $f : X \to \mathbb{R}^*$ *be a function. The following are equivalent:*

a. $\forall\, \alpha \in \mathbb{R},\ \{x : f(x) > \alpha\} \in \mathcal{A}$,
b. $\forall\, \alpha \in \mathbb{R},\ \{x : f(x) \geq \alpha\} \in \mathcal{A}$,
c. $\forall\, \alpha \in \mathbb{R},\ \{x : f(x) < \alpha\} \in \mathcal{A}$,
d. $\forall\, \alpha \in \mathbb{R},\ \{x : f(x) \leq \alpha\} \in \mathcal{A}$,
e. $\forall\, U \subseteq \mathbb{R}$, U *open*, $f^{-1}(U) \in \mathcal{A}$, *and*

$$f^{-1}(\pm\infty) \in \mathcal{A}.$$

Proof. $a \Longrightarrow d$. Clearly, $\{x : f(x) \leq \alpha\} = X \setminus \{x : f(x) > \alpha\}$, so that since $\{x : f(x) > \alpha\} \in \mathcal{A}$ and $\mathcal{A}$ is an algebra, we have $\{x : f(x) \leq \alpha\} \in \mathcal{A}$.

Similarly, $d \Longrightarrow a$ and $b \Longleftrightarrow c$.

$a \Longrightarrow b$. We have $\{x : f(x) \geq \alpha\} = \bigcap \{x : f(x) > \alpha - (1/n)\}$, so that we have the required implication, since $\mathcal{A}$ is a σ-algebra.

Similarly, $b \Longrightarrow a$, and parts a through d are equivalent.

Assume parts a–d and let $U = \bigcup I_j$, where I_j is an open interval. Then $f^{-1}(U) = \bigcup f^{-1}(I_j)$ and $f^{-1}(I_j) \in \mathcal{A}$ because of a–d. Thus, $f^{-1}(U) \in \mathcal{A}$, since $\mathcal{A}$, is a σ-algebra. Conversely, suppose we assume part e and take $U = (\alpha, \infty]$ or $U = (\alpha, \infty)$; then $f^{-1}(U) \in \mathcal{A}$, and we have part a. □

Definition 2.4.11. Measurable functions
a. An *extended real-valued function* $f : X \to \mathbb{R}^*$, defined on a measure space $(X, \mathcal{A}, \mu)$, is *measurable*, more precisely, $\mathcal{A}$*-measurable* or μ*-measurable*, if any of the conditions *a–e* in Proposition 2.4.10 holds, cf., Problem 2.36. When we introduce the notion of "almost everywhere" (*a.e.*) in Definition 2.4.15, we shall assume that measurable functions are finite μ-*a.e.*

b. A complex-valued function $f : X \to \mathbb{C}$ is *measurable* if its real and imaginary parts, $\mathrm{Re}\,(f)$ and $\mathrm{Im}\,(f)$, are measurable.

c. If $\mathcal{A}$ is $\mathcal{M}(\mathbb{R}^d)$ or $\mathcal{B}(\mathbb{R}^d)$, the corresponding measurable function is *Lebesgue* or *Borel measurable*, respectively.

d. We extend parts *a–c* of this definition in the following way. Let $(X, \mathcal{A})$ and $(Y, \mathcal{E})$ be measurable spaces. A function $f : X \to Y$ is *measurable* if

$$\forall\, E \in \mathcal{E}, \quad f^{-1}(E) \in \mathcal{A}.$$

Measurable mappings are a staple in several important subjects such as ergodic theory and the topics of Chapter 9. Perhaps, more surprisingly, they are essential in the underlying geometry of wavelet sets [48].

Note that when X is a topological space (see Appendix A.1), a real-valued function f on X is *continuous* if $f^{-1}(U)$ is open for all open sets $U \subseteq \mathbb{R}$, a fact that we proved in Section 1.3.2 using the metric definition of continuity. Whenever X is a topological space and $(X, \mathcal{A}, \mu)$ is a measure space we shall assume that $\mathcal{B}(X) \subseteq \mathcal{A}$; hence continuous functions $f : (X, \mathcal{A}, \mu) \to \mathbb{R}$ are $\mathcal{A}$-measurable. Of course, every continuous function $f : X \to \mathbb{R}$ is $\mathcal{B}(X)$-measurable.

Also, $A \in \mathcal{A}$ if and only if $\mathbb{1}_A$ is a measurable function.

Proposition 2.4.12. *Let f and g be real-valued measurable functions on a measure space $(X, \mathcal{A}, \mu)$. Then, $f \pm g$, fg, $f + c$, and cf are measurable, where $c \in \mathbb{R}$.*

Proof. We first outline the proof that $f + g$ is measurable. Let $S = \{x : f(x) + g(x) < \alpha\}$. Thus, if $x \in S$ there is $r \in \mathbb{Q}$ such that $f(x) < r < \alpha - g(x)$. Hence,

$$S = \bigcup_{r \in \mathbb{Q}} \left(\{x : f(x) < r\} \cap \{x : g(x) < \alpha - r\}\right).$$

Since f and g are measurable and the union is countable, $f + g$ is measurable.

To show that fg is measurable note that we need only prove that f^2 is measurable, since $fg = (1/2)[(f+g)^2 - f^2 - g^2]$. For $\alpha \geq 0$, $\{x : f^2(x) > \alpha\} = \{x : f(x) > \sqrt{\alpha}\} \cup \{x : f(x) < -\sqrt{\alpha}\} \in \mathcal{A}$ implies that f^2 is measurable, and for $\alpha < 0$, $\{x : f^2(x) > \alpha\}$ is the domain of f. □

Example 2.4.13. Nonmeasurable subsets of perfect symmetric sets
We shall prove that there is a perfect symmetric set $E \subseteq [0, 1]$ having positive Lebesgue measure and with a subset $N \notin \mathcal{M}(\mathbb{R})$. The proof is by

contradiction. From Example 2.2.16, choose $S \notin \mathcal{M}(\mathbb{R})$, where $S \subseteq [0,1]$; and let $\{E_n : n = 1, \ldots\}$ be a sequence of perfect symmetric nowhere-dense sets as defined in Example 1.2.8 such that

$$\forall\, n, \;\; 1 \geq m(E_n) \geq 1 - \frac{1}{n}.$$

If $S \cap E_n \in \mathcal{M}(\mathbb{R})$ for each n then

$$\bigcup_{n=1}^{\infty} (S \cap E_n) \in \mathcal{M}(\mathbb{R}).$$

Let $[0,1] \cap (\bigcup E_n)^{\sim} = F \subseteq [0,1]$. Then $m(F) = 0$ and

$$S = S \cap \left(F \cup \left(\bigcup_{n=1}^{\infty} E_n \right) \right).$$

Now $S \cap F \subseteq F$ implies $m(S \cap F) = 0$, since $(\mathbb{R}, \mathcal{M}(\mathbb{R}), m)$ is complete; and so $(S \cap F) \cup (\bigcup (S \cap E_n)) \in \mathcal{M}(\mathbb{R})$, i.e., $S \in \mathcal{M}(\mathbb{R})$, the desired contradiction. Thus, $S \cap E_n \notin \mathcal{M}(\mathbb{R})$ for some $n \in \mathbb{N}$; and we let $E = E_n$ and $N = S \cap E_n$.

Example 2.4.14. $(\mathbb{R}, \mathcal{B}(\mathbb{R}), m)$ is not complete
We shall find a set $L \in \mathcal{M}(\mathbb{R}) \setminus \mathcal{B}(\mathbb{R})$ such that $L \subseteq C$, where C is the ternary Cantor set. Since $C \in \mathcal{B}(\mathbb{R})$ and $m(C) = 0$, we can conclude that $(\mathbb{R}, \mathcal{B}(\mathbb{R}), m)$ is not a complete measure space.

i. Take the set E of Example 2.4.13 and put the contiguous intervals of C in one-to-one correspondence with those of E by the "almost linear" mapping g depicted in Figure 2.2. Thus, g is defined on $[0,1] \setminus E$, it is increasing, and it maps $[0,1] \setminus E$ onto $[0,1] \setminus C$. By the monotonicity, $g(x\pm)$ exist for all $x \in [0,1]$; and since C is nowhere dense, g can be extended to a continuous increasing surjection $g : [0,1] \to [0,1]$. Thus, g is a Borel measurable function. Take $N \subseteq E$ as in Example 2.4.13 and define $L = g(N)$. Hence $L \subseteq C$, and, since $m(C) = 0$, we conclude that $L \in \mathcal{M}(\mathbb{R})$ and $m(L) = 0$. Also $g^{-1}(L) = N$ because g is injective.

ii. Finally, observe that $L \notin \mathcal{B}(\mathbb{R})$, for, by a routine property of Borel measurable functions f, $f^{-1}(\mathcal{B}(\mathbb{R})) \subseteq \mathcal{B}(\mathbb{R})$ (Problem 2.35); cf. Problem 2.36; and this would imply that N is a Borel set if in fact L were a Borel set.

Definition 2.4.15. Almost everywhere
a. Let $(X, \mathcal{A}, \mu)$ be a measure space and let $S(x)$ be a statement about a point $x \in X$. For example, for a given function $f : X \to \mathbb{R}$, $S(x)$ could be the statement $f(x) > 0$. A statement $S(x)$ is valid *almost everywhere* if there is a set $N \in \mathcal{A}$, for which $\mu(N) = 0$, such that

$$\forall\, x \in X \setminus N, \;\; S(x) \text{ is true.}$$

In this case we write S μ-*a.e.*

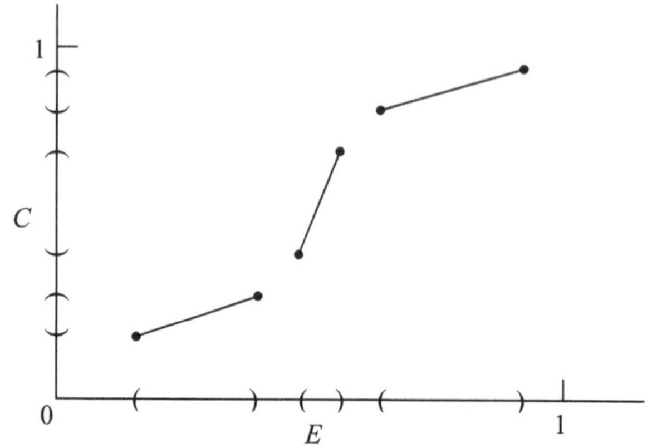

Fig. 2.2. A general Cantor function.

b. For two $\mathbb{C}$- or $\mathbb{R}^*$-valued measurable functions f and g on X, $f = g$ μ*-a.e.* signifies that $\mu(\{x : f(x) \neq g(x)\}) = 0$. Suppose we use the notation $f \sim g$ to mean $f = g$ μ*-a.e.* It is elementary to verify that $\sim$ is a well-defined *equivalence relation* on the set of $\mathbb{C}$- or $\mathbb{R}^*$-valued measurable functions on X.

c. We shall *always* assume that our measurable functions are finite μ*-a.e.*

Proposition 2.4.16. *Let $(X, \mathcal{A}, \mu)$ be a complete measure space. If f is measurable and $f = g$ μ-a.e., then g is measurable.*

Proof. Let $E = \{x : f(x) \neq g(x)\}$; and so $\mu(E) = 0$. Observe that

$$\{x : g(x) > \alpha\} = (\{x : f(x) > \alpha\} \cup \{x \in E : g(x) > \alpha\}) \setminus \{x \in E : g(x) \leq \alpha\}.$$

Since $(X, \mathcal{A}, \mu)$ is complete and $\mu(E) = 0$,

$$\mu(\{x \in E : g(x) > \alpha\}) = 0 = \mu(\{x \in E : g(x) \leq \alpha\}).$$

Thus $\{x : g(x) > \alpha\} \in \mathcal{A}$, because f is measurable. □

Example 2.4.17. Noncompleteness and nonmeasurable functions
For noncomplete measure spaces Proposition 2.4.16 does not hold. Let $L \in \mathcal{M}(\mathbb{R}) \setminus \mathcal{B}(\mathbb{R})$ have Lebesgue measure $m(L) = 0$, and let $G \supseteq L$ be a $\mathcal{G}_\delta$ set for which $m(G) = m(L)$. Define $f = \mathbb{1}_{\mathbb{R}}$ and

$$g(x) = \begin{cases} 1, & \text{if } x \in \mathbb{R} \setminus G, \\ \frac{1}{2}, & \text{if } x \in G \setminus L, \\ 0, & \text{if } x \in L. \end{cases}$$

Then, since $G \in \mathcal{B}(\mathbb{R})$, we have $f = g$ m*-a.e.* in the measure space $(\mathbb{R}, \mathcal{B}(\mathbb{R}), m)$; but $g^{-1}(0) \notin \mathcal{B}(\mathbb{R})$ and so g is not measurable on $(\mathbb{R}, \mathcal{B}(\mathbb{R}), m)$.

Remark. Recall that if $A \in \mathcal{M}(\mathbb{R})$ then $A = B \cup E$, where $B \in \mathcal{B}(\mathbb{R})$ and $m(E) = 0$. On the other hand, there are measure spaces $(\mathbb{R}, \mathcal{A}, \mu)$, for which $\mathcal{B}(\mathbb{R}) \subseteq \mathcal{A}$, where this decomposition is not true; see Problem 2.28.

Remark. WACŁAW SIERPIŃSKI [435] has given an example of a set $A \notin \mathcal{M}(\mathbb{R}^2)$ that has at most two points on each straight line. Using this set A one can find a function $f : \mathbb{R} \times \mathbb{R} \to \mathbb{R}$ whose projections are Borel measurable functions but that itself is not Lebesgue measurable. There are positive results in the opposite direction, dating from LEBESGUE in 1905, which are important for the finer study of the Fubini theorem. The basic statement of the Fubini theorem is given in Section 3.7.

We shall end this section with the following observation. We have been studying extensions of certain set functions into σ-additive set functions on σ-algebras (and such extensions are called measures). However, if we wish to construct a measure, without any prerequisites, there is another way to do so. This method can be easily extracted from our construction of Lebesgue measure on $\mathbb{R}$. To this end, recall that we defined outer measures in terms of set functions in Sections 2.2 and 2.3. Now we proceed axiomatically.

Definition 2.4.18. Outer measure
An *outer measure* is a nonnegative set function $\mu^* : \mathcal{P}(X) \to \mathbb{R}^+ \cup \{\infty\}$ with the following properties:

i. $A \subseteq B \Longrightarrow \mu^*(A) \leq \mu^*(B)$,
ii. $\mu^*(\emptyset) = 0$,
iii. $\mu^*\left(\bigcup_{n=1}^{\infty} A_n\right) \leq \sum_{n=1}^{\infty} \mu^*(A_n)$.

Theorem 2.4.19. Measures in terms of outer measures
Let $\mu^* : \mathcal{P}(X) \to \mathbb{R}^+ \cup \{\infty\}$ be an outer measure. There exist a σ-algebra $\mathcal{A} \subseteq \mathcal{P}(X)$ and a nonnegative σ-additive set function μ on $\mathcal{A}$ that is the restriction of μ^* to $\mathcal{A}$.

Proof. We use CARATHÉODORY's approach, see [87], one more time, and define $\mathcal{A}$ to be all of the sets $A \subseteq X$ with the following property:

$$\forall\, E \subseteq X, \quad \mu^*(E) = \mu^*(E \cap A) + \mu^*(E \cap A^{\sim}).$$

We also define μ to be μ^* restricted to $\mathcal{A}$. We observe that in order to complete the proof we need the analogues of Theorem 2.2.5*a* and Theorem 2.2.6*b*. Thus, we are done because the only properties of μ^* that were used in the proofs of these two results are the analogues of the properties *i*, *ii*, *iii* in Definition 2.4.18 and the nonnegativity of μ^*. □

Finally, the process of obtaining μ^* from μ and vice versa can be summarized in the following way.

Let $(X, \mathcal{A}, \mu)$ be a measure space. Following the constructions in Sections 2.2 and 2.3, define the set function $\mu^* : \mathcal{P}(X) \to \mathbb{R}^*$ to be

$$\forall\, E \in \mathcal{P}(X), \quad \mu^*(E) = \inf \left\{ \sum_{j=1}^{\infty} \mu(A_j) \right\}, \tag{2.19}$$

where the infimum is taken over all collections $\{A_j : j = 1, \ldots\} \subseteq \mathcal{A}$ that cover E, see Problem 2.20. It is not difficult to verify that μ^* is an outer measure. Moreover, the restriction of μ^* to the σ-algebra $\mathcal{A}$ coincides with the measure μ. We say that μ^* is the *outer measure associated with* μ.

If, on the other hand, we are given an outer measure μ^*, we can use Theorem 2.4.19 to generate a σ-algebra $\mathcal{A}$ and a measure μ on $\mathcal{A}$ that is a restriction of μ^*. The outer measure μ^o associated with μ does not necessarily coincide with μ^* on $\mathcal{P}(X)$. However, we have

$$\forall\, A \subseteq X, \quad \mu^*(A) \leq \mu^o(A).$$

2.5 Approximation theorems for measurable functions

Recall that our measurable functions are finite *a.e.*

Proposition 2.5.1. *Let $\{f_n : n = 1, \ldots\}$ be a sequence of $\mathbb{R}^*$-valued measurable functions on the measure space $(X, \mathcal{A}, \mu)$. The operations in the following assertions are pointwise for each $x \in X$.*

a. *For each n, $g = \sup\{f_1, \ldots, f_n\}$ and $h = \inf\{f_1, \ldots, f_n\}$ are measurable functions.*

b. *$g = \sup_{n\in\mathbb{N}}\{f_n\}$ and $h = \inf_{n\in\mathbb{N}}\{f_n\}$ are measurable functions if they are finite μ-a.e.*

c. *$g = \overline{\lim}\, f_n$ and $h = \underline{\lim}\, f_n$ are measurable functions if they are finite μ-a.e.*

Proof. We prove one case for each part.

a. $\{x : g(x) > \alpha\} = \bigcup_{j=1}^{n}\{x : f_j(x) > \alpha\} \in \mathcal{A}$, and so g is measurable.

b. A routine argument shows that

$$\{x : g(x) > \alpha\} = \bigcup_{j=1}^{\infty}\{x : f_j(x) > \alpha\} \in \mathcal{A}.$$

c. $\overline{\lim}\, f_n = \inf_n \sup_{k\geq n} f_k$, so that g is measurable because of part *b*. □

The following result is a consequence of Proposition 2.5.1*c*.

Proposition 2.5.2. *Let $\{f_n : n = 1, \ldots\}$ be a sequence of $\mathbb{C}$- or $\mathbb{R}^*$-valued measurable functions on the measure space $(X, \mathcal{A}, \mu)$, and assume that $\lim f_n(x) = f(x)$ exists pointwise for each $x \in X$. Then f is a measurable function.*

Example 2.5.3. Nonmeasurable *a.e.* limits
Naturally we would like to have Proposition 2.5.2 be true with the weaker hypothesis that $f_n \to f$ μ-*a.e.* Unfortunately, such is not the case generally unless $(X, \mathcal{A}, \mu)$ is complete, e.g., Theorem 2.5.4. As a counterexample let $(X, \mathcal{A}, \mu) = (\mathbb{R}, \mathcal{B}(\mathbb{R}), m)$, set $f_n = 0$ on $\mathbb{R}$, and set $f = \mathbb{1}_L$, where $L \in \mathcal{M}(\mathbb{R}) \setminus \mathcal{B}(\mathbb{R})$ and $m(L) = 0$. Then f is not Borel measurable, whereas $f_n \to f$ m-*a.e.* in $(\mathbb{R}, \mathcal{B}(\mathbb{R}), m)$.

The proof of the following theorem is left as an exercise; see Problem 2.34.

Theorem 2.5.4. Measurable functions as limits *a.e.*
Let $(X, \mathcal{A}, \mu)$ be a complete measure space and let $\{f_n : n = 1, \ldots\}$ be a sequence of $\mathbb{C}$- or $\mathbb{R}^*$-valued measurable functions each defined μ-a.e. on X. If f is defined μ-a.e. and $\lim f_n = f$ pointwise μ-a.e., then f is measurable.

For a given measure space $(X, \mathcal{A}, \mu)$, a function $f : X \to \mathbb{R}$ is *simple* if it can be written in the form

$$f = \sum_{j=1}^{n} a_j \mathbb{1}_{A_j}, \quad \text{where } a_j \in \mathbb{R} \text{ and } A_j \in \mathcal{A}.$$

A function $f : X \to \mathbb{C}$ is *simple* if the analogous definition holds for $a_j \in \mathbb{C}$. Clearly, simple functions are measurable. We can approximate measurable functions by simple functions in the following way.

Theorem 2.5.5. Measurable functions as limits of simple functions
Let f be a $\mathbb{C}$- or $\mathbb{R}^*$-valued measurable function on $(X, \mathcal{A}, \mu)$. There is a sequence $\{f_n : n = 1, \ldots\}$ of simple functions such that
i. $\forall\, j$ *and* $\forall\, x \in X$, $|f_j(x)| \leq |f_{j+1}(x)|$
and
ii. $\forall\, x \in X$, $\lim_{j \to \infty} f_j(x) = f(x)$.
If there exists $K > 0$ such that for all $x \in X$, $|f(x)| \leq K$, i.e., f is a bounded function on X, then the convergence in part ii is uniform.

Proof. **a.** Assume $f \geq 0$. For each $n = 1, \ldots$ and $1 \leq k \leq n2^n$ define the measurable sets,

$$A_{n,k} = \left\{ x \in X : \frac{k-1}{2^n} \leq f(x) < \frac{k}{2^n} \right\}$$

and

$$B_n = \{x \in X : f(x) \geq n\}.$$

Set

$$f_n(x) = \sum_{k=1}^{n2^n} \frac{k-1}{2^n} \mathbb{1}_{A_{n,k}}(x) + n \mathbb{1}_{B_n}(x).$$

Clearly, f_n is a simple function for which $0 \leq f_n \leq n$. We also have

$$\forall\, x \in X \text{ and } \forall\, j = 1, \ldots, \quad 0 \leq f_j(x) \leq f_{j+1}(x) \leq f(x).$$

These inequalities are a consequence of the dyadic partition of each $A_{n,k}$:

$$A_{n,k} = A_{n+1,2k-1} \cup A_{n+1,2k}.$$

Thus, if $x \in A_{n,k}$ then $x \in A_{n+1,2k-1}$ or $x \in A_{n+1,2k}$. In the former case we have $f_n(x) = f_{n+1}(x)$, and, in the latter, we have $f_{n+1}(x) > f_n(x)$.

If $f(x) = \infty$ then $x \in B_n$ for each $n = 1, \ldots$. Therefore, $f_n(x) = n$, for each n, and, so $f_n(x) \to \infty$. If, on the other hand, $f(x)$ is finite, then there is m such that, for all $n \geq m$, $x \in B_n^{\sim}$, i.e., $f(x) < n$; and thus $|f(x) - f_n(x)| < 1/2^n$ for all such n. Consequently, part *ii* is proved.

If $0 \leq f \leq K$ on X, then

$$\forall\, n \geq K, \quad \sup_{x \in X} |f(x) - f_n(x)| < \frac{1}{2^n},$$

and this yields the desired uniform convergence.

b If f is $\mathbb{R}^*$-valued, we define the nonnegative functions $f^+(x) = \max\{f(x), 0\}$ and $f^-(x) = \max\{(-f)(x), 0\}$, so that $f = f^+ - f^-$. (The same procedure is used in Definition 3.2.3.) Similarly, if f is $\mathbb{C}$-valued we may write $f = (f_1 - f_2) + i(f_3 - f_4)$, where each $f_j \geq 0$. We can then apply the technique of part *a* to obtain the result. □

Remark. The idea of the above proof is to make a finer and finer grid of $X \times [0, \infty]$ as $n \to \infty$, and then to draw the appropriate simple function on the grid. The use of dyadic partitions (or similar ideas) is more than a technical convenience; see Definition 8.6.11 and Theorem 8.6.14. Such partitions arise in any number of places. We mention three. First, the primordial constructions of wavelet theory in the 1980s, going back to ALFRÉD HAAR's thesis (1909–1910), use such partitions, e.g., [118], [348]. GEORG ZIMMERMANN's English translation of HAAR's thesis appears in [228]. Second, the deep Littlewood–Paley theory from the 1930s makes use of dyadic partitions in a fundamental way; see [183]. Third, there is a dyadic counterpart to NIKOLAI N. LUZIN's problem. In Section 3.8.3 and 4.7.5, we shall discuss LUZIN's problem, which deals with the convergence *a.e.* of Fourier series. In 1924, ANDREI N. KOLMOGOROV [289] solved a dyadic version of this problem, and the dyadic partitions inherent to his work inspired constructions of wavelet sets 75 years later [48].

Theorem 2.5.5 can be compared with the characterization of regulated functions in Problem 3.21. Also the following corollary is an elementary consequence of Theorem 2.5.5, see Problem 2.40.

Corollary 2.5.6. *Let $(X, \mathcal{A}, \mu)$ be a measure space and let $f : X \to \mathbb{R}^*$ be a function that is finite μ-a.e. or let f be $\mathbb{C}$-valued μ-a.e.*

a. *Then f is measurable if and only if f is an everywhere pointwise limit of simple functions increasing in absolute value to $|f|$.*

b. *Assume that f is a bounded function on X. Then f is measurable if and only if f is the uniform limit of simple functions increasing in absolute value to $|f|$.*

In 1911, DIMITRI F. EGOROV proved the following result for intervals.

Theorem 2.5.7. Egorov theorem
Let $(X, \mathcal{A}, \mu)$ be a finite measure space and let $\{f_n : n = 1, \ldots\}$ be a sequence of $\mathbb{C}$- or $\mathbb{R}^*$-valued measurable functions each defined μ-a.e. on X. Assume that f is defined and finite μ-a.e. on X and that $f_n \to f$ μ-a.e. Then

$$\forall\, \delta > 0,\ \exists\, A \in \mathcal{A} \text{ such that } \mu(A) < \delta \text{ and } \lim_{n\to\infty} f_n = f \text{ uniformly on } A^{\sim}.$$

Proof. Using the convenience (tradition) of tradition (convenience) we give the proof in the following two parts.

i. Given $\varepsilon > 0$ and $\delta > 0$, we shall find a set $A_{\varepsilon,\delta} \in \mathcal{A}$ and an integer N such that $\mu(A_{\varepsilon,\delta}) < \delta$ and

$$\forall\, x \notin A_{\varepsilon,\delta},\ \forall\, n \geq N, \quad |f_n(x) - f(x)| < \varepsilon.$$

Without loss of generality assume $f_n \to f$ everywhere and set

$$A_m = \{x : \exists\, n \geq m \text{ such that } |f_n(x) - f(x)| \geq \varepsilon\}.$$

Clearly $A_m \supseteq A_{m+1}$, and for $x \in X$ there is an m for which $x \notin A_m$. Therefore, $\bigcap A_m = \emptyset$, and, since $\mu(A_j) < \infty$, we have $\lim \mu(A_m) = 0$. Consequently, choose $A_{\varepsilon,\delta} = A_N$, where N is chosen so that $\mu(A_m) < \delta$ for all $m \geq N$.

ii. Given $\delta > 0$ and m, we use part i to find $B_m \in \mathcal{A}$ and N_m such that $\mu(B_m) < \delta/2^m$ and

$$\forall\, x \notin B_m, \forall\, n \geq N_m, \quad |f_n(x) - f(x)| < \frac{1}{m}.$$

Set $A = \bigcup B_m$. □

Example 2.5.8. Finite measure hypothesis for Egorov theorem
The hypothesis that $\mu(X) < \infty$ is necessary in Theorem 2.5.7. Take $f_n = \mathbb{1}_{[n,n+1]}$ on $(\mathbb{R}, \mathcal{M}(\mathbb{R}), m)$. Clearly, $f_n \to f = 0$ pointwise, whereas, if we take $A \in \mathcal{M}(\mathbb{R})$ for which $m(A) < 1$, then

$$\forall\, n\,, \exists\, x_n \in [n, n+1] \cap A^{\sim} \text{ such that } |f(x_n) - f_n(x_n)| = 1.$$

Definition 2.5.9. $L^\infty_\mu(X)$
a. Let $(X, \mathcal{A}, \mu)$ be a measure space and let $f : X \to \mathbb{C}$ be measurable. Define

$$\|f\|_\infty = \inf\{M : \mu(\{x : |f(x)| > M\}) = 0\}.$$

Notationally, we also write

$$\|f\|_\infty = \operatorname*{ess\,sup}_{x\in X} |f(x)|.$$

b. $\mathcal{L}^\infty_\mu(X)$ is the set of all measurable functions $f : X \to \mathbb{C}$ such that $\|f\|_\infty < \infty$.

c. Let $L^\infty_\mu(X)$ be the set of equivalence classes defined by the equivalence relation $\sim$ on $\mathcal{L}^\infty_\mu(X)$.

Recall from Definition 2.4.15 that $f \sim g$ signifies that $\mu(\{x : f(x) \neq g(x)\}) = 0$. We shall usually think of $F \in L^\infty_\mu(X)$ in terms of any one of its representatives $f \in \mathcal{L}^\infty_\mu(X)$. In this case, we write $f \in L^\infty_\mu(X)$. No confusion arises, either technically or conceptually, because of the following Remark, which also includes another calculation about $L^\infty_\mu(X)$.

Remark. *a.* Note that if $f, g \in \mathcal{L}^\infty_\mu(X)$ then

$$f \sim g \quad \Longrightarrow \quad \|f\|_\infty = \|g\|_\infty < \infty.$$

In fact, if $\|f\|_\infty < \|g\|_\infty$, then we can choose $\|f\|_\infty \leq M < \|g\|_\infty$. Thus,

$$\mu(\{x : |f(x)| > M\}) = 0 \text{ and } \mu(\{x : |g(x)| > M\}) > 0.$$

This contradicts the hypothesis that $f \sim g$.

b. We shall verify that

$$f \in L^\infty_\mu(X) \quad \Longrightarrow \quad |f(x)| \leq \|f\|_\infty \ \mu\text{-a.e.} \tag{2.20}$$

First note that if $f \in \mathcal{L}^\infty_\mu(X)$, then

$$\forall\, K > \|f\|_\infty, \quad \mu(\{x : |f(x)| > K\}) = 0 \tag{2.21}$$

by definition of $\|f\|_\infty < \infty$. Next, let $\{K_n\} \subseteq \mathbb{R}^+$ decrease strictly to $\|f\|_\infty$. Then we have

$$\{x : |f(x)| > \|f\|_\infty\} = \bigcup_{n=1}^\infty \{x : |f(x)| > K_n\}. \tag{2.22}$$

Each term of the union on the right side of (2.22) has measure 0 by (2.21); and so $\mu(\{x : |f(x)| > \|f\|_\infty\}) = 0$. If $x \notin \{x : |f(x)| > \|f\|_\infty\}$ then $|f(x)| \leq \|f\|_\infty$, and (2.20) is verified.

c. It is elementary to prove that $\|\ldots\|_\infty$ is a norm and that $L^\infty_\mu(X)$ is a Banach space, i.e., a complete normed vector space. For these facts and more on $L^\infty_\mu(X)$, see Section 5.5.

It is elementary to prove the following result.

Proposition 2.5.10. *Let $\{f, f_n : n = 1, \ldots\} \subseteq L^\infty_\mu(X)$ be a sequence of measurable functions on the σ-finite measure space $(X, \mathcal{A}, \mu)$. Then $\lim_{n\to\infty} \|f_n - f\|_\infty = 0$ if and only if there is a set $A \in \mathcal{A}$ such that $\mu(A) = 0$ and $f_n \to f$ uniformly on $A^\sim$.*

After Definition 2.4.11, we noted that when X is both a topological space and a measure space then continuous functions $f : X \to \mathbb{R}$ are also measurable. We close this section by looking more closely at the relation between continuous and measurable functions, especially in light of Theorems 2.5.4. 2.5.5, and 2.5.7.

Recall that $\mathbb{1}_{\mathbb{R}\setminus\mathbb{Q}}$ is not the pointwise limit of continuous functions, e.g., Problem 1.15, although, from Theorem 2.5.5, every measurable function is the pointwise limit of simple functions. On the other hand, we have the following example due to JOHANN P. G. LEJEUNE DIRICHLET.

Example 2.5.11. Dirichlet example
Set

$$g_m(x) = \lim_{n\to\infty} (1 - \cos^{2n}(m!\pi x)).$$

Clearly,

$$g_m(x) = \begin{cases} 0, & \text{if } m!x \in \mathbb{Z}, \\ 1, & \text{if } m!x \notin \mathbb{Z}, \end{cases}$$

and so

$$\forall\, x \in \mathbb{R}, \quad \lim_{m\to\infty} g_m(x) = \mathbb{1}_{\mathbb{R}\setminus\mathbb{Q}}.$$

Remark. In Problem 1.15*b* we gave necessary conditions in order that a sequence of continuous functions converge pointwise to a function on $[a, b]$. In fact there are the following necessary and sufficient conditions, e.g., [64], pages 99–102. *A sequence $\{f_n : n = 1, \ldots\}$ of continuous functions $f_n : [a, b] \to \mathbb{R}$ converges pointwise to a function f if and only if for every closed set $P \subseteq [a, b]$ without isolated points*

$$\overline{C(f) \cap P} = P.$$

Also, from Theorem 2.5.7, say, we know that if a topological space X has enough continuous functions defined on it, e.g., if $X = [a, b]$, then the following assertion is true. *Let $f : X \to \mathbb{R}^*$ be measurable, $\mu(X) < \infty$, and let $\varepsilon > 0$; then there is a continuous function $h : X \to \mathbb{R}$ such that $\mu(\{x : |f(x) - h(x)| \geq \varepsilon\}) < \varepsilon$.* A much more powerful result is the *Vitali–Luzin theorem*, commonly called the *Luzin theorem.* We shall give VITALI's original proof from 1905 [484] for two reasons: first, for historical reasons, and, second, since it is an efficient and intuitive proof. LUZIN's proof appeared in 1912 [332]; and there are standard proofs due to SIERPIŃSKI [436] and

L. COHEN [107]. All the proofs that we have seen have a similar conceptual flavor. Such a theorem, relating topological and measure-theoretic notions, was certainly thought of quite early, and VITALI suggests that BOREL and LEBESGUE might have known of it before him; in any case, VITALI published the first proof. Naturally the setting for VITALI was $([a,b], \mathcal{M}([a,b]), m)$. We shall state the result in its most convenient present-day setting, and this requires some definitions.

A Hausdorff topological space X is *locally compact* if each point has a neighborhood basis of compact sets (see Appendix A.1). Such spaces are auspicious in measure theory, since they guarantee the existence of nontrivial continuous functions, and we *do* want to integrate continuous functions. If the notion of local compactness is not in your toolkit yet you may think of X as an interval for the time being.

Definition 2.5.12. Regular Borel measures
Let X be a locally compact Hausdorff space and let $(X, \mathcal{A}, \mu)$ be a measure space for which $\mathcal{B}(X) \subseteq \mathcal{A}$. Then μ is a *Borel measure* and $(X, \mathcal{A}, \mu)$ is a *Borel measure space.* The measure μ is a *regular Borel measure* and $(X, \mathcal{A}, \mu)$ is a *regular Borel measure space* if

i. $\forall\, F \subseteq X$, compact, $\mu(F) < \infty$,

ii. $\forall\, A \in \mathcal{A}$,

$$\mu(A) = \inf \{\mu(U) : A \subseteq U \text{ and } U \text{ is open}\},$$

iii. $\forall\, U \subseteq X$, open or $\mu(U) < \infty$,

$$\mu(U) = \sup \{\mu(F) : F \subseteq U \text{ and} F \text{ is compact}\}.$$

This definition is equivalent to an apparently weaker criterion, where the supremum property over compact F is omitted in the case, $\mu(U) < \infty$; see [235], Theorem 10.30 and page 178.

Theorem 2.5.13. Vitali–Luzin theorem
Let $(X, \mathcal{A}, \mu)$ be a measure space, where X is a locally compact Hausdorff space and μ is regular. Choose $A \in \mathcal{A}$ for which $\mu(A) < \infty$ and take a measurable function $f : X \to \mathbb{R}^*$ that vanishes on $A^\sim$. For each $\varepsilon > 0$ there is a continuous function $g : X \to \mathbb{R}$ that vanishes outside of a compact set such that

$$\mu\left(\{x : f(x) \neq g(x)\}\right) < \varepsilon.$$

Proof. Without loss of generality assume that $\mu(X) < \infty$ and $A = X$. We first prove that

$$\forall\, \varepsilon, \delta > 0,\ \exists\, F \subseteq X, \text{ compact, such that } \mu(F) > \mu(X) - \varepsilon, \tag{2.23}$$

and that

$$\forall\, x \in F,\ \exists\, U_x, \text{ an open neighborhood of } x, \text{ such that} \quad \sup_{y \in U_x \cap F} f(y) - \inf_{y \in U_x \cap F} f(y) \leq \delta; \tag{2.24}$$

cf. the notion of oscillation $\omega(f, I)$ of f on I defined before Proposition 1.3.6. To do this, first set

$$A_{n,\delta} = \{x : n\delta \leq f(x) < (n+1)\delta\},$$

and note that

$$\mu(X) = \sum_{n=-\infty}^{\infty} \mu_{n,\delta},$$

where $\mu_{n,\delta} = \mu(A_{n,\delta})$. Choose $n_1, \ldots, n_k$ such that

$$\mu(X) - \sum_{j=1}^{k} \mu_{n_j,\delta} < \frac{\varepsilon}{2},$$

and let $r_j < \mu_{n_j,\delta}$, $j = 1, \ldots, k$, have the property that $\sum_{j=1}^{k} r_j < \varepsilon/2$. From the regularity of μ there are compact sets $F_j \subseteq A_{n_j,\delta}$, $j = 1, \ldots, k$, for which $\mu(F_j) > \mu_{n_j,\delta} - r_j$. Consequently, $F = \bigcup_{j=1}^{k} F_j$ is compact and

$$\mu(F) = \sum_{j=1}^{k} \mu(F_j) > \sum_{j=1}^{k} (\mu_{n_j,\delta} - r_j) > \mu(X) - \varepsilon.$$

Thus, (2.23) is obtained; and (2.24) follows from the definition of $A_{n,\delta}$.

We now use (2.23) and (2.24) countably often to obtain the result. Let $\varepsilon > 0$. Choose $\varepsilon_j, \delta_j > 0$ such that $\sum_{j=1}^{\infty} \varepsilon_j < \varepsilon$ and $\lim_{j\to\infty} \delta_j = 0$. At the first step let F_1 be compact, with measure $\mu(F_1) > \mu(X) - \varepsilon_1$, and let it also satisfy the oscillation condition (2.24) with $\delta = \delta_1$.

Choose a compact set $F_2 \subseteq F_1$ such that $\mu(F_2) > \mu(F_1) - \varepsilon_2$ and such that the oscillation condition (2.24) is satisfied with $\delta = \delta_2$. Thus, $\mu(\bigcap F_j) > \mu(X) - \varepsilon$ and f is continuous on $\bigcap F_j$. The fact that we can extend f from the compact set $\bigcap F_j$ to a continuous function g on X follows from the Urysohn lemma (Theorem A.1.3). □

Nicolas Bourbaki [69], Chapitre IV.5.1, uses the Vitali–Luzin criterion as the definition of measurable function. In fact, we have the following corollary to Theorem 2.5.13.

Corollary 2.5.14. *Let $(X, \mathcal{A}, \mu)$ be a measure space, where X is a locally compact Hausdorff space and μ is regular. A function $f : X \to \mathbb{R}^*$ that is finite μ-a.e. is measurable if and only if for every compact set $K \subseteq X$ and for every $\varepsilon > 0$,*

$$\exists\, F \subseteq K, \text{ compact, such that } \mu(K \setminus F) < \varepsilon$$

and such that f restricted to F is continuous.

A locally compact Hausdorff space is *σ-compact* if it is the countable union of compact sets.

Corollary 2.5.15. *Let $(X, \mathcal{A}, \mu)$ be a complete measure space, where X is σ-compact and μ is regular. A function $f : X \to \mathbb{R}^*$ that is finite μ-a.e. is measurable if and only if there is a sequence of continuous functions $f_n : X \to \mathbb{R}$ such that $f_n \to f$ μ-a.e.*

Proof. The sufficient conditions follow from Theorem 2.5.4. For the necessary conditions we use Theorem 2.5.13, and for each n we choose a continuous function f_n such that

$$\mu(A_n) = \mu(\{x : f(x) \neq f_n(x)\}) < \frac{1}{2^n}. \tag{2.25}$$

Setting

$$A = \bigcap_{k=1}^{\infty} \bigcup_{n=k}^{\infty} A_n,$$

we have $\mu(A) = 0$, since $\mu(\bigcup_{m=k}^{\infty} A_m) \leq 1/2^{k-1}$; cf. Theorem 2.4.4.

Also, by definition, A is precisely the set of points x that are in infinitely many A_k. Consequently, if $x \notin A$ then x is in at most finitely many A_k, and so $f_n(x) = f(x)$ for all large n. Then $f_n \to f$ μ-*a.e.* □

2.6 Potpourri and titillation

1. The *axiom of choice* (AC) states that *for every family* **A** *of disjoint nonempty sets there exists a set B that has exactly one element in common with each set in* **A**.

AC is equivalent to the following statement KZ: *If in a nonempty, partially ordered set A each linearly ordered subset has a supremum then there exists a maximal element of A.*

KZ is known as Zorn's lemma, the Zorn maximum principle, or the Kuratowski–Zorn lemma. KZ was proved independently by KAZIMIERZ KURATOWSKI [306] and MAX ZORN [523]. The formulations of the two authors were different, but they turned out to be equivalent to each other and to AC; e.g., see [278], [279], [308].

There is an unverifiable anecdote that relates that KURATOWSKI was once attending a conference at which ZORN was a speaker. During his talk, ZORN referred to Zorn's lemma, and then paused and said "ZORN is me, and over there sits Professor KURATOWSKI, who proved it first".

For a detailed treatment of AC, see [231].

2. In 1964, Robert M. Solovay proved that a stronger version of Zermelo–Fraenkel set theory (ZF), but one that still does not contain the axiom of choice, is consistent, i.e., there is a model of set theory with the property LM, which states that

$$\text{every } A \subseteq \mathbb{R} \text{ is Lebesgue measurable.}$$

The main paper in which his work appears is [446]; see [104], [494] for more recent expositions. Solovay's theorem does *not* say that

$$\text{if AC fails then LM,} \tag{2.26}$$

i.e., it does not say that AC is necessary to find $A \notin \mathcal{M}(\mathbb{R})$. In fact, in Paul J. Cohen's model for establishing certain results concerning the independence of the axiom of choice fails, and there are nonmeasurable sets. Solovay's result does say that LM is consistent with but independent of the failure of AC. If we wanted to prove that the existence of a nonmeasurable set implies the axiom of choice, then we would have to show that in every model in which AC fails every set is Lebesgue measurable. Solovay has given one such model in which this occurs. He also proves that ZF is consistent with the property that every uncountable set $X \subseteq \mathbb{R}$ contains an uncountable closed set; e.g., Problem 1.4*b*, Example 2.2.14, and Problem 2.10. Related to statement (2.26) he shows that even with AC we cannot produce a definable (from a countable sequence of ordinals) set $A \notin \mathcal{M}(\mathbb{R})$. We mention again the reference [104], which makes all of this understandable to nonexperts like ourselves.

3. Cantor's *continuum hypothesis* is the statement that

$$\forall\, A \subseteq \mathbb{R}, \quad \text{card}\, A > \aleph_0 \implies \text{card}\, A = \text{card}\, \mathbb{R},$$

i.e., card $A > \aleph_0$ implies that there is a bijection $A \to \mathbb{R}$. Explicitly using the continuum hypothesis, Banach and Kuratowski [22] and Ulam [477] proved that there is no nontrivial measure on all of $\mathcal{P}(\mathbb{R})$ that satisfies (2.2); see [440], pages 107–109, and [362], pages 24–26. In fact, the result is that there are no nontrivial continuous measures (such measures are defined in Chapters 4 and 5) defined on $\mathcal{P}(\mathbb{R})$. These are stronger conclusions than what we proved in Example 2.2.16. In 1950, Shizuo Kakutani, Kunihiko Kodaira, and Oxtoby [272], [273] proved a positive result in this area by extending Lebesgue measure to a large σ-algebra $\mathcal{A} \supsetneq \mathcal{M}(\mathbb{R})$ while preserving the translation-invariance property (2.1) on $\mathcal{A}$; see also [247]. It is not difficult to define translation-invariant measures on $\mathcal{P}(\mathbb{R})$ that do not have property (2.2), e.g., Example 2.4.2.

4. With regard to the continuum hypothesis, it can be verified that *if $F \subseteq \mathbb{R}$ is closed and uncountable then*

$$\text{card}\; F = \text{card}\; \mathbb{R}.$$

This is a consequence of the following results.

i. If card $F > \aleph_0$, *then* $F = P \cup D$, *where* D *is countable and* P *is a closed set without isolated points.*

ii. If $P \subseteq \mathbb{R}$ *is a nonempty closed set without isolated points then* card $P =$ card $\mathbb{R}$.

Part *i* is the *Cantor–Bendixson theorem.* Part *ii* is proved using CANTOR's theorem, which asserts that *if* $\{F_n : n = 1, \ldots\} \subseteq \mathcal{P}(\mathbb{R})$ *is a nested decreasing sequence of closed sets whose diameters tend to* 0 *then* $\bigcap F_n$ *is a single point.* We shall prove CANTOR's result in the proof of the Baire category theorem (Theorem A.6.1). Compare CANTOR's theorem with Theorem 1.2.10*b*.

Two of the major theorems dealing with the continuum hypothesis are due to KURT GÖDEL (late 1930s) and P. COHEN (1966). Both results assume the consistency of the traditional axioms of set theory, viz., the Zermelo–Fraenkel axioms and the axiom of choice, but conclude with apparently alarmingly diverse conclusions about the continuum hypothesis. A recent analysis of the topic, which placated one of the authors, who studied GÖDEL's Annals of Mathematics Studies in 1960–1961 as an assignment from GLEASON (see page 1), is found in [513], [514].

5. In Example 2.2.16 the countable subadditivity of m was crucial to establishing the existence of an element $A \in \mathcal{P}(\mathbb{R}) \setminus \mathcal{M}(\mathbb{R})$. For $\mathbb{R}^3$ there is the following more general result.

Theorem 2.6.1. Nonexistence of finitely additive set functions in $\mathcal{P}(\mathbb{R}^3)$
There is no nontrivial finitely additive set function $\mu : \mathcal{P}(\mathbb{R}^3) \to \mathbb{R}^+ \cup \{+\infty\}$ *such that* $\mu(B) < \infty$ *for bounded sets* $B \subseteq \mathbb{R}^3$ *and for which* $\mu(A) = \mu(B)$ *if* A *and* B *are isometric.*

(See Appendix A.1 for the definition of isometry.)

Theorem 2.6.1 is proved using the axiom of choice in the form of the *Hausdorff paradox: Let* S^2 *be the surface of the unit sphere in* $\mathbb{R}^3$*; then there is a countable set* $D \subseteq S^2$ *and there are disjoint subsets* $A, B, C \subseteq S^2 \setminus D$ *such that*

$$S^2 \setminus D = A \cup B \cup C \text{ and } A \simeq B \simeq C \simeq B \cup C,$$

where "$\simeq$" designates *congruence*, i.e., a surjective isometry. We refer to [441] for an exposition of the equivalence of AC and the Hausdorff paradox. To prove Theorem 2.6.1 from the Hausdorff paradox, assume that such a μ exists and define $\nu : \mathcal{P}(S^2) \to \mathbb{R}^+ \cup \{+\infty\}$ as follows: if $X \subseteq S^2$, $\nu(X)$ is μ of the union of all radii of the unit sphere with endpoints in X. Then, ν satisfies the same conditions as μ and we obtain a contradiction to the hypothesis of finite additivity because of FELIX HAUSDORFF's result.

The *Banach–Tarski paradox*, see [25], is a corollary of the Hausdorff paradox. We shall settle for a statement of the Banach–Tarski result and again refer to [441] for details: *Let* $A, B \subseteq \mathbb{R}^3$ *be disjoint solid spheres having the same radius; there exist* $C_1, \ldots, C_{41} \subseteq A$ *and* $D_1, \ldots, D_{41} \subseteq A \cup B$ *such that* $A = \bigcup C_j$, $A \cup B = \bigcup D_j$, $C_j \simeq D_j$ *for each* j, *and*

$$C_i \cap C_j = D_i \cap D_j = \emptyset, \quad 1 \le i < j \le 41.$$

There is also a proof of the Banach–Tarski paradox as a consequence of the Hahn–Banach theorem; see [367]. An explanation of this phenomenon is due to JOHN (JOHANN) VON NEUMANN, who showed that the Banach–Tarski paradox is impossible in $\mathbb{R}$ and $\mathbb{R}^2$, because only $\mathbb{R}^d$ for $d \ge 3$ contains free nonabelian groups; e.g., see JEAN DIEUDONNÉ's biography of VON NEUMANN in [197].

As our final remark on these matters, BANACH has shown that on $\mathbb{R}$ and $\mathbb{R}^2$ there are nonnegative finitely additive set functions on $\mathcal{P}(\mathbb{R})$ and $\mathcal{P}(\mathbb{R}^2)$ that are finite for bounded sets, translation-invariant, and equal to Lebesgue measure for Lebesgue measurable sets. It is consistent with ZF to extend BANACH's result to the σ-additive case.

6. In *The Scottish Book* [344], STEINHAUS proposed in the following problem related to finite measures.

Given three sets A_1, A_2, A_3 *in* $\mathbb{R}^3$, *each having finite Lebesgue measure, does there exist a plane cutting each of the three sets* A_1, A_2, A_3 *into two parts of equal measure? There is the analogous problem for* d *sets in* $\mathbb{R}^d$.

In $\mathbb{R}^3$ this problem was solved by STEINHAUS in 1936. His result is called the "ham sandwich theorem"; see WILLIAM A. BEYER and ANDREW ZARDECKI, Amer. Math. Monthly 111 (2004), 58–61, for an early history of the ham sandwich theorem. The idea is to start by bisecting one set ("ham") by a plane and observing that continuous changes of these bisections, which cut the "two slices of bread", provide an odd mapping, g, i.e., for all $x \in S^2 \subseteq \mathbb{R}^3$ $g(-x) = -g(x)$, from $S^2 \subseteq \mathbb{R}^3$ into $\mathbb{R}^2$. The rest of the proof follows from an application of the *Borsuk–Ulam theorem*, conjectured by ULAM and proved by KAROL BORSUK in 1932. In one of its forms the Borsuk–Ulam theorem asserts that *any continuous function from* S^d, *the* d*-dimensional sphere in* $\mathbb{R}^{d+1}$, *to* $\mathbb{R}^d$, *must send some pair of opposite (bipolar) points to the same point*; see ANDREW BROWDER, Amer. Math. Monthly 113 (2006), 935–937, for an elementary proof in dimension 2.

The Borsuk–Ulam theorem itself has led to extensive developments. For example, it implies the *Brouwer fixed-point theorem*; see FRANCIS E. SU, Amer. Math. Monthly 104 (1997), 855–859.

7. In his 1933 classic [293], KOLMOGOROV defined a *probability space* as a triple $(X, \mathcal{R}, p)$ satisfying the following conditions on the set X, the algebra $\mathcal{R}$ of subsets of X, and the function $p : \mathcal{R} \to \mathbb{R}^+$: $p(X) = 1$, p is finitely additive, and if $\{A_n : n = 1, \ldots\} \subseteq \mathcal{R}$ is decreasing and $\bigcap A_n = \emptyset$, then $\lim_{n \to \infty} p(A_n) = 0$. In this case the following theorem can be proved; cf. Theorems 2.2.6*c*, 2.3.7, 2.4.1, and 2.4.3*c*, as well as the uniqueness results in Problems 2.19 and 2.20: *Let* $(X, \mathcal{R}, p)$ *be a probability space; there is a*

unique measure on $\mathcal{A}$, the σ-algebra generated by $\mathcal{R}$, that is an extension of p. This measure is the *probability measure* on $\mathcal{A}$ and is also denoted by p.

Because of this theorem and the results proved in Chapter 3, and as mentioned prior to Theorem 2.4.6, we say that a *probability space* is any triple $(X, \mathcal{A}, p)$, where X is a set, $\mathcal{A}$ is a σ-algebra in $\mathcal{P}(X)$, and p is a measure for which $p(X) = 1$.

The definition of a probability space frequently assumes the completeness of the measure space, and, as we have seen, the Carathéodory extension procedure leads automatically to complete measures. After KOLMOGOROV's short treatise [293] there appeared several other major probability books in the 1940s and 1950s including those of ALEKSANDR Y. KHINCHIN (1948) [284], PAUL LÉVY (1948) [319], WILLIAM FELLER (1950) [171], and JOSEPH DOOB (1953) [144]. Of course, probability theory has been around forever, e.g., the work of GEROLAMO CARDANO (1501–1576) on gambling and the remarkable contribution of probability to arithmetic by BOREL in 1909; and it is an important and multifaceted field.

One such facet with extraordinary implications is prediction theory and the spectral theory of minimal sequences due to KOLMOGOROV himself [292]. (There is an English translation by GOPINATH KALLIANPUR given to one of the authors by the magisterial PESI MASANI.) It is of the KOLMOGOROV paper that HARALD CRAMÉR (1976) wrote, "The fundamental importance of this work of Kolmogorov lies in the fact that he showed how the abstract theory of Hilbert space (as well, of course, as other types of spaces) could be applied to the theory of random variables and stochastic processes". We refer to [38] for an exposition of this area that relates several topics in modern analysis including frames and positive operator-valued measures, e.g., [47].

8. Let $(X, \mathcal{A}, p)$ be a probability space. An $\mathcal{A}$-measurable function $f : X \to \mathbb{R}$ is a *random variable*. In the case $f : X \to \mathbb{C}$ then f is a *random variable* if $\operatorname{Re}(f)$ and $\operatorname{Im}(f)$ are $\mathcal{A}$-measurable. It is elementary to check that *if $f : X \to \mathbb{R}$ is a random variable and $B \in \mathcal{B}(\mathbb{R})$, then $f^{-1}(B) \in \mathcal{A}$.*

One chooses $\mathcal{B}(\mathbb{R})$ and not $\mathcal{M}(\mathbb{R})$ in the probabilistic interpretation of a random variable for the following reasons. The $\mathcal{B}(\mathbb{R})$-case gives more random variables than the $\mathcal{M}(\mathbb{R})$-case, because $\mathcal{B}(\mathbb{R}) \subseteq \mathcal{M}(\mathbb{R})$; and it is easier to check that one has a random variable f in the $\mathcal{B}(\mathbb{R})$-case, because one need only check that $f^{-1}(I) \in \mathcal{A}$ for intervals I.

This measure-theoretic setting for probability theory and random variables is a natural model for many probabilistic experiments. In particular, X represents a set of *experimental outcomes*, $\mathcal{A}$ is a class of *events*, and p is the probability assigned to these events. For example, in the experiment of rolling a die, X consists of six faces of the die.

2.7 Problems

Some of the more elementary problems in this set are Problems 2.1, 2.3, 2.4, 2.7, 2.9, 2.11, 2.13, 2.14, 2.16, 2.18, 2.27, 2.29, 2.31, 2.33, 2.34, 2.36, 2.38, 2.39.

2.1. Prove Proposition 2.2.3.

2.2. Find a function $f : [0,1] \to \mathbb{R}$ and a set $D \subseteq [0,1]$ such that $D \in \mathcal{M}([0,1])$, $m(D) = 0$, D is uncountable, f is continuous on D, and f is discontinuous on $[0,1] \setminus D$.
[*Hint.* Modify the Cantor function appropriately. Also see the example in the proof of Proposition 2.2.9.]

2.3. ***a.*** Prove Theorem 2.2.6.

b. Give an example to show that the hypothesis is necessary in Theorem 2.2.6*c*.

c. Find two disjoint sets $A_1, A_2 \subseteq [0,1]$ such that

$$m^*(A_1 \cup A_2) < m^*(A_1) + m^*(A_2).$$

d. Find $\{A_j : j = 1, \ldots\} \subseteq \mathcal{P}([0,1])$, pairwise disjoint, such that

$$m^*\left(\bigcup_{j=1}^{\infty} A_j\right) < \sum_{j=1}^{\infty} m^*(A_j).$$

e. Find $\{A_j : j = 1, \ldots\} \subseteq \mathcal{P}([0,1])$ such that $A_{j+1} \subseteq A_j$ and

$$m^*\left(\bigcap_{j=1}^{\infty} A_j\right) < \lim_{j\to\infty} m^*(A_j).$$

Obviously, $m^*(A_j) < \infty$ for each j.

2.4. Let $\mu : \mathcal{A} \to \mathbb{R}^+ \cup \{\infty\}$ be a σ-additive set function on an algebra $\mathcal{A}$. Let μ^* be the associated outer measure. If $\mu^*(A) = 0$ prove that A is measurable.

2.5. Let $(\mathbb{R}^d, \mathcal{A}, \mu)$, $\mathcal{B}(\mathbb{R}^d) \subseteq \mathcal{A}$, be a regular measure space, and let μ^* be the associated outer measure. Generalize Proposition 2.2.3 to this situation.

2.6. Let $(X, \mathcal{A}, \mu)$ be a measure space, where $\mathcal{A}$ is a σ-algebra generated by an algebra $\mathcal{R}$. Prove that μ is σ-finite on $\mathcal{R}$ if and only if μ is σ-finite on $\mathcal{A}$.

2.7. Define $\mathcal{A} \subseteq \mathcal{P}(\mathbb{R})$ by the property that either card A or card $(\mathbb{R} \setminus A)$ is countable for $A \in \mathcal{A}$. The set function $\mu : \mathcal{A} \to \mathbb{R} \cup \{\infty\}$ is given by

$$\mu(A) = \begin{cases} 0, & \text{if card } A \le \aleph_0, \\ \infty, & \text{if card } (\mathbb{R} \setminus A) \le \aleph_0. \end{cases}$$

Prove that $(\mathbb{R}, \mathcal{A}, \mu)$ is a measure space.

2.8. For each $X \subseteq \mathbb{R}$, define

$$X_d = \{x \in (0,1) : \exists\, k \in \mathbb{Z} \text{ such that } kx \in X\}.$$

Prove that

$$\forall\, N > 0 \text{ and } \forall\, \varepsilon > 0,\ \exists\, X \subseteq \mathbb{R} \text{ such that } m(X) > N \text{ and } m(X_d) < \varepsilon.$$

[*Hint.* Begin with the following special case. Take

$$X = \bigcup_{k=1}^{[n/\delta]} \left(\frac{n}{k}, \frac{n+\delta}{k}\right),$$

where $[n/\delta]$ is the greatest integer less than or equal to n/δ. Show that

$$\forall\, n,\ m(X) = \sum_{k=1}^{[n/\delta]} \frac{\delta}{k} > \delta \log\left(\frac{n}{\delta}\right)$$

and that for large n

$$m(X_d) = \delta + \sum_{k=n+1}^{[n/\delta]} \frac{\delta}{k} < \delta + 2\delta \log\left(\frac{1}{\delta}\right).$$

Consequently, "$m(X) \to \infty$" and "$m(X_d) \to 0$".]

2.9. ***a.*** Find $X \subseteq [0,1]$ such that $m(X) = 0$ and X is not of first category. [*Hint.* Consider a perfect symmetric set with perfect symmetric sets in each of its contiguous intervals, etc.; make the measures add up to 1 and look at the complementary set.]

b. For each $\varepsilon > 0$ construct an open set $U \subseteq [0,1]$ such that $\overline{U} = [0,1]$ and $m(U) = \varepsilon$.
[*Hint.* Take intervals of length $\varepsilon/2^k$ about the rationals.]

c. Find $A, B \subseteq [0,1]$ such that $A, B \in \mathcal{M}([0,1])$, $A \cap B = \emptyset$,

$$m(A) = 0 \text{ and } A \text{ is not of first category},$$

and

$$m(B) = 1 \text{ and } B \text{ is of first category}.$$

[*Hint.* Check Proposition 2.2.9. Can you find a simpler example?]

2.10. ***a.*** Prove that every $B \in \mathcal{B}(\mathbb{R})$ that is not of first category contains a closed uncountable subset.

b. Is part *a* true for uncountable Borel sets of first category?

c. Prove that FELIX BERNSTEIN's example, which is what we reference in Problem 1.4*b*, contains uncountable sets of measure 0.

2.11. Prove Theorem 2.2.12.

2.12. Let m^2 be Lebesgue measure on $\mathbb{R}^2$.

a. A *packing problem.* Let U be the open unit disk in $\mathbb{R}^2$ and let $\{U_n : n = 1, \ldots\}$ be a sequence of open disks in $\mathbb{R}^2$ such that

i. $\overline{U}_n \subseteq U$ for each n,
ii. $\{\overline{U}_n : n = 1, \ldots\}$ is pairwise disjoint,
iii. $\sum r_n < \infty$, where r_n is the radius of U_n.

Define $X = U \setminus \bigcup_{n=1}^{\infty} U_n$ and prove $m(X) > 0$; see [499].

b. Show that every set of positive Lebesgue measure in $\mathbb{R}^2$ contains the vertices of an equilateral triangle.

c. Is part *b* true for other polygons in $\mathbb{R}^2$?

Remark. There are many packing problems with many modern applications, e.g., to error-correcting codes [468]. We refer to the treatise by JOHN CONWAY and NEIL J. A. SLOANE [114] for a tantalizing and deep treatment.

Related to packing there is the following problem/theorem due to HERMAN AUERBACH, BANACH, MAZUR, and ULAM. It is called the "sack of potatoes theorem", and it appeared in *The Scottish Book* [344] in the following form: *If $\{K_n\}_{n=1}^{\infty}$ is a sequence of convex bodies in $\mathbb{R}^d$, each of diameter less than or equal to a, and let the sum of their volumes be less than or equal to b; prove that there exists a cube with diameter $c = f(a, b)$ such that one can put all the given bodies in it disjointly.* Recall that $X \subseteq \mathbb{R}^d$ is a *convex set* if, for any $x, y \in \mathbb{R}^d$, the line segment $\{z = (1-t)x + ty : 0 \le t \le 1\} \subseteq \mathbb{R}^d$ is contained in X. The *diameter* of X is defined in Definition A.1.4*f*.

The first published proof is in [300].

2.13. ***a.*** Prove that there are no numbers ε and δ, such that $0 < \varepsilon \le \delta < 1$, that have the following property: if $\{A_n : n = 1, \ldots\} \subseteq \mathcal{P}([0,1])$ is any sequence of Lebesgue measurable sets each satisfying $m(A_n) \ge \delta$ then there is a set A of measure ε for which $A \subseteq A_n$ for infinitely many n.
[*Hint.* Let A_n be the set of $x \in (0,1)$ such that the nth digit in its decimal expansion is nonzero.]

b. Let $\{A_n : n = 1, \ldots\} \subseteq \mathcal{M}([0,1])$, $A_n \subseteq [0,1]$, and assume that 1 is a limit point of $\{m(A_n) : n = 1, \ldots\}$. Prove that there is a subsequence $\{n_k : k = 1, \ldots\}$ for which

$$m\left(\bigcap_{k=1}^{\infty} A_{n_k}\right) > 0. \tag{2.27}$$

[*Hint.* $\sum_{k=1}^{\infty}(1 - m(A_{n_k})) < 1$.]

Remark. If we begin by taking an uncountable collection then (2.27) is always possible. For further results on this type of problem we refer to [195], [196].

2.14. Let $\mu : \mathcal{R} \to \mathbb{R}^+ \cup \{\infty\}$ be a σ-additive set function on a ring $\mathcal{R}$. Prove that μ satisfies the following properties for $E, E_j, F \in \mathcal{R}$:

a. $E \subseteq F \implies \mu(E) \leq \mu(F)$;
b. $E \subseteq F \implies \mu(E) + \mu(F \setminus E) = \mu(F)$;
c. $E \subseteq \bigcup_{j=1}^{\infty} E_j$ and $\mu(E) < \infty \implies \mu(E) \leq \sum_{j=1}^{\infty} \mu(E_j)$;
d. if $\bigcup_{j=1}^{\infty} E_j \subseteq E$ and if the E_j are pairwise disjoint, then

$$\sum_{j=1}^{\infty} \mu(E_j) \leq \mu(E).$$

2.15. Let $\mathcal{R} \subseteq \mathcal{P}(X)$ be a ring and let $\mu : \mathcal{R} \to \mathbb{R}^+ \cup \{+\infty\}$ be a σ-additive set function. Let $E \in \mathcal{H}(\mathcal{R})$ and let $\{A_1, \ldots, A_N\} \subseteq \mathcal{A}$ be a collection of measurable, pairwise disjoint elements of the σ-algebra $\mathcal{A}$ generated by $\mathcal{R}$. Prove the equality

$$\mu^*\left(\bigcup_{n=1}^{N} A_n \cap E\right) = \sum_{n=1}^{N} \mu^*(A_n \cap E).$$

[*Hint.* First prove it for $N = 2$ and use induction.]

2.16. Prove that a ring $\mathcal{R}$ that is closed under taking countable unions of its elements is a σ-ring.

2.17. Prove that a family of sets that is closed under countable increasing unions and under countable decreasing intersections, and that contains an algebra $\mathcal{A}$, also contains the σ-algebra generated by $\mathcal{A}$.

2.18. Let $(X, \mathcal{A}, \mu)$ be a measure space, and let $A, B \in \mathcal{A}$. Show that if $\mu(A \cup B) = \mu(A) + \mu(B)$, then $\mu(A \cap B) = 0$.

2.19. Let $(X, \mathcal{A})$ be a measurable space, and let μ and ν be measures on $(X, \mathcal{A})$. Assume that $\mu = \nu$ on some subset $\mathcal{D} \subseteq \mathcal{A}$. A reasonable problem is to consider nontrivial conditions in order that $\mu = \nu$ on $\mathcal{A}$. More specifically, prove the following result. *Assume that X and $\emptyset$ are elements of $\mathcal{D}$, that $\mathcal{D}$ is closed under finite unions and intersections, and that $\mathcal{A}$ is the smallest σ-algebra containing $\mathcal{D}$; if*

$$\exists\, \{D_n : n = 1, \ldots\} \subseteq \mathcal{D} \text{ such that } \bigcup_{n=1}^{\infty} D_n = X \text{ and } \forall\, n,\ \mu(D_n) < \infty$$

then $\mu = \nu$ on $\mathcal{A}$; cf. Problem 2.20. Consequently, we see that a measure defined on the Borel subsets of the line is uniquely determined there by its values on the half-open intervals $(a, b]$, including $(-\infty, b]$ and (a, ∞). This uniqueness issue leads to the following problem: Suppose that ν is not necessarily a measure but only finitely additive on $\mathcal{D}$, whereas the other hypotheses for the above problem are satisfied; when can we conclude that $\mu = \nu$ on $\mathcal{A}$?

2.20. Let X be a set and let $\mathcal{R} \subseteq \mathcal{P}(X)$ be an algebra. Assume that the set function $\mu : \mathcal{R} \to \mathbb{R}^+$, for which $\mu(\emptyset) = 0$, satisfies

$$\forall\, \{A_j\} \subseteq \mathcal{R}, \text{ a disjoint sequence such that } \bigcup_{j=1}^{\infty} A_j \in \mathcal{R},$$

$$\mu\left(\bigcup_{j=1}^{\infty} A_j\right) = \sum_{j=1}^{\infty} \mu(A_j).$$

Define

$$\forall\, E \in \mathcal{P}(X), \quad \mu^*(E) = \inf\left\{\sum_{j=1}^{\infty} \mu(A_j)\right\},$$

where the infimum is taken over all collections $\{A_j : j = 1, \ldots\} \subseteq \mathcal{R}$ that cover E. Define $E \in \mathcal{P}(X)$ to be *μ^*-measurable* if

$$\forall\, F \in \mathcal{P}(X), \quad \mu^*(F) \geq \mu^*(F \cap E) + \mu^*(F \cap E^{\sim});$$

cf. the development in Section 2.2.

a. Prove the *Carathéodory theorem: The μ^*-measurable sets form a σ-algebra $\mathcal{A}$ containing $\mathcal{R}$ such that $(X, \mathcal{A}, \mu^*)$ is a measure space and $\mu^* = \mu$ on $\mathcal{R}$.* Note that $\mathcal{A}$ is not the σ-algebra generated by $\mathcal{R}$, but the σ-algebra of measurable sets with respect to $\mathcal{R}$ and μ^*.

b. Prove that *if μ is σ-finite then μ^* is the unique extension of μ as a measure to the smallest σ-algebra containing $\mathcal{A}$* (this latter part is precisely Problem 2.19).

2.21. Let $f : \mathbb{R} \to \mathbb{R}$ be an increasing, right continuous function. Prove that the set function μ_f defined in Example 2.3.10 is σ-additive on the ring of disjoint unions of parallelepipeds in $\mathbb{R}^d$. The set of parallelepipeds in $\mathbb{R}^d$ is an example of a semiring; see Problem 2.22.

2.22. We say that a collection $\mathcal{S} \subseteq \mathcal{P}(X)$ is a *semiring* if

$$\forall\, A, B \in \mathcal{S}, \quad A \cap B \in \mathcal{S}$$

and

$$A \setminus B = \bigcup_{j=1}^{n} A_j, \text{ for some pairwise disjoint sequence } \{A_1, \ldots, A_n\} \subseteq \mathcal{S}.$$

If $\mathcal{S} \subseteq \mathcal{P}(X)$ is a semiring and $X \in \mathcal{S}$, then we say that $\mathcal{S}$ is a *semialgebra.*

a. If $\mathcal{R}$ is a ring generated by a semiring $\mathcal{S}$, i.e., the smallest ring that contains $\mathcal{S}$, prove that

$$\forall\, A \in \mathcal{R},\ A = \bigcup_{j=1}^{n} A_j, \text{ for some pairwise disjoint sequence } \{A_1, \ldots, A_n\} \subseteq \mathcal{S}.$$

b. Let $\mu : \mathcal{S} \to \mathbb{R}^+ \cup \{\infty\}$ be a σ-additive set function on a semiring $\mathcal{S}$. Prove that there exists a unique extension of μ to a σ-additive set function on the ring generated by $\mathcal{S}$.

Thus, if, in addition, $\mathcal{S}$ is a semialgebra and μ is σ-finite on $\mathcal{S}$, then there exists a unique extension of μ from $\mathcal{S}$ to a σ-additive set function on the σ-algebra generated by $\mathcal{S}$.
[*Hint.* Use part *a.*]

2.23. Let $f : \mathbb{R} \to \mathbb{R}$ be a Lebesgue measurable function and let $G : \mathbb{R}^2 \to \mathbb{R}$ be continuous. Show that if

$$\forall\, x, y \in \mathbb{R}, \quad |f(x+y)| \leq G(f(x), f(y)),$$

then f is bounded on bounded sets.
[*Hint.* $h(x) = |f(x)| + |f(-x)|$ is measurable, and so there is a set $X \subseteq \mathbb{R}$ such that $m(X) > 0$ and h is bounded on X; thus for all $x, y \in X$, $f(x-y)$ is bounded. Since $X - X$ is a neighborhood of 0 (a useful result with an ingenious proof due to STEINHAUS, which we shall give in Problem 3.6), f is bounded on U. By applying induction to the hypothesis we see that $f(nz)$ is bounded, for $z \in U$; and so f is bounded on any bounded set.]

2.24. ***a.*** Let $A \subseteq [0,1]$, and assume $m^*(A) > 0$. Prove that there is a nonmeasurable set $E \subseteq A$. Can you show that for any $\alpha \in (0,1)$ there is a nonmeasurable set $E \subseteq [0,1]$ for which $m^*(E) = \alpha$? For purposes of comparison note Example 2.4.13.

b. For $Y \subseteq \mathbb{R}$ let $\tau_{-\alpha} Y = \{y + \alpha : y \in Y\}$. If $S = \bigcup_{k=-\infty}^{\infty} [2k, 2k+1)$, prove that $\tau_{-\alpha} S = S^{\sim}$ when α is an odd integer.

c. Define $\tau S = \inf\, \{\alpha \in \mathbb{R}, \alpha > 0 : \tau_{-\alpha} S = S^{\sim}\}$; thus τS is 1 for S defined in part *b*. Prove that if $\tau S = 0$, then $S \notin \mathcal{M}(\mathbb{R})$.

d. The set $H \subseteq \mathbb{R}$ is a *Hamel basis* if

$$\forall\, x \in \mathbb{R}, \ \exists\, \{r_\alpha\} \subseteq \mathbb{Q} \text{ and } \exists\, \{h_\alpha\} \subseteq H, \ \text{ such that } x = \sum r_\alpha h_\alpha,$$

where the sum is finite and the representation is unique. Using Zorn's lemma, which is an equivalent form of the axiom of choice, it is easy to prove that Hamel bases exist using the following argument: let $\mathcal{F}$ be the family of all subsets $S \subseteq \mathbb{R}$ that are linearly independent over $\mathbb{Q}$; then there is a maximal element $H \in \mathcal{F}$. Prove that if H is a Hamel basis and $H \in \mathcal{M}(\mathbb{R})$ then $m(H) = 0$. Thus, if H is a Hamel basis and $m^*(H) > 0$, we conclude that $H \notin \mathcal{M}(\mathbb{R})$.

e. The existence of a Hamel basis with $m^*(H) > 0$ is a corollary of the following fact: *There is a Hamel basis H_B that intersects every closed uncountable set in* $\mathbb{R}$.

This result was first proved by C. BURSTIN in 1916, and a simpler proof of this fact is due to ALEXANDER ABIAN [1]. Assuming BURSTIN's result prove that $m^*(H_B) > 0$.

f. Let $H = \{x\} \cup \{x_\alpha : \alpha \in A,\ \text{ an index set}\} \subseteq \mathbb{R}$ be a Hamel basis, and define X to be the set of elements $y \in (0,1)$ whose Hamel expansions do not use x. Prove that $X \notin \mathcal{M}(\mathbb{R})$, e.g., [2], [434].

Remark. In [440], SIERPIŃSKI defined a set $X \subseteq \mathbb{R}$ to have the *property* S if

$$\forall\, A \subseteq \mathbb{R},\ m(A) = 0 \implies \text{card}\, A \cap X \leq \aleph_0.$$

Thus, X has the property S if and only if every uncountable subset of X is nonmeasurable. He then proved that uncountable sets with the property S exist if the continuum hypothesis is assumed; cf. [145] for a more recent development.

2.25. ***a.*** Let $(X, \mathcal{A}, \mu)$ be a measure space and let $E \in \mathcal{A}$. Suppose $M \in \mathcal{P}(X)$ has the property that the only measurable subsets of $M \cap E$ and $M^\sim \cap E$ are of measure zero. Prove that $\mu^*(M \cap E) = \mu^*(M^\sim \cap E) = \mu(E)$.

b. Let $(\mathbb{R}, \mathcal{M}(\mathbb{R}), m)$ be the Lebesgue measure space. Prove that there exists an m-measurable set $M \in \mathcal{M}(\mathbb{R})$ such that

$$\forall\, E \subseteq \mathcal{M}(\mathbb{R}), \quad m^*(M \cap E) = m^*(M^\sim \cap E) = m^*(E).$$

[*Hint.* Take any $\xi \in \mathbb{R}\backslash\mathbb{Q}$ and define $A = \{n + m\xi : n, m \in \mathbb{Z}\}$, $B = \{n + 2m\xi : n, m \in \mathbb{Z}\}$, and $C = \{n + (2m+1)\xi : n, m \in \mathbb{Z}\}$. Similarly to Example 2.2.16, let $x \sim y$ if $x, y \in \mathbb{R}$ and $x - y \in A$. Clearly, "$\sim$" is an equivalence relation. Let S be a set that contains one representative from each equivalence class and define

$$M = S + B = \bigcup_{b \in B} (S + b).$$

Observe that $(M - M) \cap C = \emptyset$, and use STEINHAUS' theorem, mentioned in Problem 2.23 and Problem 3.6, to show that every measurable subset of M must have measure zero. Also, observe that $M^\sim = M + \xi$, and conclude a similar result for $M^\sim$. To finish the problem use part *a.*]

2.26. ***a.*** Consider the situation as in Theorem 2.4.1, i.e., let $\mathcal{R} \subseteq \mathcal{P}(X)$ be an algebra, and let $\mu : \mathcal{R} \to \mathbb{R}^+ \cup \{+\infty\}$ be a σ-additive set function that extends to a measure (also denoted by μ) on the σ-algebra of measurable sets $\mathcal{A}$. Prove that this extension $(X, \mathcal{A}, \mu)$ is a complete measure space.

b. Let $(X, \mathcal{A}, \mu)$ be a σ-finite measure space, where $\mathcal{A}$ is the σ-algebra generated by an algebra $\mathcal{R}$. Define $(X, \mathcal{A}_0, \mu_0)$ to be the complete measure space corresponding to $(X, \mathcal{A}, \mu)$, as in Theorem 2.4.8. Prove that $\mathcal{A}_0 = \mathcal{M}(\mathcal{R})$, the σ-algebra of measurable sets generated by $\mathcal{R}$ and μ.

[*Hint.* Clearly, by definition, $\mathcal{A}_0 \subseteq \mathcal{M}(\mathcal{R})$. To prove the other inclusion, we first note that if $A \in \mathcal{A}_0$ and $\mu^*(A) < \infty$, then $A \in \mathcal{A}_0$. This follows since, for a given $n > 0$, we can find $F_n \in \mathcal{A}$ such that $A \subseteq F_n$ and $\mu^*(F_n \setminus A) < 1/n$; and hence $F = \bigcap F_n \in \mathcal{A}$. The general assertion, that $A \in \mathcal{M}(\mathcal{R})$ implies $A \in \mathcal{A}_0$, is a consequence of the fact that $(X, \mathcal{A}, \mu)$ is σ-finite.]

2.27. With regard to Example 2.3.11, prove that $\lim_{d\to\infty} m^d(B_d) = 0$. In fact, the rate at which $\lim_{d\to\infty} m^d(B_d) = 0$ can be quantified. For example, prove that if $d = 2n$ is even, then, not only is $m^d(B_d) = \pi^n/n!$, but also

$$\frac{\pi^n}{n!} = \frac{1}{\sqrt{2\pi n}} \left(\frac{e\pi}{n}\right)^n u_n,$$

where $u_n \to 1$.
[*Hint.* Use the Stirling formula,

$$\lim_{n\to\infty} \frac{(n/e)^n \sqrt{2\pi n}}{n!} = 1;$$

e.g., [504].]

2.28. Find an example of a measure space $(\mathbb{R}, \mathcal{A}, \mu)$, for which $\mathcal{B}(\mathbb{R}) \subseteq \mathcal{A}$, and where it is not true that if $A \in \mathcal{M}(\mathbb{R})$ then $A = B \cup E$, for $B \in \mathcal{B}(\mathbb{R})$ and $m(E) = 0$.

2.29. Let $\mu : \mathcal{R} \to \mathbb{R}^+ \cup \{\infty\}$ be a σ-additive set function on an algebra $\mathcal{R}$. Prove that the outer measure μ^* is complete on the σ-algebra $\mathcal{A}$ of measurable sets generated by $\mathcal{R}$.

2.30. Prove Theorem 2.4.9.

2.31. Prove that there is a continuous function $f : [0,1] \to \mathbb{R}$ such that $f(x) \in \mathbb{Q}$ *m-a.e.*, and for which f is not constant on any interval.
[*Hint.* At the first step form a sequence $\{f_{n,1} : n = 1, \ldots\}$ of continuous functions such that $f_{n,1} \to f_1$ uniformly and f_1 has rational values on a set of Lebesgue measure $1/2$.]

2.32. For each $x \in (0,1]$, let $f(x) = \sum a_j/j$ when $x = .a_1 \ldots$ (2), where we take the expansion of x to have an infinite number of 1s in the case that x has the form $.a_1 \ldots a_n$ (2). Are $f^{-1}(\mathbb{R}^+)$ and $f^{-1}(\infty)$ Lebesgue measurable sets?

2.33. Let f be a decreasing and bounded real-valued function on $[0,1]$. Show that there is a sequence $\{f_n : n = 1, \ldots\}$ of continuous decreasing functions such that $f_n \to f$ *m-a.e.*

2.34. Prove Theorem 2.5.4.
[*Hint.* Define $A \in \mathcal{A}$ with empty complement in terms of $\{f_n : n = 1, \ldots\}$ and f, set $g_n = f_n \mathbb{1}_A$, $g = f \mathbb{1}_A$, and use Proposition 2.4.16.]

2.35. Let $f : \mathbb{R} \to \mathbb{R}$ be Borel measurable. Prove that $f^{-1}(\mathcal{B}(\mathbb{R})) \subseteq \mathcal{B}(\mathbb{R})$.

2.36. ***a.*** Let $f : [a,b] \to \mathbb{R}$ be a Lebesgue measurable function. Prove that

$$\forall\, B \in \mathcal{B}(\mathbb{R}), \quad f^{-1}(B) \in \mathcal{M}([a,b]).$$

Thus, Lebesgue measurable functions f on $[a,b]$ are characterized by the condition, $f^{-1}(\mathcal{B}(\mathbb{R})) \subseteq \mathcal{M}([a,b])$, as Example 2.4.14 shows.
[*Hint.* Let $\mathcal{C} = \{A \subseteq \mathbb{R} : f^{-1}(A) \in \mathcal{M}([a,b])\}$ and prove that $\mathcal{C}$ is a σ-algebra.]
b. Give an example of a Lebesgue measurable function $f : [a,b] \to \mathbb{R}$ for which $f^{-1}(\mathcal{M}(\mathbb{R}))$ is not contained in $\mathcal{M}([a,b])$; cf. Figure 2.2.
c. Is it true that $f^{-1}(\mathcal{B}(\mathbb{R})) \subseteq \mathcal{B}([a,b])$ for every Lebesgue measurable function $f : [a,b] \to \mathbb{R}$?

2.37. Find an everywhere discontinuous function $f : \mathbb{R} \to \mathbb{R}$ such that f is measurable on $(\mathbb{R}, \mathcal{M}(\mathbb{R}), m)$ but not measurable on $(\mathbb{R}, \mathcal{B}(\mathbb{R}), m)$.
[*Hint.* Use the idea of Example 2.4.17.]

2.38. Let the functions $f, g : (0,1) \to \mathbb{R}$ have the following property:

$$\forall\, \alpha \in \mathbb{R}, \quad m(\{x : f(x) > \alpha\}) = m(\{x : g(x) > \alpha\}). \qquad (2.28)$$

Prove that if f and g are left continuous for all $x \in (0,1)$, as well as being decreasing, then $f = g$ on $(0,1)$.
[*Hint.* Assume $f(x_0) > g(x_0)$, let $\varepsilon = f(x_0) - g(x_0)$, and find $\delta > 0$ such that

$$m(\{x : f(x) \geq f(x_0)\}) = x_0 \quad \text{and} \quad m(\{x : g(x) \geq g(x_0)\}) = x_0 - \delta.]$$

2.39. Find a sequence $\{f_n : n = 1, \ldots\}$ of functions $[0,1] \to \mathbb{R}$ such that

$$\forall\, x \in [0,1], \quad \lim_{n\to\infty} f_n(x) = 0,$$

whereas for each $[a,b] \subseteq [0,1]$, $\{f_n : n = 1, \ldots\}$ does not converge uniformly on $[a,b]$.

2.40. Prove Corollary 2.5.6.

2.41. With respect to Problem 1.13, prove or disprove the following statement: for any fixed n there is a set $X \subseteq \mathbb{R}$, with $m(X) = 0$, such that

$$\forall\, d_1, \ldots, d_n > 0,\ \exists\, x_0, \ldots, x_n \in X \text{ for which } \forall\, j = 1, \ldots, n,$$
$$d_j = x_j - x_{j-1}.$$

2.42. Let $\alpha_n \to 0$. Find a bounded Lebesgue measurable function $f : [0,1] \to \mathbb{R}$ such that $f(x - \alpha_n)$ does not tend to $f(x)$ m-*a.e.*
[*Hint.* Let $f = \mathbb{1}_E$, where E is a perfect symmetric set of positive Lebesgue measure.]

2.43. Let $(X, \mathcal{A}, \mu)$ be a measure space, let X be a locally compact Hausdorff space, let μ be a Borel measure, and assume that μ is regular on the Borel algebra $\mathcal{B}(X)$. Prove that μ need not be regular on $\mathcal{A}$.

Remark. A refinement of this exercise is found in [212], page 230, exercise (5); the related exercise (10) on page 231 is due to DIEUDONNÉ.

2.44. Let $\mathcal{L}$ be a collection of closed line segments in $\mathbb{R}^d$ and let $E(\mathcal{L})$ be the set of all endpoints of the members of $\mathcal{L}$. Any set $X \subseteq \mathbb{R}^d$ of two or more points has the form $E(\mathcal{L})$ for some $\mathcal{L}$. A set $X \subseteq \mathbb{R}^d$ is an *endset* if $X = E(\mathcal{L})$ for a pairwise disjoint collection $\mathcal{L}$. Prove the following results.

a. In $\mathbb{R}$, if S is an endset then $\operatorname{card} S \leq \aleph_0$.

b. In $\mathbb{R}^2$, if S is a closed bounded endset then $m^2(S) = 0$.

Remark. Part *b* does not extend to $\mathbb{R}^d$ for $d \geq 3$. In fact, for $d \geq 4$ there is a compact endset of positive Lebesgue measure. For this result and a discussion of the difficult case of $d = 3$, we refer to the Amer. Math. Monthly 78 (1971), 516–518, article by ANDREW M. BRUCKNER and J. G. CEDER.

2.45. ***a.*** Let $A \in \mathcal{M}(\mathbb{R})$, where $A \subseteq [0,1]$ and $0 < m(A) < 1$. Prove that

$$\inf m(A \cap I)/m(I) = 0 \text{ and } \sup m(A \cap I)/m(I) = 1,$$

where the inf and sup are taken over all nontrivial proper subintervals I of $[0,1]$.

b. Find $A \in \mathcal{M}(\mathbb{R})$, where $A \subseteq [0,1]$, such that for each nontrivial proper subinterval $I \subseteq [0,1]$,

$$m(A \cap I) > 0 \text{ and } m(A^{\sim} \cap I) > 0.$$

Remark. In BERNSTEIN's example, mentioned in Problems 1.4*b* and 2.10*c*, $\mathbb{R}$ is decomposed as a disjoint union $A \cup B$ where neither A nor B contains uncountable closed sets F but where for each such F, $A \cap F$, $B \cap F \neq \emptyset$. It is easy to check that neither A nor B is measurable; in fact, assume that A is measurable, approximate the measure from within by compact sets, and obtain a contradiction to the decomposition properties. Compare this situation with Problem 2.45*b*. Also, see Example A.6.7.

2.46. Let $(X, \mathcal{B}(X), \mu)$ be a Borel measure space on a separable metric space X; see Appendix A.1 and A.3 for definitions.

a. If μ is a bounded measure, prove that it is also regular on the Borel algebra $\mathcal{B}(X)$.

b. If μ is a σ-finite measure, prove that

$$\forall\, A \in \mathcal{B}(X), \quad \mu(A) = \sup\{\mu(F) : F \subseteq A \text{ and } F \text{ is compact}\}.$$

c. If μ is a σ-finite measure with the property that every compact set has finite measure, prove that μ is regular.

d. Find an example of a σ-finite measure on the measurable space $(X, \mathcal{B}(X))$ that does not satisfy the following condition:

$$\forall\, A \in \mathcal{B}(X), \quad \mu(A) = \inf\{\mu(U) : A \subseteq U \text{ and } U \text{ is open}\}.$$

2.47. A *homeomorphism* $h : [0,1] \to [0,1]$ is a continuous bijection whose inverse is also continuous; cf. the definition in Appendix A.1. Note that the Borel sets are invariant under homeomorphisms $h : [0,1] \to [0,1]$, and that such is not the case for the Lebesgue measurable sets (Example 2.4.14). This is another reason why in probability theory the class of "probabilizable" sets are the Borel sets and not the Lebesgue measurable sets; see Section 2.6.8.

a. Let $E \subseteq [0,1]$ be a closed nowhere dense set. Show that there is a homeomorphism $h : [0,1] \to [0,1]$ such that $m(h(E)) = 0$.
[*Hint.* Define

$$h(x) = m([0,x] \cap ([0,1] \setminus E))/m([0,1] \setminus E).]$$

b. Let $E \subseteq [0,1]$ be a set of first category. Show that there is a homeomorphism $h : [0,1] \to [0,1]$ such that $m(h(E)) = 0$.
[*Hint.* Let $\mathcal{H}$ be the family of homeomorphisms $[0,1] \to [0,1]$ with metric $\rho(f,g) = \sup_{x\in[0,1]} |f(x) - g(x)|$. If $E = \bigcup E_n$, where E_n is nowhere dense, define

$$A_{n,k} = \{h \in \mathcal{H} : m(\overline{E}_n) < 1/k\}.$$

Prove that $A_{n,k}$ is open in $\mathcal{H}$, and, setting $A = \bigcap A_{n,k}$, note that $m(h(E)) = 0$ for each $h \in A$.]

Remark. Actually it is possible to find an uncountable set $F \subseteq [0,1]$ such that

$$\forall\, h \in \mathcal{H},\ \ m(h(F)) = 0. \tag{2.29}$$

This result makes explicit use of the continuum hypothesis and was first proved by ABRAM S. BESICOVITCH [54]. For each $h \in \mathcal{H}$ there is a corresponding Hausdorff measure μ_h (we define this notion in Chapter 9), and BESICOVITCH actually proved $\mu_h(F) = 0$ for each $h \in \mathcal{H}$; (2.29) follows from this in a straightforward manner. Relative to our remark about Borel sets and homeomorphisms at the beginning of this exercise, we know that $F \in \mathcal{M}([0,1]) \setminus \mathcal{B}([0,1])$.

BESICOVITCH's solution to (2.29) is essentially equivalent to his solution of a famous conjecture made by BOREL (1919). In order to state the conjecture we say that a set $F \subseteq [0,1]$ has the *property* C if

$$\forall\, \{a_n : a_n > 0, n = 1, \ldots\},\ \exists\, \{I_n = (c_n, d_n) : d_n - c_n = a_n, n = 1, \ldots\}$$

such that

$$F \subseteq \bigcup_{n=1}^{\infty} I_n.$$

The *Borel conjecture* was that every set with the property C had to be countable. In order to construct his counterexample to the Borel conjecture, BESICOVITCH defined the notion of a *concentrated set F in the neighborhood of a given countable set H* by the property that

$$\text{card}\,(F \setminus (U \cap F)) \leq \aleph_0$$

for each open set U containing H. Using the continuum hypothesis, BESICOVITCH was able to construct such sets F for which card $F > \aleph_0$. These concentrated sets provide the solution to BOREL's conjecture as well as to (2.29); see Section 9.3.

3 The Lebesgue Integral

3.1 Motivation

An excellent description of the motivation to develop the notion of the Lebesgue integral has been given by LEBESGUE himself in an article, "Development of the integral concept" (1926), which appears in [316].

We begin by recalling the definition of the Riemann integral for certain bounded functions $f : [a, b] \to \mathbb{R}$. For any partition

$$P : a = x_0 < x_1 < \cdots < x_n = b$$

of $[a, b]$ consider the numbers

$$S_P = \sum_{i=1}^{n} M_i(x_i - x_{i-1}) \quad \text{and} \quad s_P = \sum_{i=1}^{n} m_i(x_i - x_{i-1}), \tag{3.1}$$

where

$$M_i = \sup \{f(x) : x_{i-1} < x \leq x_i\}$$

and

$$m_i = \inf \{f(x) : x_{i-1} < x \leq x_i\}.$$

Define

$$R\overline{\int_a^b} f = \inf_P S_P \quad \text{and} \quad R\underline{\int_a^b} f = \sup_P s_P. \tag{3.2}$$

Clearly, $R\overline{\int} \geq R\underline{\int}$, and we say that f is *Riemann integrable* if $R\overline{\int} = R\underline{\int}$. In this case, the *Riemann integral* of f *over* $[a, b]$ is

$$R\int_a^b f = R\underline{\int_a^b} f = R\overline{\int_a^b} f.$$

Note that

$$R\overline{\int_a^b} f = \inf \left\{ \int_a^b \psi : \psi \geq f,\ \psi = \sum_{j=1}^{n} c_j \mathbb{1}_{(x_{j-1}, x_j]} \right\},$$

J.J. Benedetto and W. Czaja, *Integration and Modern Analysis*, Birkhäuser Advanced Texts,
DOI 10.1007/978-0-8176-4656-1_3,

where $x_0 < x_1 < \cdots < x_n$ is a partition of $[a, b]$ and

$$\int_a^b \psi = \sum_{j=1}^{n} c_j (x_j - x_{j-1}).$$

Example 3.1.1. A non-Riemann-integrable function
Define the function $f : [a, b] \to \mathbb{R}$ as

$$f(x) = \begin{cases} 0, & \text{if } x \in [a, b] \text{ is irrational,} \\ 1, & \text{if } x \in [a, b] \text{ is rational.} \end{cases}$$

Clearly,

$$R\overline{\int_a^b} f = b - a \quad \text{and} \quad R\underline{\int_a^b} f = 0,$$

so that f is not Riemann integrable.

LEBESGUE's observation goes something like this. The numbers $R\underline{\int}$ and $R\overline{\int}$ will be close if somehow there is a lot of continuity in each interval of each partition. Example 3.1.1 shows that this will not happen if f has many discontinuities. LEBESGUE's goal was to collect approximately equal values of f. He proceeded in the following way. Let $f : [a, b] \to \mathbb{R}$ be a bounded Lebesgue measurable function, and consider the partition

$$Q : \alpha = y_0 < y_1 < \cdots < y_n = \beta,$$

where

$$\alpha = \inf \{f(x) : x \in [a, b]\} \quad \text{and} \quad \beta = \sup \{f(x) : x \in [a, b]\},$$

and the *norm* of Q is

$$|Q| = \max\{|y_j - y_{j-1}| : j = 1, \ldots, n - 1\}. \tag{3.3}$$

If $A_j = \{x : y_{j-1} < f(x) \leq y_j\}$, $j = 1, \ldots, n$, $A_0 = \{x : f(x) = \alpha\}$, and $y_{-1} = \alpha$, we define

$$S_Q = \sum_{j=0}^{n} y_j m(A_j) \quad \text{and} \quad s_Q = \sum_{j=0}^{n} y_{j-1} m(A_j),$$

and

$$\overline{\int_a^b} f = \inf_Q S_Q \quad \text{and} \quad \underline{\int_a^b} f = \sup_Q s_Q.$$

A major initial result is the following.

Theorem 3.1.2. Integrability of bounded measurable functions *Let $f : [a,b] \to \mathbb{R}$ be a bounded Lebesgue measurable function. Then*

$$\underline{\int_a^b} f = \overline{\int_a^b} f,$$

and the common value is denoted by $\int_a^b f$.

Proof. i. We must prove that $\sup_Q s_Q = \inf_Q S_Q$, where Q is a partition of $[\alpha, \beta]$ and $\alpha \leq f(x) \leq \beta$ for $x \in [a,b]$. We begin by noting that $-\infty < \sup_Q s_Q, \inf_Q S_Q < \infty$; in fact, $\alpha(b-a) \leq s_Q \leq S_Q \leq \beta(b-a)$. For any $\varepsilon > 0$ we shall verify that $|\sup_Q s_Q - \inf_Q S_Q| < \varepsilon$.

ii. If $Q' \subseteq Q$, we note that

$$s_{Q'} \leq s_Q \quad \text{and} \quad S_{Q'} \geq S_Q. \tag{3.4}$$

In fact, without loss of generality, take $Q' : \alpha = y_0 < \cdots < y_n = \beta$ and let $Q = Q' \cup \{y\}$, where $y_{j-1} < y < y_j$. Define $B_j = \{x : y < f(x) \leq y_j\}$ and $C_j = \{x : y_{j-1} < f(x) \leq y\}$, so that

$$y_j\, m(B_j) + y\, m(C_j) \leq y_j(m(B_j) + m(C_j)) = y_j\, m(A_j).$$

Consequently, $S_Q \leq S_{Q'}$, with a similar calculation for s_Q. Hence (3.4) is obtained.

iii. Therefore, since $\sup s_Q$ and $\inf S_Q$ are finite, there are partitions Q_1 and Q_2 of $[\alpha, \beta]$ such that

$$\left|\sup_Q\, s_Q - s_{Q_1}\right| < \frac{\varepsilon}{3} \quad \text{and} \quad \left|\inf_Q\, S_Q - S_{Q_2}\right| < \frac{\varepsilon}{3}.$$

Moreover, because of (3.4), we may assume that if $\{y_{i,k}\} = Q_k$, $k = 1, 2$, then $y_{i,k} - y_{i-1,k} < \varepsilon/(3(b-a))$. Let $\widetilde{Q}$ be the partition formed by the points in both Q_1 and Q_2, i.e., $\widetilde{Q} = Q_1 \cup Q_2 = \{\tilde{y}_i\}$. Then, $|s_{\widetilde{Q}} - S_{\widetilde{Q}}| < \varepsilon$ because of the triangle inequality and the fact that

$$0 \leq S_{\widetilde{Q}} - s_{\widetilde{Q}} = \sum_{i=0}^{n} (\tilde{y}_{i+1} - \tilde{y}_i) m(\widetilde{A}_i) < \frac{\varepsilon}{3(b-a)} \sum_{i=0}^{n} m(\widetilde{A}_i) = \frac{\varepsilon}{3}.$$

Combining these inequalities we obtain $|\inf_Q s_Q - \sup_Q S_Q| < \varepsilon$. □

LEBESGUE's partitioning of the "$f(x)$-axis" is both an ingenious technical observation and conceptual insight, allowing the full power and subtlety of measure theory to operate; see LEBESGUE's book [313], second edition, page 136; cf. [130].

Remark. If g is a simple function

$$g = \sum_{j=1}^{k} a_j \mathbb{1}_{A_j}$$

defined on $([a,b], \mathcal{M}([a,b]), m)$, it is natural to define the integral $\int_a^b g$ as

$$\int_a^b g = \sum_{j=1}^{k} a_j\, m(A_j);$$

cf. the Remark at the beginning of Section 3.2. It is not difficult to check that this definition agrees with our definition of the Lebesgue integral in Theorem 3.1.2; see also Section 3.4. In light of Theorem 2.5.5, Theorem 3.1.2 can be restated as follows. *If $f : [a,b] \to \mathbb{R}$ is a bounded Lebesgue measurable function, then there is a sequence $\{f_n : n = 1, \ldots\}$ of simple functions such that $f_n \to f$ pointwise* (in fact, the convergence is uniform and $\|f_n\|_\infty \leq \|f\|_\infty$) *and*

$$\int_a^b f_n \to \int_a^b f. \tag{3.5}$$

In this form it is interesting to compare this result with the Lebesgue dominated convergence theorem (LDC) in Section 3.3.

Because of Theorem 3.1.2 we define the *Lebesgue integral* of a bounded Lebesgue measurable function $f : [a,b] \to \mathbb{R}$ to be $\int_a^b f$.

Example 3.1.3. $\mathbb{1}_{\mathbb{Q}}$ on $[a,b]$
a. Take f as in Example 3.1.1. Clearly f is Lebesgue measurable. For the partition $R : 0 = y_0 < y_1 < \cdots < y_n = 1$ we have

$$S_R = y_1(b-a).$$

Therefore, $\inf_Q S_Q = \int_a^b f = 0$.

b. Let $\{r_n : n = 1, \ldots\}$ be an ordering of $\mathbb{Q} \cap [a,b]$ and define the function $f_m : [a,b] \to \mathbb{R}$ as

$$f_m(x) = \begin{cases} 1, & \text{if } x = r_1, \ldots, r_m, \\ 0, & \text{if otherwise.} \end{cases}$$

Clearly $f_m \geq 0$ and $\{f_m : m = 1, \ldots\}$ increases pointwise to the function f defined in Example 3.1.1. Also,

$$\forall\, m = 1, \ldots, \quad R\int_a^b f_m = 0. \tag{3.6}$$

Even though (3.6) is true, $R\int_a^b f$ does not exist, as we showed in Example 3.1.1. One of the beauties of LEBESGUE's theory is that (on a finite interval $[a,b]$, say) if $g_n \geq 0$ is measurable and the sequence $\{g_n : n = 1, \ldots\}$ increases pointwise to a bounded function g, then $\int g$ exists and equals $\lim_{n\to\infty} \int g_n$.

Remark. Example 3.1.3 is *not* a good example of a function that is Lebesgue integrable but not Riemann integrable. In fact, f is 0 m-*a.e.* and so there is a Riemann integrable function, viz., g identically 0, in the same equivalence class as f, recalling that "*a.e.*" defines an equivalence relation. In Example 3.4.6c we shall give examples of functions g whose Lebesgue integrals exist but for which there are no Riemann integrable functions h with the property that $g = h$ m-*a.e.*

3.2 The Lebesgue integral

Let $(X, \mathcal{A}, \mu)$ be a measure space. We define the *integral* of a simple function

$$f = \sum_{j=1}^{n} a_j \mathbb{1}_{A_j}, \quad A_j \in \mathcal{A},\ a_j \in \mathbb{R}, \tag{3.7}$$

to be

$$\int_X f\, d\mu = \sum_{j=1}^{n} a_j \mu(A_j)$$

if $A_j = \{x : f(x) = a_j\}$ and $\mu(A_j) < \infty$. For each such f we write

$$\int_A f\, d\mu = \int_X \mathbb{1}_A f\, d\mu, \quad A \in \mathcal{A}.$$

Remark. We can write a simple function f in many ways. Our canonical criterion will be (3.7) with the property that $A_j = \{x : f(x) = a_j\}$. On the other hand, the operation "$\int$" is a linear mapping on the vector space of simple functions that vanish outside of a set of finite measure, and so we do not have to worry if $A_j \cap A_k \neq \emptyset$. This fact can be proved in the following way. Write a given simple function $f = \sum b_j \mathbb{1}_{B_j}$ canonically as $\sum a_j \mathbb{1}_{A_j}$; define $\int f\, d\mu$ as $\sum a_j \mu(A_j)$ and check that this is well defined; finally, calculate that $\sum a_j \mathbb{1}_{A_j} = \sum b_j \mathbb{1}_{B_j}$. All the details are routine.

Theorem 3.1.2 can be generalized in a straightforward way to the following context.

Theorem 3.2.1. Measurability criterion for bounded functions
Let $(X, \mathcal{A}, \mu)$ be a complete finite measure space, and let $f : X \to \mathbb{R}$ be a bounded function. The function f is μ-measurable if and only if

$$\begin{aligned} &\inf \left\{ \int_X h\, d\mu : f \leq h \text{ and } h \text{ is simple} \right\} \\ &= \sup \left\{ \int_X g\, d\mu : f \geq g \text{ and } g \text{ is simple} \right\}. \end{aligned} \tag{3.8}$$

The completeness hypothesis in Theorem 3.2.1 is not required to prove (3.8).

Because of Theorem 3.2.1, we define the *μ-integral* of a μ-measurable bounded function $f : X \to \mathbb{R}$, $\mu(X) < \infty$, as

$$\int_X f \, d\mu = \inf \left\{ \int_X h \, d\mu : f \leq h \text{ and } h \text{ is simple} \right\}. \tag{3.9}$$

Example 3.2.2. Step and regulated functions
We say that $f : [a, b] \to \mathbb{R}$ is a *step function* if there exists an increasing finite sequence $\{x_j : j = 0, \ldots, n\} \subseteq [a, b]$, such that $x_0 = a$, $x_n = b$, and f is constant in each of the open intervals (x_j, x_{j+1}), $j = 0, \ldots, n-1$. Also, $g : [a, b] \to \mathbb{R}$ is a *regulated function* if $\lim_{y \to x\pm} g(y)$ exist for each $x \in (a, b)$, and $\lim_{y \to a+} g(y)$ and $\lim_{y \to b-} g(y)$ exist. For each step function f there is a simple function h such that $f = h$. Further, continuous functions are regulated, as are functions of bounded variation, e.g., Chapter 4, and each regulated function is bounded m-*a.e.* A basic fact about regulated functions is the following: *f is regulated if and only if f is the uniform limit of a sequence of step functions*; cf. Theorem 2.5.5 and Example A.9.7.

Definition 3.2.3. μ-integrable function and its integral
Let $f \geq 0$ be a μ-measurable function on the measure space $(X, \mathcal{A}, \mu)$. We define

$$\int_X f \, d\mu = \sup_{h \leq f} \int_X h \, d\mu, \tag{3.10}$$

where h is a bounded μ-measurable function and $\mu(\{x : h(x) \neq 0\}) < \infty$. The function f is *μ-integrable* with *integral* $\int_X f \, d\mu$ if $\int f \, d\mu < \infty$. For $X \subseteq \mathbb{R}^d$, $\mathcal{A} = \mathcal{M}^d(X)$, and $\mu = m^d$, we write

$$\int_X f \, dm^d = \int_X f(x) \, dx = \int_X f \, dx = \int_X f.$$

This definition of a μ-integrable function for $f \geq 0$ is reasonable in light of Theorem 3.2.1. Moreover, because of Theorem 3.2.1, it agrees with our definition of an integral for bounded measurable functions in (3.9). On the other hand, there is a slight possible pathology involved if there are not enough such functions h on the given measure space $(X, \mathcal{A}, \mu)$. This point arises explicitly in Theorem 3.3.5, and we shall tacitly assume that the measure spaces with which we deal do not have this deficiency. Also, with this definition of a μ-integrable function, the innocent-looking linearity in Theorem 3.2.6*b*, and a consequence of this linearity in Theorem 3.2.7*a*, really depend on the results in Section 3.3. Since no logical problems evolve and since aesthetically the linearity should be mentioned now, we have done so.

Next, let $f : X \to \mathbb{R}^*$ be a μ-measurable function on the measure space $(X, \mathcal{A}, \mu)$, and set

$$f^+(x) = \max\{f(x), 0\} \quad \text{and} \quad f^-(x) = \max\{(-f)(x), 0\}.$$

Since f is μ-measurable, both f^+ and f^- are μ-measurable and

$$f = f^+ - f^- \quad \text{and} \quad |f| = f^+ + f^-.$$

A function f is *μ-integrable* if f^+ and f^- are μ-integrable, and we define the *μ-integral of f* as

$$\int_X f \, d\mu = \int_X f^+ \, d\mu - \int_X f^- \, d\mu. \tag{3.11}$$

Similarly, we define $\int_X f \, d\mu$ for $f : X \to \mathbb{C}$.

Proposition 3.2.4. *Let $(X, \mathcal{A}, \mu)$ be a measure space and let f be a μ-measurable function that is nonnegative μ-a.e. Then,*

$$\int_X f \, d\mu = 0 \quad \Longleftrightarrow \quad f = 0 \quad \mu\text{-a.e.}$$

Proof. Define the sets $A_n = \{x : f(x) > 1/n\}$ and $A = \{x : f(x) > 0\}$, and let $\int_X f \, d\mu = 0$. Thus, $A = \bigcup A_n$, and we need only prove that, for each n, $\mu(A_n) = 0$. If $\mu(A_n) > 0$ we set $g_n = (1/n)\mathbb{1}_{A_n}$. Hence, $g_n \leq f$ and $\int_X g_n \, d\mu = (1/n)\mu(A_n)$. Thus, $\int_X f \, d\mu > 0$, the required contradiction. Conversely, if $f = 0$ *μ-a.e.*, then its integral is 0 by definition of the integral. □

Instead of considering the space of μ-integrable functions it is more advantageous to employ the space $L^1_\mu(X)$ (defined below), each of whose elements is the collection of μ-integrable functions that are equal *μ-a.e.* There are two reasons that this is done: first, the operation of integration assigns the same value to two functions that are equal *μ-a.e.* (this is a consequence of Proposition 3.2.4 and of the linearity of the operation of taking an integral of μ-integrable functions); and, second, the natural norm topology on $L^1_\mu(X)$ is Hausdorff, a notion that essentially dispenses with the topological crises inherent in the lives of identical twins. On the other hand, there is no problem in computation if we deal with integrable functions instead of the corresponding elements in $L^1_\mu(X)$.

Definition 3.2.5. $\boldsymbol{L^1_\mu(X)}$

Let $(X, \mathcal{A}, \mu)$ be a measure space. Let $\mathcal{L}^1_\mu(X)$ be the set of all μ-measurable functions $f : X \to \mathbb{C}$ such that

$$\int_X |f| \, d\mu < \infty.$$

Let $\sim$ denote the equivalence relation $f \sim g$ if $f = g$ *μ-a.e.* We define the space $L^1_\mu(X)$ to be the collection of all equivalence classes in $\mathcal{L}^1_\mu(X)$. Moreover, if f is an element of any such equivalence class, we set

$$\|f\|_1 = \int_X |g| \, d\mu,$$

where $g = f$ μ-*a.e.*; and, with abuse of notation, we write $f \in L^1_\mu(X)$. This is well defined by Proposition 3.2.4 and the linearity of the integral, since if $f \in L^1_\mu(X)$, then $\|f\|_1 = 0$ if and only if $f = 0$ μ-*a.e.*

The next theorem contains some of the basic properties of the space $L^1_\mu(X)$. For more on this space and related $L^p_\mu(X)$ spaces see Section 5.5.

Theorem 3.2.6. Linearity, monotonicity, and additivity
Let $(X, \mathcal{A}, \mu)$ be a measure space.

a. *The set of simple functions $\sum_{j=1}^n a_j \mathbb{1}_{A_j}$, $\mu(A_j) < \infty$, is dense in $L^1_\mu(X)$ taken with the L^1-norm $\|...\|_1$.*

b. *$L^1_\mu(X)$ is a vector space, and*

$$\int_X (\alpha f + \beta g) \, d\mu = \alpha \int_X f \, d\mu + \beta \int_X g \, d\mu,$$

where $f, g \in L^1_\mu(X)$ and α and β are scalars. Thus, $L^1_\mu(X)$ is a real, respectively, complex, vector space if the elements of $L^1_\mu(X)$ are $\mathbb{R}$-valued, respectively, $\mathbb{C}$-valued.

c. *If $f, g \in L^1_\mu(X)$ and $f \leq g$ μ-a.e., then*

$$\int_X f \, d\mu \leq \int_X g \, d\mu.$$

d. *Let $A \in \mathcal{A}$ and $B \in \mathcal{A}$ be disjoint, and choose $f \in L^1_\mu(X)$. Then*

$$\int_{A \cup B} f \, d\mu = \int_A f \, d\mu + \int_B f \, d\mu.$$

Part a of Theorem 3.2.6 is a consequence of our definition of the integral, part b is a consequence of the Levi–Lebesgue theorem (Theorem 3.3.6), and the remaining parts are elementary. Theorem 3.3.6 does not depend on part b. Part a is also a consequence of Theorem 2.5.5 and the Lebesgue dominated convergence theorem (Theorem 3.3.7), which does not depend on this result.

Theorem 3.2.7. Elementary role of $\int_X |f| \, d\mu$
Let $(X, \mathcal{A}, \mu)$ be a measure space.

a. *$f \in L^1_\mu(X) \iff |f| \in L^1_\mu(X)$ and f is μ-measurable.*

b. *Assume $f \in L^1_\mu(X)$. Then $\left|\int_X f \, d\mu\right| \leq \int_X |f| \, d\mu$.*

c. *Assume $f \in L^1_\mu(X)$. Then $\left|\int_X f \, d\mu\right| = \int_X |f| \, d\mu \iff$ there is $c \in \mathbb{C}$ such that $|c| = 1$ and $cf \geq 0$ μ-a.e.*

Proof. **a.** Part a follows from Theorem 3.2.6 and the decomposition of f in terms of f^+ and f^-.

b. There is $c \in \mathbb{C}$, for which $|c| = 1$, such that

$$\left| \int_X f \, d\mu \right| = \int_X fc \, d\mu.$$

Thus, by the definition of the integral,

$$\begin{aligned}\left| \int_X fc \, d\mu \right| &= \int_X fc \, d\mu = \int_X \mathrm{Re}\,(cf) \, d\mu + i \int_X \mathrm{Im}\,(cf) \, d\mu \\ &= \int_X \mathrm{Re}\,(cf) \, d\mu \leq \int_X |cf| \, d\mu,\end{aligned}$$

where the inequality follows since $|cf| - \mathrm{Re}\,(cf) \geq 0$. Here $\mathrm{Re}\,(a)$ and $\mathrm{Im}\,(a)$ are the *real* and *imaginary parts of* $a \in \mathbb{C}$.

c. Because of part *b* we have the desired equality if and only if $\mathrm{Re}\,(cf) = |cf|$ μ-*a.e.*, and this occurs if and only if $cf \geq 0$ μ-*a.e.* □

Example 3.2.8. f nonmeasurable and $|f|$ measurable
It is necessary to assume that f is μ-measurable in Theorem 3.2.7*a*. In fact, if we consider the measure space $([0,1], \mathcal{M}([0,1]), m)$ and choose $A \notin \mathcal{M}([0,1])$, then the function

$$f(x) = \begin{cases} 1, & \text{if } x \in A, \\ -1, & \text{if } x \notin A \end{cases}$$

is not m-measurable, whereas $|f|$ is m-measurable.

Remark. As we shall see, we can find functions $f : [a,b] \to \mathbb{R}$ such that f' exists everywhere but $f' \notin L^1_m([a,b])$. There are general integration theories with the property that the fundamental theorem of calculus holds whenever f' exists everywhere on $[a,b]$; these are due to OSKAR PERRON and ARNAUD DENJOY. Such theories are important, but, as of now, they have not achieved the general success of LEBESGUE's theory; one of the reasons for this is precisely that they do not have the property of Theorem 3.2.7*a*. We refer to [370] (translated from Russian) for a modern and historical approach to the PERRON–DENJOY theories. Classically, there are some introductory remarks in [471], Chapter 11, and more full-fledged treatments in [283], [355], volume II, and [412], Chapters 6–8. Denjoy has written extensively on the subject. We should also mention the *Kempisty integral* [281] from 1925, which can be used in harmonic analysis; see [33], pages 233–234, for such an application as well as reference to its relationship to DENJOY's work.

Example 3.2.9. A recursive integral equation
Let $f_0 > 0$ on $(0,1)$ be an element of $L^1_m((0,1))$ and set

$$f_{n+1}(x) = \left(\int_0^x f_n(t) \, dt \right)^{1/2}.$$

It can be shown that $\lim_{n\to\infty} f_n(x) = x/2$.

Example 3.2.10. Improper Riemann integrability for non-Lebesgue-integrable functions

a. Let $f(x) = x^2 \sin(1/x^2)$ for $x \in (0,1]$, and set $f(0) = 0$. Then

$$f'(x) = \begin{cases} 0, & \text{if } x = 0, \\ 2x \sin\left(1/x^2\right) - \frac{2}{x}\cos\left(1/x^2\right), & \text{if } x \in (0,1]. \end{cases}$$

Note that, for $x > 0$,

$$|f'(x)| \geq \frac{2}{x}\left|\cos\left(\frac{1}{x^2}\right)\right| - 2x\left|\sin\left(\frac{1}{x^2}\right)\right| \geq \frac{2}{x}\left|\cos\left(\frac{1}{x^2}\right)\right| - 2x.$$

Define

$$I_n = \left[\left(\left(2n + \frac{1}{3}\right)\pi\right)^{-1/2}, \left(\left(2n - \frac{1}{3}\right)\pi\right)^{-1/2}\right].$$

Observe that

$$\forall\, x \in I_n, \quad \left|\cos\left(\frac{1}{x^2}\right)\right| \geq \frac{1}{2};$$

in fact, for $x = \left((2n + (1/3))\,\pi\right)^{-1/2}$,

$$\left|\cos\left(\frac{1}{x^2}\right)\right| = \left|\cos\left(2n\pi + \frac{\pi}{3}\right)\right| = \left|\cos\left(\frac{\pi}{3}\right)\right|.$$

Thus, for any $x \in I_n$, $|f'(x)| \geq (1/x) - 2x$, and so

$$\begin{aligned} \int_{I_n} |f'| &\geq \int_{I_n} \left(\frac{1}{x} - 2x\right)\, dx \\ &= \frac{1}{2}\log\left(\frac{2n + (1/3)}{2n - (1/3)}\right) - \frac{2}{3}\frac{1}{\pi(2n + (1/3))(2n - (1/3))}. \end{aligned}$$

The fact that the sequence $\{I_n : n = 1, \ldots\}$ is a disjoint family follows, since $2n+2-(1/3) > 2n+(1/3)$ implies $[(2n+2-(1/3))\pi]^{-1/2} < [(2n-(1/3))\pi]^{-1/2}$. Consequently, if we let $a_N = \sum_{n=1}^{N} 1/[(2n + (1/3))(2n - (1/3))]$ we have

$$\begin{aligned} \int_0^1 |f'| &\geq \sum_{n=1}^{N} \int_{I_n} |f'| \geq \frac{1}{2}\sum_{n=1}^{N} \log\left(\frac{2n + (1/3)}{2n - (1/3)}\right) - \frac{2a_N}{3\pi} \\ &\geq \frac{1}{2}\sum_{n=1}^{N} \log\left(1 + \frac{1}{6n}\right) - \frac{2a_N}{3\pi}. \end{aligned}$$

Because the sequence $\{a_N : N = 1, \ldots\}$ is convergent and

$$\sum_{n=1}^{N} \log\left(1 + \frac{1}{6n}\right)$$

is divergent, noting that $\sum_{n=1}^{N} 1/(6n)$ and therefore $\prod_{n=1}^{N}(1+1/(6n))$ diverge, we conclude that

$$f' \notin L_m^1([0,1]).$$

b. Consider the function f of part *a*. From the fundamental theorem of calculus for Riemann integration, e.g., Problem 1.30, we compute

$$R\int_{\varepsilon}^{1} f' = f(1) - f(\varepsilon) = \sin(1) - \varepsilon^2 \sin(1/\varepsilon^2)$$

for each $\varepsilon > 0$, and so

$$\lim_{\varepsilon \to 0} R\int_{\varepsilon}^{1} f' = \sin(1).$$

We shall see in Section 3.4 that $f' \in L_m^1([\varepsilon,1])$, since it is bounded and Riemann integrable on $[\varepsilon,1]$; consequently,

$$\lim_{\varepsilon \to 0} \int_{\varepsilon}^{1} f' = \sin(1).$$

c. We again consider the function f from part *a*. We now observe that even though f' exists everywhere on $[0,1]$, f is not a function of bounded variation; see Definition 4.1.1 and Theorem 4.3.2. In fact,

$$f(1/\sqrt{k\pi}) = 0 \quad \text{and} \quad f\left(1/\sqrt{k\pi + (\pi/2)}\right) = (-1)^k / (k\pi + (\pi/2)),$$

and so the variation of f is larger than

$$\sum_{k=1}^{\infty} \left| 0 - \frac{(-1)^k}{k\pi + (\pi/2)} \right| = \infty.$$

As we shall see in Theorem 4.6.7, if g' exists everywhere on $[a,b]$ and $g' \in L_m^1([a,b])$, then not only is g a function of bounded variation, but it is also absolutely continuous.

3.3 The Lebesgue dominated convergence theorem (LDC)

In his thesis, cf. Section 1.3, LEBESGUE notes that his dominated convergence theorem is a generalization, with simplification in the proof, of a theorem by WILLIAM F. OSGOOD [361] (1897). OSGOOD's result is Proposition 3.3.1 for the special case of a continuous function f, and Proposition 3.3.1 was originally proved by CESARE ARZELÀ [10] (1885). We shall state ARZELÀ's result before making additional remarks.

A sequence $\{f_n : n = 1, \ldots\}$ of functions $[a,b] \to \mathbb{R}$ converges *boundedly* to a function $f : [a,b] \to \mathbb{R}$ if $\{f_n : n = 1, \ldots\}$ converges pointwise to f and

$$\sup_{n \in \mathbb{N}} \|f_n\|_\infty < \infty.$$

Clearly, if $\{f_n : n = 1, \ldots\}$ is a sequence of bounded functions on $[a, b]$ and if $f_n \to f$ uniformly on $[a, b]$, then $f_n \to f$ boundedly.

Proposition 3.3.1. *Let $\{f, f_n : n = 1, \ldots\}$ be a sequence of Riemann integrable functions $f, f_n : [a, b] \to \mathbb{R}$, and assume that $f_n \to f$ boundedly. Then*

$$\lim_{n\to\infty} R \int_a^b f_n = R \int_a^b f.$$

Starting with the definition of the Riemann integral it is nontrivial to prove Proposition 3.3.1, whereas starting from the axioms of measure theory it is not difficult to prove the corresponding and more general Lebesgue dominated convergence theorem (Theorem 3.3.2). The reason for this is not so mysterious. ARZELÀ's result depends on a σ-additivity property, and, starting with RIEMANN's definition of integral, the route to proving such a property requires some effort; on the other hand, in LEBESGUE's theory the σ-additivity is essentially built into the preliminaries. The history of the elementary (that is, without LEBESGUE's theory) proofs of Proposition 3.3.1 has been recorded by WILHELMUS A. J. LUXEMBURG [334]; he also gives another elementary proof of his own that is basically a corrected version of an old proof due to HAUSDORFF (1927). The problem of "taking limits under the integral"—which, of course, has many forms—is one of the absolutely fundamental issues in analysis; consequently, new proofs of such results as Proposition 3.3.1 are valuable for providing insights on the matter of "switching limits".

ARZELÀ's original proof depended on a complicated lemma that is, in fact, an easy corollary of our Problem 2.13*b*. It was in this lemma that he derived the "countable additivity property"—mentioned in the previous paragraph—that was necessary for his theorem. Mind you, Problem 2.13*b* is straightforward to prove when one begins with the σ-additivity of Lebesgue measure. Using Problem 2.13*b*, we now give ARZELÀ's proof, properly streamlined, of Proposition 3.3.1.

Proof. (Proposition 3.3.1) Without loss of generality assume that $[a, b] = [0, 1]$, $f = 0$, and $f_n(x) \in [-1, 1]$ for all x and n. If the result is false, then $\overline{\lim}_{n\to\infty} R \int_0^1 f_n$ or $\underline{\lim}_{n\to\infty} R \int_0^1 f_n$ is not zero. Assume $\underline{\lim}_{n\to\infty} R \int_0^1 f_n = r > 0$. Define $A_n = \{x : f_n(x) \geq r/2\}$, so that $\underline{\lim}_{n\to\infty} m(A_n) > 0$. By Problem 2.13*b* we see that there is a point $y \in [0, 1]$ such that $f_n(y) \geq r/2$ for infinitely many n. This contradicts the hypothesis that $f_n(y) \to 0$. □

Observe that we assumed f to be Riemann integrable in Proposition 3.3.1. The corresponding assumption will not have to be made in Theorem 3.3.2.

Recall that $\mathbb{R}^* = \mathbb{R} \cup \{\pm\infty\}$.

Theorem 3.3.2. Special case of Lebesgue dominated convergence (LDC) theorem
Let $(X, \mathcal{A}, \mu)$ be a finite measure space and let $\{f_n : n = 1, \ldots\}$ be a sequence of measurable functions $f_n : X \to \mathbb{R}^*$ for which

$$\sup_{n \in \mathbb{N}} \|f_n\|_\infty = M < \infty.$$

If $f_n \to f$ pointwise on X then $f \in L^1_\mu(X)$, $\|f\|_\infty \leq M$, and

$$\lim_{n \to \infty} \|f_n - f\|_1 = 0;$$

in particular,

$$\lim_{n \to \infty} \int_X f_n \, d\mu = \int_X f \, d\mu. \tag{3.12}$$

Proof. Clearly f is measurable by Proposition 2.5.2, and $\|f\|_\infty \leq M$; thus $f \in L^1_\mu(X)$. Take $\varepsilon > 0$ and choose N and $A \in \mathcal{A}$, by Egorov's theorem, such that $\mu(A) < \varepsilon/(4M)$ and

$$\forall\, n \geq N,\ \forall\, x \notin A, \quad |f_n(x) - f(x)| < \frac{\varepsilon}{2\mu(X)}.$$

Thus, for all $n \geq N$,

$$\int_X |f_n - f| \, d\mu = \int_A + \int_{X \setminus A} |f_n - f| \, d\mu \leq \frac{2M\varepsilon}{4M} + \mu(X \setminus A) \frac{\varepsilon}{2\mu(X)} \leq \varepsilon. \quad \square$$

Example 3.3.3. Elementary examples for LDC
a. We can find a sequence $\{f_n : n = 1, \ldots\}$ of simple functions on $([0,1], \mathcal{M}([0,1]), m)$ such that
i. $f_n \geq 0$,
ii. $\int_0^1 f_n(x) \, dx = 1$,
iii. $f_n \to 0$ pointwise,
iv. $\overline{\lim}_{n \to \infty} \|f_n\|_\infty = \infty$.
In fact, take $f_n = n(n+1)\mathbb{1}_{[1/(n+1), 1/n]}$.
a.′ We can also find a sequence $\{f_n : n = 1, \ldots\}$ of simple functions on $([0,1], \mathcal{M}([0,1]), m)$ that satisfies
i. $f_n \geq 0$,
ii. $\int_0^1 f_n(x) \, dx = 1$,
iii'. $f_n \nrightarrow 0$ pointwise,
iv. $\overline{\lim}_{n \to \infty} \|f_n\|_\infty = \infty$.
Take $f_n = n/2\mathbb{1}_{[1/2 - 1/n, 1/2 + 1/n]}$.
b. Observe that properties *ii* and *iii* in part *a* cannot hold along with the condition that $\sup_n \|f_n\|_\infty < \infty$; for if we had *iii* and the norm boundedness we would contradict *ii* by Theorem 3.3.2.

c. Define $f_n(x) = x^n$ on $[0,1]$. Then, $f_n \to f = 0$ pointwise on $[0,1)$ but the convergence is not uniform. On the other hand, $|f_n| \leq 1$, $\int_0^1 x^n \, dx = 1/(n+1)$, and $\int_0^1 f(x) \, dx = 0$.

d. Set

$$f_n(x) = \begin{cases} 0, & \text{if } x \leq 0 \text{ and } x \geq 1/n, \\ n^2, & \text{if } 0 < x < 1/n. \end{cases}$$

Then, $f_n \to f = 0$ pointwise on $\mathbb{R}$ but once again the convergence is not uniform: $\int_{\mathbb{R}} f_n(x) \, dx = n \to \infty$ and $\int_{\mathbb{R}} f(x) \, dx = 0$.

e. Set

$$f_n(x) = \begin{cases} 1/n, & \text{if } |x| \leq n^2, \\ 0, & \text{if } |x| > n^2. \end{cases}$$

Then, $f_n \to f = 0$ uniformly on $\mathbb{R}$, $\int_{\mathbb{R}} f_n(x) \, dx = 2n \to \infty$, and $\int_{\mathbb{R}} f(x) \, dx = 0$.

We now give *Fatou's lemma* (Theorem 3.3.5) which PIERRE FATOU, a friend of LEBESGUE, published in his famous thesis [162]. In order to motivate this result we first establish some notation and make some remarks on Fourier series.

Example 3.3.4. Fourier series and Fatou
a. By definition, a function f is an element of $L^1_m(\mathbb{T})$ if f is a 1-periodic complex-valued m-measurable function defined on $\mathbb{R}$ such that, for some (and therefore for all) $r \in \mathbb{R}$,

$$\int_r^{r+1} |f| < \infty.$$

The associated measure space is designated by $(\mathbb{T}, \mathcal{M}(\mathbb{T}), m)$, and the $L^1_m(\mathbb{T})$ norm of $f \in L^1_m(\mathbb{T})$ is

$$\|f\|_1 = \int_0^1 |f|.$$

We also define $f \in L^\infty_m(\mathbb{T})$ with norm $\|f\|_\infty$ if $f \in L^1_m(\mathbb{T})$ and

$$\|f\|_\infty = \operatorname*{ess\,sup}_{x \in \mathbb{R}} |f(x)| < \infty.$$

If $f \in L^1_m(\mathbb{T})$, then the *Fourier coefficients* of f are defined by

$$\hat{f}(n) = c_n = \int_0^1 f(x) e^{2\pi i n x} \, dx, \quad n \in \mathbb{Z},$$

and the *Fourier series* of f is

$$S(f)(x) = \sum_{n \in \mathbb{Z}} c_n e^{-2\pi i n x}.$$

There are many expositions of Fourier series, from [39] to ZYGMUND's classic [524]. Also, see Appendix B.

b. Setting

$$u(r,x) = \sum_{n\in\mathbb{Z}} c_n r^n e^{-2\pi i n x}, \quad r \in (0,1),$$

FATOU proved

$$\lim_{r\to 1-} u(r,x) = f(x) \quad m\text{-}a.e. \tag{3.13}$$

This result is quite important in function theory, and in order to prove it FATOU used Theorem 3.3.5. Combining (3.13) and Theorem 3.3.5 he also obtained the *Parseval equality*

$$\int_0^1 |f(x)|^2 \, dx = \sum_{n\in\mathbb{Z}} |c_n|^2$$

for $f \in L^2_m(\mathbb{T}) = \{f : f^2 \in L^1_m(\mathbb{T})\}$, a fact that LEBESGUE initially proved only for $f \in L^\infty_m(\mathbb{T})$. Shortly thereafter, FRIGYES (FRÉDÉRIC) RIESZ used FATOU's theory to prove that $L^2_m(\mathbb{T})$ is a complete metric space with metric ρ defined by

$$\rho(f,g) = \left(\int_{\mathbb{T}} |f(x) - g(x)|^2 \, dx \right)^{1/2},$$

i.e., he proved that *$L^2_m(\mathbb{T})$ is a Hilbert space*, e.g., Appendix A.2.

Theorem 3.3.5. Fatou lemma

Let $\{f_n : n = 1, \ldots\}$ be a sequence of measurable functions $X \to \mathbb{R}^*$ defined on the measure space $(X, \mathcal{A}, \mu)$. Assume that $\{f_n : n = 1, \ldots\}$ is bounded below by some $g \in L^1_\mu(X)$, $f_n \to f$ μ-a.e., and f is μ-measurable. Then

$$\int_X f \, d\mu \le \underline{\lim}_{n\to\infty} \int_X f_n \, d\mu.$$

Proof. First note that we make no claims about integrability either in the hypothesis or conclusion, since our integrals may be unbounded.

Without loss of generality assume that g is identically 0, that $f_n \to f$ pointwise everywhere, and that $\underline{\lim} \int f_n \, d\mu < \infty$. Take a bounded measurable function h such that $0 \le h \le f$ and $h = 0$ off a set Y of finite measure; cf. the remark on pathology after Definition 3.7.6.

Define $h_n(x) = \min\{h(x), f_n(x)\}$. Thus, $0 \le h_n \le h$ and $h_n = 0$ on $Y^\sim$. Since $f_n \to f$ pointwise and $h \le f$, we have $h_n \to h$ pointwise by the definition of h_n. Consequently we apply Theorem 3.3.2 to obtain

$$\int_X h \, d\mu = \int_Y h \, d\mu = \lim_{n\to\infty} \int_Y h_n \, d\mu. \tag{3.14}$$

Now $f_n \geq h_n \geq 0$ implies $\int_X f_n \, d\mu \geq \int_Y f_n \, d\mu \geq \int_Y h_n \, d\mu$, and so

$$\varliminf_{n\to\infty} \int_X f_n \, d\mu \geq \varliminf_{n\to\infty} \int_Y h_n \, d\mu = \lim_{n\to\infty} \int_Y h_n \, d\mu.$$

This combined with (3.14) yields the result by the definition of $\int_X f \, d\mu$. □

The following result allows us to prove Theorem 3.2.6*b*. This is important since we use linearity in Theorem 3.3.7.

Theorem 3.3.6. Levi–Lebesgue theorem
Let $(X, \mathcal{A}, \mu)$ be a measure space and let $\{f_n : n = 1, \ldots\}$ be a sequence of μ-measurable $\mathbb{R}^*$-valued functions defined on X. Assume that $\{f_n : n = 1, \ldots\}$ is bounded below by some $g \in L^1_\mu(X)$ and that $\{f_n : n = 1, \ldots\}$ converges μ-a.e. to a μ-measurable function f. If

$$\forall\, n = 1, \ldots, \quad f_n \leq f \quad \mu\text{-a.e.},$$

then

$$\int_X f \, d\mu = \lim_{n\to\infty} \int_X f_n \, d\mu.$$

Proof. Since $f_n \leq f$ μ-a.e., we have $\int f_n \, d\mu \leq \int f \, d\mu$. By Theorem 3.3.5,

$$\int_X f \, d\mu \leq \varliminf_{n\to\infty} \int_X f_n \, d\mu \leq \varlimsup_{n\to\infty} \int_X f_n \, d\mu \leq \int_X f \, d\mu,$$

and we are done. □

If f_n increases to f then the conditions of Theorem 3.3.6 hold. It was in this form that the result was proved in 1906 by Beppo Levi (May 14, 1875, Torino, Italy–August 28, 1961, Rosario, Argentina). For this setting, Theorem 3.3.6 is referred to as the *monotone convergence theorem.*

The fundamental criterion for taking limits under the integral sign is the following theorem.

Theorem 3.3.7. Lebesgue dominated convergence (LDC) theorem
Let $(X, \mathcal{A}, \mu)$ be a measure space and let $\{f_n : n = 1, \ldots\}$ be a sequence of μ-measurable functions each of which is complex-valued μ-a.e. Assume that $f_n \to f$ μ-a.e., that f is μ-measurable, and that there is an element $g \in L^1_\mu(X)$ such that

$$\forall\, n = 1, \ldots, \quad |f_n| \leq g \quad \mu\text{-a.e.}$$

Then $f \in L^1_\mu(X)$ and

$$\lim_{n\to\infty} \|f - f_n\|_1 = 0.$$

In particular,

$$\int_X f \, d\mu = \lim_{n\to\infty} \int f_n \, d\mu. \tag{3.15}$$

Proof. We have $f \in L^1_\mu(X)$, since $|f| \leq g$ μ-*a.e.* Without loss of generality take f_n real-valued μ-*a.e.*, and so

$$0 \leq g - f_n \quad \mu\text{-}a.e.$$

Thus, from Theorem 3.3.5,

$$\int_X (g - f)\, d\mu \leq \varliminf_{n\to\infty} \int_X (g - f_n)\, d\mu.$$

Consequently, from properties of the $\underline{\lim}$ and $\overline{\lim}$ and since $f, f_n \in L^1_\mu(X)$,

$$\int_X g\, d\mu - \int_X f\, d\mu \leq \int_X g\, d\mu - \varlimsup_{n\to\infty} \int_X f_n\, d\mu.$$

This yields

$$\varlimsup_{n\to\infty} \int_X f_n\, d\mu \leq \int_X f\, d\mu.$$

For the opposite direction we consider $g + f_n$ and compute

$$\int_X f\, d\mu \leq \varliminf_{n\to\infty} \int_X f_n\, d\mu.$$

This proves (3.15).

Now let $g_n = |f - f_n|$, $n = 1, \ldots$. Since

$$\forall\, n = 1, \ldots, \qquad |g_n| \leq |f| + |f_n| \leq |f| + |g| \in L^1_\mu(X),$$

and, since $g_n \to 0$ μ-*a.e.*, we can apply (3.15) to the sequence $\{g_n\}$, and we obtain that

$$\lim_{n\to\infty} \int |f - f_n|\, d\mu = 0. \qquad \square$$

Remark. *Nets* (or *directed sets*) [279], page 65, or *filters* [188] arise in non-metric convergence, e.g., Appendix A.9. See [188], pages 306–307, for a brief history.

If a net of functions is indexed by a countable set then Theorem 3.3.7 holds [69], Chapter IV.3.7. We shall give an example of a convergent net for which Theorem 3.3.7 fails. Let N be the set of characteristic functions $\mathbb{1}_F$, where $F \subseteq [0, 1]$ is a finite set. By definition (which we have not yet given!), N is a *net*, since $\mathbb{1}_{F_1 \cup F_2}$ dominates the functions $\mathbb{1}_{F_1}$ and $\mathbb{1}_{F_2}$; and the net N *converges* to the function $\mathbb{1}_{[0,1]}$. On the other hand, Theorem 3.3.7 fails, since $\int \mathbb{1}_F = 0$ and $\int \mathbb{1}_{[0,1]} = 1$. The deficiency in this example is that $\mathbb{1}_F = 0$ m-*a.e.* In this context it is natural to inquire whether a genuine failure of Theorem 3.3.7 for nets is possible; for this question it is well to keep in mind the fact that measurable functions are characterized in terms of *sequences* of measurable functions; see Corollary 2.5.15.

One means of generalizing Theorem 3.3.7 is the following; see Problem 3.9.

Theorem 3.3.8. A general LDC
Let $(X, \mathcal{A})$ be a measurable space and let $\{\mu_n : n = 1, \ldots\}$ be a sequence of measures on $\mathcal{A}$ such that

$$\forall\, A \in \mathcal{A}, \quad \lim_{n\to\infty} \mu_n(A) = \mu(A), \tag{3.16}$$

where μ is a measure on $\mathcal{A}$. Assume that the sequence $\{g, g_n : n = 1, \ldots\} \subseteq L^1_\mu(X)$ satisfies the conditions that $g_n \to g$ pointwise and

$$\lim_{n\to\infty} \int_X g_n \, d\mu_n = \int_X g \, d\mu.$$

If $\{f_n : n = 1, \ldots\}$ is a sequence of functions with the properties that $f_n \to f$ pointwise, each f_n is μ_n-measurable, and

$$\forall\, n, \quad |f_n| \le g_n,$$

then $f \in L^1_\mu(X)$ and

$$\lim_{n\to\infty} \|f - f_n\|_1 = 0.$$

In particular, $\int_X f \, d\mu = \lim_{n\to\infty} \int_X f_n \, d\mu_n$.

Generally we shall refer to Theorem 3.3.2, Theorem 3.3.6, Theorem 3.3.7, and Theorem 3.3.8 as LDC (Lebesgue dominated convergence theorem) in the sequel. The hypothesis that μ in (3.16) is a measure raises the problem to find conditions such that (3.16) defines a measure. We shall study this question when we discuss weak convergence of measures in Chapter 6, and we shall do it in the context of trying to find necessary and sufficient conditions for the conclusion of LDC to hold. For now we refer to Problem 3.9*b* and the following "outline", which we shall expand on later.

Proposition 3.3.9. *Let $(X, \mathcal{A}, \mu)$ be a measure space and let $f \in L^1_\mu(X)$. Then*

$$\forall\, \varepsilon > 0,\ \exists\, \delta > 0 \text{ such that } \forall\, A \in \mathcal{A}, \text{ for which } \mu(A) < \delta, \quad \int_A |f| \, d\mu < \varepsilon.$$

Proof. The result is obvious if $\|f\|_\infty < \infty$. Define

$$f_n(x) = \begin{cases} |f|(x), & \text{if } |f(x)| \le n, \\ n, & \text{if } |f(x)| > n. \end{cases}$$

Then $\|f_n\|_\infty \le n$ and $|f_n| \to |f|$ pointwise. From LDC

$$\exists\, N \text{ such that } \forall\, n \ge N, \quad \int_X (|f| - |f_n|) \, d\mu < \varepsilon/2.$$

Letting $0 < \delta < \varepsilon/(2N)$ we have

$$\left|\int_A f\,d\mu\right| \le \int_A |f|\,d\mu = \int_A (|f| - |f_N|)\,d\mu + \int_A |f_N|\,d\mu$$
$$\le \int_X (|f| - |f_N|)\,d\mu + N\mu(A) < \varepsilon$$

if $\mu(A) < \delta$. □

Definition 3.3.10. Absolute continuity
Let $(X, \mathcal{A}, \mu)$ be a measure space.

a. A μ-measurable function f on X is *absolutely continuous* with respect to μ if

$$\forall\, \varepsilon > 0,\ \exists\, \delta > 0 \text{ such that } \forall\, A \in \mathcal{A}, \text{ for which } \mu(A) < \delta,$$
$$\int_A |f|\,d\mu < \varepsilon.$$

Thus, each element $f \in L^1_\mu(X)$ is absolutely continuous with respect to μ. In Chapter 5 we shall define a "measure ν absolutely continuous with respect to μ" and show that such measures are actually characterized by $L^1_\mu(X)$.

b. A collection $\{f_\alpha\} \subseteq L^1_\mu(X)$ is *uniformly absolutely continuous* if

$$\forall\, \varepsilon > 0,\ \exists\, \delta > 0 \text{ such that } \forall\, A \in \mathcal{A}, \text{ for which } \mu(A) < \delta, \text{ and } \forall\, \alpha,$$
$$\int_A |f_\alpha|\,d\mu < \varepsilon.$$

c. If $\mathcal{F} \subseteq \mathcal{P}(X)$ and ν is a scalar-valued function on $\mathcal{F}$, we say that ν is *Vitali continuous* if, for each decreasing sequence $\{A_n : n = 1, \ldots\} \subseteq \mathcal{F}$ for which $\bigcap A_n = \emptyset$, we can conclude that

$$\lim_{n\to\infty} \nu(A_n) = 0.$$

A sequence of Vitali continuous functions ν_m on $\mathcal{F}$ is *Vitali equicontinuous* if for each decreasing sequence $\{A_n : n = 1, \ldots\} \subseteq \mathcal{F}$, for which $\bigcap A_n = \emptyset$, we have

$$\forall\, \varepsilon > 0,\ \exists\, N \text{ such that } \forall\, n > N \text{ and } \forall\, m,$$
$$|\nu_m(A_n)| < \varepsilon.$$

d. If $\mathcal{F}$ from part c is $\mathcal{A}$ then a sequence $\{\nu_m : m = 1, \ldots\}$ of scalar-valued functions on $\mathcal{A}$ is *uniformly absolutely continuous* if

$$\forall\, \varepsilon > 0,\ \exists\, \delta > 0 \text{ such that } \forall\, A \in \mathcal{A}, \text{ for which } \mu(A) < \delta, \text{ and } \forall\, m,$$
$$|\nu_m(A)| < \varepsilon.$$

This definition obviously generalizes the above definition of uniformly absolutely continuous sets of integrable functions.

Clearly, if $\|f - f_n\|_1 \to 0$ then

$$\forall\, \varepsilon > 0, \quad \lim_{n\to\infty} \mu(\{x : |f(x) - f_n(x)| \geq \varepsilon\}) = 0. \tag{3.17}$$

We are basically interested in finding a converse to this observation in order to obtain the best possible LDC. We make the following definition. A sequence $\{f_n : n = 1, \ldots\}$ of μ-measurable functions on a measure space $(X, \mathcal{A}, \mu)$ *converges in measure* to a μ-measurable function f if (3.17) holds. VITALI made initial and deep progress in characterizing LDC with essentially the following results [486].

Theorem 3.3.11. Vitali uniform absolute continuity theorem
Let $(X, \mathcal{A}, \mu)$ be a finite measure space and choose a sequence $\{f_n : n = 1, \ldots\} \subseteq L^1_\mu(X)$. Then

$$\lim_{n\to\infty} \|f - f_n\|_1 = 0 \tag{3.18}$$

for some $f \in L^1_\mu(X)$ if and only if

i. $\{f_n : n = 1, \ldots\}$ converges in measure to a μ-measurable function f, and

ii. $\{f_n : n = 1, \ldots\}$ is uniformly absolutely continuous.

Theorem 3.3.12. Vitali equicontinuity theorem
Let $(X, \mathcal{A}, \mu)$ be a measure space and choose a sequence $\{f_n : n = 1, \ldots\} \subseteq L^1_\mu(X)$. Then (3.18) holds for some $f \in L^1_\mu(X)$ if and only if

i. $\{f_n : n = 1, \ldots\}$ converges in measure to a μ-measurable function f, and

ii. $\{\nu_n : \forall\, A \in \mathcal{A},\ \nu_n(A) = \int_A |f_n|\, d\mu\}$ is Vitali equicontinuous.

LEBESGUE proved that if $f_n \to f$ μ-a.e. on a measure space $(X, \mathcal{A}, \mu)$, where f is μ-measurable, and if $|f_n| \leq g$ μ-a.e. for some $g \in L^1_\mu(X)$, then $f_n \to f$ in measure; cf. Theorem 3.3.13*b* and Theorem 6.1.1. Further, with these hypotheses it is straightforward to deduce part *ii* in Theorem 3.3.12. Thus, LDC is a corollary of Theorem 3.3.12.

Part *a* of the following result is due to F. RIESZ, and part *b* is due to LEBESGUE; see Problem 3.26*d,e*.

Theorem 3.3.13. F. Riesz–Lebesgue theorem
Let $(X, \mathcal{A}, \mu)$ be a measure space and let $\{f, f_n : n = 1, \ldots\}$ be a sequence of measurable functions.

a. *If $f_n \to f$ in measure on $(X, \mathcal{A}, \mu)$, then there is a subsequence $\{f_{n_k} : k = 1, \ldots\}$ that converges to f pointwise μ-a.e.*

b. *Assume that $(X, \mathcal{A}, \mu)$ is a finite measure space. If $f_n \to f$ μ-a.e., then $f_n \to f$ in measure.*

Theorem 3.3.14. Strongly nonconvergent dilations
Let f be a nonconstant bounded Lebesgue measurable function defined on $\mathbb{R}$ such that

$$\forall\, x \in \mathbb{R}, \quad f(x+1) = f(x).$$

Set $f_n(x) = f(nx)$. *There is no subsequence of* $\{f_n : n = 1, \ldots\}$ *that converges* m*-a.e. on any bounded interval* $[a, b]$, *where* $b > a$.

Proof. i. Take any interval $[\alpha, \beta]$. From the periodicity of f,

$$\begin{aligned}\int_\alpha^\beta f(nx)\,dx &= \frac{1}{n}\int_{n\alpha}^{n\beta} f(x)\,dx \\ &= \frac{1}{n}\left\{\sum_{j=[n\alpha]}^{[n\beta]}\int_j^{j+1} f(x)\,dx + \int_{[n\beta]}^{n\beta} f(x)\,dx - \int_{[n\alpha]}^{n\alpha} f(x)\,dx\right\} \\ &= \frac{1}{n}(1+[n\beta]-[n\alpha])\int_0^1 f(x)\,dx + \frac{1}{n}\left(\int_{[n\beta]}^{n\beta} f(x)\,dx - \int_{[n\alpha]}^{n\alpha} f(x)\,dx\right),\end{aligned}$$

where $[x]$ is the greatest integer less than or equal to x. Clearly, the last term in the last expression is bounded by $2\|f\|_\infty/n$ and so tends to 0 as $n \to \infty$.

Now note that

$$\begin{aligned}\frac{1}{n}([n\beta]-[n\alpha]) &= \frac{1}{n}(n\beta - n\alpha + ([n\beta]-n\beta) + (n\alpha - [n\alpha])) \\ &= \beta - \alpha + \frac{[n\beta]-n\beta}{n} + \frac{n\alpha - [n\alpha]}{n}.\end{aligned}$$

Consequently,

$$\lim_{n\to\infty}\int_\alpha^\beta f_n(x)\,dx = (\beta-\alpha)\int_0^1 f(x)\,dx. \tag{3.19}$$

ii. Assume that $f(n_k x) \to g(x)$ m-a.e. on $[a, b]$ as $k \to \infty$. Take $[\alpha, \beta] \subseteq [a, b]$ and let $K = \int_0^1 f(x)\,dx$. From LDC we compute

$$\int_\alpha^\beta g(x)\,dx = \lim_{k\to\infty}\int_\alpha^\beta f(n_k x)\,dx = (\beta-\alpha)K,$$

and so

$$\int_\alpha^\beta (g(x) - K)\,dx = 0.$$

Since $[\alpha, \beta]$ is an arbitrary subinterval of $[a, b]$ we conclude that $g = K$ m*-a.e.* in $[a, b]$.

We now use $|f - g| = |f - K|$ and $[a, b]$ instead of f and $[\alpha, \beta]$ in (3.19). Thus,

$$(b-a)\int_0^1 |f(x) - K|\,dx = \lim_{k\to\infty}\int_a^b |f_{n_k}(x) - K|\,dx = 0.$$

The last equality follows by LDC, and so $f = K$ m*-a.e.* on $[0, 1]$, a contradiction. □

Example 3.3.15. Examples of strongly nonconvergent dilations
Apply Theorem 3.3.14 to the following functions.

a. $f(x) = x - [x]$ on $[0, 1]$; see Figure 3.1.

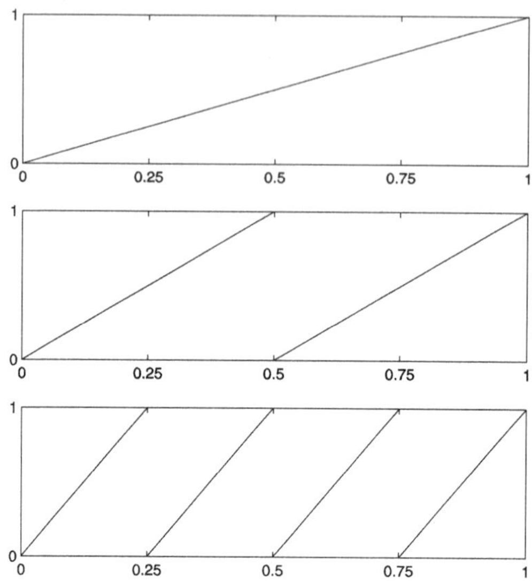

Fig. 3.1. Linear nonconvergent dilations.

b. $f(x) = \sin(2\pi x)$ on $[0, 1]$; see Figure 3.2.

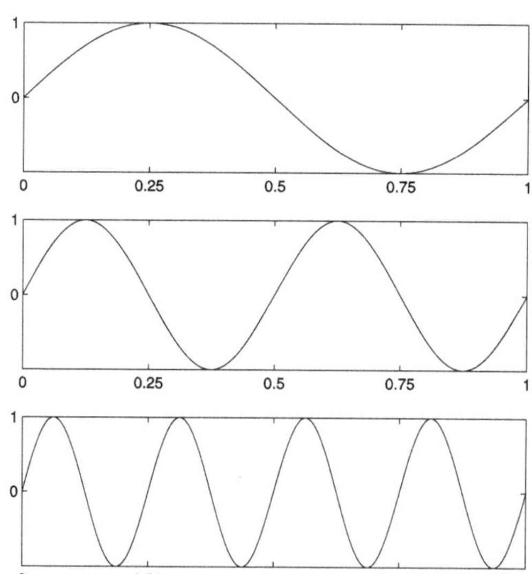

Fig. 3.2. Trigonometric nonconvergent dilations.

Remark. ***a.*** Theorem 3.3.14 is interesting in light of LIPÓT FEJÉR's theorem, e.g., Problem 3.14, which tells us, in particular, that if f is a bounded Lebesgue measurable function on $\mathbb{R}$ with period 1, then

$$\lim_{n\to\infty}\int_0^1 f(nx)\,dx = \int_0^1 f(x)\,dx;$$

cf. Section 6.3.

b. BANACH suggested the following result; see [344], Problem 162: If f is a Lebesgue measurable function on $\mathbb{R}$ with period 1, then

$$\overline{\lim_{n\to\infty}}\, f(nx) = \operatorname*{ess\,sup}_{x\in[0,1]} f(x)$$

and

$$\underline{\lim_{n\to\infty}}\, f(nx) = \operatorname*{ess\,inf}_{x\in[0,1]} f(x)$$

for m-$a.e.$ $x \in \mathbb{R}$.

3.4 The Riemann and Lebesgue integrals

We begin by proving a fact that we have implicitly assumed for a while now.

Proposition 3.4.1. *Let $f : [a,b] \to \mathbb{R}$ be a bounded function. If $R\int_a^b f$ exists, then $f \in L^1_m([a,b])$ and*

$$R\int_a^b f = \int_a^b f(x)\,dx. \tag{3.20}$$

Proof. Let g and h denote simple functions. Then

$$R\underline{\int_a^b} f \le \sup_{g\le f}\int_a^b g \le \inf_{h\ge f}\int_a^b h \le R\overline{\int_a^b} f,$$

since, for example, $R\underline{\int_a^b} f = \sup\{R\int_a^b g : g = \sum a_j \mathbb{1}_{(x_{j-1},x_j]} \le f$ and $a = x_0 < x_1 < \cdots < x_n = b\}$ and $\{g : g = \sum a_j \mathbb{1}_{(x_{j-1},x_j]}$ and $a = x_0 < x_1 < \cdots < x_n = b\}$ is a subfamily of the class of simple functions.

Equation (3.20) follows because $R\int_a^b f$ exists, by Theorem 3.2.1, and from the definition of the Lebesgue integral. □

Proposition 3.4.2. *A function $f : \mathbb{R} \to \mathbb{R}$ is continuous m-a.e. if and only if*

$$\forall\, V \subseteq \mathbb{R},\ open, \quad f^{-1}(V) = U \cup A,$$

where U is open and $A \in \mathcal{M}(\mathbb{R})$ has Lebesgue measure $m(A) = 0$.

Proof. ($\Longrightarrow$) Let $X = f^{-1}(V)$ and set $X = X_c \cup X_d$, where $X_c = X \cap C(f)$ and $X_d = X \cap D(f)$. For each $x \in X_c$ choose an open neighborhood U_x of x such that $f(U_x) \subseteq V$; we can do this since f is continuous on X_c. Clearly U_x need not be contained in X_c; for example, take the ruler function. Since $f^{-1}(V) = X$ and $U_x \subseteq f^{-1}(V)$, we have

$$X = X_c \cup X_d \subseteq \left(\bigcup_{x \in X_c} U_x \right) \cup X_d \subseteq X.$$

Let $A = X_d$ and $U = \bigcup_{x \in X_c} U_x$. Then $m(A) = 0$ by hypothesis, and U is obviously open.

($\Longleftarrow$) For each $x \in D(f)$ there is an open set V such that $f(x) \in V$, and for each open neighborhood U of x, $f(U) \not\subseteq V$. Let $f(x) \in N(s,r) = \{y \in \mathbb{R} : y \in (s-r, s+r),\ s, r \in \mathbb{Q}\} \subseteq V$. Consequently, for all such sets U, $f(U) \not\subseteq N(s,r)$. By hypothesis, $f^{-1}(N(s,r)) = U_{s,r} \cup A_{s,r}$, where $U_{s,r}$ is open and $m(A_{s,r}) = 0$. Now, because of the above observations, $x \in f^{-1}(N(s,r))$ and $x \notin U_{s,r}$. Thus, $x \in A_{s,r}$, and so $D(f) \subseteq \bigcup \{A_{s,r} : s, r \in \mathbb{Q}\}$. Hence f is continuous m-*a.e.* □

Example 3.4.3. Composition of continuous *m-a.e.* functions
We shall show that if f and g are continuous m-*a.e.* on $\mathbb{R}$, then $f \circ g$ is not necessarily continuous m-*a.e.*. Define

$$g(x) = \begin{cases} 0, & \text{if } x \text{ is irrational}, \\ 1, & \text{if } x = 0, \\ 1/|q|, & \text{if } x = p/q, \text{ where } (p,q) = 1, \end{cases}$$

and

$$f(x) = \begin{cases} 1, & \text{if } x = 1/n,\ n = 1, \ldots, \\ 0, & \text{otherwise.} \end{cases}$$

Then $f \circ g = \mathbb{1}_{\mathbb{Q}}$, and this function is not continuous anywhere. See Problem 3.20 in this regard. Note that g is the analogue on $\mathbb{R}$ of the ruler function.

We now sketch a "first approximation" to one direction of LEBESGUE's characterization of Riemann integrable functions. The proof is elementary and does not involve Lebesgue measure.

Proposition 3.4.4. *Let f be a Riemann integrable function defined on $[a,b]$. Then f is continuous on a dense subset of $[a,b]$.*

Proof. Assume that f is a real-valued function. Let $\{P_n : n = 1, \ldots\}$ be a sequence of partitions of $[a,b]$ such that each P_n divides $[a,b]$ into n segments of equal length. Thus, $S_{P_n} - s_{P_n} \to 0$. Letting "$\varepsilon = (1/2)(b-a)$" we have

$$\exists\, m \geq 4 \text{ such that } \forall\, n \geq m,\ \sum_{i=1}^{n} (M_i - m_i)(x_i - x_{i-1}) < (1/2)(b-a).$$

By the definition of $\{P_n : n = 1, \ldots\}$, and since $n \geq 4$,

$$\sum_{i=1}^{n}(M_i - m_i) < (1/2)(b-a)\frac{n}{b-a} = \frac{n}{2} \leq n-2.$$

Consequently, for each $n \geq m$ there are at least three integers, i, where $1 \leq i \leq n$, such that $M_i - m_i < 1$.

With these estimates we generate inductively a nested sequence of intervals $[a_n, b_n] \subseteq [a_{n-1}, b_{n-1}]$ such that $a_n \neq a_{n-1}$, $b_n \neq b_{n-1}$, $b_n - a_n \leq (b-a)/4^n$, and

$$\omega(f, [a_n, b_n]) < 1/n.$$

Thus, $\bigcap [a_n, b_n] = \{x_0\}$ and f is continuous at x_0. Using the same technique we obtain continuity on a dense set.

We needed three integers above to determine continuity instead of one-sided continuity. $\square$

Theorem 3.4.5. Riemann integrability and continuity *m*-a.e.
Let $f : [a, b] \to \mathbb{R}$ be a bounded function. Then $R\int_a^b f$ exists if and only if f is continuous m-a.e.

Proof. ($\Longrightarrow$) Let

$$D_k = \left\{ x : \lim_{\delta \to 0} \sup_{y,z \in [x-\delta, x+\delta)} |f(y) - f(z)| > 1/k \right\},$$

so that $D(f) = \bigcup_k D_k$. We assume that $m(D(f)) > 0$ and we shall show that $R\int_a^b f$ does not exist.

Since $m(D(f)) > 0$, there is a k for which $m(D_k) = d_k > 0$. Let P : $x_0 < \cdots < x_n$ be a partition of $[a, b]$. Then, $\bigcup(x_{j-1}, x_j)$ covers all but a finite number of points of D_k, and so

$$\sum_{j \in \dot{P}} (x_j - x_{j-1}) \geq d_k,$$

where $j \in \dot{P}$ indicates that $(x_{j-1}, x_j) \cap D_k \neq \emptyset$. From the definition of D_k, if $j \in \dot{P}$ then

$$M_j - m_j > 1/k.$$

Thus,

$$\begin{aligned} S_P - s_P = \sum (M_j - m_j)(x_j - x_{j-1}) &\geq \sum_{j \in \dot{P}} (M_j - m_j)(x_j - x_{j-1}) \\ &> \frac{1}{k} \sum_{j \in \dot{P}} (x_j - x_{j-1}) \geq \frac{d_k}{k} > 0. \end{aligned} \tag{3.21}$$

The number k is fixed and (3.21) is true for all partitions P. Consequently $R\int_a^b f$ does not exist.

($\Longleftarrow$) Let $M = \|f\|_\infty$ and consider the functions $R\underline{\int_a^x} f$ and $R\overline{\int_a^x} f$, where $R\overline{\int_a^x} f$ is defined as $R\overline{\int_a^b} f$ if $x \geq b$ and we similarly extend $R\underline{\int_a^x} f$. For each $a \leq x_1 < x_2 \leq b$ we have

$$\left| R\overline{\int_a^{x_1}} f - R\overline{\int_a^{x_2}} f \right| \leq M(x_2 - x_1). \tag{3.22}$$

Let

$$\overline{F}(x) = R\overline{\int_a^x} f.$$

It is easy to check directly, e.g., Problem 1.30 and Problem 3.22, that

$$\overline{F}'(x) = f(x) \quad m\text{-a.e.} \tag{3.23}$$

Here, of course, we use the hypothesis that f is continuous m-a.e.

Define the functions

$$f_n(x) = n\left[\overline{F}\left(x + \frac{1}{n}\right) - \overline{F}(x)\right], \quad n = 1, \ldots.$$

Each f_n is continuous on $[a, b]$, and, because of (3.22), $\{f_n : n = 1, \ldots\}$ is a uniformly bounded sequence of functions. It is also clear, from (3.23), that $f_n \to f$ m-a.e. Consequently, we apply LDC and obtain

$$\int_a^b f = \lim_{n\to\infty} \int_a^b f_n. \tag{3.24}$$

Next observe that

$$\int_a^b f_n = n\int_b^{b+(1/n)} \overline{F} - n\int_a^{a+(1/n)} \overline{F} = \overline{F}(b) - n\int_a^{a+(1/n)} \overline{F}.$$

Also,

$$\left| n\int_a^{a+(1/n)} \overline{F} \right| \leq \sup_{x\in[a,a+(1/n)]} \left| R\overline{\int_a^x} f \right| n\int_a^{a+(1/n)} 1\, dx \leq M/n.$$

Hence,

$$\lim_{n\to\infty} \int_a^b f_n = \overline{F}(b) = R\overline{\int_a^b} f.$$

Combining this with (3.24) yields

$$\int_a^b f = R\overline{\int_a^b} f;$$

and we compute similarly that

$$\int_a^b f = R\underline{\int_a^b} f. \qquad \square$$

Example 3.4.6. Lebesgue integrable non-Riemann-integrable functions

a. Let $E \subseteq [0,1]$ be a perfect symmetric set with $m(E) > 0$, and let f be its associated Volterra function defined in Example 1.3.1. Recall that the derivative of f vanishes on E, and for x just to the right of a, in the contiguous interval (a,b), it has the value

$$2(x-a)\sin\left(\frac{1}{x-a}\right) - \cos\left(\frac{1}{x-a}\right).$$

From the symmetric definition of f on (a,b) we therefore see that f' is bounded and Lebesgue measurable. Thus, $f' \in L^1_m([0,1])$. Since f' is not continuous m-$a.e.$ we can conclude from Theorem 3.4.5 that $R\int_0^1 f'$ does not exist; see part *c.ii* and Example 3.4.10.

b. For the ruler function $r : [0,1] \to \mathbb{R}$, $R\int_0^1 r$ exists since $m(D(r)) = m(\mathbb{Q} \cap [0,1]) = 0$. The function $\mathbb{1}_{[0,1]\cap\mathbb{Q}}$ is not Riemann integrable, as we saw in Example 3.1.1. For the generalized ruler function r_γ, of which r and $\mathbb{1}_{[0,1]\cap\mathbb{Q}}$ are special cases, we observed that if $\gamma_q \to 0$ then $m(D(r_\gamma)) = m(\mathbb{Q}\cap[0,1]) = 0$, and so $R\int_0^1 r_\gamma = 0$, e.g., Example 1.3.8.

c. We have seen that $f(x) = x^{-1/2}$ and $f(x) = \mathbb{1}_{[0,1]\cap\mathbb{Q}}$ are not Riemann integrable on $[0,1]$, whereas they are both Lebesgue integrable. In some sense both of these functions provide unsatisfactory examples: in the first case f is unbounded, and although the Riemann integral necessarily supposes f to be bounded, $\int_0^1 x^{-1/2}\,dx$ exists in an "improper" Riemann sense; similarly $f = \mathbb{1}_{[0,1]\cap\mathbb{Q}} = 0$ m-$a.e.$, so that even though $R\int_0^1 \mathbb{1}_{[0,1]\cap\mathbb{Q}}$ does not exist, the function g, which is identically 0 and equal to f m-$a.e.$, is Riemann integrable. We now give genuine examples of bounded Lebesgue measurable functions f defined on $[0,1]$ such that

$$\forall\, g = f \ m\text{-}a.e., \quad R\int_0^1 g \text{ does not exist.} \tag{3.25}$$

c.i. Let $E \subseteq [0,1]$ be a perfect symmetric set with Lebesgue measure $m(E) > 0$. Setting $f = \mathbb{1}_E$, we obtain (3.25) since $m(D(f)) > 0$.

c.ii. VOLTERRA's example f (part a) has the property that f' is a bounded Lebesgue measurable function and $m(D(f')) > 0$.

A complete solution to the problem implied by Example 3.4.6*b* is given in the following result, due to ISAAC J. SCHOENBERG.

Proposition 3.4.7. *The Riemann integral of the generalized ruler function r_γ on $[0,1]$ exists if and only if $\gamma_q \to 0$; in this case $R\int_0^1 r_\gamma = 0$.*

Proof. Obviously we need only check that if $R\int_0^1 r_\gamma$ exists then $\gamma_q \to 0$. Assume not. We prove that $R\int_0^1 r_\gamma$ does not exist by showing that r_γ is not continuous at any irrational point and applying Theorem 3.4.5.

Given n, consider the $\phi(n)$ integers $r_1 < \cdots < r_{\phi(n)} \leq n$ for which $(r_j, n) = 1$, and define the partition

$$P_n : 0 < r_1/n < \cdots < r_{\phi(n)}/n < 1$$

with norm

$$g_n = \sup \left\{ \frac{r_j}{n} - \frac{r_{j-1}}{n} : j = 1, \ldots, \phi(n) + 1, \ r_0 = 0, \text{ and } r_{\phi(n)+1} = n \right\}.$$

Here ϕ is the *Euler function* mentioned in Problem 1.25. We now make an act of faith and state that

$$\lim_{n \to \infty} g_n = 0. \tag{3.26}$$

The proof of (3.26) depends on a theorem due to GYÖRGY (GEORGE) PÓLYA, e.g., [374], I, problem 188, which proves that the partitions P_n are asymptotically uniformly distributed; recall Problem 1.28 and see Problem 3.29 with regard to uniform distribution.

Pick any irrational number $y \in (0, 1)$, and for each n let $k = k(n)$ be the integer for which

$$\frac{r_{k-1}}{n} < y < \frac{r_k}{n}.$$

From (3.26), $\lim_{n \to \infty} r_{k(n)}/n = y$. On the other hand $r_\gamma(y) = 0$ and $r_\gamma(r_{k(n)}/n) = \gamma_n$. Since we have assumed that $\{\gamma_n : n = 1, \ldots\}$ does not tend to 0 we conclude that r_γ is not continuous m-*a.e.* □

The following is left as an exercise (Problem 3.21).

Proposition 3.4.8. *Assume that the function $f : [a, b] \to \mathbb{R}$ has the property that*

$$\forall\, x \in (a, b), \quad \lim_{y \to x\pm} f(y) \text{ exist.}$$

Then

$$R \int_a^b f \text{ exists.}$$

For perspective we have the following proposition.

Proposition 3.4.9. *If $f \in L^1_m([a, b])$ then there is a sequence $\{\alpha_n > 0 : n = 1, \ldots\}$ such that if $\beta_n \in (0, \alpha_n), n = 1, \ldots,$ then $\lim_{n \to \infty} f(x + \beta_n) = f(x)$, m-a.e.*

This result is proved by choosing a sequence $\{\alpha_n : n = 1, \ldots\}$ for which $\sum \|\tau_{\alpha_n} f - f\|_1 < \infty$, where $\tau_\alpha f(x) = f(x - \alpha)$. It should be compared with Problem 2.42; and the details of its proof are the content of Problem 3.16.

Example 3.4.10. More on the function $f(x) = 1/x^{1/2}$
a. In Problem 3.3*b* we show that $f(x) = 1/x^{1/2} \in L^1_m([0,1])$. Using this result we now observe that if $f \geq 0$ is a Lebesgue measurable function defined on $\mathbb{R}$ and if

$$\forall\, a < b,\ \forall\, r \in \mathbb{R}, \quad m(\{(a,b) \cap \{x : f(x) \geq r\}\}) > 0,$$

then it is not necessarily true that

$$\int_{\mathbb{R}} f(x)\, dx = \infty.$$

In fact, define

$$f_k(x) = \begin{cases} (x - r_k)^{-1/2}, & \text{if } x \in (r_k, r_k + 1), \\ 0, & \text{otherwise}, \end{cases}$$

and set

$$f = \sum_{k=1}^{\infty} \frac{f_k}{2^k},$$

where $\{r_k : k = 1, \ldots\} = \mathbb{Q}$. We then compute that

$$\int_{\mathbb{R}} f(x)\, dx = \frac{2}{3}.$$

Clearly, for any $a < b$ and $r > 0$ there are $q \in \mathbb{Q} \cap (a,b)$ and c such that $(q,c) \subseteq (q, q+1)$ and $f \geq r$ on (q,c).

b. In view of the function $f(x) = 1/x^{1/2}$, $x \neq 0$, $f(0) = 0$, defined on $[0,1]$, we now define a function g, unbounded on $[0,1]$, such that $g \in L^1_m([0,1])$ and the improper Riemann integral of g does not exist. Note that for $f(x) = \sin(x)/x$ on $(0,\infty)$ the opposite phenomenon is true (Problem 3.32*d*). Let $\sum_{k=1}^{\infty} a_k$ be a convergent series of positive terms, and set

$$g(x) = \sum_{k=1}^{\infty} \frac{a_k}{|x - b_k|^{1/2}},$$

where $\{b_k : k = 1, \ldots\}$ is a dense subset of $[0,1]$. Of course we could even define some sort of improper Riemann integral in this case in terms of partial sums.

3.5 Lebesgue–Stieltjes measure and integral

In this chapter we have defined and developed a general theory of integration for arbitrary measures. For historical reasons, we would like to devote this section to a class of integrals that is closely related to the Lebesgue integral and, more importantly, that may be viewed as a nascent means of formulating the spectacular equivalence of measure-theoretic integration theory and the functional-analytic theory of Radon measures developed in Chapter 7.

We shall start by recalling Example 2.3.10, where we have considered the family $\mathcal{Q} = \{(a_1, b_1] \times \cdots \times (a_d, b_d] : a_1, \ldots, a_d, b_1, \ldots, b_d \in \mathbb{R}\}$ of half-open parallelepipeds in $\mathbb{R}^d$; these are half-open intervals $(a, b]$ in $\mathbb{R}$ if $d = 1$. This family forms a *semiring* in $\mathcal{P}(\mathbb{R}^d)$ as defined in Problem 2.22. In view of Theorem 2.3.7 and Problem 2.22, in order to find a Borel measure on $\mathbb{R}^d$ it is enough to find a σ-additive nonnegative set function on a σ-algebra that contains $\mathcal{Q}$. We shall restrict our attention to the case $d = 1$, and we begin with the following observation.

Theorem 3.5.1. σ-additive set functions on $\mathcal{Q}$
Let $f : \mathbb{R} \to \mathbb{R}$ be an increasing right-continuous function. The set function μ_f defined by

$$\mu_f((a, b]) = f(b) - f(a)$$

is a nonnegative, σ-additive set function on the semiring $\mathcal{Q}$.

Proof. Clearly, μ_f is nonnegative and finitely additive, so we need to prove only σ-additivity. Let $A = \bigcup_{j=1}^{\infty} A_j$, where $\{A = (a, b], A_j = (a_j, b_j] : j = 1, \ldots\} \subseteq \mathcal{Q}$ and where the sets A_j, $j = 1, \ldots,$ are pairwise disjoint. Since $\bigcup_{j=1}^{n} A_j \subseteq A$, it follows from the finite additivity and nonnegativity of μ_f that

$$\sum_{j=1}^{n} \mu_f(A_j) \le \mu_f(A).$$

Therefore, taking the limit as $n \to \infty$, we have

$$\sum_{j=1}^{\infty} \mu_f(A_j) \le \mu_f(A).$$

To prove the opposite inequality fix $\varepsilon > 0$, and, using the right continuity of f, choose $\delta > 0$ and $\delta_j > 0$, $j = 1, \ldots,$ such that

$$\mu_f((a, a + \delta]) = f(a + \delta) - f(a) \le \varepsilon$$

and

$$\mu_f((b_j, b_j + \delta_j]) = f(b_j + \delta_j) - f(b_j) \le \frac{\varepsilon}{2^j}.$$

Then

$$[a + \delta, b] \subseteq \bigcup_{j=1}^{\infty} (a_j, b_j + \delta_j).$$

By compactness of a closed interval there exists k such that (without loss of generality)

$$[a+\delta, b] \subseteq \bigcup_{j=1}^{k} (a_j, b_j + \delta_j).$$

Therefore, the nonnegativity and finite additivity of μ_f imply that

$$\begin{aligned}\mu_f((a+\delta, b]) &\leq \sum_{j=1}^{k} \mu_f((a_j, b_j + \delta_j]) \leq \sum_{j=1}^{\infty} \mu_f((a_j, b_j + \delta_j]) \\ &\leq \sum_{j=1}^{\infty} \mu_f((a_j, b_j]) + \sum_{j=1}^{\infty} \mu_f((b_j, b_j + \delta_j]).\end{aligned}$$

Considering our choice of δ and δ_j we obtain

$$\begin{aligned}\mu_f((a,b]) &\leq \mu_f((a, a+\delta]) + \sum_{j=1}^{\infty} \mu_f((a_j, b_j]) + \sum_{j=1}^{\infty} \mu_f((b_j, b_j + \delta_j]) \\ &\leq \varepsilon + \sum_{j=1}^{\infty} \mu_f((a_j, b_j]) + \sum_{j=1}^{\infty} \frac{\varepsilon}{2^j} \\ &\leq \sum_{j=1}^{\infty} \mu_f((a_j, b_j]) + 2\varepsilon.\end{aligned}$$

We complete the proof by letting $\varepsilon \to 0$, being technically rigorous by taking the $\overline{\lim}$ of both sides. □

Theorem 3.5.1 together with Problem 2.22 implies that μ_f extends to a σ-additive set function on $\mathcal{R}$, the ring generated by $\mathcal{Q}$. Following the developments in Section 2.3 that led us to Theorem 2.3.7, we note that μ_f further extends to a σ-additive set function on the σ-ring $\mathcal{A}_{\mathcal{Q}}$ of sets measurable with respect to μ_f and $\mathcal{Q}$. Since the elements of the ring $\mathcal{R}$ are measurable, $\mathcal{A}_{\mathcal{Q}}$ is in fact a σ-algebra; and this extension of μ_f is a measure. We shall also denote this new measure by μ_f, and it is defined to be the *Lebesgue–Stieltjes measure* associated with the function f.

The assumption in Theorem 3.5.1 that f is real-valued implies in particular that μ_f is σ-finite on $\mathbb{R}$. In view of Problem 2.20 this means that there exists a unique extension of μ_f to the smallest σ-algebra generated by $\mathcal{Q}$.

We have the following "converse" of Theorem 3.5.1.

Proposition 3.5.2. *Let $\mu : \mathcal{Q} \to \mathbb{R}^+$ be a σ-additive set function on the semiring $\mathcal{Q}$ of half-open intervals $(a, b]$ in $\mathbb{R}$. There exists an increasing right-continuous function $f : \mathbb{R} \to \mathbb{R}$ such that*

$$\forall\, a < b, \quad \mu((a,b]) = f(b) - f(a). \tag{3.27}$$

Proof. We define

$$f(x) = \begin{cases} \mu((0,x]), & \text{if } x \geq 0, \\ 0, & \text{if } x = 0, \\ -\mu((x,0]), & \text{if } x < 0. \end{cases}$$

This function satisfies (3.27) since μ is additive. Clearly this choice is not unique. We see that f is increasing, since

$$a \leq b \Longrightarrow f(b) - f(a) = \mu((a,b]) \geq 0 \Longrightarrow f(b) \geq f(a).$$

To show that f is right continuous consider a sequence $\{a_n : n = 1, \ldots\} \subseteq \mathbb{R}$, where $a_1 \geq \cdots \geq a_n$ and $a_n \to b$. We can write

$$(b, a_1] = \bigcup_{j=1}^{\infty} (a_{j+1}, a_j],$$

and use the σ-additivity of μ to obtain

$$\begin{aligned} f(a_1) - f(b) = \mu((b, a_1]) &= \sum_{j=1}^{\infty} \mu((a_{j+1}, a_j]) \\ &= \sum_{j=1}^{\infty} (f(a_j) - f(a_{j+1})) \\ &= f(a_1) - \lim_{j\to\infty} f(a_j). \end{aligned}$$

□

As a consequence, we have the following result.

Proposition 3.5.3. *Let $\mu : \mathcal{B}(\mathbb{R}) \to \mathbb{R}^+ \cup \{\infty\}$ be a measure that is finite on all bounded intervals. There exists an increasing right-continuous function $f : \mathbb{R} \to \mathbb{R}$ such that (3.27) holds.*

The integral of Section 3.2 associated with the measure μ_f is the *Lebesgue–Stieltjes integral*, and the Lebesgue–Stieltjes integral of h is naturally denoted by

$$\int_{\mathbb{R}} h \, d\mu_f.$$

Clearly, Lebesgue measure and the Lebesgue integral are a special example of the Lebesgue–Stieltjes theory for the case that $f(x) = x$ and $\mu_f = m$.

In view of the results of Section 3.4, we shall now define the *Riemann–Stieltjes integral with respect to an increasing right-continuous function*, and we shall give its relationship with the Lebesgue–Stieltjes integral.

Let $f : \mathbb{R} \to \mathbb{R}$ be an increasing right-continuous function. Let $h : [a,b] \to \mathbb{R}$ be a bounded function. Analogous to Section 3.1, for any partition

$$P : a = x_0 < x_1 < \cdots < x_n = b$$

of $[a, b]$, we let

$$S_P^f(h) = \sum_{i=1}^{n} M_i(f(x_i) - f(x_{i-1})) \quad \text{and} \quad s_P^f(h) = \sum_{i=1}^{n} m_i(f(x_i) - f(x_{i-1})),$$

where

$$M_i = \sup \{h(x) : x_{i-1} < x \leq x_i\}$$

and

$$m_i = \inf \{h(x) : x_{i-1} < x \leq x_i\},$$

for $i = 1, \ldots, n$. Define

$$\overline{\int_a^b} h \, df = \inf_P \, S_P^f(h) \quad \text{and} \quad \underline{\int_a^b} h \, df = \sup_P \, s_P^f(h). \tag{3.28}$$

As in the case for the Riemann integral, we have

$$\overline{\int_a^b} h \, df \geq \underline{\int_a^b} h \, df.$$

If $\overline{\int_a^b} h \, df = \underline{\int_a^b} h \, df$, we say that h is *Riemann–Stieltjes integrable* with respect to f. In this case, the *Riemann–Stieltjes integral* of f over $[a, b]$ is

$$\int_a^b h \, df = \underline{\int_a^b} h \, df = \overline{\int_a^b} h \, df. \tag{3.29}$$

The Riemann–Stieltjes integral played a fundamental role historically in the functional-analytic formulation of integration theory exemplified by the Riesz representation theorem; see Theorem 7.1.1.

In analogy with Theorem 3.4.5 we have the following result.

Theorem 3.5.4. Riemann–Stieltjes integrability and continuity μ_f-a.e.
Let $f : \mathbb{R} \to \mathbb{R}$ be an increasing right-continuous function and let $h : [a, b] \to \mathbb{R}$ be a bounded function. Then h is Riemann–Stieltjes integrable with respect to f if and only if h is continuous μ_f-a.e.

The proof of Theorem 3.5.4 is left as an exercise (Problem 3.23).

We shall now use Theorem 3.5.4 to prove that Lebesgue–Stieltjes integration generalizes the Riemann–Stieltjes idea.

Theorem 3.5.5. Lebesgue–Stieltjes integrals generalize Riemann–Stieltjes integrals
Let $f : \mathbb{R} \to \mathbb{R}$ be an increasing right-continuous function and let $h : [a, b] \to \mathbb{R}$ be a bounded function. If h is Riemann–Stieltjes integrable, then it is integrable with respect to the measure μ_f, and

$$\int_a^b h \, df = \int_a^b h \, d\mu_f. \tag{3.30}$$

Proof. Reformulating the definition of Riemann–Stieltjes integrability, it is not difficult to see that there are increasing sequences $\{P_k\}$ and $\{Q_k\}$ of partitions with norms $|P_k|, |Q_k| \to 0$. Associated with these partitions there are sequences $\{h_k : k = 1, \ldots\}$ and $\{H_k : k = 1, \ldots\}$ of simple functions, such that $h_k \le h_{k+1} \le h \le H_{k+1} \le H_k$ on $[a, b]$, and for which

$$\overline{\int_a^b} h \, df = \lim_{k\to\infty} \int_a^b H_k \, d\mu_f$$

and

$$\underline{\int_a^b} h \, df = \lim_{k\to\infty} \int_a^b h_k \, d\mu_f.$$

As such, we define the functions

$$m(x) = \lim_{k\to\infty} h_k(x)$$

and

$$M(x) = \lim_{k\to\infty} H_k(x).$$

Clearly, $m(x) \le h(x) \le M(x)$ on $[a, b]$.

Further, we verify that if h is continuous at $x \in [a, b]$ then $m(x) = h(x) = M(x)$. Indeed, if h is continuous at x, then for every $\varepsilon > 0$ there is $\delta > 0$ such that $|x - y| < \delta$ implies $|h(x) - h(y)| < \varepsilon$. Because we have assumed that the norms of partitions converge to 0, we obtain $m(x) = h(x) = M(x)$ by letting $\varepsilon \to 0$.

On the other hand, a converse statement is true for all but countably many $x \in [a, b]$. To see this, let $x \notin \bigcup_{k=1}^{\infty}(P_k \cup Q_k)$ and suppose $m(x) = h(x) = M(x)$. Then there exists a decreasing sequence of partition intervals $(x_k, y_k]$ containing x in their interiors such that $m_k = \inf\{h(y) : y \in (x_k, y_k]\} \to h(x)$ and $M_k = \sup\{h(y) : y \in (x_k, y_k]\} \to h(x)$. This implies the continuity of h at x.

The functions h_k and H_k are μ_f-measurable. Therefore, m and M are μ_f-measurable as limits of μ_f-measurable functions.

By Theorem 3.5.4, Riemann–Stieltjes integrability of h with respect to f implies that h is continuous μ_f-*a.e.* In view of our previous observation this means that $m(x) = h(x) = M(x)$ μ_f-*a.e.* Since both m and M are μ_f-measurable, and since, in view of Problem 2.26, the Lebesgue–Stieltjes measure is complete, it follows from Theorem 2.5.4 that f is measurable with respect to the measure μ_f on the σ-algebra of measurable sets generated by $\mathcal{Q}$. Moreover, by LDC,

$$\int_a^b h \, df = \int_a^b m \, d\mu_f = \int_a^b M \, d\mu_f = \int_a^b h \, d\mu_f,$$

and so (3.30) is obtained. □

The above discussion extends easily to $\mathbb{R}^d$ in the following way. Let $\mathcal{Q}$ be the family of all half-open parallelepipeds in $\mathbb{R}^d$ and let $f : \mathbb{R} \to \mathbb{R}$ be an increasing right-continuous function. Then there exists a nonnegative finitely additive set function μ_f^d on $\mathcal{Q}$ such that

$$\mu_f^d((a_1, b_1] \times \cdots \times (a_d, b_d]) = (f(b_1) - f(a_1)) \cdots (f(b_d) - f(a_d)).$$

We leave the verification of this assertion for the reader, as well as proving the fact that such a function μ_f^d extends to a measure on $\mathcal{B}(\mathbb{R}^d)$; see Problem 3.31. With this definition of μ_f^d, all of the results in this section can be extended to $\mathbb{R}^d$; cf. [15]. For a more general approach to Lebesgue–Stieltjes measures on $\mathbb{R}^d$, see [14].

3.6 Some fundamental applications

The following three results are variants of LDC, and they are left as exercises (Problem 3.24).

Theorem 3.6.1. The interchange of limits and integration
Let $(X, \mathcal{A}, \mu)$ be a measure space and choose an open interval $S \subseteq \mathbb{R}$. Pick $t_0 \in S$. Define the function $f(x,t)$ on $X \times S$ assuming that:

i. $\forall\, t \in S$, $f(\cdot, t)$ is a μ-measurable function,

ii. $\forall\, t_0 \in S$, $\lim_{t \to t_0} f(x,t) = f(x, t_0)$ exists μ-a.e.,

and

iii. $\exists\, g \in L^1_\mu(X)$ such that $\forall\, t \in S$, $|f(x,t)| \leq g(x)$ μ-a.e.

Then $f(\cdot, t_0) \in L^1_\mu(X)$ and

$$\lim_{t \to t_0} \int_X f(x,t)\, d\mu(x) = \int_X f(x, t_0)\, d\mu(x).$$

Theorem 3.6.1 is true for the case that S is a metric space; see Appendix A.1 for the definition of a metric space.

Theorem 3.6.2. The interchange of summation and integration
Let $(X, \mathcal{A}, \mu)$ be a measure space and assume that $\{f_n : n = 1, \ldots\}$ is a sequence of μ-measurable functions for which

$$\sum_{n=1}^{\infty} \int_X |f_n|\, d\mu < \infty.$$

Then $\sum_{n=1}^{\infty} f_n \in L^1_\mu(X)$ and

$$\int_X \left(\sum_{n=1}^{\infty} f_n \right) d\mu = \sum_{n=1}^{\infty} \left(\int_X f_n\, d\mu \right).$$

Theorem 3.6.3. The interchange of differentiation and integration
Let $(X, \mathcal{A}, \mu)$ be a σ-finite measure space and choose an open interval $S \subseteq \mathbb{R}$. Define the function $f(x,t)$ on $X \times S$ assuming that:

i. $f(\cdot, t)$ is a μ-measurable function for m-a.e. $t \in S$, and there exists $t_0 \in S$ such that $f(\cdot, t_0)$ is integrable,

ii. $\frac{d}{dt} f(x,t)$ exists and is continuous on S μ-a.e.,

and

iii. $\exists\, h \in L^1_\mu(X)$ such that $\forall\, t \in S$, $|\frac{d}{dt} f(x,t)| \leq h(x)$ μ-a.e.

Then $\int_X f(x, \cdot)\, d\mu(x)$ is continuously differentiable on S and

$$\frac{d}{dt} \int_X f(x,t)\, d\mu(x) = \int_X \frac{d}{dt} f(x,t)\, d\mu(x).$$

Theorems 3.6.1, 3.6.2, and 3.6.3 generalize classical criteria that one proves with a uniform convergence hypothesis, e.g., [7], [407].

The *Riemann–Lebesgue lemma*, which we now prove, is a fundamental result in Fourier series; we indicate another proof of it in Problem 3.14*d*.

Theorem 3.6.4. Riemann–Lebesgue lemma
For each $f \in L^1_m(\mathbb{T})$,

$$\lim_{|n| \to \infty} \hat{f}(n) = 0.$$

(The Fourier coefficients $\hat{f}$ of $f \in L^1_m(\mathbb{T})$ were defined in Example 3.3.4*a*.)

Proof. If $(a,b) \subseteq [0,1)$,

$$\lim_{|n| \to \infty} \left| \int_a^b e^{2\pi i n x}\, dx \right| = \lim_{|n| \to \infty} \left| \frac{1}{2\pi n} \left(e^{2\pi i n b} - e^{2\pi i n a} \right) \right| = 0.$$

Consequently, the result is true for characteristic functions of intervals; and the expected application of our approximation theorems yields the result for arbitrary elements of $L^1_m(\mathbb{T})$. □

Example 3.6.5. Weak convergence that is not strong convergence
There is a sequence $\{f_n : n = 1, \ldots\} \subseteq L^1_m(\mathbb{T})$ such that

$$\forall\, A \in \mathcal{M}(\mathbb{T}), \quad \lim_{n \to \infty} \int_A f_n(x)\, dx = 0$$

and

$$\overline{\lim_{n \to \infty}} \int_0^1 |f_n(x)|\, dx > 0.$$

In fact, $\lim_{n \to \infty} \int_A \sin(2\pi n x)\, dx = 0$ by Theorem 3.6.4, whereas

$$\int_0^1 |\sin(2\pi n x)|\, dx = \frac{2}{\pi}.$$

From the point of view of functional analysis this will serve as an example of weak convergence that is not strong convergence. We shall discuss this phenomenon more fully in Chapter 6.

Example 3.6.6. Unboundedness associated with a generalized sinc function on $\mathbb{Z}$
We shall construct an open set $S \subseteq [0,1)$ such that

$$\overline{\lim_{n\to\infty}} \left|n\hat{\mathbb{1}}_S(n)\right| = \infty. \tag{3.31}$$

Note that $\overline{\lim}_{n\to\infty} \hat{\mathbb{1}}_S(n) = 0$ because of Theorem 3.6.4. In particular, $\mathbb{1}_S$ is not a function of bounded variation; e.g., Problem 4.4*d*, a notion we study systematically in the next chapter. If $S = [-T,T]$, $0 < T < 1/2$, then

$$\hat{\mathbb{1}}_S(n) = \frac{1}{\pi n}\sin(2\pi T n),$$

and so it is natural to seek sets S for which (3.31) holds; see [339], [78], and [203].

For each j and each $k = 1,\dots,2j$ set

$$S_{j,k} = ((k-1/2)/(2j)!, k/(2j)!).$$

These $2j$ intervals are obviously disjoint. We define

$$S_j = \bigcup_{k=1}^{2j} S_{j,k} \quad \text{and} \quad S = \bigcup_{j=1}^{\infty} S_j,$$

noting that the sequence $\{S_j : j = 1,\dots\}$ forms a disjoint family. We shall prove that

$$\lim_{m\to\infty} (2m)!\hat{\mathbb{1}}_S((2m)!) = \infty.$$

We first calculate

$$\int_{S_{j,k}} e^{2\pi i(2m)!x}\,dx = \frac{e^{2\pi i k(2m)!/(2j)!} - e^{2\pi i(k-1/2)(2m)!/(2j)!}}{2\pi i(2m)!}. \tag{3.32}$$

Observe that the numerator in (3.32) vanishes if $j < m$ and it takes the value 2 if $j = m$. Therefore,

$$\begin{aligned}\hat{\mathbb{1}}_S((2m)!) &= \sum_{j=m}^{\infty}\int_{S_j} e^{2\pi i(2m)!x}\,dx\\ &= \sum_{j=m+1}^{\infty}\int_{S_j} e^{2\pi i(2m)!x}\,dx + \frac{2m}{\pi i(2m)!}.\end{aligned}$$

Further, $\bigcup_{j=m+1}^{\infty} S_j \subseteq (0, 1/(2m+1)!)$, and so

$$\left|\sum_{j=m+1}^{\infty}\int_{S_j} e^{2\pi i(2m)!x}\,dx\right| \le \frac{1}{(2m+1)!}.$$

Consequently,

$$\left|(2m)!\hat{\mathbb{1}}_S((2m)!)\right| \ge \frac{2m}{\pi} - \frac{1}{2m+1} \to \infty.$$

3.7 Fubini and Tonelli theorem

We now present the Fubini and Tonelli theorems. These are among the most important results in analysis. Let $(X, \mathcal{A}_1, \mu)$ and $(Y, \mathcal{A}_2, \nu)$ be σ-finite measure spaces. The Fubini–Tonelli theorems give conditions such that

$$\begin{aligned}\iint_{X\times Y} f\, d(\mu \times \nu) &= \int_X \left(\int_Y f(x,y)\, d\nu(y) \right) d\mu(x) \\ &= \int_Y \left(\int_X f(x,y)\, d\mu(x) \right) d\nu(y).\end{aligned} \tag{3.33}$$

Obviously, the formal expressions in (3.33) require explanations. We shall start by constructing a *product measure* on certain subsets of the direct product $X \times Y$ of the spaces X and Y.

For $A \in \mathcal{A}_1$ and $B \in \mathcal{A}_2$, $A \times B$ is a *measurable rectangle.* The collection $\mathcal{R}$ of all such rectangles is a *semialgebra* as defined in Problem 2.22, i.e., $X \times Y \in \mathcal{R}$, it is closed under finite intersections,

$$(A_1 \times B_1) \cap (A_2 \times B_2) = (A_1 \cap A_2) \times (B_1 \cap B_2),$$

and the complement of a rectangle is a finite union of rectangles,

$$(A \times B)^{\sim} = (A^{\sim} \times B) \cup (A \times B^{\sim}) \cup (A^{\sim} \times B^{\sim}).$$

Let $\mathcal{A}$ be the collection of all finite unions of disjoint elements of $\mathcal{R}$. Clearly $\mathcal{A}$ is an algebra, as we have seen in Section 2.3.

For any rectangle $A \times B \in \mathcal{R}$ we define

$$\omega(A \times B) = \mu(A) \cdot \nu(B),$$

with the convention that $0 \cdot \infty = 0$. This convention is consistent with the assumption that the measure spaces $(X, \mathcal{A}_1, \mu)$ and $(Y, \mathcal{A}_2, \nu)$ are σ-finite. For example, if $\mu(A) = 0$ and $\nu(B) = \infty$, then there exist sets B_n, $n = 1, \ldots,$ such that $B \subseteq \bigcup B_n$ and $\nu(B_n) < \infty$, $n = 1, \ldots$. Hence, if we want ω to be extended to a σ-additive set function, we must let $\omega(A \times B) = 0$, since $\sum \omega(A \times B_n) = 0$.

Next, let $A \times B \in \mathcal{R}$ and suppose $A \times B = \bigcup_{i\in I} A_i \times B_i$, where each $A_i \times B_i \in \mathcal{R}$ and I is countable. Then we have

$$\omega(A \times B) = \sum_{i \in I} \omega(A_i \times B_i),$$

which is a consequence of the Levi–Lebesgue theorem (Theorem 3.3.6). Because of this observation, ω has a unique extension to a σ-additive set function on the algebra $\mathcal{A}$, see Problem 3.38, and we also denote this extension by ω. In particular, $\omega : \mathcal{A} \to \mathbb{R}^+$ is a well-defined σ-additive set function with the property that for any $\bigcup_{i=1}^{N} (A_i \times B_i) \in \mathcal{A}$, a disjoint union, we have

$$\omega\left(\bigcup_{i=1}^{N}(A_i \times B_i)\right) = \sum_{i=1}^{N}\omega(A_i \times B_i).$$

Hence, following our construction of a measure in Section 2.3, or using the Carathéodory theorem (Problem 2.20), there exists an extension of ω to a σ-additive set function on the σ-algebra generated by $\mathcal{A}$. We denote this σ-algebra by $\mathcal{A}_1 \times \mathcal{A}_2 \subseteq \mathcal{P}(X \times Y)$. Moreover, since we have assumed that $(X, \mathcal{A}_1, \mu)$ and $(Y, \mathcal{A}_2, \nu)$ are σ-finite, such an extension is unique. We call this extension the *product measure* on the measurable space $(X \times Y, \mathcal{A}_1 \times \mathcal{A}_2)$, and $(X \times Y, \mathcal{A}_1 \times \mathcal{A}_2, \mu \times \nu)$ is the associated *product measure space*, where $\mu \times \nu$ is the unique extension of ω.

We combine a few simple properties of product measures in the next proposition.

Proposition 3.7.1. *Let $(X, \mathcal{A}_1, \mu)$ and $(Y, \mathcal{A}_2, \nu)$ be measure spaces.*

a. *If $(X, \mathcal{A}_1, \mu)$ and $(Y, \mathcal{A}_2, \nu)$ are finite measure spaces, then $(X \times Y, \mathcal{A}_1 \times \mathcal{A}_2, \mu \times \nu)$ is a finite measure space.*

b. *If $(X, \mathcal{A}_1, \mu)$ and $(Y, \mathcal{A}_2, \nu)$ are σ-finite measure spaces, then $(X \times Y, \mathcal{A}_1 \times \mathcal{A}_2, \mu \times \nu)$ is a σ-finite measure space.*

c. *If $(X, \mathcal{A}_1, \mu)$ and $(Y, \mathcal{A}_2, \nu)$ are complete σ-finite measure spaces, then $(X \times Y, \mathcal{A}_1 \times \mathcal{A}_2, \mu \times \nu)$ is not necessarily complete.*

For a set $E \subseteq X \times Y$ and for $x \in X$, we define the *section* $E_x \subseteq Y$ as

$$E_x = \{y \in Y : (x, y) \in E\}.$$

Similarly, for $y \in Y$, we define the section $E^y \subseteq X$ to be

$$E^y = \{x \in X : (x, y) \in E\}.$$

Proposition 3.7.2. *Let $(X, \mathcal{A}_1, \mu)$ and $(Y, \mathcal{A}_2, \nu)$ be σ-finite measure spaces, and let $E \in \mathcal{A}_1 \times \mathcal{A}_2$. Then,*

$$\forall\, x \in X,\ E_x \in \mathcal{A}_2, \quad \textit{and} \quad \forall\, y \in Y,\ E^y \in \mathcal{A}_1.$$

Proof. Since the situation is symmetric for x and y, we shall prove only the first part. Thus, it is enough to show that the collection $\mathcal{E}$ of all sets $E \in \mathcal{A}_1 \times \mathcal{A}_2$ for which $E_x \in \mathcal{A}_2$, for all $x \in X$, is a σ-algebra that contains $\mathcal{A}$.

Clearly, if $E = A \times B$ is a measurable rectangle, then, for all $x \in X$, $E_x \in \mathcal{A}_2$, since either $E_x = B$ or $E_x = \emptyset$. In particular, $X \times Y \in \mathcal{E}$. Also, if $E \in \mathcal{E}$, then

$$\forall\, x \in X, \quad ((X \times Y) \setminus E)_x = Y \setminus E_x \in \mathcal{A}_2,$$

because $\mathcal{A}_2$ is an algebra. If $\{E_n : n = 1, \ldots\} \subseteq \mathcal{E}$ is a pairwise disjoint collection of elements of $\mathcal{E}$, then, from the fact that $\mathcal{A}_2$ is a σ-algebra, we have that

$$\left(\bigcup_{n=1}^{\infty} E_n\right)_x = \bigcup_{n=1}^{\infty} (E_n)_x$$

is also an element of $\mathcal{E}$. □

For a function $f : X \times Y \to \mathbb{R}^*$ we define the *sections* f_x and f^y of f by

$$f_x(y) = f^y(x) = f(x, y).$$

Proposition 3.7.3. *Let $(X \times Y, \mathcal{A}_1 \times \mathcal{A}_2, \mu \times \nu)$ be the σ-finite product measure space associated with the σ-finite measure spaces $(X, \mathcal{A}_1, \mu)$ and $(Y, \mathcal{A}_2, \nu)$. Let $f : X \times Y \to \mathbb{R}^*$ be a $(\mu \times \nu)$-measurable function. For all $x \in X$, f_x is ν-measurable, and for all $y \in Y$, f^y is μ-measurable.*

Proof. The proof is again symmetric for x and y, and so we shall prove only the first part. For a set $I \subseteq \mathbb{R}^*$, we have

$$f_x^{-1}(I) = \left(f^{-1}(I)\right)_x.$$

Thus, the ν-measurability of f_x follows from Proposition 3.7.2 and Proposition 2.4.10. □

Theorem 3.7.4. Measurability of sections
Let $(X, \mathcal{A}_1, \mu)$ and $(Y, \mathcal{A}_2, \nu)$ be σ-finite measure spaces. For every $(\mu \times \nu)$-measurable set $E \in \mathcal{A}_1 \times \mathcal{A}_2$, the functions $\nu(E_x)$ and $\mu(E^y)$ are measurable, e.g., $\nu(E_x)$ is a μ-measurable function $X \to \mathbb{R}^*$, and

$$\int_X \nu(E_x)\, d\mu(x) = \int_Y \mu(E^y)\, d\nu(y) = (\mu \times \nu)(E). \tag{3.34}$$

Proof. The proof is symmetric for $\nu(E_x)$ and $\mu(E^y)$. We shall provide details for $\nu(E_x)$.

Assume $\mu(X)$, $\nu(Y) < \infty$. If $E = A \times B$ then $\nu(E_x) = \nu(B) \cdot \mathbb{1}_A(x)$. Thus,

$$\int_X \nu(E_x)\, d\mu(x) = \int_X \nu(B) \cdot \mathbb{1}_A(x)\, d\mu(x) = \mu(A) \cdot \nu(B).$$

For any $E = \bigcup_{n=1}^N E_n = \bigcup_{n=1}^N (A_n \times B_n) \in \mathcal{A}$, a disjoint union, we have by the additivity of ν that

$$\begin{aligned}
\int_X \nu(E_x)\, d\mu(x) &= \int_X \nu\left(\left(\bigcup_{n=1}^N E_n\right)_x\right) d\mu(x) \\
&= \int_X \nu\left(\bigcup_{n=1}^N (E_n)_x\right) d\mu(x) \\
&= \int_X \sum_{n=1}^N \nu\left((E_n)_x\right)\, d\mu(x)
\end{aligned}$$

$$= \sum_{n=1}^{N} \int_X \nu(B_n) \cdot \mathbb{1}_{A_n}(x)\, d\mu(x)$$
$$= \sum_{n=1}^{N} \mu(A_n) \cdot \nu(B_n).$$

Therefore, for all $E \in \mathcal{A}$, $\nu(E_x)$ is μ-measurable and

$$\int_X \nu(E_x)\, d\mu(x) = (\mu \times \nu)(E).$$

Let $\mathcal{E}$ denote the family of all sets $E \in \mathcal{A}_1 \times \mathcal{A}_2$ for which our hypothesis is true. We have just shown that $\mathcal{A} \subseteq \mathcal{E}$.

In view of Problem 2.17, in order to show that $\mathcal{E}$ contains $\mathcal{A}_1 \times \mathcal{A}_2$, the σ-algebra generated by $\mathcal{A}$, it is enough to prove that $\mathcal{E}$ is closed under increasing countable unions and decreasing countable intersections.

If $E = \bigcup_{n=1}^{\infty} E_n$, $E_n \subseteq E_{n+1}$, $n = 1, \ldots$, then the functions $\nu((E_n)_x)$ are measurable and they increase pointwise to the function $\nu(E_x)$. Thus, $\nu(E_x)$ is also measurable. From Theorem 3.3.6 it follows that

$$\begin{aligned}\int_X \nu(E_x)\, d\mu(x) &= \lim_{n\to\infty} \int_X \nu\left((E_n)_x\right)\, d\mu(x)\\ &= \lim_{n\to\infty} (\mu \times \nu)(E_n) = (\mu \times \nu)(E),\end{aligned}$$

where the last equality holds in view of Theorem 2.4.3*d*.

The proof that $\mathcal{E}$ is closed under countable decreasing intersections is similar, and it uses Theorem 3.3.7 and the fact that the measures are bounded.

If we now assume that μ and ν are σ-finite, then

$$X \times Y = \bigcup_{n=1}^{\infty} X_n \times Y_n,$$

where $\mu \times \nu(X_n \times Y_n) < \infty$ and $X_n \times Y_n \subseteq X_{n+1} \times Y_{n+1}$, $n = 1, \ldots$. For $E \in \mathcal{A}_1 \times \mathcal{A}_2$ and for all n, $\nu((E \cap (X_n \times Y_n))_x)$ is μ-measurable and $\mu((E \cap (X_n \times Y_n))^y)$ is ν-measurable; and hence respectively so are $\nu(E_x)$ and $\mu(E^y)$. Equation (3.34) now follows from another application of Theorem 3.3.6:

$$\begin{aligned}\int_X \nu(E_x)\, d\mu(x) &= \int_X \lim_{n\to\infty} \nu\left((E \cap (X_n \times Y_n))_x\right)\, d\mu(x)\\ &= \int_X \lim_{n\to\infty} \left(\mathbb{1}_{X_n}(x) \nu\left(E_x \cap Y_n\right)\right)\, d\mu(x)\\ &= \lim_{n\to\infty} \int_{X_n} \nu\left(E_x \cap Y_n\right)\, d\mu(x)\\ &= \lim_{n\to\infty} (\mu \times \nu)(E \cap (X_n \times Y_n)) = (\mu \times \nu)(E).\end{aligned}$$

A similar argument works for $\mu(E^y)$. □

Theorem 3.7.5. Fubini theorem for integrable functions
Let $(X, \mathcal{A}_1, \mu)$ and $(Y, \mathcal{A}_2, \nu)$ be σ-finite measure spaces, and let $h : X \times Y \to \mathbb{R}^*$ be an element of $L^1_{\mu\times\nu}(X \times Y)$.

a. *For μ-a.e. $x \in X$, $h_x \in L^1_\nu(Y)$ and $\int_Y h_x \, d\nu \in L^1_\mu(X)$.*

b. *Similarly, for ν-a.e. $y \in Y$, $h^y \in L^1_\mu(X)$ and $\int_X h^y \, d\mu \in L^1_\nu(Y)$.*

c. *Moreover,*

$$\begin{aligned}\iint_{X\times Y} h(x,y)\, d(\mu\times\nu)(x,y) &= \int_X \left(\int_Y h(x,y)\, d\nu(y)\right) d\mu(x) \\ &= \int_Y \left(\int_X h(x,y)\, d\mu(x)\right) d\nu(y).\end{aligned} \tag{3.35}$$

Proof. It follows from Theorem 3.7.4 that the theorem is true for characteristic functions $\mathbb{1}_E$ of $\mu \times \nu$-measurable sets $E \subseteq X \times Y$. Thus, it also holds for any measurable simple function $\sum_{j=1}^n a_j \mathbb{1}_{E_j}$.

Let $h \in L^1_{\mu\times\nu}(X \times Y)$. Write $h = h^+ - h^-$, with $h^+, h^- \geq 0$. There exist monotone sequences, $\{h_n^+ : n = 1, \ldots\}$ and $\{h_n^- : n = 1, \ldots\}$, of simple functions, such that $h^+ = \lim_{n\to\infty} h_n^+$ and $h^- = \lim_{n\to\infty} h_n^-$, pointwise on $X \times Y$. This implies, in particular, that $h_x^\pm = \lim_{n\to\infty} (h_n^\pm)_x$ exist, and the convergence is monotone. Thus, the functions $h_x^\pm$ are ν-measurable for all x, and it follows from Theorem 3.3.6 that

$$\int_Y h_x^\pm(y)\, d\nu(y) = \lim_{n\to\infty} \int_Y \left(h_n^\pm\right)_x(y)\, d\nu(y).$$

Consequently, $\int_Y h_x^\pm(y)\, d\nu(y)$ is μ-measurable.

Another application of Theorem 3.3.6 and Theorem 3.7.4 yields

$$\begin{aligned}\int_X \left(\int_Y h_x^\pm(y)\, d\nu(y)\right) d\mu(x) &= \int_X \lim_{n\to\infty} \left(\int_Y \left(h_n^\pm\right)_x(y)\, d\nu(y)\right) d\mu(x) \\ &= \lim_{n\to\infty} \int_X \left(\int_Y \left(h_n^\pm\right)_x(y)\, d\nu(y)\right) d\mu(x) \\ &= \lim_{n\to\infty} \iint_{X\times Y} h_n^\pm(x,y)\, d(\mu\times\nu)(x,y) \\ &= \iint_{X\times Y} h^\pm(x,y)\, d(\mu\times\nu)(x,y) < \infty.\end{aligned}$$

Therefore, $\int_Y h_x^\pm(y)\, d\nu(y) \in L^1_\mu(X)$, and, consequently, $h_x^\pm \in L^1_\nu(Y)$ for μ-*a.e.* x. Analogous arguments hold for $(h^\pm)^y$. □

Example 3.7.6. Intuitive proof of Fubini theorem
a. We outline the following alternative and intuitive proof of Theorem 3.7.5. Let $(X, \mathcal{A}_1, \mu)$ and $(Y, \mathcal{A}_2, \nu)$ be σ-finite complete measure spaces and let $\mu \times \nu$ be the associated complete product measure.

As a first step, if $Q \subseteq X \times Y$ and $Q_x \in \mathcal{A}_2$ we set

$$F(x) = \nu(Q_x). \tag{3.36}$$

Consequently, as in the case of ordinary integration, we should have

$$(\mu \times \nu)(Q) = \int_X F(x)\, d\mu(x).$$

Precisely, let $\mathcal{R}_\sigma$ denote the collection of countable unions of elements of the ring $\mathcal{R}$ of measurable rectangles in $X \times Y$, and let $\mathcal{R}_{\sigma\delta}$ denote the collection of countable intersections of elements of $\mathcal{R}_\sigma$. Take $Q \in \mathcal{R}_{\sigma\delta}$ with $(\mu\times\nu)(Q) < \infty$. Then F is μ-measurable and (3.36) is valid.

The next step extends the last statement to any $Q \in \mathcal{A}_1 \times \mathcal{A}_2$ satisfying $(\mu \times \nu)(Q) < \infty$. In order to prove this we must use the fact, which is elementary to show, that if $Q \in \mathcal{A}_1 \times \mathcal{A}_2$ then there is $P \in \mathcal{R}_{\sigma\delta}$ such that $Q \subseteq P$ and $(\mu \times \nu)(P) = (\mu \times \nu)(Q)$; cf. Proposition 2.2.3. Observe that this step clinches (3.35) for the case of

$$f(x, y) = \mathbb{1}_Q(x, y).$$

The final step in the proof of Theorem 3.7.5 for "arbitrary" functions $f(x, y)$ involves standard approximation techniques.

b. We note that, in order to use the above argument for an arbitrary $Q \in \mathcal{A}_1 \times \mathcal{A}_2$, $\mathcal{R}_{\sigma\delta}$ cannot approximate Q from within by an element $P \in \mathcal{R}_\sigma$.

Let $(X, \mathcal{A}_1, \mu)$ and $(Y, \mathcal{A}_2, \nu)$ be $([0,1], \mathcal{M}([0,1]), m)$, and choose a perfect totally disconnected set $E \subseteq [0,1]$ satisfying $m(E) > 0$. Set

$$Q = \{(x, y) : x - y \in E,\ x \in X,\ y \in Y\},$$

so that $(m \times m)(Q) > 0$; cf. Problem 3.6. Now if we take any $A \times B \in \mathcal{R}$ then $A \setminus B \subseteq E$ if $A \times B \subseteq Q$. Note that if A and B are both of positive measure then $A \setminus B$ contains an interval, e.g., Problem 3.6. This contradicts the fact that $A \setminus B \subseteq E$. Thus, $(m \times m)(A \times B) = 0$.

In general, to apply Fubini's theorem, we must know that the function h is integrable with respect to the product measure $\mu \times \nu$. It is a condition that is not always easy to check. However, the following result provides us with a large class of functions for which we do not have to worry about this assumption.

Theorem 3.7.7. Tonelli theorem for nonnegative functions
Let $(X, \mathcal{A}_1, \mu)$ and $(Y, \mathcal{A}_2, \nu)$ be σ-finite measure spaces, and let $h : X \times Y \to \mathbb{R}^+$ be a $(\mu \times \nu)$-measurable function. Then, $\int_Y h_x\, d\nu$ is μ-measurable and $\int_X h^y\, d\mu$ is ν-measurable. Moreover, the equations (3.35) are valid.

The proof of Theorem 3.7.7 is straightforward and it is the content of Problem 3.39.

Theorem 3.7.8. Tonelli theorem
Let $(X, \mathcal{A}_1, \mu)$ and $(Y, \mathcal{A}_2, \nu)$ be σ-finite measure spaces, and let $h : X \times Y \to \mathbb{R}^*$ be a $(\mu \times \nu)$-measurable function such that

$$\int_X \int_Y |h(x,y)|\, d\nu(y)\, d\mu(x) < \infty \quad \text{or} \quad \int_Y \int_X |h(x,y)|\, d\mu(x)\, d\nu(y) < \infty.$$

Then $h \in L^1_{\mu\times\nu}(X \times Y)$ and the equations (3.35) are valid.

Proof. It is enough to apply Theorem 3.7.7 to the function $|h(x,y)|$. □

GUIDO FUBINI proved his theorem in 1907, [187]. Two years later, in 1909, LEONIDA TONELLI [472] observed that it is enough to assume that a function is nonnegative to obtain the conclusions of FUBINI's result. Consequently, we often refer to one or the other of their results as the *Fubini–Tonelli theorem*.

The proof of the following result is Problem 3.40.

Theorem 3.7.9. Fubini–Tonelli theorem for complete σ-finite measure spaces
Let $(X, \mathcal{A}_1, \mu)$ and $(Y, \mathcal{A}_2, \nu)$ be complete σ-finite measure spaces, and let $(X \times Y, \overline{\mathcal{A}_1 \times \mathcal{A}_2}, \overline{\mu \times \nu})$ be the completion of the product measure space $(X \times Y, \mathcal{A}_1 \times \mathcal{A}_2, \mu \times \nu)$. Let $h : X \times Y \to \mathbb{C}$ be $(\overline{\mu \times \nu})$-measurable.

a. *If $h \in L^1_{\overline{\mu\times\nu}}(X \times Y)$ then $h_x \in L^1_\nu(Y)$ for μ-a.e. $x \in X$, $h^y \in L^1_\mu(X)$ for ν-a.e. $y \in Y$, $\int_X h^y\, d\mu \in L^1_\nu(Y)$, $\int_Y h_x\, d\nu \in L^1_\mu(X)$, and the equations (3.35) are valid.*

b. *If $h \geq 0$ then h_x is ν-measurable for μ-a.e. $x \in X$, h^y is μ-measurable for ν-a.e. $y \in Y$, $\int_X h^y\, d\mu$ is ν-measurable, $\int_Y h_x\, d\nu$ is μ-measurable, and the equations (3.35) are valid.*

In the context of the Fubini–Tonelli theorem it is natural to ask whether a function that is continuous in each variable must necessarily be measurable. This question was studied by LEBESGUE in his first published paper; see [408] for more on this matter.

Theorem 3.7.10. Lebesgue's first theorem
Let $f(x,y) : \mathbb{R}^2 \to \mathbb{R}$ be continuous in x for each $y \in \mathbb{R}$ and let it be continuous in y for each $x \in \mathbb{R}$. Then f is Lebesgue measurable.

Proof. Let $X_n = \{(k/n, y) \in \mathbb{R}^2 : k \in \mathbb{Z},\, y \in \mathbb{R}\}$, for $n = 1, \ldots$. Define f_n to be equal to f on X_n; and for (x,y) such that $k/n < x < k + 1/n$ for some $k \in \mathbb{Z}$, let $f_n(x,y)$ be the linear interpolation between $f(k/n, y)$ and $f((k+1)/n, y)$. For each n, f_n is continuous on $\mathbb{R}^2$, because f is continuous in y for each x. On the other hand, since f is continuous in x for each fixed y, we have

$$f(x,y) = \lim_{n\to\infty} f_n(x,y).$$

Thus, f is Lebesgue measurable since the limit of measurable functions is measurable. □

We close this section with a complete elementary proof of the Fubini–Tonelli theorem for a special case. We need the following standard result from advanced calculus, which itself is easy to prove.

Lemma 3.7.11. *Let $g : [a,b] \to \mathbb{C}$ be bounded and assume that $\lim_{n\to\infty} S_{P_n}$ and $\lim_{n\to\infty} s_{P_n}$ exist and are equal for every sequence $\{P_n : n = 1, \ldots\}$ of partitions for which $\operatorname{card} P_n = n+1$ and the norms $|P_n| \to 0$. Then, $R\int_a^b g$ exists.*

Theorem 3.7.12. Fubini–Tonelli theorem for Lebesgue and Riemann integrable functions

Let $(X, \mathcal{A}, \mu)$ be a measure space, and let $f : X \times [a,b] \to \mathbb{C}$. Assume that $R\int_a^b f(x,y)\,dy$ exists μ-a.e., f is μ-measurable for all $y \in [a,b]$, and that there is $F \in L^1_\mu(X)$ for which

$$\forall\,(x,y) \in X \times [a,b], \quad |f(x,y)| \leq F(x).$$

Then $\int_X f(x,t)\,d\mu(x)$ is Riemann integrable and

$$\int_X \left(R\int_a^b f(x,y)\,dy \right) d\mu(x) = R\int_a^b \left(\int_X f(x,y)\,d\mu(x) \right) dy.$$

Proof. Let $\{P_n : n = 1, \ldots\}$ be a sequence of partitions of $[a,b]$ for which the norms $|P_n| \to 0$ and $\operatorname{card} P_n = n+1$. Form

$$S_{P_n}(x) = \sum_{j=1}^{n} f(x, M_j)(y_j - y_{j-1}).$$

Then

$$|S_{P_n}(x)| \leq (b-a)F(x) \tag{3.37}$$

and

$$\lim_{n\to\infty} S_{P_n}(x) = R\int_a^b f(x,y)\,dy \quad \mu\text{-a.e.}$$

Now,

$$\int_X S_{P_n}(x)\,d\mu(x) = \sum_{j=1}^{n} \left(\int_X f(x, M_j)\,d\mu(x) \right)(y_j - y_{j-1}).$$

By Theorem 3.3.7, which we can use by (3.37),

$$\begin{aligned} \lim_{n\to\infty} \sum_{j=1}^{n} \left(\int_X f(x, M_j)\,d\mu(x) \right)(y_j - y_{j-1}) &= \lim_{n\to\infty} \int_X S_{P_n}(x)\,d\mu(x) \\ &= \int_X \left(R\int_a^b f(x,y)\,dy \right) d\mu(x). \end{aligned} \tag{3.38}$$

This is true for all such $\{P_n : n = 1, \ldots\}$ and for $s_{P_n}(x)$. Consequently, by Lemma 3.7.11,

$$R\int_a^b \left(\int_X f(x,y)\, d\mu(x) \right)\, dy$$

exists, and (3.38) completes the proof. □

If a function does not satisfy the hypotheses of the Fubini–Tonelli theorem as stated in this section (noting that a measure space can always be completed), there is probably no other useful version of Fubini–Tonelli that will help to switch the order of integration—counterexamples abound; and if it is possible to switch the order at all, it will have to be done by ad hoc, and probably ingenious, methods.

3.8 Measure zero and sets of uniqueness

3.8.1 B. Riemann

BERNHARD RIEMANN's life (September 17, 1826–July 20, 1866) was documented by his friend RICHARD DEDEKIND.

In his *Habilitationsschrift* [388] RIEMANN begins with an important historical note on Fourier series, defines the Riemann integral to provide a broader setting for an analytically precise theory of Fourier series, and develops the Riemann localization principle, which is a key technique in the study of U-sets. A set $E \subseteq [0, 1) = \mathbb{T}$ is a *set of uniqueness*, or U-set, if

$$\lim_{N\to\infty} \sum_{|n|\le N} c_n e^{2\pi i n x} = 0 \text{ off } E \implies \forall\, n \in \mathbb{Z},\ c_n = 0.$$

U-sets have Lebesgue measure 0.

The problem to determine U-sets is important, since one would like to know whether the representation of a function by a trigonometric series is unique. The first explicit results in this direction were given by CANTOR, e.g., Section 3.8.2, although the following fundamental theorem for uniqueness questions, first proved by CANTOR, was apparently known by RIEMANN [314], page 110, [388]:

$$\lim_{N\to\infty} \sum_{|n|\le N} c_n e^{2\pi i n x} = 0 \text{ on } [a,b] \implies \lim_{|n|\to\infty} c_n = 0, \tag{3.39}$$

e.g., Problem 3.34*e*.

It is interesting to observe the overlap between DINI (1845–1918) and RIEMANN. RIEMANN convalesced and toured Italy during the winter of 1862, and then came to Pisa during 1863. He became quite friendly with ENRICO BETTI and EUGENIO BELTRAMI. BETTI was director of the Scuola

Normale Superiore in Pisa from 1865 to 1892; and there is a BETTI–RIEMANN correspondence that has yet to be studied as far as we know. DINI graduated from the Scuola Normale in 1864 when he was 19 years old; he then spent a year in Paris with JOSEPH L. F. BERTRAND, and returned to the Scuola Normale, where he spent the next 52 years. DINI became one of the nineteenth-century giants in real-variable and Fourier analysis, and includes VOLTERRA and VITALI among his students. RIEMANN returned to Germany for the winter of 1864–65, but then went back to Pisa. He died and was buried at Biganzolo in the northern part of Verbania, the Italian resort town on the western banks of Lago Maggiore just 15 miles south of the Swiss border.

3.8.2 G. Cantor

GEORG CANTOR (March 3, 1845–January 6, 1918) wrote several important papers on U-sets during the early 1870s (Crelle's Journal, volumes 72 and 73, and Math. Ann., volume 5). In the first, he proved (3.39) and, using this fact, proved (in the second) that $\emptyset$ is a U-set. The subsequent papers gave simplifications of proof and extensions of the basic result, showing finally that certain countable infinite sets are U-sets. The study of special types of infinite sets in this work influenced his later research activity, and it was in 1874 that he gave his famous and controversial proof that there are only countably many algebraic numbers; cf. Problem 1.24.

The remainder of his life was devoted to the study of the infinite, not only in a mathematical milieu, but often delving into various philosophical notions of infinity due to the Greeks, the scholastic philosophers, and his contemporaries. An interesting letter in this latter regard was sent by CANTOR, in London at the time, to BERTRAND RUSSELL (at Trinity College, Cambridge); he writes: "... and I am quite an adversary of Old KANT, who, in my eyes has done much harm and mischief to philosophy, even to mankind; as you easily see by the most perverted development of metaphysics in Germany in all that followed him, as in FICHTE, SCHILLING, HEGEL, HERBART, SCHOPENHAUER, HARTMAN, NIETZSCHE, etc. etc. on to this very day. I never could understand why ... reasonable ... peoples ... could follow yonder sophistical philistine, who was so bad a mathematician."

DEDEKIND was also involved in work related to CANTOR's, and many of his set-theoretic contributions are found in their 27-year correspondence (edited by JEAN CAVAILLÈS and EMMY NOETHER and published by Hermann of Paris in 1937). FEDOR A. MEDVEDEV (1966) has studied these particular results. The marvelous DEDEKIND, by the way, taught at the Gymnasium in Brunswick for fifty years beginning in 1862.

CANTOR's U-set papers were preceded by HEINE's uniqueness theorem (Crelle's Journal, volume 71) in 1870; which assumed that the given trigonometric series were uniformly convergent off arbitrary neighborhoods of a finite number of points. HEINE was at the University of Halle with CANTOR and attributes this approach to CANTOR.

CANTOR, of course, tried to prove that all countable sets are U-sets; and this was finally achieved by F. BERNSTEIN (1908) and WILLIAM H. YOUNG (1909). Actually, BERNSTEIN proved somewhat more for the setting of compact abelian groups, showing that E is a U-set if it does not contain any non-empty closed sets without isolated points.

3.8.3 D. Menshov

DMITRI E. MENSHOV (April 18, 1892–November 25, 1988) proved a key result on U-sets in 1916 by finding a non-U-set E with Lebesgue measure $m(E) = 0$. He did this just after graduating from Moscow University, where he wrote his thesis under LUZIN. His example stimulated a great deal of study about sets of measure zero; and research about specific sets of measure zero forms a part of Fourier analysis, potential theory, and fractal geometry. Actually, on the basis of MENSHOV's example, LUZIN and NINA K. BARI defined the notion "U-set" as such. Earlier, CHARLES DE LA VALLÉE-POUSSIN had proved that if a trigonometric series converges to $f \in L^1_m(\mathbb{T})$ off a countable set E then the series is the Fourier series of f; and it was generally felt that the same would be true if $m(E) = 0$. Consequently, MENSHOV's example had a certain amount of shock value to say the least.

Since we shall be discussing CARLESON's solution to LUZIN's problem in Section 4.7.5, it is interesting to note that MENSHOV solved the analogue for measurable functions in 1940–41. LUZIN, in 1915, had noted that if f is measurable on $\mathbb{T}$ and finite m-$a.e.$ then there is a trigonometric series that converges to f by both Riemann and Abel summation. The problem was to see whether such a series exists that converges pointwise m-$a.e.$ to f; MENSHOV showed precisely this. Thus, with the Carleson–Hunt theorem and with KOLMOGOROV's example of $f \in L^1_m(\mathbb{T})$ with Fourier series diverging everywhere (1926), "all that remains" (in the broad sense) of LUZIN's problem is an investigation of the analogous situation for f measurable but taking infinite values on a set of positive measure. Actually MENSHOV has an affirmative answer to this latter problem for the case of convergence in measure instead of convergence m-$a.e.$

3.8.4 N. Bari and A. Rajchman

What with MENSHOV's example, ALEKSANDER RAJCHMAN (November 13, 1890–Sachsenhausen, 1940) "seems to have been the first to realize that for sets of measure zero that occur in the theory of trigonometric series it is not so much the metric as the arithmetic properties that matter" [414] (from ANTONI ZYGMUND's biography of RAPHAËL SALEM). RAJCHMAN [381] (1922) proved the existence of closed uncountable U-sets. He was motivated by some work of HARDY and JOHN E. LITTLEWOOD [217], [218] (1914), and later (1920) STEINHAUS (who was RAJCHMAN's adviser), on Diophantine approximation to introduce "H-sets" and proved that such sets are U-sets. RAJCHMAN, in a letter to LUZIN, thought that any U-set is contained in a countable union of H-sets, and it was only in 1952 that ILYA I. PIATETSKI–SHAPIRO proved this conjecture false. The Cantor set is H and therefore U. It is easy to verify that if $m(E) > 0$ then E is not U.

Actually, BARI had proved the existence of closed uncountable U-sets in 1921 and presented her results at LUZIN's seminar at Moscow University; they were unpublished at the time of RAJCHMAN's paper, although they were communicated to him in [381] (1923). This does not minimize the importance of RAJCHMAN's results, since he established a large class of uncountable U-sets and illustrated the need for Diophantine properties in constructing such sets.

NINA K. BARI (November 19, 1901–July 15, 1961) established her first results on U-sets as an undergraduate. Throughout her life, although she engaged in several other research areas, she was an outstanding expositor and contributor on the tricky business of uniqueness. One of her major results is that the countable union of closed U-sets is a U-set, although the problem is open for the finite union of arbitrary U-sets. Another of her results, which she proved in 1936–37 and which has an interesting sequel, e.g., Section 9.6.1, shows that if α is rational and E is the perfect symmetric set determined by $\xi_k = \alpha$, then E is a U-set if and only if $1/\alpha$ is an integer. Her theorem depends heavily on Diophantine considerations. See [347] and [299] for recent developments.

3.9 Potpourri and titillation

1. "Sensational" is the proper level of sober language to describe the discovery in 1899 of a palimpsest containing a Greek Orthodox ritual, *but* with an underlying ARCHIMEDES text, theretofore thought lost to civilization. The text, written on goatskin parchment from the tenth-century, contained ARCHIMEDES' *Method* ($E\phi o\delta o\sigma$)—one might say, *Method of Attack*. See [139], [451] for more on this riveting saga in scholarship, led by the Danish classicist JOHAN L. HEIBERG in the early twentieth-century. The palimpsest disappeared after World War I and reemerged in 1998. Since then, HEIBERGS's work has expanded under the leadership of WILLIAM NOEL and ABIGAIL QUANDT of the Walters Art Museum in Baltimore.

ARCHIMEDES (c. 287 B.C.–212 B.C.) began the *Method* with a letter to ERATOSTHENES saying that he first writes down what "became clear to us by the mechanical method" and *then* he formulates these results in the geometrical rigueur du jour. The latter, of course, with its systematic sequence of logical steps, often has the capability of obfuscating ideas, especially as mathematics weaned itself from Euclidean geometrical technology through the centuries. So, the *Method* really illustrates how ARCHIMEDES divined some of his greatest creations. In his letter he writes that "there will be some among [...] future generations who by means of the method here explained will be enabled to find other theorems which have not yet fallen to our share".

As an illustration, ARCHIMEDES tells about the 1 : 2 : 3 theorem. To state this theorem, let T be an isosceles triangle whose base is twice its height. We inscribe T in a semicircle C, which, in turn, we inscribe in a rectangle R; see Figure 3.3. We generate the corresponding cone, hemisphere, and cylinder by rotating T, C, and R about the segment AO. The respective volumes of these solids are denoted by V_1, V_2, and V_3.

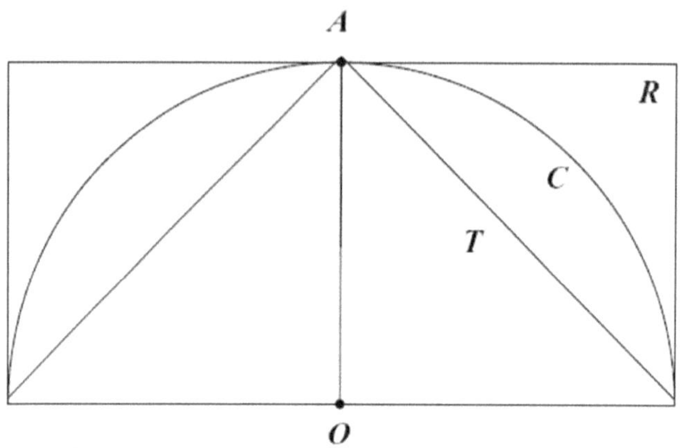

Fig. 3.3. Archimedes 1 : 2 : 3 theorem.

Theorem 3.9.1. Archimedes 1 : 2 : 3 theorem
We have

$$V_1/V_2 = 1/2 \quad \textit{and} \quad V_2/V_3 = 2/3.$$

ARCHIMEDES proved this result geometrically in *On the sphere and cylinder*. Elaborating on the contributions of DEMOCRITUS and EUDOXUS, ARCHIMEDES says that DEMOCRITUS discovered the fact that $V_1/V_3 = 1/3$, but that the part of the proof depending on the method of exhaustion was due to EUDOXUS. This proof is found in EUCLID's *Book 12*, Proposition 10; and Book 12 is still difficult to read. ARCHIMEDES completed the proof of the 1 : 2 : 3 theorem by proving that $V_1/V_2 = 1/2$.

ARCHIMEDES obviously considered the 1 : 2 : 3 theorem to be a remarkable result, since he asked that on his tombstone should be carved a sphere inscribed in a cylinder. This request was followed, and CICERO (106 B.C.–43 B.C.), the Roman orator, found the tombstone when he was quaestor of Sicily. At that time the tombstone had been neglected, and CICERO was responsible for its restoration (the tombstone was again forgotten and rediscovered in 1965). Unfortunately, CICERO more than balanced this action with

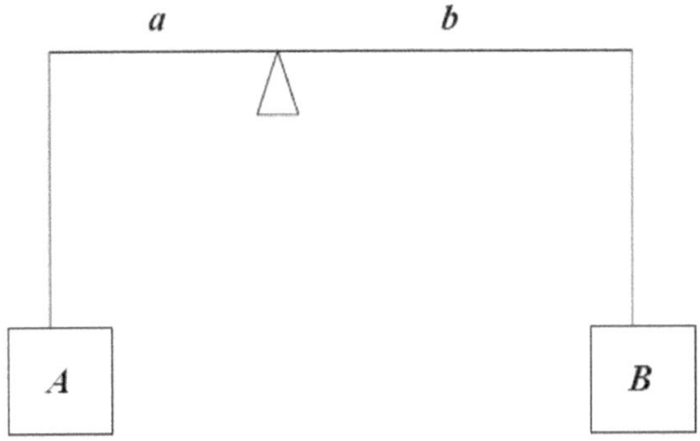

Fig. 3.4. Law of the lever.

the remark, "Among the Greeks nothing was more glorious than mathematics. But we [the Romans] have limited the usefulness of this art to measuring and calculating".

We shall verify that $V_1/V_2 = 1/2$. We shall give the mechanical proof that is in ARCHIMEDES' *Method.* This proof uses the DEMOCRITUS–EUDOXUS result that $V_3 = 3V_1$.

We begin by describing the *law of the lever*: the lever in Figure 3.4 is in equilibrium if

$$aA = bB, \tag{3.40}$$

where A and B are weights having distance a and b, respectively, from the fulcrum.

Figure 3.5 is the figure we shall use to prove $V_1/V_2 = 1/2$. Let S be a sphere of radius r and center 0; we shall work on a plane P through 0. Take perpendicular diameters AB and CD of S in P.

When we rotate the triangle ACD about the segment AO we obtain a cone inscribed in the hemisphere obtained by rotating the arc CAD about AO. We also form the rectangle $EFGH$ determined by the lines parallel to the circle at A and B, and by the points F and H, which are on the lines AC and AD. Let XY be a line intersecting AO at Z and parallel to CD. Also, XY intersects the circle at M and N and the triangle ACD at U and V. Finally, we extend AB to the point I so that the lengths $|IA|$ and $|IB|$ satisfy $|IA| = 2r$ and $|IB| = 4r$, respectively.

Note that

$$|FB| = |BH| = 2r. \tag{3.41}$$

In fact, AOC and ABF are similar triangles, and so (3.41) follows since $|AB| = 2r$, $|AO| = r$, and $|OC| = r$. Next, set $|UZ| = a$ and $|MZ| = b$, and note that $|AZ| = a$, since AUZ is a right triangle and the angle $\angle ZAU$ is $\pi/4$ radians. Consequently, the right triangle AMZ has the property that

$$|AM|^2 = a^2 + b^2. \tag{3.42}$$

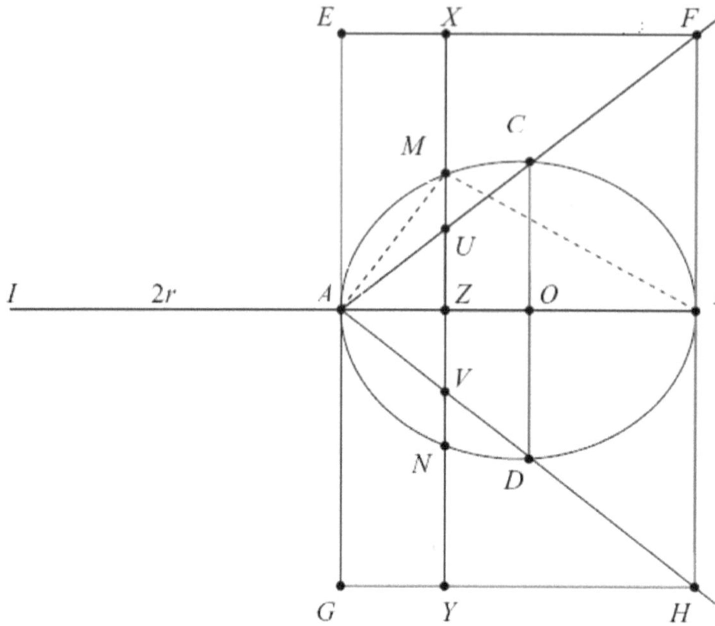

Fig. 3.5. Archimedes method.

Observe that AMB is a right triangle with right angle at M, since AB is a diameter. Thus,

$$|AM|^2 = (2r)^2 - |MB|^2. \tag{3.43}$$

From the right triangle ZMB we obtain

$$b^2 = |MB|^2 - (2r - a)^2. \tag{3.44}$$

Then (3.42) and (3.44) yield

$$|AM|^2 = a^2 + |MB|^2 - (2r)^2 + 4ra - a^2, \tag{3.45}$$

so that by adding (3.43) and (3.45) we obtain

$$|AM|^2 = 2ra. \tag{3.46}$$

Also we combine (3.42) and (3.46), and have

$$2ra = a^2 + b^2;$$

we write this as

$$2r\pi(a^2 + b^2) = a\pi(2r)^2. \tag{3.47}$$

We now rotate our figure about IB so that the circle generates S, the triangle AFH generates a cone whose base is a circle of radius $|BF| = 2r$ and whose height is $|AB| = 2r$, and the rectangle $EFGH$ generates a cylinder whose base has radius $|BF| = 2r$ and whose height is $|AB| = 2r$. In this rotation, XY determines a plane that intersects the cone in a circle C_c of radius a, the sphere in a circle C_S of radius b, and the cylinder in a circle C_C of radius $2r$. With this notation and the fact that $|AI| = 2r$, (3.47) becomes

$$|AI|(A(C_c) + A(C_S)) = |AZ|A(C_C), \tag{3.48}$$

where $A(X)$ represents the area of X.

Equation (3.48) should be compared with (3.40); in fact, we suppose IZ is a lever with fulcrum at A. Heuristically, then, we consider circular disks with weights proportional to their areas, so that (3.48) expresses the law of the lever. Consequently, if we consider n such cuts XY equidistributed by the points $A = Z_0, Z_1, \ldots, Z_n = 0$ along AO, then we can think of the sum of n areas $A(\cdot)$ as approximating the volume V of the corresponding solid. Thus, since $|AI| = 2r$, the left-hand side of (3.48) becomes

$$2r(V_1 + V_2). \tag{3.49}$$

For the right-hand side we see from the equidistribution that

$$\begin{aligned}\sum_{j=1}^{n} |AZ_j|A(C_C) &= \frac{1}{n}\widetilde{V}_3 \sum_{j=1}^{n} |AZ_j| = \frac{1}{n}\widetilde{V}_3 r\left(\frac{1}{n} + \frac{2}{n} + \cdots + \frac{n}{n}\right) \\ &= \frac{r}{n^2}\widetilde{V}_3 \frac{n(n+1)}{2},\end{aligned} \tag{3.50}$$

where $\widetilde{V}_3 = \pi(2r)^2 r$ is the volume of the cylinder whose base has radius $|AE| = 2r$ and whose height is $|AO| = r$. Since $V_3 = \pi r^3$, we have $\widetilde{V}_3 = 4V_3$. Therefore, equating (3.49) and (3.50), we compute

$$V_1 + V_2 = V_3, \tag{3.51}$$

since $\lim_{n\to\infty}(n+1)/n = 1$. We now use the DEMOCRITUS–EUDOXUS result, $V_3 = 3V_1$, to obtain ARCHIMEDES' theorem, $V_2 = 2V_1$.

The point of this illustration of ARCHIMEDES' *Method*, especially the previous paragraph, is to relate ARCHIMEDES' analysis with the definition of the Riemann integral; see his prescient comment above about "future generations".

2. Let $(X, \mathcal{A}, p)$ be a probability space and let $f : X \to \mathbb{R}$ be a random variable as in Section 2.6. If $f \in L^1_p(X)$ then

$$E(f) = \int_X f\, dp$$

is the *mean value* or *expected value* of f. The mapping

$$p_f : \mathcal{B}(\mathbb{R}) \to \mathbb{R}, \quad B \mapsto p(f^{-1}(B)),$$

is a probability measure on $\mathbb{R}$ (the range of f) called the *distribution* of f, not to be confused with the Schwartz distributions of Section 7.5 or the distribution functions of Section 8.6.

If $g : \mathbb{R} \to \mathbb{R}$ is $\mathcal{B}(\mathbb{R})$-measurable and $f \in L^1_p(X)$ then it is not difficult to prove that

$$E(g \circ f) = \int_{\mathbb{R}} g \, dp_f,$$

and so

$$E(f) = \int_{\mathbb{R}} t \, dp_f(t).$$

Also, in probabilistic language, if $\{f_n : n = 1, \ldots\}$ and f are random variables on X, then one says that $\{f_n\}$ *converges in probability* to f if $\{f_n\}$ converges in measure to f.

3.10 Problems

Some of the more elementary problems in this set include Problems 3.1, 3.2, 3.3, 3.4, 3.6, 3.8, 3.9, 3.10, 3.12, 3.13, 3.14, 3.15, 3.18, 3.19, 3.20, 3.22, 3.26, 3.27, 3.33.

3.1. Let $f_n(x) = (\pi n)^{-1/2} e^{-x^2/n}$ on $\mathbb{R}$ and consider $f(x) = \lim_{n\to\infty} f_n(x)$. Compute f, and determine whether (and why) $\int_{\mathbb{R}} f_n \to \int_{\mathbb{R}} f$.
[*Hint.* In order to prove that

$$\int_{-\infty}^{\infty} e^{-x^2} \, dx = \sqrt{\pi},$$

Pierre-Simon de Laplace considered the product

$$\left(\int_0^{\infty} e^{-x^2} \, dx \right) \left(\int_0^{\infty} e^{-y^2} \, dy \right)$$

and then used polar coordinates. Brilliant!]

3.2. Let $f \in L^1_m(\mathbb{R})$ be a function differentiable at the origin, and let $f(0) = 0$. Prove that

$$\int_{-\infty}^{\infty} \frac{f(x)}{x} \, dx$$

exists.

3.3. ***a.*** Let $f \geq 0$ on $[a,b]$ be Lebesgue measurable. Prove that

$$\int_a^b f = \lim_{\varepsilon \to 0+} \int_a^b f \mathbb{1}_{[a+\varepsilon, b]},$$

where $\int_a^b f$ may be infinite.

b. Prove that $\int_0^1 x^{\alpha-1}\, dx = 1/\alpha$ if $\alpha > 0$.

c. With respect to part *a* and Problem 1.28*a*, find a function $f : (0,1] \to \mathbb{R}$ such that

$$\forall\, \varepsilon > 0, \quad f \in L^1_m([\varepsilon, 1])$$

and $\lim_{n\to\infty}(1/n)\sum_{k \leq n} f(k/n)$ exists, but $\lim_{\varepsilon \to 0+} \int f \mathbb{1}_{[\varepsilon,1]}$ does not exist.

3.4. For real-valued functions $f \in L^\infty_m((0,\infty))$ define

$$f \odot f(x) = \int_0^x f(t) f(x-t)\, dt.$$

It is not unreasonable to expect that $f \odot f \geq 0$ on some small interval $(0, \varepsilon)$. Show that this is not necessarily the case.
[*Hint.* Take $f(t) = \sin(1/t)$.]

3.5. let $f, g \in L^1_m(\mathbb{R})$. Prove the following.

a. The function $\mathbb{R} \to L^1_m(\mathbb{R})$, $x \mapsto f(y-x)g(x)$, is an element of $L^1_m(\mathbb{R})$ for *a.e.* $y \in \mathbb{R}$. For all such y, define

$$f * g(y) = \int_{-\infty}^{\infty} f(y-x) g(x)\, dx.$$

We call $f * g$ the *convolution* of f and g.

b. If $f, g \in L^1_m(\mathbb{R})$, then $f * g \in L^1_m(\mathbb{R})$ and $\|f * g\|_1 \leq \|f\|_1 \|g\|_1$.
[*Hint.* A basic part of the proof of both parts *a* and *b* is to prove that the function $(x,y) \mapsto f(y-x)g(x)$ is appropriately measurable. Assuming this fact for the hint, we calculate

$$\int_{-\infty}^{\infty} \int_{-\infty}^{\infty} |f(y-x)g(x)|\, dy\, dx = \|f\|_1 \|g\|_1 < \infty.$$

Thus, the hypotheses of Fubini's theorem are satisfied. Hence, part *a* is obtained, $f * g \in L^1_m(\mathbb{R})$, and part *b* is completed, since

$$\begin{aligned} \|f * g\|_1 &\leq \int_{-\infty}^{\infty} \int_{-\infty}^{\infty} |f(y-x)g(x)|\, dx\, dy \\ &= \int_{-\infty}^{\infty} \int_{-\infty}^{\infty} |f(y-x)g(x)|\, dy\, dx = \|f\|_1 \|g\|_1 . \end{aligned}$$]

Remark. Because of this result, $L^1_m(\mathbb{R})$ becomes a Banach algebra with multiplication defined by convolution $*$. This is an elegant gateway to harmonic analysis; see Appendix B.

3.6. Prove STEINHAUS' result, mentioned in Problem 1.13 and in Problem 2.23: *If* $X \in \mathcal{M}(\mathbb{R})$ *and* $m(X) > 0$, *then* $X - X = \{x - y : x, y \in X\}$ *is a neighborhood of* 0.
[*Hint.* Let $f(x) = \mathbb{1}_X(x)$ and note that $f * f$ is continuous. We can also prove the result set-theoretically.]

STEINHAUS' result can be used to characterize absolute continuity, as we observe in the remark after Proposition 5.4.4 at the end of Section 5.4.

3.7. ***a.*** Let $X \subseteq \mathbb{R}$ have the property that finite linear combinations of elements from X with rational coefficients generate an interval, I. Prove that X contains a Hamel basis; see Problem 2.24 for the definition and see Problem 3.17. Thus, *the Cantor set* C *contains a Hamel basis.*
[*Hint.* Let $0 \in I$; and so

$$\forall\, x \in \mathbb{R},\ \exists\, r \in \mathbb{Q} \text{ such that } (1/r)x \in I.$$

Hence, every real number is a finite rational linear combination from X. Let $\mathcal{F}$ be the family of all subsets $Y \subseteq X$ with the same property. Apply Zorn's lemma.]

b. Take $A \in \mathcal{M}(\mathbb{R})$ for which $0 < m(A) < \infty$; prove that A contains a Hamel basis.
[*Hint.* Use part *a* and the method of proof from Problem 3.6. This is simpler to prove than Problem 2.24*e*, since we do not have to deal with the likes of BURSTIN's result; see Problem 2.24*e*. Of course, it is also a weaker result.]

3.8. ***a.*** Consider a sequence of functions $f, f_n \in L^1_m([a,b])$, $n = 1, \ldots,$ such that $f = \lim_{n\to\infty} f_n$ *m-a.e.* Prove that

$$\lim_{n\to\infty} \int_a^b |f_n - f| = 0 \quad \Longleftrightarrow \quad \lim_{n\to\infty} \int_a^b |f_n| = \int_a^b |f|;$$

cf. Theorem 6.5.3.

b. Find an example of a sequence $\{f_n\}$ of m-integrable functions on $[a,b]$ that converges *m-a.e.* to an m-integrable function f such that

$$\lim_{n\to\infty} \int_a^b f_n = \int_a^b f, \quad \text{but} \quad \lim_{n\to\infty} \int_a^b |f_n| \neq \int_a^b |f|.$$

[*Hint.* Take

$$f_n(x) = \begin{cases} n, & \text{if } x \in [0, 1/(2n)), \\ -n, & \text{if } x \in [1/(2n), 1/n), \\ 0, & \text{otherwise.]} \end{cases}$$

3.9. ***a.*** Prove Theorem 3.3.8.
[*Hint.* Generalize Theorem 3.3.5 appropriately.]

b. Show that there can be a strict inequality in Theorem 3.3.5.
[*Hint.* Let $f_n = \mathbb{1}_{[n,n+1)}$.]

c. Prove that Theorem 3.3.6 can fail if we replace "$f_n \le f$ μ-*a.e.*" by the hypothesis that

$$\forall\, x \text{ and } \forall\, n, \quad f_{n+1}(x) \le f_n(x).$$

[*Hint.* Let $f_n = \mathbb{1}_{[n,\infty)}$.]

3.10. Find a function $f : [0,1] \to \mathbb{R}$ such that f' exists on $[0,1]$, $f''(0)$ exists, and $R\int_0^b f'$ does not exist for any $b \in (0,1]$.
[*Hint.* Let g be the Volterra function defined in Example 1.3.1 and set $f(x) = x^2 g(x)$.]

3.11. Let $(X, \mathcal{A})$ and $(Y, \mathcal{E})$ be measurable spaces and let $\phi : X \to Y$ be measurable (Definition 2.4.11*d*).

a. If $(X, \mathcal{A}, \mu)$ is a measure space prove that $(Y, \mathcal{E}, \nu)$ is a measure space, where $\nu = \mu \circ \phi^{-1}$.

b. Prove that

$$\forall\, f \in L^1_\nu(Y), \quad \int_Y f\, d\nu = \int_X f \circ \phi\, d\mu.$$

Remark. The assertions of Problem 3.11, which are general and not too *difficult* to verify, are compatible with the change of variables formula, Theorem 8.7.3, for Lebesgue measure on $\mathbb{R}^d$. In fact, consider the notation and hypotheses of Theorem 8.7.3. Assume the advanced calculus form of the change of variables theorem in the setting of continuous functions on rectangles; e.g., see [7] for the role of Jordan content and the theorem itself. This result can be used to prove that if μ is defined as $|\det D(\phi)| m^d$ on U then $\mu \circ \phi^{-1} = m^d$ on $\phi(U)$, noting that $\nu = \mu \circ \phi^{-1}$ in Problem 3.11.

3.12. Let $f : \mathbb{R} \to \mathbb{R}$ be a bounded Lebesgue measurable function with the following property:

$$\exists\, C > 0 \text{ such that } \forall\, \varepsilon > 0,\ m(\{x \in \mathbb{R} : |f(x)| > \varepsilon\}) \le \frac{C}{\sqrt{\varepsilon}}.$$

Prove that $f \in L^1_m(\mathbb{R})$.

3.13. Let $(X, \mathcal{A}, \mu)$ be a measure space and let $f : X \to \mathbb{R}^*$ be a measurable function. Prove the *Chebyshev inequality*:

$$\forall\, t > 0, \quad \mu\left(\{x \in X : |f(x)| \ge t\}\right) \le \frac{1}{t} \int_X |f|\, d\mu.$$

[*Hint.* Let $A_t = \{x : |f(x)| \ge t\}$. Then, $A_t \in \mathcal{A}$ and $t\mathbb{1}_{A_t} \le |f|$ μ-*a.e.* Thus, $t\int_X \mathbb{1}_{A_t}\, d\mu \le \int_X |f|\, d\mu$.]

3.14. ***a.*** Prove that for each $f \in L^1_m(\mathbb{R})$ and $\varepsilon > 0$ there is a continuous function g such that

$$m(\{x : g(x) \ne 0\}) < \infty$$

and

$$\int_{\mathbb{R}} |f(x) - g(x)| \, dx < \varepsilon.$$

[*Hint.* Use LDC and the observation on this matter given after Example 2.5.11. Also, see Theorem 7.2.6.]

b. Prove:

$$\forall\, f \in L^1_m(\mathbb{R}), \quad \lim_{h\to 0} \int_{\mathbb{R}} |f(x+h) - f(x)| \, dx = 0. \tag{3.52}$$

[*Hint.* Use part *a* and a fact about uniformly continuous functions.]

Remark. We shall later characterize functions of bounded variation, respectively, constants, on $[a, b]$ as those elements in $L^1_m([a, b])$ that satisfy

$$\int_a^b |f(x+h) - f(x)| \, dx = O(|h|), \text{ respectively, } o(|h|), \quad h \to 0.$$

The "O" and "o" notation is standard in classical analysis: $F(x) = O(G(x))$, respectively, $o(G(x))$, as $x \to r$, means that

$$\exists\, M > 0 \text{ and } \exists\, y \text{ such that } \forall\, x \in [r - y, r + y], \quad |F(x)| \le MG(x),$$
$$\text{respectively, } \lim_{x\to r} F(x)/G(x) = 0,$$

where G is positive. For $r = \pm\infty$ there is an obvious adjustment in the definition.

c. A *trigonometric polynomial* on $\mathbb{T}$ has the form

$$t_N(x) = \sum_{|n|\le N} a_N e^{-2\pi i n x}.$$

Using part *a*, prove that

$$\forall\, \varepsilon > 0 \text{ and } \forall\, f \in L^1_m(\mathbb{T}),\ \exists\, t_N \text{ such that } \int_{\mathbb{T}} |f - t_N| < \varepsilon.$$

[*Hint.* Find a function $g \in C^2(\mathbb{R})$ in terms of indefinite integrals such that g has period 1 and $\|f - g\|_1 < \varepsilon/2$; then find N for which

$$\left\| g(x) - \sum_{|n|\le N} \hat{g}(n) e^{-2\pi i n x} \right\|_\infty < \varepsilon,$$

where $\hat{g}(n) = \int_{\mathbb{T}} g(x) e^{2\pi i n x} \, dx$.]

d. We proved the *Riemann–Lebesgue lemma* in Theorem 3.6.4. Complete the following proof, which uses (3.52):

$$|\hat{f}(n)| = \left(\frac{1}{2}\right) |(f - \tau_{-1/(2n)} f)\hat{\ }(n)| \leq \frac{1}{2} \int_0^1 |f(x) - f(x + 1/2n)| \, dx,$$

where $\tau_y f(x) = f(x - y)$.

e. Prove the *Fejér lemma:* If $f \in L^1_m(\mathbb{T})$ and $g \in L^\infty_m(\mathbb{T})$ then

$$\lim_{|n| \to \infty} \int_0^1 f(x) g(nx) \, dx = \int_0^1 f(x) \, dx \int_0^1 g(x) \, dx.$$

This result reduces to the Riemann–Lebesgue lemma for the case of $g(x) = e^{2\pi i x}$.

3.15. Let $\{f_n : n = 1, \ldots\} \subseteq L^1_m([0,1])$ be a sequence of nonnegative functions defined on $[0,1]$. Assume that $\sum_{n=1}^\infty \int_0^1 f_n < \infty$ and prove that $f_n \to 0$ *m-a.e.* This result is an often used technical device, e.g., Proposition 3.4.9. [*Hint.* $\int \sum f_n = \sum \int f_n < \infty$, and so $\sum f_n(x)$ is finite *m-a.e.*]

3.16. Prove Proposition 3.4.9.

3.17. Let $H \subseteq \mathbb{R}$ be a Hamel basis; see Problem 2.24 for the definition. Prove that

$$\forall\, a \in \mathbb{R},\ a \neq 1,\ \exists\, h \in H \text{ such that } ah \notin H.$$

[*Hint.* Take such an a and assume $aH \subseteq H$; for each $x = \sum r_\alpha h_\alpha$ define $f(x) = \sum r_\alpha$ and note that $f(ax) = f(x)$. Compute $f(x)$, where $x = h/(a-1)$ and $h \in H$.]

3.18. Let $f \in L^1_m((0,\infty))$ and define I_r to be a subset of $(0,\infty)$ parametrized by $r > 0$.

a. Prove that $\lim_{r \to \infty} \int_{I_r} f(x) \cos(rx) \, dx = 0$ if I_r is an interval.

b. Show that part *a* is not generally true if I_r is taken to be a finite union of intervals.

These results are interesting in light of the Riemann–Lebesgue lemma.

3.19. Compute

$$\lim_{n \to \infty} \int_0^n \left(1 - \frac{x}{n}\right)^n e^{x/2} \, dx.$$

[*Hint.* Transform the integral to $\int_{-n}^0$ and note, comparing like terms of the binomial expansions of $(1 + t/n)^n$ and $(1 + t/(n+1))^{n+1}$, that $(1 + t/n)^n e^{-t/2}$ increases to $e^{t/2}$; then use LDC.]

3.20. Prove that if $f : \mathbb{R} \to \mathbb{R}$ is a continuous function and $g : \mathbb{R} \to \mathbb{R}$ is continuous *m-a.e.*, then the composition $f \circ g$ is continuous *m-a.e.*; cf. Problem 4.48.

3.21. Prove Proposition 3.4.8.
[*Hint.* The characterization of regulated functions in Example 3.2.2 gives us a means of verifying Proposition 3.4.8. In fact, *if f is a regulated function then there is a function F such that $F' = f$ except for possibly a countable set.*]

3.22. Prove that if f is bounded on $[a, b]$ and continuous m-*a.e.*, then $\overline{F}' = f$ m-*a.e.*, where

$$\overline{F}(x) = R\overline{\int_a^x} f.$$

3.23. Prove Theorem 3.5.4.

3.24. Verify Theorems 3.6.1, 3.6.2, and 3.6.3.

Remark. Theorems 3.6.1 and 3.6.2 require LDC for their proofs. Theorem 3.6.3 requires not only LDC but also the fundamental theorem of calculus and the Fubini–Tonelli theorem.

3.25. Let $(X, \mathcal{A}, \mu)$ be a measure space, where X is a locally compact space (Appendix A.1) and μ is a positive measure on X. Choose an open interval $S \subseteq \mathbb{R}$ and let $f(x, t)$ be a function on $X \times S$. Assume that

i. $\forall\, t \in S$, $f(\cdot, t)$ is a μ-measurable function,
ii. $\exists\, g \in L^1_\mu(X)$ such that $\forall\, t \in S$, $|f(x, t)| \le g(x)$ μ-*a.e.*,
iii. $\exists\, t_0 \in S$ such that $f(x, \cdot)$ is continuous at t_0 μ-*a.e.*,
iv. $f(x, \cdot)$ is finite and differentiable on S μ-*a.e.*,
v. $\exists\, h \in L^1_\mu(X)$ such that $\forall\, t \in S$, $|d/dt f(x, t)| \le h(x)$ μ-*a.e.*

Prove that $\int_X f(x, t)\, d\mu(x)$ is differentiable on S and

$$\frac{d}{dt}\int_X f(x, t)\, d\mu(x) = \int_X \frac{d}{dt} f(x, t)\, d\mu(x).$$

Remark. This result obviously begs comparison with Theorem 3.6.3. For all the little differences, the important common hypothesis is the bound on the derivative by h. This hypothesis is the analogue to the uniform convergence hypothesis for the corresponding result from calculus.

3.26. ***a.*** Let $(X, \mathcal{A}, \mu)$ be a finite measure space. Let Y be the space of real-valued μ-measurable functions defined on X; and let $\widetilde{Y}$ be the space of equivalence classes of such functions where $f, g : X \to \mathbb{R}$ are defined to be equivalent if $f = g$ μ-*a.e.* Define

$$\forall\, f, g \in Y, \quad \rho(f, g) = \int_X \frac{|f - g|}{1 + |f - g|}\, d\mu.$$

Prove that ρ is a metric on $\widetilde{Y} \times \widetilde{Y}$ (see Appendix A.1 for the definition of metric).
[*Hint.* If a and b are real, then $|a + b|/(1 + |a + b|) \le |a|/(1 + |a|) + |b|/(1 + |b|)$.]

b. Consider the setting of part *a*. Prove that $\rho(f_n, f) \to 0$ if and only if $f_n \to f$ in measure.
[*Hint.* Let $A_n(\varepsilon) = \{x : |f_n(x) - f(x)| \geq \varepsilon\}$ and prove that

$$\frac{\varepsilon}{1+\varepsilon}\mu(A_n(\varepsilon)) \leq \rho(f_n, f) \leq \mu(A_n(\varepsilon)) + \varepsilon\mu(X).]$$

c. Find an example of a sequence $\{f_n : n = 1, \ldots\}$ of m-measurable functions on the measure space $([0,1], \mathcal{M}([0,1]), m)$ such that $f_n \to 0$ in measure, whereas $\{f_n : n = 1, \ldots\}$ does not converge pointwise at any point. Actually, the example can be constructed so that $\lim_{n\to\infty} \|f_n\|_1 = 0$ [69], Chapitre IV, page 236.

d. Prove Theorem 3.3.13 (F. Riesz–Lebesgue).
[*Hint.* Use the Egorov theorem for part *b*.]

e. Show by example that the hypothesis $\mu(X) < \infty$ is necessary generally in Theorem 3.3.13*b*.

f. We know that if $\|f_n - f\|_1 \to 0$, then $\{f_n\}$ converges in measure to f; and, from Theorem 3.3.13*a*, we know that if $f_n \to f$ in measure, then there is a subsequence $\{f_{n_k}\}$ that converges to f μ-*a.e.* Construct a sequence $\{f_n\} \subseteq L^1_m([0,1])$ such that $\sup_n \|f_n\|_\infty < \infty$ and $\|f_n\|_1 \to 0$ (hence, $f_n \to 0$ in measure), but such that $\{f_n\}$ does *not* converge m-*a.e.*

3.27. ***a.*** Give an equivalent description of the space $L^1_c(\mathbb{Z})$ in terms of a summability property of sequences $\{c_n\} \subseteq \mathbb{C}$; see Definition 5.5.1.

b. Show that if $f \in L^1_c(\mathbb{Z})$, where c is counting measure defined in Example 2.4.2*d*, then

$$\lim_{|n|\to\infty} f(n) = 0.$$

c. Prove that there are continuous functions $f \in L^1_m(\mathbb{R})$ such that

$$\overline{\lim_{|x|\to\infty}} |f(x)| = \infty.$$

3.28. The *Fejér kernel* $\{W_N : N = 1, \ldots\}$ is defined by

$$\forall\, x \in \mathbb{R} \text{ and } \forall\, N \geq 0, \quad W_N(x) = \sum_{|n|\leq N} \left(1 - \frac{|n|}{N+1}\right) e^{-2\pi i n x}.$$

a. Prove that if $x \in \mathbb{Z}$ then $W_N(x) = N+1$, and that if x is any other real number then

$$W_N(x) = \frac{1}{N+1} \frac{\sin^2((N+1)x/2)}{\sin^2(x/2)}.$$

b. Clearly $W_N \geq 0$. Prove that

$$\int_0^1 W_N = 1,$$

and that

$$\forall\, \delta \in (0, 1/2), \quad \lim_{N\to\infty} \int_\delta^{1-\delta} |W_N| = 0.$$

c. An *approximate identity* in $L^1_m(\mathbb{T})$ is a sequence $\{K_N : N = 1, \ldots\} \subseteq L^1_m(\mathbb{T})$ that satisfies

$$\sup_N \int_0^1 |K_N| < \infty, \quad \int_0^1 K_N = 1,$$

and

$$\forall\, \delta \in (0, 1/2), \quad \lim_{N\to\infty} \int_\delta^{1-\delta} |K_N| = 0.$$

Thus, $\{W_N : N = 1, \ldots\}$ is an approximate identity. We defined convolution in $L^1_m(\mathbb{R})$ in Problem 3.5; the same definition makes sense on $\mathbb{T}$ as long as we define our functions 1-periodically. Hence, if $f,\, g \in L^1_m(\mathbb{T})$ we have

$$f * g(x) = \int_0^1 f(x-y)g(y)\, dy.$$

Prove that if $f \in C(\mathbb{R})$ has period 1 and $\{K_N : N = 1, \ldots\}$ is an approximate identity in $L^1_m(\mathbb{T})$, then

$$\lim_{N\to\infty} \|K_N * f - f\|_\infty = 0,$$

where "$\|...\|_\infty$" designates the $L^\infty_m(\mathbb{T})$-norm.
[*Hint.*

$$\|K_N * f - f\|_\infty \le \int_0^1 |K_N(y)| \|\tau_y f - f\|_\infty\, dy = I.$$

Take $\varepsilon > 0$ and fix $0 < \delta < 1/2$ such that $\|\tau_y f - f\|_\infty < \varepsilon$ if $y \in [0, \delta] \cup [1-\delta, 1]$. Compute

$$I \le M\varepsilon + 2\|f\|_\infty \int_\delta^{1-\delta} |K_N|,$$

where M is independent of N.]

In particular, we have proved that

$$f(0) = \lim_{N\to\infty} \frac{1}{N+1} \int_0^1 f(x) \frac{\sin^2((N+1)x/2)}{\sin^2(x/2)}\, dx.$$

d. Let $f : [a, b] \times [a, b] \to \mathbb{R}$ be a function that is continuous in each variable separately. Prove that

$$\forall\, (x, y) \in [a, b] \times [a, b], \quad \lim_{N\to\infty} f_N(x, y) = f(x, y),$$

where $\{f_N : N = 1, \ldots\}$ is a sequence of continuous functions on $[a, b] \times [a, b]$.

[*Hint.* Assume without loss of generality that $[a, b] = [0, 1]$, that $\|f\|_\infty < \infty$, and that for each y, $f(0, y) = f(1, y)$. Set

$$f_N(x, y) = \sum_{|n|\leq N} \left(1 - \frac{|n|}{N+1}\right) c_n(y) e^{-2\pi inx},$$

where

$$c_n(y) = \int_0^1 f(x, y) e^{2\pi inx}\, dx.$$

From part *c*,

$$\forall\, y \in \mathbb{T}, \quad \lim_{N\to\infty} \sup_{x\in\mathbb{T}} |f_N(x, y) - f(x, y)| = 0.$$

In particular, we have the desired pointwise convergence. To prove that f_N is continuous we have only to check that each c_n is continuous, and this is clear from LDC.]

3.29. We have mentioned the concept of uniform distribution in Problem 1.28 and Proposition 3.4.7. We shall now define it. A sequence $\{x_n : n = 1, \ldots\} \subseteq \mathbb{R}$ is *uniformly distributed modulo* a, $a > 0$, if, when $y_n \in [0, a)$ and $(x_n - y_n)/a \in \mathbb{Z}$, we can conclude that

$$\forall\, I \subseteq [0, a), \quad \lim_{N\to\infty} \frac{C(I, N)}{N} = \frac{1}{a} m(I),$$

where $I = [c, d] \subseteq [0, a)$ is an arbitrary interval and where we define $C(I, N) = \text{card}\ (\{y_1, \ldots, y_N\} \cap I)$. This concept was introduced in a systematic way by WEYL in 1914; see [500] (1916) and the first Remark in Section 9.6.1. WEYL proved that *a sequence* $\{x_n : n = 1, \ldots\} \subseteq \mathbb{R}$ *is uniformly distributed modulo a if and only if*

$$\lim_{N\to\infty} \frac{1}{N} \sum_{n=1}^{N} f(y_n) = R \int_0^a f$$

for every Riemann integrable function f *on* $[0, a]$, e.g., [304]. Here $\{y_n\} \subseteq [0, a)$ corresponds to $\{x_n\}$ as in the definition.

a. Let $x \in \mathbb{R} \setminus \mathbb{Q}$. Prove that, for each Riemann integrable function g defined on $[0, 1)$,

$$\lim_{N\to\infty} \frac{1}{N} \sum_{1}^{N} g(nx \mod 1) = \hat{g}(0).$$

[*Hint.* First verify that

$$\left|\sum_{n=1}^{N} e^{2\pi imnx}\right| \leq \frac{1}{|\sin(\pi mx)|};$$

and then use the hypothesis that x is irrational to prove that if $m \neq 0$ then

$$\lim_{N \to \infty} \frac{1}{N} \sum_{n=1}^{N} e^{2\pi i m n x} = 0.$$

The rest of the demonstration follows from standard approximation results.]

b. Let $x \in \mathbb{R} \setminus \mathbb{Q}$. Prove that the sequence $\{nx : n = 1, \ldots\}$ is uniformly distributed modulo 1.

[*Hint.* Part b is clear from part a by taking the appropriate function g.]

c. let $a = 1$, let $\{x_n : n = 1, \ldots\} \subseteq \mathbb{R}$, and let $\{y_n\} \subseteq [0,1)$ correspond to $\{x_n\}$ as in the definition. For an arbitrary subset $E \subseteq [0,1)$, let $C(E,N) = \text{card}\,(\{y_1, \ldots, y_N\} \cap E)$. Construct a Lebesgue measurable subset $A \subseteq [0,1)$ such that $m(A) = 1$ and $\lim_{N \to \infty} C(A,N)/N = 0$.

[*Hint.* Let $A = [0,1) \setminus \{y_n\}$.]

3.30. For $a > 0$ and a sequence $\{x_n : n = 1, \ldots\} \subseteq \mathbb{R}$ define $y_n \in \mathbb{R}$ as in Problem 3.29 by the condition that $y_n \in [0,a)$ and $(x_n - y_n)/a \in \mathbb{Z}$. The antithesis of uniform distribution for a sequence $\{x_n : n = 1, \ldots\}$ occurs when $y_n \to 0$. Assume that the sequence $\{x_n : n = 1, \ldots\} \subseteq \mathbb{R}$ does not tend to 0 and let $A = \{a > 0 : y_n \to 0\}$. Prove that $m(A) = 0$.

[*Hint.* Without loss of generality we can assume that $\lim_{n \to \infty} |x_n| = \infty$. If $a \in A$ then $\lim_{n \to \infty} e^{2\pi i x_n / a} = 1$. Consequently, $\lim_{n \to \infty} e^{2\pi i x_n r} = 1$ for $r \in A^{-1} = \{a^{-1} : a \in A\}$. Thus,

$$\forall\, b > 0, \quad \lim_{n \to \infty} \int_{A_b^{-1}} e^{2\pi i x_n r}\, dr = m(A_b^{-1}),$$

where $A_b^{-1} = A^{-1} \cap (0,b)$, by LDC. On the other hand, from the Riemann–Lebesgue lemma,

$$\lim_{n \to \infty} \int_{A_b^{-1}} e^{2\pi i x_n r}\, dr = 0.$$

Therefore, $m(A^{-1}) = m(A) = 0$.]

Note that A can be uncountable.

3.31. Let $\mathcal{Q}$ be the family of all half-open parallelepipeds in $\mathbb{R}^d$ and let $f : \mathbb{R} \to \mathbb{R}$ be an increasing right-continuous function. Prove that there exists a nonnegative finitely additive set function μ_f^d on $\mathcal{Q}$ such that

$$\mu_f^d((a_1,b_1] \times \cdots \times (a_d,b_d]) = (f(b_1) - f(a_1)) \cdots (f(b_d) - f(a_d)).$$

3.32. Given the open interval (a,b), where b can be ∞, and a function $f : (a,b) \to \mathbb{R}$, assume that $R\int_a^r f$ exists for each $r \in (a,b)$. The *Cauchy–Riemann integral of f on (a,b)* is

$$CR\int_a^b f = \lim_{r \to b-} R\int_a^r f$$

when the right-hand side exists. Naturally we can define the "Cauchy–Lebesgue" integral if we replace $R\int_a^r$ by $\int_a^r$.

a. Give an example of an unbounded function on (a,∞) whose Cauchy–Riemann integral exists, e.g., Problem 3.27*c*.

b. Let f be a nonnegative function on the interval (a,∞) and assume that $R\int_a^r f$ exists for all $r > a$. Prove that $CR\int_a^\infty f$ exists if and only if $f \in L^1_m((a,\infty))$.

c. Prove that if $f \in L^1_m((a,\infty))$ and $R\int_a^r f$ exists for all $r > a$ then $\int_a^\infty f = CR\int_a^\infty f$.

d. Let $f(x) = \sin(x)/x$. Show that $CR\int_0^\infty f$ exists but that $f \notin L^1_m((0,\infty))$.

[*Hint.*

$$\int_0^\infty \left|\frac{\sin(x)}{x}\right| dx = \sum_{k=0}^\infty \int_{k\pi}^{(k+1)\pi} \left|\frac{\sin(x)}{x}\right| dx$$

$$\geq \sum_{k=0}^\infty \frac{1}{(k+1)\pi} \int_{k\pi}^{(k+1)\pi} |\sin(x)|\, dx.]$$

Thus, f can be integrated, whereas $|f|$ cannot be integrated; this phenomenon is referred to as *conditional convergence.*

e. Show that the Cauchy–Riemann integral does not integrate every bounded Lebesgue measurable function on $[a,b]$, $b < \infty$.

f. Find a Cauchy–Riemann integrable function f on $[a,b]$, $b < \infty$, such that $f \notin L^1_m([a,b])$.

[*Hint.* Example 3.2.10*a*.]

g. Let $f(x) = (1/x)\sin(\pi/x)$ on $(0,1)$. Is f' Cauchy–Riemann integrable on $(0,1)$? Does $f' \in L^1_m([0,1])$?

h. In light of Problem 3.27 we would like to know conditions on a function f so that we could conclude that

$$\lim_{x\to\infty} f(x) = 0 \tag{3.53}$$

when $CR\int_a^\infty f$ exists. In fact, the following is true: *Let f be continuous on $[a,\infty)$ and assume that $CR\int_a^\infty f$ exists; then (3.53) is valid if and only if f is uniformly continuous.*

Prove this assertion.

i. Let f be a twice continuously differentiable real-valued function defined on (a,∞), and assume that $CR\int_a^\infty f^2$ and $CR\int_a^\infty (f'')^2$ exist. Prove that (3.53) is valid.

[*Hint.* Integrate ff'' by parts and use part *h*.]

3.33. ***a.*** In light of Example 3.6.5, does there exist a sequence $\{f_n : n = 1,\ldots\} \subseteq L^1_m([0,1])$ such that $f_n \to 0$ pointwise and

$$\forall\, A \in \mathcal{M}([0,1]), \quad \lim_{n\to\infty} \int_0^1 f_n(x)\mathbb{1}_A(x)\, dx = 0,$$

whereas $\overline{\lim}_{n\to\infty} \int |f_n(x)|\, dx > 0$?

b. Construct a sequence $\{f_n : n = 1, \ldots\} \subseteq L^1_m(\mathbb{R})$ such that
i. f_n is continuous and $\lim_{|x|\to\infty} f_n(x) = 0$,
ii. $\sup_n \int_{\mathbb{R}} |f_n(x)|\, dx < \infty$,
iii. $f_n \to f$ uniformly on $\mathbb{R}$,
but such that $\{f_n : n = 1, \ldots\}$ contains no subsequence $\{f_{n_k} : k = 1, \ldots\}$ for which

$$\int_{\mathbb{R}} |f_{n_k}(x) - f(x)|\, dx \to 0.$$

[*Hint.* Start with $g_n = (1/n)\mathbb{1}_{[n,2n]}$.]

3.34. Let $\{f_n : n = 1, \ldots\}$ be a sequence of functions on $[a, b]$. The general problem for this exercise is to investigate pointwise convergence of subsequences on certain subsets of $[a, b]$.

a. Let $\{n_k : k = 1, \ldots\} \subseteq \mathbb{N}$ be a sequence converging to infinity. We let S_{n_k} denote the set $\{x : \lim_{k\to\infty} \sin(2\pi n_k x) \text{ exists}\} \subseteq \mathbb{R}$. Show that, even though $m(S_{n_k}) = 0$, it is possible to find subsequences $\{n_k : k = 1, \ldots\} \subseteq \mathbb{N}$ for which $\operatorname{card} S_{n_k} = \operatorname{card} \mathbb{R}$.

b. Do there exist $\delta \in (0, 1]$ and $X \subseteq [0, 1)$ such that $m(X) > 0$ and

$$\operatorname{card} \left\{ n : \sup_{x\in X} \sin(2\pi n x) \geq \delta \right\} < \infty?$$

c. Solve part *b* for any nonconstant periodic function $f \in L^\infty_m(\mathbb{R})$, where $\sin(2\pi n x)$ is replaced by $f(nx)$.

d. Let $A_{n_k} = S_{n_k} \cap [0, 1)$. Give examples to show whether A_{n_k} can be closed and infinite, countable, or finite. Find $A_{n!}$. See Remark *b* below.

e. Prove the *Cantor–Lebesgue theorem: If* $\sum_{|n|\leq N} c_n e^{-2\pi i n x} \to 0$ *pointwise on a set* $X \subseteq [0, 1)$, *as* $N \to \infty$, *and* $m(X) > 0$, *then* $\lim_{|n|\to\infty} c_n = 0$. [*Hint.* Without loss of generality consider the series $\sum_{n=0}^{\infty} r_n \cos(2\pi(nx + a_n))$, and assume that there are a $\delta > 0$ and a subsequence $\{n_k : k = 1, \ldots\} \subseteq \mathbb{N}$ such that for each k, $r_{n_k} > \delta > 0$. Thus, $\lim_{k\to\infty} \cos(2\pi(n_k x + a_{n_k})) = 0$. Now use LDC and the fact that

$$\int_X \cos^2(2\pi(kx + a))\, dx = \frac{1}{2} m(X) + \frac{1}{2} \int_X \cos(4\pi(kx + a))\, dx. \tag{3.54}$$

Obtain a contradiction by employing the Riemann–Lebesgue lemma.]

See Sections 3.8.1 and 3.8.2.

f. Suppose that $\sum_{n=1}^{\infty} a_n$ is an absolutely convergent series. By way of considering the converse situation to part *e*, prove the following: If $\sum_{n=1}^{\infty} (a_n/n) \cos(2\pi n x)$ is not constant on $[0, 1)$, then $\sum_{n=1}^{\infty} a_n \sin(2\pi n x)$ cannot converge pointwise m-*a.e.* to 0.

Remark. a. Using (3.54) and LDC we have another proof, besides Theorem 3.3.14, that $m(A_{n_k}) = 0$.

b. MAZURKIEWICZ [345] proved that if $\{f_n : n = 1, \ldots\}$ is a uniformly bounded sequence of continuous functions on $[a, b]$, then there are a closed uncountable set $F \subseteq [a,b]$ (without isolated points) and integers $n_1 < n_2 < \cdots$ such that $\{f_{n_k} : k = 1, \ldots\}$ converges uniformly on F. Using the continuum hypothesis, SIERPIŃSKI [439] proved the following result: *There is a sequence of nonmeasurable functions on $[a,b]$ that is uniformly bounded but such that no subsequence converges on an uncountable set.*

Note, with regard to Problem 3.34*d*, that if $\{f_n : n = 1, \ldots\}$ is any sequence of functions and $D \subseteq [a,b]$ is countable then there is a subsequence $\{f_{n_k} : k = 1, \ldots\}$ that converges on D.

c. Some interesting positive results on the general problem stated at the beginning of this exercise have been given by KEITH SCHRADER [422], [421]; cf. Problem 4.39 and the Remark following it on the selection principle. For example, he proved the following theorem. *Let $f_n : [a,b] \to \mathbb{R}$ be continuous; and assume that for some $N \geq 0$ the sequence $\{f_n : n = 1, \ldots\}$ has the following property: if $f_k = f_j$ for more than N values of $x \in [a,b]$ then f_k is identical to f_j on $[a,b]$. Then, there is a subsequence $\{f_{n_k} : k = 1, \ldots\}$ that converges pointwise, possibly with infinite value, on $[a,b]$.*

3.35. ***a.*** Prove that $\sum_{n=1}^{\infty} e^{2\pi i n x}/n$ converges on $[0,1)$ if and only if $x \neq 0$. [*Hint.* Use partial summation to obtain a bound on the geometric series $\sum_{n=1}^{N} e^{2\pi i n x}$.]

b. Let $\{a_n : n = 1, \ldots\}$ be a sequence decreasing to 0 and define $g(x) = \sum_{n=1}^{\infty} a_n \sin(2\pi n x)$. By an argument similar to that of part *a*, the series converges for each x. Prove that $g \in L^1_m(\mathbb{T})$ if and only if $\sum_{n=1}^{\infty} a_n/n < \infty$, e.g., [155], volume I, page 115. Note that $\sum_{n=1}^{\infty} e^{2\pi i n x}/n \in L^2_m(\mathbb{T})$ by the Parseval equality stated in Example 3.3.4.

3.36. Let $f \in L^1_m([0,1])$ and assume that

$$\int_0^1 f\phi = 0$$

for all functions $\phi \in C([0,1])$ such that $\phi(0) = \phi(1) = 0$. Prove that $f = 0$ *m-a.e.*

[*Hint.* Assume that the result is not true. Then $\int_0^1 |f| = \alpha > 0$. For any $\varepsilon > 0$ consider a set $G_\varepsilon \subseteq [0,1]$ such that $\int_{[0,1]\setminus G_\varepsilon} |f| < \varepsilon$. Using Theorem A.1.3, construct a continuous function ϕ_ε on $[0,1]$ such that $|\phi_\varepsilon| \leq 1$, $\phi_\varepsilon(0) = \phi_\varepsilon(1) = 0$, and $f\phi_\varepsilon = |f|$ on C_ε, a compact subset of G_ε such that $\int_{G_\varepsilon \setminus C_\varepsilon} |f| < \varepsilon$ and $0, 1 \notin C_\varepsilon$. Note that there exists $\delta > 0$ such that, for all $\varepsilon > 0$, $\int_{C_\varepsilon} |f| > \delta$, which yields a contradiction to the fact that $\int_0^1 f\phi_\varepsilon = 0$.]

3.37. Let $f : [0,1] \to \mathbb{R}$ be an increasing right-continuous function with the property that

$$\forall\, g \in C([0,1]), \quad \int_0^1 g\, df = 0.$$

Prove that f is a constant function.

3.38. Let $\mathcal{R}$ be a semialgebra of sets in X and let $\mu : \mathcal{R} \to \mathbb{R}^+$ be a set function that satisfies the following properties:

i. $\mu(\emptyset) = 0$, and

ii. if $A \in \mathcal{R}$ is a disjoint union of a countable family of sets $A_i \in \mathcal{R}$, $i \in I$, then $\mu(A) = \sum_{i \in I} \mu(A_i)$.

Prove that μ has a unique extension to a σ-additive set function on the algebra $\mathcal{A}$ generated by $\mathcal{R}$.

3.39. Prove Theorem 3.7.7.

3.40. Prove Theorem 3.7.9.

3.41. Let X be an uncountable set and let

$$\mathcal{A} = \{A \subseteq X : A \text{ is countable or } A^{\sim} \text{ is countable}\}.$$

Define $\mu(A) = 1$ if card $A^{\sim} \leq \aleph_0$ and $\mu(A) = 0$ if card $A \leq \aleph_0$. Prove that $(X, \mathcal{A}, \mu)$ is a measure space.

In this case, and with the discrete topology on X, we have a measure space with open covering $\{U_\alpha\}$, such that $\mu(X \cap U_\alpha) = 0$ for each α, whereas $\mu(X) = 1$; cf. Corollary 7.4.2.

3.42. Let $I \subseteq \mathbb{R}$ be an interval and given $\phi : I \to \mathbb{R}$. Then ϕ is a *convex function* on I if

$$\forall\, a, b \in I, \text{ where } a < b, \text{ and } \forall\, t \in [0,1], \quad \phi(ta+(1-t)b) \leq t\phi(a)+(1-t)\phi(b).$$

Also, ϕ is a *weak convex function* if

$$\forall\, a, b \in I, \quad \phi\left(\frac{a+b}{2}\right) \leq \frac{1}{2}(\phi(a) + \phi(b)).$$

Clearly, convex functions ϕ are weak convex, and ϕ is convex if and only if

$$\forall\, x_1 < x_2 < x_3, \quad \det \begin{pmatrix} \phi(x_1) & x_1 & 1 \\ \phi(x_2) & x_2 & 1 \\ \phi(x_3) & x_3 & 1 \end{pmatrix} \leq 0.$$

($\phi(x) = x^2$ gives rise to the Vandermonde determinant.)

a. Prove that if a weak convex function ϕ is Lebesgue measurable then it is convex. Counterexamples can be constructed for nonmeasurable functions by a Hamel basis argument.

b. Let $\phi : [a, b] \to \mathbb{R}$ be convex. Prove $\phi \in C((a, b))$ as well as the following facts for right derivatives, which are also true for left derivatives:

i. $\forall\, x \in (a, b),\ \exists\, \phi'_+(x) = \lim_{h \to 0^+} (\phi(x+h) - \phi(x))/h$;

ii. ϕ'_+ is increasing on (a, b);

iii. $\forall\, c, d$, where $a < c < d < b$, $\int_c^d \phi'_+ = \phi(d) - \phi(c)$; see Chapter 4.

3.43. *a.* Let $a < x_1 \leq x_2 \leq \cdots \leq x_n < b$, let $\{w_j > 0 : j = 1, \ldots, n\}$ satisfy $\sum_{j=1}^{n} w_j = 1$, and assume that $\phi : (a,b) \to \mathbb{R}$ is convex and continuous; cf. Problem 3.42*b*. Prove the discrete version of *Jensen's inequality* (1906):

$$\phi\left(\sum_{j=1}^{n} w_j x_j\right) \leq \sum_{j=1}^{n} w_j \phi(x_j),$$

with equality if and only if $x_1 = x_2 = \cdots = x_n$.
[*Hint (for a special case).* Assume that ϕ has a positive second derivative in (a,b), and so ϕ is convex. Set $\overline{x} = \sum_{j=1}^{n} w_j x_j$. Then, $\overline{x} \in (a,b)$ and

$$\phi(x_j) = \phi(\overline{x}) + (x_j - \overline{x})\phi'(\overline{x}) + \frac{(x_j - \overline{x})^2}{2}\phi''(\xi_j).$$

Therefore,

$$\sum_{j=1}^{n} w_j \phi(x_j) = \phi(\overline{x}) + \frac{1}{2}\sum_{j=1}^{n} w_j (x_j - \overline{x})^2 \phi''(\xi_j) > \phi(\overline{x}).]$$

b. *Jensen's inequality.* Let $(X, \mathcal{A}, \mu)$ be a nontrivial ($\mu(X) \neq 0$) measure space, and let $f \in L^1_\mu(X)$ be real-valued. Let ϕ be convex on an interval $I \subseteq \mathbb{R}$, and assume $f(X) \subseteq I$ and $\phi \circ f \in L^1_\mu(X)$. Prove that

$$\phi\left(\frac{1}{\mu(X)} \int_X f \, d\mu\right) \leq \frac{1}{\mu(X)} \int_X \phi \circ f \, d\mu.$$

[*Hint.* Let $c = (1/\mu(X)) \int_X f \, d\mu$ and note that $c \in I$. Assume $c \in \operatorname{int} I$. By the calculations used to solve Problem 3.42*b*, there is a slope s such that $\phi(t) \geq s(t-c) + \phi(c)$ on I. Let $t = f(x)$ and integrate. If c is an endpoint of I, the proof is elementary.]

c. AUGUSTIN L. CAUCHY (1821) proved the *arithmetic–geometric mean inequality*:

$$(r_1 r_2 \cdots r_n)^{1/n} \leq \frac{r_1 + \cdots + r_n}{n} \tag{3.55}$$

for $r_j > 0$. Use Jensen's inequality and the fact that $\phi(x) = e^x$ is convex to prove the following standard generalization of (3.55) (see [220] and [30] for encyclopedic classical treatments):

$$\prod_{j=1}^{n} r_j^{w_j} \leq \sum_{j=1}^{n} w_j r_j,$$

where $r_j, w_j > 0$ and $\sum_{j=1}^{n} w_j = 1$.
[*Hint.* Let $X = [0,1]$ and let

$$g = \sum_{j=1}^{n} r_j \mathbb{1}_{[w_0 + \cdots + w_{j-1}, w_0 + \cdots + w_j)},$$

where $w_0 = 0$. Write $\sum_{j=1}^{n} w_j r_j = \int_0^1 g$ and $g = e^f$, and compute.]

d. $\phi(x) = \log(x)$, $x > 0$, is *concave*. Let f be a nonnegative Lebesgue measurable function on $[0,1]$ and assume $\int_0^1 \log(f(x))\,dx$ is finite. Prove that

$$\log\left(\int_0^1 f(x)\,dx\right) \geq \int_0^1 \log(f(x))\,dx.$$

Remark. Notwithstanding CAUCHY's proof of (3.55), CARL F. GAUSS also made a fascinating observation on this topic dealing with elliptic integrals, which are fundamental in astronomy as well as in the study of solving quintic polynomial equations. Let $r > s > 0$ and define

$$r_n = \frac{r_{n-1} + s_{n-1}}{2} \quad \text{and} \quad s_n = (r_{n-1}s_{n-1})^{1/2},$$

where $r_0 = r$, $s_0 = s$, and $n = 1, \ldots$. GAUSS proved that there is a number $G(r,s)$ for which

$$\lim_{n\to\infty} r_n = \lim_{n\to\infty} s_n = G(r,s).$$

He also showed the relation of G to complete elliptic integrals by proving that

$$\frac{1}{G(1-x, 1+x)} = \frac{1}{\pi}\int_0^\pi \frac{dy}{\sqrt{1 - x^2\cos^2(y)}}.$$

4 The Relationship between Differentiation and Integration on $\mathbb{R}$

4.1 Functions of bounded variation and associated measures

A function $f : X \to \mathbb{R}$ is *increasing*, respectively, *decreasing*, if

$$\forall\, x \le y,\ x, y \in X, \quad f(x) \le f(y), \quad \text{respectively, } f(x) \ge f(y).$$

Definition 4.1.1. Bounded variation on $[a, b]$
a. A real- or complex-valued function f defined on an interval $[a, b]$ is of *bounded variation on* $[a, b]$, in which case we write $f \in BV([a, b])$, if

$$V(f, [a, b]) = \sup \left\{ \sum_{j=1}^{n} |f(x_j) - f(x_{j-1})| : a \le x_0 \le \cdots \le x_n \le b \right\} < \infty.$$

If $f \in BV([a, b])$ then $V(f, [a, b])$ is the *total variation* of f on $[a, b]$. Clearly, if f is increasing on $[a, b]$, then $V(f, [a, b]) = f(b) - f(a)$.

b. If $f : \mathbb{R} \to \mathbb{R}$ is an element of $BV([a, b])$ for each interval $[a, b] \subseteq \mathbb{R}$, we define

$$V(f)(x) = \begin{cases} f(0) + V(f, [0, x]), & \text{if } x \ge 0, \\ f(0) - V(f, [x, 0]), & \text{if } x \le 0. \end{cases}$$

We call $V(f)$ the *variation function* of f. It is an increasing function on $\mathbb{R}$. Note that the function $V(f)$ can take negative values.

c. With the hypothesis of part *b*, it is straightforward to verify that

$$\forall\, a \le b, \quad V(f)(b) - V(f)(a) = V(f, [a, b]);$$

see Problem 4.1*a*, which can be used to prove this equality.

d. If $f \in L^1_m([a, b])$, then we write $f \in BV([a, b])$ if there is a function $g \in BV([a, b])$ defined pointwise on $[a, b]$ such that $g = f$ *m-a.e.* As such, $BV([a, b]) \subseteq L^1_m([a, b])$, since each element of $BV([a, b])$ is bounded and Lebesgue measurable.

The following result is the *Jordan decomposition theorem*, which we shall prove in general measure-theoretic form in Theorem 5.1.8.

J.J. Benedetto and W. Czaja, *Integration and Modern Analysis*, Birkhäuser Advanced Texts,
DOI 10.1007/978-0-8176-4656-1_4,

Theorem 4.1.2. Jordan decomposition theorem
a. *Let the function $f : \mathbb{R} \to \mathbb{R}$ be of the form*

$$f = g_1 - g_2,$$

where each g_i is an increasing function. Then for each interval $[a, b] \subseteq \mathbb{R}$, $f \in BV([a, b])$ and

$$V(f, [a, b]) \leq g_1(b) - g_1(a) + g_2(b) - g_2(a).$$

b. *If the function $f : \mathbb{R} \to \mathbb{R}$ is an element of $BV([a, b])$ for each interval $[a, b] \subseteq \mathbb{R}$, then the following hold:*

$$P(x) = (1/2)(V(f)(x) + f(x)) \text{ and } N(x) = (1/2)(V(f)(x) - f(x)) \\ \text{are increasing functions,} \tag{4.1}$$

$$f = P - N, \quad V(f) = P + N, \\ f(0) = P(0) = V(f)(0), \text{ and } N(0) = 0. \tag{4.2}$$

Proof. **a.** Take an interval $[a, b]$ and a partition $a \leq x_0 \leq x_1 \leq \cdots \leq x_n \leq b$. Clearly,

$$\begin{aligned}
\sum_{j=1}^{n} |f(x_j) - f(x_{j-1})| &= \sum_{j=1}^{n} |g_1(x_j) - g_1(x_{j-1}) - g_2(x_j) + g_2(x_{j-1})| \\
&\leq \sum_{j=1}^{n} |g_1(x_j) - g_1(x_{j-1})| + \sum_{j=1}^{n} |g_2(x_j) - g_2(x_{j-1})| \\
&= g_1(x_n) - g_1(x_0) + g_2(x_n) - g_2(x_0) \\
&\leq g_1(b) - g_1(a) + g_2(b) - g_2(a).
\end{aligned}$$

b. (4.2) is obvious from (4.1). Using the additivity of $V(f, [a, b])$, e.g., Problem 4.1, observe, as stated in Definition 4.1.1*c*, that

$$\forall\, [a, b] \subseteq \mathbb{R}, \quad V(f)(b) - V(f)(a) = V(f, [a, b]). \tag{4.3}$$

Consequently,

$$\begin{aligned}
P(b) - P(a) &= \frac{1}{2}[V(f, [a, b]) + (f(b) - f(a))] \\
&\geq \frac{1}{2}[|f(b) - f(a)| + (f(b) - f(a))] \geq 0.
\end{aligned}$$

Thus, P is an increasing function. A similar argument works for N, and so (4.1) is obtained. □

Let $BV_{\text{loc}}(\mathbb{R})$ be the space of functions $f : \mathbb{R} \to \mathbb{R}$ such that

$$\forall\, [a,b] \subseteq \mathbb{R}, \quad f \in BV([a,b]).$$

By Theorem 4.1.2, $BV_{\text{loc}}(\mathbb{R})$ is a real vector space.

Definition 4.1.3. Bounded variation on $\mathbb{R}$
$BV(\mathbb{R}) \subseteq BV_{\text{loc}}(\mathbb{R})$ is the space of all functions $f \in BV_{\text{loc}}(\mathbb{R})$ for which

$$V(f) = V(f, \mathbb{R}) = \sup_{a \leq b} V(f, [a,b]) < \infty.$$

$BV(\mathbb{R})$ is the real vector space of functions of *bounded variation on* $\mathbb{R}$. Clearly, $f \in BV(\mathbb{R})$ implies $V(f) \in \mathbb{R}^+$, and $V(f)$ is the *total variation* of f.

Remark. Let $V(f, (-\infty, b]) = \lim_{a \to -\infty} V(f, [a,b])$ and let $V(f, [a, \infty)) = \lim_{b \to \infty} V(f, [a,b])$. The limits exist by monotonicity. If $f \in BV_{\text{loc}}(\mathbb{R})$, then for each $a \in \mathbb{R}$, it is straightforward to check that

$$V(f) = V(f, (-\infty, a]) + V(f, [a, \infty))$$

and

$$\forall\, a \leq b, \quad V(f, [a, \infty)) = V(f, [a,b]) + V(f, [b, \infty)).$$

Further, if $f \in BV(\mathbb{R})$ then f is bounded on $\mathbb{R}$ and

$$\lim_{x \to -\infty} V(f, (-\infty, x]) = 0 \quad \text{and} \quad \lim_{x \to \infty} V(f, [x, \infty)) = 0.$$

The following result is a consequence of Theorem 4.1.2.

Theorem 4.1.4. Jordan decomposition theorem for $BV(\mathbb{R})$
Given $f : \mathbb{R} \to \mathbb{R}$, then $f \in BV(\mathbb{R})$ if and only if $f = f_1 - f_2$, where f_1, f_2 are bounded increasing functions on $\mathbb{R}$. In this case, $\lim_{x \to -\infty} f(x)$ and $\lim_{x \to \infty} f(x)$ exist.

A complex-valued function $f : \mathbb{R} \to \mathbb{C}$ is defined to be an element of the complex-valued class $BV_{\text{loc}}(\mathbb{R})$ if both $\text{Re}\,(f)$ and $\text{Im}\,(f)$ are elements of $BV_{\text{loc}}(\mathbb{R})$. Similarly we define a space $BV(\mathbb{R})$ for complex-valued functions. If $g \in BV([a,b])$ then the function

$$f(x) = \begin{cases} g(x) - g(a), & \text{if } x \in [a,b], \\ 0, & \text{if } x \leq a, \\ g(b) - g(a), & \text{if } x \geq b, \end{cases}$$

is an element of $BV(\mathbb{R})$ and $V(f, \mathbb{R}) = V(g, [a,b])$. As such, an element of $BV([a,b])$, properly normalized, can be considered as an element of $BV(\mathbb{R})$.

In the case of $BV_{\text{loc}}(\mathbb{R})$ and $BV(\mathbb{R})$, there is no ambiguity between real- and complex-valued functions. In particular, the following result is also valid for complex-valued functions.

Proposition 4.1.5. *$BV_{\text{loc}}(\mathbb{R})$ is a vector space and $BV(\mathbb{R}) \subseteq BV_{\text{loc}}(\mathbb{R})$ is a linear subspace.*

Example 4.1.6. $f \notin BV([-1,1])$ and $f' \notin L^1_m([-1,1])$
Let $f(x) = x^2 \sin(1/x^2)$ if $x \neq 0$ and let $f(0) = 0$. We showed in Example 3.2.10c that $f \notin BV([-1,1])$. In this regard note Problem 4.4. Observe that f' exists on $[-1,1]$ but that $f' \notin L^1_m([-1,1])$, e.g., Theorem 4.6.7.

We denote $BV([0,1])$ by $BV(\mathbb{T})$ when we want to think of 1-periodic functions on $\mathbb{R}$; and, if $f \in BV(\mathbb{T})$, we denote $V(f,[0,1])$ by $V(f,\mathbb{T})$. We give the following clever proof due to MITCHELL TAIBLESON [464] of the following classical result; see Example 3.3.4 and Appendix B for the definition of Fourier coefficients.

Theorem 4.1.7. Fourier coefficients of $BV(\mathbb{T})$
If $f \in BV(\mathbb{T})$ then

$$\forall\, n \in \mathbb{Z}, \quad |n\hat{f}(n)| \leq V(f,\mathbb{T}).$$

Proof. Note that

$$\forall\, k \in \mathbb{Z} \text{ and } \forall\, n \in \mathbb{Z} \setminus \{0\}, \quad \int_{k/|n|}^{(k+1)/|n|} e^{2\pi i n x}\, dx = 0. \tag{4.4}$$

When $n \neq 0$, set $a_k = k/|n|$ for $k = 0, \ldots, |n|$. Let g be the function

$$g(x) = \sum_{k=1}^{n} f(a_k) \mathbb{1}_{(a_{k-1},a_k]}(x).$$

Then by (4.4),

$$\forall\, n \neq 0, \quad \hat{g}(n) = 0.$$

Thus, for $n \neq 0$,

$$\begin{aligned} |\hat{f}(n)| &= \left| \int_0^1 (f(x) - g(x)) e^{2\pi i n x}\, dx \right| \leq \sum_{k=1}^{|n|} \int_{a_{k-1}}^{a_k} |f(x) - f(a_k)|\, dx \\ &\leq \frac{1}{|n|} \sum_{k=1}^{|n|} V(f,[a_{k-1},a_k]) = \frac{1}{|n|} V(f,\mathbb{T}), \end{aligned}$$

where the final step follows from the additivity of $V(f,[a,b])$. □

Remark. The converse to Theorem 4.1.7 is not true. In fact, there is an element $f \in L^1_m(\mathbb{T})$ such that

$$|n\hat{f}(n)| = o(1), \quad |n| \to \infty.$$

In particular, $|n\hat{f}(n)| = O(1)$, $|n| \to \infty$, and $f \notin BV(\mathbb{T})$; see Problem 4.5.

Proposition 4.1.8. *Let* $(\mathbb{R}, \mathcal{B}(\mathbb{R}), \mu)$ *be a finite measure space and set*

$$f(x) = \mu((-\infty, x]) \tag{4.5}$$

for the given bounded measure μ. *Then* f *is an increasing element of* $BV(\mathbb{R})$ *that is right continuous at each point* $x \in \mathbb{R}$.

Proof. Clearly, f is increasing and

$$\mu((a,b]) = \mu((-\infty, b]) - \mu((-\infty, a]) = f(b) - f(a).$$

Because $\mu(\mathbb{R}) < \infty$ we conclude that $f \in BV(\mathbb{R})$. Finally, we must show that $f(b) = \lim_{x \to b+} f(x)$ for each $b \in \mathbb{R}$ (compare with the proof of Proposition 3.5.2). We write

$$(-\infty, b] = \bigcap_{n=1}^{\infty} (-\infty, b + 1/n],$$

so that, by the properties of measures,

$$f(b) = \mu((-\infty, b]) = \lim_{n \to \infty} \mu\left((-\infty, b + 1/n]\right) = \lim_{n \to \infty} f\left(b + 1/n\right). \qquad \square$$

Theorem 4.1.9. Equivalence of bounded measures and increasing functions
Consider the measurable space $(\mathbb{R}, \mathcal{B}(\mathbb{R}))$.

a. *If* μ *is a bounded measure on* $(\mathbb{R}, \mathcal{B}(\mathbb{R}))$, *then* f *defined by (4.5) is an increasing function of bounded variation, right continuous at each* $x \in \mathbb{R}$.

b. *If* f *satisfies the conclusions of part a, then there is a unique bounded measure* μ *on* $(\mathbb{R}, \mathcal{B}(\mathbb{R}))$ *such that*

$$\forall\, (a,b] \subseteq \mathbb{R}, \quad \mu((a,b]) = f(b) - f(a). \tag{4.6}$$

Proof. **a.** Part *a* is Proposition 4.1.8.

b. An increasing function of bounded variation must have real values. Thus, part *b* follows from Theorem 3.5.1 and the remark following it.

For the proof of uniqueness see Problem 2.19. $\square$

The reader should compare Theorem 4.1.9 with Theorem 5.4.1.

Remark. For perspective note the relation between Theorem 4.1.9 and the *Riesz representation theorem* (F. RIESZ, 1909): μ *is a continuous linear functional on* $C([a,b])$, *where* $C([a,b])$ *is taken with the* $\|...\|_\infty$ *norm* (see Definition 2.5.9 and Appendix A.2.1), *if and only if there is a function* $g \in BV([a,b])$ *such that*

$$\forall\, f \in C([a,b]), \quad \mu(f) = \int_a^b f\, dg; \tag{4.7}$$

cf. Section 7.1.

The Riesz representation theorem has been generalized by BOURBAKI in the most perfect way by becoming the definition of measure; and it has been a powerful influence on the formation of modern analysis. This metamorphosis is the subject of Chapter 7.

In (4.7), $\int_a^b f\,dg$ is the *classical Riemann–Stieltjes integral with respect to a function g of bounded variation.* Its definition is analogous to that in (3.29) of the Riemann–Stieltjes integral with respect to an increasing right-continuous function. However, here, because we consider only the integral for continuous functions f, we do not need to assume that the function g is right continuous. For a more detailed discussion of the classical Riemann–Stieltjes integral, see, e.g., [7], [355]. It should be observed that if $f \in C([a,b])$ and $g \in BV([a,b])$, then

$$\left|\int_a^b f\,dg\right| \leq \|f\|_\infty V(g,[a,b]).$$

We now give an application of (4.7) and several other results. The statement of Proposition 4.1.10 is true by a simpler proof for the cases that $|\alpha| < 1$ and α is a root of unity, and it is false for $|\alpha| > 1$; the proof need not be assimilated the first time around. The result is taken from Amer. Math. Monthly 75 (1968) 312–313, and is due to DONALD J. (DJ) NEWMAN (JOHN NASH's Gnu); cf. [358] for further developments.

Proposition 4.1.10. *Let $|\alpha| = 1$. There is a nonzero function $f \in BV(\mathbb{T})$ such that*

$$\forall\, x \in [0,1], \quad \alpha f(x) = f(x/2) + f((x+1)/2).$$

Proof. For this result let $C(\mathbb{T})$ be the space of $\mathbb{C}$-valued continuous 1-periodic functions g on $\mathbb{R}$ for which $g(0) = g(1)$. From the Sidon theorem, e.g., Problem 4.6,

$$\sup_{|z|=1}\left|\sum_{n=1}^{N} a_n z^{2^n}\right| \geq \frac{1}{4}\sum_{n=1}^{N}|a_n|. \tag{4.8}$$

We shall use this fact to prove that the set

$$S = \{\alpha g(x) - g(2x) : g \in C(\mathbb{T})\}$$

is not dense in $C(\mathbb{T})$; in fact, we shall show that

$$\inf\{\|e - f\|_\infty : f \in S\} \geq \frac{1}{4}, \tag{4.9}$$

where $e(x) = e^{2\pi i x}$. Set $\alpha g(x) - g(2x) = e(x) + \delta(x)$. Then

$$\begin{aligned}\alpha^2 g(x) - g(2^2 x) &= \alpha^2 g(x) + \alpha g(2x) - \alpha g(2x) - g(2^2 x)\\ &= \alpha(e(x) + \delta(x)) + e(2x) + \delta(2x).\end{aligned}$$

Consequently, when we iterate N times,

$$\alpha^N g(x) - g(2^N x) = \sum_{n=1}^{N} \alpha^{N-n} e(2^{n-1}x) + \sum_{n=1}^{N} \alpha^{N-n} \delta(2^{n-1}x). \tag{4.10}$$

From (4.10) we compute

$$2\|g\|_\infty \geq \left|\alpha^N g(x) - g(2^N x)\right| \geq \left|\sum_{n=1}^{N} \alpha^{N-n} e(2^{n-1}x)\right| - N\|\delta\|_\infty,$$

so that, from (4.8),

$$2\|g\|_\infty \geq \frac{1}{4}\sum_{n=1}^{N} |a_n|^{N-n} - N\|\delta\|_\infty = N((1/4) - \|\delta\|_\infty).$$

Thus,

$$\|\delta\|_\infty \geq \frac{1}{4} - \frac{2\|g\|_\infty}{N},$$

and since for any $g \in C(\mathbb{T})$ this is valid for all N, we have (4.9).

We use (4.7), (4.9), and the Hahn–Banach theorem (see Appendix A.8) to conclude that a function $h \in BV([0,1])$ exists such that

$$\forall\, g \in C(\mathbb{T}), \quad \int_0^1 (\alpha g(x) - g(2x))\, dh(x) = 0.$$

Computing

$$\int_0^1 g(2x)\, dh(x) = \int_0^1 g(x)\, d\left\{h(x/2) + h((x+1)/2)\right\},$$

we obtain

$$\forall\, g \in C(\mathbb{T}), \quad \int_0^1 g(x)\, d\left\{\alpha h(x) - h(x/2) - h((x+1)/2)\right\} = 0;$$

and thus $\alpha h(x) - h(x/2) - h((x+1)/2)$ is a constant c. Set $f = h - d$, where $d = c/(\alpha - 2)$. □

Proposition 4.1.11. *Let $f \in BV([a,b])$.*

a. *For each $c \in [a,b]$, $f(c\pm)$ exists and* card $D(f) \leq \aleph_0$.

b. *f is Lebesgue measurable and bounded.*

Proof. **a.** $f(c\pm)$ exists from Theorem 4.1.2. For each n let $D_n = \{c \in (a,b) : |f(c+) - f(c-)| > 1/n\}$. Clearly, card $D_n < \aleph_0$ since $f \in BV([a,b])$. Thus, card $D(f) \leq \aleph_0$.

b. Part b is clear from part a. □

Example 4.1.12. Increasing functions with discontinuities on $\mathbb{Q}$
We construct a bounded increasing function f on $(0,1)$ that is discontinuous precisely on the set $\mathbb{Q} \cap (0,1) = \{r_n : n = 1, \ldots\}$. Define

$$f_n(x) = \begin{cases} 0, & \text{if } x \in [0, r_n), \\ 1/2^n, & \text{if } x \in [r_n, 1). \end{cases}$$

Then $f = \sum f_n$ is an increasing function, since each f_n is increasing; and $0 \le f \le 1$ from the definition. For any fixed r_k, the fact that $\sum_{n \ne k} f_n$ is increasing implies that

$$\lim_{x \to r_k -} \sum_{n \neq k} f_n(x) \le \lim_{x \to r_k +} \sum_{n \neq k} f_n(x).$$

Consequently, from the definition of $f_k(r_k \pm)$ we have $f(r_k-) < f(r_k+)$.

An interesting discussion of total variation and related concepts is found in [97].

4.2 Decomposition into discrete and continuous parts

Let $f : \mathbb{R} \to \mathbb{R}$ be an increasing function. We shall try to find the "continuous part" of f. Recall from Problem 1.16 that an increasing function can have at most countably many points of discontinuity. If we write $D(f) = \{x_n : n = 1, \ldots\}$, define

$$s_n(x) = \begin{cases} 0, & \text{if } x < x_n, \\ f(x_n) - f(x_n-), & \text{if } x = x_n, \\ f(x_n+) - f(x_n-), & \text{if } x > x_n. \end{cases}$$

Clearly, s_n is a nonnegative increasing function. (The "s" is for "saltus", which, in turn, is for old times' sake.)

Proposition 4.2.1. *Assume that $f : \mathbb{R} \to \mathbb{R}$ is an increasing function, and let $D(f) = \{x_n : n = 1, \ldots\}$. Define*

$$g_n = f - \sum_{k=1}^{n-1} s_k.$$

Then,

a. *g_n is an increasing function;*
b. *g_n is continuous on the set $C(f) \cup \{x_1, \ldots, x_{n-1}\}$.*

Proof. **a.** If $x < y < x_1$ then $g_1(y) = f(y) \geq f(x) = g_1(x)$. If $x < x_1$ then $g_1(x_1) = f(x_1) - (f(x_1) - f(x_1-)) = f(x_1-) \geq f(x) = g_1(x)$.

For $x_1 < x$, $g_1(x) = (f(x) - f(x_1+)) + f(x_1-) \geq f(x_1-) = g_1(x_1)$.

Finally, when $x_1 < x < y$, $g_1(y) = f(y) - (f(x_1+) - f(x_1-)) \geq f(x) - (f(x_1+) - f(x_1-)) = g_1(x)$.

Consequently, g_1 is increasing; and for arbitrary g_n the result follows by induction.

b. It is sufficient to check that for $n > 1$,

$$\lim_{x \to x_1 \pm} g_n(x) = g_n(x_1).$$

Note that $g_n(x_1) = f(x_1-) - \sum_{j=2}^{n-1} s_j(x_1)$. We shall check that

$$\lim_{x \to x_1 -} g_n(x) = g_n(x_1);$$

a similar argument works for $x \to x_1+$. Clearly,

$$\lim_{x \to x_1-} g_n(x) = f(x_1-) - \lim_{x \to x_1-} s_1(x) - \lim_{x \to x_1-} \sum_{j=2}^{n-1} s_j(x).$$

In this case $s_1(x) = 0$ since $x < x_1$, and

$$\lim_{x \to x_1-} \sum_{j=2}^{n-1} s_j(x) = \sum_{j=2}^{n-1} s_j(x_1)$$

by the definition of s_j. □

Remark. Assume that $f : \mathbb{R} \to \mathbb{R}$ is an element of $BV(\mathbb{R})$. With the previous notation observe that $\sum_{j=1}^{\infty} s_j(x)$ converges. In fact, if all but finitely many x_n are greater than x then the sum is finite, and if infinitely many x_n are less than x we have convergence by the definition of s_j and the fact that f increases. For example, if $x_n < x$ for each n then

$$\sum_{j=1}^{\infty} s_j(x) = \sum_{j=1}^{\infty} (f(x_j+) - f(x_j-)),$$

and the second sum is finite since $V(f) < \infty$; cf. Example 4.1.12. Besides the pointwise convergence we also have that $\sum s_j$ converges uniformly on any closed bounded interval $[a, b]$. To verify this latter assertion note that if $s = \sum s_j$ and $\varepsilon > 0$ is given, then

$$\exists\, N \text{ such that } \forall\, n > N, \quad \left| \sum_{j=n}^{\infty} s_j(a) \right| < \varepsilon \quad \text{and} \quad \left| \sum_{j=n}^{\infty} s_j(b) \right| < \varepsilon.$$

Consequently, for all $x \in [a, b]$ and for all $n > N$,

$$-\varepsilon < \sum_{j=n}^{\infty} s_j(a) \le \sum_{j=n}^{\infty} s_j(x) = s(x) - \sum_{j=1}^{n-1} s_j(x) \le \sum_{j=n}^{\infty} s_j(b) < \varepsilon.$$

If f is an increasing element of $BV(\mathbb{R})$ then $s = \sum s_j$ is the *discrete* or *discontinuous* part of f. If $f \in BV(\mathbb{R})$ and $f = P - N$, with notation as in Theorem 4.1.2, the *discrete* part of f is

$$s = s_P - s_N,$$

where s_P, respectively, s_N, is the discrete part of P, respectively, N.

The following result is straightforward to prove; cf. Problem 4.19*b*, which concerns the differentiability of the variation function $V(f)$.

Proposition 4.2.2. *Let $f \in BV(\mathbb{R})$. If f is continuous at x then the variation function $V(f)$, and hence P and N, are continuous at x.*

From Proposition 4.2.1, Proposition 4.2.2, and the previous Remark, we conclude with the following theorem.

Theorem 4.2.3. The continuous part of $f \in BV(\mathbb{R})$
Let $f \in BV(\mathbb{R})$ have discrete part s. Then

$$g = f - s \in BV(\mathbb{R})$$

is a continuous function on $\mathbb{R}$. Further, if f is an increasing function then g is an increasing function, and if f is a continuous function then $s = 0$.

The function g in Theorem 4.2.3 is the *continuous part* of f. In this regard and taking μ and then f as in Proposition 4.1.8, note that

$$\forall\, x \in \mathbb{R}, \quad \mu(\{x\}) = \lim_{n\to\infty} \mu\left((x - 1/n, x]\right) = f(x) - f(x-),$$

and so, since f defined by (4.5) is right continuous, $f \in BV(\mathbb{R})$ *is continuous at x if and only if $\mu(\{x\}) = 0$*. Also note that the function f in Example 4.1.12 is a discrete function that is discontinuous precisely on the rational numbers in $[0, 1]$.

Example 4.2.4. Cantor–Lebesgue continuous measure
Let $E \subseteq [0, 1]$ be a perfect symmetric set determined by $\{\xi_k : k = 1, \ldots\} \subseteq (0, 1/2)$ and let C_E be its associated Cantor function; see Example 1.3.17. If $m(E) = 0$ then $C'_E = 0$ *m-a.e.*; C_E has no discrete part and the measure μ_E associated with C_E (by Theorem 4.1.9*b*) is the *Cantor–Lebesgue continuous measure* for E. In light of the following example note that C_E takes constant values on a set of positive measure, and, in particular, it is not a strictly increasing function.

Example 4.2.5. Hellinger example
We now present the *Hellinger example* (1907); cf. Example 4.6.6. This is a continuous, strictly increasing function $H : [0,1] \to \mathbb{R}$ such that $H' = 0$ m-*a.e.* The function H will, in fact, depend on a fixed $t \in (0,1)$, and so, later, we shall sometimes write $H = H_t$. We shall see that H takes the form $\lim_{n\to\infty} H_n$, and we begin by setting $H_0(x) = x$ on $[0,1]$. Assume that H_{n-1} is constructed. We shall take H_n to be continuous and linear in each dyadic interval $[k/2^n, (k+1)/2^n]$, $k = 0, \ldots, 2^n - 1$. Besides that, we set

$$H_n\left(\frac{k}{2^{n-1}}\right) = H_{n-1}\left(\frac{k}{2^{n-1}}\right), \quad k = 0, \ldots, 2^{n-1},$$

and for our fixed $t \in (0,1)$ we define

$$H_n\left(\frac{2k+1}{2^n}\right) = \frac{1-t}{2} H_{n-1}\left(\frac{k}{2^{n-1}}\right) + \frac{1+t}{2} H_{n-1}\left(\frac{k+1}{2^{n-1}}\right),$$

$k = 0, \ldots, 2^{n-1} - 1$. Note that $(2k+1)/2^n$ is the midpoint of $[k/2^{n-1}, (k+1)/2^{n-1}]$. For example, take $t = 1/2$ and observe that $H_1(1/2) = 3/4$; see Figure 4.1.

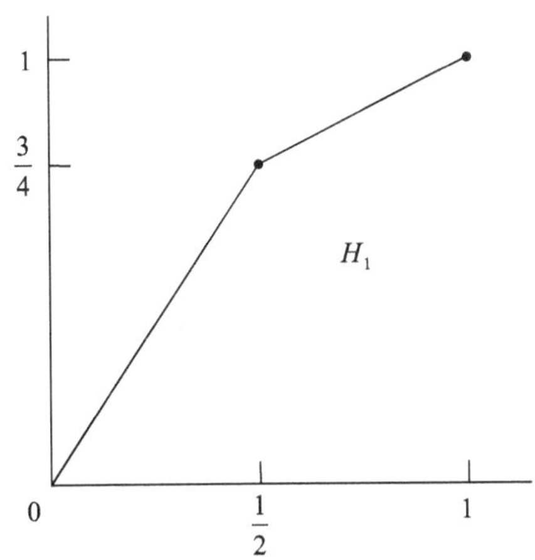

Fig. 4.1. Hellinger example.

From this construction we see that H_n is continuous and strictly increasing, and that

$$\forall\, x \in [0,1], \quad 0 \le H_n(x) \le H_{n+1}(x) \le 1.$$

Consequently, $H = \lim_{n\to\infty} H_n$ exists and is an increasing function. Now let $x < y$ and choose an n and a corresponding $k \in \{0, \ldots, 2^n - 1\}$ such that

$$x < k/2^n < y.$$

Then

$$H(x) \le H(k/2^n) = H_n(k/2^n) < H_n(y) \le H(y),$$

so that H is *strictly increasing.*

We now prove that H is continuous at each $x \in [0,1]$. Take

$$\{x\} = \bigcap_{n=1}^{\infty} [\alpha_n, \beta_n], \quad [\alpha_n, \beta_n] = [k_n/2^n, (k_n+1)/2^n],$$

where $[\alpha_{n+1}, \beta_{n+1}] \subseteq [\alpha_n, \beta_n]$. Clearly, $\alpha_{n+1} = \alpha_n$ and $\beta_{n+1} = \beta_n - (1/2^{n+1})$, or $\alpha_{n+1} = \alpha_n + (1/2^{n+1})$ and $\beta_{n+1} = \beta_n$. In the first case,

$$\begin{aligned} H(\beta_{n+1}) - H(\alpha_{n+1}) &= H_{n+1}(\beta_{n+1}) - H_{n+1}(\alpha_{n+1}) \\ &= H_{n+1}(\beta_{n+1}) - H_n(\alpha_n) \\ &= \frac{1-t}{2} H_n(\alpha_n) + \frac{1+t}{2} H_n(\beta_n) - H_n(\alpha_n) \\ &= \frac{1+t}{2} H_n(\beta_n) - \frac{1+t}{2} H_n(\alpha_n). \end{aligned} \tag{4.11}$$

For the second case, the computation yields "$(1-t)/2$" instead of "$(1+t)/2$", and so in either situation, we have

$$H(\beta_{n+1}) - H(\alpha_{n+1}) = \frac{1 \pm t}{2} (H_n(\beta_n) - H_n(\alpha_n)).$$

Continuing this process we obtain

$$H(\beta_{n+1}) - H(\alpha_{n+1}) = \prod_{j=1}^{n} \frac{1+\varepsilon_j t}{2}, \quad \varepsilon_j = \pm 1, \tag{4.12}$$

and hence

$$|H(\beta_{n+1}) - H(\alpha_{n+1})| \le \left(\frac{1+t}{2}\right)^n, \tag{4.13}$$

since $1 + \varepsilon_j t \le 1 + t$. (As a gift, (4.12) also yields another proof that H is strictly increasing.) The *continuity of H at x follows from (4.13), which in fact tells us that $H(x+) = H(x-)$.*

Now, $H'(x)$ exists *m-a.e.*, as we shall prove in Theorem 4.3.2. For any such x choose $\{[\alpha_n, \beta_n]\}$ as before, and observe that

$$\frac{H(\beta_{n+1}) - H(\alpha_{n+1})}{\beta_{n+1} - \alpha_{n+1}} = 2^{n+1} \prod_{j=1}^{n} \frac{1+\varepsilon_j t}{2} = 2 \prod_{j=1}^{n} (1 + \varepsilon_j t), \quad \varepsilon_j = \pm 1.$$

The product $\prod_{j=1}^{\infty} (1 + \varepsilon_j t)$ diverges to 0, since we are assuming that $H'(x)$ exists, and so $H'(x) = 0$ *m-a.e.*

A similar function was introduced by HERMANN MINKOWSKI in 1912 in his study of quadratic irrationals. Further examples have been given by BORGE JESSEN, SALEM, WIENER, and AUREL WINTNER, e.g., [414], pages 282–294.

Proposition 4.2.6. *Let $f : [a, b] \to \mathbb{R}$ be a continuous function. Then*

$$\lim_{|P| \to 0} \sum |f(x_j) - f(x_{j-1})| = V(f, [a, b]),$$

where $V(f, [a, b])$ may be infinite and the norm $|P|$ of a partition P is defined as $\max\{|x_j - x_{j-1}| : j = 1, \ldots, n\}$ for the partition $P : a = x_0 \leq x_1 \leq \cdots \leq x_n = b$.

In order to determine sets of continuity in Proposition 1.3.6 we dealt with the oscillations $\omega(f, I_j)$, $I_j = [x_j, x_{j+1}]$. If f is continuous on $[a, b]$ we can easily check, using Proposition 4.2.6 and the notation there, that

$$\lim_{|P| \to 0} \sum_{j=0}^{n-1} \omega(f, I_j) = V(f, [a, b]).$$

We shall now give a criterion in order that a continuous function be a function of bounded variation. Let $f : [a, b] \to \mathbb{R}$ be continuous and define

$$\forall\, y \in \mathbb{R}, \quad R_y = f^{-1}(y) = \{x \in [a, b] : f(x) = y\}.$$

The *Banach indicatrix* $B : \mathbb{R} \to \mathbb{R}^+ \cup \{\infty\}$ is defined as

$$\forall\, y \in \mathbb{R}, \quad B(y) = \text{card } R_y.$$

We want to mention that BANACH denoted the indicatrix function by N and, naturally, he did not call it the *Banach indicatrix.*

Theorem 4.2.7. Banach–Vitali theorem
Let $f : [a, b] \to \mathbb{R}$ be a continuous function.

a. *B is Lebesgue measurable and*

$$\int_c^d B(y)\, dy = V(f, [a, b]),$$

where $c = \inf\{f(x) : x \in [a, b]\}$ and $d = \sup\{f(x) : x \in [a, b]\}$.

b. *$f \in BV([a, b]) \iff B \in L^1_m([c, d])$.*

c. *If $f \in BV([a, b])$ then $m(\{y : B(y) \geq \aleph_0\}) = 0$.*

Proof. **a.** Let $P_n : a = x_0 < a + (b - a)/2^n = x_1 < \cdots < a + j(b - a)/2^n = x_j < \cdots < x_{2^n} = b$ be a partition of $[a, b]$. We have $|P_n| = (b - a)/2^n$. (The norm is defined in Section 3.1 and in the statement of Proposition 4.2.6.) For each $j = 1, \ldots, 2^n$ define

$$\forall\, y \in \mathbb{R}, \quad S_j(y) = \begin{cases} 1, & \text{if } \exists\, x \in (x_{j-1}, x_j] \text{ such that } f(x) = y, \\ 0, & \text{if } \forall\, x \in (x_{j-1}, x_j] \;\; f(x) \neq y. \end{cases}$$

Observe that S_j is a bounded Lebesgue measurable function with at most two points of discontinuity, and that

$$\int_c^d S_j(y)\, dy = \omega(f, J_j), \quad \text{where} \quad J_j = (x_{j-1}, x_j].$$

Define $B_n = \sum_{j=1}^{2^n} S_j$. From the above remarks,

$$\lim_{n\to\infty} \int_c^d B_n(y)\, dy = V(f, [a, b]).$$

Now, $\{B_n : n = 1, \ldots\}$ is an increasing sequence, so that by LDC

$$\int_c^d \widetilde{B}\, dy = \lim_{n\to\infty} \int_c^d B_n(y)\, dy,$$

where $\widetilde{B}$ is the pointwise limit of the sequence $\{B_n : n = 1, \ldots\}$.

Clearly, $\widetilde{B} \leq B$ since $B_n \leq B$, and so we shall complete the proof of part *a* once we prove that $B \leq \widetilde{B}$.

Let $m \leq B(y)$ and choose $z_1 < z_2 < \cdots < z_m$, where each $f(z_k) = y$. Take n large enough that

$$\forall\, k = 1, \ldots, n, \quad \frac{b - a}{2^n} < z_k - z_{k-1}.$$

Consequently, no two z_ks will be in the same $[x_{j-1}, x_j]$ for the partition P_n. Thus, $B_n(y) \geq m$, and so $\widetilde{B}(y) \geq m$. Since this is true for each $m \leq B(y)$ we conclude that $B(y) \leq \widetilde{B}(y)$.

b. Part *b* follows from part *a*.

c. Part *c* is clear from part *b*. □

BANACH is usually given full credit for the above result; see Théorème 1.2 in [18]. The relevant VITALI paper, with the result obtained independently, is [488]. Theorem 4.2.7 has been generalized by SERGEI M. LOZINSKII, [330], [331], to include cases of functions f that are discontinuous but for which one-sided limits exist on $[0, 1]$. For more on this subject we also refer the interested reader to the commentary of LIPIŃSKI on the work of BANACH in [21].

4.3 The Lebesgue differentiation theorem

In 1904, LEBESGUE proved the *differentiation theorem: continuous functions of bounded variation have a derivative m-a.e.* In 1907, while trying to extend the fundamental theorem of calculus (FTC) to $\mathbb{R}^2$, VITALI proved what is now known as the *Vitali covering lemma.* This became the basis for proving general differentiation theorems; see Chapter 8. In 1932, F. RIESZ gave a different proof of LEBESGUE's result using the so-called "rising sun lemma". RIESZ' proof is found in [15], [392], pages 6–11. RIESZ' approach to proving the result without integration theory was anticipated by GEORG FABER (1910) and the YOUNGS (1911). L. COHEN [109] has made a simplification of RIESZ' technique and used it to obtain the differentiation theorem for df/dg, where f and g are increasing; this is interesting, since $g(x) = x$ is the usual result and because it provides a means to compute a Radon–Nikodym derivative without using the integration theory centered about the Radon–Nikodym theorem; cf. the development in Chapter 8. VITALI's approach has been studied extensively, and we refer to [79] for an important survey. There are also proofs of the differentiation theorem by DONALD G. AUSTIN [16], GIORGIO LETTA [318], and MICHAEL W. BOTSKO [68]. None of the proofs we know is transparent, but these last three seem simpler than most. VITALI's proof is long but readable, and we shall give BANACH's proof [17] of VITALI's theorem. BANACH's technique is brilliant, and the exposition of it in [235] is perfect; as such we have no choice but to engage in a copying exercise.

A collection $\mathcal{V}$ of intervals is a *Vitali covering* of $X \subseteq \mathbb{R}$ if

$$\forall\, \varepsilon > 0 \text{ and } \forall\, x \in X,\ \exists\, I \in \mathcal{V}, \text{ for which } m(I) < \varepsilon, \text{ such that } x \in I.$$

VITALI's result below does not preclude the possibility that if $X \subseteq \bigcup_{n=1}^N I_n$ then $m^*\left(\bigcup_{n=1}^N I_n \setminus X\right)$ is large.

Theorem 4.3.1. Vitali covering lemma
Let $X \subseteq \mathbb{R}$ with $m^*(X) < \infty$, and let $\mathcal{V}$ be a Vitali covering of X. Then for each $\varepsilon > 0$ there is a disjoint family $\{I_n : n = 1, \ldots, N\} \subseteq \mathcal{V}$ such that

$$m^*\left(X \setminus \bigcup_{n=1}^N I_n\right) < \varepsilon. \tag{4.14}$$

Proof. i. Without loss of generality we can take each interval $I \in \mathcal{V}$ to be closed, since

$$m^*\left(X \setminus \bigcup_{n=1}^N I_n\right) = m^*\left(X \setminus \bigcup_{n=1}^N \overline{I}_n\right).$$

Also without loss of generality let U be an open set, with $m(U) < \infty$, such that

$$\forall\, I \in \mathcal{V}, \quad I \subseteq U.$$

We can do this since $m^*(X) < \infty$ and $\mathcal{V}$ is a Vitali covering, even though we may be throwing away some of our original elements from $\mathcal{V}$.

ii. We now choose a disjoint family $\{I_n : n = 1, \ldots\} \subseteq \mathcal{V}$ that "almost" covers X in the following way. Take any $I_1 \in \mathcal{V}$ and assume that a disjoint family $\{I_1, \ldots, I_n\} \subseteq \mathcal{V}$ has been chosen. If $X \subseteq \bigcup_{j=1}^N I_j$ we stop the procedure. Assume otherwise. Define $r_n = \sup\{m(I) : I \cap I_j = \emptyset,\ j = 1, \ldots, n,\ I \in \mathcal{V}\}$, noting that

$$\forall\, n, \quad r_n \leq m(U) < \infty.$$

We see that r_n is positive, since $\mathcal{V}$ is a Vitali covering and $\bigcup_{j=1}^N I_j$ does not cover X. Thus, we can take $I_{n+1} \in \mathcal{V}$ for which

$$m(I_{n+1}) > \frac{1}{2} r_n$$

and

$$\forall\, j = 1, \ldots, n, \quad I_{n+1} \cap I_j = \emptyset.$$

Obviously, without loss of generality, we can choose $\{m(I_j) : j = 1, \ldots\}$ to be a decreasing sequence.

iii. Using part *ii* we see that

$$\exists\, N > 0 \text{ such that } \sum_{j=N+1}^{\infty} m(I_j) < \frac{\varepsilon}{5};$$

this follows since $\{I_j : j = 1, \ldots\}$ is a disjoint family and $m(U) < \infty$. We shall prove that

$$m^*\left(X \setminus \bigcup_{j=1}^{N} I_j\right) < \varepsilon.$$

iv. If $y \in X \setminus \bigcup_{j=1}^N I_j$ we can choose $I_y \in \mathcal{V}$ such that $y \in I_y$ and

$$\forall\, j = 1, \ldots, N, \quad I_y \cap I_j = \emptyset; \tag{4.15}$$

this follows from the definition of $\mathcal{V}$, including the hypothesis that each of its elements is a closed interval.

Next observe that $I_y \cap I_n \neq \emptyset$ for some n. To see this, take any n (larger than N in light of (4.15)) and assume that $I_y \cap I_j = \emptyset$ for each $j \leq n$. Then, from the definition of r_n and the construction of $\{I_j : j = 1, \ldots\}$,

$$m(I_y) \leq r_n < 2m(I_{n+1}). \tag{4.16}$$

Since $\sum m(I_j) < \infty$, we have $m(I_j) \to 0$; and thus $I_y \cap I_j \neq \emptyset$ for some j.

v. Let $n = n(y)$ be the smallest integer ($n > N$) for which $I_y \cap I_n \neq \emptyset$, where $y \in X \setminus \bigcup_{j=1}^N I_j$, and let x_n be the midpoint of I_n. We shall prove that

$$|y - x_n| \leq \frac{5}{2} m(I_n). \tag{4.17}$$

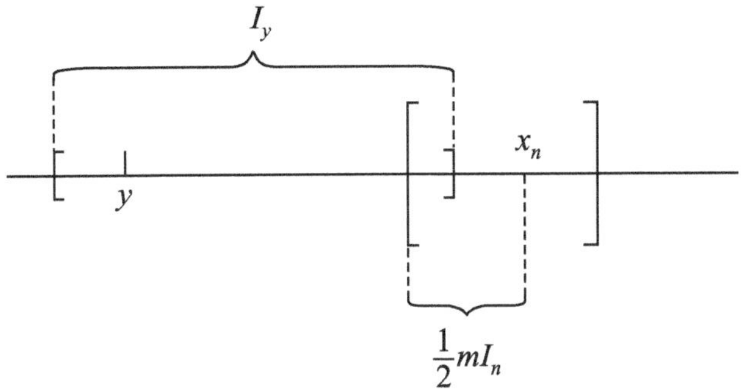

Fig. 4.2. Vitali covering estimate.

Clearly (from Figure 4.2, for example),

$$|y - x_n| \leq m(I_y) + \frac{1}{2}m(I_n). \tag{4.18}$$

Because of (4.16) and the fact that $\{m(I_j) : j = 1, \ldots\}$ decreases, we obtain (4.17) from (4.18).

vi. For each n let J_n be the closed interval with center x_n and length five times that of I_n. From (4.17), $y \in J_n$. Since $n = n(y) > N$ and y is arbitrary in $X \setminus \bigcup_{j=1}^{N} I_j$, we have

$$X \setminus \bigcup_{j=1}^{N} I_j \subseteq \bigcup_{n=N+1}^{\infty} J_n$$

and

$$m\left(\bigcup_{n=N+1}^{\infty} J_n\right) \leq \sum_{n=N+1}^{\infty} m(J_n) \leq 5 \sum_{n=N+1}^{\infty} m(I_n) < \varepsilon.$$

This yields (4.14). □

The *Dini derivates* of a function $f : \mathbb{R} \to \mathbb{R}$ are

$$D^+f(x) = \overline{\lim_{h\to 0+}} \frac{f(x+h) - f(x)}{h},$$

$$D_+f(x) = \underline{\lim_{h\to 0+}} \frac{f(x+h) - f(x)}{h},$$

$$D^-f(x) = \overline{\lim_{h\to 0-}} \frac{f(x+h) - f(x)}{h} = \overline{\lim_{h\to 0+}} \frac{f(x) - f(x-h)}{h},$$

$$D_-f(x) = \underline{\lim_{h\to 0-}} \frac{f(x+h) - f(x)}{h} = \underline{\lim_{h\to 0+}} \frac{f(x) - f(x-h)}{h}.$$

These numbers always exist, and it is obvious that

i. $D^+f(x) \geq D_+f(x), \quad D^-f(x) \geq D_-f(x)$,

ii. $\exists\, f'(x) \iff D^+f(x) = D_+f(x) = D^-f(x) = D_-f(x) \neq \pm\infty$.

Now for the differentiation theorem formulated by LEBESGUE in 1904; see [392], pages 5–11, for background *and* foreground.

Theorem 4.3.2. Lebesgue differentiation theorem
Let $f \in BV([a,b])$. Then,

a. *f' exists m-a.e.*

b. *$f' \in L^1_m([a,b])$.*

c. *If f is increasing then $\int_a^b f' \leq f(b) - f(a)$.*

Proof. Without loss of generality we assume that f is increasing.

a. *i.* To prove part a we must show that the set of points where any two derivatives are unequal has Lebesgue measure 0. We shall do the computation for

$$A = \{x : D^+f(x) > D_-f(x)\},$$

proving that

$$r = m^*(A_{p,q}) = 0, \tag{4.19}$$

where

$$A_{p,q} = \{x : D^+f(x) > p > q > D_-f(x)\}, \quad p,\, q \in \mathbb{Q}.$$

It is sufficient to prove (4.19), since $A = \bigcup\{A_{p,q} : p,\, q \in \mathbb{Q}\}$. In order to verify (4.19) we assume $r \neq 0$ and show that

$$\forall\, \varepsilon > 0, \quad b(r+\varepsilon) > a(r - 2\varepsilon); \tag{4.20}$$

this tells us that $q \geq p$, the desired contradiction.

ii. We construct a Vitali covering $\mathcal{V}$ of $A_{p,q}$.

Let U be an open set containing $A_{p,q}$ such that $m(U) < r + \varepsilon$. For each $x \in A_{p,q}$ choose $h > 0$ for which

$$[x-h, x] \subseteq U, \quad f(x) - f(x-h) < bh; \tag{4.21}$$

this can be done from the definition of $A_{p,q}$ and because U is open.

The set $\mathcal{V}$ is defined to be the collection of all possible intervals $[x-h,x]$, where $x \in A_{p,q}$ and h satisfies (4.21). Obviously, $\mathcal{V}$ is a Vitali covering of $A_{p,q}$, and so, from Theorem 4.3.1, there is a disjoint family $\{I_j = [x_j - h_j, x_j] \in \mathcal{V} : j = 1, \ldots, N\}$ satisfying

$$m^*\left(A_{p,q} \setminus \bigcup_{j=1}^{N} I_j\right) < \varepsilon.$$

iii. Let

$$B = A_{p,q} \cap \left(\bigcup_{j=1}^{N} I_j \right),$$

and observe that

$$\left(A_{p,q} \setminus \bigcup_{j=1}^{N} I_j \right) \cup B = A_{p,q}.$$

Thus, from part *ii*,

$$m^*(A_{p,q}) < \varepsilon + m^*(B). \tag{4.22}$$

We now proceed to define a Vitali covering $\mathcal{U}$ of B in terms of $\{I_1, \ldots, I_N\}$. For each $y \in B$ there is $j \in \{1, \ldots, N\}$ such that $y \in I_j$; choose $k \in \mathbb{R}^+$ such that $[y, y+k] \subseteq I_j$ and

$$f(y+k) - f(y) > pk. \tag{4.23}$$

Inequality (4.23) is possible by the definition of $A_{p,q}$.

The set $\mathcal{U}$ is defined to be the collection of intervals $[y, y+k]$, for $y \in B$ and $k \in \mathbb{R}$, for which $[y, y+k]$ is contained in some $I_j, j = 1, \ldots, N$, and (4.23) is satisfied. Obviously, $\mathcal{U}$ is a Vitali covering of B, and so, from Theorem 4.3.1, there is a disjoint family $\{J_j = [y_j, y_j + k_j] : j = 1, \ldots, M\}$ satisfying

$$m^* \left(B \setminus \bigcup_{j=1}^{M} J_j \right) < \varepsilon.$$

iv. Let

$$D = B \cap \left(\bigcup_{j=1}^{M} J_j \right)$$

and observe that

$$B = \left(B \setminus \bigcup_{j=1}^{M} J_j \right) \cup D;$$

thus

$$m^*(B) < \varepsilon + m^*(D),$$

which, when combined with (4.22) and (4.23), yields

$$p(r - 2\varepsilon) < pm^*(D) \leq pm^* \left(\bigcup_{j=1}^{M} J_j \right) = p \sum_{j=1}^{M} k_j < \sum_{j=1}^{M} (f(y_j + k_j) - f(y_j)). \tag{4.24}$$

v. From the definition of $\mathcal{U}$, each J_j, $j = 1, \ldots, M$, is contained in some $I_i, i = 1, \ldots, N$. Hence, for each fixed $n = 1, \ldots, N$,

$$\sum_{J_j \subseteq I_n} (f(y_j + k_j) - f(y_j)) \leq f(x_n) - f(x_n - h_n),$$

since f is increasing and $\{J_j : j = 1, \ldots, M\}$ is a disjoint family.

This observation, combined with (4.24) and (4.21), implies (4.20):

$$p(r - 2\varepsilon) \leq q \sum_{j=1}^{N} h_j \leq q\, m(U) < q(r + \varepsilon).$$

Part a is complete.

b. and ***c.*** Define

$$g_n(x) = n\left[f(x + (1/n)) - f(x)\right],$$

where we define $f(x) = f(b)$ if $x \geq b$. From part a, $g_n \to f'$ pointwise m-$a.e.$ Also, g_n, f', and $|f'|$ are measurable since f is measurable. By Fatou's lemma and the fact that f is increasing,

$$\begin{aligned}\int_a^b |f'| &\leq \underline{\lim}_{n\to\infty} \int_a^b |g_n| = \underline{\lim}_{n\to\infty} \left[n \int_b^{b+(1/n)} f - n \int_a^{a+(1/n)} f \right] \\ &\leq f(b) - f(a) < \infty.\end{aligned}$$

□

Theorem 4.3.3. A measure-theoretic result about Diophantine approximation

Let $P : \mathbb{N} \to \mathbb{R}^+$ be a function and let A be a set of irrational numbers x such that

$$\left| x - \frac{p}{q} \right| < \frac{1}{qP(q)}, \quad (p, q) = 1,$$

for infinitely many integers $q > 1$. If

$$\sum_{q=1}^{\infty} \frac{1}{P(q)} < \infty$$

then $m(A) = 0$.

Proof. Take $A \subseteq [0, 1]$ and define

$$f(x) = \begin{cases} 0, & \text{if } x \text{ is irrational in } [0,1], \\ 1/(qP(q)), & \text{if } x = p/q \in [0,1], \text{ where } (p,q) = 1. \end{cases}$$

We first prove that $f \in BV([0, 1])$. Observe that for each $q > 1$ there are fewer than q rational numbers $p/q \in [0, 1]$, where $(p, q) = 1$. Take any partition $0 = x_0 < x_1 < \cdots < x_n = 1$, and observe that if each $x_j = p/q$ then $\sum_{j=1}^{n} |f(x_j)| < 1/P(q)$. This is the "worst" possible situation in the sense that no matter what the x_j are,

$$\sum_{j=1}^{n} |f(x_j) - f(x_{j-1})| \leq 2\sum_{j=0}^{n} |f(x_j)| \leq 2\sum_{q=1}^{m} \frac{1}{P(q)} \leq 2\sum_{q=1}^{\infty} \frac{1}{P(q)} < \infty.$$

Thus, $f \in BV([0,1])$.

Since $f \in BV([0,1])$ we can apply the Lebesgue differentiation theorem, and hence f' exists m-$a.e.$ Consequently, f' exists for almost all irrationals. Take such an irrational x. Then $f'(x) = 0$ since in the difference quotient we can approximate x by rationals. Observe that in this case

$$\frac{f(p/q) - f(x)}{(p/q) - x} = \frac{1}{qP(q)((p/q) - x)},$$

and so for $\varepsilon = 1$ there is a (largest possible) $\delta(1)$ for which

$$\left|\frac{p}{q} - x\right| < \delta(1) \Longrightarrow \frac{1}{qP(q)} < \left|\frac{p}{q} - x\right|. \tag{4.25}$$

If we assume that there exist infinitely many integers $q > 1$ such that

$$\left|x - \frac{p}{q}\right| < \frac{1}{qP(q)},$$

then (4.25) implies that for such q,

$$\left|\frac{p}{q} - x\right| \geq \delta(1).$$

Thus, we obtain a contradiction, since, for infinitely many q,

$$\frac{1}{qP(q)} \geq \left|\frac{p}{q} - x\right| \geq \delta(1) > 0,$$

and the left-hand side tends to 0 as $q \to \infty$.

Therefore, $x \notin A$, and so an irrational number $x \in [0,1]$ is in A only if $f'(x)$ does not exist; and this latter possibility can occur only on a set of Lebesgue measure 0. Thus, $m(A) = 0$. □

4.4 Fundamental Theorem of Calculus I

If $f \in L^1_m([a,b])$, we set

$$\forall\, x \in [a,b], \quad F(x) = \int_a^x f$$

for the next two results.

Proposition 4.4.1. *If $f \in L^1_m([a,b])$ then F is a continuous function of bounded variation.*

Proof. The fact that F is continuous follows immediately from Proposition 3.3.9. For the bounded variation observe that

$$\sum_{j=1}^{n} |F(x_j) - F(x_{j-1})| \le \sum_{j=1}^{n} \int_{x_{j-1}}^{x_j} |f| = \int_a^b |f| < \infty. \qquad \square$$

Once we define absolute continuity of point functions we shall see immediately (using Proposition 3.3.9) that F is absolutely continuous; this notion is stronger than that of bounded variation and continuity.

Proposition 4.4.2. *Let $f \in L^1_m([a,b])$ and assume that F is identically 0. Then $f = 0$ m-a.e.*

Proof. Without loss of generality assume that $f > 0$ on a set A of positive Lebesgue measure, $m(A) > 0$. By the properties of Lebesgue measure, there is a closed set $K \subseteq A$ such that $m(K) > 0$. Let $U = [a,b] \setminus K$. Then

$$0 = \int_a^b f = F(b) = \int_K f + \int_U f,$$

and so

$$\int_U f = -\int_K f < 0.$$

The set $U = \bigcup_{n=1}^{\infty} (a_n, b_n)$ is a disjoint union of open intervals, so that, by LDC,

$$0 \ne \int_U f = \sum_{n=1}^{\infty} \int_{a_n}^{b_n} f,$$

and consequently

$$\int_{a_n}^{b_n} f \ne 0$$

for some n. Thus, either $\int_a^{a_n} f$ or $\int_a^{b_n} f$ is nonzero, and this contradicts the hypothesis on F. $\square$

Theorem 4.4.3. Fundamental Theorem of Calculus (FTC) I
Let $f \in L^1_m([a,b])$ and take $r \in \mathbb{R}$. Define the function $F : [a,b] \to \mathbb{R}$ as

$$F(x) = r + \int_a^x f$$

(so that $F(a) = r$). Then

$$F' = f \quad m\text{-a.e.}$$

Proof. We consider the case that $\|f\|_\infty < \infty$. The unbounded case is left as an exercise (Problem 4.21). We have $F \in BV([a,b])$ from Proposition 4.4.1, and so F' exists m-*a.e.* because of Theorem 4.3.2. Define $F(x) = F(b)$ for $x \geq b$ and set

$$f_n(x) = \frac{F(x + (1/n)) - F(x)}{1/n} = n \int_x^{x+(1/n)} f.$$

Consequently, $f_n \to F'$ m-*a.e.* Since $\|f\|_\infty < \infty$ we can employ LDC to obtain

$$\begin{aligned}\int_a^c F' &= \lim_{n\to\infty} \int_a^c f_n = \lim_{h\to 0} \frac{1}{h} \int_a^c (F(x+h) - F(x))\, dx \\ &= \lim_{h\to 0} \frac{1}{h} \left[\int_c^{c+h} F - \int_a^{a+h} F \right] = F(c) - F(a) = \int_a^c f\end{aligned}$$

for each $c \in [a,b]$. The penultimate equality follows by the continuity of F and the observation that

$$\left| F(c) - \frac{1}{h} \int_c^{c+h} F \right| = \left| \frac{1}{h} \int_c^{c+h} (F(c) - F) \right| \leq \sup_{x\in[c,c+h]} |F(c) - F(x)|.$$

Thus,

$$\forall\, c \in [a,b], \qquad \int_a^c (F' - f) = 0,$$

and we can apply Proposition 4.4.2 to conclude that $F' = f$ m-*a.e.* □

Example 4.4.4. The primitive of sin(1/x)
Let $f(x) = \sin(1/x)$ for $x > 0$ and let $f(0) = 0$. Define the function

$$F(x) = \begin{cases} 0, & \text{if } x = 0, \\ x^2 \cos(1/x) - 2 \int_0^x t \cos(1/t)\, dt, & \text{if } x > 0. \end{cases}$$

Then $F' = f$ in $[0, \infty)$.

Let us look at the differentiation of indefinite integrals as an averaging procedure. Thus, from Theorem 4.4.3, if $f \in L^1_m([a,b])$ and

$$F(x) = \int_a^x f,$$

we have $F' = f$ m-*a.e.*, and so

$$\lim_{h\to 0} \frac{1}{h} \int_x^{x+h} (f(t) - f(x))\, dt = 0 \quad m\text{-}a.e. \tag{4.26}$$

Now (4.26) could be valid due to the cancellation caused by change of sign in the integration, or the stronger result

$$\lim_{h\to 0}\frac{1}{h}\int_x^{x+h}|f(t)-f(x)|\,dt=0 \quad m\text{-}a.e. \tag{4.27}$$

could be true. In fact, (4.27) is true and is a corollary of the following even more general theorem.

Theorem 4.4.5. Lebesgue's extension of FTC-I
Let $f\in L^1_m([a,b])$ be real-valued. There is a subset $L\subseteq[a,b]$, for which $m(L)=b-a$, such that

$$\forall\, r\in\mathbb{R} \text{ and } \forall\, x\in L,\quad \lim_{h\to 0}\frac{1}{h}\int_x^{x+h}|f(t)-r|\,dt=|f(x)-r|.$$

Proof. Let $\{r_n : n=1,\dots\}\subseteq\mathbb{R}$ be dense, and define the elements $g_n\in L^1_m([a,b])$ as

$$g_n(x)=|f(x)-r_n|.$$

Because of Theorem 4.4.3, for each n there is a subset $L_n\subseteq[a,b]$, for which $m(L_n)=b-a$, such that

$$\forall\, x\in L_n,\quad \lim_{h\to 0}\frac{1}{h}\int_x^{x+h}g_n=g_n(x). \tag{4.28}$$

Setting $L=\bigcap L_n$, we have $m(L^{\sim})=m(\bigcup L_n^{\sim})\leq\sum m(L_n^{\sim})=0$, and so $m(L)=b-a$.

Let $\varepsilon>0$ and $r\in\mathbb{R}$. Choose $n=n(\varepsilon,r)$ such that $|r-r_n|<\varepsilon/3$. Then

$$\forall\, x\in[a,b],\quad \big||f(x)-r|-|f(x)-r_n|\big|<\frac{\varepsilon}{3};$$

and so, for all $x\in[a,b]$,

$$\left|\frac{1}{h}\int_x^{x+h}|f(t)-r|\,dt-|f(x)-r|\right|<\frac{2\varepsilon}{3}+\left|\frac{1}{h}\int_x^{x+h}(g_n(t)-g_n(x))\,dt\right|.$$

If $x\in L$ then x is in L_n, where $n=n(\varepsilon,r)$; consequently, from (4.28) there is $h_\varepsilon>0$ for which

$$\left|\frac{1}{h}\int_x^{x+h}(g_n(t)-g_n(x))\,dt\right|<\frac{\varepsilon}{3}$$

for all $|h|<h_\varepsilon$. □

We rewrite Theorem 4.4.5 as follows; also, see Theorem 8.4.7 for a version for regular Borel measures on $\mathbb{R}^d$.

Corollary 4.4.6. *Let $f \in L^1_m([a,b])$. There is a subset $L \subseteq [a,b]$, for which $m(L) = b-a$, such that*

$$\forall\, x \in L, \qquad \int_0^h |f(x+t) - f(x)|\, dt = o(h), \quad h \to 0. \tag{4.29}$$

The largest set of points $x \in [a,b]$ for which (4.29) holds, for a given $f \in L^1_m([a,b])$, is the *Lebesgue set $L(f)$ of f*. It is clear from (4.29) that

$$C(f) \subseteq L(f)$$

and that

$$m(L(f)) = b - a.$$

Remark. We can reformulate the notion of Lebesgue set in the following more elegant way; see [264], pages 458–459 and pages 491–494. Let $f \in L^1_m([a,b])$. Then $x \in (a,b)$ is an element of the *Lebesgue set $L(f)$* of $f \in L^1_m([a,b])$ if

$$\exists\, c_x \in \mathbb{C} \text{ such that } \int_0^h |f(x+t) - c_x|\, dt = o(h), \quad h \to 0. \tag{4.30}$$

Because of Corollary 4.4.6, c_x is unique for any given x and f. If g is a point function representative of the equivalence class $f \in L^1_m([a,b])$, then $g(x) = c_x$ in (4.30) for almost every $x \in L(f)$.

Example 4.4.7. The Fejér kernel and the Lebesgue set
The Fejér kernel $\{W_N : N = 1, \ldots\}$ and convolution $*$ on $\mathbb{T}$ were defined in Problem 3.28, and we saw that $\|W_N * f - f\|_\infty \to 0$ for 1-periodic functions $f \in C(\mathbb{R})$, i.e., $f \in C(\mathbb{T})$. The convolution $W_N * f$ is the arithmetic mean of the first $N+1$ partial sums of the Fourier series for f; see Problem 4.28. In particular,

$$\lim_{N\to\infty} |W_N * f(x) - f(x)| = 0 \tag{4.31}$$

for each $x \in [0,1)$. The question arises whether (4.31) has any meaningful generalization for arbitrary elements $f \in L^1_m(\mathbb{T})$. The answer is contained in LEBESGUE's result: *If $f \in L^1_m(\mathbb{T})$ and $x \in L(f)$ then (4.31) is valid*, see [524] for a proof as well as Sections 3.8.3 and 4.7.5.

4.5 Absolute continuity and Fundamental Theorem of Calculus II

The problem now is to find a converse to Theorem 4.4.3; that is, given a continuous function $F : [a,b] \to \mathbb{R}$ of bounded variation, is it true that

$$F(x) - F(a) = \int_a^x F'(t)\, dt?$$

Recall that $F' \in L^1_m([a,b])$ if $F \in BV([a,b])$.

Example 4.5.1. The integral of the derivative of the Cantor function
Note that the Cantor function $C_C \in BV([0,1])$ is continuous and $C_C' = 0$ m-*a.e.* Hence

$$\int_0^1 C_C' = 0,$$

whereas $C_C(1) - C_C(0) = 1$. This fact tells us that C_C is not *absolutely continuous*; see Definition 4.5.2.

Thus, the above question is answered negatively, and so, if a converse is to hold, we need F to satisfy a requirement that is stronger than both bounded variation and continuity. Such a notion exists and was given explicitly by VITALI [482], although CARL G. A. HARNACK had used a similar notion in a different context during the late 1890s. VITALI's concept is absolute continuity, and with it he characterized the relation between differentiation and integration in the context of LEBESGUE's theory. This characterization is the most general form of FTC, i.e., Theorem 4.5.5. LEBESGUE's role in this matter is important. His major work was available to VITALI (cf. Section 5.6.3 for a historical remark on this point), and although VITALI published the first proof of Theorem 4.5.5 in [482], LEBESGUE produced a quite efficient proof shortly thereafter in 1907.

Definition 4.5.2. Absolute continuity
A function $F : [a,b] \to \mathbb{R}$ is *absolutely continuous on* $[a,b]$ if

$\forall\, \varepsilon > 0,\ \exists\, \delta > 0$ such that $\forall\, \{(x_j, y_j) \subseteq [a,b] : j = 1, \ldots, n\}$, a disjoint family,

$$\sum_{j=1}^{n} (y_j - x_j) < \delta \Longrightarrow \sum_{j=1}^{n} |F(y_j) - F(x_j)| < \varepsilon.$$

In light of the following proposition this definition of absolute continuity is equivalent to Definition 3.3.10*a*.

Proposition 4.5.3. a. *Let* $f \in L^1_m([a,b])$ *and set*

$$F(x) = r + \int_a^x f, \quad x \in [a,b].$$

Then $r = F(a)$ *and* F *is absolutely continuous on* $[a,b]$.

b. *If* $G : [a,b] \to \mathbb{R}$ *is absolutely continuous on* $[a,b]$, *then* G *is a continuous function of bounded variation, and, in particular,* G' *exists* m-*a.e. and is an element of* $L^1_m([a,b])$.

Proof. **a.** This is immediate from Proposition 3.3.9, as was Proposition 4.4.1.

b. Let $\varepsilon = 1$ and choose a corresponding δ from the definition of absolute continuity. Then

$$V(G,[a,b]) \le \frac{1 + b - a}{\delta}. \qquad \square$$

Remark. Thus, FTC–I says that each $f \in L^1_m([a,b])$ is the derivative m-*a.e.* of an absolutely continuous function. We know that the converse of this is true: *If* $F : [a,b] \to \mathbb{R}$ *is absolutely continuous then* $F' \in L^1_m([a,b])$. LUZIN [333] proved: *Each Lebesgue measurable function* $f : [a,b] \to \mathbb{R}$ *is the derivative* m-*a.e. of a continuous function.* What sort of converse would you have in mind for this?

With regard to C_E and the Hellinger example H, we note the following fact.

Proposition 4.5.4. *Let* F *be absolutely continuous on* $[a,b]$ *and assume that* $F' = 0$ m-*a.e. Then* F *is a constant.*

Proof. Take $c \in (a,b]$. To prove $F(c) = F(a)$, by hypothesis there is a subset $A \subseteq (a,c)$ such that $m(A) = c - a$ and $F' = 0$ on A. Let $\varepsilon > 0$. Since $F' = 0$ on A we have that

$$\forall\, x \in A,\ \exists\, [x,y] \subseteq (a,c) \text{ such that } |F(y) - F(x)| < \frac{\varepsilon(y-x)}{2(c-a)}.$$

From the Vitali covering lemma we can find

$$\{[x_j, y_j] : j = 1, \ldots, n \text{ and } x_j \in A\},$$

a disjoint family, for which

$$|F(y_j) - F(x_j)| < \frac{\varepsilon(y_j - x_j)}{2(c-a)} \tag{4.32}$$

and

$$m\left(A \setminus \left(\bigcup_{j=1}^{n} [x_j, y_j]\right)\right) < \delta, \tag{4.33}$$

where $\delta = \delta(\varepsilon/2)$ is determined from the absolute continuity of F.

Without loss of generality (a labeling problem) take $x_j < x_{j+1}$. Using (4.33), we have

$$\sum_{j=0}^{n} |x_{j+1} - y_j| < \delta, \quad x_{n+1} = c, \ \ y_0 = a,$$

so that from the absolute continuity

$$\sum_{j=1}^{n} |F(x_{j+1}) - F(y_j)| < \frac{\varepsilon}{2};$$

and, because of (4.32),

$$\sum_{j=1}^{n} |F(y_j) - F(x_j)| < \frac{\varepsilon}{2(c-a)} \sum_{j=1}^{n} |y_j - x_j|.$$

Combining the last two inequalities gives

$$|F(c) - F(a)| < \varepsilon.$$

□

Theorem 4.5.5. Fundamental Theorem of Calculus (FTC) II
A function $F : [a,b] \to \mathbb{R}$ is absolutely continuous on $[a,b]$ if and only if there is an element $f \in L^1_m([a,b])$ such that

$$\forall\, x \in [a,b], \quad F(x) - F(a) = \int_a^x f.$$

Proof. ($\Longleftarrow$) This is Proposition 4.5.3*a*.

($\Longrightarrow$) Define the function $H = F - G$, where F is absolutely continuous on $[a,b]$ and

$$G(x) = \int_a^x F'.$$

Note that $F' \in L^1_m([a,b])$ from Theorem 4.3.2. By hypothesis and Proposition 4.5.3, H is absolutely continuous; and

$$H' = F' - G' = 0 \;\; m\text{-}a.e.$$

Consequently, $F = G + r$ m-*a.e.* by Proposition 4.5.4, and so

$$F(x) - r = \int_a^x F'.$$

Thus, $r = F(a)$ and we set $f = F'$. □

Part *b* of the following is a useful consequence of FTC; see Problems 4.37 and 4.38 for hints to the proof of both parts.

Theorem 4.5.6. Integral formula for total variation
a. *If $F \in BV([a,b])$ then $V(F,[a,b]) \geq \int_a^b |F'|$.*

b. *If $F : [a,b] \to \mathbb{C}$ is absolutely continuous, then*

$$V(F,[a,b]) = \int_a^b |F'|.$$

Example 4.5.7. Fourier series and absolute continuity
We prove that if f is a continuous function in $BV(\mathbb{T})$ and

$$\overline{\lim_{|n|\to\infty}}\, |n\hat{f}(n)| > 0,$$

then f is not absolutely continuous. Thus, *if f is absolutely continuous on $\mathbb{T}$ then $n\hat{f}(n) = o(1)$, $|n| \to \infty$*. In fact, if f is absolutely continuous then $f' \in L^1_m(\mathbb{T})$ and

$$\widehat{(f')}(n) = -2\pi i n \hat{f}(n); \tag{4.34}$$

and so an application of the Riemann–Lebesgue lemma (Theorem 3.6.4) to f' concludes the proof. Theorem 4.6.3, which follows from FTC, is used to prove (4.34). See Problem 4.5 and Theorem 4.1.7 for perspective.

Remark. ***a.*** In broad terms, FTC–I asserts that

$$D \int = Id,$$

and FTC–II asserts that

$$\int D = Id,$$

where D, $\int$, Id are the differentiation, integration, and identity operators, respectively.

b. The Green and Stokes theorems can be considered as generalizations of FTC to $\mathbb{R}^d$ in that they assert the equality of the integral over a region $A \in \mathcal{M}(\mathbb{R}^d)$ of a derivative with the integral of the original function over the boundary $\partial(A)$ of A. For example, formally, we have the familiar two-dimensional version,

$$\iint_A \left[\frac{\partial Q}{\partial x}(x,y) - \frac{\partial P}{\partial y}(x,y)\right] \,dx\,dy = \int_{\partial(A)} P(x,y)\,dy + Q(x,y)\,dx.$$

c. Generalizations of FTC to $\mathbb{R}^d$ and to quite arbitrary measure spaces constitute one of the most important facets of the theory, and lead, through the Radon–Nikodym theorem, to some of the most significant properties of measures; see Chapter 8.

With the new tools that we have developed in this section, we now come back to the discussion of bounded variation.

Remark. The comments in this remark began in Problem 3.14*b*. As a convention we set $f(x+h) = 0$ if f is defined on $[a,b]$, $x \in [a,b]$, and $x+h \notin [a,b]$. For all $f \in L^1_m([a,b])$,

$$\|\tau_{-h}f - f\|_1 = o(1), \quad |h| \to 0. \tag{4.35}$$

We can show further, e.g., Problem 4.17, that if $f \in L^1_m([a,b])$ and

$$\|\tau_{-h}f - f\|_1 = o(|h|), \quad |h| \to 0, \tag{4.36}$$

then f is a constant k m-*a.e.* Consider the condition,

$$\|\tau_{-h}f - f\|_1 = O(|h|), \quad |h| \to 0. \tag{4.37}$$

Clearly,

$$(4.37) \Longrightarrow (4.36) \Longrightarrow (4.35).$$

The following characterization of bounded variation in terms of (4.37) is due to HARDY and LITTLEWOOD (1928).

Theorem 4.5.8. Equivalence of bounded variation and L^1 boundedness of difference quotients
Let $f \in L^1_m([a,b])$. Then $f \in BV([a,b])$ if and only if (4.37) is valid.

Proof. ($\Longrightarrow$) Let $f = g_1 - g_2$, where each g_j is increasing and bounded, and $f = 0$ on $[a,b]^\sim$. For $x > b$, without loss of generality, we let $g_1(x) = g_2(x) = \max(g_1(b), g_2(b))$. Thus, for each $h > 0$,

$$\begin{aligned}\int_a^b |f(x+h) - f(x)|\,dx &\leq \int_a^b (g_1(x+h) - g_1(x))\,dx \\ &\quad + \int_a^b (g_2(x+h) - g_2(x))\,dx \\ &= \int_b^{b+h} g_1 - \int_a^{a+h} g_1 + \int_b^{b+h} g_2 - \int_a^{a+h} g_2 \\ &\leq 2h \max(g_1(b), g_2(b)).\end{aligned}$$

($\Longleftarrow$) Assume (4.37) and let K be the constant associated with the "O" hypothesis. Set

$$f_n(x) = n \int_x^{x+(1/n)} f(t)\,dt.$$

Then, using the Fubini–Tonelli theorem (Theorem 3.7.7),

$$\begin{aligned}\int_a^b |f_n(x+h) - f_n(x)|\,dx &= n \int_a^b \left| \int_0^{1/n} (f(x+h+t) - f(x+t))\,dt \right| dx \\ &\leq n \int_0^{1/n} \left(\int_a^b |f(x+h+t) - f(x+t)|\,dx \right) dt \\ &\leq K|h|.\end{aligned}$$

Thus,

$$\|\tau_{-h} f_n - f_n\|_1 = O(|h|), \quad |h| \to 0, \text{ uniformly in } n. \tag{4.38}$$

Because of the differentiation theory we shall develop in Sections 4.3 and 4.4, especially Theorem 4.4.3, f_n' exists m-a.e. and $f_n' \in L^1_m([a,b])$. Also, by Fatou's lemma,

$$\int_a^b |f_n'| \leq \underline{\lim}_{h \to 0} \int_a^b \left| \frac{f_n(x+h) - f_n(x)}{h} \right| dx;$$

and from (4.38) the right-hand side is uniformly bounded in n as h tends to 0. Hence, there is a constant $M > 0$ such that for each n and for each finite collection $\{(x_j, x_j + h_j) : j = 1, \ldots, m\}$ of nonoverlapping intervals,

$$\sum_{j=1}^{m} |f_n(x_j + h_j) - f_n(x_j)| \leq M; \tag{4.39}$$

this follows since

$$\sum_{j=1}^{m} |f_n(x_j + h_j) - f_n(x_j)| = \sum_{j=1}^{m} \left| \int_{x_j}^{x_j+h_j} f_n' \right| \leq \int_a^b |f_n'|$$

by Theorem 4.5.5. Using Theorem 4.4.3 again, we see that $f_n \to f$ m-*a.e.* Consequently, aside from a set S of Lebesgue measure $m(S) = 0$ to which the points x_j, $x_j + h_j$ cannot belong, we have

$$\sum_{j=1}^{m} |f(x_j + h_j) - f(x_j)| \leq M.$$

Now define functions $V_S(f)$, P_S, and N_S analogous to the functions $V(f)$, P, and N, but excluding values from S in the partitions. Then we can extend P_S and N_S to increasing functions on $[a, b]$, and in this way we form an element of $BV([a, b])$ that is equal to f m-*a.e.* □

4.6 Absolutely continuous functions

We have seen that every absolutely continuous function is a continuous function of bounded variation; but the converse is not true; for example the Cantor function C_C is continuous and increasing on $[0, 1]$, but it is not absolutely continuous (Section 4.5). Theorem 4.6.2, below, gives necessary and sufficient conditions in order that a continuous element of $BV([a, b])$ be absolutely continuous. In this regard, recall that the Banach–Vitali theorem, Theorem 4.2.7, characterizes those continuous functions that have bounded variation. With regard to the criterion of Theorem 4.6.2, compare Problem 2.47.

The following lemma is Problem 4.32, and a proof can be found in [412], page 226, or in Dale E. Varberg's interesting exposition on absolutely continuous functions [480].

Lemma 4.6.1. *Let $F : [a, b] \to \mathbb{R}$ be a function and let $A \subseteq [a, b]$ be any subset on which F' exists. If*

$$\exists\, K \text{ such that } \forall\, x \in A, \quad |F'(x)| \leq K,$$

then

$$m^*(F(A)) \leq K m^*(A). \tag{4.40}$$

The next result was proved independently by BANACH [18] and MOISEJ A. ZARETSKY [520] in 1925.

Theorem 4.6.2. Banach–Zaretsky theorem
Let $F \in BV([a,b])$ be a continuous function. Then F is absolutely continuous on $[a,b]$ if and only if

$$m(A) = 0 \Longrightarrow m(F(A)) = 0.$$

Proof. ($\Longrightarrow$) This is the easy direction. Without loss of generality take a subset $A \subseteq (a,b)$ with Lebesgue measure $m(A) = 0$. Let $\varepsilon > 0$. We shall prove that $m^*(F(A)) < \varepsilon$. From the hypothesis of absolute continuity there is $\delta > 0$ such that, no matter what disjoint family $\{(a_k, b_k) : k = 1, \ldots, n\}$ we take,

$$\sum_{k=1}^{n} (b_k - a_k) < \delta \Longrightarrow \sum_{k=1}^{n} (M_k - m_k) < \varepsilon,$$

where $M_k = \sup\{F(x) : x \in (a_k, b_k)\}$ and $m_k = \inf\{F(x) : x \in (a_k, b_k)\}$, e.g., Problem 4.35.

Choose an open set $U = \bigcup_{k=1}^{n} (a_k, b_k)$, a disjoint union, for which $A \subseteq U \subseteq (a,b)$ and $m(U) < \delta$. Note that since

$$F(A) \subseteq F(U) = \bigcup_{k=1}^{n} F((a_k, b_k)),$$

we have

$$m^*(F(A)) \leq \sum_{k=1}^{n} m^*([m_k, M_k]) < \varepsilon.$$

($\Longleftarrow$) *i.* We first use Lemma 4.6.1 to prove that if F' exists on $A \in \mathcal{M}([a,b])$ then

$$m^*(F(A)) \leq \int_A |F'|. \tag{4.41}$$

Assume that there is an integer p such that

$$\forall\, x \in A, \quad |F'(x)| < p,$$

and define

$$A_{k,n} = \left\{ x \in A : \frac{k-1}{2^n} \leq |F'(x)| < \frac{k}{2^n} \right\},$$

where $k = 1, \ldots, p2^n$ and $n = 1, \ldots$. Using Lemma 4.6.1 we compute

$$\begin{aligned} m^*(F(A)) &= m^*\left(\bigcup_{k=1}^{p2^n} F(A_{k,n}) \right) \leq \sum_{k=1}^{p2^n} m^*(F(A_{k,n})) \leq \sum_{k=1}^{p2^n} \frac{k}{2^n} m(A_{k,n}) \\ &= \sum_{k=1}^{p2^n} \frac{k-1}{2^n} m(A_{k,n}) + \frac{1}{2^n} \sum_{k=1}^{p2^n} m(A_{k,n}). \end{aligned}$$

From the definition of the Lebesgue integral and since $m^*(F(A))$ is independent of n, we have (4.41).

The case for an unbounded function F' uses the bounded case and sets of the form

$$A_k = \{x \in A : k - 1 \leq |F'(x)| < k\}$$

in the expected manner.

ii. Let $\{(a_k, b_k) : k = 1, \ldots, n\} \subseteq [a, b]$ be a disjoint family and let

$$A_k = \{x \in [a_k, b_k] : \exists\, F'(x)\}.$$

Since F' exists m-*a.e.* we have that $m([a_k, b_k] \setminus A_k) = 0$; by hypothesis, then, $m(F([a_k, b_k])) = m(F(A_k))$. Using the continuity of F and the result from part i we compute

$$\begin{aligned}\sum_{k=1}^{n} |F(b_k) - F(a_k)| &\leq \sum_{k=1}^{n} m(F([a_k, b_k])) = \sum_{k=1}^{n} m(F(A_k)) \\ &\leq \sum_{k=1}^{n} \int_{A_k} |F'| = \sum_{k=1}^{n} \int_{a_k}^{b_k} |F'|.\end{aligned} \tag{4.42}$$

Since $F' \in L^1_m([a, b])$, if $\sum_{k=1}^{n}(b_k - a_k)$ is small then the last sum in (4.42) is small. □

We now prove the *integration by parts* formula.

Theorem 4.6.3. Integration by parts formula
Let f, $g \in L^1_m([a, b])$ with corresponding functions

$$F(x) = r + \int_a^x f \quad \textit{and} \quad G(x) = s + \int_a^x g.$$

Then

$$\int_a^b fG + \int_a^b gF = F(b)G(b) - F(a)G(a).$$

Proof. Note that

$$\begin{aligned}|F(x)G(x) - F(y)G(y)| &= |F(x)(G(x) - G(y)) + G(y)(F(x) - F(y))| \\ &\leq \|F\|_\infty |G(x) - G(y)| + \|G\|_\infty |F(x) - F(y)|.\end{aligned}$$

Thus, FG is absolutely continuous on $[a, b]$, and so FG is differentiable m-*a.e.* and $(FG)' = F'G + G'F$ m-*a.e.* From FTC–I, $F' = f$ and $G' = g$ m-*a.e.* Therefore,

$$\int_a^b (FG)' = \int_a^b fG + \int_a^b gF.$$

On the other hand, from the absolute continuity of FG and FTC–II,

$$\int_a^b (FG)' = F(b)G(b) - F(a)G(a).$$

□

If $F \in BV([a,b])$ and $F' = 0$ m-a.e. then F is *singular*. It is obvious that the discrete part of $F \in BV([a,b])$ is singular. What is more interesting is that there are functions $F \in BV([a,b])$ with singular continuous parts, e.g., C_C. With regard to the following result, compare Theorem 5.4.2*b*. The sufficient conditions in Theorem 4.6.4 are found in [431].

Theorem 4.6.4. Measure-theoretic characterization of 0-derivative
Let $F : [a,b] \to \mathbb{R}$ be a function and assume that F' exists on $A \subseteq [a,b]$. Then $F' = 0$ m-a.e. on A if and only if $m(F(A)) = 0$.

Proof. ($\Longrightarrow$) Let $A_k = \{t \in A : k-1 \leq |F'(t)| \leq k\}$, $k = 1, \ldots$. Then, from (4.40) and the hypothesis,

$$m^*(F(A)) \leq \sum_{k=0}^{\infty} m^*(F(A_k)) \leq \sum_{k=0}^{\infty} k m^*(A_k) = 0.$$

($\Longleftarrow$) Let $B = \{t \in A : |F'(t)| > 0\}$ and define

$$B_n = \{t \in B : |F(s) - F(t)| \geq |s-t|/n \text{ for } |s-t| < 1/n\}.$$

Then $B = \bigcup_{n=1}^{\infty} B_n$ and we must prove that $m(B_n) = 0$ for each n. Fix an n and let I be an interval with Lebesgue measure $m(I) < 1/n$; set $D = I \cap B_n \subseteq A$. Thus, it is sufficient to prove that

$$\forall\, \varepsilon > 0, \quad m^*(D) \leq n\varepsilon. \tag{4.43}$$

We use our hypothesis now to choose a sequence $\{I_j : j = 1, \ldots\}$ of intervals such that $F(D) \subseteq \bigcup_{j=1}^{\infty} I_j$ and $\sum_{j=1}^{\infty} m(I_j) < \varepsilon$. Set $D_j = F^{-1}(I_j) \cap D$. From the definition of B_n and the fact that $m^*(D) \leq \sum_{j=1}^{\infty} m^*(D_j)$, we compute (4.43). □

Relative to the decomposition earlier in this chapter we have the following result. Because of the Lebesgue differentiation theorem and the point of view of Chapter 7, the Radon–Nikodym theorem can be considered a generalization of Theorem 4.6.5.

Theorem 4.6.5. Decomposition into absolutely continuous and singular parts
If $F \in BV([a,b])$ then $F = F_a + F_s$, where F_a is absolutely continuous and F_s is singular.

Proof. Set $F_a(x) = \int_a^x F'$. We know that F_a is absolutely continuous on $[a,b]$ and that $F_a' = F'$ m-a.e. Consequently, the proof will be complete when we define $F_s = F - F_a$. □

Recall that the Hellinger example, Example 4.2.5, is a strictly increasing continuous singular function $H = H_t$, where $t \in (0,1)$ is fixed. A simpler construction is the following; cf. Problem 4.33.

Example 4.6.6. A simpler example of Hellinger type
Let $[a,b] \subseteq [0,1]$ be an interval and consider a corresponding positive constant $k_{a,b}$. Then

$$k_{a,b} C_C \left(\frac{x-a}{b-a} \right), \quad x \in [a,b], \tag{4.44}$$

is continuous on $[a,b]$ and increases from 0 to $k_{a,b}$ there. Enumerate all the intervals $I_n = [a_n, b_n]$, where a_n, $b_n \in \mathbb{Q} \cap [0,1]$ and $a_n < b_n$; and define f_n on I_n, analogous to (4.44), such that f_n is continuous, increasing from 0 to $(1/2)^n$, and $f_n' = 0$ m-a.e. on I_n. We set $f = \sum_{n=1}^{\infty} f_n$ and see that f is continuous and strictly increasing. From the Fubini differentiation theorem, given in Problem 4.19, $f' = 0$ m-a.e.

Remark. Part of the technical difficulty arising in the construction of H_t, $t \in (0,1)$, is compensated for by the following fact, which will be developed in Chapter 5. Not only is H_t, $t \in (0,1)$, singular "with respect to Lebesgue measure", but, if μ_t corresponds to H_t, where the correspondence is in the sense of Theorem 4.1.9, we have a continuum of continuous singular measures μ_t such that if $t \neq s$ then μ_t and μ_s are "mutually singular", i.e., μ_t and μ_s are concentrated on disjoint sets. We shall give precise definitions of these notions in Chapter 5, but, for now, we say that a measure space $(X, \mathcal{A}, \mu)$ is *concentrated on* $A \subseteq X$ if

$$\forall\, B \in \mathcal{A}, \quad \mu(B) = \mu(A \cap B).$$

As we have seen, for C_C, say, we *cannot* conclude that a function F is absolutely continuous if

$$F \text{ is continuous}, \ \exists\, F' \ m\text{-a.e., and } F' \in L^1_m. \tag{4.45}$$

As far as positive results on this question go, we can conclude that *a function* $F : [a,b] \to \mathbb{R}$ *is absolutely continuous if it satisfies* (4.45) *and either*

$$\operatorname{card} \{t : \nexists\, F'(t)\} \leq \aleph_0 \tag{4.46}$$

or

$$m(A) = 0 \Longrightarrow m(F(A)) = 0. \tag{4.47}$$

The latter theorem, i.e., "(4.47) and (4.45) imply absolute continuity", generalizes the Banach–Zaretsky theorem—since we assumed that $F \in BV([a,b])$ there—and it has the same proof, since the only use of bounded variation in Theorem 4.6.2 was to obtain $F' \in L^1_m([a,b])$.

For the proof of the first theorem, i.e., "(4.46) and (4.45) imply absolute continuity", it is sufficient to prove that (4.46) implies (4.47). Let $m(A) = 0$ and $A = B \cup D$, where $\operatorname{card} D \leq \aleph_0$ and F' exists on B. We must prove that $m(F(B)) = 0$. Set

$$B_{n,k} = \left\{ t \in B : \forall\, J \text{ with } m(J) < \frac{1}{k} \text{ and } t \in J, \text{ we have } m(F(J)) < n m(J) \right\},$$

where J is an open interval. Since F' exists on B,

$$B = \bigcup_{n,k=1}^{\infty} B_{n,k},$$

and so we have only to check that

$$\forall\, n,k \text{ and } \forall\, \varepsilon > 0, \quad m(F(B_{n,k})) < \varepsilon. \tag{4.48}$$

To verify (4.48) we cover $B_{n,k} \subseteq A$ by nonoverlapping intervals each with length less than $1/k$ and with total length less than ε/n. This does it.

As a special case of this discussion we can state the following theorem; see [199] for a short, direct, transparent proof that uses the Vitali covering lemma.

Theorem 4.6.7. Everywhere differentiability criteria for absolute continuity
Let $F : [a,b] \to \mathbb{R}$ be an everywhere differentiable function. If $F' \in L^1_m([a,b])$ then F is absolutely continuous on $[a,b]$.

The requirement $F' \in L^1_m([a,b])$ is necessary in Theorem 4.6.7, as the example

$$F(x) = \begin{cases} 0, & \text{if } x = 0, \\ x^2 \sin(1/x^2), & \text{if } x \in (0,1] \end{cases}$$

shows; cf. Problem 4.4*a*.

An interesting application of some of our previous results and an amusing source of counterexample attempts is the following result.

Theorem 4.6.8. Elementary surprise
Let $f, f' \in L^1_m(\mathbb{R})$ and assume that

$$\forall\, x \in \mathbb{R}, \quad f'(x) \text{ exists.}$$

Then,

$$\int_{\mathbb{R}} f' = 0.$$

Proof. From Theorem 4.6.7, f is absolutely continuous on each finite interval. Take $\{F_n : n = 1, \ldots\} \subseteq C^\infty(\mathbb{R})$ such that $\sup_n \|F_n'\|_\infty \leq K$, $0 \leq F_n \leq 1$, $F_n = 1$ on $[-n,n]$, and $F_n = 0$ off $[-(n+1), n+1]$. Note that each F_n is absolutely continuous on finite intervals. Now $F_n f' \to f'$ pointwise and $|F_n f'| \leq |f'| \in L^1_m(\mathbb{R})$. Consequently, by LDC, $\int f' = \lim_{n\to\infty} \int F_n f'$.

Since each F_n and f are absolutely continuous, we use Theorem 4.6.3 to compute

$$\int_{-(n+1)}^{n+1} F_n f' = -\int_{-(n+1)}^{n+1} F_n' f.$$

Observe that $F_n'f \to 0$ pointwise and $|F_n'f| \le K|f| \in L_m^1(\mathbb{R})$, so that by LDC, $\lim_{n\to\infty} \int F_n'f = 0$. Therefore,

$$\int_{\mathbb{R}} f' = \lim_{n\to\infty} \int_{\mathbb{R}} F_n f' = 0. \qquad \square$$

There are functions $f \in L_m^1(\mathbb{R})$ that are absolutely continuous on $\mathbb{R}$ but for which $f' \notin L_m^1(\mathbb{R})$. For an elementary example see the Amer. Math. Monthly, 114 (2007), 356–357, by GILBERT HELMBERG.

4.7 Potpourri and titillation

1. HUGO STEINHAUS (January 14, 1887–February 25, 1972) studied in Göttingen and received his doctorate under the supervision of DAVID HILBERT in 1911. However, his mathematical interests were mostly influenced by LEBESGUE and not significantly by anyone in Göttingen. STEINHAUS was determined to return to Poland, where he wanted to create a group of mathematicians that would work together. Walking one day in 1916 in Cracow (Kraków), he overheard two young men talking about *Lebesgue measure*—a subject that even today is not common in our streets. This is how he met BANACH and OTTO NIKODYM. Soon thereafter they created the Mathematical Society of Cracow, which later became the Polish Mathematical Society.

Around 1920, STEINHAUS moved to Lwów (Lvov), where he was eventually a professor. Together with BANACH they became the nucleus of what is now known as the Lwów School of Mathematics—a vital part of the famous Polish School of Mathematics. Among the members of this group were MAZUR, WŁADYSŁAW ORLICZ, and JULIUSZ P. SCHAUDER—all deep mathematicians. The group was devoted to the study of functional analysis, real variables, and probability theory. STEINHAUS published the first rigorous exposition of the theory of coin tossing based on measure theory. He was also the first to define the notion of a strategy in game theory. His name is associated with many results in real and functional analysis, e.g., Problem 2.23 and Problem 3.6, for starters. However, he himself used to say that his greatest mathematical discovery was BANACH.

An interesting feature of this group was the style in which they worked on mathematics. ULAM describes it as "an enthusiastic collaboration"; see [479]. After short seminars they would go to a nearby café to discuss problems further, often sitting there until closing time. First it was *Café Roma* and later, across the street, *Café Szkocka* (*Scottish Café*). In order not to lose too much of the mathematical content of their discussions there, BANACH decided that they should be making notes. He bought a notebook that was kept in the café, and that was brought to them whenever it was needed. It was in this book that they recorded many ideas, open questions, and, frequently, answers.

The names of authors in this book include, besides the "locals", LEBESGUE, VON NEUMANN, SERGEI L. SOBOLEV, and many others. Miraculously, the book survived World War II and was popularized by STEINHAUS and ULAM as the *Scottish Book* [344]. Nowadays there are many followers worldwide in the business of writing and keeping "Scottish books". In particular, in the library of the Department of Mathematics at the University of Wrocław, where STEINHAUS moved after the war, there is the *New Scottish Book* that he started.

2. STEFAN BANACH (March 30, 1892–August 31, 1945) attended schools in Cracow and studied engineering in Lwów, never receiving his degree. He returned to Cracow with little formal mathematical education. There, with his friends NIKODYM and WITOLD WILKOSZ, he was "discovered" by STEINHAUS [452] (see the anecdote above about STEINHAUS' walk and Lebesgue measure), who interested them in some open mathematical problems. This was the start of a long and fruitful cooperation between BANACH and STEINHAUS. Soon thereafter (1920), BANACH moved to Lwów, where he obtained his doctorate in mathematics. An exception had to be made for him, since, as we noted, he lacked a mathematics, or any other, degree. In his dissertation he defined what is now called a *Banach space*. The idea was already in the air, e.g., with work by WIENER (1920), but no one had developed it into a systematic theory. MAURICE R. FRÉCHET's generalized theory of limits and differentials was WIENER's inspiration to define Banach spaces; but, as WIENER pointed out in 1955, "these spaces are quite justly named after BANACH alone" [509], page 60.

In 1927 BANACH became a full professor at the Jan Kazimierz University in Lwów. Besides many scientific publications, he also wrote several high school texts and academic textbooks for calculus and mechanics. In 1939, just before the war, he was elected president of the Polish Mathematical Society, of which he was a cofounder. For more details about the life of BANACH and his dazzling virtuosity see, e.g., [105], [274], [452].

3. MOISEJ A. ZARETSKY (1903–1930) is not a household name in mathematics, which is why we feature him now because of his beautiful result, independent of BANACH, of the Banach–Zaretsky theorem (1925) (Theorem 4.6.2). ZARETSKY was a Russian, and he died in the Caucasian resort of Batumi. The English translation of the Russian ISIDOR P. NATANSON's text uses the spelling "Zarecki" [355].

ZARETSKY published the Banach–Zaretsky theorem in conjunction with his new proof [520] of GRIGORI M. FICHTENHOLZ' theorem [174]: *Let F and G be two real-valued absolutely continuous functions defined on closed intervals $[a,b]$ and $[c,d]$, respectively, such that the values of $G(x)$ are in the closed interval $[a,b]$ on which F is defined; then the composition $F \circ G$ is absolutely continuous on $[a,b]$ if and only if $F \circ G \in BV([a,b])$.*

The condition

$$m(A) = 0 \Longrightarrow m(F(A)) = 0 \tag{4.49}$$

is due to LUZIN. In the setting of functions $F : [a, b] \to \mathbb{R}$, (4.49) *characterizes* the property that if E is Lebesgue measurable then $F(E)$ is Lebesgue measurable. For perspective, recall that F is continuous if and only if

$$\forall\, U \subseteq \mathbb{R}, \text{ open}, \quad F^{-1}(U) \text{ is open.}$$

Going back to ZARETSKY, he also proved the following result dealing with absolute continuity: *Let* $F : [a, b] \to \mathbb{R}$ *be continuous and strictly increasing; F is absolutely continuous if and only if* $m(F(A)) = 0$, *where* $A = \{x \in [a, b] : F'(x) = +\infty\}$.

4. Notwithstanding the FTC formula, remarkably characterized in terms of absolute continuity, we mention a more general problem that LEBESGUE formulated, and which we shall refer to as the *Lebesgue–Dini problem.* The problem is to determine when a function f can be recovered from one of its Dini derivatives in terms of an FTC formula of the form

$$f(b) - f(a) = \int_a^b D^+ f. \tag{4.50}$$

LEBESGUE's contributions to solving the Lebesgue–Dini problem, along with those of LUDWIG SCHEEFER, are exposited in [313], 2nd edition, pages 79–90. (One must distinguish between a solution in terms of (4.50) and an existential solution, which asserts only that f can be *determined* in terms of $D^+ f$.) In order to obtain (4.50) there must be an offsetting of the paucity of information in $D^+ f$ (compared with f') and the notion of integral used in (4.50) as well as the restrictions on f. The following succinct statement is typical of the type of solution one has in mind; see [210]: *Let* $f : [c, d] \to \mathbb{R}$ *be continuous and have finite Dini derivate* $D^+ f$ *at each* $x \in [c, d]$*; then (4.50) is valid on any interval* $[a, b] \subseteq [c, d]$ *if* $D^+ f$ *is Riemann integrable, Lebesgue integrable, or Denjoy–Perron integrable on* $[c, d]$.

We do not define the *Denjoy–Perron integral*, also known by the more recent names of Henstock and Kurzweil; see [260], Chapter 6, for references and perspective; cf. DENJOY's Harvard lectures [132], Chapitre I. DENJOY's lectures, based on research beginning in 1912, addressed the problem of computing the coefficients c_k of trigonometric series $\sum c_k e^{2\pi i k x}$ representing non-Riemann- and/or non-Lebesgues integrable functions. One of his main innovations is an operator, the *Denjoy integral operator* $\mathcal{D} \int$, with the property that if f is continuous on $[a, b]$ and f' exists on $[a, b]$, then $\mathcal{D} \int_a^x f' = f(x) - f(a)$; see Theorems 4.6.2 (Banach–Zaretsky) and 4.6.7. The Denjoy integral operator involves a countably infinite number of steps, e.g., [261].

5. In his dissertation of 1915 (actually he published an announcement at Paris in 1913 on the relevant material), LUZIN gave necessary and sufficient conditions that $f \in L^2_m(\mathbb{T})$ have a Fourier series convergent *m-a.e.*; and at the same time essentially posed the problem, the *Luzin problem*, as to whether every $f \in L^2_m(\mathbb{T})$ has its Fourier series converging *m-a.e.* ($L^2_m(\mathbb{T})$

was defined in Example 3.3.4*b*.) FATOU (1906), F. JEROSCH and WEYL (1908), WEYL (1909), W. YOUNG (1912), ERNEST W. HOBSON (1913), MICHEL PLANCHEREL (1913), and HARDY (1913) had in fact also worked on such issues. Refined "log-estimates" by KOLMOGOROV and GLEB A. SELIVERSTOV (1925), ABRAM I. PLESSNER (1926), and LITTLEWOOD and RAYMOND E. A. C. PALEY (1931) were introduced to address the problem.

Finally, in 1966, CARLESON [91] proved that *if* $f \in L^2_m(\mathbb{T})$ *then its Fourier series converges* m*-a.e. to* f; and, in 1968, using the method of CARLESON's proof and the theory of interpolation of operators, RICHARD A. HUNT extended CARLESON's theorem to $L^p_m(\mathbb{T})$, $p > 1$. (Recall KOLMOGOROV's example mentioned in Section 3.8.3 for $p = 1$.) CHARLES L. FEFFERMAN [166] later proved CARLESON's theorem along the lines initiated by KOLMOGOROV–SELIVERSTOV. An important lemma for CARLESON, which he proved using properties of harmonic functions, is the following ([91], Lemma 5, page 140): *Let* $\{I_k : k = 1, \ldots\}$ *be a disjoint cover of* $(0,1)$ *by open intervals, where* $m(I_k) = d_k$ *and the center of* I_k *is* t_k*, and define*

$$D(x) = \sum_{k=1}^{\infty} \frac{d_k^2}{(x - t_k)^2 + d_k^2}, \quad x \in (0,1),$$

and

$$U_M = \{x \in (0,1) : D(x) > M\};$$

then there are $k, K > 0$ *such that for all* M

$$m(U_M) \leq ke^{-KM}.$$

See Section 8.8.1 for more on CARLESON's theorem.

ZYGMUND [525] proved CARLESON's lemma using a technique in the spirit of FTC, but which is essentially deeper than the differentiability of the integral. He obtains not only CARLESON's lemma but other substantial results in classical harmonic analysis. The technique is due to JÓZEF MARCINKIEWICZ, and ZYGMUND takes the opportunity to describe several of MARCINKIEWICZ' results in the area. MARCINKIEWICZ was taken as a prisoner of war in 1939. He was executed, together with a few thousand Polish officers, in a Soviet prison camp (possibly Katyń) in 1940 at the age of 30; see, e.g., [127].

4.8 Problems

Some of the more elementary problems in this set are Problems 4.1, 4.2, 4.3, 4.4, 4.13, 4.14, 4.15, 4.16, 4.19, 4.20, 4.21, 4.31, 4.37.

4.1. Let $f : \mathbb{R} \to \mathbb{R}$ be an element of $BV(\mathbb{R})$.

a. Prove that if $a \leq c \leq b$ then

$$V(f,[a,c]) + V(f,[c,b]) = V(f,[a,b]);$$

cf. Theorem 5.3.6*a*.

b. Let $f = g_1 - g_2$, where each g_j is increasing. Prove that

$$\forall\, a \leq b, \quad 0 \leq P(b) - P(a) \leq g_1(b) - g_1(a),$$
$$0 \leq N(b) - N(a) \leq g_2(b) - g_2(a),$$

cf. Theorem 5.1.9*b*.

Part *b* is a uniqueness or optimality condition for the representation of $f \in BV(\mathbb{R})$ as the difference of increasing functions.

4.2. Let $f, g : \mathbb{R} \to \mathbb{R}$ be elements of $BV(\mathbb{R})$. Prove that, for all $a \leq b$,

$$V(f + g, [a, b]) \leq V(f, [a, b]) + V(g, [a, b]).$$

4.3. ***a.*** Let $f \in BV([a, b])$ be a right-continuous function. Prove that $V(f)(x) = V(f, [a, x])$ is also right continuous.

b. Let $f : \mathbb{R} \to \mathbb{R}$ be a right-continuous function. Prove that if $f \in BV_{\text{loc}}(\mathbb{R})$, then there exist increasing, right-continuous functions P and N such that $f = P - N$ and $V(f) = P + N$.

4.4. ***a.*** Let $f : [a, b] \to \mathbb{R}$ be everywhere differentiable and assume that $\|f'\|_\infty < \infty$. Prove that f is absolutely continuous; cf. Theorem 4.6.7.

b. Let $\alpha, \beta > 0$ and define the function $f : [0, 1] \to \mathbb{R}$ as

$$f(x) = \begin{cases} 0, & \text{if } x = 0, \\ x^\alpha \sin(1/x^\beta), & \text{if } x \in (0, 1]. \end{cases}$$

Prove that $f \in BV([0, 1])$ if and only if $\alpha > \beta$.

c. It is elementary to find functions that are both continuous and of bounded variation; in this exercise, for example, let $(\alpha, \beta) = (2, 1)$. Find f, as in part *b*, that is discontinuous at a point and that is not an element of $BV([0, 1])$ (respectively, that is continuous and that is not an element of $BV([0, 1])$).

[*Hint.* Take $(\alpha, \beta) = (0, 1)$ (respectively, $(\alpha, \beta) = (1, 1)$).]

d. Prove that the function $\mathbb{1}_S$ in Example 3.6.6 is not a function of bounded variation.

e. Show that if $f(x) = a_n x^n + a_{n-1} x^{n-1} + \cdots + a_0$ then $f \in BV([a, b])$. Compute $V(f, [a, b])$ given the roots of f'.

[*Hint.* Use part *a*.]

4.5. ***a.*** With regard to the Remark after Theorem 4.1.7, prove that the Fourier series (see Definition B.5.1)

$$S(f)(x) = \sum_{n=2}^{\infty} \frac{1}{n[\log(n)]} e^{-2\pi i n[\log(n)]x}$$

and the function

$$g(x) = x\sin(1/x)$$

are elements of $L^1_m(\mathbb{T})\backslash BV(\mathbb{T})$, and that $n\hat{f}(n)$, $n\hat{g}(n) = o(1)$, $|n| \to \infty$. Here "$[\log(n)]$" denotes the largest integer $k \leq \log(n)$. Note that g is continuous on $\mathbb{R}$ when defined on $[-1/2, 1/2)$ and extended 1-periodically. Also, compare this with (1.15).

b. We now give an example of a continuous function $f \in BV(\mathbb{T})$ such that $n\hat{f}(n) \neq o(1)$, $|n| \to \infty$. Construct the Cantor function C_C so that $C_C(0) = 0$ and $C_C(1) = 1$. Setting $f(x) = C_C(x) - x$ on $[0,1]$, f can be defined as a continuous periodic function on $\mathbb{R}$ with $f(0) = f(1) = 0$ and period 1. Prove that

$$\forall\, n, \quad \left|3^n \hat{f}(3^n)\right| = \left|\hat{f}(1)\right| > 0.$$

For a more extensive treatment of finding and analyzing Fourier coefficients of singular measures we refer to [32], [271]. Further remarks and examples are given in Chapter 5 and Appendix B.
[*Hint.* The calculation for the equality is routine, and to see that $|\hat{f}(1)| \neq 0$ it is sufficient to compute the real part.]

4.6. Prove the *Sidon theorem*: *Let f be a bounded Lebesgue measurable function with Fourier series expansion*

$$S(f)(x) = \sum_{k=1}^{\infty} a_k \cos(2\pi n_k x),$$

where $\{n_k : k = 1, \ldots\}$ is a Hadamard set; see Example 1.3.25 for the definition of a Hadamard set. Then $\sum_{k=1}^{\infty} |a_k| \leq 2\|f\|_\infty$, e.g., [524].

Thus, we have a criterion in order that a bounded Lebesgue measurable function have an absolutely convergent Fourier series.

4.7. ***a.*** Construct a continuous function $f : [a,b] \to \mathbb{R}$ such that for each interval $[c,d] \subseteq [a,b]$, $f \notin BV([c,d])$. This is immediate from Theorem 4.3.2 and the existence of everywhere continuous nowhere differentiable functions, but you should be able to give much simpler examples.

b. Let $f : [a,b] \to \mathbb{R}$. For $x \in [a,b]$ and $\delta > 0$ define

$$V(x,\delta) = 1/(1 + V(f, [x-\delta, x+\delta])), \quad v(x) = \lim_{\delta \to 0} V(x,\delta),$$

and

$$S_{BV} = \{x \in [a,b] : \exists\, \delta > 0 \text{ such that } V(f, [x-\delta, x+\delta]) < \infty\}.$$

Show that S_{BV} is an open set. Further, prove that

i. if f is continuous at $x \in S_{BV}$ then $v(x) = 1$,

ii. if $v(x) = 1$ then f is continuous at x.

Observe that if f is continuous at x then x need not be in S_{BV}.

4.8. With regard to Example 3.6.6 and Problem 4.5, find an open set $S \subseteq [0,1)$, with $m([0,1) \setminus S) > 0$, such that $\mathbb{1}_S \notin BV([0,1))$, whereas $\sup_n |n\hat{\mathbb{1}}_S(n)| < \infty$; cf. [78].

4.9. VAN DER WAERDEN's everywhere continuous nowhere differentiable function W was defined in Section 1.3.4 and in Problem 1.32. Let

$$W_n(x) = \sum_{j=0}^{n-1} 10^{-j} f(10^j x), \quad x \in (0,1),$$

and

$$f(x) = |x - k|, \quad x \in [k - (1/2), k + (1/2)] \text{ and } k \in \mathbb{Z}.$$

From Theorem 4.3.2, $V(W, [0,1]) = \infty$. On the other hand there is the following control on the variation:

$$\lim_{n\to\infty} \frac{1}{\sqrt{n}} V(W_n, [0,1]) = \sqrt{2/\pi}. \tag{4.51}$$

Prove (4.51).
[*Hint.* Calculate that

$$\lim_{n\to\infty} \frac{1}{\sqrt{n}} V(W_n, [0,1]) = \lim_{n\to\infty} \sum_{m=0}^{n} \left| \frac{m - (n/2)}{\sqrt{n/2}} \right| \binom{n}{m} \frac{1}{2^n}$$
$$= \frac{1}{\sqrt{2\pi}} \int_{-\infty}^{\infty} |x| e^{-x^2/2}\, dx = \sqrt{2/\pi}.]$$

4.10. ***a.*** Prove that L'Hospital's rule is not necessarily true for complex-valued functions.
[*Hint.* Let $f(x) = x$, $g(x) = xe^{-i/x}$, and take the limit as $x \to 0$.]

b. Let functions $f, g : (0,a) \to \mathbb{C}$ have the following properties:
i. f', g' are continuous on $(0,a)$,
ii. $\lim_{x\to 0} f(x) = \lim_{x \to 0} g(x) = 0$,
iii. $\forall\, x \in (0,a)$, $g'(x) \neq 0$,
iv. $\lim_{x\to 0} f'(x)/g'(x) = c$.

Define $\delta(x) = -c + f'(x)/g'(x)$, and assume that $|g(x)|$ is monotone and $\operatorname{Re}\delta$, $\operatorname{Im}\delta \in BV((0,a))$. Prove that $\lim_{x\to 0} f(x)/g(x) = c$.
[*Hint.* Define

$$R(x) = \frac{\int_0^x \delta(y) g'(y)\, dy}{g(x)}$$

and compute that

$$\lim_{x\to 0} R(x) = 0.$$

Use this fact and an integration of

$$g'(x)\delta(x) = f'(x) - cg'(x)$$

to complete the proof.]

4.11. Prove Proposition 4.2.6, and show by example that the continuity hypothesis is necessary.

4.12. Let $f : [a, b] \to \mathbb{R}$ and $g : [c, d] \to \mathbb{R}$ be continuous functions of bounded variation. Assume that $g([c, d]) \subseteq [a, b]$. Then the composition $f \circ g$ is continuous. Find conditions to ensure that $f \circ g \in BV([c, d])$, e.g., [379].

4.13. Compute P, N, and the variation function $V(f)$ for $f(x) = e^{-x^2}$.

4.14. Let $\{a_n : n = 1, \ldots\}, \{b_n : n = 1, \ldots\} \subseteq \mathbb{R}$ and define the function $f_{a,b} = \sum_{n=1}^{\infty} f_n$, where

$$f_n(x) = \begin{cases} b_n, & \text{if } x \geq a_n, \\ 0, & \text{if } x < a_n. \end{cases}$$

a. Fix a sequence $\{b_n : n = 1, \ldots\} \subseteq \mathbb{R}$, and find necessary and sufficient conditions such that for each sequence $\{a_n : n = 1, \ldots\} \subseteq \mathbb{R}$ that is bounded below,

$$f_{a,b} \in BV_{\text{loc}}(\mathbb{R}).$$

b. Do there exist sequences $\{a_n : n = 1, \ldots\}, \{a'_n : n = 1, \ldots\}, \{b_n : n = 1, \ldots\} \subseteq \mathbb{R}$ for which $f_{a,b} \in BV_{\text{loc}}(\mathbb{R})$, whereas $f_{a',b} \notin BV_{\text{loc}}(\mathbb{R})$?

4.15. Define the function $f : \mathbb{R} \to \mathbb{R}$ as

$$f(x) = \begin{cases} 0, & \text{if } x \text{ is irrational or } x = 0, \\ 1/(p^2 q^2), & \text{if } x = p/q \text{ and } (p, q) = 1. \end{cases}$$

a. Prove that $f \in BV(\mathbb{R})$, and so f' exists m-*a.e.*
b. Prove that f' does not exist on a dense subset of $\mathbb{R}$.

4.16. Construct a continuous real-valued function $f \in BV([0, 1])$ such that

$$\forall\, q \in \mathbb{Q} \cap [0, 1], \quad f'(q) \text{ does not exist.}$$

[*Hint.* Let $\mathbb{Q} \cap [0, 1] = \{r_n : n = 1, \ldots\}$ and define

$$f_n(x) = \begin{cases} 0, & \text{if } x \in [0, r_n], \\ (x - r_n)/(2^n(1 - r_n)), & \text{if } x \in (r_n, 1]. \end{cases}$$]

4.17. Assume that $f \in L^1_m([0, 1])$ satisfies the condition that

$$\|\tau_{-h} f - f\|_1 = o(|h|), \quad |h| \to 0.$$

Prove that $f = k$, a constant, m-*a.e.*
[*Hint.* Use Corollary 4.4.6 or the Fourier uniqueness theorem; see Theorem B.3.5*c*.]

Remark. Note that if $f \in L^1_m([0,1])$ and

$$\|\tau_{-h}f - f\|_1 = o(|h|^2), \quad |h| \to 0,$$

then $f = 0$ *m-a.e.*

4.18. Construct three essentially different Vitali coverings of $[a,b] \subseteq \mathbb{R}$.

4.19. ***a.*** Prove the *Fubini differentiation theorem: Let* $\{f_n : n = 1, \ldots\}$ *be a sequence of monotone increasing functions* $f_n : [a,b] \to \mathbb{R}$*, and assume that*

$$f(x) = \sum_{n=1}^{\infty} f_n(x)$$

exists for each $x \in [a,b]$*. Then,* $f' = \sum_{n=1}^{\infty} f_n'$ *exists m-a.e.*

b. Let $f \in BV([a,b])$. Prove that $(V(f))' = |f'|$ *m-a.e.*, where $V(f)$ is the variation function of f.
[*Hint.* Use part *a*.]

Remark. Thus, if $f \in BV([a,b])$, then not only does f' exist *m-a.e.* but also $(V(f))'$ exists *m-a.e.* The sets of points where f' and $(V(f))'$ exist need not be the same, e.g., $f(x) = |x|$.

4.20. If $f(x) = e^{-n^2x^2}$, then $V(f, \mathbb{R}) = 2$; see Problem 4.13. Let $f_n(x) = e^{-n^2x^2}/n^2$ and set $g(x) = \sum_{n=1}^{\infty} f_n(x)$. Prove that $g \in BV_{\mathrm{loc}}(\mathbb{R})$, whereas $\sum_{n=1}^{\infty} f_n'$ does not converge uniformly on $[-1,1]$. Consequently, g' exists *m-a.e.*, but this fact cannot be proved by the standard advanced calculus criterion. Compute g'.

4.21. Prove FTC–I for the case of unbounded f.

4.22. Let $f, f_n \in BV(\mathbb{R})$.

a. Assume that

$$\forall\, x \in \mathbb{R}, \quad \lim_{n\to\infty} V(f - f_n)(x) = 0. \tag{4.52}$$

Prove that there is a subsequence $\{f_{n_k} : k = 1, \ldots\} \subseteq \{f_n : n = 1, \ldots\}$ such that

$$\lim_{k\to\infty} f_{n_k}' = f' \quad m\text{-}a.e.$$

b. Does part *a* hold when (4.52) is weakened to almost everywhere pointwise convergence?

c. Show that the conclusion of part *a* cannot be replaced in general by

$$\lim_{n\to\infty} f_n' = f' \quad m\text{-}a.e.$$

4.23. ***a.*** In light of Example 4.2.5 find a simple example of a strictly increasing function f such that $f' = 0$ m-a.e.
[*Hint.* Use Example 4.1.12 and Problem 4.19. Note that we are not asking that f be everywhere continuous.]

b. Does there exist a continuous function $f : [a, b] \to \mathbb{R}$ such that $\|f\|_\infty < \infty$, $f \notin BV([a, b])$, and $f' = 0$ m-a.e.?

4.24. ***a.*** Let $A \subseteq \mathbb{R}$. Prove that $m(A) = 0$ if and only if there is a countably infinite sequence $\{I_n = (a_n, b_n) : n = 1, \ldots\}$, that satisfies $\sum m(I_n) < \infty$ and $A \subseteq \bigcup I_n$, and such that for each $x \in A$ there is a countably infinite subsequence $\{I_{n_k} : k = 1, \ldots\} \subseteq \{I_n : n = 1, \ldots\}$ for which

$$\forall\, k, \quad x \in I_{n_k}.$$

b. Let $A \subseteq [a, b]$ be a set with $m(A) = 0$. Does there exist a function $f \in BV([a, b])$ such that f' exists precisely on $A^{\sim}$?
[*Hint.* Use part *a.*]

c. Let C_E be the Cantor function for a perfect symmetric set E. Prove that C'_E does not exist for boundary points of contiguous intervals; cf. [64], pages 128–134.

d. Let C_E be the Cantor function for a perfect symmetric set E. Does there exist a point x for which $C'_E(x)$ exists and $C'_E(x) \neq 0$?

4.25. With regard to Problem 4.11 (Proposition 4.2.6) we would like to find conditions on the partitions P such that if $f \in BV([a, b])$ then

$$\lim_{|P| \to 0} \sum_P |f(x_j) - f(x_{j-1})| = V(f, [a, b]).$$

Let f have jump discontinuities at $\{y_j : j = 1, \ldots\} \subseteq [a, b]$ and let $P(N, \delta)$ be a partition including $y_1, \ldots, y_N$ and with norm $|P(N, \delta)| < \delta$ (see (3.3)). Prove that

$$\forall\, \varepsilon > 0,\ \exists\, N \text{ and } \exists\, \delta, \text{ such that } \forall\, P(N, \delta),$$
$$\sum_{P(N,\delta)} |f(x_j) - f(x_{j-1})| > V(f, [a, b]) - \varepsilon.$$

4.26. *Change of variable.* Define the functions $f : [a, b] \to \mathbb{R}$, $\phi : [c, d] \to \mathbb{R}$, and $F(x) = \int_a^x f$, and assume that $\phi([c, d]) \subseteq [a, b]$.

a. Suppose that $f \in L^1_m([a, b])$ and that ϕ and $F \circ \phi$ are differentiable m-a.e. on their respective domains. Prove that

$$(F \circ \phi)' = (f \circ \phi)\phi' \quad m\text{-a.e.}$$

b. Assume that $f \in L^1_m([a, b])$ and that ϕ' exists m-a.e. Prove that $F \circ \phi$ is absolutely continuous on $[c, d]$ if and only if

i. $(f \circ \phi)\phi' \in L^1_m([c, d])$;

ii.

$$\forall\, [\alpha, \beta] \subseteq [c, d], \quad \int_{\phi(\alpha)}^{\phi(\beta)} f = \int_\alpha^\beta (f \circ \phi)\phi'. \tag{4.53}$$

[*Hint.* The sufficient conditions for absolute continuity are clear, and we use part *a* for the necessary conditions.]

c. Prove (4.53) assuming any of the following sets of conditions:

i. $f \in L^1_m([a,b])$ and ϕ is an increasing absolutely continuous function;

ii. $\|f\|_\infty < \infty$ and ϕ is absolutely continuous;

iii. ϕ is absolutely continuous, $f \in L^1_m([a,b])$, and $(f \circ \phi)\phi' \in L^1_m([c,d])$.

[*Hint.* For parts *i* and *ii* show that $F \circ \phi$ is absolutely continuous; and for part *iii*, approximate by bounded functions, apply part *ii*, and use LDC.]

Remark. The change of variable formula can be proved in a more general setting; see Section 8.7.

4.27. Find f_n, f, $g : [a,b] \to \mathbb{R}$, $n = 1, \ldots$, where g and each f'_n are continuous on $[a,b]$, such that

$$\lim_{n\to\infty} f_n = f \;\; m\text{-}a.e. \quad \text{and} \quad \forall\, x \in [a,b], \;\; \lim_{n\to\infty} f'_n(x) = g(x),$$

whereas $f' \neq g$ m-$a.e.$ on $[a,b]$.

[*Hint.* Consider a sequence $\{f_n : n = 1, \ldots\}$ of smooth functions on $[0,1]$, that converges to 1 on $\mathbb{Q} \cap (0,1)$ and that converges to 0 on a set of Lebesgue measure 1, such that the sequence f'_n converges to 0 in measure. Choose a subsequence $\{f_{n_k} : k = 1, \ldots\}$ that converges to 0 m-$a.e.$ Adjust this subsequence to be smooth and convergent to 0 pointwise everywhere.]

Remark. It is an interesting question to find out what happens in Problem 4.27 if we assume that the sequence $\{f_n : n = 1, \ldots\}$ converges pointwise everywhere in $[a,b]$.

4.28. Let $\{W_N : N = 1, \ldots\}$ be the Fejér kernel defined in Problem 3.28 and let $f \in C(\mathbb{R})$ be a 1-periodic function, i.e., $f \in C(\mathbb{T})$. Prove that

$$\forall\, x \in \mathbb{R}, \quad W_N * f(x) = \frac{1}{N+1} \sum_{n=0}^{N} \sum_{m=-n}^{n} \hat{f}(m) e^{-2\pi i m x},$$

where convolution $*$ on $\mathbb{T}$ was also defined in Problem 3.28; see Example 4.4.7.

4.29. Let $f : [a,b] \to \mathbb{R}$ be a differentiable function and suppose $f' \in BV([a,b])$. Prove that $f' \in C([a,b])$.

[*Hint.* Use the intermediate value theorem for derivatives: let $f : [a,b] \to \mathbb{R}$ be a differentiable function and suppose $f'(a) < \lambda < f'(b)$; then there is $x \in (a,b)$ for which $f'(x) = \lambda$.]

4.30. Consider the function f of Problem 1.26 that is increasing, continuous on the irrationals, and discontinuous on the rationals. Define

$$F(x) = \int_0^x f.$$

Prove that $F' = f$ at each irrational, and that F' does not exist at any rational. Compare this with Problem 4.15 and Problem 4.16.

4.31. A function $f : [a, b] \to \mathbb{R}$ is *locally recurrent* if

$$\forall\, x \in [a, b] \text{ and } \forall\, \varepsilon > 0,\ \exists\, y \text{ such that } f(y) = f(x) \text{ and } 0 < |y - x| < \varepsilon.$$

Although it is not apparent, there are continuous, nonconstant, locally recurrent functions. Prove that if f is continuous, nonconstant, and locally recurrent on $[a, b]$ then $f \notin BV([a, b])$.
[*Hint.* Use Theorem 4.6.2.]

4.32. Prove Lemma 4.6.1.

4.33. ***a.*** Let $A \subseteq [0, 1]$ be a set for which $m(A) = 0$. Does there exist an increasing absolutely continuous function $f : [0, 1] \to \mathbb{R}$ such that f' exists on $A^{\sim}$ and

$$\forall\, x \in A, \quad \lim_{h \to 0} \frac{f(x + h) - f(x)}{h} = \infty?$$

[*Hint.* Cf. Problem 4.24.]

b. Find a continuous function f on $[0, 1]$ that is absolutely continuous on each interval $[\varepsilon, 1]$, $\varepsilon > 0$, but that is not absolutely continuous on $[0, 1]$.
[*Hint.* Consider $x \sin(1/x)$.]

c. Along with the hypotheses of part *b* assume that $f \in BV([0, 1])$. Prove that f is absolutely continuous on $[0, 1]$.
[*Hint.* Use the Banach–Zaretsky theorem.]

d. Construct an increasing continuous function f on $[0, 1]$ such that for each interval $[\alpha, \beta] \subseteq [0, 1]$, f is not absolutely continuous on $[\alpha, \beta]$.
[*Hint* with a capital H!]

e. Find a continuous and strictly increasing function f on $[0, 1]$ and a set $A \subseteq [0, 1]$ such that $m(A) = 0$ and $m(f(A)) = 1$.

f. Find an absolutely continuous function f on $[0, 1]$ such that f is not monotonic on any interval in $[0, 1]$.
[*Hint.* Find a set $E \subseteq [0, 1]$ such that, for each interval I, $m(E \cap I) > 0$ and $m(E^{\sim} \cap I) > 0$, as in Problem 2.45; set $f(x) = \int_0^x (\mathbb{1}_E - \mathbb{1}_{E^{\sim}})$.]

4.34. For a function $f : \mathbb{R} \to \mathbb{R}$ assume that

$$\forall\, x \in \mathbb{R},\ \exists\, N_x \subseteq \mathbb{R}, \text{ for which } m(N_x) = 0, \text{ such that } \forall\, y \notin N_x,\ f(y) = f(y + x).$$

Prove that there is a constant c such that $f = c$ *m-a.e.*

4.35. Prove that the following statements are equivalent:

i. F is absolutely continuous on $[a, b]$.

ii. For each $\varepsilon > 0$ there is $\delta > 0$ such that for all disjoint families $\{(a_j, b_j) \subseteq [a, b] : j = 1, \ldots, n\}$,

$$\sum_{j=1}^{n}(b_j - a_j) < \delta \Longrightarrow \left|\sum_{j=1}^{n}(F(b_j) - F(a_j))\right| < \varepsilon.$$

iii. For each $\varepsilon > 0$ there is $\delta > 0$ such that for all disjoint families $\{(a_j, b_j) \subseteq [a,b] : j = 1, \ldots, n\}$,

$$\sum_{j=1}^{n}(b_j - a_j) < \delta \Longrightarrow \sum_{j=1}^{n}\omega(F, [a_j, b_j]) < \varepsilon.$$

4.36. A function $f : [a,b] \to \mathbb{R}$ satisfies a *Lipschitz condition of order* $\alpha > 0$ if there exists a constant $C > 0$ such that

$$\forall\, x,\, y \in [a,b], \quad |f(x) - f(y)| \le C|x - y|^{\alpha};$$

cf. the definition of Lipschitz functions in Problem 1.31. We often say that f is a *Hölder function of order* α.

a. Prove that Lipschitz functions of order 1 are absolutely continuous.

b. Prove that the Cantor function C_C, defined in Example 1.2.7*d*, is Lipschitz of order $\log_3(2)$.

c. Find an example of a continuous function of bounded variation that is not Lipschitz of order 1.

d. Show that for any $\alpha > 1$, the set of Lipschitz functions of order α consists only of constant functions.

4.37. Let $f \in L^1_m([a,b])$ and set $F(x) = \int_a^x f$; prove that $V(F, [a,b]) = \|f\|_1$. [*Hint.* The inequality $V(F, [a,b]) \le \int_a^b |f|$ is clear as in the proof of Proposition 4.4.1. For the opposite inequality, first recall the definition of the sgn function: sgn $f = f/|f|$ if $f(x) \neq 0$ and sgn $f = 0$ if $f(x) = 0$. Choose simple functions s_j such that $s_j \to$ sgn f *m-a.e.* and $|s_j| \le 1$ (such an s should have the form $\sum a_j \mathbb{1}_{I_j}$, $I_j = (c_j, d_j)$). Check that

$$\left|\int_a^b f s_j\right| \le V(F, [a,b])$$

and apply LDC.]

Remark. This result is important because of the bijection between $BV([a,b])$ and the bounded measures on $[a,b]$, i.e., the continuous linear functionals on $C([a,b])$; the bijection is given by taking the first distributional derivative, e.g., Chapter 7. We shall see that the total variation of $F \in BV([a,b])$ is the canonical Banach space norm of the corresponding measure. The bijection identifies the absolutely continuous functions on $[a,b]$ (as a subset of $BV([a,b])$) with $L^1_m([a,b])$ (as a subset of the bounded measures). The above Banach space norm on the measures reduces to $\|...\|_1$ for $L^1_m([a,b])$.

4.38. *a.* With regard to Problem 4.37 and the Remark there, prove the following: if $F \in BV([a,b])$ then

$$V(F,[a,b]) \geq \int_a^b |F'|. \tag{4.54}$$

[*Hint.* First prove that if G is increasing and $A = \{x : \exists\, G'(x)\}$ then

$$\int_a^b G' = m^*(G(A)).$$

Next calculate $\int_a^b |F'|$ in terms of $(V(F))'$, observing that the variation function $V(F)$ is increasing.]

b. Related to (4.54), prove that

$$\forall\, B \in \mathcal{M}([a,b]), \quad m^*(V(F)(B)) \geq \int_B |F'|, \tag{4.55}$$

when $F \in BV([a,b])$. Naturally there is equality in both (4.54) and (4.55) when F is absolutely continuous.

4.39. Let the sequence $\{g_n : n = 1, \ldots\} \subseteq BV([a,b])$ have the property that $\{g_n : n = 1, \ldots\}$ converges pointwise to a function $g : [a,b] \to \mathbb{C}$ and that

$$\exists\, M > 0 \text{ such that } \forall\, n = 1, \ldots, \quad V(g_n,[a,b]) \leq M. \tag{4.56}$$

Prove that $g \in BV([a,b])$ with $V(g,[a,b]) \leq M$ and that

$$\forall\, f \in C([a,b]), \quad \lim_{n\to\infty} \int_a^b f\, dg_n = \int_a^b f\, dg.$$

[*Hint.* The fact that $g \in BV([a,b])]$ follows from its definition, the hypothesis of pointwise convergence, and a $\overline{\lim}$ argument adding and subtracting the g_n. Next write

$$\int_a^b f\, dg_n - \int_a^b f\, dg = \sum_{k=1}^m \int_{x_{k-1}}^{x_k} (f - f(x_{k-1}))\, dg_n$$
$$- \sum_{k=1}^m \int_{x_{k-1}}^{x_k} (f - f(x_{k-1}))\, dg + \sum_{k=1}^m f(x_{k-1}) \int_{x_{k-1}}^{x_k} d(g_n - g),$$

where, without loss of generality, we have taken $x_0 = a$ and $x_m = b$. The required convergence follows from another $\overline{\lim}$ calculation.]

Remark. This result was proved independently by HUBERT E. BRAY (1919) and EDUARD HELLY (1921). It is natural to ask whether any pointwise convergence can be deduced based on the assumption (4.56). In fact, HELLY (1921) proved that, in this case, *there are a strictly increasing sequence*

$\{n_j : j = 1, \ldots\} \subseteq \mathbb{N}$ and $g \in BV([a,b])$ such that $g_{n_j} \to g$ pointwise on $[a,b]$. The proof is based on the Jordan decomposition theorem and the following *selection principle*, which itself has an elementary and natural proof. *Given $\{c_{m,n} : m, n = 1, \ldots\} \subseteq \mathbb{C}$, assume that $\{c_{m,n}\}$ is uniformly bounded. Then there are a strictly increasing sequence $\{n_j : j = 1, \ldots\} \subseteq \mathbb{N}$ and a sequence $\{c_m : m = 1, \ldots\} \subseteq \mathbb{C}$ such that*

$$\forall\, m = 1, \ldots, \qquad \lim_{j \to \infty} c_{m,n_j} = c_m.$$

These techniques can be formulated in terms of complex measures, which we shall do in Chapters 5–7, but it is important to note that they were inspired by classical moment and transform (Fourier, Laplace, and Stieltjes) problems; see [503] for a definitive treatment.

4.40. ***a.*** Let $f : [a,b] \to \mathbb{R}$ be an m-measurable function and take $\varepsilon, \delta > 0$. Prove that there are an absolutely continuous function G and a set $A \in \mathcal{M}([a,b])$ such that $m(A) < \delta$ and

$$\sup_{x \in [a,b] \setminus A} |f(x) - G(x)| < \varepsilon.$$

b. If $f \in L^\infty_m([a,b])$ is an increasing, nonnegative function and $g \in L^1_m([a,b])$, then prove that there is $\xi \in (a,b)$ such that

$$\int_a^b fg = f(b-) \int_\xi^b g.$$

We also refer to [235], page 420, on this matter.

4.41. The *functional equation* for a function $f : \mathbb{R} \to \mathbb{R}$ is

$$\forall\, x,\ y \in \mathbb{R}, \quad f(x+y) = f(x) + f(y).$$

a. If f is continuous on $\mathbb{R}$ and satisfies the functional equation prove that $f(x) = f(1)x$ on $\mathbb{R}$.
[*Hint.* Check first that $f(nx) = nf(x)$ if $n \in \mathbb{Z}$ and then that $f(rx) = rf(x)$ if $r \in \mathbb{Q}$; the conclusion follows since f is continuous.]

b. Prove that if f is continuous at some point of $\mathbb{R}$ and the functional equation is satisfied, then $f(x) = f(1)x$ on $\mathbb{R}$.

c. Let $S \subseteq \mathbb{R}$ have the property that $S - S$ contains an interval; cf. STEINHAUS' theorem in Problem 3.6. If f is bounded on S and f satisfies the functional equation, prove that $f(x) = f(1)x$ on $\mathbb{R}$. In particular, if $f = 0$ on the Cantor set $C \subseteq [0,1]$, then f is identically 0 on $\mathbb{R}$. This latter statement follows directly from Problem 1.13 and the facts that $y = nx$, $x \in [0,1]$, and $f(y) = nf(x)$ for a given $y \in \mathbb{R}$.

d. Let f be a measurable function on $\mathbb{R}$ that satisfies the functional equation. Prove that $f(x) = f(1)x$ on $\mathbb{R}$.

[*Hint.* Let $g = e^{if}$ and use Theorem 4.4.3 and the hypothesis to compute that

$$\forall\, x \in \mathbb{R}, \quad g'(x) = g'(0)g(x).$$

Solving the differential equation, we find that

$$f(x) = Ax + 2\pi n(x),$$

where $n : \mathbb{R} \to \mathbb{Z}$ and $n(x+y) = n(x) + n(y)$. Use the fact that the graph of n is not dense in $\mathbb{C}$ to obtain that $n(x) = Bx$, and, from this, conclude that $B = 0$.]

e. The question now arises as to whether the functional equation has any discontinuous solutions. Prove that, in fact, it does.
[*Hint.* Let H be a Hamel basis in $\mathbb{R}$. Then for $x, y \in \mathbb{R}$,

$$x = \sum r_i(x)h_i(x), \quad y = \sum r_i(y)h_i(y),$$
$$x + y = \sum r_i(x+y)h_i(x+y),$$

where "h"$\in H$ and "r"$\in \mathbb{Q}$. By the uniqueness of representation, $r_i(x) + r_i(y) = r_i(x+y)$. If r_i were a continuous function then we would have $r_i(x) = Cx$ for each $x \in \mathbb{R}$, and this contradicts the fact that $r_i(x) \in \mathbb{Q}$.]

f. Let f satisfy the functional equation and assume that $f = 0$ on some nonmeasurable set. Does $f = 0$ on $\mathbb{R}$?
[*Hint.* Use the material about Hamel bases in Problem 2.24*d*–*f*.]

4.42. Let $f, f' \in L^1_m(\mathbb{R})$.

a. Prove that if f is absolutely continuous on each interval $[a, b] \subseteq \mathbb{R}$ then $\lim_{|x|\to\infty} f(x) = 0$.
[*Hint.* Use FTC over finite intervals and obtain a contradiction to the hypothesis that $f \in L^1_m(\mathbb{R})$.]

b. Give an example to show that the hypothesis of absolute continuity is necessary in part *a*.
[*Hint.* "Put the Cantor set C in $I_n = [n, n+(1/2^n)]$". Then define the function f_n to have value 0 outside of I_n and to be a modified Cantor function ranging from 0 to 1 to 0 in I_n; set $f = \sum f_n$. If we do not require f to remain continuous then $f = \mathbb{1}_{\mathbb{Z}}$ provides a trivial counterexample.]

4.43. Let $z(x, y)$ be a function that is absolutely continuous on every straight line parallel to the axes of the coordinate system. Let $f, g : [0,1] \to (0,1)$ be two absolutely continuous functions on $[0,1]$. Is the function $z(f(t), g(t))$ absolutely continuous on $[0,1]$?

This problem was proposed by MEIER EIDELHEIT in [344], Problem 188.1 (1940).

4.44. Let $f \in L^1_m([0,1])$. Prove that $\sum |n\hat{f}(n)|^2 < \infty$ if and only if f is absolutely continuous on $[0,1]$, $f' \in L^2_m([0,1])$, and $f(0) = f(1)$.

4.45. Suppose that $f \in L^1_m(\mathbb{R})$ and $g \in L^\infty_m(\mathbb{R})$. Find general conditions on f and g so that

$$\lim_{x\to\infty} f * g'(x) = 0 \quad \text{when} \quad \lim_{x\to\infty} f * g(x) = 0.$$

The convolution $f * g$ was defined in Problem 3.5.

4.46. Let f and g be continuous elements of $L^1_m(\mathbb{R})$. Assume that f is continuously differentiable and vanishes outside of some finite interval. Prove that $(f * g)'$ exists and that

$$(f * g)' = f' * g. \tag{4.57}$$

To what extent can the conditions on f and g be relaxed so that (4.57) is still valid?

4.47. Let $Y \subseteq \mathbb{R}$ have the property that $m(\mathbb{R}\backslash Y) = 0$ and let $f : \mathbb{R}^2 \to \mathbb{R}$ be a function such that for each $y \in Y$, $D_x f(x,y)$ exists m-*a.e.* Give reasonable hypotheses and a rigorous proof to the *Leibniz rule* for differentiating an integral:

$$D_x \int_{g(x)}^{h(x)} f(x,y)\, dy = [f(x,h(x))h'(x) - f(x,g(x))g'(x)] + \int_{g(x)}^{h(x)} D_x f(x,y)\, dy.$$

If $g(x) = a$ and $h(x) = b$ then we are in the setting of Theorem 3.6.3. If $f(x,y)$ is of the form $f(y)$ and if $g(x) = a$ and $h(x) = x$, then we are in the framework of Theorem 4.4.3.

4.48. Prove that the composition of two absolutely continuous functions need not be absolutely continuous or even of bounded variation; see Problem 4.12 and Section 4.7.3 on ZARETSKY.

4.49. Let $f : (0,\infty) \to \mathbb{R}$ be a positive strictly increasing function that is absolutely continuous on each finite interval $[a,b] \subseteq (0,\infty)$. Prove that if $f(x) = O(x^2)$, as $x \to \infty$, then $\int_0^\infty 1/f'(x)\, dx$ diverges.
[*Hint.* For each $s > r > 0$ we have

$$(s-r)^2 = \left(\int_r^s dx\right)^2 \le (f(s) - f(r)) \int_r^s \frac{1}{f'(x)}\, dx,$$

and so there is a constant A such that for each $r > 0$,

$$A \le \int_r^\infty \frac{1}{f'(x)}\, dx.]$$

4.50. In Proposition 4.9 we used LDC to prove that $F(x) = \int_a^x f$ is continuous if $f \in L^1_m([a,b])$. Now prove the following continuity property: let $f : [a,b] \to \mathbb{R}^+$ be Lebesgue measurable and pick any $y \in [0, \int_a^b f]$; then there is $A \in \mathcal{M}([a,b])$ for which $y = \int_A f$.

[*Hint. i.* Assume $\int_a^b f < \infty$. Then F is continuous, so that by the intermediate value theorem there is t for which $F(t) = y$; hence, set $A = [a,t]$.

ii. Assume $\int_a^b f = \infty$, $A_n = \{x \in [a,b] : f(x) < n\} \in \mathcal{M}([a,b])$, $A_n \subseteq A_{n+1}$, $\bigcup A_n = [a,b]$, and $\lim \int_{A_n} f = \infty$. If $y = \infty$ let $A = [a,b]$. If $y < \infty$ let $\int_{A_n} f > y$ for some n and apply part *i* to $f_n = \mathbb{1}_{A_n} f \in L^1_m(A_n)$ to obtain $B \in \mathcal{M}([a,b])$ for which $\int_B f_n = y$; therefore, let $A = B \cap A_n$.]

4.51. Let $f, g : [a,b] \to \mathbb{R}$ be increasing continuous functions. Define F and G as in Theorem 4.6.3 and let μ_G be the Lebesgue–Stieltjes measure associated with G. Prove the following *integration by parts* formula for Lebesgue–Stieltjes integrals:

$$\int_a^b fG\, dm = F(b)G(b) - \int_a^b F\, d\mu_G.$$

Can the assumptions about f and g be relaxed?

5 Spaces of Measures and the Radon–Nikodym Theorem

5.1 Signed and complex measures, and the basic decomposition theorems

Let $(X, \mathcal{A})$ be a measurable space. A function

$$\mu : \mathcal{A} \to \mathbb{R}^*, \text{ respectively, } \mathbb{C},$$

is a *signed measure*, respectively, *complex measure*, if $\mu(\emptyset) = 0$ and

$$\mu\left(\bigcup_{n=1}^{\infty} A_n\right) = \sum_{n=1}^{\infty} \mu(A_n)$$

for every disjoint sequence $\{A_n : n = 1, \ldots\} \subseteq \mathcal{A}$. If μ is a signed, respectively, complex, measure we shall refer to the symbol $(X, \mathcal{A}, \mu)$ as a *signed*, respectively, *complex, measure space*. Note that every measure on $(X, \mathcal{A})$ is a signed measure.

Proposition 5.1.1. *Let $(X, \mathcal{A}, \mu)$ be a signed measure space. Then μ cannot take both $+\infty$ and $-\infty$ as values.*

Remark. If $(X, \mathcal{A}, \mu)$ is a measure space and $f \in L^1_\mu(X)$ then

$$\forall\, A \in \mathcal{A}, \quad \rho(A) = \int_A f \, d\mu$$

defines a complex measure ρ, and this canonical representation is one of the initial reasons we study signed and complex measures. The *Radon–Nikodym theorem*, which we designate by R–N, is the means to obtain a converse of this observation as well as being a major step in determining an FTC for abstract measure spaces. In fact we shall characterize "absolutely continuous measures" in terms of the above formula.

Definition 5.1.2. μ^+, μ^-, and the total variation of signed measures Let $(X, \mathcal{A}, \mu)$ be a signed measure space. A set $A \in \mathcal{A}$ is *nonnegative*, respectively, *nonpositive* if

$$\forall\, E \subseteq A, \text{ for which } E \in \mathcal{A}, \text{ we have } \mu(E) \geq 0, \text{ respectively, } \mu(E) \leq 0.$$

J.J. Benedetto and W. Czaja, *Integration and Modern Analysis*, Birkhäuser Advanced Texts, DOI 10.1007/978-0-8176-4656-1_5,

For each $A \in \mathcal{A}$ we define

$$\mu^+(A) = \sup\{\mu(E) : E \subseteq A, E \in \mathcal{A}\}$$

and

$$\mu^-(A) = \sup\{-\mu(E) : E \subseteq A, E \in \mathcal{A}\}.$$

Here μ^+, respectively, μ^-, is the *positive*, respectively, *negative*, *variation* of μ.

For any signed measure space $(X, \mathcal{A}, \mu)$, the *total variation* $|\mu|$ is defined as

$$\forall\, A \in \mathcal{A}, \quad |\mu|(A) = \mu^+(A) + \mu^-(A).$$

Theorem 5.1.3. $\mu^\pm$ are measures
Let $(X, \mathcal{A}, \mu)$ be a signed measure space. Then, $(X, \mathcal{A}, \mu^+)$ and $(X, \mathcal{A}, \mu^-)$ are measure spaces.

Proof. Clearly, $\mu^\pm(\emptyset) = 0$ since $\mu(\emptyset) = 0$. For each $A \in \mathcal{A}$, $\emptyset \subseteq A$, and so $\mu^\pm(A) \geq 0$. Let $\{A_n : n = 1, \ldots\} \subseteq \mathcal{A}$ be a disjoint sequence. We shall prove that $\mu^+\left(\bigcup A_n\right) = \sum \mu^+(A_n)$; the same argument works for μ^-.

We need to know that if $A \subseteq B$, where $A, B \in \mathcal{A}$, then

$$\mu^+(A) \leq \mu^+(B).$$

Take $E \subseteq A$, where $E \in \mathcal{A}$. Then $\mu(E) \leq \mu^+(B)$ by the definition of $\mu^+(B)$ and since $A \subseteq B$. Because this holds for all such E we conclude that $\mu^+(A) \leq \mu^+(B)$.

If $\mu^+(A_n) = \infty$ for some n then $\mu^+(A_n) \leq \mu^+\left(\bigcup A_j\right)$ implies that $\mu^+\left(\bigcup A_j\right) = \infty$. Consequently, $\mu^+ \geq 0$ and $\mu^+(A_n) = \infty$ yield the σ-additivity of μ^+. Therefore, assume that $\mu^+(A_n) < \infty$ for each n.

Let $\varepsilon > 0$ and choose $E_n \subseteq A_n$, where $E_n \in \mathcal{A}$, such that $\mu(E_n) \geq \mu^+(A_n) - \varepsilon/2^n$. Then $\{E_n : n = 1, \ldots\}$ is a disjoint sequence since $\{A_n : n = 1, \ldots\}$ is disjoint. Thus, by the σ-additivity of μ,

$$\mu\left(\bigcup_{n=1}^{\infty} E_n\right) = \sum_{n=1}^{\infty} \mu(E_n) \geq \sum_{n=1}^{\infty} \mu^+(A_n) - \varepsilon;$$

and so

$$\mu^+\left(\bigcup_{n=1}^{\infty} A_n\right) \geq \sum_{n=1}^{\infty} \mu^+(A_n),$$

since $\mu^+(\bigcup A_n) \geq \mu(\bigcup E_n)$ and ε is arbitrary.

For the opposite inequality take any $\lambda < \mu^+\left(\bigcup A_n\right)$. It is sufficient to prove that $\lambda < \sum \mu^+(A_n)$.

Pick $E \subseteq \bigcup A_n$ for which $\mu(E) > \lambda$. Since μ is σ-additive,

$$\mu(E) = \sum_{n=1}^{\infty} \mu(E \cap A_n).$$

Because $\mu(E \cap A_n) \leq \mu^+(A_n)$ we conclude that

$$\lambda < \mu(E) \leq \sum_{n=1}^{\infty} \mu^+(A_n).$$ □

Theorem 5.1.4. Properties of $\mu^\pm$ and $|\mu|$
Let $(X, \mathcal{A}, \mu)$ be a signed measure space.

a. *$(X, \mathcal{A}, |\mu|)$ is a measure space.*

b. *The nonnegative, respectively, nonpositive, sets are closed under countable unions.*

c. *If $A \in \mathcal{A}$ is nonnegative, respectively, nonpositive, then $\mu(A) = \mu^+(A)$, respectively, $\mu(A) = -\mu^-(A)$.*

d. *If $\mu(A) = \mu^+(A) < \infty$, respectively, $-\mu(A) = \mu^-(A) < \infty$, then A is nonnegative, respectively, nonpositive.*

Proof. Part *a* is clear from Theorem 5.1.3, and part *b* is immediate.

c. Let A be nonnegative, so that $\mu(B) \geq 0$ for each $B \in \mathcal{A}$ contained in A. For such B we have $\mu(B) \leq \mu(A)$ since $A \setminus B \subseteq A$, and so $\mu(A) = \mu(B)+\mu(A\setminus B) \geq \mu(B)$. Consequently, by the definition of μ^+, $\mu^+(A) \leq \mu(A)$. Because A is such a B, $\mu^+(A) = \mu(A)$.

d. If A is not nonnegative there is $B \subseteq A$ for which $B \in \mathcal{A}$ such that $\mu(B) < 0$. Because μ is additive and $A = (A\setminus B)\cup B$, $\mu(A) = \mu(B)+\mu(A\setminus B)$. Thus, if $\mu(A) < \infty$ (our hypothesis) we have $\mu(A) < \mu(A\setminus B)$ since $\mu(B) < 0$. On the other hand, $A \setminus B \subseteq A$ implies $\mu^+(A) \geq \mu(A \setminus B)$, and so $\mu^+(A) > \mu(A)$, which is contrary to our hypothesis. □

Proposition 5.1.5. *Let $(X, \mathcal{A}, \mu)$ be a signed measure space. For each $A \in \mathcal{A}$ there is $E \subseteq A$ for which $E \in \mathcal{A}$ such that*

$$\mu(E) = \mu^+(E) \geq \mu(A).$$

Proof. i. For the first case let $\mu^+(A) < \infty$ (and so $\mu(A) < \infty$). We shall inductively define a sequence $\{A_n : n = 1, \ldots\} \subseteq \mathcal{A}$. Let $A_0 = A$ and note that

$$\mu^+(A_0) \geq \mu(A_0) \geq \mu(A).$$

If $\mu^+(A_0) = \mu(A_0)$ take $E = A_0$ and we are done. If $\mu^+(A_0) \neq \mu(A_0)$ then there is $A_1 \subseteq A_0$ such that

$$\mu^+(A_0) \geq \mu(A_1) > \mu(A_0). \tag{5.1}$$

Further, take A_1 such that $\mu(A_1) > \mu^+(A_0) - 1$; we can do this by the definition of μ^+ and since $\mu^+(A_0) < \infty$. Since $\mu^+(A_1) \geq \mu(A_1)$,

$$\mu^+(A_1) \geq \mu(A_1) > \mu(A_0) \tag{5.2}$$

by (5.1). If $\mu^+(A_1) = \mu(A_1)$ take $E = A_1$ and we are done by (5.2). If $\mu^+(A_1) > \mu(A_1)$, there is $A_2 \subseteq A_1$ such that

$$\mu^+(A_2) \geq \mu(A_2) > \mu(A_1), \quad \mu(A_2) > \mu^+(A_1) - 1,$$

and again, since $\mu^+(A_2) \geq \mu(A_2)$,

$$\mu^+(A_2) \geq \mu(A_2) > \mu(A_1) > \mu(A_0).$$

Proceeding in this way we are either done at the nth step or else we determine a decreasing sequence $\{A_n : n = 1, \ldots\} \subseteq \mathcal{A}$ such that

$$\mu^+(A_n) \geq \mu(A_n) > \mu(A) \tag{5.3}$$

and

$$\mu(A_n) > \mu^+(A_{n-1}) - \frac{1}{n-1}. \tag{5.4}$$

Set $E = \bigcap A_n$. Since μ^+ is a measure and $\mu^+(A_0) < \infty$ we have

$$\mu^+(E) = \lim_{n\to\infty} \mu^+(A_n) \geq \mu(A).$$

Thus, since $\mu^+(E) \geq \mu(E)$ by definition, it remains to show that $\mu(E) \geq \mu^+(E)$.

If E is nonnegative we are done by Theorem 5.1.4*c*. Assume not and take $F \subseteq E$ satisfying $\mu(F) < 0$. We shall obtain a contradiction. Choose n for which $\mu(F) < -1/n$. Note that

$$\mu(A_{n+1} \setminus F) = \mu(A_{n+1}) - \mu(F) > \mu(A_{n+1}) + 1/n. \tag{5.5}$$

To prove this observe that $A_{n+1} = (A_{n+1} \setminus F) \cup F$, $\mu(A_{n+1}) = \mu(F) + \mu(A_{n+1} \setminus F)$, and $\mu(A_{n+1}) \in \mathbb{R}$ by (5.3) and the fact that $\mu(A_n) \leq \mu^+(A_{n+1}) \leq \mu^+(A) < \infty$. Consequently, $\mu(F) \in \mathbb{R}$ and we can perform the algebra to obtain (5.5).

Next we observe from the monotonicity of $\{A_n : n = 1, \ldots\}$ that

$$\mu^+(A_n) \geq \mu^+(A_{n+1}) \geq \mu^+(A_{n+1} \setminus F) \geq \mu(A_{n+1} \setminus F).$$

Thus, by (5.5), $\mu^+(A_n) > \mu(A_{n+1}) + 1/n$, and this contradicts (5.4). Consequently, the result is true if $\mu^+(A) < \infty$.

ii. For the second case let $\mu^+(A) = \infty$. As in the first case we form $\{A_n : n = 1, \ldots\} \subseteq \mathcal{A}$, a decreasing sequence, such that $A_n \subseteq A$ and

$$\begin{aligned} &\mu^+(A_n) > \mu(A_{n+1}) \geq \mu(A_n) \geq \mu(A), \\ &\mu(A_{n+1}) \geq n + 1. \end{aligned} \tag{5.6}$$

Further, $\mu^+(A_n) = \infty$, for otherwise we can argue as in the first part. Again, set $E = \bigcap A_n$. For each n,

$$A_n = E \cup \bigcup_{j=n}^{\infty} (A_j \setminus A_{j+1}),$$

so that, from (5.6),

$$n \leq \mu(A_n) = \mu(E) + \sum_{j=n}^{\infty} \mu(A_j \setminus A_{j+1}) < \infty. \tag{5.7}$$

Thus, $\mu(E)$ is a fixed finite number, and, since $\mu(A_n) \to \infty$, we have

$$\lim_{n\to\infty} \sum_{j=n}^{\infty} \mu(A_j \setminus A_{j+1}) = \infty.$$

On the other hand, $\mu(A_j \setminus A_{j+1}) \leq 0$ by (5.6) and the fact that $\mu(A_j) = \mu(A_j \setminus A_{j+1}) + \mu(A_{j+1})$, where $\mu(A_j)$ is finite. This gives a contradiction. Hence, $\mu^+(A_n) < \infty$ for some n and we apply the first part. □

We shall use Proposition 5.1.5 to prove the *Hahn decomposition theorem.* There are several proofs, including one in which it is a corollary of the R–N theorem. HANS HAHN's proof of 1928 [211] and SIERPIŃSKI's [437] are quite efficient and do not use R–N. R. FRANCK [182] has given a proof using transfinite numbers. HAHN's original proof was given in 1921.

Theorem 5.1.6. Hahn decomposition theorem
Let $(X, \mathcal{A}, \mu)$ be a signed measure space. There is a nonnegative set $P \in \mathcal{A}$ such that $N = X \setminus P$ is nonpositive.

Proof. Assume $\mu^+(X) < \infty$. The case that $\mu^+(X) = \infty$ is Problem 5.2. Set $\lambda = \sup\{\mu^+(A) : A \in \mathcal{A} \text{ and } \mu^-(A) = 0\}$. The condition that $\mu^-(A) = 0$ means that if $E \subseteq A$ then $\mu(E) \geq 0$, and so A is nonnegative. There is no problem about the existence of λ since $\mu^-(\emptyset) = 0$.

Since μ^+ is a measure and $\mu^+(X) < \infty$ we have $0 \leq \lambda < \infty$. Pick a sequence $\{A_n : n = 1, \ldots\} \subseteq \mathcal{A}$ such that $\mu^-(A_n) = 0$ and $\mu^+(A_n) \to \lambda$. Define $P = \bigcup A_n$.

Note that $\mu^-(P) = 0$ since $\mu^-(P) \leq \sum \mu^-(A_n)$. Hence, P is a nonnegative set. Because $A_n \subseteq P$ we have $\mu^+(P) \geq \mu^+(A_n)$, and so $\mu^+(P) \geq \lambda$. On the other hand, $P \in \mathcal{A}$ and $\mu^-(P) = 0$ imply that $\lambda \geq \mu^+(P)$. Thus, $\lambda = \mu^+(P)$.

Our final step is to prove that $N = X \setminus P$ is nonpositive. Assume not, so that there is $B \subseteq N$ for which $\mu(B) > 0$. By Proposition 5.1.5 there is $E \subseteq B$ such that $\mu^+(E) = \mu(E) \geq \mu(B) > 0$. Consequently, since $\mu^+(X) < \infty$, we can apply Theorem 5.1.4*d* to ascertain that E is nonnegative.

Because $E \cap P = \emptyset$,

$$\mu^+(P \cup E) = \mu^+(P) + \mu^+(E) = \lambda + \mu^+(E) > \lambda.$$

Also, from the fact that $\mu^-(P) = 0$ (as we observed above),

$$\mu^-(P \cup E) = \mu^-(P) + \mu^-(E) = \mu^-(E) = 0,$$

where the last equality follows since E is nonnegative; in fact, $F \subseteq E$ implies $\mu(F) \geq 0$ and

$$0 \leq \mu^-(E) = \sup\{-\mu(F) : F \subseteq E\} \leq 0.$$

Thus, we have a contradiction to the definition of λ by taking $A = P \cup E$. Hence, N is nonpositive. □

The sets P and N are a *Hahn decomposition* of the signed measure space $(X, \mathcal{A}, \mu)$.

There is a difference between the notions of a set of signed measure 0 and of a set of measure 0. The following remark introduces a generalization of the latter in the context of signed measure spaces.

Remark. Let $(X, \mathcal{A}, \mu)$ be a signed measure space. If $A \in \mathcal{A}$ satisfies

$$-\mu^-(A) = \mu(A) = \mu^+(A)$$

then A is a *null set.* A set $A \in \mathcal{A}$ is null if and only if

$$\forall\, B \subseteq A,\ B \in \mathcal{A}, \quad \mu(B) = 0.$$

Thus, if A is null then $\mu(A) = 0$, but the converse is not true. It is a routine exercise to prove that the Hahn decomposition is not generally unique, whereas it is unique up to null sets; e.g., Problem 5.6.

Remark. We shall give the *Jordan decomposition* of a signed measure space $(X, \mathcal{A}, \mu)$ in Theorem 5.1.8. There are two points to be made.

First, the Jordan decomposition is intimately related to the decomposition of a bounded variation function as a difference of increasing functions. Recall that increasing functions give rise to measures, and, in fact, bounded variation functions give rise to signed measures. The Riesz representation theorem (see Chapter 7) gives a further relation between functions of bounded variation and signed measures.

The second point is the relation between the Hahn and Jordan decompositions. Given a Hahn decomposition P, N of a signed measure space $(X, \mathcal{A}, \mu)$, it is easy to compute that

$$\mu = \nu^+ - \nu^-,$$

where

$$\forall\, A \in \mathcal{A}, \quad \nu^+(A) = \mu(A \cap P), \quad \text{and} \quad \nu^-(A) = -\mu(A \cap N).$$

On the other hand, we shall prove the Jordan decomposition,

$$\mu = \mu^+ - \mu^-,$$

independent of HAHN's result, and then show that $\mu^\pm = \nu^\pm$.

Proposition 5.1.7. *Let $(X, \mathcal{A}, \mu)$ be a signed measure space.*

a. *If $\mu^+(X) = \infty$ then there is $A \in \mathcal{A}$ such that $\mu(A) = \infty$.*

b. *If $\mu^-(X) = \infty$ then there is $B \in \mathcal{A}$ such that $\mu(B) = -\infty$.*

Proof. By the definition of μ^+ we can choose a sequence $\{A_n : n = 1, \ldots\} \subseteq \mathcal{A}$ such that $\mu(A_n) > n$ for each n. From Proposition 5.1.5 choose $E_n \subseteq A_n$ for which $\mu^+(E_n) = \mu(E_n) \geq \mu(A_n) > n$. We shall prove that

$$\mu\left(\bigcup_{n=1}^{\infty} E_n\right) = \mu^+\left(\bigcup_{n=1}^{\infty} E_n\right).$$

This is sufficient since $\mu^+(A) \geq \mu^+(E_n) > n$ for $A = \bigcup E_n$, and so $\mu^+(A) = \infty$.

Without loss of generality take $n < \mu(E_n) = \mu^+(E_n) < \infty$ for each n. From Theorem 5.1.4*d*, E_n is nonnegative. Consequently, if $F \subseteq \bigcup E_n$ and $\bigcup E_n = \bigcup D_n$, where $\{D_n : n = 1, \ldots\}$ is a disjoint sequence and $D_n \subseteq E_n$, we have $\mu(F \cap D_n) \geq 0$ and

$$\mu(F) = \sum_{n=1}^{\infty} \mu(F \cap D_n) \geq 0.$$

Thus, $\bigcup E_n$ is nonnegative. Finally, we apply Theorem 5.1.4*c* to conclude that $\mu(A) = \mu^+(A)$. □

We could have proved Proposition 5.1.7 using the Hahn decomposition; cf. Problem 5.2: $\mu^+(X) = \mu^+(P) + \mu^+(N) = \mu^+(P) = \mu(P)$.

Theorem 5.1.8. Jordan decomposition theorem
Let $(X, \mathcal{A}, \mu)$ be a signed measure space. Then $\mu = \mu^+ - \mu^-$. In particular, either μ^+ or μ^- is bounded.

Proof. i. Let $A \in \mathcal{A}$ have the property that $|\mu(A)| < \infty$. Take $B \subseteq A$. Then

$$\mu(A) = \mu(B) + \mu(A \setminus B), \tag{5.8}$$

and so $|\mu(B)|, |\mu(A \setminus B)| < \infty$. Therefore,

$$\mu(B) = \mu(A) - \mu(A \setminus B) \leq \mu(A) + \mu^-(A), \tag{5.9}$$

and, consequently,

$$\mu^+(A) \leq \mu(A) + \mu^-(A)$$

since (5.9) is true for all $B \subseteq A$.

To prove that

$$\mu(A) \geq \mu^+(A) - \mu^-(A) \tag{5.10}$$

we must show that $\mu^-(A) < \infty$ when $|\mu(A)| < \infty$. If $\mu^-(A) = \infty$, then from Proposition 5.1.7 there is $E \subseteq A$ for which $\mu(E) = -\infty$. Choose B above satisfying $E = A \setminus B$. Because $|\mu(A)| < \infty$ we conclude that $|\mu(B)|, |\mu(E)| < \infty$ from (5.8). Thus, we have a contradiction and (5.10) follows.

Using (5.8) again we have $-\mu(A \setminus B) = \mu(B) - \mu(A)$ and therefore

$$\mu^-(A) \leq \mu^+(A) - \mu(A).$$

This fact combined with (5.10) yields $\mu(A) = \mu^+(A) - \mu^-(A)$ for the case that $|\mu(A)| < \infty$.

If $\mu(A) = -\infty$, then $\mu^-(A) = \infty$, and so we must prove that $|\mu^+(A)| < \infty$. If $\mu^+(A) = \infty$ then there is $B \subseteq A$ for which $\mu(B) = \infty$ by Proposition 5.1.7, and this contradicts Proposition 5.1.1 since we assigned $\mu(A) = -\infty$.

The case $\mu(A) = \infty$ is treated similarly.

ii. We shall now show that both μ^+ and μ^- cannot be unbounded. If this were the case we would have a sequence $\{A_n : n = 1, \ldots\} \subseteq \mathcal{A}$ such that $\lim_{n\to\infty} \mu^+(A_n) = \infty$. Let $E_n = A_n \cap P$, $n = 1, \ldots$. Thus,

$$\begin{aligned}\mu^+(A_n) &= \mu^+((A_n \cap P) \cup (A_n \cap N)) = \mu^+(A_n \cap P) + \mu^+(A_n \cap N) \\ &= \mu^+(A_n \cap P) = \mu^+(E_n).\end{aligned}$$

Let $G_n = \bigcup_{j=1}^n E_j$. Because $E_n \subseteq G_n$, we have $\mu^+(G_n) \geq \mu^+(E_n)$, and so

$$\lim_{n\to\infty} \mu^+(G_n) = \infty.$$

This, in turn, implies that

$$\mu^+\left(\bigcup_{n=1}^{\infty} G_n\right) = \infty,$$

and so we can use Proposition 5.1.7 to deduce that there exists $G \in \mathcal{A}$ such that $\mu^+(G) = \infty$.

Analogously, we show that if μ^- is unbounded then there exists a set $H \in \mathcal{A}$ such that $\mu^-(H) = -\infty$.

These two observations yield a contradiction in view of Proposition 5.1.1.

□

The formula $\mu = \mu^+ - \mu^-$ is the *Jordan decomposition* of μ.

Theorem 5.1.9. Consequences of Hahn and Jordan decompositions *Let $(X, \mathcal{A}, \mu)$ be a signed measure space.*

a. *If P, N is a Hahn decomposition of X, then*

$$\forall\, A \in \mathcal{A}, \quad \mu^+(A) = \mu(A \cap P) \quad \textit{and} \quad \mu^-(A) = -\mu(A \setminus P) = -\mu(A \cap N).$$

b. *If $\mu = \mu_1 - \mu_2$, where the μ_j are measures, then $\mu_1 \geq \mu^+$ and $\mu_2 \geq \mu^-$.*

c. *For each $A \in \mathcal{A}$, denote*

$$\sup\left\{\sum_{k=1}^{n}|\mu(A_k)| : \{A_1,\dots,A_n\}\subseteq \mathcal{A} \text{ is a finite decomposition of } A\right\}$$

by μ_A. *Then*

$$\mu_A = |\mu|(A).$$

(A set $\{A_1,\dots,A_n\}\subseteq\mathcal{A}$ *is a finite decomposition of* A *if* $A=\bigcup_{j=1}^n A_j$ *and* $A_j\cap A_k=\emptyset$ *if* $j\neq k$.*)*

d. *If for each* $A\in\mathcal{A}$, $|\mu(A)|<\infty$, *then* $K_\mu=\max\{\mu^+(X),\mu^-(X)\}<\infty$ *and*

$$\forall\, A\in\mathcal{A},\quad |\mu(A)|\leq K_\mu.$$

Proof. **a.** From the Jordan decomposition of μ,

$$\forall\, A\in\mathcal{A},\quad \mu(A\cap P)=\mu^+(A\cap P)-\mu^-(A\cap P).$$

Since P is nonnegative, $A\cap P$ is nonnegative, and so $\mu(A\cap P)=\mu^+(A\cap P)\geq 0$ (and hence $\mu^-(A\cap P)=0$). Also,

$$0\leq\mu^+(A\setminus P)\leq\mu^+(X\setminus P)=\mu^+(N)=0$$

from the fact that N is nonpositive. Thus,

$$\mu^+(A)=\mu^+(A\cap P)+\mu^+(A\setminus P)=\mu^+(A\cap P)=\mu(A\cap P).$$

b. If $\mu=\mu_1-\mu_2$ then $\mu_1\geq\mu$, and so

$$\forall\, A\in\mathcal{A},\quad \mu^+(A)=\mu(A\cap P)\leq\mu_1(A\cap P)\leq\mu_1(A).$$

Hence, $\mu_1\geq\mu^+$.

If $|\mu(A)|<\infty$ then $\mu^-(A)=\mu^+(A)-\mu(A)\leq\mu_1(A)-\mu(A)=\mu_2(A)$. The case $|\mu(A)|=\infty$ is Problem 5.5.

c. Let $A\in\mathcal{A}$ and let $\{A_1,\dots,A_n\}$ be a finite decomposition of A. Since $|\mu|$ is a measure and $\{A_j : j=1,\dots,n\}$ is disjoint,

$$\begin{aligned}\sum_{j=1}^n|\mu(A_j)| &= \sum_{j=1}^n|\mu^+(A_j)-\mu^-(A_j)| \leq \sum_{j=1}^n\left(\mu^+(A_j)+\mu^-(A_j)\right)\\ &= \sum_{j=1}^n|\mu|(A_j)=|\mu|(A).\end{aligned}$$

Thus, $|\mu|(A)\geq\mu_A$.

For the opposite inequality take a Hahn decomposition P, N of X, so that

$$A=(A\cap P)\cup(A\cap N).$$

Then

$$\mu_A \geq |\mu(A \cap P)| + |\mu(A \cap N)| = \mu^+(A) + \mu^-(A) = |\mu|(A).$$

d. If $\mu^+(X) = \infty$ then we obtain a contradiction to our hypothesis, since, by Proposition 5.1.7, there is $A \in \mathcal{A}$ for which $\mu(A) = \infty$. Hence, $\mu^+(X) < \infty$.

For each $A \in \mathcal{A}$ we choose E as in Proposition 5.1.5 and therefore

$$\mu(A) \leq \mu(E) = \mu^+(E) \leq \mu^+(X) < \infty.$$

Using part a we compute for any $A \in \mathcal{A}$ that

$$\mu(A) = \mu^+(A) - \mu^-(A) = \mu(A \cap P) + \mu(A \cap N) \geq \mu(A \cap N) = -\mu^-(A).$$

Consequently,

$$\mu(A) \geq -\mu^-(A) \geq -\mu^-(X) > -\infty$$

by Proposition 5.1.7 again. Hence,

$$\forall\, A \in \mathcal{A}, \quad |\mu(A)| \leq \max\{\mu^+(X), \mu^-(X)\}.$$ □

We expand on the first remark in this section concerning the construction of signed measures, and R–N as a converse, in the following result.

Proposition 5.1.10. *Let* $(X, \mathcal{A}, \mu)$ *be a measure space, let*

$$f : X \to \mathbb{R}^*$$

be a μ*-measurable function, and assume that*

$$\int_X f \, d\mu$$

exists. Define

$$\forall\, A \in \mathcal{A}, \quad \nu(A) = \int_A f \, d\mu.$$

Then $(X, \mathcal{A}, \nu)$ *is a signed measure space,*

$$\nu^{\pm}(A) = \int_A f^{\pm} \, d\mu, \quad \textit{and} \quad |\nu|(A) = \int_A |f| \, d\mu. \tag{5.11}$$

(f^+ *and* f^- *are defined after Definition 3.2.3.)*

Also, $P = \{x : f(x) > 0\}$, $N = X \setminus P$ *is a Hahn decomposition for* ν*, and if neither* ν^+ *nor* ν^- *is identically zero then*

$$\exists\, A \in \mathcal{A} \quad \textit{such that} \quad |\nu|(A) > |\nu(A)|.$$

Let μ be a complex measure with real and imaginary parts μ_r and μ_i, respectively. Clearly, μ_r and μ_i are signed measures with values in $\mathbb{R}$. From Theorem 5.1.8, we have

$$\forall\, A \in \mathcal{A}, \quad \mu(A) = \mu_r^+(A) - \mu_r^-(A) + i\mu_i^+(A) - i\mu_i^-(A).$$

We would like to define the total variation of μ. It is tempting to write it as

$$\mu_r^+ + \mu_r^- + \mu_i^+ + \mu_i^-.$$

In fact, we shall be most interested in results that depend on certain decompositions of X; and so we take our cue from Theorem 5.1.9c and make the following definition.

Definition 5.1.11. The total variation of complex measures
Let $(X, \mathcal{A}, \mu)$ be a complex measure space. For each $A \in \mathcal{A}$ denote

$$\sup\left\{\sum_{j=1}^{n} |\mu(A_j)| : \{A_1, \ldots, A_n\} \text{ is a finite decomposition of A}\right\}$$

by $|\mu|(A)$. Then, $|\mu| : \mathcal{A} \to \mathbb{R}^+$ is the *total variation* $|\mu|$ of μ.

Remark. The boundedness of $|\mu|$ in Theorem 5.1.12b can also be proved quite independently of our measure-theoretic arguments using the following fact:

$$\forall\, z_1, \ldots, z_n \in \mathbb{C},\ \exists\, S \subseteq \{1, \ldots, n\} \quad \text{such that} \quad \left|\sum_{j \in S} z_j\right| \geq \frac{1}{\pi}\sum_{j=1}^{n} |z_j|.$$

BOURBAKI (1955) observed that the constant $1/\pi$ is the best possible but that it cannot be achieved [71], Chapitre VIII, § 3, Ex. 1, p. 126; cf. [62], [277]. There is an interesting refinement due to GRAHAME BENNETT; see Amer. Math. Monthly 79 (1972), 905 and 80 (1973), 1139–1141. We shall employ this sort of result in Theorem 6.2.1 (Schur's lemma).

Theorem 5.1.12. The total variation measure
Let $(X, \mathcal{A}, \mu)$ be a complex measure space.

a. $\forall\, A \in \mathcal{A},\ |\mu|(A) \leq \mu_r^+(A) + \mu_r^-(A) + \mu_i^+(A) + \mu_i^-(A)$.

b. *$(X, \mathcal{A}, |\mu|)$ is a finite measure space.*

c. $\sup\{|\mu(B)| : B \in \mathcal{A}, B \subseteq A \in \mathcal{A}\} \leq |\mu|(A) \leq 4\sup\{|\mu(B)| : B \in \mathcal{A}, B \subseteq A \in \mathcal{A}\}$.

Proof. ***a.*** Let $\{A_j : j = 1, \ldots, n\}$ be a finite decomposition of $A \in \mathcal{A}$. Then

$$\sum_{j=1}^{n} |\mu(A_j)| = \sum_{j=1}^{n} \left|\mu_r^+(A_j) - \mu_r^-(A_j) + i\mu_i^+(A_j) - i\mu_i^-(A_j)\right|$$

$$\leq \sum_{j=1}^{n} \left(\mu_r^+(A_j) + \mu_r^-(A_j) + \mu_i^+(A_j) + \mu_i^-(A_j)\right)$$
$$= \mu_r^+(A) + \mu_r^-(A) + \mu_i^+(A) + \mu_i^-(A),$$

and we are done by taking the sup on the left-hand side.

b. From Theorem 5.1.9*d*, for each $A \in \mathcal{A}$,

$$|\mu|(A) \leq \mu_r^+(A) + \mu_r^-(A) + \mu_i^+(A) + \mu_i^-(A)$$
$$\leq \mu_r^+(X) + \mu_r^-(X) + \mu_i^+(X) + \mu_i^-(X) \leq 2(K_{\mu_r} + K_{\mu_i}) < \infty,$$

and so $|\mu|$ is bounded.

Now we prove that $|\mu|$ is a measure. Clearly, $|\mu|(\emptyset) = 0$ by the definition of $|\mu|$. Let $\{A_n : n = 1, \ldots\} \subseteq \mathcal{A}$ be a disjoint family and set $A = \bigcup_{n=1}^{\infty} A_n$. Take $\alpha < |\mu|(A)$. Note that for any finite decomposition $\{B_j : j = 1, \ldots, k\} \subseteq \mathcal{A}$ of A that satisfies $\alpha < \sum_{j=1}^{k} |\mu(B_j)|$ (which we can do by the definition of $|\mu|$), we have

$$\alpha < \sum_{n=1}^{\infty} \sum_{j=1}^{k} |\mu(B_j \cap A_n)|. \tag{5.12}$$

Then, since $\sum_{j=1}^{k} |\mu(B_j \cap A_n)| \leq |\mu|(A_n)$ and $\alpha < |\mu|(A)$ is arbitrary,

$$|\mu|(A) \leq \sum_{n=1}^{\infty} |\mu|(A_n). \tag{5.13}$$

Take $\varepsilon > 0$. For each j let $\{B_{j,1}, \ldots, B_{j,n_j}\} \subseteq \mathcal{A}$ be a finite decomposition of A_j for which

$$\sum_{k=1}^{n_j} |\mu(B_{j,k})| > |\mu|(A_j) - \frac{\varepsilon}{2^j}.$$

For each m,

$$\sum_{j=1}^{m} |\mu|(A_j) < \sum_{j=1}^{m} \frac{\varepsilon}{2^j} + \sum_{j=1}^{m} \sum_{k=1}^{n_j} |\mu(B_{j,k})| < \varepsilon + |\mu|(A) \tag{5.14}$$

by the definition of $|\mu|$ and since $\bigcup\{B_{j,k} : k = 1, \ldots, n_j, j = 1, \ldots, m\} \subseteq A$ is a disjoint union. Consequently,

$$\sum_{j=1}^{\infty} |\mu|(A_j) < \varepsilon + |\mu|(A),$$

and so

$$\sum_{j=1}^{\infty} |\mu|(A_j) \leq |\mu|(A).$$

This, coupled with (5.13), yields the desired σ-additivity.

c. The first inequality is clear. For the second note that

$$|\mu_r^+(A)| = |\mu_r(A \cap P_r)| \leq |\mu(A \cap P_r)|,$$

and apply part *a.* □

Definition 5.1.13. $M_b(X)$
a. Let $(X, \mathcal{A})$ be a measurable space and let $M_b(X)$, respectively, $SM(X)$, be the set of complex, respectively, signed, measures. Then $M_b(X)$ is a complex vector space where

$$\forall\, \alpha, \beta \in \mathbb{C} \quad \text{and} \quad \forall\, A \in \mathcal{A}, \quad (\alpha\mu + \beta\nu)(A) = \alpha\mu(A) + \beta\mu(A);$$

and it is a normed vector space with norm

$$\|\mu\| = \|\mu\|_1 = |\mu|(X);$$

cf. Problem 5.22. We shall develop the theory for $M_b(X)$, noting that analogous results hold for $SM(X)$. We shall usually refer to the space "$M_b(X)$" without making special note of the corresponding measurable space $(X, \mathcal{A})$.

b. Let X be a locally compact Hausdorff space, and consider the measurable space $(X, \mathcal{B}(X))$. We say that a complex measure μ defined on $\mathcal{B}(X)$ is *regular* if $|\mu|$ is regular. In this case $M_b(X)$ will denote the space of *complex regular measures* defined on $\mathcal{B}(X)$, and we refer to $M_b(X)$ as the space of *complex regular Borel measures.* There is a similar convention used to define *signed regular Borel measures.*

The space $M_b(X)$ is a normed vector space with norm $\|\mu\| = \|\mu\|_1 = |\mu|(X)$. As a consequence of the results in Chapter 7, we shall see that $M_b(X)$ is a Banach space.

The following result is elementary to verify.

Theorem 5.1.14. Measures and linear functionals
Let $\mu \in M_b(X)$.

a. *A function $f : X \to \mathbb{C}$ defined $|\mu|$-a.e. is in $L^1_{|\mu|}(X)$ if and only if*

$$f \in L^1_{\mu_r^+}(X) \cap L^1_{\mu_r^-}(X) \cap L^1_{\mu_i^+}(X) \cap L^1_{\mu_i^-}(X).$$

b. *For each $f \in L^1_{|\mu|}(X)$,*

$$\mu(f) = \int_X f\, d\mu = \int_X f\, d\mu_r^+ - \int_X f\, d\mu_r^- + i\int_X f\, d\mu_i^+ - i\int_X f\, d\mu_i^-$$

is well defined, and $\mu : L^1_{|\mu|}(X) \to \mathbb{C}$ is a linear functional.

5.2 Discrete and continuous measures, absolutely continuous and singular measures

In this section we consider the problems for measures that were initiated for functions of bounded variation in Section 4.2.

Definition 5.2.1. Discrete, continuous, absolutely continuous, and singular measures
Let $(X, \mathcal{A})$ be a measurable space.

a. A complex measure $\mu \in M_b(X)$ is *continuous*, denoted by $\mu \in M_c(X)$, if

$$\forall\, x \in X, \quad \mu(\{x\}) = 0;$$

and $\mu \in M_b(X)$ is *discrete*, denoted by $\mu \in M_d(X)$, if

$$\mu = \sum_{x \in D} a_x \delta_x,$$

where $D \subseteq X$ is countable, $\sum_{x\in D} |a_x| < \infty$, and

$$\forall\, A \in \mathcal{A}, \quad \delta_x(A) = \begin{cases} 0, & \text{if } x \notin A, \\ 1, & \text{if } x \in A. \end{cases}$$

Clearly, a nonzero continuous, respectively, discrete, measure is not discrete, respectively, continuous.

b. We say that $\mu \in M_b(X)$ is *absolutely continuous* with respect to $\nu \in M_b(X)$, denoted by $\mu \ll \nu$, if

$$\forall\, A \in \mathcal{A}, \quad |\nu|(A) = 0 \Longrightarrow \mu(A) = 0.$$

Observe that $\nu \ll \mu$ in Proposition 5.1.10. Notationally we write

$$M_{ac}(X, \nu) = \{\mu \in M_b(X) : \mu \ll \nu\}.$$

c. To define a singular measure we say (as we did earlier after Example 4.6.6) that $\eta \in M_b(X)$ is *concentrated* on $A \in \mathcal{A}$ if

$$\forall\, B \in \mathcal{A}, \quad \eta(B) = \eta(A \cap B).$$

Concentration sets are not unique, but they are unique up to sets of measure 0.

d. A complex measure $\mu \in M_b(X)$ is *singular* with respect to $\nu \in M_b(X)$, denoted by $\mu \perp \nu$, if there are disjoint sets $C_\mu, C_\nu \in \mathcal{A}$ such that μ, respectively, ν, is concentrated on C_μ, respectively, C_ν. Obviously, $\mu \perp \nu$ if and only if $\nu \perp \mu$, and we say that μ and ν are *mutually singular*. Notationally, we write

$$M_s(X, \nu) = \{\mu \in M_b(X) : \mu \perp \nu\}.$$

Remark. Our project is to decompose a given μ in terms of the various types of measures defined above and to characterize all the absolutely continuous μ with respect to a fixed ν. The Radon–Nikodym theorem is the key result for this purpose.

Remark. Before proceeding in some generality let us recall the classical situation given in Chapter 4. Consider the measurable space $(\mathbb{R}, \mathcal{B}(\mathbb{R}))$, $F \in BV(\mathbb{R})$, and the corresponding complex measure $\mu \in M_b(\mathbb{R})$ as in Theorem 4.1.9. The space $M_b(\mathbb{R})$ is determined by $(\mathbb{R}, \mathcal{B}(\mathbb{R}))$. Then, the following assertions are valid:

i. F is a continuous function if and only if for each $x \in \mathbb{R}$, $\mu(\{x\}) = 0$ (as defined after Theorem 4.2.3).

ii. F is an absolutely continuous function if and only if for each $A \in \mathcal{A}$, for which $m(A) = 0$, we have $m(F(A)) = 0$ (Theorem 4.6.2).

iii. F is a singular function if and only if there is $A \subseteq \mathbb{R}$, for which $m(\mathbb{R} \setminus A) = 0$, such that $m(F(A)) = 0$ (Theorem 4.6.4).

iv. F is a discrete function if and only if there is $A \subseteq \mathbb{R}$ such that card $(\mathbb{R} \setminus A) \leq \aleph_0$ and $m(F(A)) = 0$ (as in the Remark after Proposition 4.2.1).

Clearly, $F(A)$ corresponds to C_μ in parts *iii* and *iv*. Note that $m(\mathbb{R} \setminus C_m) = 0$. Also,

$$F \text{ is absolutely continuous} \implies F \text{ is continuous},$$

and

$$F \text{ is discrete} \implies F \text{ is singular}.$$

The following was proved for functions of bounded variation in Theorem 4.2.3.

Theorem 5.2.2. Continuous and discrete decomposition of $M_b(X)$ *Let $\mu \in M_b(X)$. There is a unique decomposition,*

$$\mu = \mu_c + \mu_d,$$

where $\mu_c \in M_c(X)$ and $\mu_d \in M_d(X)$.

Proof. Because of the Jordan decomposition theorem we take $\mu \geq 0$ without loss of generality. Also, $\mu(X) < \infty$ since $\mu \in M_b(X)$.

For each $\varepsilon > 0$ there are at most finitely many points $x \in X$ for which $\mu(\{x\}) > \varepsilon$. Let Y_m be the finite collection of such points corresponding to $\varepsilon_m = 1/m$. Set $Y = \bigcup Y_m = \{x_1, \ldots\}$. For $A \in \mathcal{A}$ define

$$\mu_d(A) = \sum_{j=1}^{\infty} \mu(\{x_j\}) \delta_{x_j}(A),$$

noting that $\mu(\{x_j\}) > 0$. Clearly,

$$\mu_d(A) \leq \sum_{j=1}^{\infty} \mu(\{x_j\}) \leq \mu(X) < \infty,$$

and hence $\mu_d \in M_d(X)$, where we have used the fact that $\mu(X) < \infty$.

Define $\mu_c = \mu - \mu_d$. If $x \notin Y$ then $\mu(\{x\}) = 0 = \mu_d(\{x\})$ by the definition of Y. If $x \in Y$ we choose $A = \{x\}$ and compute $\mu_d(A) = \mu(\{x\})$ by the definition of μ_d. In either case $\mu_c(\{x\}) = 0$, and so $\mu_c \in M_c(X)$.

The uniqueness is elementary to verify. □

Example 5.2.3. Continuous measures that are not absolutely continuous with respect to Lebesgue measure

a. There are continuous measures μ such that $\mu \not\ll m^2$, where m^2 is Lebesgue measure on $X = [0,1]\times[0,1]$. In fact, for each Lebesgue measurable set $A \subseteq X$ define

$$\mu(A) = m^2(A \cap ([0,1] \times \{0\})).$$

b. Let μ_C be the Cantor–Lebesgue continuous measure on $[0,1]$, e.g., Section 4.2. Then $m(C) = 0$ and $\mu_C(C) = 1$, so that μ_C is continuous (by its definition) but not absolutely continuous with respect to m; cf. Example 4.5.1.

Proposition 5.2.4. *Let $(X, \mathcal{A}, \nu)$ be a measure space and let $\mu \in M_b(X)$. If $\mu \ll \nu$ and $\mu \perp \nu$, then $\mu = 0$.*

Proof. Since $\mu \perp \nu$ we have $\nu(C_\mu) = 0$.

The relation $\mu \ll \nu$ implies $\mu(A) = 0$ for $A \in \mathcal{A}$ and $A \subseteq C_\mu$ because $\nu(C_\mu) = 0$. On the other hand, $\mu(A) = 0$ if $A \cap C_\mu = \emptyset$ by the definition of C_μ. □

Proposition 5.2.5. *Let $\mu \in SM(X)$. Then $\mu^+ \perp \mu^-$, and if P, N is a Hahn decomposition for μ then we can take $C_{\mu^+} = P$ and $C_{\mu^-} = N$.*

Proof. From Theorem 5.1.9*a*, $\mu^+(A) = \mu(A \cap P)$ and $\mu^-(A) = -\mu(A \cap N)$. On the other hand, since P is nonnegative, Theorem 5.1.4*c* implies that $\mu^+(A \cap P) = \mu(A \cap P)$.

A similar argument works for μ^- and we are done. □

We now prove the *Lebesgue decomposition theorem.* It can also be proved alongside R–N or as a corollary of R–N; cf. (5.21).

Theorem 5.2.6. Lebesgue decomposition theorem

Let $(X, \mathcal{A}, \nu)$ be a measure space, and assume that μ satisfies one of the following conditions:

i. $(X, \mathcal{A}, \mu)$ is a σ-finite measure space,

ii. $\mu \in M_b(X)$.

Then there is a unique pair, μ_1 and μ_2, of σ-finite measures in the case of condition i or elements in $M_b(X)$ in the case of condition ii such that

$$\mu = \mu_1 + \mu_2, \quad \mu_1 \perp \nu, \quad \textit{and} \quad \mu_2 \ll \nu.$$

Proof. We shall prove the result for μ a bounded measure and refer to Problem 5.18 for the remaining cases.

Assume that the result is false. Thus, suppose that whenever μ_1 and μ_2 are bounded measures for which $\mu = \mu_1 + \mu_2$ and $\mu_1 \perp \nu$, then there is $A \in \mathcal{A}$ for which $\mu_2(A) > 0$ and $\nu(A) = 0$.

We shall first prove that

$$\forall\, B \in \mathcal{A} \text{ such that } \nu(X \setminus B) = 0,\ \exists\, F \subseteq B, \text{ where } F \in \mathcal{A}, \text{ for which}$$
$$\nu(F) = 0 \quad \text{and} \quad \mu(F) > 0. \tag{5.15}$$

In order to do this we first define the measures

$$\forall\, E \in \mathcal{A}, \quad \mu_B(E) = \mu(E \cap B),$$

and

$$\forall\, E \in \mathcal{A}, \quad \mu_{X\setminus B}(E) = \mu(E \cap (X \setminus B)).$$

Then $\mu = \mu_B + \mu_{X\setminus B}$, so that by hypothesis there is $A \in \mathcal{A}$ for which $\nu(A) = 0$ and $\mu_B(A) > 0$. Consequently, setting $F = A \cap B$, we have $\nu(A) = 0$ and we compute

$$\mu(F) = \mu_B(A \cap B) + \mu_{X\setminus B}(A \cap B) = \mu_B(A \cap B) = \mu(A \cap B) = \mu_B(A).$$

Let $\mathcal{G} = \{F \in \mathcal{A} : \mu(F) > 0 \text{ and } \nu(F) = 0\}$. Thus $\mathcal{G} \neq \emptyset$ for we can take $B = X$. Define $S = \{\mu(F) : F \in \mathcal{G}\} \subseteq \mathbb{R}$. Clearly S is bounded below by 0 and above by $\mu(X) < \infty$. It is easy to see that $\sup\{r \in S\} = \mu(F)$ for some $F \in \mathcal{G}$. In fact, let $\mu(F_n) \to \sup\{r \in S\}$ and set $F = \bigcup F_n$. Then

$$0 \leq \nu(F) \leq \sum_{n=1}^{\infty} \nu(F_n) = 0 \quad \text{and} \quad \mu(F) \geq \mu(F_n) \Longrightarrow \mu(F) \geq \sup\{r \in S\},$$

and so (5.15) is obtained.

Fix this F. Then $\nu((X \setminus (X \setminus F)) = 0$ since $\nu(F) = 0$, so that, by the above reasoning (for "$B = X \setminus F$"), there is $D \subseteq X \setminus F$ satisfying $\nu(D) = 0$ and $\mu(D) > 0$. Thus, $F \cup D \in \mathcal{G}$ and $\mu(F \cup D) = \mu(F) + \mu(D) > \mu(F)$ since $D \subseteq X \setminus F$. This contradicts the fact that $\mu(F) = \sup\{r \in S\}$, and hence the theorem is proved. □

The following result tells us that the absolutely continuous measures are "continuous" in a reasonable sense. Theorem 5.2.8 generalizes the fact that "indefinite integrals", as in Proposition 5.1.10, are absolutely continuous, as well as giving a converse to Theorem 5.2.7. In this regard, we also refer to Example 5.2.11.

Theorem 5.2.7. Classical continuity of absolutely continuous measures

Let $(X, \mathcal{A}, \nu)$ be a measure space and let $\mu \in M_b(X)$. If $\mu \ll \nu$, then

$$\forall\, \varepsilon > 0,\ \exists\, \delta > 0 \text{ such that } |\nu(A)| < \delta \Longrightarrow |\mu(A)| < \varepsilon.$$

Proof. Let $\mu \geq 0$ and assume that the result does not hold. Then there is $\varepsilon > 0$ such that for all n we can find $A_n \in \mathcal{A}$ for which

$$|\nu(A_n)| < \frac{1}{2^n} \quad \text{and} \quad |\mu(A_n)| = \mu(A_n) \geq \varepsilon.$$

Setting $B_n = \bigcup_{k=n+1}^{\infty} A_k$ we have

$$\nu(B_n) \leq \sum_{k=n+1}^{\infty} \nu(A_k) < \frac{1}{2^n}.$$

Since $\mu \geq 0$, $\mu(B_n) \geq \mu(A_{n+1}) \geq \varepsilon$. Consequently,

$$\mu\left(\bigcap_{n=1}^{\infty} B_n\right) = \lim_{n\to\infty} \mu(B_n) \geq \varepsilon,$$

noting that $\mu(B_1) < \infty$.

Also, $\nu(B_1) < \infty$, and so

$$\nu\left(\bigcap_{n=1}^{\infty} B_n\right) = \lim_{n\to\infty} \nu(B_n) \leq \lim_{n\to\infty} \frac{1}{2^n} = 0.$$

This last statement and the assumed absolute continuity contradict the fact that $\mu\left(\bigcap_{n=1}^{\infty} B_n\right) \geq \varepsilon$.

According to Theorem 5.1.12*a*, $|\mu| \leq |\mu_r| + |\mu_i|$. Now, if μ is a signed measure and $\mu = \mu^+ - \mu^-$, then the hypothesis that $\mu \ll \nu$ implies that $|\mu| \ll \nu$. Therefore, from the first part,

$$\forall\, \varepsilon > 0,\ \exists\, \delta > 0 \text{ such that } |\nu(A)| < \delta \Longrightarrow |\mu|(A) < \varepsilon.$$

Thus, since $|\mu|(A) \geq |\mu(A)|$, we obtain that

$$\forall\, \varepsilon > 0,\ \exists\, \delta > 0 \text{ such that } |\nu(A)| < \delta \Longrightarrow |\mu_r(A)|,\ |\mu_i(A)| < \varepsilon,$$

which concludes the proof. □

Theorem 5.2.8. Classical continuity and a sufficient condition for absolutely continuous measures

Let $(X, \mathcal{A}, \nu)$ be a finite measure space and let

$$\mu : \mathcal{A} \to \mathbb{C}$$

be a finitely additive measure. Assume that

$$\forall\, \varepsilon > 0,\ \exists\, \delta > 0 \textit{ such that } |\nu(A)| < \delta \Longrightarrow |\mu|(A) < \varepsilon.$$

Then $\mu \in M_b(X)$ and $\mu \ll \nu$.

Proof. First we prove σ-additivity. Let $\{A_n : n = 1, \ldots\} \subseteq \mathcal{A}$ be a disjoint family. From the finite additivity,

$$\mu\left(\bigcup_{n=1}^{\infty} A_n\right) = \sum_{j=1}^{k} \mu(A_j) + \mu\left(\bigcup_{j=k+1}^{\infty} A_j\right),$$

and so it is sufficient to prove that

$$\lim_{k\to\infty} \left|\mu\left(\bigcup_{j=k+1}^{\infty} A_j\right)\right| = 0. \tag{5.16}$$

Since $\nu\left(\bigcup_{j=k}^{\infty} A_j\right) = \sum_{j=k}^{\infty} \nu(A_j)$ is valid and finite for each k, and since $\lim_{k\to\infty} \sum_{j=k}^{\infty} \nu(A_j) = 0$, we conclude (5.16) by our hypothesis.

Clearly, $\mu \ll \nu$, for if $\nu(A) = 0$, then, for each $\varepsilon > 0$, $|\mu(A)| < \varepsilon$.

This fact also tells us that $\mu(\emptyset) = 0$, which is also clear from σ-additivity. Thus, $\mu \in M_b(X)$. □

We can drop the hypothesis that ν is bounded in Theorem 5.2.8 if we assume that $\mu \in M_b(X)$.

Example 5.2.9. Perfect sets in compact spaces

a. Let X be a compact space (Appendix A.1) and assume that there is a nonzero Borel measure μ on X, e.g., m on $[0, 1]$. We shall show that there is a *perfect set* (i.e., closed, nonempty, and without isolated points) $P \subseteq X$, and so card $P > \aleph_0$, such that $\mu(P) = 0$. For example, we could take the Cantor set $C \subseteq [0, 1]$ in the case of m on $[0, 1]$. To do this we follow RICHARD B. DARST [117] and use the following two results by WALTER RUDIN [403]. Let Q be a compact space without any perfect subsets; Q need not be countable!

R1. If $f \in C(Q)$ then card $f(Q) \leq \aleph_0$.

R2. If $\mu \in M_c(Q)$ then $\mu(Q) = 0$.

Assertion R2 implies that there is a perfect set $A \subseteq X$, since otherwise μ is trivial. From Urysohn's lemma (Theorem A.1.3) there is a continuous surjection $f : A \to [0, 1]$. Let $\{C_\lambda\}$ be an uncountable collection of pairwise disjoint perfect subsets of $[0, 1]$. Set $A_\lambda = f^{-1}(C_\lambda)$, so that $\{A_\lambda\}$ is an uncountable disjoint family of closed sets in A. From R1 each A_λ contains a perfect subset P_λ. Now, if each $\mu(P_\lambda) \neq 0$ we have $\mu\left(\bigcup P_\lambda\right) = \infty$, and this is a contradiction. Consequently, let $P = P_\lambda$ for some P_λ for which $\mu(P_\lambda) = 0$.

b. Part *a* is really not unexpected—in this subject nothing is. More spectacular is the following example, due to GUSTAVE CHOQUET [102]. Let $X = [0, 1] \times [0, 1]$ and let $\mathcal{B}(X) \subseteq \mathcal{P}(X)$ be the Borel algebra. Then there is an uncountable set $N \subseteq X$, measurable for each measure on $\mathcal{B}(X)$, such that

$$\forall\, \mu \in M_c(X), \quad \mu(N) = 0,$$

and for which card $K \leq \aleph_0$ when $K \subseteq N$ is compact.

Remark. For a given $\mu \in M_b(X)$ we have defined C_μ. When X is locally compact we shall also be able to define the support of μ, denoted by supp μ. There is a relationship between these two notions that we shall discuss in Section 7.4.

Example 5.2.10. Continuous measures on perfect totally disconnected sets
a. Let $E \subseteq [0,1]$ be of the form $\bigcap E_k$, $E_k = \bigcup_{j=1}^{2^k} E_j^k$, where $\{E_j^k : j = 1,\ldots,2^k\}$ is disjoint and each E_j^k is a closed interval. Let $\gamma_{j,k}$ be the midpoint of E_j^k and define the measure

$$\mu_k = \sum_{j=1}^{2^k} \frac{1}{2^k}\delta_{\gamma_{j,k}}.$$

Then each $\|\mu_k\|_1 = |\mu_k|([0,1]) = 1$, so that, by the Banach–Alaoglu theorem (Theorem A.9.5), $\{\mu_k : k = 1,\ldots\}$ has a weak* limit μ, noting that each $\mu_k \in C([0,1])'$, i.e., each $\mu_k : C([0,1]) \to \mathbb{C}$ is a continuous linear functional with the $\|\ldots\|_\infty$ norm on $C([0,1])$. Clearly, $\mu \in M_c([0,1])$, $\|\mu\|_1 = 1$, and $\mu \geq 0$. This procedure can be extended to any perfect totally disconnected metric space E.

b. Suppose we adjust the above procedure in the following way. Let $F_n \subseteq [0,1]$ consist of n points and define the measure

$$\mu_n = \sum_{\gamma \in F_n} \frac{1}{n}\delta_\gamma.$$

Then $\|\mu_n\|_1 = 1$ and $\mu_n \geq 0$. Set $F = \overline{\bigcup F_n}$. By the Banach–Alaoglu theorem, there are a subsequence $\{\mu_{m_n} : n = 1,\ldots\} \subseteq \{\mu_n : n = 1,\ldots\}$ and $\mu \in M_b(F)$ such that $\mu_{m_n} \to \mu$ in the weak* topology, where $\mu \geq 0$, and $\|\mu\|_1 = 1$. Taking $F_n = \{1, 1/2, \ldots, 1/n\}$ we have $F = \{0, 1, 1/2, \ldots\}$ and $\mu(\{\gamma\}) = 0$ for each $\gamma \in (0,1]$. Also, $\mu(\{0\}) = 1$, and so $\mu = \delta_0$.

Example 5.2.11. $\mu \ll \nu$ and ν bounded does not imply μ bounded
We shall prove that if $\nu \in M_b(X)$ and $\mu \ll \nu$, then μ need not be bounded. In the process we shall show the necessity of the hypothesis that μ be bounded in Theorem 5.2.7. Let ν be Lebesgue measure m on $(0,1]$ and set

$$\mu(A) = \int_A \frac{1}{x}\,dx.$$

Clearly μ is unbounded and $\mu \ll \nu$. As far as Theorem 5.2.7 is concerned note that

$$\forall\, \varepsilon > 0,\ \exists\, N \text{ such that } \forall\, n > N, \quad \nu\left(\left(0, \frac{1}{n}\right]\right) = \frac{1}{n} < \varepsilon,$$

whereas $\mu\left(\left(0, \frac{1}{n}\right]\right) = \infty$ for all n.

5.3 The Vitali–Lebesgue–Radon–Nikodym theorem

We have indicated the roles of VITALI and LEBESGUE in the development of FTC in Section 4.3. Essentially, the *Radon–Nikodym theorem* (R–N) can be considered as the natural generalization of FTC for more general spaces, and, in this context, we refer to the historical remarks in Section 5.6.4 and to the *Epilogue* in THOMAS HAWKINS' book [226]. The former studies the contribution of VITALI and LEBESGUE, whereas the latter examines RADON's paper and THOMAS J. STIELTJES' influence, among other things.

We first prove R–N for the case that μ and ν are bounded measures. We then discuss extensions of the result to other "measures". The proof of Theorem 5.3.1 actually works if μ is a real-valued signed measure.

Theorem 5.3.1. Radon–Nikodym theorem (R–N) for bounded measures
Let μ and ν be bounded measures on a measurable space $(X, \mathcal{A})$. Assume that $\mu \ll \nu$. There is a unique element $f \in L^1_\nu(X)$ such that

$$\forall\, A \in \mathcal{A}, \quad \mu(A) = \int_A f\, d\nu. \tag{5.17}$$

Proof. For each $r \in \mathbb{R}$ consider the real-valued signed measure $\mu - r\nu$, and let P_r, N_r be a Hahn decomposition of X for $\mu - r\nu$.

a. Let $r < s$. We shall first check that

i. $\mu(A) \le r\nu(A) \le s\nu(A)$ if $A \subseteq N_r$,

ii. $s\nu(A) \le \mu(A)$ if $A \subseteq P_s$,

iii. $\nu(A) = \mu(A) = 0$ if $A \subseteq N_r \setminus N_s$.

Property *iii* follows from *i* and *ii* since $N_r \setminus N_s \subseteq X \setminus N_s = P_s$. For the first inequality of *i* we need only the definition of N_r. For the second inequality we use the facts that ν is a measure and $r < s$. Assertion *ii* is immediate from the definition of P_s.

b. Set

$$E = \bigcup_{s\in\mathbb{Q}} \bigcup_{r<s, r\in\mathbb{Q}} (N_r \setminus N_s), \quad R_r = N_r \setminus E,$$

and

$$G = X \setminus \left(\bigcup_{r\in\mathbb{Q}} R_r \setminus \bigcap_{s\in\mathbb{Q}} R_s \right).$$

We shall prove that

i. $\nu(E) = 0$,

ii. $r < s \Longrightarrow R_r \subseteq R_s$,

iii. $\nu\left(\bigcap_{r\in\mathbb{Q}} R_r\right) = 0$,

iv. $\nu\left(X \setminus \bigcup_{r\in\mathbb{Q}} R_r\right) = \nu\left(\bigcap_{r\in\mathbb{Q}} R_r^\sim\right) = 0$,

v. $\nu(G) = 0$.

Statement *i* is clear from part *a.iii* and the fact that E is a countable union. Assertion *ii* follows from an easy set-theoretic manipulation using the definitions of E and R_r. To prove *iii* we first observe that

$$\forall\, s \in \mathbb{Q}, \quad \bigcap_{r\in\mathbb{Q}} R_r \subseteq \bigcap_{r\in\mathbb{Q}} N_r \subseteq N_s,$$

and so

$$(\mu - s\nu)\left(\bigcap_{r\in\mathbb{Q}} R_r\right) \leq 0.$$

Consequently, if $\nu\left(\bigcap R_r\right) > 0$ and we let $s_n \to -\infty$, we conclude that $\mu\left(\bigcap R_r\right) = -\infty$, and this contradicts our hypothesis that μ is bounded.

For *iv* we compute

$$X \setminus \bigcup_{r\in\mathbb{Q}} R_r = \bigcap_{r\in\mathbb{Q}} (X \setminus (N_r \setminus E)) = \bigcap_{r\in\mathbb{Q}} ((X \setminus N_r) \cup E) = E \cup \left(\bigcap_{r\in\mathbb{Q}} N_r^{\sim}\right);$$

see Figure 5.1. Thus, it is sufficient to prove that $\nu\left(\bigcap P_r\right) = 0$. We use the technique of the proof of *iii*: $\bigcap P_r \subseteq P_s$, assume $\nu\left(\bigcap P_r\right) > 0$, and obtain a contradiction.

To obtain *v* we compute

$$\begin{aligned} G = \left(\bigcup_{r\in\mathbb{Q}} R_r \setminus \bigcap_{s\in\mathbb{Q}} R_s\right)^{\sim} &= \left[\left(\bigcup_{r\in\mathbb{Q}} R_r\right) \cap \left(\left(\bigcap_{s\in\mathbb{Q}} R_s\right)^{\sim}\right)\right]^{\sim} \\ &= \left[\left(\bigcup_{r\in\mathbb{Q}} R_r\right) \cap \left(\bigcup_{s\in\mathbb{Q}} R_s^{\sim}\right)\right]^{\sim} \\ &= \left(\bigcap_{r\in\mathbb{Q}} R_r^{\sim}\right) \cup \left(\bigcap_{s\in\mathbb{Q}} R_s\right); \end{aligned}$$

see Figure 5.2. Thus, $\nu(G) = 0$ by *iii* and *iv*.

c. Define

$$f(x) = \begin{cases} 0, & \text{if } x \in G, \\ \sup\{s \in \mathbb{Q} : x \notin R_s\}, & \text{if } x \notin G. \end{cases}$$

We shall show that f is ν-measurable. In fact, if $\alpha \leq 0$ then

$$\{x : f(x) < \alpha\} = \bigcup_{r<\alpha, r\in\mathbb{Q}} R_r,$$

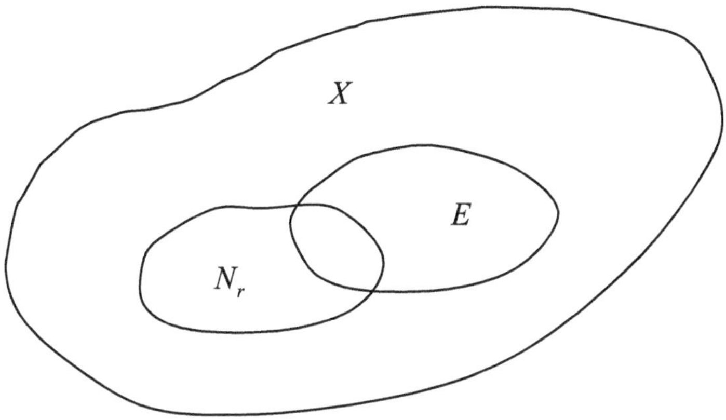

Fig. 5.1. Radon–Nikodym diagram 1.

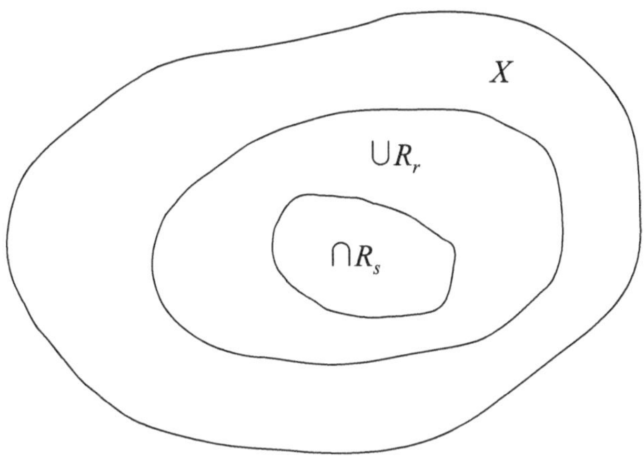

Fig. 5.2. Radon–Nikodym diagram 2.

and if $\alpha > 0$ then

$$\{x : f(x) < \alpha\} = G \cup \left(\bigcup_{r<\alpha, r\in\mathbb{Q}} R_r \right).$$

These equalities are elementary to verify.

d. Let $r < s$ and take $A \subseteq R_s \setminus R_r$; cf. *b.ii.* We shall prove that

$$\left| \int_A f \, d\nu - \mu(A) \right| \leq (s - r)\nu(A). \tag{5.18}$$

First, observe that $R_s \setminus R_r \subseteq (\bigcup R_t) \setminus (\bigcap R_u)$, and so $A \cap G = \emptyset$. Now, if $x \notin R_t$, then $f(x) \geq t$. On the other hand, if $f(x) > t$, then $x \notin R_t$. Thus,

$$r \leq f(A) \leq s,$$

and so

$$r\nu(A) \le \int_A f\,d\nu \le s\nu(A). \tag{5.19}$$

Further, $R_s \setminus R_r = (N_s \setminus E)\setminus(N_r \setminus E)$ and hence $\mu(A) \le s\nu(A)$ since $A \subseteq N_s$. Also $A \subseteq P_r$, since $R_s \setminus R_r \subseteq (N_s \setminus E) \setminus N_r \subseteq X \setminus N_r$, and so $\mu(A) \ge r\nu(A)$. Thus, $-s\nu(A) \le -\mu(A) \le -r\nu(A)$, which, when combined with (5.19), yields (5.18).

e. We next show that if $A \subseteq R_{n+1} \setminus R_n$ then

$$\mu(A) = \int_A f\,d\nu. \tag{5.20}$$

For $p \in \mathbb{N}$ set

$$A_j = A \cap \left(R_{n+(j/p)} \setminus R_{n+[(j-1)/p]}\right), \quad j = 1, \ldots, p.$$

Clearly, $\{A_j : j = 1, \ldots, p\}$ is a disjoint family and $A = \bigcup A_j$. Hence, from part *d*,

$$\left|\mu(A) - \int_A f\,d\nu\right| \le \sum_{j=1}^{p} \left|\mu(A_j) - \int_{A_j} f\,d\nu\right| \le \sum_{j=1}^{p} \frac{1}{p}\nu(A_j) = \frac{1}{p}\nu(A),$$

and so (5.20) follows since p is arbitrary.

f. For $A \in \mathcal{A}$ we write

$$A = (A \cap G) \cup \left(\bigcup_{j=1}^{\infty} A_j\right),$$

where $\{A_j : j = 1, \ldots\}$ is a disjoint family, and, for each j, there is an $n = n(j)$ for which $A_j \subseteq R_{n+1} \setminus R_n \subseteq G^{\sim}$. Thus,

$$\mu(A) = \mu(A \cap G) + \mu\left(\bigcup_{j=1}^{\infty} A_j\right),$$

so that, from (5.20),

$$\mu(A) = \mu(A \cap G) + \int_A f\,d\nu \tag{5.21}$$

(this is the Lebesgue decomposition of μ!). The fact that $\nu(G) = 0$, e.g., part *b.v*, and the hypothesis that $\mu \ll \nu$ yield

$$\mu(A) = \int_A f\,d\nu.$$

Because f is ν-measurable and $\int_A f\,d\nu$ exists for each $A \in \mathcal{A}$ we conclude that $f \in L^1_\nu(X)$. □

The proof of the following extension of Theorem 5.3.1 is straightforward; see Problem 5.37 for this extension and related material.

Theorem 5.3.2. Radon–Nikodym theorem (R–N)
Let $(X, \mathcal{A})$ be a measurable space, let $\mu \in M_b(X)$, and assume that $(X, \mathcal{A}, \nu)$ is σ-finite measure space. If $\mu \ll \nu$, then there is a unique element $f \in L^1_\nu(X)$ such that

$$\forall\, A \in \mathcal{A}, \quad \mu(A) = \int_A f\, d\nu.$$

Remark. A necessary and sufficient condition that $f \in L^1_\nu(X)$ in any statement of R–N is that $\mu \in M_b(X)$. On the other hand, if, for example, $(X, \mathcal{A}, \nu)$ is a σ-finite measure space, $\mu \in SM(X)$, and $\mu \ll \nu$, then there is a unique ν-measurable function f that satisfies (5.17). In this case, $f \in L^1_\nu(B)$ whenever $\nu(B) < \infty$; cf. Definition 5.5.1*c*.

Recall the notation that if $\nu \in M_b(X)$ then

$$M_{ac}(X, \nu) = \{\mu \in M_b(X) : \mu \ll \nu\}.$$

We rewrite Theorem 5.3.1, Theorem 5.3.2, and the Remark in the following way.

Theorem 5.3.3. R–N as a bijection on $L^1_\nu(X)$
Let $(X, \mathcal{A}, \nu)$ be a σ-finite measure space. There is a natural bijective mapping

$$L^1_\nu(X) \to M_{ac}(X, \nu),$$
$$f \mapsto \mu_f,$$

where

$$\forall\, A \in \mathcal{A}, \quad \mu_f(A) = \int_A f\, d\nu.$$

The isomorphism of Theorem 5.3.3 is also described in other terms.

Definition 5.3.4. Radon–Nikodym derivative
Consider a statement of R–N with corresponding elements μ, ν, and f. We call f the *R–N derivative* of μ with respect to ν. As such, R–N becomes a natural generalization of FTC: FTC *characterizes absolutely continuous functions g as those for which*

$$g(x) - g(a) = \int_a^x g';$$

and R–N characterizes absolutely continuous measures $\mu \ll \nu$ as those for which

$$\mu(A) = \int_A f\, d\nu.$$

The fact that the R–N derivative can be viewed as a difference quotient is discussed in Section 8.4.

Part *b* of the following should be compared with Proposition 5.1.10.

Theorem 5.3.5. R–N and total variation
a. *Let $(X, \mathcal{A})$ be a measurable space and let $\nu \in M_b(X)$. There is a $|\nu|$-integrable function h, $|h| = 1$, such that*

$$\forall\, f \in L^1_{|\nu|}(X), \quad \int_X f\, d\nu = \int_X f h\, d|\nu|.$$

b. *Let $(X, \mathcal{A})$ be a measurable space, let ν be a measure, and let $g \in L^1_\nu(X)$. Define*

$$\forall\, A \in \mathcal{A}, \quad \mu(A) = \int_A g\, d\nu.$$

Then $\mu \in M_b(X)$ and

$$\forall\, A \in \mathcal{A}, \quad |\mu|(A) = \int_A |g|\, d\nu.$$

Proof. **a.** Clearly, $\nu \ll |\nu|$, so that by R–N there is $h \in L^1_{|\nu|}(X)$ such that

$$\forall\, A \in \mathcal{A}, \quad \nu(A) = \int_A h\, d|\nu|.$$

Consequently, since we can approximate the elements of $L^1_{|\nu|}(X)$ by simple functions, it is sufficient to prove that $|h| = 1$ $|\nu|$-*a.e.*

We first prove $|h| \geq 1$ ν-*a.e.* Let $A_r = \{x : |h(x)| < r\}$, $r > 0$, and let $A_r = \bigcup_{j=1}^J A_j$, a finite disjoint union of elements from $\mathcal{A}$. Then

$$\sum_{j=1}^J |\nu(A_j)| = \sum_{j=1}^J \left| \int_{A_j} h\, d|\nu| \right| < r \sum_{j=1}^J |\nu|(A_j) = r|\nu|(A_r).$$

Taking the supremum over all finite decompositions of A_r gives $|\nu|(A_r) \leq r|\nu|(A_r)$. Thus, $|\nu|(A_r) = 0$ for each $r \in (0, 1)$, and so $|h| \geq 1$ $|\nu|$-*a.e.*

We now prove that $|h| \leq 1$ $|\nu|$-*a.e.* Note that if $|\nu|(A) > 0$ then

$$\left| \frac{1}{|\nu|(A)} \int_A h\, d|\nu| \right| = \left| \frac{\nu(A)}{|\nu|(A)} \right| \leq 1. \tag{5.22}$$

Take any closed disk $S \subseteq \{z \in \mathbb{C} : |z| \leq 1\}^\sim$ with center ζ and radius r, and set $B = h^{-1}(S)$. We must show that $|\nu|(B) = 0$. If $|\nu|(B) > 0$ then

$$\left| \zeta - \frac{1}{|\nu|(B)} \int_B h\, d|\nu| \right| \leq \frac{1}{|\nu|(B)} \int_B |h(x) - \zeta|\, d|\nu|(x) \leq r \tag{5.23}$$

by the definition of S. However, the left side of (5.23) is greater than r by (5.22) and the fact that the closed unit disk and S are disjoint. This is the desired contradiction.

b. Clearly, from the definition of $|\mu|$,

$$\forall\, A \in \mathcal{A}, \quad |\mu|(A) \leq \int_A |g|\, d\nu.$$

In particular, $\mu \in M_b(X)$.

The opposite inequality can be proved in an elementary way by approximating $|g|\mathbb{1}_A$ by gs_n, where $\{s_n : n = 1, \ldots\}$ is a sequence of simple functions satisfying

$$\left| \int_X g s_n \, d\nu \right| \leq |\mu|(A).$$

We shall give another proof of this (opposite) inequality that uses R–N. Without loss of generality let $(X, \mathcal{A}, \nu)$ be σ-finite since $g \in L^1_\nu(X)$; e.g., Problem 5.31. By hypothesis we write $\mu = g\nu$, and, since $\mu \in M_b(X)$, we have

$$\int f\, d\mu = \int f g\, d\nu.$$

We use part *a* to assert that $\mu = h|\mu|$, where $|h| = 1$ $|\mu|$-*a.e.* It is at this point that we use the fact that $(X, \mathcal{A}, \nu)$ is σ-finite, because part *a* requires R–N, which, in turn, we have stated only for σ-finite $(X, \mathcal{A}, \nu)$. Thus,

$$|\mu| = \frac{g}{h} \nu.$$

Clearly, $g/h \geq 0$ ν-*a.e.* since $|\mu| \geq 0$ and ν is a measure. Therefore, $g/h = |g|$ ν-*a.e.* □

Theorem 5.3.6. The decomposition $\mu = \mu_a + \mu_s + \mu_d$

a. *Let $(X, \mathcal{A})$ be a measurable space. If $\mu, \nu \in M_b(X)$ and $\mu \perp \nu$, then*

$$|\mu + \nu| = |\mu| + |\nu|.$$

b. *Let $(X, \mathcal{A})$ be a measurable space. If $\mu \in M_b(X)$ and ν is a measure on X, then*

$$\mu = \mu_a + \mu_s + \mu_d,$$

where $\mu_d \in M_d(X)$, $\mu_a \ll \nu$, $\mu_s \perp \nu$, and $\mu_a, \mu_s \in M_c(X)$. Also, $(\mu_a + \mu_s) \perp \mu_d$ and

$$|\mu| = |\mu_a| + |\mu_s| + |\mu_d|.$$

Proof. ***a.*** Let $\mu = h|\mu|$, where $|h| = 1$ and $h \in L^1_{|\mu|}(X)$, respectively, let $\nu = g|\nu|$. Then

$$\mu + \nu = \left(h\mathbb{1}_{C_\mu} + g\mathbb{1}_{C_\mu^\sim} \right) (|\mu| + |\nu|).$$

Observe that $\left|h\mathbb{1}_{C_\mu} + g\mathbb{1}_{C_\mu^\sim}\right| = 1$. From Theorem 5.3.5*b*, we obtain

$$|\mu+\nu|(A) = \int_A \left|h\mathbb{1}_{C_\mu} + g\mathbb{1}_{C_\mu^\sim}\right| \, d(|\mu|+|\nu|) = |\mu|(A) + |\nu|(A).$$

b. First, we know that $\mu = \mu_c + \mu_d$; and, from the Lebesgue decomposition (Theorem 5.2.6), we have $\mu_c = \mu_a + \mu_s$, where $\mu_a \ll \nu$ and $\mu_s \perp \nu$.

It is easy to see that $\mu_c \perp \mu_d$, and so $|\mu| = |\mu_c| + |\mu_d|$. Also, in proving that $\mu \ll \nu$ and $\mu \perp \nu$ imply $\mu = 0$ (Proposition 5.2.4), we really showed that $\mu \ll \nu$ and $\lambda \perp \nu$ imply $\mu \perp \lambda$; see Problem 5.35. Thus, $\mu_a \perp \mu_s$ and we apply part *a* again. □

Remark. IRVING E. SEGAL (1951) has characterized spaces for which R–N is valid as those that can be "localized". JOHN L. KELLEY [280] has streamlined SEGAL's work, making the characterization in terms of so-called Dedekind complete set functions on certain algebras of sets. These results show the equivalence of the Hahn decomposition theorem, the Lebesgue decomposition theorem, R–N, and a form of the Riesz representation theorem.

Remark. Let X be a locally compact Hausdorff space and consider the sup-norm Banach space $C_0(X)$; see Chapter 7 and Appendix A.1. R–N establishes an isomorphism between $M_{ac}(X,\nu)$ and those continuous linear functionals $C_0(X) \to \mathbb{C}$ that are elements of $L^1_\nu(X)$. The Riesz representation theorem in Chapter 7 will yield an isomorphism between $(C_0(X))'$ and $M_b(X)$, where $(C_0(X))'$ is the space of continuous linear functionals $C_0(X) \to \mathbb{C}$.

5.4 The relation between set and point functions on $\mathbb{R}$

We restate Theorem 4.1.9 as follows.

Theorem 5.4.1. Equivalence of $SM(\mathbb{R})$ and $BV_{loc}(\mathbb{R})$
Consider the measurable space $(\mathbb{R}, \mathcal{B}(\mathbb{R}))$ and a function $f : \mathbb{R} \to \mathbb{R}$, right continuous at each point, for which $\lim_{x\to-\infty} f(x) = 0$. Then $f \in BV_{\mathrm{loc}}(\mathbb{R})$ if and only if

$$\mu_f((a,b]) = f(b) - f(a) \tag{5.24}$$

can be uniquely extended to an element of $SM(\mathbb{R})$. In this situation f is continuous at x_0 if and only if $\mu_f(\{x_0\}) = 0$.

Proof. ($\Longrightarrow$) Since $f \in BV_{\mathrm{loc}}(\mathbb{R})$ is right continuous, there exist increasing right-continuous functions P and N such that $f = P - N$ (see Problem 4.3); and so $\mu_f = \mu_P - \mu_N$ on half-open intervals. From Theorem 3.5.1 it follows that the nonnegative set functions μ_P and μ_N have unique extensions to nonnegative measures on $(\mathbb{R}, \mathcal{B}(\mathbb{R}))$. Thus, μ_f extends to a σ-additive set function on the σ-algebra $\mathcal{B}(\mathbb{R})$.

The uniqueness assertion can be proved using the fact that Borel measures on $\mathbb{R}$ taking finite values on compact sets are regular; cf. Problem 2.46*c*.

($\Longleftarrow$) Since $\mu \in SM(\mathbb{R})$ we deduce from Theorem 5.1.4 that $(\mathbb{R}, \mathcal{B}(\mathbb{R}), |\mu|)$ is a regular measure space. Moreover, Theorem 5.1.8 implies that there exist two measures, μ^+ and μ^-, such that $\mu = \mu^+ - \mu^-$. From Proposition 3.5.2 we conclude that there are nonnegative right-continuous functions f^+ and f^- such that $\mu^\pm((a, b]) = f^\pm(b) - f^\pm(a)$. Let $f = f^+ - f^-$. Clearly, f is right continuous, and Theorem 4.1.2*a* implies that $f \in BV_{\mathrm{loc}}(\mathbb{R})$. □

An analogous result can be proven for $\mu \in M_b(\mathbb{R})$.

We now complete the second Remark prior to Theorem 5.2.2, and relate the results of Chapter 4 and the notions introduced in this chapter for the special case of $\mathbb{R}$. A natural example to keep in mind is the Cantor–Lebesgue function and its corresponding measure.

Theorem 5.4.2. Equivalence of subspaces of $SM(\mathbb{R})$ and $BV(\mathbb{R})$
Let $f \in BV(\mathbb{R})$ be real-valued and right continuous, and define μ_f as in (5.24).

a. *$\mu_f \ll m \Longleftrightarrow f$ is absolutely continuous on $\mathbb{R}$.*

b. *$\mu_f \perp m \Longleftrightarrow f' = 0$ m-a.e.*

c. *If μ_f is discrete then $f' = 0$ m-a.e.*

d. *$f(x) = g(x) + \int_{-\infty}^{x} f'$, where $\mu_g \perp m$; and $g = 0 \Longleftrightarrow \mu_f \ll m$.*

Proof. **a.**($\Longrightarrow$) $\mu_f \ll m$ implies that

$$\forall\, \varepsilon > 0,\ \exists\, \delta > 0 \text{ such that } A \in \mathcal{M}(\mathbb{R}) \text{ and } m(A) < \delta \Longrightarrow |\mu_f(A)| < \varepsilon.$$

Thus, choose $\varepsilon > 0$ and let $A = \bigcup (a_k, b_k]$ satisfy $m(A) < \delta$. Then

$$\left| \sum_{k=1}^{\infty} (f(b_k) - f(a_k)) \right| = \left| \sum_{k=1}^{\infty} \mu_f((a_k, b_k)) \right| = |\mu_f(A)| < \varepsilon.$$

($\Longleftarrow$) Let $A \in \mathcal{M}(\mathbb{R})$, $m(A) = 0$. Without loss of generality take $A \subseteq [-c, c]$ and let f be an increasing function. For $\varepsilon > 0$ choose $\delta > 0$ by the definition of the absolute continuity of f.

Now A is contained in a union $\bigcup (a_k, b_k]$, for which $\sum (b_k - a_k) < \delta$, since $m(A) = 0$. By the choice of δ,

$$\left| \sum_{k=1}^{\infty} (f(b_k) - f(a_k)) \right| < \varepsilon;$$

and so, since f is increasing,

$$|\mu_f(A)| \leq \left| \mu_f \left(\bigcup_{k=1}^{\infty} (a_k, b_k] \right) \right| < \varepsilon.$$

b. ($\Longleftarrow$) Without loss of generality let f be an increasing function. Assume that μ_f and m are not mutually singular. From the Lebesgue decomposition, $\mu_f = \mu_1 + \mu_2$, where $\mu_1 \ll m$, $\mu_2 \perp m$, and $\mu_1 \neq 0$. We have that μ_1 and μ_2 are measures since μ is a measure.

As remarked after Theorem 5.4.1, μ_f is regular, and for the same reason, so also are μ_1 and μ_2.

We define the function $f_1(x) = \mu_1((-\infty, x])$. Thus, by part a and because $\mu_1 \ll m$, we conclude that f_1 is absolutely continuous. Moreover, f_1 is an increasing function since μ_1 is a measure. From FTC, we have

$$\mu_1((a,b]) = f_1(b) - f_1(a) = \int_a^b f_1'.$$

Consequently, since $\mu_1 \neq 0$ and μ_1 is regular, $f_1' > 0$ on some $A \in \mathcal{B}(\mathbb{R})$, where $m(A) > 0$.

If we set $f_2(x) = \mu_2((-\infty, x])$ then f_2 is characterized by μ_2 from Theorem 5.4.1. Hence, $f = f_1 + f_2$, and f_2 is increasing since μ_2 is a measure. Therefore,

$$f' = f_1' + f_2' \geq f_1',$$

and so $f' \neq 0$ on a set of positive measure.

($\Longrightarrow$) Assume $f' > 0$ on $A \in \mathcal{B}(\mathbb{R})$, where $m(A) > 0$. Again, without loss of generality, take f to be increasing and thus $f' \geq 0$ m-*a.e.* Since f is m-measurable, f' is m-measurable, and we define

$$\forall\, B \in \mathcal{B}(\mathbb{R}), \quad \nu(B) = \int_B f'.$$

Clearly $\nu \ll m$ and $\nu(A) > 0$. Also,

$$\nu((a,b]) = \int_a^b f' \leq f(b) - f(a) = \mu_f((a,b]),$$

where the inequality follows from Theorem 4.3.2. It is easily verified that ν is regular, so that, by standard measure-theoretic manipulation,

$$\forall\, B \in \mathcal{B}(\mathbb{R}), \quad \nu(B) \leq \mu_f(B).$$

Consequently, $\mu_f(A) > 0$.

Note that $\nu \ll m$ and $\mu_f \perp m$ imply $\nu \perp \mu_f$, e.g., Theorem 5.3.6b, and this contradicts the fact that $\mu_f(A), \nu(A) > 0$. Hence, $f' = 0$ m-*a.e.*

c. For this part we use part b and can take $C_m = \mathbb{R} \setminus C_{\mu_f}$ where card $C_{\mu_f} \leq \aleph_0$.

d. Part d is just a succinct way of writing some of our previous results.

□

We wind up this peroration on the decomposition of measures with the following theorem (Problem 5.45), which is mostly an encapsulation of previous results.

Theorem 5.4.3. Equivalence of decompositions of measures and elements of $BV(\mathbb{R})$
Let $(\mathbb{R}, \mathcal{B}(\mathbb{R}), \mu)$ be a regular measure space. Then μ has the unique decomposition

$$\mu = \mu_a + \mu_s + \mu_d$$

into regular Borel measures, where μ_d is discrete, $\mu_a \ll m$, $\mu_s \perp m$, and μ_s is continuous. Further, there are unique increasing functions f, f_a, f_s, f_d, corresponding to μ, μ_a, μ_s, μ_d, such that

$$f = f_a + f_s + f_d,$$

where f_a is absolutely continuous on every compact interval, f_s is continuous and $f_s' = 0$ m-a.e., $f_d' = 0$ m-a.e., and f_d has at most countably many discontinuities x_n, $n = 1, \ldots,$ each of the form $f_d(x_n+) - f_d(x_n-)$. Finally,

$$\forall\, B \in \mathcal{B}(\mathbb{R}), \quad \mu_a(B) = \int_B f_a,$$

and

$$\forall\, y \geq x, \quad f_a(y) - f_a(x) = \int_x^y f'.$$

Remark. Theorems 5.4.1, 5.4.2, and 5.4.3 have a Schwartz distributional formulation in terms of the Riesz representation theorem (Theorem 7.5.10): *$M_b(\mathbb{R})$ is the subspace of Schwartz distributions T for which there is a function of bounded variation whose first distributional derivative is T.*

In Chapter 4 we gave several examples of increasing and strictly increasing continuous functions f for which $f' = 0$ m-a.e. An amusing application of some of our previous results, including Theorem 5.3.6*a* and a characterization of this phenomenon, is the following result.

Proposition 5.4.4. *Let f be an increasing continuous function on $[0,1]$ with $f(0) = 0$ and $f(1) = \alpha$.*

a. *Let L be the length of the graph of f. Then*

$$L = 1 + \alpha \Longleftrightarrow f' = 0 \; m\text{-a.e.}$$

b. *Let f be strictly increasing with inverse function g. Then*

$$f' = 0 \; m\text{-a.e. on } [0,1] \Longleftrightarrow g' = 0 \; m\text{-a.e. on } [0,\alpha].$$

Proof. **a.** Take $\alpha > 0$ (the $\alpha = 0$ case is clear). Denote the graph of f by $G(x) = x + if(x)$, and so $L = V(G, [0,1])$.

($\Longleftarrow$) $\mu_f \perp m$ and $\mu_{if} \perp m$ by Theorem 5.4.2*b* and the hypothesis that $f' = 0$ m-a.e. Note that if $g(x) = x$ then $\mu_g = m$. Thus,

$$L = V(x + if(x), [0,1]) = \|m + \mu_{if}\|_1 = \|m\|_1 + \|\mu_{if}\|_1$$
$$= 1 + \|\mu_f\|_1 = 1 + \alpha,$$

where $\|\mu_f\|_1 = \alpha$ since f is increasing.

($\Longrightarrow$) Assume that $L = 1 + \alpha$ and let $f = f_1 + f_2$, where f_1 is absolutely continuous and $f_2' = 0$ m-*a.e.* Then $\mu_{f_1} \perp \mu_{f_2}$ since $\mu_{f_1} \ll m$ and $\mu_{f_2} \perp m$. Thus,

$$\|\mu_f\|_1 = \|\mu_{f_1}\|_1 + \|\mu_{f_2}\|_1 = V(f_1, [0,1]) + V(f_2, [0,1]).$$

Consequently,

$$V(x + if_1(x), [0,1]) = 1 + V(f_1, [0,1]) \tag{5.25}$$

because

$$1 + \alpha = L = V(G, [0,1]) \le V(x + if_1(x), [0,1]) + V(if_2(x), [0,1])$$
$$\le 1 + \|\mu_{f_1}\|_1 + \|\mu_{f_2}\|_1 = 1 + \|\mu_f\|_1 = 1 + \alpha,$$

where the fact that f is increasing yields the last equality.

Since f_1 is absolutely continuous we can use the integral formula for the arc length, $L_1 = V(x + if_1(x), [0,1])$, of the graph of f_1. Hence,

$$V(x + if_1(x), [0,1]) = \int_0^1 \left(1 + (f_1')^2\right)^{1/2}. \tag{5.26}$$

On the other hand, $V(x+if_1(x), [0,1]) = 1 + V(f_1, [0,1])$ from (5.25), and so

$$V(x + if_1(x), [0,1]) = 1 + \|\mu_{f_1}\|_1 = 1 + \mu_{f_1}([0,1])$$
$$= 1 + f_1(1) - f_1(0) = 1 + \int_0^1 f_1'.$$

Combining this with (5.26) yields

$$\int_0^1 \left(1 + (f_1')^2\right)^{1/2} = \int_0^1 (1 + f_1'). \tag{5.27}$$

Now, by squaring the integrands and noting that f_1 is increasing, we have $(1 + (f_1')^2)^{1/2} \le 1 + f_1'$ m-*a.e.*, so that we can actually conclude that

$$\left(1 + (f_1')^2\right)^{1/2} = 1 + f_1' \quad m\text{-}a.e.$$

from (5.27). Therefore, $f_1' = 0$ m-*a.e.*, and consequently $f' = 0$ m-*a.e.*, since $f_2' = 0$ m-*a.e.*

b. Part b is Problem 5.49. □

Remark. Finally in this section we note the intimate relation between STEINHAUS' theorem given in Problem 3.6 and the notion of absolute continuity: *Let* $\mu \in M_b(\mathbb{R})$; *then* $\mu \ll m$ *if and only if the implication*

$$|\mu|(K) > 0 \Longrightarrow \text{int}\,(K - K) \neq \emptyset$$

is valid for each compact set $K \subseteq \mathbb{R}$. This result is due to S. M. SIMMONS [442].

5.5 $L^p_\mu(X)$, $1 \le p \le \infty$

We defined the space $L^1_\mu(X)$ in Section 3.2, and we defined $L^\infty_\mu(X)$ in Section 2.5. We now generalize this notion.

Definition 5.5.1. $L^p_\mu(X)$
a. Let $(X, \mathcal{A}, \mu)$ be a measure space and fix $1 \le p < \infty$. Let $\mathcal{L}^p_\mu(X)$ be the set of all μ-measurable functions $f : X \to \mathbb{C}$ such that

$$\int_X |f|^p \, d\mu < \infty.$$

We write $f \sim g$ for $f, g \in \mathcal{L}^p_\mu(X)$ if $f = g$ μ-*a.e.*; and we note that $\sim$ is an equivalence relation. We define the space $L^p_\mu(X)$ to be the collection of all equivalence classes in $\mathcal{L}^p_\mu(X)$. Moreover, we set

$$\|f\|_p = \left(\int_X |g|^p \, d\mu \right)^{1/p},$$

where $g = f$ μ-*a.e.*

b. We defined counting measure c in Example 2.4.2*d*. As such we can consider the space $L^p_c(X)$, $1 \le p \le \infty$; and, in fact, we dealt with $L^1_c(\mathbb{Z})$ in Problem 3.27. In line with almost universal convention, we shall use the notation

$$\ell^p(X) = L^p_c(X).$$

c. Let X be a locally compact Hausdorff space and let μ be a regular Borel measure. In particular, $(X, \mathcal{A}, \mu)$ is a measure space for which $\mathcal{B}(X) \subseteq \mathcal{A}$. An $\mathcal{A}$-measurable $f : X \to \mathbb{C}$ is *locally in* $L^p_\mu(X)$ if

$$\forall\, K \subseteq X, \ \text{compact}, \quad f \mathbb{1}_k \in L^p_\mu(X).$$

In the case $X \subseteq \mathbb{R}^d$ and $\mu = m^d$, the space of functions that are locally in $L^p_{m^d}(X)$ is denoted by $L^p_{\text{loc}}(X)$.

For $p \ge 1$, q will denote $p/(p-1)$; in particular, $p = 1$ if and only if $q = \infty$. We shall not consider L^p, $p < 1$, not because it is uninteresting (on the contrary) but because it has a trivial duality theory.

The L^p spaces were introduced by F. Riesz in 1910 in the celebrated paper [390]. He considered functions of one variable defined on a closed interval and the integral with respect to Lebesgue measure m.

Theorem 5.5.2. Hölder and Minkowski inequalities for the Banach space $L^p_\mu(X)$, $1 \le p \le \infty$
Let $(X, \mathcal{A}, \mu)$ be a measure space and take $1 \le p \le \infty$.

a. *$L^p_\mu(X)$ is a Banach space.*

b. *If $f \in L^p_\mu(X)$ and $g \in L^q_\mu(X)$, then $fg \in L^1_\mu(X)$ and*

$$\left|\int_X fg\, d\mu\right| \leq \int_X |fg|\, d\mu \leq \|f\|_p \|g\|_q.$$

c. *Let $\alpha_1, \ldots, \alpha_n > 0$ and $\sum \alpha_j = 1$. If $f_1, \ldots, f_n \in L^1_\mu(X)$ then*

$$\left|\int_X f_1^{\alpha_1} \cdots f_n^{\alpha_n}\, d\mu\right| \leq \int_X |f_1^{\alpha_1} \cdots f_n^{\alpha_n}|\, d\mu \leq \prod_{j=1}^n \|f_j\|_1^{\alpha_j}.$$

Proof. **a.** We shall consider $1 \leq p < \infty$. The triangle inequality,

$$\forall\, f, g \in L^p_\mu(X), \quad \|f+g\|_p \leq \|f\|_p + \|g\|_p,$$

is called the *Minkowski inequality.* This inequality is elementary for $L^1_\mu(X)$, and it is verified in the following way, using part *b*, for $p > 1$: First note that

$$\begin{aligned}\int_X |f+g|^p\, d\mu &\leq \int_X |f+g|^{p-1}|f|\, d\mu + \int_X |f+g|^{p-1}|g|\, d\mu \\ &\leq (\|f\|_p + \|g\|_p)\, \|f+g\|_p^{p/q};\end{aligned}$$

and then divide both sides of this inequality by $\|f+g\|_p^{p/q}$.

The completeness of $L^p_\mu(X)$, $1 \leq p < \infty$, is sometimes referred to as the *Riesz–Fischer theorem* and it can be proved using Proposition A.1.10; see Section 5.6.2. In fact, letting $\{f_n : n = 1, \ldots\} \subseteq L^p_\mu(X)$ be absolutely convergent and $g_n = \sum_{k=1}^n |f_k|$, we see that g_n increases to an element $g \in L^p_\mu(X)$ by using the Minkowski inequality and Fatou's lemma. It is then straightforward to check, using LDC, that

$$\lim_{n\to\infty} \left\| \sum_{k=n}^{\infty} f_k \right\|_p = 0.$$

Consequently, $\sum_{k=1}^\infty f_k$ is convergent in $L^p_\mu(X)$ and we apply Proposition A.1.10.

The $p = \infty$ case is straightforward using an elementary analysis of sets of the form $\{x : |f(x)| > M\}$, as in the definition of $L^\infty_\mu(X)$.

b. Part *b* is called the *Hölder inequality*, and its proof does not depend on the assertion of part *a*. Its verification is elementary for $p = 1$. If $p, q > 1$, then the Hölder inequality is proved by the following calculation. Set

$$h(t) = \frac{t^p}{p} + \frac{t^{-q}}{q}, \quad t > 0,$$

and observe that

$$\forall\, t > 0, \quad h(t) \geq h(1) = 1.$$

The proof is completed by choosing

$$t = \left(\frac{|f|}{\|f\|_p}\right)^{1/q} \Big/ \left(\frac{|g|}{\|g\|_q}\right)^{1/p}.$$

c. Part *c* follows from part *b* by induction. □

Theorem 5.5.3. ***L^p*-approximation by simple functions**
Let $(X, \mathcal{A}, \mu)$ be a measure space and let $1 \le p \le \infty$. The set of simple functions $\sum_{j=1}^n a_j \mathbb{1}_{A_j}$, $\mu(A_j) < \infty$, is dense in $L^p_\mu(X)$ taken with the L^p-norm $\|\ldots\|_p$. In the case $1 \le p < \infty$, each $\mu(A_j) < \infty$.

Proof. **a.** As in Theorem 3.2.6 for the case of $L^1_\mu(X)$, the proof for $1 \le p < \infty$ essentially follows from our definition of the Lebesgue integral, or more formally, from Theorem 2.5.5 and LDC.

b. For $L^\infty_\mu(X)$ the proof also follows from Theorem 2.5.5 and can be verified in the following way. Let $f \in \mathcal{L}^\infty_\mu(X)$, so that $|f(x)| \le \|f\|_\infty$ μ-*a.e.* (see the Remark after Definition 2.5.9), and let g be μ-measurable, bounded on X, and $g = f$ μ-*a.e.* As in Theorem 2.5.5, first assume $g \ge 0$ and construct simple functions

$$f_n(x) = \sum_{k=1}^{n2^n} \frac{k-1}{2^n} \mathbb{1}_{A_{n,k}}, \quad n \ge \|f\|_\infty,$$

where

$$A_{n,k} = \left\{x : \frac{k-1}{2^n} \le g(x) < \frac{k}{2^n}\right\}.$$

Then $0 \le f_n \le f_{n+1} \to g$ uniformly on X.

Using the triangle inequality from Theorem 5.5.2, we have $\|f - f_n\|_\infty \le \|g - f_n\|_\infty$, and the right side is

$$\inf\{M : \mu(\{x : |g(x) - f_n(x)| > M\}) = 0\}.$$

Since $|g(x) - f_n(x)| \le 1/2^n$ on X, this infimum is bounded above by $1/2^n$. The result is obtained. □

If $L^p_\mu(X)$ is considered as a Banach space, then $(L^p_\mu(X))'$ denotes the space of continuous linear functionals $L^p_\mu(X) \to \mathbb{C}$ on $L^p_\mu(X)$; e.g., Appendix A.8. The space $(L^p_\mu(X))'$ is called the *dual* of $L^p_\mu(X)$. We shall now characterize the dual of $L^p_\mu(X)$. If $1 < p < \infty$ this can be done without R–N, but not without some effort except in the case $p = 2$; cf. Problem 5.39. In any case we shall use R–N to provide this characterization in Theorems 5.5.5 and 5.5.7.

Proposition 5.5.4. *Let $(X, \mathcal{A}, \mu)$ be a finite measure space, let $1 \le p < \infty$, and take $g \in L^1_\mu(X)$. Assume that there is a constant $M = M_g > 0$ such that, for all simple functions s,*

$$\left| \int_X sg \, d\mu \right| \le M \|s\|_p.$$

Then $g \in L^q_\mu(X)$.

Proof. First assume that $p > 1$ and take a sequence $\{s_n : n = 1, \ldots\}$ of nonnegative simple functions that increases pointwise to $|g|^q$. Define

$$\phi_n = s_n^{1/p} \overline{\text{sgn } g},$$

where

$$\text{sgn } g = \begin{cases} 0, & \text{if } g(x) = 0, \\ \overline{g(x)}/|g(x)|, & \text{if } g(x) \neq 0. \end{cases}$$

Clearly,

$$\|\phi_n\|_p = \left(\int_X s_n \, d\mu \right)^{1/p}. \tag{5.28}$$

Observe that

$$0 \le s_n = s_n^{1/p} s_n^{1/q} \le s_n^{1/p} |g| = \phi_n g,$$

and so,

$$0 \le \int_X s_n \, d\mu \le \int_X \phi_n g \, d\mu \le M \|\phi_n\|_p.$$

Consequently, from (5.28),

$$\int_X s_n \, d\mu \le M^q,$$

and therefore

$$\int_X |g|^q \, d\mu \le M^q$$

by B. LEVI's theorem, i.e., LDC.

Now let $p = 1$ and assume, without loss of generality, that $g \ge 0$. Assume $g \notin L^\infty_\mu(X)$, so that there is a sequence $\{A_n : n = 1, \ldots\} \subseteq \mathcal{A}$ for which $g \ge n$ on A_n and $\mu(A_n) > 0$. Then, for $s_n = \mathbb{1}_{A_n}$,

$$\frac{1}{\|s_n\|_1} \int_X s_n g \, d\mu \ge \frac{n\mu(A_n)}{\mu(A_n)} = n.$$

This contradicts our hypothesis. □

If $1 < p < \infty$ the following is true for any measure space $(X, \mathcal{A}, \mu)$ (Problem 5.50*b*).

Theorem 5.5.5. $(L^p_\mu(X))' = L^q_\mu(X)$, $1 \le p < \infty$

Let $(X, \mathcal{A}, \mu)$ *be a* σ*-finite measure space and let* $1 \le p < \infty$. *Then* $F \in (L^p_\mu(X))'$ *if and only if there is a unique* $g \in L^q_\mu(X)$ *such that*

$$\forall\, f \in L^p_\mu(X), \quad F(f) = \int_X fg\, d\mu. \tag{5.29}$$

In this case, $\|F\| = \|g\|_q$*; cf. Problem 4.37. (*$\|F\|$ *is defined as* $\|F\| = \sup_{\|f\|_p \leq 1} |F(f)|$*; see Appendix A.8.)*

Proof. ($\Longleftarrow$) Assuming (5.29) we have $F \in (L^p_\mu(X))'$ by Theorem 5.5.2*b*. With (5.29) we prove that $\|F\| = \|g\|_q$. The easier direction is to show

$$\|F\| = \sup_{\|f\|_p \leq 1} |F(f)| \leq \|g\|_q.$$

This is immediate by Theorem 5.5.2*b*.

For the opposite inequality we first consider the $1 < p < \infty$ case, and define

$$f = |g|^{q-1}\overline{\text{sgn}\, g}.$$

Then

$$f^p = |g|^{(q-1)p}\overline{(\text{sgn}\, g)}^p = |g|^q\overline{(\text{sgn}\, g)}^p,$$

and so $f \in L^p_\mu(X)$. Also

$$\|f\|_p = \left(\int_X |g|^q\, d\mu\right)^{1/p} = \|g\|_q^{q/p} = \|g\|_q^{q-1}. \tag{5.30}$$

For this f and using (5.30),

$$F(f) = \int_X fg\, d\mu = \int_X |g|^q\, d\mu = \|g\|_q^q = \|f\|_p \|g\|_q.$$

Thus,

$$\|F\| = \sup_{h \in L^p_\mu(X), h \neq 0} \frac{|F(h)|}{\|h\|_p} \geq \|g\|_q.$$

Now we must prove that $\|F\| \geq \|g\|_\infty$ when $p = 1$ and when we assume (5.29). Take $\|g\|_\infty > \varepsilon > 0$ and choose a subset $A \subseteq \{x \in X : |g(x)| > \|g\|_\infty - \varepsilon\}$ for which $0 < \mu(A) < \infty$. Define

$$f = \frac{1}{\mu(A)} \mathbb{1}_A \overline{\text{sgn}\, g}.$$

Clearly, $f \in L^1_\mu(X)$ since $\|f\|_1 = 1$. Further,

$$F(f) = \frac{1}{\mu(A)} \int_A g\, \overline{\text{sgn}\, g}\, d\mu = \frac{1}{\mu(A)} \int_A |g|\, d\mu \geq \|g\|_\infty - \varepsilon,$$

and so $\|F\| \geq \|g\|_\infty$.

($\Longrightarrow$) Assume that $(X, \mathcal{A}, \mu)$ is a finite measure space, and let $F \in (L^p_\mu(X))'$. Since $(X, \mathcal{A}, \mu)$ is finite, $L^\infty_\mu(X) \subseteq L^p_\mu(X)$, and thus

$$\forall\, A \in \mathcal{A}, \quad \nu(A) = F(\mathbb{1}_A)$$

exists.

i. We use R–N to show that there is a unique element $g \in L^1_\mu(X)$ for which

$$\forall\, A \in \mathcal{A}, \quad \nu(A) = \int_A g \, d\mu. \tag{5.31}$$

To do this we first prove that $\nu \in M_b(X)$. Let $\{A_n : n = 1, \ldots\} \subseteq \mathcal{A}$ be a disjoint family and set $A = \bigcup A_n$. Since $(X, \mathcal{A}, \mu)$ is finite it is easy to check that

$$\lim_{n \to \infty} \left\| \sum_{j=n+1}^{\infty} \mathbb{1}_{A_j} \right\|_p = 0,$$

so that, by the hypothesis that F is continuous,

$$\sum_{n=1}^{\infty} \nu(A_n) = \sum_{n=1}^{\infty} F(\mathbb{1}_{A_n}) = F(\mathbb{1}_A) = \nu(A).$$

Thus, $\nu \in M_b(X)$. Also $\nu \ll \mu$ from the definition of ν, and hence (5.31) follows by R–N.

ii. If s is simple then the linearity of F and (5.31) yield

$$F(s) = \int_X sg \, d\mu. \tag{5.32}$$

This and the continuity of F imply that

$$\left| \int_X sg \, d\mu \right| \leq \|s\|_p \|F\|.$$

Consequently, from Proposition 5.5.4, $g \in L^q_\mu(X)$.

iii. Define

$$\forall\, f \in L^p_\mu(X), \quad G(f) = \int_X fg \, d\mu,$$

so that $G \in (L^p_\mu(X))'$ by Theorem 5.5.2*b*. Therefore, $G - F \in (L^p_\mu(X))'$ and $G - F = 0$ on the set of simple functions. By Theorem 5.5.3, these functions are dense in $L^p_\mu(X)$, and so $F = G$.

It remains to check that $g \in L^q_\mu(X)$ is unique. This is easy, for if $g_1 \in L^q_\mu(X)$ yielded $F(f) = \int f g_1 \, d\mu$ for each $f \in L^p_\mu(X)$ and if $\mu(\{x : g_1(x) \neq g(x)\}) > 0$ we could construct $f \in L^p_\mu(X)$ in the obvious way to obtain

$$\int_X f(g - g_1) \, d\mu \neq 0.$$

The extension to σ-finite measure spaces is not difficult (Problem 5.50*a*). □

We shall calculate $(L^\infty_\mu(X))'$ in Theorem 5.5.7.

Example 5.5.6. $(L^\infty_m([0,1]))' \neq L^1_m([0,1])$
Observe that $L^1_m([0,1]) \subseteq (L^\infty_m([0,1]))'$. We shall show that the inclusion is proper. Note that $C([0,1])$ is a closed subspace of $L^\infty_m([0,1])$. Define $F : C([0,1]) \to \mathbb{C}$ as

$$\forall\, g \in C([0,1]), \quad F(g) = g(0), \tag{5.33}$$

so that F is continuous on $C([0,1])$ with $\|F\| = 1$. By the Hahn–Banach theorem (Theorem A.8.3), F extends to an element of $(L^\infty_m([0,1]))'$. We now prove that there is no $f \in L^1_m([0,1])$ such that

$$\forall\, g \in C([0,1]), \quad F(g) = \int_0^1 fg.$$

In fact, (5.33) defines the discrete Dirac measure $\delta = \delta_0$. If such an f existed, then choose a sequence $\{g_n : n = 1, \ldots\} \subseteq C([0,1])$ such that $g_n \to 0$ pointwise on $(0,1]$, $g_n(0) = 1$, and $|g_n| \leq 1$. Then, from LDC, $F(g_n) \to 0$, whereas $F(g_n) = g_n(0) = 1$, a contradiction.

Remark. ***a.*** It is not difficult to prove that if $f \in \bigcap_{p\geq 1} L^p_m(\mathbb{T})$ then $f \in L^\infty_m(\mathbb{T})$ if and only if $\lim_{p\to\infty} \|f\|_p$ exists, in which case $\lim_{p\to\infty} \|f\|_p = \|f\|_\infty < \infty$. It turns out that

$$L^\infty_m(\mathbb{T}) \subsetneq \bigcap_{p\geq 1} L^p_m(\mathbb{T}). \tag{5.34}$$

b. The phenomenon reflected by (5.34) is not in the category of being an amusing anecdote, but brings to the stage two important spaces in harmonic analysis.

We first define the *Hardy space*,

$$H^1(\mathbb{T}) = \{f \in L^1_m(\mathbb{T}) : \forall\, n < 0, \hat{f}(n) = 0\}, \tag{5.35}$$

taken with the induced norm topology from $L^1_m(\mathbb{T})$. (For $H^1(\mathbb{T}^d)$, see [385].) Hardy spaces are naturally defined in the context of analytic function theory, and it is a theorem of F. and M. Riesz (1916) that allows us to write (5.35); see [239], [376], [152]. Second, we define the space $BMO(\mathbb{T})$ of functions of *bounded mean oscillation*. A 1-periodic function $f \in L^1_{\text{loc}}(\mathbb{R})$ is in $BMO(\mathbb{T})$ if

$$\exists\, C > 0 \text{ such that } \forall\, I \subseteq \mathbb{R}, \quad \frac{1}{m(I)} \int_I |f - f_I| \leq C,$$

where $I \subseteq \mathbb{R}$ is a bounded interval and

$$f_I = \frac{1}{m(I)} \int_I f.$$

C. FEFFERMAN proved the basic theoretical result that

$$(H^1(\mathbb{T}))' = BMO(\mathbb{T});$$

see [165], [190], Chapter I.9. With regard to (5.34), it is elementary to verify that

$$L_m^\infty(\mathbb{T}) \subsetneq BMO(\mathbb{T}) \subsetneq \bigcap_{p\geq 1} L_m^p(\mathbb{T}).$$

c. Besides their intrinsic and broad mathematical interest, Hardy spaces are used significantly in several important applications, e.g., in Kolmogorov–Wiener prediction theory; see [152], [338], [38], and Section 2.6.7.

An elementary but fundamental use of Hardy spaces is in the context of *causal* linear systems in signal processing. To fix ideas, let X and Y be two vector spaces of functions defined on $\mathbb{Z}$, and let $L : X \to Y$ be linear with the translation-invariance property that if $x \in X$, $n_0 \in \mathbb{Z}$, and $L(x) = y$, then $L(\tau_{n_0}(x)) = \tau_{n_0}(y)$, where $\tau_{n_0}(x)(n) = x(n - n_0)$. By definition, $x \in X$ is *causal* if $x(n) = 0$ for $n < 0$; see (5.35). We call L a *causal system* if causal input x results in causal output y, e.g., [365].

Let $(X, \mathcal{A}, \mu)$ be a measure space and let $F(X) = F(X, \mathcal{A}, \mu)$ be the set of finitely additive set functions

$$\nu : \mathcal{A} \to \mathbb{C}$$

that vanish on the set $\{A \in \mathcal{A} : \mu(A) = 0\}$ and that have the property that

$$\sup\{|\nu(A)| : A \in \mathcal{A}\} < \infty.$$

We call $F(X)$ the space of *complex-valued finitely additive bounded set functions on* $\mathcal{A}$. We define

$$|\nu|(A) = \sup \sum_{j=1}^{n} |\nu(A_j)|,$$

where the supremum is taken over all disjoint decompositions of $A = \bigcup_{j=1}^n A_j$. It is easy to check that $|\nu| \in F(X)$, and $F(X)$ is a normed vector space with norm,

$$\|\nu\| = |\nu|(X). \tag{5.36}$$

Further, if $f \in L_\mu^\infty(X)$ and $\nu \in F(X)$,

$$\int_X f \, d\nu$$

will denote

$$\lim_{n\to\infty} \int_X s_n \, d\nu,$$

where s_n is simple, $\|s_n - f\|_\infty \to 0$, and

$$\int_X s\, d\nu = \sum_{j=1}^{n} a_j \nu(A_j) \text{ for } \quad s = \sum_{j=1}^{n} a_j \mathbb{1}_{A_j}.$$

It is easy to see that $\int_X f\, d\nu$ is well defined, and the following facts are checked in a routine fashion:

i. $\forall\, f, g \in L^\infty_\mu(X)$, $\forall\, \alpha, \beta \in \mathbb{C}$, and $\forall\, \nu \in F(X)$,

$$\int_X (\alpha f + \beta g)\, d\nu = \alpha \int_X f\, d\nu + \beta \int_X g\, d\nu;$$

ii. $\forall\, f \in L^\infty_\mu(X)$ and $\forall\, \nu \in F(X)$, $\left|\int_X f\, d\nu\right| \le \int_X |f|\, d|\nu|$;

iii. $F(X)$ is normed by (5.36);

iv. If $f \in L^\infty_\mu(X)$, $\nu \in F(X)$, and $\mu(\{x : f(x) \ne 0\}) = 0$, then

$$\int_X f\, d\nu = 0.$$

Theorem 5.5.7. The dual of $L^\infty_\mu(X)$

Let $(X, \mathcal{A}, \mu)$ be a measure space. There is a surjective isometric isomorphism

$$F(X) \to (L^\infty_\mu(X))',$$
$$\nu \mapsto F_\nu,$$

where

$$\forall\, f \in L^\infty_\mu(X), \quad F_\nu(f) = \int_X f\, d\nu. \tag{5.37}$$

Also, $F(X)$ is a Banach space.

Proof. If $\nu \in F(X)$ is given and F_ν is defined by (5.37), then property *ii* above yields $\|F_\nu\| \le \|\nu\|$, where F_ν is considered as an element of $(L^\infty_\mu(X))'$. To prove that $\|\nu\| \le \|F_\nu\|$, let $\varepsilon > 0$ and choose a finite decomposition $\{A_1, \ldots, A_n\} \subseteq \mathcal{A}$ of X such that

$$\sum_{j=1}^{n} |\nu(A_j)| > \|\nu\| - \varepsilon.$$

Define

$$g(x) = \sum_{j=1}^{n} a_j \mathbb{1}_{A_j}, \quad \text{where } a_j = \operatorname{sgn} \nu(A_j).$$

Clearly, $\|g\|_\infty \le 1$ and

$$|F_\nu(g)| = \left|\int_X g\, d\nu\right| = \left|\sum_{j=1}^{n} a_j \nu(A_j)\right| = \sum_{j=1}^{n} |\nu(A_j)| > \|\nu\| - \varepsilon.$$

Consequently, $\|\nu\| = \|F_\nu\|$, and so the mapping $\nu \mapsto F_\nu$ is an isometry. It remains to prove the surjectivity.

Let $L \in (L^\infty_\mu(X))'$ and define ν on $\mathcal{A}$ as

$$\nu(A) = L(\mathbb{1}_A).$$

Note that

$$\sup\{|\nu(A)| : A \in \mathcal{A}\} \leq \sup\{|L(g)| : \|g\|_\infty \leq 1,\ g \in L^\infty_\mu(X)\} = \|L\|.$$

Also, if $\mu(A) = 0$ then $\mathbb{1}_A = 0 \in L^\infty_\mu(X)$, and so $0 = L(0) = \nu(A)$. Thus, to verify that $\nu \in F(X)$ we take $A, B \in \mathcal{A}$, for which $A \cap B = \emptyset$, and observe that

$$\nu(A \cup B) = L(\mathbb{1}_{A\cup B}) = L(\mathbb{1}_A) + L(\mathbb{1}_B) = \nu(A) + \nu(B).$$

Our final task then is to prove that if $g \in L^\infty_\mu(X)$ then

$$L(g) = \int_X g\, d\nu.$$

Let $\{s_n : n = 1, \ldots\}$ be a sequence of simple functions such that $\|s_n - g\|_\infty \to 0$, and note that

$$L(s_n) = \sum_{j=1}^{m_n} a_{j,n} L(\mathbb{1}_{A_{j,n}}) = \sum_{j=1}^{m_n} a_{j,n}\nu(A_{j,n}) = \int_X s_n\, d\nu.$$

Since L is continuous,

$$L(g) = \lim_{n\to\infty} L(s_n) = \lim_{n\to\infty} \int_X s_n\, d\nu = \int_X g\, d\nu,$$

where the last equality is the definition of the integral. □

Theorem 5.5.7 was proved independently in 1934 by T. H. Hildebrandt for $(\ell^\infty)'$ [236] and by Fichtenholz and Leonid V. Kantorovich for $(L^\infty_m([a,b]))'$ [175]; also see [150], pages 296–297, and [383].

We note, e.g., [32], Chapter 2, that generally the condition

$$\sup\{|\nu(A)| : A \in \mathcal{A}\} < \infty \tag{5.38}$$

for $\nu \in F(X)$ cannot be replaced by the condition that

$$\forall\, A \in \mathcal{A}, \quad |\nu(A)| < \infty. \tag{5.39}$$

Recall that if ν is σ-additive then (5.39) implies (5.38).

5.6 Potpourri and titillation

1. As mentioned before Theorem 5.5.2, F. RIESZ introduced the L^p_m spaces, $p \neq 2$, in 1910, and he also proved the L^p-duality theorem, Theorem 5.5.5; see also Example A.8.4*b*. Using an argument with *Lagrange multipliers*, he then proved the following theorem. *Let $p > 1$, and let $\{f_n\} \subseteq L^p_m(\mathbb{R})$ and $\{c_n\} \subseteq \mathbb{C}$ be given sequences; then*

$$\exists\, f \in L^q_m(\mathbb{R}) \text{ such that} \,\forall\, n = 1, \ldots, \quad c_n = \int_{\mathbb{R}} f(x)\overline{f_n(x)}\, dx \tag{5.40}$$

if and only if there is $C > 0$ such that for all finite sequences $\{\lambda_n : n = 1, \ldots, N\} \subseteq \mathbb{C}$,

$$\left| \sum_{n=1}^{N} \lambda_n c_n \right| \leq C \left\| \sum_{n=1}^{N} \lambda_n f_n \right\|_p . \tag{5.41}$$

The proof that (5.40) implies (5.41) is clear. If $p = 2$ and $\{f_n\}$ is orthonormal, then RIESZ' theorem is equivalent to the *Riesz–Fischer theorem*; see Section 5.6.2.

The proof of (5.40) from (5.41) is a consequence of the Hahn–Banach theorem as follows. Assume (5.41) and let $V = \text{span}\,\{f_n\} \subseteq L^p(\mathbb{R})$. Define $K : V \to \mathbb{C}$ by the rule that $K(\sum \lambda_n f_n) = \sum \lambda_n c_n$ for any finite sequence $\{\lambda_n\} \subseteq \mathbb{C}$. Clearly, K is linear, and it is continuous on V by (5.41). Thus, by the Hahn–Banach theorem and Theorem 5.5.5, there is $L = f \in L^q(\mathbb{R})$ for which $L = K$ on V. Hence, $c_n = \int f(x)\overline{f_n(x)}\, dx$ for each n.

In 1912, the Austrian mathematician EDUARD HELLY (1884–1943) (who spent his last years in the United States as an actuary) gave an elementary proof that (5.41) is a sufficient condition for the validity of (5.40). In doing so, he proved a result that is *equivalent* to the Hahn–Banach theorem (Theorem A.8.3); see [138].

2. The original statements of the *Riesz–Fischer theorem* were the following; see, e.g., [393], [394], [396], [395] and [177], [176], for the works of F. RIESZ and ERNST FISCHER, respectively.

Theorem 5.6.1. Fischer theorem
The normed space $L^2_m([a,b])$ is complete in the sense that if $\{f_n : n = 1, \ldots\} \subseteq L^2_m([a,b])$ is a Cauchy sequence, then there is $f \in L^2_m([a,b])$ for which $\lim_{n\to\infty} \|f_n - f\|_2 = 0$.

Theorem 5.6.2. F. Riesz theorem
Let $\{f_n : n = 1, \ldots\} \subseteq L^2_m([a,b])$ be an orthonormal sequence. For any sequence $\{c_n : n = 1, \ldots\} \subseteq \mathbb{C}$ with the property that $\sum_{n=1}^{\infty} |c_n|^2 < \infty$, there exists a function $f \in L^2_m([a,b])$ such that $c_n = \langle f, f_n \rangle$ for each $n = 1, \ldots$.

The question of the equivalence of Theorem 5.6.1 and Theorem 5.6.2, as well as its historical context, have been analyzed by JOHN HORVÁTH

[244]. Since both versions are true, equivalence is meant in the sense that one statement implies the other in a straightforward way. HORVÁTH deals with this problem in the context of inner product spaces H (see Appendix A.1). The following is a natural question for such spaces. Does RIESZ' theorem, Theorem 5.6.2, for H imply the completeness of the inner product space H?

3. Besides the L^p results mentioned earlier in Section 5.6.1, F. RIESZ also proved the following theorem [390], pages 466–467: *Let* $q \in (1, \infty]$ *and let* $\{g_n : n = 1, \ldots\} \subseteq L^q_m([a, b])$ *be* L^q_m*-norm bounded; there are* $g \in L^q_m([a, b])$ *and a sequence* $\{n_k : k = 1, \ldots\} \subseteq \mathbb{N}$ *such that*

$$\forall\, f \in L^p_m([a, b]), \quad \lim_{k \to \infty} \int_a^b f(x) g_{n_k}(x)\, dx = \int_a^b f(x) g(x)\, dx,$$

and

$$\|g\|_q \leq \underline{\lim}_{k \to \infty} \|g_{n_k}\|_q.$$

This theorem easily extends to more general measure spaces. Further, it is a result in weak* compactness. (Technically, it is also weak compactness for $q \in (1, \infty)$; cf. the theory developed in Chapter 6.) In fact, this theorem is an immediate consequence of Theorem A.9.8, which is due to BANACH [19], page 123, and which establishes that the closed unit ball of the dual of a separable normed vector space is weak* sequentially compact.

These first three items of Section 5.6 are lopsided with contributions of F. RIESZ, some of which were proved while he was a high-school teacher from 1905–1912. He was one of the first to understand the importance of LEBESGUE's theory, and he went on to create profound mathematics depending on the symbiosis of the then nascent fields of Lebesgue integration and functional analysis. There is a clear outline of F. RIESZ' contributions in [115], which, as the article is aptly named, is also a *panorama* of twentieth-century Hungarian analysis following RIESZ.

4. In [313], 1st edition, 1904, page 94, LEBESGUE considered the following definition:

A bounded function f is *integrable* if there is a function F
with bounded derived numbers such that $F' = f$ *m-a.e.* (5.42)
The *integral* of f in (a, b) is $F(b) - F(a)$.

Such a definition generalizes the integrals of RIEMANN and JEAN-MARIE-CONSTANT DUHAMEL. LEBESGUE introduced (5.42) by saying, "Je ne m'occuperai pas, pour le moment du moins, de la suivante". And he keeps his word (and reticence) until on the very last page of text (page 129), where in a footnote no less, we read, "Pour qu'une fonction soit intégrable indéfinie, il faut de plus que sa variation totale dans une infinité dénombrable d'intervalles de longueur totale L tende vers zéro avec L. Si, dans l'énoncé de la page 94 (i.e., (5.42) above), on n'assujettit pas f à être borneé, ni F à être à nombres dérivés bornés, mais seulement à la condition précédente, on a une definition de l'intégrale équivalente à celle développée dans ce Chapitre et applicable à toutes les fonctions sommables, bornées ou non". Thus, in an obtuse presentation and as a footnote and without proof, we are handed the fundamental theorem of calculus!

As we have seen, VITALI defined the notion of absolute continuity in 1904 and went on to state and prove FTC. His next step in this business is [485]. He begins by proving the Vitali covering lemma and uses the covering theorem to prove an FTC in $\mathbb{R}^2$. He also deduces FTC on $\mathbb{R}$ with his covering theorem. Because of the importance of set functions in the development of integration theory we note that in Section 5 of [485] VITALI considers families of rectangles and the formula R–N for A, a "rettangolo coordinato".

Essentially, in his major work of 1910 [315], LEBESGUE relies on the Vitali covering lemma and "les travaux de M. VOLTERRA, à définir la dérivée de la fonction $F(A)$ en un point P comme la limite du rapport $F(A)/m(A)$, A étant un ensemble contenant P et dont on fait tendre toutes les dimensions vers zéro" [315], page 361.

In [315], LEBESGUE begins by quoting VITALI's FTC in $\mathbb{R}^2$, and notes that an "inadvertence" by VITALI in his proof is corrected by considering a *regular* family of rectangles in $\mathbb{R}^2$. This constraint to define Radon–Nikodym derivatives by taking limits over regular families is necessary. Further, by proving FTC in $\mathbb{R}^2$ in terms of set functions (not depending on rectangular coordinates as VITALI had done), LEBESGUE set the stage for the synthesis and generalization of RADON in 1913, where he (RADON) incorporated Stieltjes integrals into the scheme of things.

Concerning the above-mentioned footnote in [313], 1st edition, LEBESGUE [315], page 365, writes, "J'avais, dans mes Leçons, tout à fait incidentment sans demonstration, fait connaître" the FTC. On the other hand, in [313], 2nd edition 1928, page 188, LEBESGUE does give VITALI his due by writing that VITALI "a montré la simplicité et la clarté que prend toute la théorie quand on met cette notion (i.e., absolute continuity) à sa base". Naturally, this too is a footnote! For other testimony by LEBESGUE to VITALI's priority and role in fundamental results of the Lebesgue theory, see [313], 2nd edition 1928, pages 29 and 131.

A perusal of [315] indicates the crucial dependence of LEBESGUE on the Vitali covering lemma ("un théorème capital" [315], page 390) and VITALI's original proof for LEBESGUE's setting ("La démonstration qu'on lira plus loin est presque copiée sur celle de M. VITALI" [315], page 390).

5. Russian born HERMANN MINKOWSKI (1864–1909) had a dazzling mathematical career in Germany and Switzerland, and he died of appendicitis in his prime. His mathematical depth was astonishing, from profound results on quadratic forms to deep insights about space and time in the context of ALBERT EINSTEIN's special theory of relativity.

Let $(X, \mathcal{A}, \mu)$ be a measure space and let $p \geq 1$. An elementary form of the *Minkowski inequality*,

$$\forall\, f, g \in L^p_\mu(X), \quad \|f + g\|_p \leq \|f\|_p + \|g\|_p, \tag{5.43}$$

was proved in Theorem 5.5.2*a*. If $p = 1$, then there is equality in (5.43) if and only if there is a positive μ-measurable function P on X such that

$$fP = g, \quad \mu\text{-}a.e. \text{ on } \{x \in X : f(x)g(x) \neq 0\}.$$

If $1 < p < \infty$, then there is equality in (5.43) if and only if $af = bg$, for $a, b \geq 0$ for which $a^2 + b^2 > 0$. It is interesting to note that if $0 < p < 1$ then

$$\|f + g\|_p \geq \|f\|_p + \|g\|_p$$

for nonnegative $f, g \in L^p_\mu(X)$.

More generally than (5.43), we have the following form of the *Minkowski inequality* for $p \geq 1$:

$$\left(\int \left| \int f(x,y)\, d\mu(x) \right|^p d\mu(y)\right)^{1/p} \leq \int \left(\int |f(x,y)|^p\, d\mu(y)\right)^{1/p} d\mu(x),$$

i.e., the L^p-norm of a "sum" is less than or equal to the "sum" of the L^p-norms.

The proof of the Minkowski inequality is in his book *Geometrie der Zahlen* (1896), pages 115–117. See the classic book *Inequalities* [220] for generalizations and geometrical interpretation.

5.7 Problems

Some of the more elementary problems in this set are Problems 5.1, 5.2, 5.5, 5.6, 5.7, 5.12, 5.14, 5.21, 5.27, 5.28, 5.29, 5.31, 5.36, 5.42, 5.48, 5.49, 5.50.

5.1. Prove Proposition 5.1.1.

5.2. Give the proof of Theorem 5.1.6 for the case that $\mu^+(X) = \infty$.
[*Hint.* Use Proposition 5.1.7.]

5.3. Let $(X, \mathcal{A}, \mu)$ be a measure space where $X = [1, \infty)$, $\mathcal{B}(X) \subseteq \mathcal{A}$, and $\mu(X) < \infty$. Define $\phi(t) = \int_X \cos(xt)\, d\mu(x)$. Prove that ϕ has a zero in $[0, \pi]$, and that π cannot be replaced by a smaller number.
[*Hint.* Set $I = \int_0^\pi \sin(t\phi(t))\, dt$ so that by Fubini's theorem,

$$I = \int_1^\infty \frac{1 + \cos(\pi x)}{1 - x^2}\, d\mu(x) \leq 0.$$

Since $\phi(0) \geq 0$, ϕ must vanish on $[0, \pi]$. For the second part we construct μ with the property that ϕ's "first" zero is arbitrarily close to π. Take $\varepsilon > 0$ and $0 < \rho < 1$. Set

$$\mu = \sum_{k=0}^{\infty} \rho^k \delta_{k(2+\varepsilon)+1},$$

so that

$$\phi(t) = \frac{\cos(t) - \rho \cos((1+\varepsilon)t)}{1 + \rho^2 - 2\rho \cos((2+\varepsilon)t)}.]$$

5.4. Let f be an increasing function defined on $[0, 1]$. Consider the following condition: $\lim_{n\to\infty} \int_0^1 f_n \, df = 0$ for all sequences $\{f_n : n = 1, \ldots\} \subseteq C([0, 1])$ for which

$$\forall\, g \in L^1_m([0,1]), \quad \lim_{n\to\infty} \int_0^1 f_n g = 0. \tag{5.44}$$

Prove that f is absolutely continuous if and only if (5.44) holds.

Taking X to be a locally compact Hausdorff space and using the notation of Chapter 7, it is not difficult to prove the following generalization: *let* $\nu \in C'(X)$, $\mu \in M_b(X)$; *then* $\mu \ll \nu$ *if and only if for every sequence* $\{f_n : n = 1, \ldots\} \subseteq C_c(X)$ *such that* $\lim_{n\to\infty} \int f_n g \, d\nu = 0$ *for all* $g \in L^1_\nu(X)$ *we have* $\lim_{n\to\infty} \int f_n h \, d\mu = 0$ *for each* $h \in L^1_{|\mu|}(X)$.

5.5. Prove the case that $|\mu(A)| = \infty$ in Theorem 5.1.9*b*.

5.6. Prove by example that a Hahn decomposition of a signed measure μ need not be unique; but show that it is unique except for possibly a null set.

5.7. With regard to Theorem 5.1.4*d* find a signed measure space $(X, \mathcal{A}, \mu)$ and $A \in \mathcal{A}$ such that $\mu(A) = \infty$ and A is not nonnegative.

5.8. Let $f \in L^1_m(\mathbb{R})$, $g \in L^\infty_m(\mathbb{R})$. Prove the following assertions.

a. $\forall\, r$, $\lim_{x\to\infty} \int_{-\infty}^r f(x-y) g(y) \, dy = 0$.

b. If $\lim_{x\to\infty} g(x) = G$, then

$$\lim_{x\to\infty} f * g(x) = \lim_{x\to\infty} \int_{-\infty}^{\infty} f(x-y) g(y) \, dy = G \int_{-\infty}^{\infty} f(x) \, dx.$$

Convolution $f * g$ for $f, g \in L^1_m(\mathbb{R})$ was defined in Problem 3.5; see Problems 4.45 and 4.46. In this case of $f \in L^1_m(\mathbb{R})$, $g \in L^\infty_m(\mathbb{R})$, the definition of $f * g$ is easier to make well defined.

Remark. An important result in harmonic analysis is WIENER's *Tauberian theorem: Let* $f \in L^1_m(\mathbb{R})$, $g \in L^\infty_m(\mathbb{R})$, *and assume*

$$\forall\, \gamma \in \mathbb{R}, \quad \int_{-\infty}^{\infty} f(x) e^{-2\pi i x \gamma} \, dx \neq 0;$$

if

$$\lim_{x \to \infty} f * g(x) = r \int_{-\infty}^{\infty} f(x)\, dx,$$

then

$$\forall\, h \in L_m^1(\mathbb{R}), \quad \lim_{x \to \infty} h * g(x) = r \int_{-\infty}^{\infty} h(x)\, dx.$$

Algebraic formulations of WIENER's Tauberian theorem, as well as many applications, are found in [33]. WIENER's Tauberian theorem is restated in Appendix B.12, where its role in WIENER's Generalized Harmonic Analysis is explained.

5.9. In Problem 3.5 the exercise was to prove that if $f, g \in L_m^1(\mathbb{R})$, then $\|f * g\|_1 \le \|f\|_1 \|g\|_1$.

a. Prove that if $f \in L_m^1(\mathbb{R})$ and $g \in L_m^p(\mathbb{R})$, $1 < p < \infty$, then

$$f * g(y) = \int_{-\infty}^{\infty} f(y - x) g(x)\, dx$$

is well defined m-*a.e.*, and $\|f * g\|_p \le \|f\|_1 \|g\|_p$.
[*Hint.* Let $h \in L_m^q(\mathbb{R})$, and, using Hölder's inequality, begin with the estimate,

$$\int_{-\infty}^{\infty} \int_{-\infty}^{\infty} |f(y - x) h(y)|\, dx\, dy \le \|f\|_1 \|g\|_p \|h\|_q < \infty.]$$

b. Prove that if $f \in L_m^p(\mathbb{R})$ and $g \in L_m^q(\mathbb{R})$, $1 \le p < \infty$, then $f * g$ is uniformly continuous on $\mathbb{R}$, and $\|f * g\|_\infty \le \|f\|_p \|g\|_q$. In the case $p > 1$, prove $f * g \in C_0(\mathbb{R})$.

Remark. In this spirit, W. YOUNG proved the following result, which depends on the generalized Hölder inequality stated in Theorem 5.5.2*c*. Given $p > 1$, $r > 1$, and $s \in \mathbb{R}$ such that $1/p + 1/r - 1 = 1/s > 0$, let $f \in L_m^p(\mathbb{R})$ and $g \in L_m^r(\mathbb{R})$; then $f * g \in L_m^s(\mathbb{R})$ and $\|f * g\|_s \le \|f\|_q \|g\|_r$.

5.10. The proof of Hölder's inequality in Theorem 5.5.2 may seem magical. There are many proofs, mostly based on the same idea, but the following has a geometrical flavor.

a. Let $\phi : [0, \infty) \to [0, \infty)$ be a continuous strictly increasing surjection with continuous strictly increasing surjective inverse $\psi : [0, \infty) \to [0, \infty)$. Prove W. YOUNG's inequality (1912),

$$\forall\, a, b \ge 0, \quad ab \le \int_0^a \phi + \int_0^b \psi,$$

with equality if and only if $b = \phi(a)$.
[*Hint.* The result is *geometrically* clear if one draws the curve $y = \phi(x)$ and the lines $x = 0$, $x = a$, and $y = b$, and then considers the various areas bounded by them; also, see [220], pages 111–113.]

b. Let $p > 1$. Prove

$$\forall\, a, b \geq 0, \quad ab \leq \frac{a^p}{p} + \frac{b^q}{q},$$

with equality if and only if $a^p = b^q$.
[*Hint.* Let $\phi(x) = x^{p-1}$.]

c. Using part *b*, prove Hölder's inequality, $\|fg\|_1 \leq \|f\|_p \|g\|_q$, given the hypotheses of Theorem 5.5.2.
[*Hint.* Let $a = |f(x)|/\|f\|_p$ and $b = |g(x)|/\|g\|_q$.]

5.11. We proved Jensen's inequality in Problem 3.43. This problem states two consequences. Let $(X, \mathcal{A}, \mu)$ be a finite measure space for which $\mu(X) = 1$.

a. Let $f \in L^p_\mu(X)$, $0 < p < \infty$. Prove that

$$\exp\left(\int_X \log(|f(x)|)\, d\mu(x)\right) \leq \|f\|_p.$$

Here $\exp(t) = e^t$, and $\|...\|_p$, $0 < p < 1$, is not a norm as defined in Definition A.1.9. This range, $0 < p < 1$, is also considered in Theorem 8.6.4.
[*Hint.* Use the form of Jensen's inequality given in Problem 3.43*d*.]

b. Let $f \in L^r_\mu(X)$ for some $r > 0$. Prove that

$$\lim_{p \to 0} \|f\|_p = \exp\left(\int_X \log(|f(x)|)\, d\mu(x)\right).$$

[*Hint.* Use LDC and Jensen's inequality in conjunction with the fact that $(x^p - 1)/p$ decreases to $\log(x)$, $x > 0$, as $p \to 0$.]

5.12. Let $(X, \mathcal{A}, \mu)$ be a signed measure space where card $X \geq \aleph_0$, $\mathcal{A} = \mathcal{P}(X)$, and

$$\forall\, x \in X, \quad \mu(\{x\}) = 1 \text{ or } \mu(\{x\}) = -1. \tag{5.45}$$

If P, N is a Hahn decomposition of X prove that the cardinality of either P or N is finite.

5.13. Let $X = [-1, 1]$, let $\mathcal{A}$ be the algebra of all finite disjoint unions of intervals $[a, b) \subseteq X$, and define

$$f(x) = \begin{cases} \frac{1}{x}, & \text{if } x \neq 0, \\ 0, & \text{if } x = 0. \end{cases}$$

Define $\mu : \mathcal{A} \to \mathbb{R}$ as

$$\mu\left(\bigcup_{k=1}^n [a_k, b_k)\right) = \sum_{k=1}^n [f(b_k) - f(a_k)].$$

Prove that μ is a well-defined finitely additive set function but that $\mu \neq \mu^+ - \mu^-$. What can you say about a Hahn decomposition for μ?

5.14. Let $(X, \mathcal{A}, \mu)$ be a signed measure space. Prove the following assertions.

a. $\forall\, A, B \in \mathcal{A}$, with $|\mu(A)| < \infty$ and $B \subseteq A$, $|\mu(B)| < \infty$.

b. $\forall\, \{A_n : n = 1, \ldots\} \subseteq \mathcal{A}$, increasing, $\lim_{n\to\infty} \mu(A_n) = \mu\left(\bigcup A_n\right)$.

c. $\forall\, \{A_n : n = 1, \ldots\} \subseteq \mathcal{A}$, decreasing and satisfying $|\mu(A_1)| < \infty$, $\lim_{n\to\infty} \mu(A_n) = \mu\left(\bigcap A_n\right)$.

5.15. Let $([0,1], \mathcal{A}, \mu_\alpha)$ be a family of measure spaces and let $f_n : [0,1] \to \mathbb{R}^*$, $n = 1, \ldots,$ have the property that for each α, $\{f_n : n = 1, \ldots\} \subseteq L^1_{\mu_\alpha}([0,1])$ is $L^1_{\mu_\alpha}([0,1])$-Cauchy.

Prove that there is a function $f \in \bigcap L^1_{\mu_\alpha}([0,1])$ for which

$$\forall\, \alpha, \quad \|f - f_n\|_\alpha \to 0,$$

where $\|\ldots\|_\alpha$ is the $L^1_{\mu_\alpha}([0,1])$ norm.
[*Hint.* For the case "α" is not indexed by a sequence, the notion of an absolute ultrafilter is used.]

5.16. Prove Proposition 5.1.10.

5.17. Let $(X, \mathcal{A}, \mu)$ be a complex measure space. With regard to Theorem 5.1.12a, for which sets $A \in \mathcal{A}$ does the equality

$$|\mu|(A) = \mu_r^+(A) + \mu_r^-(A) + \mu_i^+(A) + \mu_i^-(A)$$

hold?

5.18. ***a.*** Prove the Lebesgue decomposition as a corollary to R–N.

b. Prove the uniqueness of the Lebesgue decomposition in Theorem 5.2.6.

c. Prove the extension parts of Theorem 5.2.6: extend from μ a bounded measure to μ σ-finite, respectively, $\mu \in M_b(X)$.

d. Show that if μ is counting measure on $(0,1)$ and $\nu = m$ on $(0,1)$ then μ has no Lebesgue decomposition with respect to m.

5.19. Let $(X, \mathcal{A})$ be a measurable space and let μ, ν be measures on $\mathcal{A}$. Prove that there are measures ν_1 and ν_2 such that

$$\nu_1 \ll \mu \quad \text{and} \quad \nu = \nu_1 + \nu_2,$$

where ν_2 satisfies the following singularity condition with respect to μ:

$$\forall\, A \in \mathcal{A},\ \exists\, B \subseteq A \quad \text{such that} \quad \nu_2(A) = \nu_2(B) \text{ and } \mu(B) = 0.$$

Also show that ν_2 is always unique and that ν_1 is unique if ν is σ-finite.

5.20. Consider the measurable space $(\mathbb{T}, \mathcal{B}(\mathbb{T}))$. The Fourier coefficients of $\mu \in M(\mathbb{T})$ are defined as

$$\forall\, n \in \mathbb{Z}, \quad \hat{\mu}(n) = \int_0^1 e^{2\pi i n x}\, d\mu(x).$$

a. If $k \geq 1$ is fixed, $k \in \mathbb{Z}$, and $\mu = \delta_{1/k}$, verify that

$$\forall\, n \in \mathbb{Z}, \quad \hat{\mu}(n+k) = \hat{\mu}(n). \tag{5.46}$$

Can you characterize the subclass of $M(\mathbb{T})$ that satisfies (5.46)?

b. Assume $\hat{\mu}(n) \to 0$ as $n \to \infty$. Prove that $\hat{\mu}(n) \to 0$ as $n \to -\infty$.
[*Hint.* By R–N, there is a bounded μ-measurable function f, $|f| = 1$, such that for each n

$$|\hat{\mu}(n)| = \left| \int_0^1 e^{2\pi i n x} f(x)\, d|\mu|(x) \right|.$$

Thus, for $g(x) = \overline{f(x)}/f(x)$,

$$\overline{|\hat{\mu}(n)|} = \left| \int_0^1 e^{-2\pi i n x} g(x)\, d\mu(x) \right|;$$

and so it is sufficient to prove that

$$\lim_{n\to\infty} \left| \int_0^1 e^{2\pi i n x} g(x)\, d\mu(x) \right| = 0. \tag{5.47}$$

To do this observe first and trivially that if $g(x)$ has the form $t_N(x) = \sum_{|k|\leq N} a_k e^{2\pi i k x}$ then (5.47) is immediate. By the Stone–Weierstrass theorem (Theorem B.7.3) and the fact that each bounded μ-measurable function h on $\mathbb{T}$ is the pointwise m-*a.e.* limit of a bounded sequence of continuous functions, we can compute

$$\lim_{n\to\infty} \int_0^1 e^{2\pi i n x} h(x)\, d\mu(x) = 0$$

by LDC and Jordan decomposition.]

5.21. Let $(X, \mathcal{A}, \mu)$ be a signed measure space with values in $\mathbb{R}$. Prove that $\mu^{\pm} = (1/2)(|\mu| \pm \mu)$.

5.22. Let $(X, \mathcal{A})$ be a measurable space, assume that X is a locally compact Hausdorff space, and take $M_b(X)$ with norm $\|\mu\| = |\mu|(X)$. Prove that $M_b(X)$ is complete, i.e., $M_b(X)$ is a Banach space.

5.23. Let $(X, \mathcal{A}, \mu)$ be a complex measure space. Prove the following assertions.

a. $|\mu|(X) < \infty$.

b. $|\mu(A)| \leq |\mu|(X)$.

[*Hint.* Use Theorem 5.1.12*c*.]

c. $\mu_r^+(A), \mu_r^-(A), \mu_i^+(A), \mu_i^-(A) \leq |\mu(A)|$.

[*Hint.* Use Theorem 5.1.12*c*.]

5.24. Let X be a locally compact Hausdorff space, and let $(X, \mathcal{A}, \mu)$ be a signed regular Borel measure space. Prove that the following assertions are equivalent.

i. $|\mu|$ is a regular Borel measure.

ii. μ^+, μ^- are regular Borel measures.

[*Hint.* Use Theorem 5.1.12*c*.]

5.25. Let X be a locally compact Hausdorff space, and let $(X, \mathcal{A}, \mu)$ be a signed measure space. Assume $|\mu(A)| < \infty$ for each $A \in \mathcal{A}$. Prove that $|\mu|$ is a bounded measure and that

$$\forall\, A \in \mathcal{A}, \quad |\mu|(A) = \sup\{|\mu(B)| : B \in \mathcal{A},\ B \subseteq A\}.$$

5.26. ***a.*** Prove Theorem 5.1.14.

b. For $\mu \in M_b(X)$, prove that

$$\left| \int_X f \, d\mu \right| \leq \int_X |f| \, d|\mu|.$$

5.27. Find a measure μ such that $\mu \neq \mu_c + \mu_d$.

[*Hint.* Take μ to be counting measure.]

5.28. Find a discrete measure that is absolutely continuous with respect to ν for some ν.

5.29. Let $(X, \mathcal{A}, \mu)$ be an unbounded measure space. Prove that X contains measurable sets of arbitrarily large bounded measure or else it contains a nonempty set A for which

$$\forall\, B \subseteq A,\ B \in \mathcal{A}, \quad \mu(B) = 0 \text{ or } \mu(B) = \infty.$$

5.30. Let $\mu \in M_b([0,1])$, where $M_b([0,1])$ is determined by the measurable space $([0,1], \mathcal{A})$ and $\mathcal{B}([0,1]) \subseteq \mathcal{A}$. Prove that μ is discrete if and only if for each net $\{f_\alpha\}$, e.g., [279], page 65, of bounded μ-measurable functions that decrease pointwise to 0 we can conclude that $\mu(f_\alpha) \to 0$; see Chapter 7 for notation.

[*Hint.* Assume that μ has a continuous part ν. Set $f_\alpha = \mathbb{1}_{E_\alpha}$, where $E_\alpha \subseteq [0,1]$ and card $E_\alpha^\sim \leq \aleph_0$. The ordering $E > F$ for the net $\{f_\alpha\}$ is defined by $E \subseteq F$.]

5.31. Let $(X, \mathcal{A})$ be a measurable space. Let $\mu \in SM(X)$ or $\mu \in M_b(X)$, and let $f \in L^1_\mu(X)$. Prove that f vanishes outside of a σ-finite set.

5.32. Let $(X, \mathcal{A})$ be a measurable space. Let $\mu, \nu \in SM(X)$ or $\mu, \nu \in M_b(X)$. Find conditions on μ and ν such that

$$\mu \ll \nu \Longrightarrow \mu \in M_c(X);$$

cf. Theorem 5.2.7.

5.33. Let $(X, \mathcal{A})$ be a measurable space. Let $\mu, \nu \in M_b(X)$. Prove that the following assertions are equivalent.

i. $\mu \ll \nu$.
ii. $|\mu| \ll \nu$.
iii. $\mu^+, \mu^- \ll \nu$.

5.34. Let $(X, \mathcal{A})$ be a measurable space. Let $\mu, \nu \in M_b(X)$. Prove that the following assertions are equivalent.

i. $\mu \perp \nu$.
ii. $|\mu| \perp \nu$.
iii. $\mu^+ \perp \nu$ and $\mu^- \perp \nu$.

5.35. Let $(X, \mathcal{A})$ be a measurable space. Let $\mu, \nu, \lambda \in M_b(X)$. Prove that if $\mu \ll \nu$ and $\lambda \perp \nu$, then $\mu \perp \lambda$.

5.36. Let $(X, \mathcal{A})$ be a measurable space. Let $\mu \in M_b(X)$ and assume $|\mu|(X) = \mu(X) = 1$. Prove that μ is a measure; cf. the proof of Theorem B.9.2. [*Hint.* Use Theorem 5.3.5.]

5.37. ***a.*** Let $(X, \mathcal{A}, \mu)$ be a measure space and let f be an $\mathbb{R}^*$- or $\mathbb{C}$-valued μ-measurable function on X. Define $\nu(A) = \int_A f \, d\mu$ and let g be an $\mathbb{R}^*$- or $\mathbb{C}$-valued ν-measurable function on X. Prove that

$$\forall\, A \in \mathcal{A}, \quad \int_A g \, d\nu = \int_A fg \, d\mu,$$

and that $g \in L^1_{|\nu|}(X)$ if and only if $fg \in L^1_\mu(X)$. This result can be used to prove an extension of R–N from the case of bounded measures.

b. Prove Theorem 5.3.2.

c. Prove part *b* with the assumption that $\nu \in M_b(X)$.

d. Prove that Theorem 5.3.2 is not valid for all measure spaces $(X, \mathcal{A}, \nu)$. [*Hint.* Consider the measurable space $([0,1], \mathcal{M}([0,1]))$, let ν be counting measure c, and let $\mu = m$.]

5.38. Provide the details for the following alternative proof to Theorem 5.3.1 (R–N). Let

$$\Phi = \left\{ f \in L^1_\nu(X) : f \geq 0 \text{ and } \forall\, A \in \mathcal{A},\ \mu(A) \geq \int_A f \, d\nu \right\}.$$

Then $f = 0$ is in Φ, and so $\Phi \neq \emptyset$. Set $r = \sup \left\{ \int_X f \, d\nu : f \in \Phi \right\}$ and check that $r = \int_X g \, d\nu$ for some $g \in \Phi$. Define

$$\forall\, A \in \mathcal{A}, \quad \lambda(A) = \mu(A) - \int_A g \, d\nu,$$

and obtain a contradiction by assuming that $\lambda(A) > 0$ for some A. The following lemma is used: *let $a \ll b$ for bounded measures a and b, and* *assume* $a(A) \neq 0$*; then there is a nonnegative function* $h \in L^1_b(X)$ *such that* $\int h\, db > 0$ *and*

$$a(B) \geq \int_B h\, db$$

for each b*-measurable set* B. To prove the lemma set $h = \varepsilon \mathbb{1}_Q$, where $a(A) > \varepsilon b(A)$ and Q is a nonnegative set with respect to $a - \varepsilon b$ for which $(a - \varepsilon b)(Q) > 0$. We then apply the lemma to the case $\lambda = a$, $\nu = b$, and $g + h \in \Phi$; and we contradict the definition of r.

5.39. In 1940 VON NEUMANN proved R–N for $\mu \ll \nu$, where μ and ν are bounded measures on the measurable space $(X, \mathcal{A})$, using "only" the fact that $(L^2_{\mu+\nu}(X))' = L^2_{\mu+\nu}(X)$. The proof is extremely slick, and, as we have seen in Theorem 5.5.5, the converse also holds. The proof that $(L^2_{\mu+\nu}(X))' = L^2_{\mu+\nu}(X)$ depends on the completeness of L^2, which, we recall from Theorem 5.5.2, is called the *Riesz–Fischer theorem*; see Section 5.6.2 as well as perspective on the Riesz–Fischer theorem in [39], pages 192–195. The Riesz–Fischer theorem can be proved elegantly and relatively simply in the context of Hilbert spaces.

VON NEUMANN's proof of R–N is given in several texts, e.g., [212], page 178, [235], pages 313 ff., [402], [405], pages 123–124. We outline the proof, and the content of this problem is to fill in the details.

i. Set $\lambda = \mu + \mu$ and prove that

$$\exists\, g \in L^2_\lambda(X) \text{ such that } \forall\, f \in L^2_\lambda(X), \quad \int_X f\, d\mu = \int_X fg\, d\lambda. \tag{5.48}$$

ii. Show that $0 \leq g \leq 1$ λ*-a.e.* by taking $f = \mathbb{1}_A$, $\lambda(A) > 0$, in (5.48).

iii. Compute

$$\forall\, A \in \mathcal{A}, \quad \int_A \left(1 - g^{n+1}\right)\, d\mu = \int_A g\left(1 + g + \cdots + g^n\right)\, d\nu. \tag{5.49}$$

iv. Prove that the left-hand side of (5.49) converges to $\mu(A)$ as $n \to \infty$.

5.40. SOLOMON BOCHNER [65] proved a version of R–N for the case that μ and ν are finitely additive. His original proof used the countably additive version of R–N as well as LEBESGUE's theorem on the differentiation of monotone functions. There have been simplifications of proof during the years. We mention those due to LESTER E. DUBINS [146] and C. FEFFERMAN [164]. Here is a statement of the result: *Let* $\mathcal{A} \subseteq \mathcal{P}(X)$ *be an algebra, let*

$$\mu, \nu : \mathcal{A} \to \mathbb{C}$$

be finitely additive, and assume

$$\|\nu\| = \sup_{A \in \mathcal{A}} |\nu(A)| < \infty \tag{5.50}$$

as well as the condition,

$$\forall\, \varepsilon > 0,\ \exists\, \delta > 0 \text{ such that } |\nu(A)| < \delta \Longrightarrow |\mu(A)| < \varepsilon; \tag{5.51}$$

then there is a sequence $\{s_n : n = 1, \ldots\}$ *of simple functions such that*

$$\forall\, A \in \mathcal{A}, \quad \lim_{n\to\infty} \int_A s_n \, d\nu = \mu(A).$$

Prove this theorem.

Statement (5.50) is compared to absolute continuity in Theorem 5.2.7 and Theorem 5.2.8. The above result can be strengthened to read, *If* $\|\nu\| < \infty$ *then the set* $\{s\nu : s$ *simple and* $(s\nu)(g) = \int gs\, d\nu\}$ *of finitely additive set functions is dense under the norm* $\|...\|$ *of (5.50) in the space of finitely additive set functions* μ *that satisfy (5.51).*

The importance of (5.51) has already been made clear in the proof of Theorem 5.3.1. For the finitely additive case, (5.50) is crucial.

5.41. Give examples of $\mu \in M_c(\mathbb{T})$, respectively, $\mu \in M_d(\mathbb{T})$, for which

$$\overline{\lim_{|n|\to\infty}} |\hat{\mu}(n)| > 0;$$

cf. Problem 4.5.

5.42. Let $f \in L^p_m((0,\infty))$, $p \in [1,\infty)$, and assume that f is uniformly continuous on $(0,\infty)$. Prove that $\lim_{x\to\infty} f(x) = 0$.

5.43. Fix $\mu \in M_b(X)$ and let $Y \subseteq X$ be the intersection of all possible concentration sets C_μ of μ. Prove that $Y = \{x : |\mu|(x) > 0\}$.

5.44. Consider the measure space $([0,1], \mathcal{M}([0,1]), m)$. Is it true that for each uncountable set $A \in \mathcal{M}([0,1])$, $m(A) = 0$, there is a measure $m_A \in M_c([0,1])$ defined on $\mathcal{B}([0,1])$, $m_A(A) = 1$, such that

$$D \subseteq B \subseteq A \Longrightarrow m_A(D) = m_A(B)m_B(D)?$$

See [478], pages 77–78 for a precise statement and discussion. Compare this with the problem of measure, mentioned in the remark after the definition of "measure" in Chapter 2, and which is answered by the statement that there are no continuous nonzero measures defined on $\mathcal{P}(X)$, card X = card $\mathbb{R}$. Also see Example 5.2.9*b*.

5.45. Prove Theorem 5.4.3.

5.46. Let $(X, \mathcal{A}, \mu)$ be a σ-finite measure space. Assume that $L^p_\mu(X)$ has a Hausdorff locally convex topology $\mathcal{T}$, so that with $\mathcal{T}$ on $L^p_\mu(X)$, $(L^p_\mu(X))' = L^{p/(p-1)}_\mu(X)$. Prove that $L^p_\mu(X)$ is $\mathcal{T}$-sequentially complete.

[*Hint.* First prove the result for the topology $\mathcal{T}_w = \sigma(L^p_\mu(X), L^{p/(p-1)}_\mu(X))$. For an arbitrary $\mathcal{T}$ choose a $\mathcal{T}$-Cauchy sequence $\{f_n : n = 1, \ldots\}$, so that, a fortiori, $\{f_n : n = 1, \ldots\}$ is $\mathcal{T}_w$-Cauchy. Thus, $f_n \to f$ in $\mathcal{T}_w$. Take a convex neighborhood V of 0, so that, by the Hahn–Banach theorem, V is $\mathcal{T}_w$ closed if it is $\mathcal{T}$ closed. Hence, eventually, $f_n \in f + V$. The notation "$(L^p_\mu(X))'$" requires some explanation in this exercise. And the problem should be attempted only if you have had some exposure to the theory of topological vector spaces.]

5.47. Let $\alpha, \beta > 0$, $\alpha\beta < 1$. If $f \in L^{1+\alpha}_\mu(X)$, $g \in L^{1+\beta}_\mu(X)$, and $f^{1+\alpha}g^{1+\beta} \in L^1_\mu(X)$, then prove that

$$\left|\int_X fg\, d\mu\right|^{(1+\alpha)(1+\beta)/(1-\alpha\beta)} \leq \left(\int_X |f|^{1+\alpha}|g|^{1+\beta}\, d\mu\right) \times \left(\int_X |f|^{1+\alpha}\, d\mu\right)^{\beta(1+\alpha)/(1-\alpha\beta)} \left(\int_X |g|^{1+\beta}\, d\mu\right)^{\alpha(1+\beta)/(1-\alpha\beta)}.$$

5.48. Let $g \in L^1_m([0,1])$ be nonnegative and take $r \geq 1$. Prove that

$$\left(\int_0^1 g\right)^r \leq \int_0^1 g^r.$$

5.49. Prove Proposition 5.4.4*b*.

5.50. ***a.*** Prove the part of Theorem 5.5.5 that extends the result from the finite to σ-finite case, and which was omitted in the text.

b. Prove Theorem 5.5.5 for any measure space $(X, \mathcal{A}, \mu)$ in the $1 < p < \infty$ case.

5.51. Let $\{\mu, \mu_n : n = 1, \ldots\} \subseteq M_b(X)$ and assume

$$\forall\, A \in \mathcal{A}, \quad \mu_n(A) \to \mu(A).$$

Can you find nontrivial situations to conclude that $\|\mu_n\|_1 \to \|\mu\|_1$; cf. Chapter 6?
[*Hint. i.* Take each μ_n to be a measure.

ii. Assume that $\mu_n \ll \nu$, where ν is a bounded measure, and that f_n converges in measure, where $f_n \in L^1_\nu(X)$ corresponds to μ_n.]

5.52. ***a.*** Consider the measure space $([0,1], \mathcal{B}([0,1]), m)$ and assume $\mu \ll m$. We define $f(x)$ in the manner of Theorem 5.4.1 and so $\mu_f = \mu$. Prove that f' is the R–N derivative with respect to m, cf., Section 8.4.

b. Prove that if g is absolutely continuous on $[0,1]$ and $f \in C([0,1])$, then

$$\int_0^1 f\, dg = \int_0^1 fg',$$

where $\int_0^1 f\, dg$ is the Riemann–Stieltjes integral.

5.53. Let $\nu \in M_b(X)$. Prove that $M_d(X)$, respectively, $M_c(X)$, $M_{ac}(X,\nu)$, $M_s(X,\nu)$, is closed in $M_b(X)$, where $M_b(X)$ is considered as a normed space with the total variation norm $\|\ldots\|_1$. Recall from Problem 5.22 that $M_b(X)$ is a Banach space.

5.54. Consider the measure space $(\mathbb{R}^d, \mathcal{M}^d(\mathbb{R}^d), m^d)$, where m^d is Lebesgue measure on $\mathbb{R}^d$ and $\mathcal{M}^d(\mathbb{R}^d)$ is the family of Lebesgue measurable sets in $\mathbb{R}^d$. If $f : \mathbb{R}^d \to \mathbb{C}$ is m^d-measurable and $A \in \mathcal{M}^d(\mathbb{R}^d)$ is fixed, define

$$\forall\, y > 0, \quad w(y) = m^d\left(\{x \in A : |f(x)| > y\}\right).$$

Prove that

$$\forall\, f \in L^1_{m^d}(A), \quad \int_A |f| = \int_0^\infty w(y)\, dy,$$

so that *the Lebesgue integral in* $\mathbb{R}^d$ *can be viewed as a one-dimensional improper integral*; see Theorem 8.6.4.
[*Hint.* Note that if $m^d(A) < \infty$ then

$$w(y) \le \frac{1}{y} \int_A |f|.]$$

5.55. Prove WIENER's theorem,

$$\forall\, \mu \in M(\mathbb{T}), \quad \sum_{x \in \mathbb{T}} |\mu(\{x\})|^2 = \lim_{N\to\infty} \frac{1}{2N+1} \sum_{n=-N}^{N} |\hat{\mu}(n)|^2.$$

As a consequence we obtain WIENER's characterization of continuous measures (1924): $\mu \in M_c(\mathbb{T})$ *if and only if*

$$\lim_{N\to\infty} \frac{1}{2N+1} \sum_{n=-N}^{N} |\hat{\mu}(n)|^2 = 0,$$

e.g., [505], pages 146–149, and [33] for applications with pseudomeasures. *In particular, if* $\mu \in M(\mathbb{T})$ *and* $|\hat{\mu}| = 1$ *then* $\mu \notin M_c(\mathbb{T})$.

5.56. Let $\mathcal{A} \subseteq \mathcal{P}(X)$, card $\mathcal{A} \ge \aleph_0$, be an algebra. Prove that there is a finitely additive function $\nu : \mathcal{A} \to \mathbb{R}$ such that

$$\sup_{A \in \mathcal{A}} |\nu(A)| = \infty.$$

[*Hint.* First prove that there is an infinite disjoint family $\{A_n : n = 1, \ldots\} \subseteq \mathcal{A}$. Choose $x_n \in A_n$ and define

$$\mu = \sum_{n=1}^{\infty} \frac{\pi^n}{4^n} \delta_{x_n}.$$

Since $\{\pi^n/4^n : n = 1, \ldots\}$ is linearly independent over $\mathbb{Q}$ it can be embedded in a Hamel basis H for $\mathbb{R}$. Define the appropriate function $f : H \to \mathbb{R}$, extend linearly, and set

$$\forall\, A \in \mathcal{A}, \quad \nu(A) = f(\mu(A)).]$$

Remark. Using a set-theoretic argument it can be shown that there is a nontrivial finitely additive set function $\nu : \mathcal{P}(\mathbb{N}) \to [0,1]$ such that $\nu(F) = 0$ if $F \subseteq \mathbb{N}$ is a finite set.

5.57. Decompose μ_E, the Cantor–Lebesgue measure for the perfect symmetric set E, as in Theorem 5.3.6 for $\nu = m$. Do not forget the case $m(E) > 0$. With regard to this and Problem 4.24*c* we note the paper by Lipiński [325], which completely characterizes the set of points in which a singular function can have an infinite derivative.

5.58. Find a nonnegative function $f \in L^2_m([1,\infty))$ such that

$$\int_1^\infty \frac{f(x)}{\sqrt{x}}\, dx$$

diverges.
[*Hint.* Let $f = \sum_{n=1}^\infty k_n \mathbb{1}_{[n,n+1)}$ and set

$$\frac{1}{n \log n} = k_n \int_n^{n+1} \frac{dx}{\sqrt{x}}.]$$

5.59. Let $1 < p < \infty$ and let $1/p + 1/q = 1$. Suppose that $\{b_n : n = 1, \ldots\} \notin \ell^q(\mathbb{N})$. Prove that there is $\{a_n : n = 1, \ldots\} \in \ell^p(\mathbb{N})$ such that $\sum_{n=1}^\infty |a_n b_n| = \infty$.
[*Hint.* Use the Uniform Boundedness Principle.]

The sequence can actually be constructed; see Helmberg, Amer. Math. Monthly, 111 (2004), 518–520.

6 Weak Convergence of Measures

6.1 Vitali theorem

We shall now prove VITALI's theorems, Theorem 3.3.11 and Theorem 3.3.12. As we noted in the Remark after the statement of Theorem 3.3.12, VITALI's results give nontrivial necessary and sufficient conditions in order that

$$\lim_{n\to\infty} \int_X |f_n - f| \, d\mu = 0, \tag{6.1}$$

where $(X, \mathcal{A}, \mu)$ is a measure space and $\{f_n : n = 1, \ldots\} \subseteq L^1_\mu(X)$. The following result was mentioned after the statement of Theorem 3.3.12 and is used to deduce LDC from VITALI's theorem.

Theorem 6.1.1. Lebesgue characterization of convergence in measure
Let $(X, \mathcal{A}, \mu)$ be a measure space and let $\{f_n : n = 1, \ldots\}$ be a sequence of μ-measurable functions that converges μ-a.e. to a μ-measurable function f. If there is $g \in L^1_\mu(X)$ such that

$$\forall\, n = 1, \ldots, \quad |f_n| \leq g \quad \mu\text{-a.e.},$$

then $f_n \to f$ in measure.

Proof. Without loss of generality assume that

$$\forall\, n = 1, \ldots \text{ and } \forall\, x \in X, \quad |f_n(x)|, |f(x)| \leq g(x).$$

Then, for each $\varepsilon > 0$, we define

$$A_{n,\varepsilon} = \bigcup_{j=n}^{\infty} \{x : |f_j(x) - f(x)| \geq \varepsilon\} \subseteq \{x : |g(x)| \geq \varepsilon/2\},$$

so that by the integrability of g, $\mu(A_{n,\varepsilon}) < \infty$ for $n = 1, \ldots$. Since $f_n \to f$ μ-*a.e.* we have

$$\mu\left(\bigcap_{n=1}^{\infty} A_{n,\varepsilon}\right) = 0. \tag{6.2}$$

J.J. Benedetto and W. Czaja, *Integration and Modern Analysis*, Birkhäuser Advanced Texts,
DOI 10.1007/978-0-8176-4656-1_6,

Thus, for each $\varepsilon > 0$,

$$\overline{\lim_{n\to\infty}}\ \mu(\{x : |f_n(x) - f(x)| \geq \varepsilon\}) \leq \overline{\lim_{n\to\infty}}\ \mu(A_{n,\varepsilon}) = 0,$$

where the equality follows from Theorem 2.4.3*c*, from (6.2), and because $\mu(A_{1,\varepsilon}) < \infty$ and each $A_{n,\varepsilon} \subseteq A_{n-1,\varepsilon}$. □

Theorem 6.1.2. Vitali uniform absolute continuity theorem (Theorem 3.3.11)
Let $(X, \mathcal{A}, \mu)$ be a finite measure space and let $\{f_n : n = 1, \ldots\} \subseteq L^1_\mu(X)$. Equation (6.1) is valid for some $f \in L^1_\mu(X)$ if and only if

i. $\{f_n : n = 1, \ldots\}$ converges in measure to a μ-measurable function g, and

ii. $\{f_n : n = 1, \ldots\}$ is uniformly absolutely continuous.

Proof. ($\Longleftarrow$) Without loss of generality take each f_n to be real-valued.

a. Given condition *ii* we first prove that $\{|f_n| : n = 1, \ldots\}$ is uniformly absolutely continuous. Let $\varepsilon > 0$. Choose $\delta > 0$ such that

$$\forall\, A \in \mathcal{A}, \text{ for which } \mu(A) < \delta, \text{ and } \forall\, n, \quad \left| \int_A f_n \, d\mu \right| < \frac{\varepsilon}{2}. \tag{6.3}$$

For each such A we define the disjoint sets

$$A_n^+ = \{x \in A : f_n(x) \geq 0\} \quad \text{and} \quad A_n^- = \{x \in A : f_n(x) < 0\}.$$

The sets A_n^+ and A_n^- are measurable since f_n is measurable, and

$$\forall\, n = 1, \ldots, \quad \mu(A_n^+) < \delta \quad \text{and} \quad \mu(A_n^-) < \delta,$$

since $\mu(A) < \delta$. By definition,

$$\int_{A_n^+} |f_n| \, d\mu = \left| \int_{A_n^+} f_n \, d\mu \right| < \frac{\varepsilon}{2} \quad \text{and} \quad \int_{A_n^-} |f_n| \, d\mu = \left| \int_{A_n^-} f_n \, d\mu \right| < \frac{\varepsilon}{2}.$$

Consequently, for each n,

$$\int_A |f_n| \, d\mu = \int_{A_n^+ \cup A_n^-} |f_n| \, d\mu = \int_{A_n^+} |f_n| \, d\mu + \int_{A_n^-} |f_n| \, d\mu < \varepsilon,$$

and so $\{|f_n| : n = 1, \ldots\}$ is uniformly absolutely continuous.

b. Using the uniform absolute continuity of $\{|f_n|\}$ we shall now prove that $\{f_n : n = 1, \ldots\}$ is $L^1_\mu(X)$-Cauchy.

For each $\sigma > 0$, define

$$A_{m,n}(\sigma) = \{x : |f_m(x) - f_n(x)| \geq \sigma\}$$

and

$$B_{m,n}(\sigma) = \{x : |f_m(x) - f_n(x)| < \sigma\}.$$

Then

$$\begin{aligned}\int_X |f_m - f_n|\, d\mu &= \int_{A_{m,n}(\sigma)} + \int_{B_{m,n}(\sigma)} |f_m - f_n|\, d\mu \\ &\leq \int_{A_{m,n}(\sigma)} |f_m - f_n|\, d\mu + \sigma\mu\left(B_{m,n}(\sigma)\right) \qquad (6.4)\\ &\leq \int_{A_{m,n}(\sigma)} |f_m|\, d\mu + \int_{A_{m,n}(\sigma)} |f_n|\, d\mu + \sigma\mu(X).\end{aligned}$$

For $\varepsilon > 0$ there is $\sigma_0 > 0$ such that

$$\forall\, \sigma \leq \sigma_0, \quad \sigma\mu(X) < \frac{\varepsilon}{3} \qquad (6.5)$$

(here we use the fact that $\mu(X) < \infty$). From the first part of the proof there is $\delta > 0$ for which

$$\int_A |f_n|\, d\mu < \frac{\varepsilon}{3} \qquad (6.6)$$

if $A \in \mathcal{A}$ satisfies $\mu(A) < \delta$. From *i* we can choose N, where

$$\forall\, m, n > N, \quad \mu\left(A_{m,n}(\sigma_0)\right) < \delta. \qquad (6.7)$$

Thus, given $\varepsilon > 0$ we have picked σ_0 and δ as above, i.e., (6.5) and (6.6), and then used *i* to find N as in (6.7). We apply (6.5) and (6.6) to (6.4) to obtain

$$\forall\, m, n > N, \quad \int_X |f_m - f_n|\, d\mu < \varepsilon.$$

c. Since $L^1_\mu(X)$ is complete, we obtain (6.1). Moreover, g in part *i* equals f in (6.1) μ-*a.e.* by the following argument. By Theorem 3.3.13*a* there exists a subsequence $\{f_{n_k}\}$ that converges pointwise to g μ-*a.e.* On the other hand, $\{f_{n_k}\}$ converges in $L^1_\mu(X)$ to f, so we can choose a subsequence of $\{f_{n_k}\}$ that converges pointwise to f μ-*a.e.* (see Theorem 3.3.13).

($\Longrightarrow$) Fix $\varepsilon > 0$. Let

$$A_n = \{x : |f_n(x) - f(x)| \geq \varepsilon\}.$$

Then

$$\int_X |f_n - f|\, d\mu \geq \int_{A_n} |f_n - f|\, d\mu \geq \varepsilon\mu(A_n),$$

so that $\mu(A_n) \to 0$ from (6.1), and we have proved part *i*.

Also, by (6.1), for any $\varepsilon > 0$ we can choose N for which

$$\forall\, m, n \geq N, \quad \int_X |f_m - f_n|\, d\mu < \frac{\varepsilon}{3}.$$

Since each $f_n \in L^1_\mu(X)$ we can find $\delta > 0$ such that if $\mu(A) < \delta$ for $A \in \mathcal{A}$, then

$$\forall\, n = 1, \ldots, N, \quad \int_A |f_n|\, d\mu < \frac{\varepsilon}{3}.$$

Consequently, for $n > N$ and $\mu(A) < \delta$, $A \in \mathcal{A}$, we have

$$\left|\int_A f_n\, d\mu\right| \leq \int_A |f_n - f_N|\, d\mu + \int_A |f_N|\, d\mu < \varepsilon.$$

Hence, for each n, if $\mu(A) < \delta$, then

$$\left|\int_A f_n\, d\mu\right| < \varepsilon;$$

and we have proved part *ii.* □

Remark. One is tempted to prove directly that $f \in L^1_\mu(X)$ using the Fatou lemma, the uniform absolute continuity of $\{|f_n| : n = 1, \ldots\}$, and the fact (from *i*) that $f_{n_k} \to f$ μ-*a.e.* for some subsequence of $\{f_n : n = 1, \ldots\}$. By this method,

$$\int_A |f|\, d\mu \leq \underline{\lim}_{k\to\infty} \int_A |f_{n_k}|\, d\mu \leq \sup_{n=1,\ldots} \int_A |f_n|\, d\mu < \varepsilon$$

for $\mu(A) < \delta$. Thus, $f \in L^1_\mu(A)$, so that since $\mu(X)$ is finite we write $X = \bigcup_{j=1}^n A_j$, $\mu(A_j) < \delta$, and we have $f \in L^1_\mu(X)$. This procedure works except when there are not enough sets of small enough measure to guarantee that $X = \bigcup_{j=1}^n A_j$.

The notions of uniform absolute continuity and Vitali equicontinuity are obviously closely related. The following result quantifies this assertion.

Proposition 6.1.3. *Let $(X, \mathcal{A}, \mu)$ be a measure space, let $M_b(X)$ be the space of complex measures associated with the measurable space $(X, \mathcal{A})$, and let $\{\nu_n : n = 1, \ldots\} \subseteq M_b(X)$.*

a. *Assume that for each n, ν_n is a measure and $\nu_n \ll \mu$. If $\{\nu_n : n = 1, \ldots\}$ is Vitali equicontinuous then it is uniformly absolutely continuous.*

b. *Assume $\mu(X) < \infty$. If $\{\nu_n : n = 1, \ldots\}$ is uniformly absolutely continuous then it is Vitali equicontinuous.*

Proof. ***a.*** We use the argument of Theorem 5.2.7. Assume that $\{\nu_n : n = 1, \ldots\}$ is not uniformly absolutely continuous. Then choose $\varepsilon > 0$, $\{B_k : k = 1, \ldots\} \subseteq \mathcal{A}$, and integers $\{n_k : k = 1, \ldots\}$ such that

$$\mu(B_k) < \frac{1}{2^{k+1}} \quad \text{and} \quad \nu_{n_k}(B_k) \geq \varepsilon. \tag{6.8}$$

Set $E_k = \bigcup_{j=k}^{\infty} B_j$, so that $E_{k+1} \subseteq E_k$. Also

$$\mu(E_k) \le \sum_{j=k}^{\infty} \frac{1}{2^{j+1}} = \frac{1}{2^k}.$$

We write $A_k = E_k \setminus E_0$, where $E_0 = \bigcap_{k=1}^{\infty} E_k$. Thus, $A_{k+1} \subseteq A_k$ and $\bigcap_{k=1}^{\infty} A_k = \emptyset$.

By Vitali equicontinuity, there exists $N \in \mathbb{N}$ such that for all $k > N$

$$|\nu_{n_k}(A_k)| < \varepsilon. \tag{6.9}$$

Since $\nu_{n_k} \ll \mu$ and $\mu(E_k) = 2^{-k}$, we have

$$\nu_{n_k}(E_0) = 0. \tag{6.10}$$

Then (6.9) and (6.10) imply that

$$|\nu_{n_k}(E_k)| < \varepsilon.$$

However, from (6.8), $\nu_{n_k}(E_k) \ge \nu_{n_k}(B_k) \ge \varepsilon$. This is the desired contradiction.

b. Fix $\varepsilon > 0$ and take $\{A_n : n = 1, \ldots\} \subseteq \mathcal{A}$ decreasing to $\emptyset$. By hypothesis there is $\delta > 0$ such that if $A \in \mathcal{A}$, $\mu(A) < \delta$, then $|\nu_m(A)| < \varepsilon$ for each m. Since $\mu(X) < \infty$, we have $\mu(A_n) \to 0$, and so there is N for which $\mu(A_n) < \delta$ if $n > N$. Consequently, $|\nu_m(A_n)| < \varepsilon$ for all $n > N$ and for all m. □

In light of Proposition 6.1.3 and the first part of the proof of the sufficient conditions in Theorem 6.1.2 we have the following result.

Proposition 6.1.4. *Let $(X, \mathcal{A}, \mu)$ be a finite measure space and let $\{f_n : n = 1, \ldots\} \subseteq L^1_\mu(X)$. Define ν_n by*

$$\forall\, A \in \mathcal{A}, \quad \nu_n(A) = \int_A |f_n|\, d\mu.$$

Then $\{\nu_n : n = 1, \ldots\}$ is Vitali equicontinuous if and only if $\{f_n : n = 1, \ldots\}$ is uniformly absolutely continuous.

The "only if" part of Proposition 6.1.4 does not require that $\mu(X) < \infty$.

Theorem 6.1.5. Vitali equicontinuity theorem (Theorem 3.3.12) *Let $(X, \mathcal{A}, \mu)$ be a measure space and choose a sequence $\{f_n : n = 1, \ldots\} \subseteq L^1_\mu(X)$. Then (6.1) is valid for some $f \in L^1_\mu(X)$ if and only if*

i. $\{f_n : n = 1, \ldots\}$ converges in measure (to a μ-measurable function g), and

ii. $\left\{\nu_n : \forall\, A \in \mathcal{A},\ \nu_n(A) = \int_A |f_n|\, d\mu\right\}$ is Vitali equicontinuous.

Proof. ($\Longleftarrow$) If $\mu(X) < \infty$ the result follows by Theorem 6.1.2 and Proposition 6.1.4. Note that $\bigcup_n \{x : f_n(x) \neq 0\}$ is σ-finite (Problem 5.31) so that, without loss of generality, we take X to be σ-finite.

Let $\{B_m : m = 1, \ldots\} \subseteq \mathcal{A}$ be an increasing sequence of sets that forms a cover of X and assume that each $\mu(B_m) < \infty$. If we set $A_m = B_m^{\sim}$ then $A_{m+1} \subseteq A_m$ and $\bigcap A_k = \emptyset$.

We shall prove that $\{f_n : n = 1, \ldots\}$ is $L^1_\mu(X)$-Cauchy. Take $\varepsilon > 0$. From *ii* there is k such that

$$\forall\, m = 1, \ldots, \quad \int_{A_k} |f_m| \, d\mu < \frac{\varepsilon}{4};$$

and so, for each m and n,

$$\int_{A_k} |f_m - f_n| \, d\mu < \frac{\varepsilon}{2}. \tag{6.11}$$

Using Theorem 6.1.2 and the restriction measure space $(B_k, \mathcal{A}, \mu)$, we see that there is $N > 0$ such that

$$\forall\, m, n \geq N, \quad \int_{B_k} |f_m - f_n| \, d\mu < \frac{\varepsilon}{2}. \tag{6.12}$$

Then (6.11) and (6.12) yield the result.

($\Longrightarrow$) The proof of *i* from (6.1) is the same as in Theorem 6.1.2. To prove *ii* pick any $\varepsilon > 0$ and take N such that

$$\forall\, n \geq N, \quad \int_X |f_n - f| \, d\mu < \frac{\varepsilon}{2}. \tag{6.13}$$

Let $\{A_n : n = 1, \ldots\} \subseteq \mathcal{A}$ decrease to $\emptyset$. Since $f_n, f \in L^1_\mu(X)$, $n = 1, \ldots, N$, and since $\mathbb{1}_{A_m} \to 0$ μ-*a.e.*, we invoke LDC to assert that there is $M > 0$ for which

$$\int_{A_m} |f| \, d\mu < \frac{\varepsilon}{2} \quad \text{and} \quad \int_{A_m} |f_n - f| \, d\mu < \frac{\varepsilon}{2} \tag{6.14}$$

if $m > M$ and $n = 1, \ldots, N$. Consequently, for each $m > M$ and for all n,

$$\int_{A_m} |f_n| \, d\mu \leq \int_{A_m} |f_n - f| \, d\mu + \int_{A_m} |f| \, d\mu < \varepsilon,$$

where the last inequality follows from (6.13) or the second part of (6.14), depending on whether $n > N$ or $n \leq N$. $\square$

In Sections 6.2 and 6.3 we shall see that VITALI's results are intimately related to subsequent important work by HAHN, NIKODYM, STANISŁAW SAKS, DIEUDONNÉ, and ALEXANDRE GROTHENDIECK in the area of weak convergence of measures.

A point of information and a corresponding word of caution are in order. We use "weak convergence" in the topological sense defined in Definition A.9.2; also see Definition 6.3.1. Modern probabilists also have a notion of weak convergence of measures on complete separable metric spaces X. They say that $\mu_n \to \mu$ *weakly* if, for every bounded continuous function f on X, $\mu_n(f) \to \mu(f)$. In this setting our weak sequential convergence generates a stronger topology than "probabilistic weak sequential convergence"; see Section 6.6.5.

6.2 Nikodym and Hahn–Saks theorems

In this section the work of HAHN dates from 1922 and that of SAKS from 1933 [411], [410]. NIKODYM announced his results in 1931 and published them in 1933. Our presentation is due to JAMES K. BROOKS [75].

We use the following version of Schur's lemma [76]; cf. Theorem A.7.3.

Theorem 6.2.1. Schur lemma
Let $\{c_{i,j} : i, j = 1, \ldots\} \subseteq \mathbb{C}$ have the following properties:

i.

$$\forall\, i = 1, \ldots, \quad \sum_{j=1}^{\infty} |c_{i,j}| < \infty,$$

ii.

$$\forall\, S \subseteq \mathbb{N}, \quad \exists \lim_{i\to\infty} \sum_{j\in S} c_{i,j}.$$

Then there is $\{c_j : j = 1, \ldots\} \subseteq \mathbb{C}$ such that $\sum |c_j| < \infty$ and

$$\lim_{i\to\infty} \sum_{j=1}^{\infty} |c_{i,j} - c_j| = 0. \tag{6.15}$$

Proof. a. We first prove that if we are given $\{z_1, \ldots, z_n\} \subseteq \mathbb{C}$, then there is $S \subseteq \{j : 1 \leq j \leq n\}$ such that

$$\sum_{j=1}^{n} |z_j| \leq 4\sqrt{2} \left| \sum_{j\in S} z_j \right|. \tag{6.16}$$

To begin, divide the complex plane into its four "diagonal" quadrants Q_1, Q_2, Q_3, Q_4; see Figure 6.1.

Assume that

$$\sum_{z_j\in Q_1} |z_j| \geq \frac{1}{4} \sum_{j=1}^{n} |z_j|,$$

and let $S = \{j : z_j \in Q_1\}$. Noting that $Q_1 = \{z = x + iy : |y| \leq x, x \geq 0\}$, we have

$$\frac{|z|}{\sqrt{2}} \leq \operatorname{Re}(z),$$

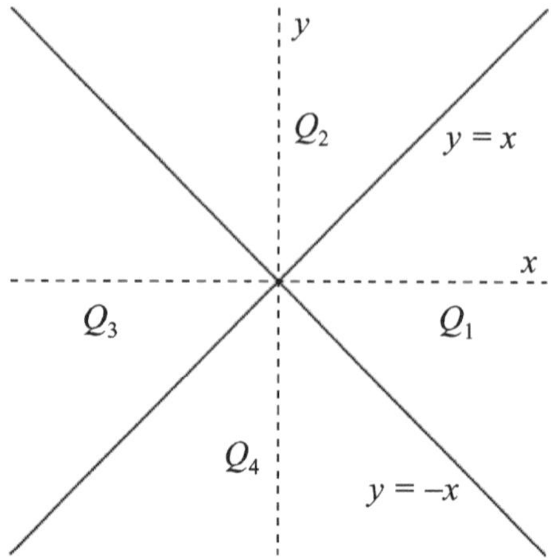

Fig. 6.1. Schur lemma diagram.

since, for $w = a + ib \in Q_1$,

$$\sqrt{2}\mathrm{Re}\,(w) = a\sqrt{2} = |a|\sqrt{2} = |a + ia| \geq |a + ib| = |w|.$$

Thus,

$$\left|\sum_{j\in S} z_j\right| \geq \sum_{j\in S} \mathrm{Re}\,(z_j) \geq \frac{1}{\sqrt{2}} \sum_{j\in S} |z_j| \geq \frac{1}{4\sqrt{2}} \sum_{j=1}^{n} |z_j|,$$

which is (6.16). Inequality (6.16) is standard, and we refer as well to [405], page 119, and our Remark prior to Theorem 5.1.12.

b. We now prove (6.15) using the additional hypothesis that the limit in *ii* is 0 for any S. Thus, in this case, we are taking each $c_j = 0$ in (6.15). We assume that (6.15) is false and shall obtain a contradiction. Since (6.15) is false, there are an $\varepsilon > 0$ and a subsequence $\{i_k : k = 1, \ldots\}$ such that

$$\forall\, k = 1, \ldots, \quad \sum_{j=1}^{\infty} |c_{i_k,j}| > \varepsilon > 0.$$

Observe that there are strictly increasing sequences $\{p_k : k = 1, \ldots\} \subseteq \mathbb{N}$ and $\{n_k : k = 1, \ldots\} \subseteq \{i_k : k = 1, \ldots\}$ for which

$$\forall\, k = 1, \ldots, \quad \sum_{j=1}^{p_k} |c_{n_k,j}| < \frac{\varepsilon}{r}, \tag{6.17}$$

and

$$\forall\, k = 1, \ldots, \quad \sum_{j=p_{k+1}+1}^{\infty} |c_{n_k,j}| < \frac{\varepsilon}{r}, \tag{6.18}$$

where $r > 2 + 8\sqrt{2}$. To see this, start by taking any p_1. Then, using (6.16), observe that for each n there is $Y_n \subseteq \{0, \ldots, p_1\}$ such that

$$\sum_{j=1}^{p_1} |c_{n,j}| \leq 4\sqrt{2} \left| \sum_{j \in Y_n} c_{n,j} \right| .$$

Since there are only finitely many possible subsets Y_n we obtain (6.17) for some n_1 from *ii*. Then, using *i*, choose p_2 for which (6.18) is true. Continue this process. From (6.17) and (6.18) we compute

$$\begin{aligned} \sum_{j=p_k+1}^{p_{k+1}} |c_{n_k,j}| &= \sum_{j=1}^{\infty} |c_{n_k,j}| - \sum_{j=1}^{p_k} |c_{n_k,j}| - \sum_{j=p_{k+1}+1}^{\infty} |c_{n_k,j}| \\ &> \varepsilon - \sum_{j=1}^{p_k} |c_{n_k,j}| - \sum_{j=p_{k+1}+1}^{\infty} |c_{n_k,j}| > \varepsilon\left(1 - \frac{2}{r}\right). \end{aligned} \tag{6.19}$$

We now use (6.16) in the following way. For each k choose $S_k \subseteq \{j : p_k + 1 \leq j \leq p_{k+1}\}$ such that

$$4\sqrt{2} \left| \sum_{j \in S_k} c_{n_k,j} \right| \geq \sum_{j=p_k+1}^{p_{k+1}} |c_{n_k,j}|. \tag{6.20}$$

Letting $S = \bigcup S_k$ we combine (6.19) and (6.20), and obtain for each k that

$$\begin{aligned} \left| \sum_{j \in S} c_{n_k,j} \right| &\geq \left| \sum_{j \in S_k} c_{n_k,j} \right| - \sum_{j=1}^{p_k} |c_{n_k,j}| - \sum_{j=p_{k+1}+1}^{\infty} |c_{n_k,j}| \\ &\geq \varepsilon\left(\frac{1}{4\sqrt{2}} - \frac{1}{2r\sqrt{2}} - \frac{r}{2} \right) > 0. \end{aligned}$$

This is the desired contradiction (of *ii*).

c. We now reduce the general case to the setting of part *b*. Let $\{m_i : i = 1, \ldots\} \subseteq \mathbb{N}$ increase to infinity and define

$$a_{i,j} = c_{m_{i+1},j} - c_{m_i,j}.$$

Then $\{a_{i,j} : i, j = 1, \ldots\}$ satisfies the hypotheses of part *b*, so that we can conclude that

$$\lim_{i \to \infty} \sum_{j=1}^{\infty} \left| c_{m_{i+1},j} - c_{m_i,j} \right| = 0. \tag{6.21}$$

Since $\{m_i : i = 1, \ldots\}$ is arbitrary, (6.21) tells us that $\{c_i : c_i = \{c_{i,1}, c_{i,2}, \ldots\}\}$ is a Cauchy sequence in $\ell^1(\mathbb{N})$. The proof is finished by the completeness of $\ell^1(\mathbb{N})$. □

We cannot replace hypothesis *ii* in Theorem 6.2.1 with the condition that

$$\forall\, S \subseteq \mathbb{N}, \quad \exists \lim_{i\to\infty} \left|\sum_{j\in S} c_{i,j}\right|.$$

In fact, let $c_{m,n} = (1/2^m)e^{in\pi/2}$.

Theorem A.7.3, which states the equivalence of weak and norm sequential convergence in $\ell^1(\mathbb{N})$, is an immediate consequence of Theorem 6.2.1. In fact, if $\{c_i : i = 1,\ldots\} \subseteq \ell^1(\mathbb{N})$, with $c_i = \{c_{i,1},\ldots\}$, converges weakly to 0 then

$$\forall\, S \subseteq \mathbb{Z}, \quad \mathbb{1}_S \in \ell^\infty(\mathbb{N}) \quad \text{and} \quad \lim_{i\to\infty} \int \mathbb{1}_S c_i = 0.$$

Thus, the hypotheses of Theorem 6.2.1 are satisfied.

Theorem 6.2.2. Nikodym theorem
Let $(X,\mathcal{A})$ be a measurable space and let $\{\mu_n : n = 1,\ldots\} \subseteq M_b(X)$. Assume that

$$\forall\, A \in \mathcal{A}, \quad \lim_{n\to\infty} \mu_n(A) = \mu(A)$$

exists. Then $\mu \in M_b(X)$ and $\{\mu_n : n = 1,\ldots\}$ is Vitali equicontinuous.

Proof. *a.* Clearly μ is finitely additive on $\mathcal{A}$. Let $\{A_j : j = 1,\ldots\} \subseteq \mathcal{A}$ decrease to $\emptyset$. We shall prove

$$\lim_{j\to\infty} \mu(A_j) = 0. \tag{6.22}$$

a.i. Set $E_k = A_k \setminus A_{k+1}$, so that $\{E_k : k = 1,\ldots\}$ is a disjoint family and

$$A_k = (A_k \setminus A_{k+1}) \cup (A_{k+1} \setminus A_{k+2}) \cup \cdots = \bigcup_{j=k}^{\infty} E_j.$$

Our immediate task is to verify (6.23) below.

Define

$$c_{i,j} = \mu_i(E_j).$$

We shall check that $\{c_{i,j} : i,j = 1,\ldots\}$ satisfies the hypotheses of Theorem 6.2.1. For each $i = 1,\ldots,$

$$\sum_{j=1}^{\infty} |c_{i,j}| = \sum_{j=1}^{\infty} |\mu_i(E_j)| \le \sum_{j=1}^{\infty} |\mu_i|(E_j) = |\mu_i|\left(\bigcup_{j=1}^{\infty} E_j\right) = |\mu_i|(A_1) < \infty,$$

and condition *i* of Theorem 6.2.1 is satisfied. For condition *ii*, let $S \subseteq \mathbb{N}$ and note that

$$\lim_{i\to\infty}\sum_{j\in S} c_{i,j} = \lim_{i\to\infty} \mu_i\left(\bigcup_{j\in S} E_j\right) = \mu\left(\bigcup_{j\in S} E_j\right) \in \mathbb{C}.$$

By assumption, we have $\lim_{i\to\infty}\mu_i(E_j) = \mu(E_j)$. Combining this hypothesis with the existence of $\{c_j\}$ from Theorem 6.2.1, we see that $c_j = \mu(E_j)$ for all $j = 1, \ldots$. Consequently, from Theorem 6.2.1, we have

$$\lim_{i\to\infty}\sum_{j=1}^{\infty} |\mu_i(E_j) - \mu(E_j)| = 0. \tag{6.23}$$

a.ii. We now show that

$$\lim_{n\to\infty}\sum_{j=n}^{\infty} \mu(E_j) = 0. \tag{6.24}$$

To this end we calculate

$$\begin{aligned}\sum_{j=n}^{\infty} |\mu(E_j)| &\le \sum_{j=n}^{\infty} |\mu_i(E_j) - \mu(E_j)| + \sum_{j=n}^{\infty} |\mu_i(E_j)| \\ &\le \sum_{j=1}^{\infty} |\mu_i(E_j) - \mu(E_j)| + \sum_{j=n}^{\infty} |\mu_i(E_j)|.\end{aligned}$$

Thus,

$$\overline{\lim_{n\to\infty}}\sum_{j=n}^{\infty} |\mu(E_j)| \le \sum_{j=1}^{\infty} |\mu_i(E_j) - \mu(E_j)| + \overline{\lim_{n\to\infty}}\, |\mu_i|\left(\bigcup_{j=n}^{\infty} E_j\right),$$

so that by our hypothesis on $\{A_n : n = 1, \ldots\}$ and the fact that μ_i is bounded,

$$\overline{\lim_{n\to\infty}}\sum_{j=n}^{\infty} |\mu(E_j)| \le \sum_{j=1}^{\infty} |\mu_i(E_j) - \mu(E_j)|.$$

Therefore, (6.24) follows from (6.23).

a.iii. Our next step is to prove

$$\lim_{j\to\infty}\left(\mu_i(A_j) - \sum_{k=j}^{\infty} \mu(E_k)\right) = 0, \quad \text{uniformly in } i. \tag{6.25}$$

Equation (6.25) is certainly true for each i (not uniformly) by (6.24) and the hypothesis on $\{A_j : j = 1, \ldots\}$. Given $\varepsilon > 0$, use (6.23) to choose $I > 0$ such that

$$\forall\, i > I, \quad \sum_{j=1}^{\infty} |\mu_i(E_j) - \mu(E_j)| < \varepsilon.$$

Next, take $J > 0$ with the property that

$$\forall\, i = 1, \ldots, I, \quad \sum_{j=J}^{\infty} |\mu_i(E_j) - \mu(E_j)| < \varepsilon.$$

Hence,

$$\forall\, j > J \text{ and } \forall\, i = 1, \ldots, \quad \left|\mu_i(A_j) - \sum_{k=j}^{\infty} \mu(E_k)\right| = \left|\sum_{k=j}^{\infty}(\mu_i(E_k) - \mu(E_k))\right|$$
$$\leq \sum_{k=J}^{\infty} |\mu_i(E_k) - \mu(E_k)| < \varepsilon,$$

and this is (6.25).

a.iv. Finally, to obtain (6.22), we first use the Moore–Smith theorem (Theorem A.4.1), in conjunction with *a.iii*, and the fact,

$$\lim_{i\to\infty} \left(\mu_i(A_j) - \sum_{k=j}^{\infty} \mu(E_k)\right) = \mu(A_j) - \sum_{k=j}^{\infty} \mu(E_k),$$

to compute

$$\lim_{j\to\infty} \left(\mu(A_j) - \sum_{k=j}^{\infty} \mu(E_k)\right) = \lim_{i\to\infty} \lim_{j\to\infty} \left(\mu_i(A_j) - \sum_{k=j}^{\infty} \mu(E_k)\right). \tag{6.26}$$

The right-hand side of (6.26) is 0, as we observed after (6.25). Consequently, we can apply (6.24) (again) to the left-hand side of (6.26), and (6.22) follows.

b. The desired Vitali equicontinuity is entailed by (6.24) and (6.25). To prove that μ is σ-additive, let $\{B_n : n = 1, \ldots\} \subseteq \mathcal{A}$ be disjoint and set $A_j = \bigcup_{n=j}^{\infty} B_n$. Thus, $\{A_j : j = 1, \ldots\}$ decreases to $\emptyset$ and

$$\mu\left(\bigcup_{n=1}^{\infty} B_n\right) = \sum_{n=1}^{j} \mu(B_n) + \mu(A_{j+1}).$$

The σ-additivity follows by (6.22). □

Remark. Consider the measurable space $(\mathbb{R}, \mathcal{B}(\mathbb{R}))$, and define

$$\forall\, n = 1, \ldots \quad \text{and} \quad \forall\, A \in \mathcal{B}(\mathbb{R}), \quad \mu_n(A) = m(A \cap [n, \infty)).$$

Then, each μ_n is a measure. Since $\mu_n \geq \mu_{n+1}$ on $\mathcal{B}(\mathbb{R})$, we define μ as

$$\forall\, A \in \mathcal{B}(\mathbb{R}), \quad \lim_{n\to\infty} \mu_n(A) = \mu(A), \tag{6.27}$$

and note that each $\mu(A)$ exists with possibly infinite value. Note that μ is *not* a measure. In fact,

$$\mu\left(\bigcup_{n=1}^{\infty}[n,n+1)\right)=\infty \quad \text{and} \quad \sum_{n=1}^{\infty}\mu([n,n+1))=0.$$

It is interesting to note that if $\{\mu_n : n = 1,\ldots\}$ is a sequence of measures on a measurable space $(X,\mathcal{A})$ and $\mu_n \le \mu_{n+1}$, then the limit μ in (6.27) is a measure.

We use Theorem 6.2.2 to prove the following result.

Theorem 6.2.3. Hahn–Saks theorem
Let $(X,\mathcal{A})$ be a measurable space, let ν be a measure, and let $\{\mu_n : n = 1,\ldots\} \subseteq M_b(X)$. Assume
i. $\forall\, A \in \mathcal{A}$, $\lim_{n\to\infty}\mu_n(A) = \mu(A)$ *exists,*
ii. $\forall\, n = 1,\ldots,\ \mu_n \ll \nu$.
Then, $\{\mu_n : n = 1,\ldots\}$ is uniformly absolutely continuous.

Proof. Assume that $\{\mu_n : n = 1,\ldots\}$ is not uniformly absolutely continuous. Then there exist $\varepsilon > 0$, a subsequence $\{n_m : m = 1,\ldots\}$ of $\mathbb{N}$, and $\{A_m : m = 1,\ldots\} \subseteq \mathcal{A}$ such that, for each $m = 1,\ldots,$

$$\nu(A_m) < \frac{1}{2^m} \tag{6.28}$$

and

$$|\mu_{n_m}(A_m)| \ge \varepsilon. \tag{6.29}$$

For simplification of notation write $n_m = m$, so that (6.29) is replaced by

$$\forall\, m = 1,\ldots, \quad |\mu_m(A_m)| \ge \varepsilon. \tag{6.30}$$

Using this we shall obtain a contradiction.

a. For each $k \ge 0$ we first prove that there is a strictly increasing subsequence $\{n_i^k : i = 1,\ldots\} \subseteq \mathbb{N}$ such that $\{n_i^{k+1} : i = 1,\ldots\}$ is a subsequence of $\{n_i^k : i = 1,\ldots\}$,

$$n_1^{k+1} > n_2^k, \quad n_2^{k+1} > n_3^k, \quad n_3^{k+1} > n_4^k, \quad \ldots,$$

and

$$\sum_{i=1}^{\infty}\left|\mu_{n_1^k}\right|\left(A_{n_i^{k+1}}\right) < \frac{\varepsilon}{2}, \tag{6.31}$$

where the elements of the sum decrease to 0.

In order to do this, first observe that (6.28) and the hypothesis $\mu_1 \ll \nu$ yield

$$\lim_{m\to\infty}|\mu_1|(A_m) = 0.$$

Consequently, there is a strictly increasing sequence $\{n_i^1 : i = 1, \ldots\}$, for which $|\mu_1|(A_{n_i^1})$ is decreasing, $n_i^1 \to \infty$, and

$$|\mu_1|(A_{n_i^1}) < \frac{\varepsilon}{2^{i+1}}.$$

Now since $\left|\mu_{n_1^1}\right| \ll \nu$ we use (6.28) again to obtain a strictly increasing sequence $\{n_i^2 : i = 1, \ldots\} \subseteq \{n_i^1 : i = 1, \ldots\}$ such that

$$\forall\, j = 1, \ldots, \quad n_{j+1}^1 < n_j^2, \quad \text{and} \quad \forall\, i = 1, \ldots, \quad \left|\mu_{n_1^1}\right| \left(A_{n_i^2}\right) < \frac{\varepsilon}{2^{i+1}},$$

and $\left\{\left|\mu_{n_1^1}\right| \left(A_{n_i^2}\right)\right\}$ decreases as $n_i^2 \to \infty$. Next we examine $\left|\mu_{n_1^2}\right|$ and construct $\{n_i^3 : i = 1, \ldots\}$ in the same way, and continue the process.

b. Define $\nu_i = \mu_{n_1^i}$ and $B_i = A_{n_1^i}$. Observe that when we chose $\{n_j^{k+1} : j = 1, \ldots\}$ from $\{n_j^k : j = 1, \ldots\}$ we had

$$n_1^{k+1} > n_2^k > n_1^k.$$

Thus, $n_1^{k+1} \geq n_1^k + 1$, and so $n_1^k \geq k$. Therefore, by (6.28),

$$\nu(B_i) < \frac{1}{2^i}. \tag{6.32}$$

Also, we compute

$$\sum_{j=m+1}^{\infty} |\nu_m|(B_j) = \sum_{j=1}^{\infty} |\nu_m|(B_{m+j}) \leq \sum_{j=1}^{\infty} |\nu_m| \left(A_{n_j^{m+1}}\right) < \frac{\varepsilon}{2}, \tag{6.33}$$

where the last two inequalities follow from *a*. The latter inequality is a consequence of (6.31). For the former, note that for fixed m, $\{|\nu_m| \left(A_{n_j^{m+1}}\right) : j = 1, \ldots\}$ is decreasing and

$$n_j^{m+1} < \cdots < n_2^{m+j-1} < n_1^{m+j}.$$

c. Set $C_n = \bigcup_{j=n}^{\infty} B_j$ and $D = \bigcap_{n=1}^{\infty} \bigcup_{j=n}^{\infty} B_j = \bigcap_{n=1}^{\infty} C_n$.

c.i. Clearly, $C_n \subseteq C_{n-1}$, so that $\{C_n : n = 1, \ldots\}$ decreases to D. Consequently, $E_n = C_n \setminus D$ decreases to $\emptyset$. Using (6.32) we compute $\nu(D) = 0$ since

$$\nu(D) \leq \nu(C_{n+1}) \leq \sum_{j=n+1}^{\infty} \nu(B_j) < \sum_{j=n+1}^{\infty} \frac{1}{2^j} = \frac{1}{2^n}.$$

Thus, by hypothesis *ii*,

$$\forall\, i = 1, \ldots, \quad \nu_i(D) = 0.$$

We therefore conclude that

$$\forall\, i = 1, \ldots \text{ and } \forall\, n = 1, \ldots, \quad \nu_i(E_n) = \nu_i(C_n). \tag{6.34}$$

c.ii. From hypothesis *i* and Theorem 6.2.2 we have

$$\lim_{n\to\infty} \nu_m(E_n) = 0, \quad \text{uniformly in } m. \tag{6.35}$$

Applying (6.34) to (6.35), we obtain

$$\lim_{n\to\infty} \nu_m(C_n) = 0, \quad \text{uniformly in } m.$$

Hence, choose $N > 0$ such that

$$\forall\, n \geq N \text{ and } \forall\, m = 1, \ldots, \quad |\nu_m(C_n)| < \frac{\varepsilon}{2}. \tag{6.36}$$

c.iii. Note that

$$C_n = B_n \cup \left(B_n^{\sim} \cap \left(\bigcup_{j=n+1}^{\infty} B_j \right) \right),$$

where B_n and $B_n^{\sim} \cap \left(\bigcup_{j=n+1}^{\infty} B_j\right)$ are disjoint.

d. For $n \geq N$ and $m = n$ we use (6.30) and *c.iii* to compute

$$\varepsilon \leq |\nu_n(B_n)| \leq |\nu_n(C_n)| + \left| \nu_n \left(B_n^{\sim} \cap \left(\bigcup_{j=n+1}^{\infty} B_j \right) \right) \right|,$$

so that, from (6.36),

$$\varepsilon < \frac{\varepsilon}{2} + |\nu_n| \left(\bigcup_{j=n+1}^{\infty} B_j \right) \leq \frac{\varepsilon}{2} + \sum_{j=n+1}^{\infty} |\nu_n|(B_j). \tag{6.37}$$

We obtain a contradiction ($\varepsilon < \varepsilon$) using (6.33) in (6.37). □

Remark. The Hahn–Saks theorem (Theorem 6.2.3) was originally proved by Saks independent of Theorem 6.2.2. He then deduced Theorem 6.2.2 in the following way. Given $\{\mu_n : n = 1, \ldots\} \subseteq M_b(X)$ and the hypothesis of Theorem 6.2.2, define

$$\forall\, A \in \mathcal{A}, \quad \nu(A) = \sum_{n=1}^{\infty} \frac{1}{2^n \|\mu_n\|_1} |\mu_n|(A). \tag{6.38}$$

Then ν is a bounded measure and $\mu_n \ll \nu$ for each n. Consequently, we can apply Theorem 6.2.3 to obtain the uniform absolute continuity of $\{\mu_n : n = 1, \ldots\}$ with respect to ν. Thus, using Proposition 6.1.3*b* and the uniform absolute continuity, we conclude that $\mu \in M_b(X)$ by a double limit argument. Also, $\mu \ll \nu$ (this is the μ defined in the hypothesis of Theorem 6.2.2).

6.3 Weak sequential convergence

Definition 6.3.1. Weak sequential convergence in $L^p_\mu(X)$

a. Let $(X, \mathcal{A}, \mu)$ be a measure space and let $1/p + 1/q = 1$. Consider a sequence $\{f_n : n = 1, \ldots\} \subseteq L^p_\mu(X)$ and $f \in L^p_\mu(X)$, $1 \leq p < \infty$. We say that $\{f_n : n = 1, \ldots\}$ *converges weakly* to f, and we write $f_n \to f$ weakly, if

$$\forall\, g \in L^q_\mu(X), \quad \lim_{n\to\infty} \int_X f_n g \, d\mu = \int_X fg \, d\mu.$$

b. Let $(X, \mathcal{A}, \mu)$ be a σ-finite measure space. A sequence $\{f_n : n = 1, \ldots\} \subseteq L^\infty_\mu(X)$ *converges weakly* to $f \in L^\infty_\mu(X)$ if for each set function ν, finitely additive on $\mathcal{A}$ and vanishing on $\{A \in \mathcal{A} : \mu(A) = 0\}$, we have

$$\lim_{n\to\infty} \int_X f_n \, d\nu = \int_X f \, d\nu.$$

As we pointed out in Example 3.6.5, weak convergence does not necessarily imply norm convergence; cf. Theorem A.7.3. In fact, weak convergence does not imply pointwise *a.e.* convergence, uniform convergence, or convergence in measure.

The weak sequential convergence defined above is actually sequential convergence for a certain topology, called the *weak topology*. We discuss the weak topology generally in Appendix A.9, but for now we consider some special results. In particular, we have the following characterization of weak sequential convergence in $L^1_\mu(X)$.

Theorem 6.3.2. Characterization of weak sequential convergence
Let $(X, \mathcal{A}, \mu)$ be a measure space and consider a sequence $\{f_n : n = 1, \ldots\} \subseteq L^1_\mu(X)$ and $f \in L^1_\mu(X)$. Then $f_n \to f$ weakly if and only if

i. $\forall\, A \in \mathcal{A}$, $\int_A (f_n - f)\, d\mu \to 0$,

ii. $\sup_{n \in \mathbb{N}} \|f_n\|_1 = K < \infty$.

Proof. ($\Longrightarrow$) Part *i* is immediate by definition of weak convergence. To prove *ii* we first define $F_n : L^\infty_\mu(X) \to \mathbb{C}$ by

$$\forall\, g \in L^\infty_\mu(X), \quad F_n(g) = \int_X (f_n - f) g \, d\mu.$$

Each F_n is continuous by LDC. Since $F_n(g) \to 0$ for each $g \in L^\infty_\mu(X)$ by hypothesis, we know that

$$\forall\, g \in L^\infty_\mu(X),\ \exists\, N_g \text{ such that } \forall\, n \geq N_g, \quad |F_n(g)| \leq 1;$$

and so

$$\forall\, g \in L^\infty_\mu(X), \quad \exists\, M_g = \max\{1, |F_1(g)|, \ldots, |F_{N_g - 1}(g)|\}$$

such that

$$\forall\, n = 1, \ldots, \quad |F_n(g)| \le M_g.$$

Therefore, we can apply the Uniform Boundedness Principle (Theorem A.7.1 or Theorem A.8.6) and conclude that

$$\exists\, M \text{ such that } \forall\, n = 1, \ldots \text{ and } \forall\, g \in L^\infty_\mu(X), \quad |F_n(g)| \le M \|g\|_\infty.$$

Clearly,

$$\|f_n\|_1 = \int_X f_n g_n \, d\mu,$$

for some $g_n \in L^\infty_\mu(X)$, where, without loss of generality, we take $|g_n| = 1$ on X. Therefore, *ii* follows.

($\Longleftarrow$) Take any $\varepsilon > 0$. We prove that for any fixed $g \in L^\infty_\mu(X)$,

$$\overline{\lim_{n\to\infty}} \left| \int_X (f_n - f) g \, d\mu \right| \le \varepsilon. \tag{6.39}$$

Choose a simple function h such that $\|g - h\|_\infty \le \varepsilon/(2\max(K, \|f\|_1))$; see Theorem 2.5.5 and Definition 2.5.9. Then

$$\left| \int_X (f_n - f) g \, d\mu \right| \le \|g - h\|_\infty \|f_n - f\|_1 + \left| \int_X (f_n - f) h \, d\mu \right|,$$

and this gives (6.39). □

The incredible thing is that we can drop condition *ii* in Theorem 6.3.2, even in the setting of quite general measure spaces. This is the content of Theorem 6.3.7.

Proposition 6.3.3. *Let $(X, \mathcal{A}, \mu)$ be a measure space and let $\{f_n : n = 1, \ldots\} \subseteq L^1_\mu(X)$ have the property that*

$$\forall\, A \in \mathcal{A}, \quad \lim_{n\to\infty} \int_A f_n \, d\mu$$

exists and is finite. Then

$$\forall\, \varepsilon > 0,\ \exists\, \delta > 0 \text{ such that } \forall\, n = 1, \ldots, \quad \mu(A) < \delta \Longrightarrow \int_A |f_n| \, d\mu < \varepsilon.$$

Proof. We define the following family of measures μ_n:

$$\forall\, A \in \mathcal{A}, \quad \mu_n(A) = \int_A f_n \, d\mu.$$

Thus, according to Theorem 5.3.3,

$$\forall\, n = 1, \ldots, \quad \mu_n \in M_{ac}(X, \mu).$$

Moreover, by our hypothesis, we have

$$\forall\, A \in \mathcal{A}, \quad \exists \lim_{n\to\infty} \mu_n(A).$$

These two statements are the assumptions of Theorem 6.2.3. Therefore, we conclude that $\{\mu_n : n = 1, \ldots\}$ is uniformly absolutely continuous on X, i.e.,

$$\forall\, \varepsilon > 0,\ \exists\, \delta > 0, \text{ such that } \forall\, A \in \mathcal{A},$$
$$\mu(A) < \delta \Longrightarrow \forall\, n = 1, \ldots, \quad \left| \int_A f_n \, d\mu \right| < \varepsilon.$$

In other words, the sequence $\{f_n : n = 1, \ldots\}$ is uniformly absolutely continuous. Therefore, to finish the proof, we only need to observe that in the proof of the "if" part of Theorem 6.1.2 we have, in particular, established that $\{f_n : n = 1, \ldots\}$ is uniformly absolutely continuous if and only if $\{|f_n| : n = 1, \ldots\}$ is uniformly absolutely continuous. □

Remark. It is an interesting observation that in the proof of Proposition 6.3.3 we may assume without loss of generality that

$$\forall\, A \in \mathcal{A}, \quad \lim_{n\to\infty} \int_A f_n \, d\mu = 0. \tag{6.40}$$

To see this we proceed as follows.

First, we shall prove that for any $\varepsilon > 0$ there exist $\delta > 0$ and $k \in \mathbb{N}$ such that

$$\forall\, m, n \geq k \text{ and } \forall\, A \in \mathcal{A}, \text{ for which } \mu(A) < \delta, \quad \int_A |f_m - f_n| \, d\mu < \frac{\varepsilon}{2}. \tag{6.41}$$

In fact, if (6.41) were not true then there would exist $\varepsilon > 0$ and sequences $\{m_i : m_i \geq i, i = 1, \ldots\}$, $\{n_i : n_i \geq i, i = 1, \ldots\}$, and $\{A_i : \mu(A_i) < 1/i, i = 1, \ldots\}$ such that

$$\forall\, i = 1, \ldots, \quad \int_{A_i} |f_{m_i} - f_{n_i}| \, d\mu \geq \frac{\varepsilon}{2}. \tag{6.42}$$

On the other hand, the sequence $\{f_{m_i} - f_{n_i} : i = 1, \ldots\} \subseteq L^1_\mu(X)$ satisfies the assumption of our proposition with $\lim_{i\to\infty} \int_A (f_{m_i} - f_{n_i}) \, d\mu = 0$ for all sets $A \in \mathcal{A}$. Thus, if the conclusion of Proposition 6.3.3 were true for 0 limits, it would yield a contradiction with (6.42). Hence, (6.41) is established.

Second, we note that for any $\varepsilon > 0$ there exists $\delta' > 0$ such that

$$\forall\, A \in \mathcal{A}, \text{ for which } \mu(A) < \delta', \text{ and } \forall\, j \leq k, \quad \int_A |f_j| \, d\mu < \frac{\varepsilon}{2}. \tag{6.43}$$

This is a consequence of Proposition 3.3.9.

Equation (6.40) follows from (6.41) and (6.43) because

$$\forall\, \varepsilon > 0,\ \exists\, \delta, \delta' > 0 \text{ such that } \forall\, n > k,$$

$$\mu(A) < \min(\delta, \delta') \Longrightarrow \int_A |f_n|\, d\mu \le \int_A |f_n - f_k|\, d\mu + \int_A |f_k|\, d\mu < \varepsilon.$$

Using Proposition 6.3.3 we may strengthen Theorem 6.3.2.

Theorem 6.3.4. Sufficient conditions for weak sequential convergence in $L^1_\mu(X)$
Let $(X, \mathcal{A}, \mu)$ be a measure space. Consider the sequence $\{f_n : n = 1, \ldots\} \subseteq L^1_\mu(X)$. If
i. $\forall\, A \in \mathcal{A}$, $\lim_{n\to\infty} \int_A f_n$ exists and is finite,
and
ii. $\sup_{n\in\mathbb{N}} \|f_n\|_1 = K < \infty$,
then $f_n \to f$ weakly for some $f \in L^1_\mu(X)$.

Proof. Fix $g \in L^\infty_\mu(X)$. We first show that $\{\int_X f_n g\, d\mu : n = 1, \ldots\}$ is a Cauchy sequence. Let $\varepsilon > 0$. There exists a simple function h such that $\|g - h\|_\infty < \varepsilon/(3K)$; see, e.g., Theorem 2.5.5. Thus,

$$\forall\, n = 1, \ldots, \quad \left| \int_X f_n g\, d\mu - \int_X f_n h\, d\mu \right| \le \|g - h\|_\infty \|f_n\|_1 < \frac{\varepsilon}{3} \tag{6.44}$$

and

$$\lim_{n\to\infty} \int_X f_n h\, d\mu$$

exists and is finite. By hypothesis, choose M such that for all $i, j \ge M$,

$$\left| \int_X f_i h\, d\mu - \int_X f_j h\, d\mu \right| < \frac{\varepsilon}{3}. \tag{6.45}$$

Then, by (6.44) and (6.45), we have

$$\forall\, i, j \ge M, \quad \left| \int_X f_i g\, d\mu - \int_X f_j g\, d\mu \right| < \varepsilon,$$

and so $\{\int_X f_n g\, d\mu\}$ is a Cauchy sequence. Thus, we can assert that

$$\forall\, g \in L^\infty_\mu(X), \quad \lim_{n\to\infty} \int_X f_n g\, d\mu$$

exists and is finite. We write

$$\lambda(g) = \lim_{n\to\infty} \int_X f_n g\, d\mu,$$

for $g \in L^\infty_\mu(X)$. Because of assumption *ii*, we conclude that λ is a bounded linear functional on $L^\infty_\mu(X)$. Also, $\lambda_n(A) = \int_A f_n\, d\mu$ defines $\lambda_n \in M_b(X)$, and hence $\lambda \in M_b(X)$ by Theorem 6.2.2. If X is a locally compact Hausdorff space, this assertion is also a consequence of the Riesz representation theorem proved in Chapter 7, since λ is a bounded linear functional on $C_0(X)$.

Proposition 6.3.3, together with Theorem 5.2.8 and the observation following it, yields that $\lambda \ll \mu$. Thus, according to the Radon–Nikodym theorem (Theorem 5.3.1), there exists $f \in L^1_\mu(X)$ such that

$$\forall\, A \in \mathcal{A}, \quad \lambda(A) = \int_A f\, d\mu.$$

Hence, $\lambda(h) = \int_X hf\, d\mu$ for all simple functions h. Since these functions are dense in $L^\infty_\mu(X)$ (see, e.g., Theorem 2.5.5), and since λ is bounded on $L^\infty_\mu(X)$, it follows that

$$\forall\, g \in L^\infty_\mu(X), \quad \lim_{n\to\infty} \int_X f_n g\, d\mu = \int_X fg\, d\mu.$$

Indeed, fix $g \in L^\infty_\mu(X)$ and $\varepsilon > 0$. Choose a simple function h such that $\|g - h\|_\infty < \varepsilon/(2\max(K, \|f\|_1))$. Then

$$\overline{\lim_{n\to\infty}} \left| \int_X f_n g\, d\mu - \int_X fg\, d\mu \right| \le \overline{\lim_{n\to\infty}} \left| \int_X f_n g\, d\mu - \int_X f_n h\, d\mu \right|$$
$$+ \overline{\lim_{n\to\infty}} \left| \int_X f_n h\, d\mu - \int_X fh\, d\mu \right| + \left| \int_X fh\, d\mu - \int_X fg\, d\mu \right| < \varepsilon.$$

Therefore, f_n converges weakly to f. □

The following result is due to DIEUDONNÉ [137] in his work on Köthe spaces (a term DIEUDONNÉ introduced), extending work of GOTTFRIED KÖTHE and his coauthors from the setting of bilinear forms; cf. [149] and Theorem 6.3.8.

Proposition 6.3.5. *Let X be a locally compact Hausdorff space, let $(X, \mathcal{A}, \mu)$ be a measure space, and let μ be regular. Assume that the sequence $\{f_n : n = 1, \ldots\} \subseteq L^1_\mu(X)$ has the property that*

$$\forall\, A \in \mathcal{A}, \quad \lim_{n\to\infty} \int_A f_n\, d\mu$$

exists and is finite. Then

$$\forall\, \varepsilon > 0,\ \exists\, K \subseteq X,\ \textit{compact, such that}\ \ \forall\, n = 1, \ldots, \quad \int_{K^\sim} |f_n|\, d\mu < \varepsilon.$$

Proof. As in the Remark following Proposition 6.3.3 we may assume without loss of generality that the limits $\lim_{n\to\infty} \int_A f_n\, d\mu$ are 0.

i. We start with the observation that, for every $f \in L^1_\mu(X)$, there exists a countable collection $\{A_n : n = 1, \ldots\} \subseteq \mathcal{A}$ of measurable sets with each $\mu(A_n) < \infty$ such that

$$\forall\, x \in \left(\bigcup_{n=1}^{\infty} A_n\right)^{\sim}, \quad f(x) = 0.$$

Thus, there exists a measurable set $B \in \mathcal{A}$ such that $B = \bigcup_{n=1}^{\infty} B_n$, where $\mu(B_n) < \infty$, for all $n = 1, \ldots$, and where

$$\forall\, x \in B^{\sim} \text{ and } \forall\, n = 1, \ldots, \quad f_n(x) = 0.$$

Since the measure μ is regular, we can write $B = N \cup (\bigcup K_j)$, where each $K_j, j = 1, \ldots$, is compact, $K_j \subseteq K_{j+1}$, and $\mu(N) = 0$.

Our goal is to show that

$$\forall\, \varepsilon > 0,\ \exists\, j \in \mathbb{N}, \text{ such that } \forall\, n = 1, \ldots, \quad \int_{K_j^{\sim}} |f_n|\, d\mu < \varepsilon.$$

ii. Let $H \subseteq L^1_\mu(X)$ be the set consisting of characteristic functions of measurable subsets of B. For each $j = 1, \ldots$, set

$$\rho_j(g_1, g_2) = \int_{K_j} |g_1 - g_2|\, d\mu$$

for any two elements $g_1, g_2 \in H$. We note that if $\rho_j(g_1, g_2) = 0$ for every $j = 1, \ldots$, then $g_1 - g_2 = 0$ μ-*a.e.* in B and therefore in X. Thus, we may define a metrizable uniform structure on H; see Definition A.1.8.

Let us begin by showing that this metrizable uniform space H is complete.

First, if $\{g_n : n = 1, \ldots\}$ is a Cauchy sequence in H, then it is a Cauchy sequence for each metric space (H, ρ_j), $j = 1, \ldots$. This, in turn, implies that, for each $j = 1, \ldots$, $\{g_n \mathbb{1}_{K_j} : n = 1, \ldots\}$ is a Cauchy sequence in $L^1_\mu(X)$, and so it has a limit $g^j \in L^1_\mu(X)$.

Second, from the sequence $\{g_n \mathbb{1}_{K_1} : n = 1, \ldots\}$, we can choose a subsequence $\{g_{m_n} \mathbb{1}_{K_1} : n = 1, \ldots\}$ that converges pointwise μ-*a.e.* to g^1. In particular, we observe that g^1 equals 0 or 1 μ-*a.e.* Next, the sequence $\{g_{m_n} \mathbb{1}_{K_2} : n = 1, \ldots\}$ converges in norm to g^2, and so we can choose a subsequence that converges pointwise μ-*a.e.* to g^2, and g^2 is also a characteristic function of a measurable set in B.

Finally, continuing this diagonal process, we find a subsequence of $\{g_n : n = 1, \ldots\} \subseteq H$ that converges pointwise μ-*a.e.* to g^j on each K_j, $j = 1, \ldots$. If we let

$$g(x) = \begin{cases} g^j(x), & x \in K_j, \\ 0, & x \in \left(\bigcup_{j=1}^{\infty} K_j\right)^{\sim}, \end{cases}$$

then $g \in H$ and

$$\forall\, j = 1, \ldots, \quad \lim_{n \to \infty} \rho_j(g, g_n) = 0.$$

This proves that g is the limit of $\{g_n : n = 1, \ldots\}$ on B in the topology of the uniform structure, and we have proved that H is complete.

iii. Let $f \in L^1_\mu(X)$ and suppose $f = 0$ on $B^\sim$. The linear mapping

$$\begin{aligned} T_f : H &\to \mathbb{C}, \\ g &\mapsto \int_X gf \, d\mu, \end{aligned}$$

is continuous. In fact, for each $\varepsilon > 0$ there exists $j \in \mathbb{N}$ such that

$$\int_{K_j^\sim} |f| \, d\mu < \varepsilon; \tag{6.46}$$

and there exists $n \in \mathbb{N}$ such that if $F_n = \{x : |f(x)| \leq n\} \subseteq X$ then

$$\int_{F_n^\sim} |f| \, d\mu < \varepsilon. \tag{6.47}$$

If $g_1, g_2 \in H$ one has, by (6.46), that

$$\left| \int_{K_j^\sim} (g_1 - g_2) f \, d\mu \right| < 2\varepsilon,$$

and, by (6.47), that

$$\left| \int_{F_n^\sim} (g_1 - g_2) f \, d\mu \right| < 2\varepsilon.$$

Therefore, we can conclude that

$$\begin{aligned} \left| \int_B (g_1 - g_2) f \, d\mu \right| &\leq \left| \int_{K_j \cap F_n} (g_1 - g_2) f \, d\mu \right| \\ &\quad + \left| \int_{K_j \cap F_n^\sim} (g_1 - g_2) f \, d\mu \right| + \left| \int_{K_j^\sim} (g_1 - g_2) f \, d\mu \right| \\ &\leq n \int_{K_j} |g_1 - g_2| \, d\mu + \left| \int_{K_j^\sim} (g_1 - g_2) f \, d\mu \right| \\ &< n\rho_j(g_1, g_2) + 2\varepsilon + 2\varepsilon = n\rho_j(g_1, g_2) + 4\varepsilon; \end{aligned}$$

and so the continuity of T_f is established thanks to the property of convergence in the uniform structure.

iv. We shall apply the Baire category theorem (Theorem A.6.1 and Theorem A.6.2) to the complete metric space H to establish the following fact. For every $\varepsilon > 0$, there exist $m, n_0 \in \mathbb{N}$, a measurable set $A_0 \subseteq B$, and $\delta > 0$ such that, for each measurable set $A \in \mathcal{A}$ satisfying

$$\int_{K_m} |\mathbb{1}_A - \mathbb{1}_{A_0}| \, d\mu < \delta, \tag{6.48}$$

we have

$$\forall\, n \geq n_0, \qquad \left| \int_A f_n \, d\mu \right| < \varepsilon. \tag{6.49}$$

Indeed, let

$$G_n = \left\{ g \in H : \forall\, m \geq n, \left| \int_X g f_m \, d\mu \right| \leq \varepsilon \right\} \subseteq H,$$

for $n = 1, \ldots$. Then $H = \bigcup_{n=1}^{\infty} G_n$, where $G_n \subseteq G_{n+1}$ for each $n = 1, \ldots,$ since we are assuming $\lim_{n\to\infty} \int_A f_n \, d\mu = 0$. Also, each G_n is closed in H by the continuity of the T_{f_m}s and the fact that the intersection of closed sets is closed. We use these properties to apply the Baire category theorem. Consequently, there exists $n_0 \in \mathbb{N}$ such that G_{n_0} contains a nonempty open set $U \subseteq G_{n_0}$. Thus, G_{n_0} contains an open ball of the form defined in (6.48). Therefore, by the definition of G_{n_0}, the elements of this ball satisfy (6.49).

v. If we now take $A = A_0 \cap K_m$, we have

$$\int_{K_m} |\mathbb{1}_{A_0 \cap K_m} - \mathbb{1}_{A_0}| \, d\mu = 0.$$

Therefore, (6.48) holds for $A = A_0 \cap K_m$, and, hence,

$$\forall\, n \geq n_0, \qquad \left| \int_{A_0 \cap K_m} f_n \, d\mu \right| < \varepsilon.$$

Let $M \subseteq K_m^{\sim}$ be any measurable set. If $B = M \cup (A_0 \cap K_m)$, one has

$$\int_{K_m} |\mathbb{1}_B - \mathbb{1}_{A_0}| \, d\mu = 0,$$

which again implies that

$$\forall\, n \geq n_0, \qquad \left| \int_B f_n \, d\mu \right| < \varepsilon;$$

and so

$$\forall\, n \geq n_0, \qquad \left| \int_M f_n \, d\mu \right| < 2\varepsilon. \tag{6.50}$$

Without loss of generality, assume that each f_n is real-valued. Let $M_{m,n}^+ = \{x \in K_m^{\sim} : f_n(x) \geq 0\}$ and $M_{m,n}^- = \{x \in K_m^{\sim} : f_n(x) < 0\}$. Since $K_m^{\sim} = M_{m,n}^+ \cup M_{m,n}^-$, we conclude from (6.50) that

$$\forall\, n \geq n_0, \qquad \int_{K_m^{\sim}} |f_n| \, d\mu < 4\varepsilon.$$

On the other hand, for each $n < n_0$, there exists m_n such that

$$\int_{K_{m_n}^{\sim}} |f_n| \, d\mu < \varepsilon,$$

by the integrability of f_n and the facts that $\{K_j\}$ is increasing and $B = N \cup (\bigcup K_j)$. Thus, for $j = \max(m, m_1, \ldots, m_{n_0 - 1})$,

$$\forall\, \varepsilon > 0,\ \exists\, j \in \mathbb{N}, \text{ such that } \forall\, n = 1, \ldots, \quad \int_{K_j^{\sim}} |f_n| \, d\mu < \varepsilon. \qquad \square$$

Using the theorems in Sections 6.1 and 6.2 it is now not difficult to prove Theorem 6.3.7, taken from [137], page 89. To this end we need the following result.

Proposition 6.3.6. *Let X be a locally compact Hausdorff space, let $(X, \mathcal{A}, \mu)$ be a measure space, and let μ be continuous and regular. Then, for all $0 \le t < \mu(X)$, there exists $A \in \mathcal{A}$ such that*

$$\mu(A) = t.$$

Proof. First, by the regularity of μ, we may assume without loss of generality that X is compact.

Let $0 \le t < \mu(X)$. We need to construct a decreasing sequence of measurable sets $A_n \subseteq X$, $n = 1, \ldots,$ with the property that

$$\forall\, n = 1, \ldots, \quad t \le \mu(A_n) < t + \frac{\mu(X)}{2^n}.$$

We show that this is possible using induction. Assume that $A_1, \ldots, A_n$ have been constructed, and consider a finite open covering of X by sets M_j, $j = 1, \ldots, N$, where each $\mu(M_j) \le \mu(X)/2^{n+1}$. This is accomplished by the hypotheses that X is compact and μ is continuous. By dividing these open sets appropriately, we may assume that $M_1, \ldots, M_N$ is a finite disjoint covering (no longer an open covering) of X and that each $\mu(M_j) \le \mu(X)/2^{n+1}$.

Define $B_{n,j} = A_n \cap M_j$, $j = 1, \ldots, N$. We can choose a subfamily B_{n,j_k}, $k = 1, \ldots, K$, for which

$$t \le \sum_{k=1}^{K} \mu(B_{n,j_k}) \le t + \frac{\mu(X)}{2^{n+1}}.$$

Take $A_{n+1} = \bigcup_{k=1}^{K} B_{n,j_k}$ and set $A = \bigcap_{n=1}^{\infty} A_n$. $\square$

Theorem 6.3.7. Characterization of weak sequential convergence of functions
Let X be a locally compact Hausdorff space, let $(X, \mathcal{A}, \mu)$ be a measure space, and let μ be regular. A sequence $\{f_n : n = 1, \ldots\} \subseteq L^1_\mu(X)$ converges weakly to some $f \in L^1_\mu(X)$ if and only if

$$\forall\, A \in \mathcal{A}, \quad \lim_{n\to\infty} \int_A f_n \, d\mu$$

exists and is finite; cf. Theorem 6.3.2.

Proof. $(\Longrightarrow)$ For $A \in \mathcal{A}$ let $g = \mathbb{1}_A \in L^\infty_\mu(X)$. Thus,

$$\lim_{n\to\infty} \int_A f_n \, d\mu = \lim_{n\to\infty} \int_X f_n g \, d\mu = \int_X fg \, d\mu = \int_A f \, d\mu.$$

$(\Longleftarrow)$ According to Theorem 6.3.4, it is enough to show that

$$\sup_{n\in\mathbb{N}} \|f_n\|_1 < \infty.$$

In view of Proposition 6.3.5, there exists a compact set $K \subseteq X$ such that

$$\forall\, n = 1, \ldots, \quad \int_{K^\sim} |f_n| \, d\mu < 1. \tag{6.51}$$

Since μ is regular, we have $\mu(K) < \infty$. Thus, there exists at most a countable collection of points $D = \{a_j : j = 1, \ldots\} \subseteq K$ with the property that $\mu(\{a_j\}) > 0$ for all $j = 1, \ldots$. Moreover,

$$\forall\, \delta > 0, \ \exists D_\delta \subseteq D, \text{ a finite set, for which } \mu(D \setminus D_\delta) \le \delta. \tag{6.52}$$

On the other hand, for any $\delta > 0$ we can partition the set $K \setminus D$ into a finite collection of measurable subsets K_i, $i = 1, \ldots, m$, such that

$$\forall\, i = 1, \ldots, m, \quad \mu(K_i) \le \delta. \tag{6.53}$$

This is a consequence of Proposition 6.3.6 in the following way. Choose $K_1 \in \mathcal{A}$ such that $K_1 \subseteq K \setminus D$ and $\mu(K_1) = \delta$. If $\mu((K \setminus D) \setminus K_1) \le \delta$ then let $K_2 = (K \setminus D) \setminus K_1$ and we are done. If $\mu((K \setminus D) \setminus K_1) > \delta$ then choose a K_2 analogous to our choice of K_1. The procedure is complete in a finite number of steps since $\mu(K \setminus D) < \infty$.

From Proposition 6.3.3 and for $\varepsilon = 1$, there is $\delta_1 > 0$ such that

$$\forall\, n = 1, \ldots, \quad \mu(A) \le \delta_1 \quad \Longrightarrow \quad \int_A |f_n| \, d\mu \le 1. \tag{6.54}$$

Let $\delta = \delta_1$ in (6.52) and (6.53). Thus, taking $A = K_i$ and $A = D \setminus D_{\delta_1}$ in (6.54) we have

$$\forall\, i = 1,\ldots,m \text{ and } \forall\, n = 1,\ldots,\quad \int_{K_i} |f_n|\, d\mu \;\text{ and }\; \int_{D\setminus D_{\delta_1}} |f_n|\, d\mu \le 1. \tag{6.55}$$

Finally, we observe, by hypothesis, that $\lim_{n\to\infty} f_n(a_j)$ exists and is finite for each $a_j \in D_{\delta_1}$. Since there are only finitely many such points,

$$\exists\, N > 0 \text{ such that } \forall\, n = 1,\ldots,\quad \int_{D_{\delta_1}} |f_n|\, d\mu \le N. \tag{6.56}$$

Combining (6.51), (6.55), and (6.56) yield that

$$\exists\, M > 0 \text{ such that } \forall\, n = 1,\ldots,\quad \int_X |f_n|\, d\mu \le M. \qquad \square$$

Remark. Since $2\mathbb{1}_A = \mathbb{1}_X + (\mathbb{1}_A - \mathbb{1}_{A^\sim})$ it is clear that the assumption that the limits,

$$\forall\, A \in \mathcal{A},\quad \lim_{n\to\infty} \int_A f_n\, d\mu,$$

exist and are finite is equivalent to the assumption that the limits,

$$\forall\, A \in \mathcal{A},\quad \lim_{n\to\infty} \left(\int_A f_n\, d\mu - \int_{A^\sim} f_n\, d\mu \right),$$

exist and are finite. This observation is important in view of the result of John Rainwater (nom de plume) [380] that *a bounded sequence* $\{x_n : n = 1,\ldots\} \subseteq B$ *of elements of a Banach space converges weakly to an element* $x \in B$ *if and only if* $\lim_{n\to\infty} f(x_n) = f(x)$ *for each extremal point* f *on the unit sphere in the dual Banach space* B'. *Extremal points* of a convex set are the points that are not interior points of any line segment contained in this set.

The following theorem is a corollary of Theorem 6.3.4.

Theorem 6.3.8. Dunford–Pettis theorem
Let X *be a locally compact Hausdorff space, let* $(X, \mathcal{A}, \mu)$ *be a measure space, and let* μ *be regular. A sequence* $\{f_n : n = 1,\ldots\} \subseteq L^1_\mu(X)$ *converges weakly to some* $f \in L^1_\mu(X)$ *if and only if*

$$\forall\, \varepsilon > 0,\ \exists\, \delta > 0,\ \textit{such that}\ \ \mu(A) < \delta \Longrightarrow \forall\, n = 1,\ldots,\quad \int_A |f_n|\, d\mu < \varepsilon,$$

$$\forall\, \varepsilon > 0,\ \exists\, K \subseteq X,\ \textit{compact, such that}\ \forall\, n = 1,\ldots,\quad \int_{K^\sim} |f_n|\, d\mu < \varepsilon,$$

$$\forall\, U \subseteq X,\ \textit{open},\quad \lim_{n\to\infty} \int_U f_n\, d\mu\ \ \textit{exists and is finite},$$

and

$$\sup_{n\in\mathbb{N}} \|f_n\|_1 = K < \infty.$$

Proof. ($\Longrightarrow$) This direction is a consequence of Theorem 6.3.7, Proposition 6.3.3, Proposition 6.3.5, and Theorem 6.3.2.

($\Longleftarrow$) We shall invoke Theorem 6.3.4 to prove weak sequential convergence. Let $A \in \mathcal{A}$. It is sufficient to prove that $\lim_{n\to\infty} \int_A f_n \, d\mu$ exists and is finite. Let $\varepsilon > 0$. By hypothesis there is K, compact, such that

$$\int_{K^\sim} |f_n| \, d\mu < \varepsilon, \quad \text{uniformly in } n \in \mathbb{N}. \tag{6.57}$$

By regularity, for each $\delta > 0$ there is open $U_\delta \subseteq K$ such that $A \cap K \subseteq U_\delta$ and $\mu(U_\delta \setminus (A \cap K)) < \delta$. Choose δ from the first of the necessary conditions to obtain

$$\forall\, n = 1, \ldots, \quad \int_{U_\delta \setminus (A\cap K)} |f_n| \, d\mu < \varepsilon. \tag{6.58}$$

By hypothesis, $\lim_{n\to\infty} \int_{U_\delta} f_n \, d\mu$ exists and is finite. This fact, combined with (6.57), (6.58), and a straightforward interchanging of limits argument, yields the desired result. □

The fundamental result of NELSON DUNFORD and B. J. PETTIS [149], pages 376–378, is their characterization of weakly compact subsets of $L^1_\mu(X)$ for a regular measure space $(X, \mathcal{A}, \mu)$. A version of their theorem, due to DIEUDONNÉ [137], pages 93–94, is the following: $H \subseteq L^1_\mu(X)$ *is relatively weakly compact if and only if*

i. $H \subseteq L^1_\mu(X)$ *is bounded;*

ii. $\forall\, \varepsilon > 0, \exists \delta > 0$ *such that*

$$\forall\, A \in \mathcal{A}, \quad \mu(A) < \delta \Longrightarrow \forall\, f \in H, \quad \int_A |f| \, d\mu < \varepsilon;$$

iii. $\forall\, \varepsilon > 0, \exists Y \subseteq X$ *compact, such that*

$$\forall\, f \in H, \quad \int_{Y^\sim} |f| \, d\mu < \varepsilon.$$

6.4 Dieudonné–Grothendieck theorem

The Dieudonné–Grothendieck theorem, Theorem 6.4.2, characterizes the weak convergence of a sequence $\{\mu_n : n = 1, \ldots\} \subseteq M_b(X)$, where X is a locally compact Hausdorff space. We require locally compact Hausdorff spaces because of their property that if K_1 and K_2 are disjoint compact sets, then there are disjoint open sets V_1 and V_2 such that $K_1 \subseteq V_1$ and $K_2 \subseteq V_2$. By our convention in Definition 5.1.13, $M_b(X)$ consists of the *complex regular Borel measures on* X. An analogous characterization of weak sequential convergence was established in Section 6.3 for sequences $\{f_n : n = 1, \ldots\} \subseteq L^1_\mu(X)$. By definition of the weak topology in Appendix A.9 and the fact that $L^1_\mu(X)' = L^\infty_\mu(X)$, the proof of this latter characterization might seem relatively concrete, compared with the complexity of $M_b(X)'$. However, employing the technique that SAKS used to prove NIKODYM's theorem (see Section 6.2), we can transfer weak convergence of $\{\mu_n\}$ to that of a sequence of functions. As such we rewrite Theorem 6.3.7 in the following way.

Theorem 6.4.1. Characterization of weak sequential convergence of measures
Let X be a locally compact Hausdorff space, and consider the measurable space $(X, B(X))$. A sequence $\{\mu_n : n = 1, \ldots\} \subseteq M_b(X)$ converges weakly to some $\mu \in M_b(X)$ if and only if

$$\forall\, A \in \mathcal{B}(X), \quad \lim_{n\to\infty} \mu_n(A)$$

exists and is finite.

Proof. We begin as suggested in the Remark after the proof of Theorem 6.2.3, i.e., for a given sequence $\{\mu_n : n = 1, \ldots\} \subseteq M_b(X)$, we define

$$\forall\, A \in \mathcal{B}(X), \quad \nu(A) = \sum_{n=1}^{\infty} \frac{1}{2^n \|\mu_n\|_1} |\mu_n|(A).$$

Then $(X, \mathcal{B}(X), \nu)$ is a measure space and ν is regular. Also, note that $\mu_n \ll \nu$ for all n. Thus, we may use R–N to find a sequence of functions $h_n \in L^1_\nu(X)$, $n = 1, \ldots,$ such that $\mu_n = h_n \nu$. This last equality can be used to define a linear injective isometric isomorphism $i : L^1_\nu(X) \to M_b(X)$; cf. Theorem 5.3.3.

We now assert that the weak topology on $L^1_\nu(X)$ is equivalent to the topology on $L^1_\nu(X)$ induced by the weak topology on $M_b(X)$ by means of the mapping i. The space $i(L^1_\nu(X))$, in both induced weak topologies, from $M_b(X)$ and $L^1_\nu(X)$, has topological bases of closed convex sets. Theorem A.9.4 implies that these topologies on $i(L^1_\nu(X))$ are identical. This gives our assertion by the injectivity of i.

Consequently, $\mu_n \to \mu$ weakly in $M_b(X)$ if and only if $\{h_n\}$ converges weakly in $L^1_\nu(X)$. Our result now follows by a direct application of Theorem 6.3.7. □

The transference technique mentioned before Theorem 6.4.1 does not imply that Theorem 6.4.2 is not a significant strengthening of Theorem 6.4.1 – on the contrary. Theorem 6.4.2 is due to DIEUDONNÉ [136], pages 35–36, and GROTHENDIECK [206], pages 146–150; cf. [137].

Theorem 6.4.2. Dieudonné–Grothendieck theorem
Let X be a locally compact Hausdorff space and let $\{\mu_n : n = 1, \ldots\} \subseteq M_b(X)$ be a $\|\ldots\|_1$-norm bounded sequence. Then, $\{\mu_n : n = 1, \ldots\}$ converges weakly to some $\mu \in M_b(X)$ if and only if

$$\forall\, U \subseteq X, \textit{ open}, \quad \lim_{n\to\infty} \mu_n(U)$$

exists and is finite.

Proof. It is clear that we need to show the sufficiency of the condition only for weak convergence. We shall assume without loss of generality that the measures μ_n are real-valued.

i. Let $\{A_n : n = 1, \ldots\} \subseteq \mathcal{B}(X)$ be a disjoint sequence of open subsets of X and let $J \subseteq \mathbb{N}$ be a set of positive integers. Observe that, by our assumption,

$$\sum_{j\in J} \mu_n(A_j) = \mu_n\left(\bigcup_{j\in J} A_j\right)$$

converges to a finite limit as $n \to \infty$. We also note that for each n, $\{\mu_n(A_k) : k = 1, \ldots\} \in \ell^1(\mathbb{N})$, and that

$$\int_J \mu_n(A_j)\, dc(j) = \sum_{j\in J} \mu_n(A_j).$$

Thus, in particular, we may use Theorem 6.3.7 for $(\mathbb{N}, \mathcal{P}(\mathbb{N}), c)$ to deduce that for each n, $\{\mu_n : n = 1, \ldots\}$ is weakly convergent in $\ell^1(\mathbb{N})$.

Similarly, using Proposition 6.3.5, we conclude that

$$\forall\, \varepsilon > 0,\ \exists\, k \in \mathbb{N}, \text{ such that } \forall\, n = 1, \ldots, \quad \sum_{j>k} |\mu_n(A_j)| \leq \varepsilon;$$

and so, for all $j > k$, we have

$$\forall\, n = 1, \ldots, \quad |\mu_n(A_j)| \leq \varepsilon.$$

Thus, for each disjoint sequence $\{A_j : j = 1, \ldots\}$ of open sets in X,

$$\lim_{j\to\infty} \mu_n(A_j) = 0, \quad \text{uniformly for } n \in \mathbb{N}. \tag{6.59}$$

ii. We shall now prove that the condition (6.59) implies that

$$\begin{gathered}\forall\, \varepsilon > 0,\ \forall\, K \subseteq X, \text{ compact},\ \exists\, U \text{ open},\ K \subseteq U, \text{ such that},\\ \forall\, n = 1, \ldots, \quad |\mu_n|(U \setminus K) < \varepsilon.\end{gathered} \tag{6.60}$$

Indeed, assume that (6.60) is not true. Then

$$\exists\, \varepsilon > 0 \text{ and } \exists\, K \subseteq X, \text{ compact, such that } \exists\, n \text{ for which } |\mu_n|(U \setminus K) \geq \varepsilon. \tag{6.61}$$

For this $\varepsilon > 0$ and compact set $K \subseteq X$, we shall prove that there are a decreasing sequence $\{A_m : m = 1, \ldots\}$ of open sets containing K, a subsequence $\{\mu_{n_m} : m = 1, \ldots\}$, and a sequence $\{B_m : m = 1, \ldots\}$ of open sets whose closures are compact (i.e., relatively compact sets) such that

$$\forall\, m = 1, \ldots, \quad \overline{B}_m \subseteq A_{m-1} \setminus A_m \quad \text{and} \quad |\mu_{n_m}(B_m)| > \frac{\varepsilon}{2}. \tag{6.62}$$

To this end, assume that the construction for (6.62) has been done up to step m. By our assumption (6.61), there exists $\mu_{n_{m+1}}$ such that $|\mu_{n_{m+1}}|(A_m \setminus K) > \varepsilon/2$. Thus, there is a compact set $F \subseteq A_m \setminus K$ such that $|\mu_{n_{m+1}}|(F) > \varepsilon/2$ by the regularity of $\mu_{n_{m+1}}$. This, in turn, implies that there exists an open, relatively compact set V such that $F \subseteq V \subseteq \overline{V} \subseteq A_m \setminus K$ and $|\mu_{n_{m+1}}|(V) > \varepsilon/2$, since $|\mu_{n_{m+1}}|$ is a positive measure. Because $\mu_{n_{m+1}}$ is real-valued, there exists $A \in \mathcal{B}(X)$, $A \subseteq V$, such that $|\mu_{n_{m+1}}(A)| > \varepsilon/2$; see Problem 5.25. This inequality and the regularity of $\mu_{n_{m+1}}$ imply that there exists an open relatively compact set B_{m+1} such that

$$A \subseteq B_{m+1} \subseteq V \quad \text{and} \quad \left|\mu_{n_{m+1}}(B_{m+1})\right| > \varepsilon/2. \tag{6.63}$$

To verify (6.63), first note that $\mu_{n_{m+1}} = \mu^+_{n_{m+1}} - \mu^-_{n_{m+1}}$, where $\mu^+_{n_{m+1}}, \mu^-_{n_{m+1}}$ are regular positive measures; see Problem 5.24. Let $B^{\pm}$ be open relatively compact sets for which $A \subseteq B^{\pm}$ and for which

$$0 < \mu^{\pm}_{n_{m+1}}(B^{\pm}) - \mu^{\pm}_{n_{m+1}}(A) < \left|\mu_{n_{m+1}}(A)\right| - \frac{\varepsilon}{2}.$$

Then set $B_{m+1} = B^+ \cap B^-$ to obtain (6.63). Finally, we take A_{m+1} to be any open neighborhood of K contained in A_m and disjoint from $\overline{B}_{m+1}$. Condition (6.62) is obtained; see Figure 6.2 for an illustration of this construction.

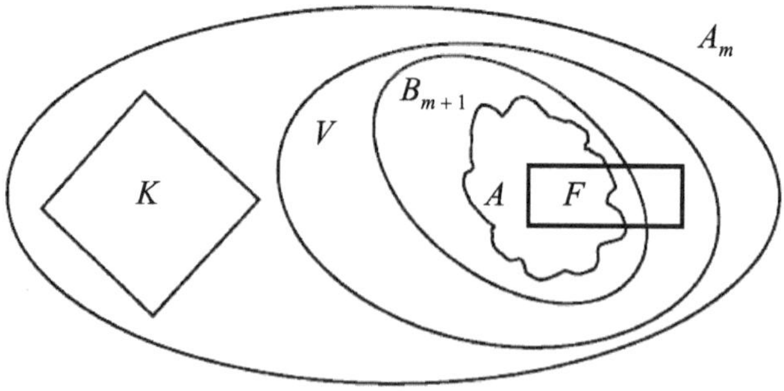

Fig. 6.2. Dieudonné–Grothendieck diagram.

It is now elementary to see that this construction for (6.62) contradicts (6.59).

iii. Analogous to part *ii*, condition (6.59) also implies that

$$\forall\, \varepsilon > 0,\ \exists\, K \subseteq X,\ \text{compact, such that } \forall\, n = 1, \ldots,\quad |\mu_n|(K^\sim) < \varepsilon. \tag{6.64}$$

To verify (6.64) assume that it is not true. Then

$$\exists\, \varepsilon > 0 \text{ such that } \forall\, K \subseteq X,\ \text{compact},\ \exists\, n \text{ such that } |\mu_n|(K^\sim) \geq \varepsilon. \tag{6.65}$$

For any such K_1 in (6.65) choose n_1 for which $|\mu_{n_1}|(K_1^\sim) \geq \varepsilon$. In part *ii*, we showed that if $A \in \mathcal{B}(X)$ and $|\mu|(A) > \varepsilon$, then there is an open set $V \supseteq A$ such that $|\mu(V)| > \varepsilon$. Symmetrically, in this part *iii*, we use the regularity of μ_{n_1} to assert the existence of a compact set $K_2 \subseteq K_1^\sim$ such that $|\mu_{n_1}(K_2)| > \varepsilon/2$. Then, as in part *ii*, we proceed by induction to obtain a contradiction to (6.59).

iv. As with the argument to prove Theorem 6.4.1, there exists a measure ν such that, for all $n \in \mathbb{N}$, $\mu_n \ll \nu$; and the question of weak convergence of $\{\mu_n\}$ in $M_b(X)$ can be reduced to the question of weak convergence of a sequence $\{h_n : n = 1, \ldots\}$ of functions in $L^1_\nu(X)$, where each $\mu_n = h_n \nu$.

According to Theorem 6.3.8, to prove that $\{\mu_n : n = 1, \ldots\}$ is weakly convergent it is enough to show that (6.60) and (6.64) imply

$$\forall\, \varepsilon > 0,\ \exists\, \delta > 0, \text{ such that } \mu(A) \leq \delta \Longrightarrow \forall\, n = 1, \ldots, \int_A |h_n|\, d\nu < \varepsilon \tag{6.66}$$

and

$$\forall\, \varepsilon > 0,\ \exists\, K \subseteq X,\ \text{compact, such that } \forall\, n = 1, \ldots, \int_{K^\sim} |h_n|\, d\nu < \varepsilon. \tag{6.67}$$

Since ν is nonnegative, then, according to Theorem 5.3.5*b*, (6.64) is equivalent to (6.67).

To prove (6.66), we first observe that, because of the regularity of ν, it is sufficient to prove it for open sets.

To this end, we assume that (6.66) does not hold for open sets, and we shall obtain a contradiction to (6.60), as written in (6.70). Thus, there exist $\varepsilon > 0$ and a sequence of open sets A_n such that $\nu(A_n) < 2^{-n}$ and

$$\int_{A_n} |h_n|\, d\nu > \varepsilon.$$

Let $B_n = \bigcup_{m=n}^{\infty} A_m$, so that

$$\int_{B_n} |h_n|\, d\nu > \varepsilon. \tag{6.68}$$

Assumption (6.64) allows us to assume without loss of generality that X is compact. Thus, (6.60) implies that for each open set B and for each $\varepsilon > 0$ there exists a compact set $F \subseteq B$ such that

$$\forall\, n = 1, \ldots, \quad \int_{B \setminus F} |h_n|\, d\nu < \varepsilon.$$

In fact, let $B = K^{\sim}$ and $F = U^{\sim}$ in (6.60). Therefore, we can find a sequence of compact sets $F_n \subseteq B_n$ with the property that

$$\forall\, m = 1, \ldots, \quad \int_{B_n \setminus F_n} |h_m|\, d\nu < \frac{\varepsilon}{2^{n+1}}.$$

Let $L_n = \bigcap_{m=1}^{n} F_m$. Then (6.68) implies that

$$\int_{L_n} |h_n|\, d\nu = \int_{B_n} |h_n|\, d\nu - \int_{B_n \setminus L_n} |h_n|\, d\nu > \varepsilon - \int_{B_n \setminus L_n} |h_n|\, d\nu.$$

Moreover, because $B_n \setminus L_n = \bigcup_{m=1}^{n} (B_n \setminus F_m)$, we have

$$\int_{B_n \setminus L_n} |h_n|\, d\nu \le \sum_{m=1}^{n} \int_{B_m \setminus F_m} |h_n|\, d\nu \le \sum_{m=1}^{n} \frac{\varepsilon}{2^{m+1}} < \frac{\varepsilon}{2}.$$

The last two inequalities imply that

$$\int_{L_n} |h_n|\, d\nu > \frac{\varepsilon}{2}. \tag{6.69}$$

Let $F = \bigcap_{n=1}^{\infty} F_n$. Then (6.60) implies that there exists an open set $U \supseteq F$ such that

$$\forall\, n = 1, \ldots, \quad \int_{U \setminus F} |h_n|\, d\nu < \frac{\varepsilon}{2}.$$

Recall that the sets B_n form a decreasing sequence with ν-measure approaching 0. Since $F_n \subseteq B_n$, we conclude that $\nu(F) = 0$, and so

$$\forall\, n = 1, \ldots, \quad \int_{U} |h_n|\, d\nu < \frac{\varepsilon}{2}. \tag{6.70}$$

On the other hand, because the L_ns are compact and $\{L_n\}$ is decreasing, and since $F = \bigcap_{n=1}^{\infty} L_n$, if U contains F, then there exists n_0 such that $L_{n_0} \subseteq U$. This implies that

$$\int_{U} |h_{n_0}|\, d\nu \ge \int_{L_{n_0}} |h_{n_0}|\, d\nu > \frac{\varepsilon}{2},$$

which contradicts (6.70). □

The following example should caution us from reading too much into Theorem 6.4.2.

Example 6.4.3. Convergence for open intervals but not all open sets
Let $f_n : [0,1] \to \mathbb{R}^+$, $n = 1, \ldots$, have the form

$$f_n(x) = \sum_{k=0}^{2^n-1} \Delta_k(x),$$

where Δ_k is an "isosceles triangle function" having its base of length $1/2^{2n+1}$ centered in $[k/2^n, (k+1)/2^n]$, Δ_j is congruent to Δ_k, and $\int_0^1 f_n = 1$. It is easy to see that for all $(a,b) \subseteq [0,1]$,

$$\lim_{n\to\infty} \int_a^b f_n \, dx = \int_a^b dx.$$

In light of Theorem 6.3.7 we would like to conclude that the sequence $\{f_n : n = 1, \ldots\}$ converges weakly to 1 (cf. Theorem 6.4.2, noting that, after all, open sets in $[0,1]$ are just countable unions of open intervals). Such is not the case. Let $A = \{x : \forall\, n, f_n(x) = 0\}$. Since $m(\{x : f_n(x) > 0\}) = 1/2^{n+1}$, then $m(\{x : \exists\, n \text{ for which } f_n(x) > 0\}) \leq 1/2$; and thus $m(A) \geq 1/2$. Consequently,

$$\lim_{n\to\infty} \int_A f_n \, dx \neq \int_A dx \geq \frac{1}{2}$$

because $\int_A f_n = 0$.

An interesting problem with applications to harmonic analysis (e.g., [32], Chapters 2 and 7, and [78]) is to find other families of sets, besides the open sets, for which Theorem 6.4.2 is true. The best results for compact Hausdorff spaces are due to BENJAMIN B. WELLS [498]. JOHANN PFANZAGL [371] has proved the analogue of Theorem 6.4.2 for the case of regular complex measures on any Hausdorff space; cf. Problem 2.19.

Because of Theorem 6.4.2 and the Riesz representation theorem in Chapter 7, it is interesting to investigate the relation betwen weak and weak* convergence in $M_b(X)$. GROTHENDIECK proved that *if X is compact and the closure of every open set is open then any weak* convergent sequence in $M_b(X)$ is weak convergent*; cf. Theorem A.9.6. This has been generalized by GALEN L. SEEVER and HELMUT SHAEFER [418].

Example 6.4.4. $\mathcal{B}([0,1])$
Consider the measurable space $([0,1], \mathcal{B}([0,1]))$. Choose a sequence $\{\mu_n : n = 1, \ldots\} \subseteq M_b([0,1])$ such that $\mu_n \to m$ in the weak* topology, card supp $\mu_n < \infty$, and supp $\mu_n \subseteq \mathbb{Q}$. For example, we could take $\mu_n = \sum_{k=1}^{n} (1/n)\delta_{k/n}$. Let $A \subseteq [0,1]$ be a closed set of irrationals with positive Lebesgue measure,

e.g., Problem 1.4*a*. If $f = \mathbb{1}_A$ then $\int f\,d\mu_n = 0$ and $\int f = m(A)$. Obviously, $f \in (M_b([0,1]))'$. Consequently, weak* convergence does not entail weak convergence in $M_b([0,1])$. Also, we have proved that, generally, the following statement is *false*: *Let X be a compact space, let $A \subseteq X$ be closed, and assume that for $\mu_n, \mu \in M_b(X)$, $\mu_n \to \mu$ in the weak* topology; then*

$$\forall\, f \in C(X), \quad \int_A f\,d\mu_n \to \int_A f\,d\mu.$$

Remark. The proof of the Dieudonné–Grothendieck theorem (Theorem 6.4.2) depends on the structure of the weakly compact sets in $M_b(X)$; cf. the Dunford–Pettis theorem (Theorem 6.3.8). Besides those already referenced, we mention work, *relevant to the point of view of this chapter*, of ROBERT G. BARTLE, DUNFORD, and JACOB T. SCHWARTZ (1955), DARST (1967), FLEMMING TOPSØE (1970), WOLFGANG ADAMSKI, PETER GÄNSSLER, and SIGURD KAISER (1976), ENNIO DEGIORGI and LETTA (1977), BROOKS and RAFAEL V. CHACON (1980), BROOKS (1980), T. V. PANCHAPAGESAN (1998), BROOKS, KAZUYUKI SAITÔ, and J. D. MAITLAND WRIGHT (2003). We emphasize that this general topic is also of interest in probability theory; see Sections 6.6.4 and 6.6.5.

6.5 Norm and weak sequential convergence

In terms of weak convergence, some of the results of Sections 6.1 and 6.2 can be rephrased as follows.

Theorem 6.5.1. Norm and weak sequential convergence
Let $(X, \mathcal{A}, \mu)$ be a measure space and assume that the sequence $\{f_n : n = 1, \ldots\} \subseteq L^1_\mu(X)$ converges weakly to $f \in L^1_\mu(X)$. Then (6.1) is valid if and only if $f_n \to f$ in measure on each $A \in \mathcal{A}$ satisfying $\mu(A) < \infty$; cf. Example 3.3.15 and Example 3.6.5.

Proof. ($\Longrightarrow$) This is immediate from Theorem 6.1.5.

($\Longleftarrow$) The characterization of weak sequential convergence in Theorem 6.3.2 yields that the assumptions of Theorem 6.2.2 are satisfied. Theorem 6.2.2 implies, in turn, that condition *ii* in Theorem 6.1.5 holds. The result follows now from Theorem 6.1.5 by noting that in its "if" part one needs only that $\{f_n : n = 1, \ldots\}$ converges in measure on sets of finite measure. □

Proposition 6.5.2. *Let $(X, \mathcal{A}, \mu)$ be a measure space and let $\{f_n, f : n = 1, \ldots\} \subseteq L^1_\mu(X)$. Assume $f_n \to f$ μ-a.e. and $\|f_n\|_1 \to \|f\|_1$. Then*

$$\forall\, A \in \mathcal{A}, \quad \lim_{n\to\infty} \int_A |f_n|\,d\mu = \int_A |f|\,d\mu.$$

Proof. Take $A \in \mathcal{A}$. From Fatou's lemma,

$$\underline{\lim}_{n\to\infty} \int_A |f_n|\, d\mu \geq \int_A |f|\, d\mu \geq \int_X |f|\, d\mu - \underline{\lim}_{n\to\infty} \int_{A^\sim} |f_n|\, d\mu. \tag{6.71}$$

Since $\|f_n\|_1 \to \|f\|_1$ we have

$$\int_X |f|\, d\mu \geq \underline{\lim}_{n\to\infty} \int_{A^\sim} |f_n|\, d\mu + \overline{\lim}_{n\to\infty} \int_A |f_n|\, d\mu. \tag{6.72}$$

Combining (6.71) and (6.72) gives the result. □

Theorem 6.5.3*b* is due to RADON (1913) and F. RIESZ (1928).

Theorem 6.5.3. Radon–Riesz theorem
Let $(X, \mathcal{A}, \mu)$ be a measure space and let $\{f_n, f : n = 1, \ldots\} \subseteq L^p_\mu(X)$, $1 \leq p < \infty$. Assume $\|f_n\|_p \to \|f\|_p$.

a. *If $1 \leq p < \infty$ and $f_n \to f$ μ-a.e., then*

$$\lim_{n\to\infty} \|f_n - f\|_p = 0.$$

b. *If $1 < p < \infty$ and $f_n \to f$ weakly, then*

$$\lim_{n\to\infty} \|f_n - f\|_p = 0.$$

Proof. **a.** For any $a, b \geq 0$, $(a+b)^p \leq 2^p(a^p + b^p)$. Let $|f_n| = a$ and $|f| = b$. Then the nonnegative functions,

$$g_n = 2^p(|f_n|^p + |f|^p) - |f_n - f|^p, \quad n = 1, \ldots,$$

converge μ-*a.e.* to $2^{p+1}|f|^p$. Using Fatou's lemma we have

$$\begin{aligned} 2^{p+1} \int_X |f|^p\, d\mu &\leq \underline{\lim}_{n\to\infty} \int_X g_n\, d\mu \\ &= 2^{p+1} \int_X |f|^p\, d\mu - \overline{\lim}_{n\to\infty} \int_X |f_n - f|^p\, d\mu, \end{aligned}$$

where the equality follows since $\|f_n\|_p \to \|f\|_p$. Therefore,

$$\lim_{n\to\infty} \int_X |f_n - f|^p\, d\mu = 0.$$

b. There are different proofs of part *b*, e.g., in [235], page 233, or [392], pages 78–80. We shall follow the latter proof.

We start by noting that it is sufficient to prove this result for real-valued functions $\{f_n : n = 1, \ldots\}$ that converge weakly to a real-valued function f in the space of real-valued functions.

i. Assume that $p \geq 2$. We shall consider the following function:

$$h(y) = \frac{|1+y|^p - 1 - py}{|y|^p} = \frac{g(y)}{|y|^p}.$$

Since $p \geq 2$, g'' exists on $\mathbb{R}$ and

$$\forall\, y \in \mathbb{R}, \quad g''(y) = p(p-1)|1+y|^{p-2}.$$

Thus, $g'' > 0$ on $\mathbb{R} \setminus \{-1\}$. (If $p = 2$ then $g'' > 0$ on $\mathbb{R}$, and if $p > 2$ then $g''(-1) = 0$.) Using this fact and applying the mean value theorem two times, we see that $g > 0$ on $\mathbb{R} \setminus \{0\}$. Further, $g(0) = 0$. Therefore, $h > 0$ on $\mathbb{R} \setminus \{0\}$. If $p = 2$ then $\lim_{|y|\to 0} h(y) = 1$ and if $p > 2$ then $\lim_{|y|\to 0} h(y) = \infty$. Clearly, $\lim_{|y|\to\infty} h(y) = 1$. Combining these facts we see that

$$\exists\, C > 0 \text{ such that } \forall\, y \in \mathbb{R}, \quad |1+y|^p \geq 1 + py + C|y|^p. \tag{6.73}$$

In particular, (6.73) is valid when C is replaced by $\min(C, 1)$.

We now replace y in (6.73) by

$$\frac{f_n(x) - f(x)}{f(x)}$$

on $P = \{x \in X : f(x) \neq 0\}$. Multiplying both sides of (6.73), with this substitution, by $|f(x)|^p$, we obtain

$$|f_n(x)|^p \geq |f(x)|^p + p|f(x)|^{p-1}\frac{|f(x)|}{f(x)}(f_n(x) - f(x)) + C|f_n(x) - f(x)|^p \tag{6.74}$$

on P. However, (6.74) is also true for $f(x) = 0$ in the case $0 < C \leq 1$. For any such C, we integrate (6.74) over X to obtain

$$\int_X |f_n|^p\, d\mu \geq \int_X |f|^p\, d\mu + p\int_X |f|^{p-1}\frac{|f|}{f}(f_n - f)\, d\mu + C\int_X |f_n - f|^p\, d\mu.$$

Since $|f|^{p-1}\frac{|f|}{f} \in L^q_\mu(X)$, where $1/p + 1/q = 1$, it follows that

$$\lim_{n\to\infty} \int_X |f|^{p-1}\frac{|f|}{f}(f_n - f)\, d\mu = 0,$$

by the weak convergence of $\{f_n : n = 1, \ldots\}$ to f. Combining this observation with the assumption that $\|f_n\|_p \to \|f\|_p$, we obtain

$$0 \geq \overline{\lim_{n\to\infty}} \int_X |f_n - f|^p\, d\mu.$$

Consequently, part *b* of the theorem is proved for $p \geq 2$.

ii. Let $1 < p < 2$. Consider the function

$$F(y) = \begin{cases} (|1+y|^p - 1 - py)/(|y|^p), & |y| \geq 1, \\ (|1+y|^p - 1 - py)/(|y|^2), & |y| < 1. \end{cases} \tag{6.75}$$

By arguments similar to those of part *i*,

$$\exists\, C \in (0,1] \text{ such that } \forall\, y \in \mathbb{R}, \quad F(y) \geq C. \tag{6.76}$$

Let

$$X_n = \{x \in X : |f_n(x) - f(x)| \geq |f(x)|\}.$$

Setting $y = (f_n(x) - f(x))/f(x)$ on P and multiplying by $|f|^p$, (6.76) becomes

$$|f_n(x)|^p \geq |f(x)|^p + p|f(x)|^{p-1}\frac{|f(x)|}{f(x)}(f_n(x) - f(x)) + C|f_n(x) - f(x)|^p \tag{6.77}$$

for $x \in P \cap X_n$. As in (6.73) we can take $0, C \leq 1$. For $x \in P$, write $g(x) = |f(x)|/f(x)$, and extend it to a μ-measurable function g on X for which $|g| \leq 1$. Then (6.77) is also valid for $x \in X_n \setminus P$ in the case $0 < C \leq 1$.

For $x \in P \cap X_n^{\sim}$, (6.76) becomes

$$|f_n(x)|^p \geq |f(x)|^p + p|f(x)|^{p-1}g(x)(f_n(x) - f(x)) + C\frac{|f_n(x) - f(x)|^2}{|f(x)|^2}|f(x)|^p. \tag{6.78}$$

Since $P^{\sim} \cap X_n^{\sim} = \emptyset$, then (6.78) is valid for all $x \in X_n^{\sim}$.

We now integrate (6.77) and (6.78) over X_n and $X_n^{\sim}$, respectively, to obtain

$$\int_X |f_n|^p \, d\mu \geq \int_X |f|^p \, d\mu + p \int_X |f|^{p-1} g(f_n - f) \, d\mu$$
$$+ C\left(\int_{X_n} |f_n - f|^p \, d\mu + \int_{X_n^{\sim}} \left|\frac{f_n - f}{f}\right|^2 |f|^p \, d\mu\right).$$

Combining this observation with our assumptions yields

$$\lim_{n\to\infty} \int_{X_n} |f_n - f|^p \, d\mu = 0 = \lim_{n\to\infty} \int_{X_n^{\sim}} |f_n - f|^2 |f|^{p-2} \, d\mu. \tag{6.79}$$

Moreover, we have that

$$\begin{aligned} \int_{X_n^{\sim}} |f_n - f|^p \, d\mu &\leq \int_{X_n^{\sim}} |f_n - f||f|^{p-1} \, d\mu \\ &\leq \left(\int_{X_n^{\sim}} |f_n - f|^2 |f|^{p-2} \, d\mu\right)^{1/2} \left(\int_{X_n^{\sim}} |f|^p \, d\mu\right)^{1/2} \\ &\leq \left(\int_{X_n^{\sim}} |f_n - f|^2 |f|^{p-2} \, d\mu\right)^{1/2} \left(\int_X |f|^p \, d\mu\right)^{1/2}. \end{aligned} \tag{6.80}$$

Finally (6.79) together with (6.80) yields the conclusion of part b for $1 < p < 2$. □

Corollary 6.5.4. *Let $(X, \mathcal{A}, \mu)$ be a measure space, and let $\{f_n, f : n = 1, \dots\} \subseteq L^p_\mu(X)$, $1 \le p < \infty$. Assume $f_n \to f$ μ-a.e. Then $\|f_n\|_p \to \|f\|_p$ if and only if $\|f_n - f\|_p \to 0$.*

Theorem 6.5.5. Pointwise and weak sequential convergence
Let $(X, \mathcal{A}, \mu)$ be a measure space, and let $\{f_n : n = 1, \dots\} \subseteq L^p_\mu(X)$, $1 < p < \infty$. Assume $f_n \to f$ μ-a.e., f is μ-measurable, and $\sup_n \|f_n\|_p = K < \infty$. Then $f \in L^p_\mu(X)$ and $f_n \to f$ weakly in $L^p_\mu(X)$; cf. Theorem A.9.11.

Proof. From Fatou's lemma, for the case $|f_n|^p$ and $|f|^p$, we have

$$\|f\|_p^p \le K^p,$$

and so $f \in L^p_\mu(X)$. Take $\varepsilon > 0$ and $g \in L^q_\mu(X)$. We shall find N such that

$$\forall\, n \ge N, \quad \left| \int_X (f - f_n) g \, d\mu \right| < \varepsilon. \tag{6.81}$$

Since $|g|^q \in L^1_\mu(X)$ there is $\delta > 0$ such that

$$\forall\, A \in \mathcal{A} \text{ for which } \mu(A) < \delta, \quad \left(\int_A |g|^q \, d\mu \right)^{1/q} < \frac{\varepsilon}{6K}; \tag{6.82}$$

see Proposition 3.3.9. Also, from the integrability of $|g|^q$ there is $B \in \mathcal{A}$ such that $\mu(B) < \infty$ and

$$\left(\int_{B^\sim} |g|^q \, d\mu \right)^{1/q} < \frac{\varepsilon}{6K}. \tag{6.83}$$

Because of Egorov's theorem (Theorem 2.5.7) there is $E \subseteq B$, such that $E \in \mathcal{A}$ and $\mu(E) > 0$, for which

$$\mu(B \cap E^\sim) < \delta \quad \text{and} \quad f_n \to f \text{ uniformly on } E.$$

From the uniform convergence we can choose $N \in \mathbb{N}$ such that

$$\forall\, n \ge N \text{ and } \forall\, x \in E, \quad |f(x) - f_n(x)| < \frac{\varepsilon}{3\|g\|_q (\mu(E))^{1/p}}.$$

Consequently,

$$\forall\, n \ge N, \quad \left(\int_E |f - f_n|^p \, d\mu \right)^{1/p} \|g\|_q < \frac{\varepsilon}{3}. \tag{6.84}$$

Thus, taking $A = B \cap E^\sim$, and using the Hölder and Minkowski inequalities, we obtain

$$\int_X |f_n - f||g|\, d\mu = \int_A + \int_{B^\sim} + \int_E |f_n - f||g|\, d\mu < \varepsilon,$$

which, in turn, gives (6.81). In fact, $\int_A |f_n - f||g|\, d\mu < \varepsilon/3$ by Hölder, Minkowski, and (6.82); $\int_{B^\sim} |f_n - f||g|\, d\mu < \varepsilon/3$ by Hölder, Minkowski, and (6.83); and $\int_E |f_n - f|\, d\mu < \varepsilon/3$ by Hölder and (6.84). □

In Theorem 6.3.2 we noted that weak convergence in $L^1_\mu(X)$ implies norm boundedness. This phenomenon is rather general because of the Uniform Boundedness Principle (Theorem A.8.6 and Corollary A.8.7). Thus, we can restate Theorem 6.5.5 as follows.

Corollary 6.5.6. *Let $(X, \mathcal{A}, \mu)$ be a measure space, and let $\{f_n : n = 1, \ldots\} \subseteq L^p_\mu(X)$, $1 < p < \infty$. Assume $f_n \to f$ μ-a.e. and assume that f is μ-measurable. Then $f \in L^p_\mu(X)$, and $f_n \to f$ weakly in $L^p_\mu(X)$ if and only if $\sup_{n \in \mathbb{N}} \|f_n\|_p < \infty$.*

Example 6.5.7. Failure of Theorem 6.5.5 for $p = 1$
Theorem 6.5.5 is false when $p = 1$. Let $f_n : [0, 1] \to \mathbb{R}^+$ be 0 on $[1/n, 1]$, n at $x = 0$, and linear on $[0, 1/n]$. Thus, $f_n \to 0$ m-a.e., and $\|f_n\|_1 = 1/2$, whereas $f \not\to 0$ weakly.

Because of Theorem A.9.4 we know that if $f_n \to f$ weakly in $L^p_\mu(X)$ then certain linear combinations of the f_ns converge to f in the $L^p_\mu(X)$ topology. The *Banach–Saks theorem* [23], [392], pages 78–81, which we now state, is much finer in that instead of rather arbitrary linear combinations we can use arithmetic means.

Theorem 6.5.8. Banach–Saks theorem
Let $(X, \mathcal{A}, \mu)$ be a measure space, and let $\{f_n, f : n = 1, \ldots\} \subseteq L^p_\mu(X)$, $1 \le p < \infty$. Assume $f_n \to f$ weakly. Then, there is a subsequence $\{f_{n_k} : k = 1, \ldots\}$ whose arithmetic means $(1/m) \sum_{k=1}^m f_{n_k}$ converge in the $L^p_\mu(X)$-norm topology to f.

Banach and Saks proved Theorem 6.5.8 for only the $1 < p < \infty$ cases. The result was proved for $L^1_\mu(X)$ in 1965 by Wieslaw Szlenk. Otto Schreier (1930) showed that $C([0, 1])$ does not have the property of the Banach–Saks theorem.

Because of Problem 3.26*d* we see that if $\|f_n - f\|_1 \to 0$ then $f_{n_k} \to f$ μ-a.e., for some subsequence $\{f_{n_k} : k = 1, \ldots\}$. We then ask whether $f_n \to f$ weakly yields the same result. Generally the answer is negative, as we have seen in Theorem 3.3.14, Example 3.3.15, Example 3.6.5, Problem 3.33, and Problem 3.34. Note that if we take $\{f_n : n = 1, \ldots\}$ as in Theorem 3.3.14 then $f_n \to \alpha$ weakly, where $\alpha = \int_0^1 f$; cf. [519], pages 87–88.

We close this section by noting that $L^1_\mu(X)$ and $M_b(X)$ are weakly sequentially complete. This is proved using Schur's lemma (Theorem 6.2.1) and the results of Sections 6.1 and 6.2; see, e.g., [137]. On the other hand, these spaces are never weakly complete as uniform spaces. In fact, no infinite-dimensional normed space is weakly complete.

6.6 Potpourri and titillation

1. Giuseppe Vitali (August 26, 1875–February 29, 1932), the oldest of five children, was born in Ravenna. After graduating from the "liceo" in Ravenna, he studied mathematics at the University of Bologna in 1895. He then received a scholarship to the Scuola Normale Superiore; cf. Section 3.8.1. Dini and Luigi Bianchi were there at the time. He graduated in 1899 and in his thesis he extended a theorem of Gösta Mittag-Leffler to Riemann surfaces. His next work was devoted to Abelian integrals. In 1901 he was Dini's assistant, and then he began his teaching career in a "liceo". After two brief appointments in Sassari and Voghera he taught at the Liceo Colombo in Genova from 1904 to 1922. This, of course, does not match Dedekind's record of under-undergraduate teaching. Also, Weierstrass' "defenders" would point out that his (Weierstrass) service in teaching penmanship and gymnastics, besides mathematics, should count for something extra. Finally, in 1923, Vitali received a position at the University of Modena (a weak counter-example to one of the fundamental theorems of life that "you can keep a good person down"). In 1924 he went to the University of Padova, where, at the end of 1926, he was struck with hemiplegia, a paralysis resulting from injury to the motor center of the brain. His intellectual powers were unaffected. He left Padova in 1930 for the University of Bologna. For more, see Vitali's obituary [474].

2. The Arzelà–Ascoli theorem (Theorem A.4.3) provides criteria for characterizing compact subsets of the Banach space $C([0,1])$; the Dunford–Pettis theorem (Theorem 6.3.8) gives criteria for characterizing weakly compact subsets of the Banach space $L^1_\mu(X)$; and the Banach–Alaoglu theorem (Theorem A.9.5) asserts that the closed unit ball of the dual of a normed vector space is weak* compact.

It is natural to ask whether there are workable criteria for characterizing compact subsets of $L^1_\mu(X)$. In fact, the relatively compact sets of $L^p_{m^d}(\mathbb{R}^d)$, $p \geq 1$, are characterized in the following result, whose proof uses the Arzelà–Ascoli theorem.

Theorem 6.6.1. Kolmogorov compactness theorem
Let $p \geq 1$. A subset $K \subseteq L^p_{m^d}(\mathbb{R}^d)$ is relatively compact in the L^p-norm topology if and only if the following conditions hold:

i. $\sup_{f\in K} \|f\|_p < \infty$;

ii. $\lim_{T\to\infty} \int_{|x|\geq T} |f(x)|^p \, dx = 0$ *uniformly for* $f \in K$;

iii. $\forall\, \varepsilon > 0,\ \exists\, r = r(\varepsilon) > 0$ *such that* $\forall\, y \in \overline{B(0,r)}$ *and* $\forall\, f \in K$,

$$\int_{\mathbb{R}^d} |f(x+y) - f(x)|^p \, dx < \varepsilon.$$

Here $\overline{B(0,r)} \subseteq \mathbb{R}^d$ is the closed Euclidean ball centered at $0 \in \mathbb{R}^d$ with radius r.

The conditions *i*, *ii*, *iii* are norm boundedness of K, uniform "small tails" at infinity, and uniform translation continuity, respectively. KOLMOGOROV [290], [291] (1931) proved the result for $L^p_m([a,b])$, $p>1$, and with the $f(x+y)$ in condition *iii* replaced by

$$f_y(x) = \frac{1}{2y}\int_{x-y}^{x+y} f, \quad y>0. \tag{6.85}$$

In fact, he proved that, for each $y>0$, the set $\{f_y : f \in K\}$ is equicontinuous. His proof required $p>1$. A. N. TULAIKOV [476] (1933) dealt with the case $p=1$. MARCEL RIESZ [397] (1933) introduced condition *iii*. Accessible proofs of the Kolmogorov compactness theorem can be found in [356], pages 212–216, [326], pages 41–44, [275], pages 292–299, and [517], pages 274–277. The statement for $\mathbb{R}^d$ can be generalized to any locally compact group using the same proof.

With either *iii* as written or using the mean (6.85), we are reminded of the FTC.

An ingenious application of Theorem 6.6.1 to the Heisenberg uncertainty principle for *classes* of functions is due to HAROLD SHAPIRO [432] (1991). For example, he used Theorem 6.6.1 to prove that *if* $\{f_n\} \subseteq L^2_m(\mathbb{R})$ *is an orthonormal sequence, then it is not possible that all four of the sequences* $\{t_n\}$, $\{\gamma_n\}$, $\{\sigma^2 f_n\}$, *and* $\{\sigma^2 \hat{f}_n\}$ *are bounded, where the expected values* t_n *and* γ_n *are*

$$t_n = \int_{\mathbb{R}} t|f_n(t)|^2\,dt \quad \textit{and} \quad \gamma_n = \int_{\widehat{\mathbb{R}}} \gamma|\hat{f}_n(\gamma)|^2\,d\gamma,$$

where the variances $\sigma^2 f_n$ *and* $\sigma^2 \hat{f}_n$ *are*

$$\sigma^2 f_n = \int_{\mathbb{R}} (t-t_n)^2|f_n(t)|^2\,dt \quad \textit{and} \quad \sigma^2 \hat{f}_n = \int_{\widehat{\mathbb{R}}} (\gamma-\gamma_n)^2|\hat{f}_n(\gamma)|^2\,d\gamma,$$

and where the Fourier transform $\hat{f}_n$ *defined on* $\widehat{\mathbb{R}} = \mathbb{R}$ *is defined in* Appendix B. See [45], Chapter 7, [227], Chapter 5, [116], [240], and [181] for perspective on the Heisenberg uncertainty principle. This definition of expected value can be formulated in terms of the probabilistic definition given in Section 3.9.2.

3. Let $(X, \mathcal{A}, u)$ be a measure space, and let $p \geq 2$ be fixed. The *Clarkson inequality* is the following:

$$\forall\, f,g \in L^p_\mu(X), \quad \left\|\frac{f+g}{2}\right\|_p^p + \left\|\frac{f-g}{2}\right\|_p^p \leq \frac{1}{2}\,\|f\|_p^p + \frac{1}{2}\,\|g\|_p^p.$$

For $1<p<2$ this inequality holds in the reverse sense. JAMES A. CLARKSON also proved several other related inequalities; see [106].

The Clarkson inequality can be used to prove that if $F \in (L^p_\mu(X))'$, as a continuous linear functional, then

$$\exists\, f \in L^p_\mu(X) \text{ such that } \|f\|_p = 1 \text{ and } F(f) = \sup_{\|g\|_p \le 1} |F(g)|.$$

This, in turn, can be used to give an alternative proof of the necessary condition in Theorem 5.5.5 in order that $F \in (L^p_\mu(X))'$.

IRVING L. GLICKSBERG used the Clarkson inequality to give a short proof of the Radon–Riesz theorem (Theorem 6.5.3), e.g., [235], page 233.

One of our reasons for recording the Clarkson inequality is to advertise the ingenious approach due to SADAHIRO SAEKI (1991, personal communication) based on establishing the inequality on a two-point probability space with probability measure ν weighted $1/2$ at each point. In this case, for $p \ge 2$, we have

$$\begin{aligned}\int \left(\left|\frac{f+g}{2}\right|^p + \left|\frac{f-g}{2}\right|^p \right)^{p/p} d\nu &\le \int \left(\left|\frac{f+g}{2}\right|^2 + \left|\frac{f-g}{2}\right|^2 \right)^{p/2} d\nu \\ &= \int \left(\frac{|f|^2 + |g|^2}{2} \right)^{p/2} d\nu \le \int \frac{|f|^p + |g|^p}{2}\, d\nu.\end{aligned}$$

4. In "The Scottish Book", STEINHAUS asked the following question; see [344] Problem 126: *Does there exist a family $\mathcal{F}$ of measurable functions on a measure space X such that*

i. $\forall\, f \in \mathcal{F}$, $|f| = 1$, *and*

ii. $\forall\, \{f_n : n = 1, \ldots\} \subseteq \mathcal{F}$, *the sequence*

$$\frac{1}{m} \sum_{k=1}^{m} f_k(x)$$

is divergent for a.e. $x \in X$?

Related to this problem, JÁNOS KOMLÓS [294] showed that every bounded sequence of integrable random variables $\{\xi_n : n = 1, \ldots\}$ has a subsequence $\{\xi_{n_k} : k = 1, \ldots\}$ such that there exists an integrable random variable η for which

$$\frac{\xi_{n_1} + \cdots + \xi_{n_m}}{m}$$

converges to η with probability 1. In the language of analysis, we can restate a stronger version of this result as follows.

Theorem 6.6.2. Komlós theorem

Let $(X, \mathcal{A}, \mu)$ be a finite measure space and let $\{f_n : n = 1, \ldots\} \subseteq L^1_\mu(X)$ be a bounded sequence of integrable functions $f_n : X \to \mathbb{R}$. Then there exist a subsequence $\{f_{n_k} : k = 1, \ldots\}$ and a function $f \in L^1_\mu(X)$ such that

$$\frac{1}{m}\sum_{l=1}^{m} f_{n_{k_l}} \to f \ \ \mu\text{-a.e.}, \tag{6.86}$$

for each subsequence $\{f_{n_{k_l}} : l = 1, \ldots\}$.

The convergence of arithmetic means in (6.86) is also called *Cesàro summability* or *Cesàro convergence.* The theorem, both hypotheses and conclusion, should be compared with the Banach–Saks theorem (Theorem 6.5.8).

The theorem of KOMLÓS has many generalizations and applications; see, e.g., [134] or [409] for characterizations of weak compactness by means of KOMLÓS' theorem. In other developments, a recent paper of HEINRICH VON WEIZSÄCKER [493] studies the question of necessity of the integrability assumption in Theorem 6.6.2 for sequences of nonnegative functions that are allowed to assume the value $+\infty$. In this case, the following converse to Komlós' theorem holds.

Theorem 6.6.3. Converse Komlós theorem
Let $(X, \mathcal{A}, \mu)$ *be a finite measure space and let* $\{f_n : n = 1, \ldots\}$ *be a sequence of measurable functions* $f_n : X \to \mathbb{R}^+ \cup \{+\infty\}$. *If there exists a function* $f \in L^\infty_\mu(X)$ *such that*

$$\frac{1}{m}\sum_{l=1}^{m} f_n \to f \ \ \mu\text{-a.e.}, \tag{6.87}$$

then there exist a subsequence $\{f_{n_k} : l = 1, \ldots\}$ *and an equivalent measure* ν *such that* $\{f_{n_k} : l = 1, \ldots\}$ *is bounded in* $L^1_\nu(X)$.

5. Let $M_1^+(\mathbb{R}^d) = \{\mu \in M_b(\mathbb{R}^d) : \mu \geq 0 \text{ and } \mu(\mathbb{R}^d) = 1\}$. This space is defined equivalently, but in slightly different terms, in the proof of Theorem B.9.2.

We have proved fundamental properties of weak sequential convergence of measures in Section 6.4. We now compare these ideas with the following notion for $\{\mu_n, \mu : n = 1, \ldots\} \subseteq M_1^+(\mathbb{R}^d)$: $\{\mu_n : n = 1, \ldots\}$ *converges to* μ *in the sense of Bernoulli* if

$$\forall\, f \in C_b(\mathbb{R}^d), \ \lim_{n\to\infty} \mu_n(f) = \mu(f).$$

When dealing with random variables in a probabilistic setting this is also referred to as *convergence in distribution.* It is *not difficult* to prove the following result; cf. with Theorems 6.4.1 and 6.4.2 and the final comment in Section 6.1 concerning "probabilistic weak sequential convergence".

Theorem 6.6.4. Bernoulli convergence theorem
Let $\{\mu_n, \mu : n = 1, \ldots\} \subseteq M_1^+(\mathbb{R}^d)$. *The following are equivalent.*

- ***a.*** $\{\mu_n : n = 1, \ldots\}$ *converges to* μ *in the sense of Bernoulli.*
- ***b.*** $\forall\, U \subseteq \mathbb{R}^d$, *open,* $\underline{\lim}_{n\to\infty} \mu_n(U) \geq \mu(U)$.

c. $\forall\, A \in \mathcal{B}(\mathbb{R}^d)$, *for which* $\mu(\overline{A}) = \mu(\operatorname{int} A)$,

$$\lim_{n\to\infty} \mu_n(A) = \mu(A).$$

Proof. The implication $c \Longrightarrow a$ is the most interesting. Let $f \in C_b(\mathbb{R}^d)$ be real-valued, and assume $-M \leq f(x) \leq M$ on $\mathbb{R}^d$. Given $\varepsilon > 0$, clearly $D = \{y \in \mathbb{R} : \mu(\{x : f(x) = y\}) > 0\}$ is countable; and, in fact, $\{x \in \mathbb{R}^d : \{f(x) = y : y \in \mathbb{R}\}\}$ is a disjoint union of closed sets. Choose $\{y_j : j = 0, \ldots, J\} \subseteq \mathbb{R} \setminus D$ such that $y_0 < -M < y_1 < \cdots < y_{J-1} < M < y_J$ and each $y_j - y_{j-1} < \varepsilon$. Define $A_j = \{x : y_{j-1} < f(x) \leq y_j\}$ and $g = \sum_{j=1}^{J} y_j \mathbb{1}_{A_j}$. We have $\mu(\overline{A_j}) = \mu(\operatorname{int} A_j)$ since

$$\begin{aligned} 0 &\leq \mu(\overline{A_j}) - \mu(\operatorname{int} A_j) \\ &= \mu(\overline{A_j} \setminus \operatorname{int} A_j) \\ &\leq \mu(\{x : f(x) = y_{j-1}\}) + \mu(\{x : f(x) = y_j\}) = 0. \end{aligned}$$

Therefore, by part c,

$$\exists\, N_\varepsilon \text{ such that } \forall\, n > N_\varepsilon, \quad |\mu_n(g) - \mu(g)| < \frac{\varepsilon}{3}.$$

On the other hand, by our definition of $\{A_j\}$, $\|f-g\|_\infty < \varepsilon/3$, and so $|\mu_n(f) - \mu_n(g)| < \varepsilon/3$ and $|\mu(f) - \mu(g)| < \varepsilon/3$. Thus, for $n > N_\varepsilon$, $|\mu_n(f) - \mu(f)| < \varepsilon$. □

A sequence $\{\mu_n : n = 1, \ldots\} \subseteq M_1^+(\mathbb{R}^d)$ is *tight* if

$$\forall\, \varepsilon > 0, \exists\, K_\varepsilon \subseteq \mathbb{R}^d, \text{ compact, such that } \forall\, n = 1, \ldots, \quad \mu_n(K_\varepsilon) > 1 - \varepsilon.$$

Theorem 6.6.5. Prohorov theorem
Given $\{\mu_n : n = 1, \ldots\} \subseteq M_1^+(\mathbb{R}^d)$.

a. *If* $\{\mu_n : n = 1, \ldots\}$ *converges to* $\mu \in M_1^+(\mathbb{R}^d)$ *in the sense of Bernoulli, then* $\{\mu_n : n = 1, \ldots\}$ *is tight.*

b. *If* $\{\mu_n : n = 1, \ldots\}$ *is tight, then there are* $\mu \in M_1^+(\mathbb{R}^d)$ *and a subsequence* $\{\mu_{n_k} : k = 1, \ldots\} \subseteq \{\mu_n : n = 1, \ldots\}$ *such that* $\{\mu_{n_k} : k = 1, \ldots\}$ *converges to* μ *in the sense of Bernoulli.*

Part a is a consequence of the first two equivalences of Theorem 6.6.4 and the regularity of the measures. Part b depends on the Helly selection principle described in the Remark after Problem 4.39; also, see Section 5.6.1. Clearly, especially in light of Theorem 6.6.5b, the Helly selection principle can be viewed as a form of the Banach–Alaoglu theorem (Theorem A.9.5 and Theorem A.9.9).

These ideas are the basis of significant development in probability theory, e.g., [61], [51], [335], [342], [461]. In this regard, see Example B.10.4.

7 Riesz Representation Theorem

7.1 Original Riesz representation theorem

Let $C([a,b])$ be the real Banach space of continuous functions $f : [a,b] \to \mathbb{R}$ with norm

$$\forall\, f \in C([a,b]), \quad \|f\|_\infty = \sup_{x\in[a,b]} |f(x)|,$$

e.g., Definition 2.5.9. From the theory of Riemann–Stieltjes integration we know that if $g \in BV([a,b])$, then $L_g : C([a,b]) \to \mathbb{R}$, defined by

$$L_g(f) = \int_a^b f\, dg,$$

is an element of $(C([a,b]))'$. In fact,

$$|L_g(f)| \leq \|f\|_\infty V(g,[a,b]).$$

The converse is F. Riesz' representation theorem [389]; cf. (4.7) and Proposition 4.1.10. In order to state and prove this result, we define the norm of a continuous linear functional $L \in (C([a,b]))'$ as follows:

$$\|L\| = \sup\{|L(f)| : f \in C([a,b]) \text{ and } \|f\|_\infty \leq 1\};$$

see equation (A.7) and Example A.8.9.

Theorem 7.1.1. Riesz representation theorem (F. Riesz, 1909) *If $L \in (C([a,b]))'$ then there is $g \in BV([a,b])$ such that $\|L\| = V(g,[a,b])$ and*

$$\forall\, f \in C([a,b]), \quad L(f) = \int f\, dg. \tag{7.1}$$

Proof. For the sake of simple subscripts let $[a,b] = [0,1]$. Since $L \in (C([0,1]))'$ we use the Hahn–Banach theorem (Theorem A.8.3) to conclude the existence of $K \in (L^\infty_m([0,1]))'$ such that $L = K$ on $C([0,1])$ and $\|L\| = \|K\|$. Let $\mathbb{1}_x = \mathbb{1}_{[0,x]}$ and write

$$g(x) = K(\mathbb{1}_x). \tag{7.2}$$

J.J. Benedetto and W. Czaja, *Integration and Modern Analysis*, Birkhäuser Advanced Texts,
DOI 10.1007/978-0-8176-4656-1_7,

We shall verify (7.3), thereby proving that $g \in BV([0,1])$. Consider a partition,

$$0 = x_0 \le x_1 \le \cdots \le x_n = 1,$$

and set $t_i = \text{sgn}\,[g(x_i) - g(x_{i-1})]$, $i = 1, \ldots, n$. Thus,

$$\begin{aligned}\sum_{i=1}^{n} |g(x_i) - g(x_{i-1})| &= \sum_{i=1}^{n} t_i (g(x_i) - g(x_{i-1})) \\ &= K\left(\sum_{i=1}^{n} t_i (\mathbb{1}_{x_i} - \mathbb{1}_{x_{i-1}})\right) = K\left(\sum_{i=1}^{n} t_i \mathbb{1}_{(x_{i-1}, x_i]}\right).\end{aligned}$$

Note that $|t_i|$ is 0 or 1, so that

$$\left\| \sum_{i=1}^{n} t_i \mathbb{1}_{(x_{i-1}, x_i]} \right\|_\infty \le 1.$$

Hence,

$$V(g, [0,1]) \le \|K\| = \|L\|. \tag{7.3}$$

Now, for $f \in C([0,1])$, define

$$f_n(x) = \sum_{j=1}^{n} f\left(\frac{j}{n}\right) \left(\mathbb{1}_{j/n}(x) - \mathbb{1}_{(j-1)/n}(x)\right).$$

Therefore, if $j = 1, \ldots, n$, then

$$\forall\, x \in \left(\frac{j-1}{n}, \frac{j}{n}\right), \quad f_n(x) = f\left(\frac{j}{n}\right).$$

Since f is continuous, $\|f - f_n\|_\infty \to 0$. Thus,

$$\lim_{n \to \infty} K(f_n) = K(f) = L(f). \tag{7.4}$$

Using (7.2) we see that

$$K(f_n) = \sum_{j=1}^{n} f\left(\frac{j}{n}\right) \left(g\left(\frac{j}{n}\right) - g\left(\frac{j-1}{n}\right)\right),$$

from which we conclude

$$\lim_{n \to \infty} K(f_n) = \int_0^1 f\, dg. \tag{7.5}$$

Equations (7.4) and (7.5) yield (7.1).

Fix $\varepsilon > 0$, and, letting $\|L\| \le |L(f)| + \varepsilon$ for some $f \in C([0,1])$ that satisfies $\|f\|_\infty \le 1$, we apply (7.1) and compute

$$\|L\| \le V(g, [0,1]) + \varepsilon.$$

Since ε is arbitrary we have $\|L\| \le V(g, [0,1])$, which, when combined with (7.3), yields $\|L\| = V(g, [0,1])$. □

Part of the motivation that led to Theorem 7.1.1 came from the *moment problem*: Let $\{a_n : n = 1, \ldots\} \subseteq \mathbb{C}$ be a sequence of moments of a distribution of electric charges and let $\{f_n : n = 1, \ldots\} \subseteq C([a,b])$ be a sequence of functions; the problem is to find $\mu \in M_b([a,b])$ such that

$$\forall\, n = 1, \ldots, \qquad \int f_n \, d\mu = a_n.$$

It is interesting that what is sometimes known as the Riesz representation theorem for $L^2([a,b])$ was discovered independently by FRÉCHET; and both F. RIESZ' and FRÉCHET's $L^2([a,b])$ results appeared in the same issue of the Comptes Rendus; see [395] and [184]. Of course, the L^2 result, dealt with in Chapter 5, has limited impact in measure theory, as opposed to the extraordinary importance of Theorem 7.1.1, as seen in the remaining sections of this chapter.

7.2 Riesz representation theorem (RRT)

Let X be a locally compact Hausdorff space and let $C_0(X)$ be the complex vector space of continuous functions $f : X \to \mathbb{C}$ with the property that

$$\forall\, \varepsilon > 0,\ \exists\, K_{f,\varepsilon} \subseteq X, \text{ compact, such that } \forall\, x \notin K_{f,\varepsilon},\ |f(x)| < \varepsilon,$$

i.e., f "vanishes at infinity". Then $C_0(X)$ is a Banach space with norm $\|f\|_\infty = \sup_{x \in X} |f(x)|$.

Let $C_c(X)$ denote the space of continuous functions with compact support, i.e., all continuous functions $f : X \to \mathbb{C}$ for which

$$\exists\, K, \text{ compact, such that } \forall\, x \notin K, \quad f(x) = 0.$$

$C_c(X)$ is a normed vector space with norm $\|\ldots\|_\infty$, and it is not a Banach space. In fact, $\overline{C_c(X)} = C_0(X)$.

By $C_0^+(X)$ and $C_c^+(X)$ we denote the subsets of $C_0(X)$ and $C_c(X)$, respectively, consisting of functions with range contained in $\mathbb{R}^+$.

In this section, when we refer to $L : C_c(X) \to \mathbb{C}$ as a *continuous* functional, we are taking the $\|\ldots\|_\infty$-norm topology on $C_c(X)$. If $L : C_c(X) \to \mathbb{C}$ is a linear functional for which $L(f) \geq 0$, whenever $f \in C_c^+(X)$, then L is a *positive linear functional.*

Let us introduce the following notation:

$$\phi \wedge \psi(x) = \min(\phi(x), \psi(x)) \quad \text{and} \quad \phi \vee \psi(x) = \max(\phi(x), \psi(x)).$$

We shall say that a family Φ of functions in $C_c^+(X)$ is *directed* if, for any two functions $\phi, \psi \in \Phi$, there exists a function $\eta \in \Phi$ such that $\phi \vee \psi \leq \eta$.

Proposition 7.2.1. *Let X be a locally compact Hausdorff space. Let $\phi \in C_c^+(X)$, let $\{\phi_\alpha : \alpha \in A\} \subseteq C_c^+(X)$ be directed, and let L be a positive linear functional on $C_c(X)$. If $\phi \leq \sup_{\alpha \in A} \phi_\alpha$, then $L(\phi) \leq \sup_{\alpha \in A} L(\phi_\alpha)$.*

Proof. It is enough to consider the case $\phi = \sup_{\alpha \in A} \phi_\alpha$. In fact, if $\phi \leq \sup_{\alpha \in A} \phi_\alpha$ then $\phi = \sup_{\alpha \in A} \phi \wedge \phi_\alpha$. Let $\psi_\alpha = \phi \wedge \phi_\alpha \in C_c^+(X)$, and observe that, since $\psi_\alpha \vee \psi_\beta = \phi \wedge (\phi_\alpha \vee \phi_\beta)$, $\{\psi_\alpha : \alpha \in A\}$ is a directed family. Thus, if the statement of our proposition is true for ϕ and the family $\{\psi_\alpha : \alpha \in A\}$, which we shall prove, then the positivity of L implies that

$$L(\phi) \leq \sup_{\alpha \in A} L(\phi \wedge \phi_\alpha) \leq \sup_{\alpha \in A} L(\phi_\alpha).$$

Fix $\varepsilon > 0$. For any $x \in X$ there exists $\alpha = \alpha(x, \varepsilon)$ such that $\phi(x) < \psi_\alpha(x) + \varepsilon$. Since ϕ and ψ_α are continuous, this inequality holds on some open neighborhood $U_\alpha(x)$ of x. Thus, for all $x \in X$, $\phi(y) < \psi_\alpha(y) + \varepsilon$ for all $y \in U_\alpha(x)$.

Since supp ϕ is compact, there exist $\alpha_1, \ldots, \alpha_m$ such that

$$\forall\, x \in \text{supp}\, \phi, \quad \phi(x) < \psi_{\alpha_1} \vee \cdots \vee \psi_{\alpha_m}(x) + \varepsilon.$$

Because we have assumed that the family Φ is directed, there exists $\beta \in A$ for which

$$\forall\, x \in \text{supp}\, \phi, \quad \phi(x) < \psi_\beta(x) + \varepsilon.$$

This, in turn, implies that

$$\|\phi - \psi_\beta\|_\infty < \varepsilon.$$

By Urysohn's lemma (Theorem A.1.3), there is $u \in C_c^+(X)$ with the property that $u = 1$ on supp ϕ. For any $\psi \in C_c^+(X)$ satisfying supp $\psi \subseteq$ supp ϕ, we have $0 \leq \psi \leq \|\psi\|_\infty u$, and hence

$$0 \leq L(\psi) \leq \|\psi\|_\infty L(u)$$

by the positivity of L. Thus, since $\phi - \psi_\beta \in C_c^+(X)$ and supp $(\phi - \psi_\beta) \subseteq$ supp ϕ, there exists $C = L(u)$ such that

$$|L(\phi - \psi_\beta)| < C\|\phi - \psi_\beta\|_\infty < C\varepsilon.$$

In particular,

$$L(\phi) < L(\psi_\beta) + C\varepsilon.$$

Consequently,

$$L(\phi) < \sup_{\alpha \in A} L(\psi_\alpha) + C\varepsilon.$$

Since this inequality is valid for all $\varepsilon > 0$, our result is proved. □

Now consider the family

$$S(X) = \{f : X \to \mathbb{R}^+ \cup \{\infty\} \text{ such that } f = \sup_{\phi \in \Phi} \phi,$$
$$\text{for some directed family } \Phi = \Phi_f \subseteq C_c^+(X)\}.$$

The elements of $S(X)$ need not be continuous. In fact, they are *lower semicontinuous.* This means that the real-valued elements $f \in S(X)$ have the property that $\{x : f(x) > r\}$ is open for each $r \in \mathbb{R}$; see the Remark after Theorem 7.3.8. Note that for any $f \in C_0^+(X)$, the family $\{\phi : \phi \leq f, \phi \in C_c^+(X)\} \subseteq C_c^+(X)$ is directed, and thus $C_0^+(X) \subseteq S(X)$.

Let L be a positive linear functional on $C_c(X)$. For $f \in S(X)$, we define the extension $\tilde{L}$ of L:

$$\tilde{L}(f) = \sup_{\phi \in \Phi} L(\phi), \tag{7.6}$$

where $f, \Phi \subseteq C_c^+(X)$, and $\phi \in \Phi$ are related as in the definition of $S(X)$. In order to verify that this extension is well defined, let Φ and Ψ be two directed families for which

$$f = \sup_{\phi \in \Phi} \phi = \sup_{\psi \in \Psi} \psi.$$

Thus, for any $\phi \in \Phi$, $\phi \leq f = \sup_{\psi \in \Psi} \psi$, and Proposition 7.2.1 implies that $L(\phi) \leq \sup_{\psi \in \Psi} L(\psi)$. Then, taking the supremum over all $\phi \in \Phi$, we obtain

$$\sup_{\phi \in \Phi} L(\phi) \leq \sup_{\psi \in \Psi} L(\psi).$$

The opposite inequality follows analogously, and so we have

$$\sup_{\phi \in \Phi} L(\phi) = \sup_{\psi \in \Psi} L(\psi).$$

If $f \in S(X)$ is bounded on X, it is possible that $\tilde{L}(f) = \infty$.

Proposition 7.2.2. *Let X be a locally compact Hausdorff space, let L be a positive linear functional on $C_c(X)$, and let $\tilde{L}$ be the extension of L to $S(X)$ defined in (7.6). Then $\tilde{L}$ has the following properties:*

a. $\forall\, \phi \in C_c^+(X)$, $\tilde{L}(\phi) = L(\phi)$;
b. $\forall\, f, g \in S(X)$, $f \leq g \Longrightarrow \tilde{L}(f) \leq \tilde{L}(g)$;
c. $\forall\, f \in S(X)$ *and* $\forall\, \alpha \geq 0$, $\tilde{L}(\alpha f) = \alpha \tilde{L}(f)$;
d. $\forall\, f, g \in S(X)$, $\tilde{L}(f + g) = \tilde{L}(f) + \tilde{L}(g)$;
e. $\forall\, \{f_n : n = 1, \ldots\} \subseteq S(X)$,

$$f = \sum_{n=1}^{\infty} f_n \in S(X) \quad \textit{and} \quad \tilde{L}(f) = \sum_{n=1}^{\infty} \tilde{L}(f_n);$$

f. $\forall\, \{g_n : n = 1, \ldots\} \subseteq S(X)$, *such that* $g_1 \leq g_2 \leq \cdots$,

$$\tilde{L}\left(\lim_{n \to \infty} g_n\right) = \lim_{n \to \infty} \tilde{L}(g_n).$$

Proof. In this proof we shall use the convention that $f = \sup_{\phi \in \Phi} \phi$ and $g = \sup_{\psi \in \Psi} \psi$.

a. Take $\Phi = \{\phi\}$, a one-element directed family.

b. Property b follows since $f \leq g$ implies $\phi \leq \sup_{\psi \in \Psi} \psi$, and we use Proposition 7.2.1 to conclude that $L(\phi) \leq \sup_{\psi \in \Psi} L(\psi) = \tilde{L}(g)$. In particular, it implies that $\tilde{L}(f) = \sup_{\phi \in \Phi} L(\phi) \leq \tilde{L}(g)$.

c. Part c follows from definition.

d. In order to prove property d we note that if Φ and Ψ are directed families, then $\Phi + \Psi = \{\phi + \psi : \phi \in \Phi, \psi \in \Psi\}$ is also a directed family. Therefore, if $f, g \in S(X)$, then $f + g \in S(X)$. Moreover,

$$\begin{aligned}\tilde{L}(f+g) &= \sup_{\phi+\psi \in \Phi+\Psi} L(\phi+\psi) = \sup_{\phi \in \Phi, \psi \in \Psi} (L(\phi) + L(\psi)) \\ &= \sup_{\phi \in \Phi} L(\phi) + \sup_{\psi \in \Psi} L(\psi).\end{aligned}$$

e. For part e we first need to show that $f = \sum_{n=1}^{\infty} f_n \in S(X)$. We can assume without loss of generality that $0 \in \Phi_n$, $n = 1, \ldots$. Further, let $f_n = \sup_{\phi \in \Phi_n} \phi$. Let $\Psi_m = \Phi_1 + \cdots + \Phi_m$ and let $\Phi = \bigcup_{m=1}^{\infty} \Psi_m$. Each Ψ_m is a directed family and $\{\Psi_m\}$ is increasing since $0 \in \Psi_m$; hence, Φ is a directed family. Moreover, since each $f_n \geq 0$,

$$f = \sup_{m \in \mathbb{N}} \sum_{n=1}^{m} f_n = \sup_{m \in \mathbb{N}} \sup_{\phi \in \Psi_m} \phi = \sup_{\phi \in \Phi} \phi.$$

Therefore, we can use part d to write

$$\begin{aligned}\tilde{L}(f) &= \sup_{\phi \in \Phi} L(\phi) = \sup_{m \in \mathbb{N}} \sup_{\phi \in \Psi_m} L(\phi) = \sup_{m \in \mathbb{N}} \tilde{L}\left(\sum_{n=1}^{m} f_n\right) \\ &= \sup_{m \in \mathbb{N}} \sum_{n=1}^{m} \tilde{L}(f_n) = \sum_{n=1}^{\infty} \tilde{L}(f_n).\end{aligned}$$

f. Since $\{g_n : n = 1, \ldots\}$ is a monotone family, we can deal with a supremum instead of a limit. Let Ψ_n be a directed family associated with g_n, $n = 1, \ldots$. This means that there is a directed family $\Psi_n \subseteq C_c^+(X)$ such that $g_n = \sup_{\phi \in \Psi_n} \phi$. Define $\Phi_n = \{\max\{\phi_1, \ldots, \phi_n\} : \phi_i \in \Psi_i, i = 1, \ldots, n\}$. Each Φ_n is a directed family, and, since we can assume without loss of generality that $0 \in \Psi_n$ for all n, we have that $\{\Phi_n\}$ is increasing. Thus, $\Psi = \bigcup_{n=1}^{\infty} \Phi_n$ is also a directed family. Moreover,

$$\lim_{n \to \infty} g_n = \sup_{n \in \mathbb{N}} \sup_{\psi \in \Psi_n} \psi = \sup_{n \in \mathbb{N}} \left(\max_{j=1,\ldots,n} \sup_{\psi \in \Psi_j} \psi \right) = \sup_{n \in \mathbb{N}} \sup_{\phi \in \Phi_n} \phi = \sup_{\psi \in \Psi} \psi.$$

Therefore,

$$\tilde{L}\left(\lim_{n \to \infty} g_n\right) = \sup_{\psi \in \Psi} L(\psi) = \sup_{n \in N} \sup_{\psi \in \Phi_n} L(\psi) = \sup_{n \in \mathbb{N}} \tilde{L}(g_n). \qquad \square$$

We now introduce one more extension of a positive linear functional L on $C_c(X)$. For any function $F : X \to \mathbb{R}^+ \cup \{\infty\}$ we define

$$L^*(F) = \inf_{f \geq F, f \in S(X)} \tilde{L}(f). \tag{7.7}$$

The set of functions $f \geq F$ is a directed family, possibly consisting of a single element identically equal to ∞. The functional L^* has properties analogous to those of $\tilde{L}$.

Proposition 7.2.3. *Let X be a locally compact Hausdorff space and let L be a positive linear functional on $C_c(X)$. Then L^*, defined by (7.7), is a well-defined extension of L to all nonnegative bounded functions on X, and L^* has the following properties:*

a. $\forall\, f \in S(X)$, $L^*(f) = \tilde{L}(f)$;

b. $\forall\, F, G$, *nonnegative*, $F \leq G \Longrightarrow L^*(F) \leq L^*(G)$;

c. $\forall\, F$, *nonnegative, and* $\forall\, \alpha \geq 0$, $L^*(\alpha F) = \alpha L^*(F)$;

d. $\forall\, F, G$, *nonnegative*, $L^*(F + G) \leq L^*(F) + L^*(G)$;

e. $\forall\, \{F_n : n = 1, \ldots\}$, *where each F_n is nonnegative,*

$$F = \sum_{n=1}^{\infty} F_n \Longrightarrow L^*(F) \leq \sum_{n=1}^{\infty} L^*(F_n);$$

f. $\forall\, \{F_n : n = 1, \ldots\}$, *where each F_n is nonnegative and where* $F_1 \leq F_2 \leq \cdots$,

$$L^*\left(\lim_{n\to\infty} F_n\right) = \lim_{n\to\infty} L^*(F_n).$$

Proof. **a.** By definition, $L^*(f) = \inf\{\tilde{L}(g) : g \in S(X), g \geq f\}$. In view of Proposition 7.2.2*b* we have $\tilde{L}(f) \leq \tilde{L}(g)$ for all $g \in S(X)$ for which $g \geq f$. Thus, the infimum equals $\tilde{L}(f)$.

b. Part *b* follows from Proposition 7.2.2*b*.

c. Part *c* follows from definition.

d. We first assume that both $L^*(F)$ and $L^*(G)$ are finite. Fix $\varepsilon > 0$, and take $f \geq F$ and $g \geq G$ such that $\tilde{L}(f) < L^*(F) + \varepsilon$ and $\tilde{L}(g) < L^*(G) + \varepsilon$. Then $F + G \leq f + g$ implies

$$L^*(F + G) \leq \tilde{L}(f + g) = \tilde{L}(f) + \tilde{L}(g) < L^*(F) + L^*(G) + 2\varepsilon.$$

Since this inequality is valid for all $\varepsilon > 0$, our result is proved.

In the case that either $L^*(F)$ or $L^*(G)$ is infinite, the result is immediate.

e. As we have done in the proof of part *d*, we can assume without loss of generality that all functions F_n satisfy $L^*(F_n) < \infty$. Hence, for each n, we can choose f_n such that $F_n \leq f_n$ and $\tilde{L}(f_n) \leq L^*(F_n) + \varepsilon/2^n$. Then $F \leq \sum_{n=1}^{\infty} f_n \in S(X)$. Thus it follows from Proposition 7.2.2*e* that

$$L^*(F) \leq \tilde{L}\left(\sum_{n=1}^{\infty} f_n\right) = \sum_{n=1}^{\infty} \tilde{L}(f_n) \leq \sum_{n=1}^{\infty} L^*(f_n) + \varepsilon.$$

Since this inequality is valid for all $\varepsilon > 0$, our result is proved.

f. From part b we have $\lim_{n\to\infty} L^*(F_n) \le L^*\left(\lim_{n\to\infty} F_n\right)$. Therefore, we need to prove only the opposite inequality. Without loss of generality we can assume that $L^*(F_n) < \infty$ for all n. Let $\varepsilon > 0$. Choose functions $f_n \in S(X)$ such that $f_n \ge F_n$ and $\tilde{L}(f_n) < L^*(F_n) + \varepsilon/2^n$. We define the functions $g_n = \max(f_1, \ldots, f_n)$, $n = 1, \ldots$. Clearly, $\{g_n : n = 1, \ldots\}$ forms an increasing sequence in $S(X)$. Thus, we may apply Proposition 7.2.2*f* and parts *a*,*b* to obtain

$$\begin{aligned}\lim_{n\to\infty} \tilde{L}(g_n) &= \tilde{L}\left(\lim_{n\to\infty} g_n\right) = L^*\left(\lim_{n\to\infty} g_n\right) \\ &= L^*\left(\sup_{n\in\mathbb{N}} f_n\right) \ge L^*\left(\lim_{n\to\infty} F_n\right).\end{aligned} \tag{7.8}$$

Observe now that

$$g_{n+1} + g_n \wedge f_{n+1} = g_n + f_{n+1}.$$

All of the functions in this expression are in $S(X)$, and so

$$\tilde{L}(g_{n+1}) + \tilde{L}(g_n \wedge f_{n+1}) = \tilde{L}(g_n) + \tilde{L}(f_{n+1}).$$

Moreover, since $\{F_n\}$ is increasing, $g_n \wedge f_{n+1} \ge F_n$ implies $\tilde{L}(g_n \wedge f_{n+1}) \ge L^*(F_n)$. Thus,

$$\tilde{L}(g_{n+1}) \le \tilde{L}(g_n) + \tilde{L}(f_{n+1}) - L^*(F_n) < \tilde{L}(g_n) + L^*(F_{n+1}) + \frac{\varepsilon}{2^{n+1}} - L^*(F_n).$$

Summing this inequality from 1 to m and using the cancellation property of telescoping sums, we have

$$\tilde{L}(g_{m+1}) < \tilde{L}(f_1) + L^*(F_{m+1}) - L^*(F_1) + \frac{\varepsilon}{2} < L^*(F_{m+1}) + \varepsilon.$$

Since these inequalities are valid for all $\varepsilon > 0$, we obtain

$$\lim_{n\to\infty} L^*(F_n) \ge \lim_{n\to\infty} \tilde{L}(g_n). \tag{7.9}$$

Now (7.8) and (7.9) prove part *f*. □

We now extend Theorem 7.1.1 to the case of positive linear functionals. Our proof may appear lengthy, but, because of the importance of the result, we have included many details that are not transparent. We begin the proof with an outline of the steps we take.

Theorem 7.2.4. RRT for positive linear functionals
Let X be a locally compact Hausdorff space and let L be a positive linear functional on $C_c(X)$. There exists a (nonnegative) regular Borel measure μ on X such that

$$\forall\, f \in C_c(X), \quad L(f) = \int_X f\, d\mu. \tag{7.10}$$

Proof. In part *i* we define an outer measure μ^* corresponding to L. In part *ii* we invoke Theorem 2.4.19 to obtain a measure μ associated with μ^*. Part *iii* shows that μ^* is finite on compact sets. Two technical properties of μ^* are proved in part *iv*; and these are used to prove that μ is a Borel measure in part *v*. In part *vi* we prove that μ is regular, and we verify the integral representation (7.10) in part *vii*.

i. The first step is to define an appropriate outer measure μ^*. This is done by considering the extension L^* of the functional L, and defining

$$\forall\, A \subseteq X, \quad \mu^*(A) = L^*(\mathbb{1}_A).$$

We shall verify the following properties of μ^*:

a. $\mu^*(\emptyset) = 0$;
b. $0 \leq \mu^*(A) \leq \infty$;
c. $A \subseteq B \Longrightarrow \mu^*(A) \leq \mu^*(B)$;
d. $\mu^*\left(\bigcup_{n=1}^{\infty} A_n\right) \leq \sum_{n=1}^{\infty} \mu^*(A_n)$.

Assertion *a* follows from Proposition 7.2.3*a*, and assertion *b* follows from the positivity of L. Proposition 7.2.3*b* implies *c*; and *d* is a consequence of the fact that $\mathbb{1}_{\bigcup_{n=1}^{\infty} A_n} \leq \sum_{n=1}^{\infty} \mathbb{1}_{A_n}$ and Proposition 7.2.3*b*,*e*.

ii. Now we invoke Theorem 2.4.19 to conclude that there exist a σ-algebra $\mathcal{M} \subseteq \mathcal{P}(X)$ and a nonnegative σ-additive set function μ on $\mathcal{M}$, i.e., μ is a measure. Further, μ is the restriction of μ^* to $\mathcal{M}$.

iii. Next, we show that if $A \subseteq X$ is a compact set, then

$$0 \leq \mu^*(A) < \infty.$$

We use Urysohn's lemma (Theorem A.1.3) to find a continuous function $f : X \to [0,1]$ such that $f = 1$ on A and $f = 0$ on $U^\sim$ for some open set U containing A for which $\overline{U}$ is compact. Thus, $\mathbb{1}_A \leq f \in C_c^+(X)$, and so

$$\mu^*(A) \leq L^*(f) = L(f) < \infty.$$

iv. In order to prove that μ is a Borel measure we shall need the following two statements:

$$\forall\, U \subseteq X, \text{ open}, \ \mu^*(U) = \sup\{\mu^*(V) : V \subseteq \overline{V} \subseteq U,\ V \text{ open},\ \overline{V} \text{ compact}\}, \tag{7.11}$$

and

$$\forall\, A \subseteq X, \quad \mu^*(A) = \inf\{\tilde{L}(\mathbb{1}_U) : A \subseteq U,\ U \text{ open}\}. \tag{7.12}$$

To verify (7.11), for an open set $U \subseteq X$, we first note that

$$\mu^*(U) = \sup\{L(\phi) : \phi \in C_c^+(X), \phi \leq \mathbb{1}_U\}. \tag{7.13}$$

Equation (7.13) is a consequence of the fact that $\mathbb{1}_U \in S(X)$, which itself follows since $\mathbb{1}_U = \sup\{\phi : \phi \in C_c^+(X), \phi \leq \mathbb{1}_U\}$ and $\{\phi : \phi \in C_c^+(X), \phi \leq \mathbb{1}_U\}$ is a directed family, and of the definitions of μ^* and L^*. Also, for any $\phi \in C_c^+(X)$ for which $\phi \leq \mathbb{1}_U$, let $V_{\phi,0} = \phi^{-1}((0,\infty))$, and let $V_{\phi,\delta} = \phi^{-1}((\delta,\infty))$, for $\delta > 0$. Fix $\varepsilon > 0$ and let $\phi \in C_c^+(X)$ satisfy $L(\phi) \geq \mu^*(U) - \varepsilon$; such a ϕ exists because of (7.13). Since $\phi \leq \mathbb{1}_{V_{\phi,0}}$, we have

$$L(\phi) \leq \mu^*(V_{\phi,0}).$$

On the other hand, by Proposition 7.2.3*f*,

$$\mu^*(V_{\phi,0}) = L^*\left(\mathbb{1}_{V_{\phi,0}}\right) = L^*\left(\lim_{\delta\to 0} \mathbb{1}_{V_{\phi,\delta}}\right) = \lim_{\delta\to 0} L^*\left(\mathbb{1}_{V_{\phi,\delta}}\right).$$

These observations together yield the inequality,

$$\forall\, \varepsilon > 0,\ \exists\, \delta > 0 \text{ such that } \quad \mu^*(U) - \varepsilon \leq \mu^*(V_{\phi,\delta}).$$

This, in turn, implies that

$$\mu^*(U) \leq \sup_{\delta>0} \mu^*(V_{\phi,\delta}). \tag{7.14}$$

Combining (7.13) and (7.14) with the facts that each $V_{\phi,\delta}$ is open, $V_{\phi,\delta} \subseteq \overline{V_{\phi,\delta}}$, $\overline{V_{\phi,\delta}}$ is closed, and $\overline{V_{\phi,\delta}} \subseteq V_{\phi,0} \subseteq U$, we compute

$$\begin{aligned}\mu^*(U) &\leq \sup\{L(\phi) : \phi \in C_c^+(X), \phi \leq \mathbb{1}_U\} \leq \mu^*(V_{\phi,\delta}) \\ &\leq \sup\{\mu^*(V) : V \subseteq \overline{V} \subseteq U, V \text{ open}, \overline{V} \text{ compact}\}.\end{aligned}$$

The opposite inequality follows from Proposition 7.2.3*b*, and so (7.11) is verified.

To verify (7.12) we write by definition that

$$\mu^*(A) = \inf\{\tilde{L}(f) : f \in S(X), f \geq \mathbb{1}_A\}.$$

Since, as we noted in the verification of (7.11), $\mathbb{1}_U \in S(X)$ for any open set U, we need to show only that $\mu^*(A) \geq \inf\{\tilde{L}(\mathbb{1}_U) : A \subseteq U, U \text{ open}\}$. Clearly, the opposite inequality is true, and so we can assume without loss of generality that $\mu^*(A) < \infty$. Fix $\varepsilon > 0$, and let $f \in S(X)$ satisfy $f \geq \mathbb{1}_A$ and $\tilde{L}(f) < \mu^*(A) + \delta$, for some $\delta > 0$ for which $\delta < \varepsilon/(1 + \varepsilon + \mu^*(A))$. Let $U = \{x : f(x) > 1 - \delta\}$. The set U is open, since f is lower semicontinuous as an element of $S(X)$, and $A \subseteq U$. We can write $f > (1-\delta)\mathbb{1}_U$. Therefore, we have

$$\tilde{L}(\mathbb{1}_U) \leq \tilde{L}\left(\frac{1}{1-\delta} f\right) \leq \frac{1}{1-\delta}(\mu^*(A) + \delta) < \mu^*(A) + \varepsilon.$$

Since these inequalities are valid for all $\varepsilon > 0$, we obtain

$$\mu^*(A) = \inf\{\tilde{L}(\mathbb{1}_U) : A \subseteq U, U \text{ open}\},$$

and so (7.12) is verified.

v. We can now prove that μ is a Borel measure. In order to check that $\mathcal{M}$ contains all Borel subsets of X, it is enough to show that $\mathcal{M}$ contains all open sets $U \subseteq X$, since $\mathcal{M}$ is a σ-algebra. In view of the Carathéodory criterion for the construction of $\mathcal{M}$ and μ, we need to verify the following for any open set U:

$$\forall\, E \subseteq X, \quad \mu^*(E) = \mu^*(E \cap U) + \mu^*(E \cap U^{\sim}).$$

Clearly, $\mu^*(E) \leq \mu^*(E \cap U) + \mu^*(E \cap U^{\sim})$.

It remains to prove the opposite inequality. Fix $\varepsilon > 0$ and use (7.12) to choose an open set $V_1 \subseteq X$, $E \subseteq V_1$, such that $\mu^*(V_1) < \mu^*(E) + \varepsilon$, and an open set $V_2 \subseteq X$, $V_1 \cap U^{\sim} \subseteq V_2$, such that $\mu^*(V_2) < \mu^*(V_1 \cap U^{\sim}) + \varepsilon$. Moreover, we use (7.11) to find an open set $V_3 \subseteq X$ for which $\overline{V}_3$ is compact, $V_3 \subseteq \overline{V_3} \subseteq V_1 \cap U$, and $\mu^*(V_1 \cap U) < \mu^*(V_3) + \varepsilon$. Let $W = V_1 \cap V_2 \cap \overline{V}_3^{\sim}$. Then W and V_3 are open disjoint sets. Thus, we have

$$V_1 \cap U^{\sim} \subseteq W \subseteq V_2,$$

and so

$$\begin{aligned} &|(\mu^*(V_3) + \mu^*(W)) - (\mu^*(V_1 \cap U) + \mu^*(V_1 \cap U^{\sim}))| \\ &\quad \leq |\mu^*(W) - \mu^*(V_1 \cap U^{\sim})| + |\mu^*(V_3) - \mu^*(V_1 \cap U)| \\ &\quad \leq (\mu^*(V_2) - \mu^*(V_1 \cap U^{\sim})) + \varepsilon \leq 2\varepsilon. \end{aligned} \tag{7.15}$$

Therefore,

$$\begin{aligned} \mu^*(E \cap U) + \mu^*(E \cap U^{\sim}) &\leq \mu^*(V_1 \cap U) + \mu^*(V_1 \cap U^{\sim}) \\ &\leq \mu^*(V_3) + \mu^*(W) + 2\varepsilon = \mu^*(V_3 \cup W) + 2\varepsilon \\ &\leq \mu^*(V_1) + 2\varepsilon \leq \mu^*(E) + 3\varepsilon. \end{aligned}$$

The first and last inequalities are a consequence of the choice of V_1. The second inequality is a consequence of (7.15). The equality follows from Proposition 7.2.2*d* and the disjointness of V_3 and W, since $\mu^*(V_3 \cup W) = \tilde{L}(\mathbb{1}_{V_3 \cup W}) = \tilde{L}(\mathbb{1}_{V_3} + \mathbb{1}_W)$, and since $\mathbb{1}_{V_3}, \mathbb{1}_W \in S(X)$ (as was noted earlier) because they are characteristic functions of open sets. The penultimate inequality is clear since $V_3, W \subseteq V$; and the last inequality follows by our choice of V_1.

Thus, all Borel sets are μ-measurable, and so μ is a Borel measure.

vi. We are now ready to prove that μ is regular. We have already shown in part *iii* that μ is finite on compact sets and in part *iv* that

$$\forall\, A \subseteq X, \quad \mu(A) = \inf\{\mu(U) : A \subseteq U,\ U \text{ open}\},$$

i.e., (7.12). It remains to prove that

$$\forall\, U \subseteq X, \text{ open}, \quad \mu^*(U) = \sup\{\mu^*(F) : F \subseteq U, F \text{ compact}\} \tag{7.16}$$

and

$$\forall\, A \in \mathcal{M},\ \mu(A) < \infty, \quad \mu^*(A) = \sup\{\mu^*(F) : F \subseteq A, F \text{ compact}\}. \quad (7.17)$$

To prove (7.16), we begin with (7.13),

$$\mu^*(U) = \sup\{L(\phi) : \phi \in C_c^+(X), \phi \leq \mathbb{1}_U\}.$$

For any $\phi \in C_c^+(X)$, such that $\phi \leq \mathbb{1}_U$, we have $\phi \leq \mathbb{1}_{\operatorname{supp}\phi} \leq \mathbb{1}_U$, and so, since $\operatorname{supp}\phi$ is compact,

$$L^*(\phi) \leq \mu^*(\operatorname{supp}\phi) \leq \sup\{\mu^*(F) : F \subseteq U, F \text{ compact}\}.$$

To prove the opposite inequality, we use Urysohn's lemma (Theorem A.1.3), and so, for any compact set $F \subseteq U$, we find $\phi \in C_c^+(X)$ for which $\mathbb{1}_F \leq \phi \leq \mathbb{1}_U$.

For the proof of (7.17), we begin by noting that we need to prove only the inequality $\mu^*(A) \leq \sup\{\mu^*(F) : F \subseteq A, F \text{ compact}\}$. To do this we proceed as follows. Fix $\varepsilon > 0$. Use (7.12) to find an open set $U \supseteq A$ such that $\mu^*(U \setminus A) < \varepsilon$. Moreover, using (7.12) again, choose open $V \supseteq U \setminus A$ such that $\mu^*(V) < \varepsilon$. Let $W = U \cap V$. Next, use (7.16) to find $K \subseteq U$, compact and such that $\mu^*(U \setminus K) < \varepsilon$. Define $F = K \setminus W = K \cap W^{\sim}$. Thus, F is compact, $F \subseteq A$, and $\mu^*(A \setminus F) = \mu^*\left((A \setminus K) \cap (A \cap W)\right) \leq \mu^*(U \setminus F) + \mu^*(K) < 2\varepsilon$.

Thus, μ is regular.

vii. Last to prove, but certainly not least, is that (7.10) holds. Let $f \in C_c^+(X)$. Recall from Theorem 2.5.5 that any measurable function may be represented as a pointwise limit of a sequence of simple functions s_n, $n = 1, \ldots$. In the case that the function f is nonnegative, we can take $\{s_n\}$ to be increasing and such that each s_n is nonnegative. Proposition 7.2.3*f* implies that

$$L^*(f) = L^*\left(\lim_{n\to\infty} s_n\right) = \lim_{n\to\infty} L^*(s_n). \quad (7.18)$$

We shall now prove that L^* is linear on the vector space of simple measurable functions. Let $F = a\mathbb{1}_A + b\mathbb{1}_B$, where A and B are disjoint measurable sets and $a, b > 0$. Clearly, because of Proposition 7.2.3*d*, we need to show only that

$$L^*(F) \geq a\mu^*(A) + b\mu^*(B). \quad (7.19)$$

Assume without loss of generality that $L^*(F) < \infty$ and that A and B are compact. (We can make this assumption, since we have proved that μ is regular, and so, for any measurable set A, $\mu(A) = \sup\{\mu(K) : K \subseteq A, K \text{ compact}\}$ and hence the standard approximation arguments work.) There are open sets U and V such that $A \subseteq U$, $B \subseteq V$, and $U \cap V = \emptyset$. Fix $\varepsilon > 0$. Let $f \in S(X)$ satisfy the conditions that $f \geq F$ and $\tilde{L}(f) < L^*(F) + \varepsilon$. Define $U' = U \cap \{x : f(x) > a - \varepsilon\}$ and $V' = V \cap \{x : f(x) > b - \varepsilon\}$. The sets U' and V' are open (since f is lower semicontinuous), disjoint, $A \subseteq U'$, $B \subseteq V'$, and $f > (a - \varepsilon)\mathbb{1}_{U'} + (b - \varepsilon)\mathbb{1}_{V'}$. Thus,

$$\begin{aligned} L^*(F)+\varepsilon > \tilde{L}(f) &\geq \tilde{L}((a-\varepsilon)\mathbb{1}_{U'} + (b-\varepsilon)\mathbb{1}_{V'}) \\ &= (a-\varepsilon)\tilde{L}(\mathbb{1}_{U'}) + (b-\varepsilon)\tilde{L}(\mathbb{1}_{V'}) \\ &\geq a\mu^*(A) + b\mu^*(B) - \varepsilon(\tilde{L}(\mathbb{1}_U) + \tilde{L}(\mathbb{1}_V)), \end{aligned}$$

where the equality follows from Proposition 7.2.2*d*. We obtain (7.19) since these inequalities are valid for all $\varepsilon > 0$.

As a consequence of this "restricted linearity" of L^*, we obtain, for a simple nonnegative measurable function $s = \sum_{j=1}^{n} a_j \mathbb{1}_{A_j}$, that

$$L^*(s) = \sum_{j=1}^{n} a_j \mu(A_j) = \int_X s \, d\mu. \tag{7.20}$$

Then (7.18), (7.20), and Theorem 3.3.6 yield (7.10) for nonnegative functions $f \in C_c^+(X)$. Since L and the integral with respect to μ are both linear, we have (7.10) for all $f \in C_c(X)$. □

Theorem 7.2.5. RRT for positive continuous linear functionals
Let X be a locally compact Hausdorff space and let L be a positive continuous linear functional on $C_c(X)$ with norm $\|\ldots\|_\infty$. There exists a (nonnegative) bounded regular Borel measure μ on X such that (7.10) holds.

Proof. The existence of a regular Borel measure μ that satisfies (7.10) follows from Theorem 7.2.4. We need to show only that the measure μ is bounded, provided that L is continuous on $C_c(X)$. Indeed, $\mathbb{1}_X \in S(X)$ since $\mathbb{1}_X = \sup_{\phi \in \Phi} \phi$ for the directed family $\Phi = \{\phi \in C_c^+(X) : \phi \leq 1\}$. Thus,

$$\begin{aligned} L^*(\mathbb{1}_X) = \tilde{L}(\mathbb{1}_X) &= \sup_{\phi \in \Phi} L(\phi) \\ &\leq \sup\{|L(\psi)| : \psi \in C_c(X) \text{ and } \|\psi\|_\infty \leq 1\} = \|L\|, \end{aligned}$$

and $\|L\|$ is finite since L is continuous; see Appendix A.8. □

Since $C_c(X)$ is dense in $C_0(X)$ in the sup-norm topology and since L is assumed to be continuous on $C_c(X)$ in Theorem 7.2.5, we could have taken $C_0(X)$ instead of $C_c(X)$ in its statement.

Combining the fact that simple functions are dense in $L^p_\mu(X)$ (Theorem 5.5.3) with the Vitali–Luzin theorem (Theorem 2.5.13), we obtain the following result.

Theorem 7.2.6. $\overline{C_c(X)} = L^p_\mu(X)$, $1 \leq p < \infty$
Let X be a locally compact Hausdorff space and let μ be a regular Borel measure on X. Then $C_c(X)$ is dense in $L^p_\mu(X)$, $1 \leq p < \infty$.

In the case $X = \mathbb{R}^d$ and $\mu = m^d$, we have $\overline{C_c^\infty(\mathbb{R}^d)} = L^p_{m^d}(\mathbb{R}^d)$, $1 \leq p < \infty$. This follows from Theorem 7.2.6 and a convolution argument.

Recall that for locally compact Hausdorff spaces X, $M_b(X)$ denotes the space of complex regular Borel measures on X; see Definitions 2.5.12 and 5.1.13. We know from Problem 5.22 that $M_b(X)$ is a Banach space with norm $\|\mu\|_1 = |\mu|(X)$. Another extension of Theorem 7.1.1 is the following result.

Theorem 7.2.7. RRT for complex measures
Let X be a locally compact Hausdorff space. Then $M_b(X)$ and $(C_0(X))'$ are isometrically isomorphic. In fact, for each $L \in (C_0(X))'$ there exists a unique $\mu \in M_b(X)$ such that

$$\forall\, f \in C_0(X), \quad L(f) = \int_X f\, d\mu. \tag{7.21}$$

Proof. i. We start by showing that for each $\mu \in M_b(X)$, $L_\mu(f) = \int_X f\, d\mu$ defines a continuous linear functional on $C_0(X)$, where $\|L_\mu\| = \|\mu\|_1$ (see Appendix A.8 for the norm of a functional). It is clear that L_μ is a linear functional, e.g., Theorem 5.1.14. Also,

$$|L_\mu(f)| \le \int_X |f|\, d|\mu| \le \|f\|_\infty |\mu|(X) = \|f\|_\infty \|\mu\|_1.$$

Hence, L_μ is continuous, i.e.,

$$\|L_\mu\| \le \|\mu\|_1. \tag{7.22}$$

By Theorem 5.3.5, let h be a $|\mu|$-integrable function for which $|h| = 1$ and such that

$$\forall\, f \in L^1_{|\mu|}(X), \quad \int_X f\, d\mu = \int_X fh\, d|\mu|.$$

Thus, we can write

$$\|\mu\|_1 = |\mu|(X) = \int_X |h|^2\, d|\mu| = \int_X \overline{h}h\, d|\mu| = \int_X \overline{h}\, d\mu. \tag{7.23}$$

By Theorem 7.2.6,

$$\forall\, \varepsilon > 0,\ \exists\, g \in C_c(X) \text{ such that } \quad \|g\|_\infty \le \|h\|_\infty \quad \text{and} \quad \int_X |\overline{h} - g|\, d|\mu| < \varepsilon. \tag{7.24}$$

Now (7.23) and (7.24) yield

$$\|\mu\|_1 \le \left|\int_X g\, d\mu\right| + \left|\int_X (\overline{h} - g)\, d\mu\right| \le |L_\mu(g)| + \varepsilon \le \|L_\mu\| \|g\|_\infty + \varepsilon \le \|L_\mu\| + \varepsilon.$$

Since these inequalities are valid for all $\varepsilon > 0$ we have $\|\mu\|_1 \le \|L_\mu\|$.

Combining this inequality with (7.22) gives $\|L_\mu\| = \|\mu\|_1$.

ii. Let L be a continuous linear functional on $C_0(X)$ as defined in the statement of the theorem. We introduce an induced mapping on $C_0^+(X)$:

$$\forall\, f \in C_0^+(X), \quad L^+(f) = \sup\{|L(g)| : g \in C_0(X), |g| \le f\}.$$

It is not difficult to see that L^+ is a positive additive mapping on $C_0^+(X)$ such that $L^+(\alpha f) = \alpha L^+(f)$ for $\alpha \ge 0$. We can extend L^+ to all of $C_0(X)$ by setting

$$L^+(f) = L^+((\operatorname{Re} f)^+) - L^+((\operatorname{Re} f)^-) + iL^+((\operatorname{Im} f)^+) - iL^+((\operatorname{Im} f)^-).$$

An elementary calculation shows that such an extended mapping L^+ is linear over $\mathbb{C}$ on $C_0(X)$. By definition it is also positive. Thus, by definition (see Appendix A.8), we have

$$\begin{aligned} \|L^+\| &= \sup\{|L^+(f)| : f \in C_0(X), \|f\|_\infty \le 1\} \\ &\le \sup\{L^+(|f|) : f \in C_0(X), \|f\|_\infty \le 1\} \\ &= \sup\{|L(f)| : f \in C_0^+(X), \|f\|_\infty \le 1\} \le \|L\|, \end{aligned} \tag{7.25}$$

where the second inequality follows since L is a continuous linear functional on $C_0(X)$. On the other hand, for any $\varepsilon > 0$ there exists $g \in C_0(X)$ such that $\|g\|_\infty \le 1$ and $\|L\| \le |L(g)| + \varepsilon$. Therefore,

$$\|L\| \le |L(g)| + \varepsilon \le L^+(|g|) + \varepsilon \le \|L^+\| + \varepsilon. \tag{7.26}$$

Hence, $\|L\| \le \|L^+\|$ since (7.26) is valid for all $\varepsilon > 0$. Combining this with (7.25) yields $\|L^+\| = \|L\|$.

iii. Part *ii* was the first step in proving the surjectivity of the mapping $M_b(X) \to (C_0(X))'$. To complete the proof of surjectivity, we proceed in the following way. We invoke Theorem 7.2.5 for the positive continuous linear functional L^+ to obtain a (nonnegative) bounded regular Borel measure μ such that

$$\forall\, f \in C_c(X), \quad L^+(f) = \int_X f\, d\mu.$$

Clearly, $\|\mu\|_1$ is bounded by $\|L\|$:

$$\|\mu\|_1 = \mu(X) = \sup\{L^+(f) : f \in C_c^+(X), f \le 1\} \le \|L^+\| = \|L\|.$$

Since $C_c(X)$ is a dense subset of $L^1_\mu(X)$, and since L is bounded on $C_c(X)$ with respect to the $L^1_\mu(X)$-norm, there is an extension of L to a bounded linear functional $\overline{L}$ on $L^1_\mu(X)$. We can apply Theorem 5.5.5 to find a function $g \in L^\infty_\mu(X)$ such that $\|g\|_\infty = \|\overline{L}\|$ and $\overline{L}(f) = \int_X fg\, d\mu$ for any $f \in L^1_\mu(X)$. Then we use Theorem 5.3.3 to associate with this function g a complex measure μ_g that is absolutely continuous with respect to μ:

$$\forall\, A \in \mathcal{A}, \quad \mu_g(A) = \int_A g\, d\mu,$$

where $\mathcal{A}$ is the σ-algebra of μ-measurable sets. Thus, since the elements of $C_0(X)$ are integrable with respect to μ and μ_g, we have

$$\forall\, f \in C_0(X), \quad \int_X f\, d\mu_g = \int_X fg\, d\mu = \overline{L}(f); \tag{7.27}$$

see Problem 5.37.

We need to verify that the measure μ_g is regular, knowing that μ is regular. By definition of μ_g, μ_g is bounded on compact sets since $\|g\|_\infty < \infty$ and μ is a bounded measure. Fix $\varepsilon > 0$ and let $A \in \mathcal{A}$. There exist a compact set F and an open set U such that $F \subseteq A \subseteq U$ and $\mu(U \setminus F) < \varepsilon$. Using Theorem 5.1.12*c* we can write

$$\begin{aligned}
|\mu_g|(U \setminus F) &\le 4 \sup\{|\mu(B)| : B \in \mathcal{A}, B \subseteq U \setminus F\} \\
&\le 4 \sup \left\{ \left| \int_B g\, d\mu \right| : B \in \mathcal{A}, B \subseteq U \setminus F \right\} \\
&\le 4 \sup\{\|g\|_\infty \mu(B) : B \in \mathcal{A}, B \subseteq U \setminus F\} \\
&\le 4\|g\|_\infty \mu(U \setminus F) \le 4\varepsilon \|g\|_\infty,
\end{aligned}$$

which proves the regularity of $|\mu_g|$ and so also of μ_g.

Since $\overline{L} = L$ on $C_c(X)$, and since $\overline{L}$ and L are continuous in $L^1_\mu(X)$-norm on $C_0(X)$, we obtain

$$\forall\, f \in C_0(X), \quad L(f) = \overline{L}(f). \tag{7.28}$$

Now (7.27) and (7.28) yield (7.21).

iv. Part *i* showed that the mapping $\mu \mapsto L_\mu$ is a linear isometry of $M_b(X)$ *into* $(C_0(X))'$. Part *ii* proved that if $L \in (C_0(X))'$ then $\|L\| = \|L^+\|$. Using part *ii*, part *iii* showed that the mapping $\mu \to L_\mu$ is in fact *onto*. Thus, $\mu \mapsto L_\mu$ must be one-to-one, which yields the uniqueness in the statement of Theorem 7.2.7. □

7.3 Radon measures

RRT (Theorem 7.2.7) gives a spectacular identity between the dual space $(C_0(X))'$ and the space $M_b(X)$ of complex regular Borel measures on a locally compact Hausdorff space X. As such we shall write $M_b(X) = (C_0(X))'$, and we deal with measures as set functions or functionals, depending on what is more convenient, either theoretically or computationally. At the moment our euphoria is tempered by the fact that the elements of $M_b(X)$ are bounded and, in particular, $m \notin M_b(\mathbb{R})$. In this section we solve this dilemma by defining Radon measures, which also serve as a natural way to introduce (SCHWARTZ) distribution theory.

Proposition 7.3.1. *Let X be a locally compact Hausdorff space that has a countable basis. For every open set $U \subseteq X$ there exists a sequence $\{K_n : n = 1, \ldots\} \subseteq X$ of compact sets such that*

$$U = \bigcup_{n=1}^{\infty} K_n, \quad \textit{where} \quad K_n \subseteq \operatorname{int} K_{n+1}.$$

Proof. The fact that a topological space X has a countable basis is called the *second countability* of X. For a locally compact Hausdorff topological space X, second countability is equivalent to the fact that there exist a metric ρ on X and a sequence of compact sets F_n such that $X = \bigcup_{n=1}^{\infty} F_n$, $F_n \subseteq \operatorname{int} F_{n+1}$; see [278], Theorem I.5.3.

Fix $U \subseteq X$, an open set. Let $E_n = \{x \in X : \rho(x, U^{\sim}) \geq 1/n\}$. Each such E_n is closed. Let $K_n = E_n \cap F_n$. Thus, K_n is compact for $n = 1, \ldots,$ and

$$K_n = E_n \cap F_n \subseteq (\operatorname{int} E_{n+1}) \cap (\operatorname{int} F_{n+1}) \subseteq \operatorname{int} K_{n+1}.$$

Clearly,

$$U = \bigcup_{n=1}^{\infty} K_n.$$

□

Proposition 7.3.2. *Let $(X, \mathcal{B}(X), \mu)$ be a measure space, where X is a locally compact Hausdorff space that has a countable basis and where μ is a (nonnegative) measure that is finite for all compact sets in X. Then μ is a regular Borel measure.*

Proof. From the fact that second countability of X implies that there exists a sequence $\{F_n : n = 1, \ldots\}$ of compact sets such that $X = \bigcup_{n=1}^{\infty} F_n$, $F_n \subseteq F_{n+1}$, and from Theorem 2.4.3*d*, it follows that

$$\sup\{\mu(F) : F \subseteq A,\ F \text{ compact}\} = \sup\{\mu(F) : F \subseteq A,\ F \text{ closed}\}$$

for any Borel set $A \subseteq X$. Consider the collection $\mathcal{A}$ of all the Borel sets $A \subseteq X$ for which

$$\mu(A) = \sup\{\mu(F) : F \subseteq A,\ F \text{ closed}\} \tag{7.29}$$

and

$$\mu(A) = \inf\{\mu(U) : A \subseteq U,\ U \text{ open}\}. \tag{7.30}$$

From Proposition 7.3.1 and Theorem 2.4.3*d*, (7.29) is satisfied by all open sets A. Since (7.30) is naturally satisfied by any open set, we conclude that $\mathcal{A}$ contains all open sets.

It is also not difficult to see that $\mathcal{A}$ is closed under completions and under countable unions, and so $\mathcal{A} = \mathcal{B}(X)$. Regularity follows from this fact and from the assumption that μ is finite on compact sets. □

Theorem 7.3.3. Uniqueness for RRT
Let $(X, \mathcal{B}(X))$ be a measurable space, where X is a locally compact Hausdorff space. Let μ and ν be two regular measures on X such that

$$\forall\, f \in C_c(X), \quad \int_X f\, d\mu = \int_X f\, d\nu.$$

Then $\mu = \nu$ on $\mathcal{B}(X)$.

Proof. i. Fix a compact set $K \subseteq X$. By the regularity of μ and ν, there exist collections $\{U_n^\mu : n = 1, \ldots\}$ and $\{U_n^\nu : n = 1, \ldots\}$ of open sets such that $\mu(U_1^\mu)$, $\nu(U_1^\nu) < \infty$, $K \subseteq U_n^\mu$, $K \subseteq U_n^\nu$, $n = 1, \ldots$, and $\mu(K) = \lim_{n\to\infty} \mu(U_n^\mu)$ and $\nu(K) = \lim_{n\to\infty} \nu(U_n^\nu)$. Define sets $U_n = \bigcap_{j=1}^n (U_j^\mu \cap U_j^\nu)$, $n = 1, \ldots$. Then, $U_1 \supseteq U_2 \supseteq \cdots \supseteq K$, and $\mu(K) = \lim_{n\to\infty} \mu(U_n)$ and $\nu(K) = \lim_{n\to\infty} \nu(U_n)$. By Urysohn's lemma, for each n, let f_n be a continuous nonnegative function supported in $\overline{U}_n$ and equal to 1 on K such that $f_n(x) \leq 1$ for all $x \in X$. We define a monotone sequence $\{g_n\}$, where $g_n = \min(f_1, \ldots, f_n)$, and observe that $g_n \to \mathbb{1}_K$ pointwise μ-*a.e.* as $n \to \infty$. From the Levi–Lebesgue theorem (Theorem 3.3.6) and from our assumptions on μ and ν, it follows that

$$\mu(K) = \int_X \mathbb{1}_K\, d\mu = \lim_{n\to\infty} \int_X g_n\, d\mu = \lim_{n\to\infty} \int_X g_n\, d\nu = \int_X \mathbb{1}_K\, d\nu = \nu(K).$$

ii. Since $\mu(K) = \nu(K)$ for all compact sets, we also have that $\mu(U) = \nu(U)$ for all open sets in X, because of the regularity of μ and ν. Consequently, by regularity again, we obtain $\mu(B) = \nu(B)$ for all Borel sets. □

Theorem 7.3.4. Equivalence of positive functionals and measures
Let $(X, \mathcal{B}(X))$ be a measurable space, where X is a locally compact Hausdorff space that has a countable basis. There is a bijection between positive linear functionals on $C_c(X)$ and (nonnegative) Borel measures that are finite on compact subsets of X.

Proof. From Theorem 7.2.4 and Theorem 7.3.3 it follows that on any locally compact Hausdorff space X there is a bijection between positive linear functionals on $C_c(X)$ and regular Borel measures on X.

If we assume additionally that X is second countable, then it follows from Proposition 7.3.2 that any Borel measure that is finite on compact subsets of X is regular, and the proof is complete. □

Let X be a locally compact Hausdorff space. There is a Hausdorff locally convex topology $\mathcal{T}$ on $C_c(X)$ such that sequential convergence $f_n \to f$, $\{f_n, f : n = 1, \ldots\} \subseteq C_c(X)$, is defined by the following properties:

i. $\exists\, K \subseteq X$, compact, such that $\forall\, n = 1, \ldots,\ \operatorname{supp} f_n \subseteq K$,

ii. $\lim_{n\to\infty} \|f_n - f\|_\infty = 0$;

e.g., [69], [242], [428]; cf. Example A.6.5. Further, for each compact set $K \subseteq X$ let $C_c(K) = \{f \in C_c(X) : \operatorname{supp} f \subseteq K\}$. It turns out that $\mathcal{T}$ is characterized by the property that a linear functional $\mu : C_c(X) \to \mathbb{C}$ is continuous if and only if μ is continuous on each sup-normed space $C_c(K)$. We shall not write out the elements of $\mathcal{T}$ nor prove its characterization in terms of the spaces $C_c(K)$, K compact (but see [69], [242], [428]).

Definition 7.3.5. Radon measures
Let X be a locally compact Hausdorff space. Let $\mathcal{T}$ be the topology on $C_c(X)$ just described for sequential convergence. The space of continuous linear functionals $\mu : C_c(X) \to \mathbb{C}$ is the space of *Radon measures*, and it is denoted by $\mathcal{C}'(X)$. This notation indicates that the Radon measures form a special subset of $\mathcal{D}'(X)$, the space of Schwartz distributions on X, e.g., [428] and Definition 7.5.2. We say that $\mu \in \mathcal{C}'(X)$ is a *positive Radon measure* if $\mu(f) \geq 0$ whenever $f \in C_c^+(X)$.

Example 7.3.6. The sup norm and $\mathcal{T}$ topologies on $C_c(\mathbb{R})$
Let $f_n \in C_c(\mathbb{R})$ be defined by

$$f_n(x) = \begin{cases} \frac{1}{n}, & \text{if } |x| \leq n, \\ 0, & \text{if } |x| > n+1, \end{cases}$$

and assume that for all $x \in \mathbb{R}$, $|f(x)| \leq 1/n$. Thus, $\|f_n\|_\infty \to 0$, whereas $f_n \not\to 0$ in the topology of $C_c(\mathbb{R})$.

Clearly, then, $\mathcal{T}$ is a finer topology than the usual sup norm topology on $C_c(X)$, and so $M_b(X) \subseteq \mathcal{C}'(X)$ since $\overline{C_c(X)} = C_0(X)$, where the closure is taken in the sup norm; cf. Appendix A.11. Since $M_b(X) \subseteq \mathcal{C}'(X)$, we refer to $M_b(X)$ as the space of *bounded Radon measures*. If we define a functional μ_m by

$$\forall\, f \in C_c(\mathbb{R}), \quad \mu_m(f) = \int_{\mathbb{R}} f\, dm,$$

we see that $\mu_m \in \mathcal{C}'(\mathbb{R}) \setminus M_b(\mathbb{R})$. Consequently, Lebesgue measure m is an unbounded Radon measure. The following are other examples of unbounded Radon measures.

Example 7.3.7. Unbounded Radon measures
a. Define

$$\mu = \sum_{n=-\infty}^{\infty} a_n \delta_n, \tag{7.31}$$

where $\{a_n : n = 1, \ldots\} \subseteq \mathbb{C}$. δ_x was defined as a Borel measure in Example 2.4.2; see Section 7.6.3. Clearly, $\mu \in \mathcal{C}'(\mathbb{R})$. If we put $a_0 = 0$ and $a_n = 1/n$ for $n \neq 0$, and if we interpret the sum in (7.31) as

$$\mu = \sum_{n=1}^{\infty} \frac{1}{n}(\delta_n - \delta_{-n}),$$

then $\mu \in \mathcal{C}'(\mathbb{R}) \setminus M_b(\mathbb{R})$. To prove this, let f_k have the form of the function in Figure 7.1, where each triangle has height $1/\log(k)$. Then $\|f_k\|_\infty \to 0$, whereas

$$\mu(f_k) \sim 2, \quad k \to \infty.$$

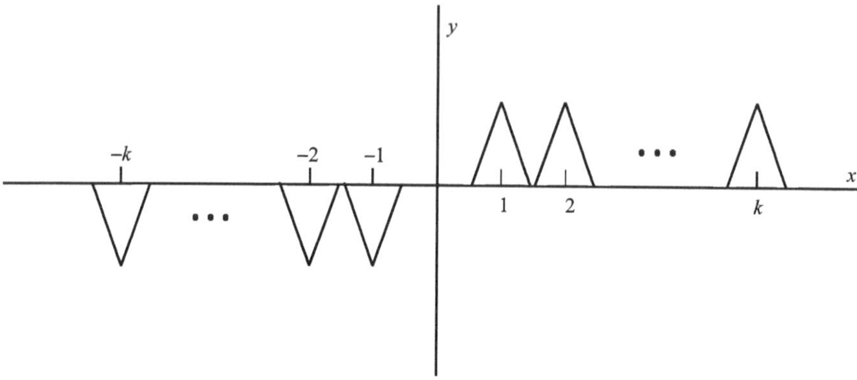

Fig. 7.1. Unbounded Radon measures.

b. Note that if X is a compact Hausdorff space, then $\mathcal{C}'(X) = M_b(X)$. On the other hand, if $X = (-\pi, \pi)$, so that $\overline{X}$ is compact, we have $\mathcal{C}'(X) \setminus M_b(X) \neq \emptyset$. For example, define

$$\forall\, A \in \mathcal{B}((-\pi, \pi)), \quad \mu(A) = m(\tan(A)).$$

For positive linear functionals on $C_c(X)$ we have the following theorem, which does not require the functionals to be continuous, e.g., [69], Chapitre III.1.5.

Theorem 7.3.8. Positive linear functionals are Radon measures
Let X be a locally compact Hausdorff space.

a. Let $L : C_c(X) \to \mathbb{C}$ be a positive linear functional. Then, L is a positive Radon measure μ, and, in particular, $\mu \in \mathcal{C}'(X)$.

b. Let $\mu \in \mathcal{C}'(X)$. There exist positive linear functionals $\{\mu_i : i = 1, \ldots, 4\}$ such that

$$\mu = \mu_1 - \mu_2 + i\mu_3 - i\mu_4.$$

Proof. To prove part *a*, let $K \subseteq X$ be compact and let $g \in C_c(X)$ be a real-valued function for which $\operatorname{supp} g \subseteq K$. By Urysohn's lemma (Theorem A.1.3), let $f \in C_c(X)$ have the properties that $f(X) \subseteq [0, 1]$ and $f = 1$ on K. Thus,

$$-\|g\|_\infty f \leq g \leq \|g\|_\infty f$$

on X. By hypothesis,

$$-\|g\|_\infty L(f) \leq L(g) \leq \|g\|_\infty L(f).$$

Hence

$$|L(g)| \leq C_K \|g\|_\infty \quad (C_K = \mu(f))$$

independent of $g \in C_c(X)$ for which supp $g \subseteq K$. Consequently, $L \in \mathcal{C}'(X)$.

The proof of part b is more involved. In fact, part b is the Jordan decomposition theorem; cf. Theorem 5.1.8. □

Remark. Let us come back to the proof of Theorem 7.2.4 and discuss its meaning in view of the following *extension problem*. Let $\mu \in \mathcal{C}'(X)$. We wish to find a large superspace $Y \supseteq C_c(X)$ such that μ is defined on Y and the canonical properties of integration theory hold. The extension from $C_c(X)$ to Y is the *Lebesgue extension*. This extension is closely related to the extension of measures that has been described in Chapter 2.

LEBESGUE's approach is based on the following notions: $f : X \to \mathbb{R}^*$ is *lower semicontinuous* at $x_0 \in X$ if $f(x_0) = -\infty$ or

$$\forall\, \varepsilon > 0,\ \exists\, V \subseteq X, \text{ an open set containing } x_0, \text{ such that } \forall\, x \in V,$$
$$f(x) > f(x_0) - \varepsilon;$$

cf. the definition of a directed family. The notions of *upper semicontinuity* at $x_0 \in X$ and *lower* and *upper semicontinuity on* X are defined in analogous ways. Clearly, $f : X \to \mathbb{R}$ is continuous at x if and only if it is both lower and upper semicontinuous at x. Also, $f : X \to \mathbb{R}^*$ is lower semicontinuous on X if and only if

$$f = \sup\{g : \text{ real-valued } g \leq f,\ g \in C_c(X)\}.$$

We let $J^+(X)$ denote the nonnegative elements of the class of lower semicontinuous functions. Thus, a nonnegative function $f : X \to \mathbb{R}^*$ has the form

$$f = \sup\{g : g \leq f,\ g \in C_c^+(X)\} \tag{7.32}$$

if and only if $f \in J^+(X)$; cf. the definition of $S(X)$ in Section 7.2.

The fundamental idea of LEBESGUE, in terms of functional analysis, was to complete $C_c(X)$ in terms of a given $\mu \in \mathcal{C}'(X)$, considered as a positive linear functional on $C_c(X)$, when the space $C_c(X)$ has a suitable topology. This completion is the *Lebesgue extension*, and we wish to compare it with the method of proof of Theorem 7.2.4. In order to describe intuitively this "suitable topology" (which turns out to be $\mathcal{T}$) on $C_c(X)$, we note that

$$\|f\|_\mu = \mu(|f|) \tag{7.33}$$

defines a seminorm on $C_c(X)$. (A *seminorm* has all the properties of a norm except that $\|x\|$ can be 0 for some nonzero x.) Because of (7.32) we define

$$\forall\, f \in J^+(X), \quad \mu^*(f) = \sup\{\mu(g) : g \le f,\ g \in C_c^+(X)\}; \tag{7.34}$$

cf. the definition of $\tilde{L}$ for a given positive linear functional L on $C_c(X)$. Then, for any nonnegative $\mathbb{R}^*$-valued function f on X, we set

$$\mu^*(f) = \inf\{\mu^*(g) : g \ge f,\ g \in J^+(X)\}; \tag{7.35}$$

cf. the definition of L^* in Section 7.2. Equation (7.35) is well defined since $g = +\infty$ on X is in $J^+(X)$, and (7.34) and (7.35) are compatible for $f \in J^+(X)$. Thus, with an eye to (7.33), we let $\mathcal{F}_\mu(X)$ be the set of $\mathbb{R}^*$-valued functions f on X for which

$$\|f\|_\mu = \mu^*(|f|) < \infty. \tag{7.36}$$

Then $\mathcal{F}_\mu(X)$ is a seminormed space and $\mathcal{L}^1_\mu(X)$, as defined in Definition 3.2.5, is the closure of $C_c(X)$ in $\mathcal{F}_\mu(X)$. We note that μ has a unique continuous linear extension to $\mathcal{L}^1_\mu(X)$, which we again designate by μ, and $\mu(f)$ is the integral of $f \in \mathcal{L}^1_\mu(X)$.

We call $\mathcal{L}^1_\mu(X)$ the *Lebesgue extension* of $\mu \in \mathcal{C}'(X)$. This process was first accomplished by LEBESGUE, when he extended integration theory from the Riemann integrable functions on $[a,b]$ to $\mathcal{L}^1_m([a,b])$.

7.4 Support and the approximation theorem

Let X be a locally compact Hausdorff space and let $\mu \in \mathcal{C}'(X)$. We shall define the *support* of μ, denoted by supp μ. Naturally we want the notion of supp μ to extend that of supp f, where $f \in C_c(X)$. To make this definition it is necessary to use the following partition of unity result [69], Chapitre III.2.1.

Theorem 7.4.1. Construction of measures in terms of "local measures"
Let $\{V_\alpha : \alpha \in I\}$ be an open covering of X and let $\mu_\alpha \in \mathcal{C}'(V_\alpha)$, noting that V_α is also locally compact. Assume that for each α and β and for each $f \in C_c(X)$, for which supp $f \subseteq V_\alpha \cap V_\beta$, *we have*

$$\mu_\alpha(f) = \mu_\beta(f).$$

Then there is a unique $\mu \in \mathcal{C}'(X)$ such that

$$\forall\, \alpha \in I \text{ and } \forall\, f \in C_c(V_\alpha), \quad \mu(f) = \mu_\alpha(f).$$

The following result is a consequence of Theorem 7.4.1; cf. Problem 3.41.

Corollary 7.4.2. *Let $\{V_\alpha : \alpha \in I\}$ be a family of open sets in X and let $\mu \in \mathcal{C}'(X)$. Assume that*

$$\forall\, \alpha \in I \text{ and } \forall\, f \in C_c(V_\alpha), \quad \mu(f) = 0.$$

Then

$$\forall\, f \in C_c\left(\bigcup_{\alpha\in I} V_\alpha\right), \quad \mu(f) = 0.$$

Definition 7.4.3. Support of $\mu \in \mathcal{C}'(X)$

a. Let X be a locally compact Hausdorff space. A Radon measure $\mu \in \mathcal{C}'(X)$ is *zero on an open set* $V \subseteq X$ if

$$\forall\, f \in C_c(V), \quad \mu(f) = 0.$$

($C_c(V) = \{f \in C_c(X) : \text{supp}\, f \subseteq V\}$.)

For a given $\mu \in \mathcal{C}'(X)$, let $\mathcal{V}$ be the family of open sets $V_\alpha \subseteq X$ such that μ is zero on V_α. Thus, if $V = \bigcup\{V_\alpha : V_\alpha \in \mathcal{V}\}$, then, by Corollary 7.4.2, μ is zero on V.

b. Hence we can define the *support of* $\mu \in \mathcal{C}'(X)$, denoted by supp μ, as the complement of the union of the open sets $V \subseteq X$ for which μ is zero on V. The set supp μ is the smallest closed set outside of which μ is zero, or, equivalently, considering μ as a set function, supp μ is the smallest closed set on which μ is concentrated.

Note that if supp $f \cap$ supp $\mu = \emptyset$, where $f \in C_c(X)$ and $\mu \in \mathcal{C}'(X)$, then $\mu(f) = 0$. In fact, just take an open set $V \supseteq$ supp f with $V \subseteq (\text{supp}\, \mu)^\sim$, so that μ is zero on V and hence $\mu(f) = 0$. This does not tell us that if $f = 0$ on supp μ then $\mu(f) = 0$. Nevertheless, such a result is true, as the following theorem shows.

Theorem 7.4.4. Continuous $f = 0$ on supp μ implies $\mu(f) = 0$

Let X be a locally compact Hausdorff space and let $\mu \in \mathcal{C}'(X)$. Assume that $f \in C_c(X)$ vanishes on supp μ. *Then $\mu(f) = 0$.*

Proof. Let $K =$ supp f, $E =$ supp μ. By definition of the topology $\mathcal{T}$ on $C_c(X)$, there is $M_K > 0$ such that for each $g \in C_c(X)$ with supp $g \subseteq K$ we have

$$|\mu(g)| < M_K \|g\|_\infty.$$

Fix $\varepsilon > 0$. We shall show that $|\mu(f)| < \varepsilon$. Let $V = \{x \in X : |f(x)| < \varepsilon/(2M_K)\}$. Then V is open since f is continuous, and $E \subseteq V$. Clearly, $E^\sim$ is an open set containing the closed set $V^\sim$. By Urysohn's lemma (Theorem A.1.3), there is a continuous function $h : X \to [0,1]$ such that $h = 1$ on $V^\sim$ and supp $h \subseteq E^\sim$. Note that $E \cap$ supp $(fh) = \emptyset$, and hence $\mu(fh) = 0$. Furthermore, $f = fh$ on $K \cap V^\sim$ and $|fh| \leq |f|$ on X. Consequently, since $f = 0$ on $K^\sim$,

$$\|f - fh\|_\infty = \sup_{x \in V\cap K} |f(x)(1 - h(x))| \leq 2 \sup_{x\in V} |f(x)| \leq \frac{\varepsilon}{M_K}.$$

Therefore, noting that supp $(f - fh) \subseteq K$,

$$|\mu(f)| = |\mu(f - fh)| < M_K \|f - fh\|_\infty < \varepsilon. \qquad \square$$

Theorem 7.4.4 and the Hahn–Banach theorem are used to prove the following approximation theorem.

Theorem 7.4.5. The density of finitely supported measures in $\mathcal{C}'(X)$ *Let X be a locally compact Hausdorff space and let $\mu \in \mathcal{C}'(X)$. Set*

$$V_\mu = \text{span}\,\{\delta_x : x \in \text{supp}\,\mu\}.$$

Then μ is a weak limit point of V_μ. (In terms of nets, mentioned but not fully developed in the Remark after Theorem 3.3.7, this means that there is a net $\{\mu_\alpha\} \subseteq V_\mu$ such that*

$$\forall\, f \in C_c(X), \quad \lim_\alpha \mu_\alpha(f) = \mu(f),$$

where each μ_α is a finite sum,

$$\mu_\alpha = \sum_{x \in F_\alpha} c_x(\alpha)\delta_x, \quad x \in \text{supp}\,\mu \quad \textit{and} \quad c_x(\alpha) \in \mathbb{C}.)$$

Proof. We begin by noting that the dual of $\mathcal{C}'(X)$, taken with the weak* topology $\sigma(\mathcal{C}'(X), C_c(X))$, is $C_c(X)$. See Theorem A.9.3, which is stated for normed vector spaces but is true for any Hausdorff locally convex topological vector space (LCTVS), of which $C_c(X)$ with the topology $\mathcal{T}$ is an example.

Suppose μ does not belong to the weak* closure of V_μ. Then, by the Hahn–Banach theorem (Appendix A.8) formulated for Hausdorff LCVTSs, there is $f \in C_c(X)$ such that $\mu(f) \neq 0$ and

$$\forall \nu \in V_\mu, \quad \nu(f) = 0. \tag{7.37}$$

However, (7.37) implies $f = 0$ on supp μ. Thus, $\mu(f) = 0$ by Theorem 7.4.4. This is the desired contradiction, and we can conclude that μ is a weak* limit point of V_μ. □

Remark. Clearly, Theorem 7.4.5 can be formulated for the pair $C_0(X)$, $M_b(X)$ instead of $C_c(X)$, $\mathcal{C}'(X)$, respectively. In this case, Theorem A.9.3 is applied directly to the normed vector space $C_0(X)$. More important, the elements of the net $\{\mu_\alpha\}$ in the statement of Theorem 7.4.5 can be chosen to have the property that

$$\forall\, \alpha, \quad \|\mu_\alpha\|_1 = \|\mu\|_1;$$

see [69], Chapitre III, Section 2, number 4, pages 71–72.

Example 7.4.6. Finitely supported measures converging to μ
Consider the measurable space $(\mathbb{R}, \mathcal{B}(\mathbb{R}))$ and take real-valued $f \in BV(\mathbb{R})$. Assume that f increases from 0 to 1 and define the corresponding measure μ_f as in (5.24). In particular, $\mu_f(\mathbb{R}) = 1$. We write

$$f_n(x) = \frac{1}{n}[nf(x)],$$

where $[y]$ is the largest integer less than or equal to y. Then f_n is a monotone step function. The measure μ_n, which corresponds to f_n as in (5.24), has finite support contained in supp μ and $\mu_n(f) \to \mu(f)$ for each $f \in C_0(\mathbb{R})$.

Also, for perspective, take

$$\mu = \sum_{\gamma \in D} a_\gamma \delta_\gamma, \quad \sum_{\gamma \in D} |a_\gamma| < \infty,$$

where D is a countable dense subset of the Cantor set $C \subseteq [0,1]$. Then μ is concentrated on both C and D, whereas supp $\mu = C$.

7.5 Distribution theory

The purpose of the theory of distributions is to provide a unified setting and *calculus* for many of the objects arising in analysis. These objects include the customary functions, viz., the elements of the space $L^1_{\text{loc}}(\mathbb{R})$ of locally integrable functions. They also include impulses (Dirac measures), dipoles, and other notions from the sciences whose role and mathematical identity could not be assimilated by the "ordinary" calculus. A key feature of the theory of distributions is that all of these objects can be differentiated in a natural way inspired by the integration by parts formula. Further, some of the important results from real analysis, allowing for switching of operations, such as summation and differentiation, are true without hypotheses in the case of distributions. At the same time we shall see the close connection of distributions with the theory we have developed so far, especially with RRT.

Let $C^\infty(\mathbb{R})$ be the vector space of infinitely differentiable functions $f : \mathbb{R} \to \mathbb{C}$, and let

$$C_c^\infty(\mathbb{R}) = \{f : f \in C_c(\mathbb{R}) \cap C^\infty(\mathbb{R})\}.$$

Clearly, $C_c^\infty(\mathbb{R})$ is a vector space. At this time it is also convenient to define the space $C^n(\mathbb{R})$ of n-times continuously differentiable functions $f : \mathbb{R} \to \mathbb{C}$.

Example 7.5.1. $C_c^\infty(\mathbb{R})$

The space $C_c^\infty(\mathbb{R})$ is not the trivial space $\{0\}$. This statement is not so absurd, since there are no analytic functions in $C_c^\infty(\mathbb{R})$. In fact, $C_c^\infty(\mathbb{R})$ is an infinite-dimensional space, so let us write down at least one element.

a. Let

$$\phi(t) = \begin{cases} e^{1/t}, & \text{if } t < 0, \\ 0, & \text{if } t \geq 0, \end{cases}$$

and define $f(t) = c\phi(|t|^2 - 1)$. Clearly, $f(t) = 0$ if $|t| \geq 1$, so that supp $f \subseteq [-1,1]$. The constant c is chosen so that $\int f(t)\, dt = 1$, and it is a straightforward calculation to show that f is infinitely differentiable.

b. To generate other examples of elements of $C_c^\infty(\mathbb{R})$, take any $g \in L^1_m(\mathbb{R})$ having compact support. Then $f * g \in C_c^\infty(\mathbb{R})$. To see this we must verify that supp $f * g$ is compact and that $f * g$ is infinitely differentiable. The first fact is routine to prove, and the second involves checking conditions to switch the operations of differentiation and integration; see, for example, Theorem 3.6.3 for conditions and [173], pages 72 and 93, for motivation.

Definition 7.5.2. Distributions on $\mathbb{R}$
A linear functional,

$$T : C_c^\infty(\mathbb{R}) \to \mathbb{C},$$
$$f \mapsto T(f),$$

is a *distribution* or *generalized function* if $\lim_{n\to\infty} T(f_n) = 0$ for every sequence $\{f_n : n = 1, \ldots\} \subseteq C_c^\infty(\mathbb{R})$ satisfying the following properties:

i. $\exists\, K \subseteq \mathbb{R}$, compact, such that $\forall\, n = 1, \ldots,$ supp $f_n \subseteq K$,

ii. $\forall\, k \geq 0$, $\lim_{n\to\infty} \|f_n^{(k)}\|_\infty = 0$.

The space of all distributions on $\mathbb{R}$ is denoted by $\mathcal{D}'(\mathbb{R})$, and it is clearly a vector space over $\mathbb{C}$. We denote the evaluation of $T \in \mathcal{D}'(\mathbb{R})$ at $f \in C_c^\infty(\mathbb{R})$ by $T(f)$. If $T_1, T_2 \in \mathcal{D}'(\mathbb{R})$, we say that T_1 *equals* T_2 if

$$\forall\, f \in C_c^\infty(\mathbb{R}), \quad T_1(f) = T_2(f).$$

In particular, $T = 0$ if $T(f) = 0$ for all $f \in C_c^\infty(\mathbb{R})$.

This notion of equality can be explained in functional-analytic terms. Intuitively, however, the idea is clear: $T(f) = 0$ for all f in the domain $C_c^\infty(\mathbb{R})$ of T implies that T is the 0-distribution, just as $g(t) = 0$ for all $t \in \mathbb{R}$ implies g is the 0-function.

Example 7.5.3. $L^1_{\text{loc}}(\mathbb{R}) \subseteq \mathcal{D}'(\mathbb{R})$
a. Let $g \in L^1_{\text{loc}}(\mathbb{R})$ and let the functional $T_g : C_c^\infty(\mathbb{R}) \to \mathbb{C}$ be defined by

$$\forall\, f \in C_c^\infty(\mathbb{R}), \quad T_g(f) = \int f(x) g(x)\, dx.$$

If there is no confusion we sometimes write $g(f)$ instead of $T_g(f)$. It is easy to check that $T_g \in \mathcal{D}'(\mathbb{R})$, and the mapping,

$$L^1_{\text{loc}}(\mathbb{R}) \to \mathcal{D}'(\mathbb{R}),$$
$$g \mapsto T_g,$$

allows us to identify $L^1_{\text{loc}}(\mathbb{R})$ with a subspace of $\mathcal{D}'(\mathbb{R})$.

To see this, we must show that this mapping is injective, i.e., that if $T_g = 0$ in $\mathcal{D}'(\mathbb{R})$, then $g = 0$ *m-a.e.*, where $g \in L^1_{\text{loc}}(\mathbb{R})$. By defnition, $T_g = 0$ implies that $T_g(f) = 0$ for all $f \in C_c^\infty(\mathbb{R})$. This, in turn, implies that $T_g(f) = 0$ for all bounded, measurable, compactly supported functions f. Next, fix $R > 0$ and set $f(x) = 0$ if $|x| > R$ or if $g(x) = 0$. Otherwise, set $f(x) = e^{i\theta(x)}$, where $g(x) = |g(x)|e^{-i\theta(x)}$. This yields

$$\int_{\mathbb{R}} fg\, dm = \int_{-R}^{R} |g|\, dm = 0,$$

and, consequently, $g = 0$ m-$a.e.$ on $[-R, R]$. Since R is arbitrary, $g = 0$ m-$a.e.$

b. The *Dirac measure* δ, also called the Dirac δ function, is a mapping from $C_c^\infty(\mathbb{R})$ to $\mathbb{C}$ defined as

$$\delta(f) = f(0).$$

It is also a distribution, and we hasten to point out that it is not an element of $L^1_{\text{loc}}(\mathbb{R})$ in the sense of part *a*. In light of the previous sections in this chapter and Example 2.4.2*b*, note that δ is a Borel measure on $(\mathbb{R}, \mathcal{A})$, $\mathcal{B}(\mathbb{R}) \subseteq \mathcal{A}$, defined by $\delta(A) = 0$ if $0 \notin A$ and $\delta(A) = 1$ if $0 \in A$, $A \in \mathcal{A}$.

Remark. LAURENT SCHWARTZ received the Fields medal in 1950 for developing the theory of distributions. His classic book is Théorie des Distributions [426], [428]. The first edition was published in two volumes in 1950 and 1951. These volumes are a compendium of diverse accomplishments, a unification of technologies, an original formulation of ideas both new and old, and a research manual leading to new mathematics and applications.

Two other monumental contributions are ISRAEL M. GELFAND and SHILOV's Generalized Functions [191], and LARS V. HÖRMANDER's The Analysis of Linear Partial Differential Equations [241]. SCHWARTZ himself wrote an exquisite chapter on distributions and partial differential equations in [427].

The origins of distribution theory are based in the operational calculus from engineering and the concepts of "turbulent" and "weak" solutions of partial differential equations from physics. SCHWARTZ' Introduction and GELFAND and SHILOV's bibliographic notes give a significant scientific overview. For an overview that is personal, expository, nontechnical, historical, and authoritative, one can do no better than SCHWARTZ' autobiography [429], Chapter VI; cf. the obituary in the Notices of the Amer. Math. Soc., 50 (2003), 1072–1084.

A great number of books have been written on the theory of distributions, with a panorama from presentations of pure topological vector spaces [242] to applications in optics and supersonic wing theory [131]; see [39], Chapter 2, [426], and [457] for workaday theoretically sound presentations at three different levels.

The duality between the small space $C_c^\infty(\mathbb{R})$ and the large space $\mathcal{D}'(\mathbb{R})$, allowing us to define so many objects $T \in \mathcal{D}'(\mathbb{R})$, can be coupled with the integration by parts formula, see Theorem 4.6.3, to provide a definition of the distributional derivative of T. Let $f \in C_c^\infty(\mathbb{R})$ and let g be sufficiently smooth, e.g., let $g \in C^1(\mathbb{R})$, the space of continuously differentiable functions on $\mathbb{R}$. Then

$$\int_{\mathbb{R}} g'(x) f(x)\, dx = -\int_{\mathbb{R}} g(x) f'(x)\, dx. \tag{7.38}$$

By Theorem 4.6.3, we could even have taken g to be locally absolutely continuous on $\mathbb{R}$. The integration by parts formula in (7.38) is the distributional "duality formula",

$$\forall\, f \in C_c^\infty(\mathbb{R}), \quad g'(f) = -g(f'). \tag{7.39}$$

Since the right side of (7.39) is well defined when g is replaced by $T \in \mathcal{D}'(\mathbb{R})$, we are motivated to make the following definition.

Definition 7.5.4. Differentiation of distributions
The *distributional derivative* $D(T)$ of $T \in \mathcal{D}'(\mathbb{R})$ is defined by the formula,

$$\forall\, f \in C_c^\infty(\mathbb{R}), \quad D(T)(f) = -T(f'). \tag{7.40}$$

To establish the viability of (7.40) as an effective definition of the notion of derivative we must prove that for any $T \in \mathcal{D}'(\mathbb{R})$ we also have $D(T) \in \mathcal{D}'(\mathbb{R})$ and, moreover, that

$$\forall\, g, \text{ locally absolutely continuous on } \mathbb{R}, \quad D(T_g) = T_{g'}, \tag{7.41}$$

where g' denotes the ordinary pointwise derivative of g. The verification of (7.41) and the fact that $D(T) \in \mathcal{D}'(\mathbb{R})$ are routine calculations.

For each $T \in \mathcal{D}'(\mathbb{R})$ we define $D^n(T)$, the *nth distributional derivative* of T, as the distribution defined by the formula,

$$\forall\, f \in C_c^\infty(\mathbb{R}), \quad D^n(T)(f) = (-1)^n T(f^{(n)}).$$

In general, we cannot expect (7.41) to hold for arbitrary elements $g \in L^1_{\text{loc}}(\mathbb{R})$, even though T_g is a well-defined distribution. In fact, if g is infinitely differentiable on $\mathbb{R} \setminus \{x_0\}$ in the ordinary pointwise sense and if g' is the ordinary pointwise derivative of g defined everywhere except at x_0, then, in general, $D(T_g)$ and $T_{g'}$ are distributions, but $D(T_g) \neq T_{g'}$, e.g., Example 7.5.5.

Example 7.5.5. $H' = \delta$
$H = \mathbb{1}_{[0,\infty)}$ is the *Heaviside function.* The ordinary pointwise derivative H' of H exists and takes the value 0 on $\mathbb{R} \setminus \{0\}$. Thus, $H' \in L^1_{\text{loc}}(\mathbb{R})$ and H' is the 0 distribution.

On the other hand, the distributional derivative $D(H)$ is evaluated as follows. Choose $f \in C_c^\infty(\mathbb{R})$ and compute

$$D(H)(f) = -H(f') = -\int_0^\infty f'(x)\, dx = f(0) = \delta(f).$$

Since $D(H)(f) = \delta(f)$ for all $f \in C_c^\infty(\mathbb{R})$, we can conclude that

$$D(H) = \delta \neq 0.$$

It turns out that the space of distributions $\mathcal{D}'(\mathbb{R})$ is sometimes too large to define naturally the Fourier transform of each of its elements. As such, we shall introduce an important subspace of $\mathcal{D}'(\mathbb{R})$ in Definition 7.5.7.

Definition 7.5.6. The Schwartz space
a. An infinitely differentiable function $f : \mathbb{R} \to \mathbb{C}$ is an element of the *Schwartz space* $\mathcal{S}(\mathbb{R})$ if

$$\forall\, n \in \mathbb{N}, \quad \|f\|_{(n)} = \sup_{0 \le j \le n} \sup_{x \in \mathbb{R}} (1 + |x|^2)^n \left| f^{(j)}(x) \right| < \infty.$$

b. An important example of a Schwartz function is a Gaussian, $g(x) = e^{-x^2}$. We also note that

$$C_c^\infty(\mathbb{R}) \subseteq \mathcal{S}(\mathbb{R}) \subseteq L^1(\mathbb{R}) \cap L^2(\mathbb{R}).$$

c. The space $\mathcal{S}(\mathbb{R})$ is given a metrizable topology (see Example 7.5.9), where convergence of a sequence $\{f_n : n = 1, \ldots\} \subseteq \mathcal{S}(\mathbb{R})$ to $0 \in \mathcal{S}(\mathbb{R})$ is defined as follows:

$$\forall\, k, m \ge 0, \quad \lim_{n \to \infty} \|x^m f_n^{(k)}(x)\|_\infty = 0. \tag{7.42}$$

Definition 7.5.7. Tempered distributions
A linear functional,

$$T : \mathcal{S}(\mathbb{R}) \to \mathbb{C},$$
$$f \mapsto T(f),$$

is a *tempered distribution* if $\lim_{n\to\infty} T(f_n) = 0$ for every sequence $\{f_n : n = 1, \ldots\} \subseteq \mathcal{S}(\mathbb{R})$ converging to 0 in the sense of Definition 7.5.6*c*.

The space of all tempered distributions on $\mathbb{R}$ is denoted by $\mathcal{S}'(\mathbb{R})$.

Example 7.5.8. $C_c^\infty(\mathbb{R})$ and $\mathcal{S}(\mathbb{R})$
a. We observe that if $\{f_n : n = 1, \ldots\} \subseteq C_c^\infty(\mathbb{R})$ and "$f_n \to 0$" in the sense of Definition 7.5.2, then "$f_n \to 0$" in the sense of (7.42). In fact, from Definition 7.5.2*i*, there exists a compact set $K \subseteq \mathbb{R}$ such that for all n and k, supp $f_n^{(k)} \subseteq K$; and so there exists $0 \le C < \infty$ such that for all $x \in K$, $|x^m| \le C^m$. Thus, we need to show only that $\lim_{n\to\infty} \|f_n^{(k)}(x)\|_\infty = 0$, which, in turn, follows from Definition 7.5.2*ii*. Therefore, the mapping

$$C_c^\infty(\mathbb{R}) \to \mathcal{S}(\mathbb{R}),$$
$$f \mapsto f,$$

is continuous.

b. Moreover,

$$\overline{C_c^\infty(\mathbb{R})} = \mathcal{S}(\mathbb{R}). \tag{7.43}$$

To prove (7.43), for any function $f \in \mathcal{S}(\mathbb{R})$, we need to find a sequence $\{f_n : n = 1, \ldots\} \subseteq C_c^\infty(\mathbb{R})$ such that "$f_n - f \to 0$" in the sense of (7.42). To this end, let $g \in C_c^\infty(\mathbb{R})$ be a function for which $g(x) = g(-x)$, supp $g = [-1,1]$, and $g(0) = 1$. Then, for any $n = 1, \ldots$, let f_n be defined as

$$f_n(x) = \begin{cases} f(x), & \text{if } x \in [-n, n], \\ 0, & \text{if } x \in (-\infty, -n-1] \cup [n+1, \infty), \\ f(x)g(x-n), & \text{if } x \in [n, n+1], \\ f(x)g(x+n), & \text{if } x \in [-n-1, -n]. \end{cases}$$

Clearly, $f_n \in C_c^\infty(\mathbb{R})$, and the verification that "$f_n - f \to 0$" in the sense of (7.42) follows with the help of the Leibniz formula.

Combining Example 7.5.8 with the *embedding theorem* (Theorem A.11.2) in Appendix A.11, we obtain

$$\mathcal{S}'(\mathbb{R}) \subseteq \mathcal{D}'(\mathbb{R}).$$

Example 7.5.9. Metrizability of $\mathcal{S}(\mathbb{R})$
Let $\rho : \mathcal{S}(\mathbb{R}) \times \mathcal{S}(\mathbb{R}) \to \mathbb{R}^+$ be the function defined as

$$\forall\, f, g \in \mathcal{S}(\mathbb{R}), \quad \rho(f,g) = \sum_{n=0}^{\infty} \frac{1}{2^n} \frac{\|f-g\|_{(n)}}{1 + \|f-g\|_{(n)}}.$$

It is not difficult to verify that "$f_n \to 0$" in the sense of (7.42) if and only if

$$\lim_{n\to\infty} \rho(f_n, 0) = 0.$$

We shall now combine the notions introduced in Section 3.5 and Section 4.1 with distributions to reformulate RRT in a tantalizing manner, especially as regards generalization.

Remark. As a preliminary motivation we let $F \in BV_{\text{loc}}(\mathbb{R})$ be real-valued and right continuous, and we define two distributions, T_F and S_F, associated with F as follows:

$$\forall\, f \in C_c^\infty(\mathbb{R}), \quad T_F(f) = \int_{\mathbb{R}} f(x)F(x)\, dx, \tag{7.44}$$

and

$$\forall\, f \in C_c^\infty(\mathbb{R}), \quad S_F(f) = \int_{\mathbb{R}} f(x)\, dF(x). \tag{7.45}$$

Since $BV_{\text{loc}}(\mathbb{R}) \subseteq L^1_{\text{loc}}(\mathbb{R})$, T_F is a distribution as defined in Example 7.5.3.

On the other hand, in view of Theorem 5.4.1, the real-valued right-continuous elements of $BV_{\text{loc}}(\mathbb{R})$ define signed Lebesgue–Stieltjes measures, which, in fact, are signed regular Borel measures on $\mathbb{R}$. Clearly, the functions in $C_c^\infty(\mathbb{R})$ are integrable with respect to all such measures, and so (7.45) is well defined.

Next, we use the integration by parts formula for Lebesgue–Stieltjes integrals, e.g., Problem 4.51, to derive the connection between S_F and T_F in (7.46). Indeed,

$$\forall\, f \in C_c^\infty(\mathbb{R}), \quad \int_{\mathbb{R}} f(x)\, dF(x) = -\int_{\mathbb{R}} f'(x)F(x)\, dx,$$

can be rewritten as

$$S_F = D(T_F) = D(F). \tag{7.46}$$

Theorem 7.5.10. RRT in terms of distributions
$(C_0(\mathbb{R}))' = \{D(F) \in \mathcal{D}'(\mathbb{R}) : F \in BV(\mathbb{R})$ and F is right continuous$\}$, i.e., $M_b(\mathbb{R})$ is the space of all distributions T for which there exists right-continuous $F \in BV(\mathbb{R})$ such that $D(F) = T$.

Proof. The RRT, Theorem 7.2.7, tells us that $(C_0(\mathbb{R}))'$ and $M_b(\mathbb{R})$ are isometrically isomorphic.

Let $F \in BV(\mathbb{R})$ be right continuous. In order to prove that $D(F) \in (C_0(\mathbb{R}))'$, we begin by writing $F = F_r + iF_i$, where $F_r, F_i \in BV(\mathbb{R})$ are real-valued. We apply Theorem 5.4.1, which was formulated for real-valued functions (F_r and F_i), and we combine real and imaginary parts of the conclusion to form a complex regular Borel measure μ_F corresponding to F. From the discussion preceding the statement of Theorem 7.5.10, we can identify $D(F)$ with a Lebesgue–Stieltjes integral. Moreover, the functional defined by this integral is bounded in sup norm, since $|\mu_F|(\mathbb{R}) < \infty$; see Theorem 5.1.12 and Problem 5.24. Because of this and the density of $C_c(\mathbb{R})$ in $C_0(\mathbb{R})$, we can extend the functional, and hence $D(F)$, to a bounded linear functional on $C_0(\mathbb{R})$, i.e., $D(F) \in (C_0(\mathbb{R}))'$.

Conversely, to show that each $\mu \in M_b(\mathbb{R})$ can be represented by the distributional derivative of a function $F \in BV(\mathbb{R})$, we take the Jordan decomposition of $\mu \in M_b(\mathbb{R})$, which we write as $\mu = \mu_1 - \mu_2 + i\mu_3 - i\mu_4$. From Theorem 5.1.12 and Problem 5.24 we deduce that $\mu \in M_b(\mathbb{R})$ if and only if $\mu_i \in M_b(\mathbb{R})$, $i = 1, \ldots, 4$. Moreover, μ is a complex measure if and only if each μ_i, $i = 1, \ldots, 4$, is a bounded measure. By Theorem 4.1.9, for each μ_i we can associate a right-continuous increasing function $F_i \in BV(\mathbb{R})$. It follows from Theorem 4.1.2a that $F = F_1 - F_2 + iF_3 - iF_4 \in BV(\mathbb{R})$, and it is right continuous. Furthermore, $\mu = \mu_F$ in the notation of (5.24).

Then, on the one hand, we have

$$\forall\, f \in C_c(\mathbb{R}), \quad D(F)(f) = \int f\, d\mu_F,$$

and, on the other, we have

$$\forall\, f \in C_0(\mathbb{R}), \quad \mu_F(f) = \int f\, d\mu_F.$$

Thus, these two functionals agree on $C_c(\mathbb{R})$, and since they are both continuous in sup norm, $D(F)$ extends to μ_F on $C_0(\mathbb{R})$. □

Remark. From Theorem 7.5.10 we see how RRT has evolved from a *theorem* (Theorem 7.1.1), associating certain linear functionals with elements of $BV(\mathbb{R})$ by means of the Riemann–Stieltjes integral, to the *definition of integral* in terms of such functionals, e.g., [69].

For perspective with Theorem 7.5.10 we record the following d-dimensional result due to HORVÁTH [243].

Theorem 7.5.11. $\mu = D_1 D_2 \cdots D_d(f)$
Let $\mu \in M_b(\mathbb{R}^d)$, $d > 1$. *Then there is* $f \in L^\infty_{\text{loc}}(\mathbb{R}^d)$ *such that* $\mu = D_1 D_2 \cdots D_d(f)$, *where* D_j *is the distributional derivative in the direction of the* j*th coordinate.*

This result is not fine enough to relate bounded variation in $\mathbb{R}^d$ with $\mu \in M_b(\mathbb{R}^d)$; see Chapter 8.

We say that $T \in \mathcal{D}'(\mathbb{R})$ is a *positive distribution* if $T(f) \geq 0$ whenever $f \in C_c^\infty(\mathbb{R})$ is nonnegative on $\mathbb{R}$. Besides the fact that there is a natural embedding $\mathcal{C}'(\mathbb{R}) \subseteq \mathcal{D}'(\mathbb{R})$, the following result establishes a fundamental property of positive distributions.

Theorem 7.5.12. Positive distributions as positive Radon measures
If $T \in \mathcal{D}'(\mathbb{R})$ *is a positive distribution, then* T *is a positive Radon measure.*

Proof. i. We first prove that $T \in \mathcal{C}'(\mathbb{R})$. Let $K \subseteq \mathbb{R}$ be compact and let $\{f_n\} \subseteq C_c^\infty(\mathbb{R})$ have the property that supp $f_n \subseteq K$ and $\|f_n\|_\infty \to 0$. It is sufficient to show that $T(f_n) \to 0$.

Let $f \in C_c^\infty(\mathbb{R})$ be nonnegative on $\mathbb{R}$ and assume $f \geq 1$ on K. Then there is a positive sequence $\{\varepsilon_n\}$ such that $\varepsilon_n \to 0$ and

$$\forall\, x \in \mathbb{R} \text{ and } \forall\, n = 1, \ldots, \quad |f_n(x)| < \varepsilon_n f(x).$$

We write $f_n = g_n + i h_n$, where $g_n, h_n \in C_c^\infty(\mathbb{R})$ are real-valued. Further,

$$\forall\, x \in \mathbb{R} \text{ and } \forall\, n = 1, \ldots, \quad -\varepsilon_n f(x) \leq g_n(x) \leq \varepsilon_n f(x)$$
$$\text{and } -\varepsilon_n f(x) \leq h_n(x) \leq \varepsilon_n f(x).$$

Since T is a positive distribution, we have

$$\forall\, n = 1, \ldots, \quad -\varepsilon_n T(f) \leq T(g_n) \leq \varepsilon_n T(f)$$
$$\text{and } -\varepsilon_n T(f) \leq T(h_n) \leq \varepsilon_n T(f).$$

Consequently, $|T(f_n)| \leq 2\varepsilon_n T(f)$ for each n, and so $T \in \mathcal{C}'(\mathbb{R})$.

ii. Now we must prove that $T \in \mathcal{C}'(\mathbb{R})$ is a positive Radon measure knowing that $T(f) \geq 0$ for every nonnegative $f \in C_c^\infty(\mathbb{R})$. Let $g \in C_c^+(\mathbb{R})$. We can choose $\{g_n\} \subseteq C_c^\infty(\mathbb{R})$ such that each $g_n \geq 0$, each supp $g_n \subseteq [-1, 1]$, and $\|g * g_n - g\|_\infty \to 0$; see Example 7.5.1 and Definition B.3.2; cf. [428], Chapitre I, Théorème I. Noting that each $g * g_n \in C_c^\infty(\mathbb{R})$, each $g * g_n$ is nonnegative on $\mathbb{R}$, and supp $g * g_n$ is contained in a fixed compact set independent of n, we obtain $T(g * g_n) \geq 0$ and $T(g * g_n) \to T(g)$. Thus, $T(g) \geq 0$, and so T is a positive Radon measure. □

7.6 Potpourri and titillation

1. The first distributional derivative of δ, $D(\delta)$, is the *dipole* at the origin, and the distributions $D^n(\delta)$, $n \geq 1$, are the *multipoles* that arise in several important applications. For example, in fluid mechanics, a dipole is the limiting case in fluid flow of a source and a sink of equal strength approaching each other under the constraint that the product of the distance between them and their strength is constant.

To quantify the fact that multipoles arise in applications, let us consider the case of electromagnetism and the potential due to point charges $q_1, \ldots, q_n$ at $v_1, \ldots, v_n \in \mathbb{R}^3$. Then the electric field $F : \mathbb{R}^3 \to \mathbb{R}^3$ due to these charges can be formulated in terms of CHARLES A. COULOMB's law and linear superposition. For conservative fields there is a potential function $V : \mathbb{R}^3 \to \mathbb{R}$ whose gradient is F. In the case the charges are close to the origin and $r = |x|$ is much larger than any $|v_j|$, then $V(x)$ is of the form

$$V(x) = c\left\{\frac{1}{r}\sum_{j=1}^{n} q_j + \frac{1}{r^2}\left[\frac{x_1}{r}\sum_{j=1}^{n} q_j v_j^1 + \frac{x_2}{r}\sum_{j=1}^{n} q_j v_j^2 + \frac{x_3}{r}\sum_{j=1}^{n} q_j v_j^3\right] + \ldots\right\},$$

where $x = (x_1, x_2, x_3)$ and $v_j = (v_j^1, v_j^2, v_j^3)$. Here the first sum, $\sum_{j=1}^{n} q_j$, is the total charge or *monopole moment* of the charge distribution. The next three sums are components of the *dipole moment*, and we have omitted writing the remaining multipole moments.

Consider the particular example of two charges on the x-axis, viz., q at $\varepsilon/2$ and $-q$ at $-\varepsilon/2$. The monopole moment vanishes, and the dipole moment is $q(\varepsilon)$. If $q = q(\varepsilon) = -1/\varepsilon$, then it is natural to formulate the notion of the *dipole of moment* 1 *at the origin* to be

$$-\lim_{\varepsilon \to 0} \frac{1}{\varepsilon}(\delta_{\varepsilon/2} - \delta_{-\varepsilon/2}),$$

which, in turn, is $D(\delta)$.

It should be pointed out that SCHWARTZ chose the name "distribution" because "if μ is a measure, i.e., a particular kind of distribution, it can be considered as a distribution of electric charges in the universe" [429], page 238. So, besides being a generalization of the notion of function, it is also "a generalization of the notion of the distribution of electric charges", e.g., the aforementioned multipoles. The total charge of a distribution is $T(1)$, and so the charge of a dipole is zero.

The classical notions of potential and moments that guided SCHWARTZ have twenty-first-century progeny including fast multipole methods in scientific computation [205], [245], [207], applications dealing with the Schrödinger equation and function spaces, e.g., [359], and frame potentials [44], which have a variety of applications including modeling for Internet erasure problems. Frame potentials are used to characterize finite unit norm tight frames; see

Appendix A.14. This concept was inspired by examples of ZIMMERMANN [522] and GÖTZ PFANDER (oral communication).

2. The *spectral synthesis problem* for L^1 was an important force in harmonic analysis in the middle half of the twentieth-century. It leaves a legacy of deep, open, creative problems awaiting new technologies. We shall describe a facet of the spectral synthesis problem appropriate to the setting of Theorem 7.4.4 and to a distributional calculation; see [33], [271], [269], [276].

The space $A(\widehat{\mathbb{R}})$ of absolutely convergent Fourier transforms $F = \hat{f}$, developed in Appendix B, is a Banach algebra under pointwise multiplication with Banach space norm $\|F\|_A = \|f\|_1$. The dual space $A'(\widehat{\mathbb{R}})$ of $A(\widehat{\mathbb{R}})$ is the space of *pseudomeasures*. We have the inclusions

$$M_b(\widehat{\mathbb{R}}) \subseteq A'(\widehat{\mathbb{R}}) \subseteq \mathcal{S}'(\widehat{\mathbb{R}}).$$

The Banach space $A'(\widehat{\mathbb{R}})$ is a Banach algebra under convolution, and $M_b(\widehat{\mathbb{R}})$ is a proper subalgebra. For each closed set $E \subseteq \widehat{\mathbb{R}}$, let $A'(E)$ be those elements $T \in A'(\widehat{\mathbb{R}})$ with support contained in E, and let $A'_S(E)$ be those elements $T \in A'(E)$ such that $T(F) = 0$ for all $F \in A(\widehat{\mathbb{R}})$ vanishing on E. We call E a *spectral synthesis set*, or *S-set*, if $A'(E) = A'_S(E)$.

The goal is to characterize all S-sets. By Theorem 7.4.4, we know that if $T \in A'(E)$ is an element of $M_b(\widehat{\mathbb{R}})$ and if $F \in A(\widehat{\mathbb{R}})$ vanishes on E, then $T(F) = 0$; cf. [242], pages 342–343, for the annihilation property $T(F) = 0$ for various test functions and subspaces of distributions. Question: Can some elements of $A'(\widehat{\mathbb{R}}) \setminus M_b(\widehat{\mathbb{R}})$ have nonannihilation properties, i.e., are there non-S-sets? Answer: Yes!

ARNE BEURLING originally explicitly posed the spectral synthesis problem, but one could argue that it was "in the air", whether because of WIENER's work in harmonic analysis or potential theory. In fact, WIENER's Tauberian theorem (1930), which preceded BEURLING, can be interpreted as the assertion that the empty set is an S-set. This result was extended significantly by VITALII A. DITKIN, SHILOV, IRVING KAPLANSKY, SEGAL, and HENRY HELSON, and their general Wiener theorem tells us, in particular, that if E is closed and has a countable boundary then E is an S-set. Using a different idea, CARL HERZ (1958) proved that the Cantor set is an S-set. See Problem 5.8 for one version of WIENER's Tauberian theorem.

In the other direction, SCHWARTZ (1948) showed that the surface S^{d-1} of the unit sphere in $\mathbb{R}^d$, $d \geq 3$, is a non-S-set; and PAUL MALLIAVIN (1959) proved that every nondiscrete locally compact abelian group has non-S-sets.

SCHWARTZ' proof that $S^{d-1} \subseteq \mathbb{R}^d$ is a non-S-set fits in naturally with the material of Section 7.4 in the following way. Let $\mu \in M_b(\mathbb{R}^d)$ be the unit mass, uniformly distributed over S^{d-1}, i.e., let μ be surface measure on S^{d-1}. (Surface measures are defined in Chapters 8 and 9.) SCHWARTZ considered $T = D_1(\mu)$, the distributional derivative of μ in the "x-direction". He proved that $T \in A'(S^{d-1})$, which comes down to proving that the Fourier transform of T is an element of $L^\infty_{m^d}(\mathbb{R}^d)$. This conclusion is valid only for $d \geq 3$. He then constructed $F \in A(\widehat{\mathbb{R}}^d)$, vanishing on S^{d-1}, for which $T(F) \neq 0$; cf. Theorem 7.4.4. Thus, $T \notin A'_S(S^{d-1})$, and SCHWARTZ' result follows. SCHWARTZ' proof is eminently readable [425]. Details are also provided in many places, e.g., [33], pages 185–186, [233], volume II, pages 533–537, [404], pages 165–166. One of the interesting consequences of SCHWARTZ' counterexample is NICHOLAS VAROPOULOS' observation [481] (1967–1968) that *primary ideals are not necessarily maximal ideals in the Banach algebra of radial functions on* $\mathbb{R}^d$, $d \geq 3$; cf. [384], as well as WARNER's result [496] for the fascinating union problem.

The spectral synthesis problem is inspiring, and we refer to [33] for extensive perspective and "relevance", ranging from the most profound notions of harmonic analysis to the Nullstellensatz in algebraic geometry, to number theory and geometry and applicable mathematics.

3. In the Remark after Example 5.5.6, we mentioned the notion of a causal system. We shall examine causality further in the context of saying more about convolution.

Formally, motivated by the definition of convolution in $L^1_m(\mathbb{R})$, we define the convolution $S * T$ of $S, T \in \mathcal{D}'(\mathbb{R})$ by

$$\forall\, f \in C_c^\infty(\mathbb{R}), \quad (S * T)(f) = T(u)\,(S(v)(f(u+v))), \tag{7.47}$$

where, although T is not necessarily a point function on $\mathbb{R}$, we write $T(u)$ to indicate its dependence on the u variable. Statement (7.47) is not well defined for all $S, T \in \mathcal{D}'(\mathbb{R})$, and we refer to [428], Chapitre VI, and [242], pages 382–388, for the most useful criteria for the existence of distributional convolution. If we define the Dirac measure δ_x at $x \in \mathbb{R}$ by

$$\forall\, f \in C_c^\infty(\mathbb{R}), \quad \delta_x(f) = f(x),$$

which is compatible with Example 2.4.2*b*, then it is natural to define the *translation of* $T \in \mathcal{D}'(\mathbb{R})$ *by* $x \in \mathbb{R}$ as

$$\tau_x(T) = \delta_x * T.$$

Let X be a linear subspace of $\mathcal{S}'(\mathbb{R})$ with the properties that $\delta \in X$ and $\tau_x(T) \in X$ for each $T \in X$ and $x \in \mathbb{R}$. A *linear translation-invariant* (LTI) *system* is a linear function $L : X \to X$. The *impulse response* of L is $L(\delta) = h \in X$. The *filter* (or *frequency response* or *transfer function*) corresponding to L is $\hat{h}$. An LTI system L is *causal* if, whenever $x_0 \in \mathbb{R}$ and $T \in X$ vanishes on $(-\infty, x_0)$, then $L(T)$ vanishes on $(-\infty, x_0)$. We first discussed causal systems in the Remark after Example 5.5.6.

Adjusting the previous setup, let $L : C_c^\infty(\mathbb{R}) \to \mathcal{D}'(\mathbb{R})$ be a causal LTI system, without necessarily assuming continuity of L. Then *there is a unique distribution* (impulse response) $h \in \mathcal{D}'(\mathbb{R})$ *such that* $L(f) = h * f$ *for all* $f \in C_c^\infty(\mathbb{R})$, e.g., [4] and [428], pages 197–198.

4. The most celebrated problem in analytic number theory is to settle the question of the validity of the *Riemann hypothesis.*

The *Riemann zeta function* $\zeta(s)$ is defined as

$$\zeta(s) = \sum_{n=1}^{\infty} \frac{1}{n^s}, \quad \mathrm{Re}\,(s) > 1,$$

and it has an analytic continuation, whereby it is analytic on $\mathbb{C} \setminus \{1\}$ and has a simple pole at $s = 1$. The *Riemann hypothesis* is the statement that the complex zeros of $\zeta(s)$ all have real part equal to $1/2$, e.g., [154], [470], [251].

The *Weil distribution* $W \in \mathcal{D}'(\mathbb{R})$ is defined as

$$\forall\, f \in C_c^{\infty}(\mathbb{R}), \quad W(f) = \sum_{\rho} \Phi_f(\rho),$$

where $\zeta(\rho) = 0$ for $0 \le \mathrm{Re}\,(\rho) \le 1$, and where

$$\Phi_f(s) = \int_{\mathbb{R}} f(t) e^{(s-\frac{1}{2})t}\, dt.$$

This integral is the bilateral *Laplace transform* of $f(t)e^{-t/2}$. Tempered distributions arise in the following result: *The Riemann hypothesis is valid if and only if* $W \in \mathcal{S}'(\mathbb{R})$ [34], [265], page 6.

The Riemann hypothesis can be considered a strong form of the prime number theorem; and WIENER's Tauberian theorem, already mentioned with regard to spectral synthesis, is an indispensable tool in this type of analysis, e.g., [33], Section 2.3.

5. Let $(\mathbb{R}, \mathcal{B}(\mathbb{R}), p)$ be a probability space. The function $F : \mathbb{R} \to [0,1]$ defined by

$$\forall\, x \in \mathbb{R}, \quad F(x) = p(\{t \in \mathbb{R} : -\infty < t \le x\}) \tag{7.48}$$

is the *distribution function* of p. It is easy to see that F satisfies the following conditions:

i. F is increasing on $\mathbb{R}$;

ii. F is right continuous on $\mathbb{R}$;

iii. $\lim_{x\to-\infty} F(x) = 0$ and $\lim_{x\to\infty} F(x) = 1$.

Conversely, it is straightforward to prove that *if a function* $F : \mathbb{R} \to [0,1]$ *satisfies conditions i, ii, iii, then there exists a unique probability measure* $p : \mathcal{B}(\mathbb{R}) \to \mathbb{R}^+$ *related to* F *by* (7.48). In fact, to construct p we first define the algebra consisting of all finite unions of half-open intervals $(a, b]$, $-\infty \le a < b \le \infty$, and set

$$p((a,b]) = F(b) - F(a).$$

The remainder of the proof is routine; cf. Example 2.3.10 and Section 3.5.

The wonderful fact is that probability spaces $(\mathbb{R}, \mathcal{B}(\mathbb{R}), p)$ and corresponding distribution functions are characterized by the Schwartz distributional formula,

$$D(F) = p; \tag{7.49}$$

for example, let p be the Cantor–Lebesgue continuous measure μ_E and let F be the associated Cantor function C_E, for a given perfect symmetric set $E \subseteq [0,1]$ (Example 4.2.4). Equation (7.49) is a special case of Theorem 7.5.10. (Unfortunately, the words "distribution" and "distributional" have two different meanings in this description.)

Notwithstanding the spectacular structural beauty of Theorem 7.5.10, we note again, as we did in the Preface, that there are inherent limitations. In fact, in order to formulate probability theory at the level of Brownian motion and measures on infinite-dimensional spaces such as $C([0,1])$, Radon measure theory for locally compact Hausdorff spaces is inadequate; see [429], pages 163–165.

6. To complete our appetizer on probability theory, suppose we are given a probability space $(\mathbb{R}, \mathcal{B}(\mathbb{R}), p)$ and a random variable $f : (\mathbb{R}, \mathcal{B}(\mathbb{R}), p) \to \mathbb{R}$. The *cumulative distribution function* F_f defined on $f(\mathbb{R})$ is given by

$$\forall\, t \in \mathbb{R}, \quad F_f(t) = \int \mathbb{1}_{S(t)}(x)\, dp(x),$$

where integration is over the probability space $(\mathbb{R}, \mathcal{B}(\mathbb{R}), p)$ and

$$S(t) = \{x \in \mathbb{R} : p(x) \leq t\};$$

$F_f : \mathbb{R} \to [0,1]$ is increasing. We have Figure 7.2, where $D(F) = p$.

$$\begin{array}{ccc} \mathbb{R} & \xrightarrow{F} & [0,1] \\ {\scriptstyle f}\big\downarrow & & \\ \mathbb{R} & \xrightarrow{F_f} & [0,1] \end{array}$$

Fig. 7.2. Distribution functions.

We note that $D(F_f)$ is a probability measure on $\mathbb{R}$ containing the range of f. We say that f has a *probability density function* g_f if $D(F_f) = g_f \in L^1_m(\mathbb{R})$, i.e., g is the probability density function of a random variable f if $g \geq 0$, $\int_{\mathbb{R}} f = 1$, and F_f can be written as

$$F_f(t) = \int_{-\infty}^{t} g.$$

8 Lebesgue Differentiation Theorem on $\mathbb{R}^d$

8.1 Introduction

In Chapter 4, we saw that FTC–I (Theorem 4.4.3) depends essentially on the Lebesgue differentiation theorem (Theorem 4.3.2), which gives the existence and integrability of f' for $f \in BV([a,b])$. Because of this dependence, and because of the setting in this chapter, we shall use the appellation *Lebesgue differentiation theorem* to designate expressions of the form,

$$\lim_{j \to \infty} \frac{1}{\mu(A_j)} \int_{A_j} f \, d\mu = f(x),$$

where $\{A_j\}$ decreases to x in some prescribed way. In Definition 5.3.4, we saw that R–N can be viewed as a generalization of FTC. We now unfold more deeply the intimate relationship between FTC, R–N, and the Lebesgue differentiation theorem. This process, although natural at the level of comparing the ideas that are at the heart of these theorems, requires some sophisticated technology.

Our setting for this chapter is usually that of a Borel measure space on $\mathbb{R}^d$, although some of the results can be extended to general measure spaces. We first prove the Lebesgue differentiation theorem on $\mathbb{R}^d$ (Theorem 8.2.4) by means of an elementary generalization to $\mathbb{R}^d$ (Theorem 8.2.1) of the Vitali covering lemma for $\mathbb{R}$ (Theorem 4.3.1), combined with the Hardy–Littlewood lemma (Theorem 8.2.3). We then extend and refine VITALI's covering ideas to give a conceptually different proof of the Lebesgue differentiation theorem on $\mathbb{R}^d$ (Theorems 8.4.6 and 8.4.7) by means of the Besicovitch covering theorem (Theorems 8.3.2 and 8.3.3). It is this latter approach that establishes the connection with R–N and allows us to differentiate one measure with respect to another (Theorem 8.4.4).

Generalizations of FTC–II (Theorem 4.5.5) to $\mathbb{R}^d$, e.g., Theorem 8.5.6, as related to Euclidean notions of bounded variation (Section 8.5) originally arising in the Lebesgue differentiation theorem on $\mathbb{R}$, are based on classical FTC formulas on $\mathbb{R}^d$. These FTC formulas equate the integral of a derivative of a function defined on a region A with the integral of the function itself over the boundary of A. They are rooted in the analysis of physical phenomena.

Example 8.1.1 gives some FTC formulas on $\mathbb{R}^3$ from the nineteenth century. Hypotheses for the validity of the formulas are not given.

J.J. Benedetto and W. Czaja, *Integration and Modern Analysis*, Birkhäuser Advanced Texts,
DOI 10.1007/978-0-8176-4656-1_8,

Example 8.1.1. FTC formulas on $\mathbb{R}^3$

a. Green–Ostrogradsky formula:

$$\begin{aligned}\iiint_A U\Delta(V)\,dx\,dy\,dz + \iint_S U\frac{\partial V}{\partial n}\,d\sigma \\ = \iiint_A V\Delta(U)\,dx\,dy\,dz + \iint_S V\frac{\partial U}{\partial n}\,d\sigma,\end{aligned} \tag{8.1}$$

where $A \subseteq \mathbb{R}^3$ and $S = \partial(A)$ is the boundary of A, where $U, V : \mathbb{R}^3 \to \mathbb{R}$, where

$$\Delta(V) = \frac{\partial^2 V}{\partial x^2} + \frac{\partial^2 V}{\partial y^2} + \frac{\partial^2 V}{\partial z^2}$$

is the *Laplacian* of V, where σ represents surface measure (Section 8.7), and where n denotes the normal to S.

b. Gauss–Ostrogradsky formula:

$$\begin{aligned}\iiint_A \left(\frac{\partial P}{\partial x} + \frac{\partial Q}{\partial y} + \frac{\partial R}{\partial z}\right)\,dx\,dy\,dz \\ = \iint_S (P\cos(\lambda) + Q\cos(\mu) + R\cos(\nu))\,d\sigma,\end{aligned} \tag{8.2}$$

where $P, Q, R : \mathbb{R}^3 \to \mathbb{R}$. If $F = (P, Q, R) : \mathbb{R}^3 \to \mathbb{R}^3$ is the corresponding vector field, then the vector formulation of (8.2) is

$$\iiint_A \nabla \cdot F\,dx\,dy\,dz = \iiint_A \mathrm{div}\,(F)\,dx\,dy\,dz = \iint_S F \cdot n\,d\sigma, \tag{8.3}$$

where ∇ is the gradient operator and $\mathrm{div}(F)$ denotes $F \cdot \nabla$.

c. Stokes formula:

$$\begin{aligned}\iint_S \left(\lambda\left(\frac{\partial R}{\partial y} - \frac{\partial Q}{\partial z}\right) + \mu\left(\frac{\partial P}{\partial z} - \frac{\partial R}{\partial x}\right) + \nu\left(\frac{\partial Q}{\partial x} - \frac{\partial P}{\partial y}\right)\right)\,d\sigma \\ = \int_C \left(P\frac{\partial x}{\partial s} + Q\frac{\partial y}{\partial s} + R\frac{\partial z}{\partial s}\right)\,ds,\end{aligned} \tag{8.4}$$

where $C = \partial(S)$ is the curve bounding the surface S and where the notation for the parametrization $r : \mathbb{R} \to \mathbb{R}^3$ of C is $s \mapsto (x(s), y(s), z(s))$. The vector formulation of (8.4) is

$$\iint_S \mathrm{curl}\,(F) \cdot n\,d\sigma = \int_C F \cdot dr, \tag{8.5}$$

where

$$\mathrm{curl}\,(F) = \nabla \times F = \left(\frac{\partial R}{\partial y} - \frac{\partial Q}{\partial z}\right)i + \left(\frac{\partial P}{\partial z} - \frac{\partial R}{\partial x}\right)j + \left(\frac{\partial Q}{\partial x} - \frac{\partial P}{\partial y}\right)k,$$

where "$\times$" is the cross product, and where i, j, k are the unit vectors on the x, y, z-axes.

Remark. GEORGE F. GREEN was a baker and a self taught scientist. In 1828, at the age of 35, he published "An Essay on the Application of Mathematical Analysis to the Theories of Electricity and Magnetism" [204]. In Section 3 of this paper he proved the integration by parts formula (8.1).

Equation (8.1) allowed GREEN to justify calculations of LAPLACE and DENIS POISSON. It was proved independently by MIKHAIL V. OSTROGRADSKY in 1831. The *potential equation* $\Delta(V) = 0$ arises naturally in this context; and it was GREEN who introduced the term *potential* in its current usage.

If $V(x, y, z, t)$ denotes the temperature at a point $(x, y, z) \in \mathbb{R}^3$ within a solid A at time t, then

$$k\Delta(V) = \frac{\partial}{\partial t}V$$

is an elementary form of the *heat* or *diffusion equation* and k is the so-called *thermal diffusivity* of the material A. In order to analyze the heat equation, OSTROGRADSKY (1831) made use of (8.2). Equations (8.2) and (8.3) are the classical *divergence theorem*. OSTROGRADSKY announced his version (8.2) as early as 1826. See Theorem 8.5.6 for a twentieth-century update.

SIR GEORGE GABRIEL STOKES' formula (8.4) was first stated by SIR WILLIAM THOMSON (LORD KELVIN) in 1850 in a letter to STOKES. In fact, (8.4) was "published" by STOKES as an examination question for the Smith Prize at Cambridge in 1854; and by the time of STOKES' death it was widely known as Stokes' formula.

Stokes' formula on $\mathbb{R}^2$, stated at the end of Section 4.5, is usually called Green's theorem. Stokes' formula is naturally generalized to compact d-dimensional C^2-manifolds with smooth boundaries; see [95], [327], [447].

8.2 Maximal function and Lebesgue differentiation theorem

Let $X \subseteq \mathbb{R}^d$. Then $\mathcal{C} \subseteq \mathcal{P}(\mathbb{R}^d)$ is a *covering* of X if

$$\forall\, x \in X,\ \exists\, C \in \mathcal{C} \text{ such that } \quad x \in C.$$

An open *cube* $Q \subseteq \mathbb{R}^d$, of side $2\delta > 0$ and center $x = (x_1, \ldots, x_d) \in \mathbb{R}^d$, is the set $Q = Q(x) = \{y \in \mathbb{R}^d : |y_i - x_i| < \delta, i = 1, \ldots, d\}$. By adjusting the inequality with δ we refer to such sets as *cubes*.

Theorem 8.2.1. Elementary Vitali covering lemma
Let $(\mathbb{R}^d, \mathcal{M}(\mathbb{R}^d), m^d)$ be the Lebesgue measure space, and let $A \in \mathcal{M}(\mathbb{R}^d)$ satisfy $m^d(A) < \infty$. For any covering of A by a family of cubes with sides parallel to the axes, there exists a finite disjoint subfamily $\{Q_1, \ldots, Q_n\}$ such that

$$m^d(A) \leq 3^d \sum_{j=1}^{n} m^d(Q_j).$$

Proof. Without loss of generality we assume that the cubes are open and that A is compact. Thus, using the Heine–Borel theorem we deduce that there exists a finite cover of A: $\widetilde{Q}_1, \ldots, \widetilde{Q}_N$. We define Q_1 to be the element of $\{\widetilde{Q}_j : j = 1, \ldots, N\}$ with largest diameter. We define Q_2 to be the cube with largest diameter of all those elements in $\{\widetilde{Q}_j : j = 1, \ldots, N\}$ that are disjoint from Q_1. We proceed inductively until we exhaust all the $\widetilde{Q}_j$s. Thus, we have defined a family of disjoint cubes $Q_1, \ldots, Q_n$, $n \leq N$. Let Q_j^* be the cube that is concentric with Q_j and that has diameter 3 times that of Q_j. We observe that

$$A \subseteq \bigcup_{j=1}^{N} \widetilde{Q}_j \subseteq \bigcup_{j=1}^{n} Q_j^*,$$

and so

$$m^d(A) \leq \sum_{j=1}^{n} m^d(Q_j^*) = 3^d \sum_{j=1}^{n} m^d(Q_j). \qquad \square$$

Remark. Theorem 8.2.1 is also valid for rotated cubes as well as balls. The proof is essentially the same.

Definition 8.2.2. Hardy–Littlewood maximal function
Let $f \in L^1_{\mathrm{loc}}(\mathbb{R}^d)$. The *Hardy–Littlewood maximal function* $M(f)$ of f is defined as

$$\forall\, x \in \mathbb{R}^d, \quad M(f)(x) = \sup_{Q(x)} \frac{1}{m^d(Q(x))} \int_{Q(x)} |f| \, dm^d, \tag{8.6}$$

where the supremum is taken over all open cubes $Q(x)$ centered at x. The function M was introduced by HARDY and LITTLEWOOD in 1930 [219]. Maximal functions had a profound influence on the development of classical harmonic analysis in the twentieth-century; e.g., see [448], [202].

Theorem 8.2.3. Hardy–Littlewood lemma
Let $(\mathbb{R}^d, \mathcal{M}(\mathbb{R}^d), m^d)$ be the Lebesgue measure space. Then, for any $f \in L^1_{\mathrm{loc}}(\mathbb{R}^d)$,

$$\forall\, \alpha > 0, \quad m^d\left(\{x \in \mathbb{R}^d : M(f)(x) > \alpha\}\right) \leq \frac{3^d}{\alpha} \int_{\mathbb{R}^d} |f| \, dm^d.$$

Proof. Because of the regularity of Lebesgue measure it is enough to investigate the boundedness of measures of compact subsets K of $\{x \in \mathbb{R}^d : M(f)(x) > \alpha\}$. Fix $\alpha > 0$. For each $x \in K$ there exists a cube $Q(x)$ such that

$$\frac{1}{m^d(Q(x))} \int_{Q(x)} |f| \, dm^d > \alpha.$$

The collection of all such cubes forms a cover of K, and so, by Theorem 8.2.1, there exists a finite family of disjoint cubes $Q_1, \ldots, Q_n$ such that

$m^d(K) \leq 3^d \sum_{j=1}^{n} m^d(Q_j)$. Therefore, because of the disjointness of the Q_js, we have

$$m^d(K) \leq 3^d \sum_{j=1}^{n} \frac{1}{\alpha} \int_{Q_j} |f| \, dm^d \leq \frac{3^d}{\alpha} \int_{\mathbb{R}^d} |f| \, dm^d.$$

Thus, by regularity of Lebesgue measure, we obtain

$$m^d\left(\{x \in \mathbb{R}^d : M(f)(x) > \alpha\}\right) \leq \frac{3^d}{\alpha} \int_{\mathbb{R}^d} |f| \, dm^d. \qquad \square$$

If $g \in L^1_{m^d}(\mathbb{R}^d)$ and g is continuous on $\mathbb{R}^d$, then

$$\forall\, x \in \mathbb{R}^d, \quad \lim_{\delta \to 0} \int_{Q(x)} g \, dm^d = g(x), \tag{8.7}$$

since

$$\forall\, \varepsilon > 0, \ \exists\, Q(x) > 0 \ \text{ such that } \ \forall\, y \in Q(x), \quad |g(y) - g(x)| < \varepsilon.$$

This raises the question as to what extent (8.7) is valid for every $f \in L^1_{m^d}(\mathbb{R}^d)$. The question is answered by the Lebesgue differentiation theorem.

Theorem 8.2.4. Lebesgue differentiation theorem on $(\mathbb{R}^d, \mathcal{M}(\mathbb{R}^d), m^d)$
Let $(\mathbb{R}^d, \mathcal{M}(\mathbb{R}^d), m^d)$ be the Lebesgue measure space, and let $f \in L^1_{m^d}(\mathbb{R}^d)$. Then

$$\lim_{\delta \to 0} \frac{1}{m^d(Q(x))} \int_{Q(x)} f \, dm^d = f(x), \quad m^d\text{-a.e. } x \in \mathbb{R}^d,$$

where $Q(x) = \{y \in \mathbb{R}^d : |y_i - x_i| < \delta, i = 1, \ldots, d\}$ is an open cube of side $2\delta > 0$, centered at x.

Proof. Fix $\varepsilon > 0$ and let $g \in L^1_{m^d}(\mathbb{R}^d)$ be a continuous function for which

$$\int_{\mathbb{R}^d} |f - g| \, dm^d < \varepsilon;$$

cf. Problem 3.14*a*. Since g is continuous, we have

$$\forall\, x \in \mathbb{R}^d, \quad \lim_{\delta \to 0} \frac{1}{m^d(Q(x))} \int_{Q(x)} g \, dm^d = g(x)$$

by (8.7). We choose $\alpha > 0$ and write

$$\begin{aligned}
&\overline{\lim_{\delta\to 0}} \left| \frac{1}{m^d(Q(x))} \int_{Q(x)} (f(y) - f(x))\, dm^d(y) \right| \\
&= \overline{\lim_{\delta\to 0}} \left| \frac{1}{m^d(Q(x))} \left(\int_{Q(x)} (f(y) - g(y) + g(y) - g(x))\, dm^d(y) \right) + g(x) - f(x) \right| \\
&\le M(f-g)(x) + |f(x) - g(x)|.
\end{aligned}$$

From this inequality it follows that

$$\begin{aligned}
A_\alpha &= \left\{ x \in \mathbb{R}^d : \overline{\lim_{\delta\to 0}} \left| \frac{1}{m^d(Q(x))} \int_{Q(x)} (f(y) - f(x))\, dm^d(y) \right| > \alpha \right\} \\
&\subseteq \left\{ x \in \mathbb{R}^d : M(f-g)(x) > \alpha/2 \right\} \cup \left\{ x \in \mathbb{R}^d : |f(x) - g(x)| > \alpha/2 \right\}.
\end{aligned}$$

Using the Chebyshev inequality (Problem 3.13), we have

$$\frac{\alpha}{2} m^d \left(\left\{ x \in \mathbb{R}^d : |f(x) - g(x)| > \alpha/2 \right\} \right) < \varepsilon.$$

On the other hand, by the Hardy–Littlewood lemma (Theorem 8.2.3),

$$m^d \left(\left\{ x \in \mathbb{R}^d : M(f-g)(x) > \alpha/2 \right\} \right) \le \frac{3^d}{\alpha/2} \int_{\mathbb{R}^d} |f - g|\, dm^d < \frac{2\varepsilon 3^d}{\alpha}.$$

Hence,

$$m^d(A_\alpha) \le \frac{2\varepsilon}{\alpha} + \frac{2\varepsilon 3^d}{\alpha}.$$

Thus, for every fixed $\alpha > 0$, we have $m^d(A_\alpha) = 0$, since $\varepsilon > 0$ is arbitrary. Let $\{\alpha_n > 0\}$ decrease to 0. Then $\{A_{\alpha_n}\}$ is increasing, and $m^d(\bigcup A_{\alpha_n}) = 0$ by Theorem 2.4.3*d*. Consequently, by the definition of A_α, we obtain the result. □

It is possible to generalize Theorem 8.2.4 to general measure spaces; see, e.g., [232]. In Section 8.4 we provide another approach to generalizing Theorem 8.2.4 to other measure spaces.

8.3 Coverings

Following the definition of a Vitali covering in $\mathbb{R}$, from Section 4.3, we say that a collection $\mathcal{Q}$ of closed balls in $\mathbb{R}^d$ is a *Vitali covering* of a set $X \subseteq \mathbb{R}^d$ if

$$\forall\, \varepsilon > 0 \text{ and } \forall\, x \in X,\ \exists\, B \in \mathcal{Q}, \text{ such that } \quad x \in B \quad \text{and} \quad \operatorname{diam}(B) < \varepsilon.$$

Remark. We could define Vitali coverings for cubes instead of balls.

We choose to deal with balls, instead of cubes, in Sections 8.3, 8.4, 8.5, because of some interesting connections that arise between the Besicovitch covering theorem (Theorems 8.3.2 and 8.3.3) stated in terms of balls and sphere-packing problems.

The following theorem, which depends on Theorem 8.2.1 and generalizes Theorem 4.3.1, is not used in the sequel, but it is included to round out our coverage of VITALI's important ideas in this area.

Theorem 8.3.1. Vitali covering lemma
Let $(\mathbb{R}^d, \mathcal{M}(\mathbb{R}^d), m^d)$ be the Lebesgue measure space, and let $A \in \mathcal{M}(\mathbb{R}^d)$ satisfy $0 < m^d(A) < \infty$. If $\mathcal{Q}$ is a Vitali covering of A then, for each $\varepsilon > 0$, there exists a sequence $\{B_j : j = 1, \dots\} \subseteq \mathcal{Q}$ of disjoint balls such that

$$m^d\left(A \setminus \bigcup_{j=1}^{\infty} B_j\right) = 0 \quad \text{and} \quad \sum_{j=1}^{\infty} m^d(B_j) < (1+\varepsilon)m^d(A). \tag{8.8}$$

Proof. Let $\varepsilon > 0$. By the regularity of Lebesgue measure, we can choose an open set U_1 such that $A \subseteq U_1$ and $m^d(U_1) < (1+\varepsilon)m^d(A)$. Let $\mathcal{Q}_1 \subseteq \mathcal{Q}$ consist of the balls in $\mathcal{Q}$ that are contained in U_1. Note that $\mathcal{Q}_1$ is a Vitali covering of A. We shall construct a disjoint sequence $\{B_j\} \subseteq \mathcal{Q}_1$ satisfying (8.8). In fact, the second assertion of (8.8) is clear for any disjoint sequence $\{B_j\} \subseteq \mathcal{Q}_1$. We shall construct such a sequence that also satisfies the first assertion.

From Theorem 8.2.1 we know that there exist a finite disjoint collection $\{B_1, \dots, B_{N_1}\} \subseteq \mathcal{Q}_1$ and a constant $0 < \beta < 1$ ($\beta = 3^{-d}$) such that

$$\sum_{j=1}^{N_1} m^d(B_j) \geq \beta m^d(A).$$

Therefore,

$$\begin{aligned} m^d\left(A \setminus \bigcup_{j=1}^{N_1} B_j\right) &\leq m^d\left(U_1 \setminus \bigcup_{j=1}^{N_1} B_j\right) \\ &= m^d(U_1) - \sum_{j=1}^{N_1} m^d(B_j) \\ &< (1+\varepsilon-\beta)m^d(A). \end{aligned}$$

We assume without loss of generality that $\varepsilon < \beta/2$, and so we have

$$m^d\left(A \setminus \bigcup_{j=1}^{N_1} B_j\right) < \left(1 - \frac{\beta}{2}\right) m^d(A).$$

We now apply Theorem 8.2.1 to the subset of A that is not covered by $\{B_1, \ldots, B_{N_1}\}$. Choose an open set $U_2 \subseteq U_1$ that contains $A \setminus \bigcup_{j=1}^{N_1} B_j$ and such that

$$m^d(U_2) \leq (1+\varepsilon) m^d \left(A \setminus \bigcup_{j=1}^{N_1} B_j \right). \tag{8.9}$$

Let $\mathcal{Q}_2 \subseteq \mathcal{Q}_1$ consist of the balls in $\mathcal{Q}_1$ that are contained in U_2. Note that $\mathcal{Q}_2$ is a Vitali covering of $A \setminus \bigcup_{j=1}^{N_1} B_j$. Thus, there exists a disjoint collection $\{B_{N_1+1}, \ldots, B_{N_2}\} \subseteq \mathcal{Q}_2$ such that

$$\sum_{j=N_1+1}^{N_2} m^d(B_j) \geq \beta m^d \left(A \setminus \bigcup_{j=1}^{N_1} B_j \right). \tag{8.10}$$

Therefore, proceeding as above, but using (8.9) and (8.10), we obtain

$$\begin{aligned} m^d \left(A \setminus \bigcup_{j=1}^{N_2} B_j \right) &= m^d \left(\left(A \setminus \bigcup_{j=1}^{N_1} B_j \right) \setminus \bigcup_{j=N_1+1}^{N_2} B_j \right) \\ &\leq m^d(U_2) - \sum_{j=N_1+1}^{N_2} m^d(B_j) \\ &< m^d \left(A \setminus \bigcup_{j=1}^{N_1} B_j \right) \left(1 - \frac{\beta}{2} \right) \\ &< \left(1 - \frac{\beta}{2} \right)^2 m^d(A). \end{aligned}$$

Repeating this process we obtain a disjoint sequence $\{B_j : j = 1, \ldots\} \subseteq \mathcal{Q}$ that satisfies the first assertion in (8.8) because $(1 - \beta/2)^k \to 0$ as $k \to \infty$. Of course, the process could stop after a finite number of steps, in which case the first assertion is immediate. □

Theorem 8.3.1 is valid for cubes as well as for balls.

As with Theorem 8.2.1, Theorem 8.3.1 can be used in the analysis of Lebesgue measure on $\mathbb{R}^d$. The next theorem is useful in the analysis of arbitrary Borel measures. The version with closed balls is due to BESICOVITCH, [57], [58]. A reformulation of Theorem 8.3.2 for open balls is straightforward; see, e.g., [252]. There is also a version of this result, due to ANTHONY P. MORSE [349], where balls are replaced with cubes whose sides are parallel to the axes.

Theorem 8.3.2. Besicovitch covering theorem
Let $\mathcal{Q}$ be a collection of closed balls in $\mathbb{R}^d$ with the property that

$$\rho = \sup_{B \in \mathcal{Q}} \operatorname{diam}(B) < \infty.$$

If A denotes the set of centers of balls in $\mathcal{Q}$, then there exist countable disjoint subcollections $\mathcal{Q}_1, \ldots, \mathcal{Q}_{N(d)} \subseteq \mathcal{Q}$ such that

$$A \subseteq \bigcup_{k=1}^{N(d)} \bigcup_{B \in \mathcal{Q}_k} B.$$

The number $N(d)$ of subcollections of disjoint balls depends only on the dimension d.

Proof. i. We shall assume first that the set A is bounded. Our immediate task is to construct a countable subset of $\mathcal{Q}$ that covers A. The first ball $B_1 = \overline{B}(a_1, r_1)$ is chosen so that $r_1 > (3/8)\rho$ (here $\overline{B}(a, r)$ denotes a closed ball centered at $a \in \mathbb{R}^d$ and of radius $r > 0$). The next balls are chosen inductively. Let $A_k = A \setminus \left(\bigcup_{j=1}^{k-1} B_j \right)$. If $A_k \neq \emptyset$ then let $B_k = \overline{B}(a_k, r_k)$ be a ball such that $a_k \in A_k$ and $r_k \geq (3/4) \sup\{r : \overline{B}(a, r) \in \mathcal{Q},\ a \in A_k\}$. If the process stops at some point we have a finite cover of A. Otherwise, we proceed letting $k \to \infty$.

For any such k and for $j < k$ we have

$$r_k \leq \sup\{r : \overline{B}(a, r) \in \mathcal{Q},\ a \in A_j\} \leq \frac{4}{3} r_j. \tag{8.11}$$

We use (8.11) to deduce that the collection $\{\overline{B}(a_j, r_j/3) : j = 1, \ldots\}$ of balls is mutually disjoint. Indeed, for $j < k$, $a_k \notin B_j$. Moreover,

$$|a_j - a_k| > r_j = \frac{r_j}{3} + \frac{2r_j}{3} \geq \frac{r_j}{3} + \frac{2}{3}\frac{3}{4} r_k > \frac{r_j}{3} + \frac{r_k}{3}.$$

To see that the collection $\{B_j : j = 1, \ldots\}$ of balls is a covering of the set A, consider any $a \in A \setminus \left(\bigcup_{j=1}^{\infty} B_j \right)$. By definition of the set A there exists $B = \overline{B}(a, r) \in \mathcal{Q}$. Because A is assumed to be bounded it is not difficult to deduce from the disjointness of balls $\{\overline{B}(a_j, r_j/3) : j = 1, \ldots\}$ that $\lim_{j \to \infty} r_j = 0$. Thus, there exists k such that

$$r_k < \frac{3}{4} r.$$

This yields the contradiction with the definition of the ball B_k, since we require that

$$r_k \geq \frac{3}{4} \sup \left\{ r : \overline{B}(a, r) \in \mathcal{Q},\ a \in \left(\bigcup_{j=1}^{k-1} B_j \right)^{\sim} \right\} \geq \frac{3}{4} r.$$

Therefore, we have constructed a countable collection of balls from $\mathcal{Q}$ that covers the bounded set A.

ii. We continue to assume that A is bounded.

Given $k \geq 1$, the next step is to estimate the number of the sets B_j, $j < k$, that have a nonempty intersection with B_k. The estimate is completed in (8.16), and the number is, as claimed in the statement of the theorem, independent of the covering and depends only on the dimension d.

Let R_k denote the set of indices of all balls B_j, $1 \leq j < k$, such that $B_j \cap B_k \neq \emptyset$ and $r_j \leq 3r_k$. If $x \in \overline{B}(a_j, r_j/3)$ then

$$|x - a_k| \leq |x - a_j| + |a_j - a_k| \leq \frac{r_j}{3} + r_j + r_k \leq 5r_k.$$

Thus, for each such j, $\overline{B}(a_j, r_j/3) \subseteq \overline{B}(a_k, 5r_k)$. Since the balls $\{\overline{B}(a_j, r_j/3) : j = 1, \dots\}$ are disjoint, we use (8.11) to obtain

$$m^d(\overline{B}(a_k, 5r_k)) \geq \sum_{j \in R_k} m^d(\overline{B}(a_j, r_j/3)) \geq \sum_{j \in R_k} m^d(\overline{B}(a_j, r_k/4)).$$

Therefore, we conclude that

$$\operatorname{card} R_k \leq 20^d. \tag{8.12}$$

Similarly, for any k, let S_k denote the set of indices of balls B_j, $1 \leq j < k$, such that $B_j \cap B_k \neq \emptyset$ and $r_j > 3r_k$. To estimate the cardinality of S_k we consider $i, j \in S_k$, $i \neq j$, and we let $\alpha_{i,j}$ denote the angle between $a_i - a_k$ and $a_j - a_k$. We shall show that there exists $\alpha_0 > 0$ such that for each such pair $i, j \in S_k$, $\alpha_{i,j} > \alpha_0 > 0$, i.e., (8.15).

To obtain (8.15) we start with two elementary observations: since $i, j < k$, we have that $a_k \notin B_i \cup B_j$ and so $r_i < |a_i - a_k|$ and $r_j < |a_j - a_k|$; and, since $B_i \cap B_k \neq \emptyset$ and $B_j \cap B_k \neq \emptyset$, we have $|a_i - a_k| \leq r_i + r_k$ and $|a_j - a_k| \leq r_j + r_k$. We may also assume without loss of generality that $|a_i - a_k| \leq |a_j - a_k|$.

Next, we prove that if $\cos(\alpha_{i,j}) \geq 5/6$ then $a_i \in B_j$. First, if $|a_i - a_j| \leq |a_i - a_k|$ and $a_i \notin B_j$, then, by the law of cosines and the fact that $r_j < |a_i - a_j|$, we compute

$$\begin{aligned}
\cos(\alpha_{i,j}) &= \frac{|a_i - a_k|^2 + |a_j - a_k|^2 - |a_i - a_j|^2}{2|a_i - a_k||a_j - a_k|} \\
&= \frac{|a_i - a_k|}{2|a_j - a_k|} + \frac{(|a_j - a_k| + |a_i - a_j|)(|a_j - a_k| - |a_i - a_j|)}{2|a_i - a_k||a_j - a_k|} \\
&\leq \frac{1}{2} + \frac{2(|a_j - a_k| - |a_i - a_j|)|a_j - a_k|}{2|a_i - a_k||a_j - a_k|} \\
&\leq \frac{1}{2} + \frac{(r_j + r_k) - r_j}{r_i} = \frac{1}{2} + \frac{r_k}{r_i} < \frac{5}{6}.
\end{aligned}$$

Second, if, $|a_i - a_j| > |a_i - a_k|$ and $a_i \notin B_j$, then

$$\cos(\alpha_{i,j}) \leq \frac{|a_i - a_k|^2}{2|a_i - a_k||a_j - a_k|} \leq \frac{1}{2} < \frac{5}{6}.$$

Thus, as asserted above, if $\cos(\alpha_{i,j}) \geq 5/6$ then $a_i \in B_j$.

We shall now show that even when $\cos(\alpha_{i,j}) \geq 5/6$ it cannot get too close to 1. Since $a_i \in B_j$, we have $i < j$ and so $a_j \notin B_i$, which implies that $|a_i - a_j| > r_i$. Then

$$\begin{aligned}\frac{|a_i - a_j| + |a_i - a_k| - |a_j - a_k|}{|a_j - a_k|} &\leq \frac{|a_i - a_j|^2 - (|a_j - a_k| - |a_i - a_k|)^2}{|a_j - a_k||a_i - a_j|} \\ &= \frac{2|a_i - a_k|(1 - \cos(\alpha_{i,j}))}{|a_i - a_j|} \\ &\leq \frac{2(r_i + r_k)(1 - \cos(\alpha_{i,j}))}{r_i} \\ &\leq \frac{8(1 - \cos(\alpha_{i,j}))}{3}.\end{aligned} \tag{8.13}$$

Also, remembering that $|a_i - a_j| > r_i$ and $r_j \leq (4/3)r_i$,

$$\begin{aligned}|a_i - a_j| + |a_i - a_k| - |a_j - a_k| &\geq r_i + r_i - (r_j + r_k) \\ &\geq \frac{3}{2}r_j - (r_j + r_k) \geq \frac{1}{6}r_j \\ &\geq \frac{1}{8}(r_j + r_k) \geq \frac{1}{8}|a_j - a_k|.\end{aligned} \tag{8.14}$$

Combining (8.13) and (8.14) yields

$$\frac{1}{8} \leq \frac{8}{3}(1 - \cos(\alpha_{i,j})),$$

which implies that $\cos(\alpha_{i,j}) \leq 61/64$, or, equivalently, that

$$\alpha_{i,j} > \alpha_0 = \arccos(61/64) > 0. \tag{8.15}$$

We shall now use this fact to complete our estimate of the cardinality of the set S_k. The essential point is that we have given a lower bound for the possible angles between any two vectors with indices in S_k. We use this estimate to show that there is an upper bound on the number of such vectors. In fact, let C_j denote the cone with vertex at a_k in the direction of $a_j - a_k$ and with angle $\alpha_0/2$. If $i \neq j$ for $i, j \in S_k$, then

$$\big(C_j \cap \overline{B}(a_k, 1)\big) \cap \big(C_i \cap \overline{B}(a_k, 1)\big) = \emptyset.$$

Since the volume of $C_j \cap \overline{B}(a_k, 1)$ is independent of j, it is not difficult to see that there exists $K_d > 0$ such that the number of such cones, and hence the cardinality of S_k, cannot exceed K_d, i.e.,

$$\text{card } S_k \leq K_d.$$

Combining this fact with (8.12) we obtain

$$\text{card}\,\{B_j : 1 \leq j < k,\ B_j \cap B_k \neq \emptyset\} = \text{card}\, R_k + \text{card}\, S_k \leq 20^d + K_d. \quad (8.16)$$

iii. We let $N(d) = 20^d + K_d + 1$, and we shall now define the collections $\mathcal{Q}_1, \ldots, \mathcal{Q}_{N(d)}$ as follows. For $k = 1, \ldots, N(d)$, let $B_k \in \mathcal{Q}_k$. For $k > N(d)$ we proceed by induction, assigning B_k to the collection $\mathcal{Q}_l$, where l is chosen so that $B_k \cap B_j = \emptyset$ for all $B_j \in \mathcal{Q}_l$, $j < k$. This is possible since we know that B_k can intersect at most $N(d) - 1$ sets B_j with $j < k$.

The method of construction guarantees that the families $\mathcal{Q}_j$, $j = 1, \ldots, N(d)$, consist of disjoint balls.

iv. If A is an unbounded set we decompose it into a union of bounded sets $A_k = A \cap \{x \in \mathbb{R}^d : 3\rho(l-1) \leq |x| < 3\rho l\}$, $l = 1, \ldots$. For each of these sets there exist families $\mathcal{Q}_1^l, \ldots, \mathcal{Q}_{N(d)}^l$ with the properties described in parts i, ii, iii. We let

$$\mathcal{Q}_k = \bigcup_{l=1}^{\infty} \mathcal{Q}_k^{2l-1}$$

and

$$\mathcal{Q}_{k+N(d)} = \bigcup_{l=1}^{\infty} \mathcal{Q}_k^{2l},$$

for $k = 1, \ldots, N(d)$. Thus,

$$A \subseteq \bigcup_{k=1}^{2N(d)} \bigcup_{B \in \mathcal{Q}_k} B. \qquad \square$$

We now reformulate Theorem 8.3.2 as a measure-theoretic covering result in $\mathbb{R}^d$. This particular result we give is due to JOHN M. SULLIVAN [462]. It was obtained as a corollary of his work on estimating the optimal constant in the Besicovitch covering theorem. SULLIVAN showed that the optimal constant is connected to a sphere-packing problem; cf. Problem 2.12 and the Remark following it.

Theorem 8.3.3. Measure-theoretic Besicovitch covering theorem *Let $(\mathbb{R}^d, \mathcal{A}, \mu)$ be a Borel measure space. Let $\mathcal{Q}$ be a collection of closed balls in $\mathbb{R}^d$ and let A denote the set of centers of balls in $\mathcal{Q}$. Assume that $A \in \mathcal{A}$, $\mu(A) < \infty$, and that, for each $a \in A$,*

$$\inf_{\overline{B}(a,r) \in \mathcal{Q}} r = 0.$$

If $U \subseteq \mathbb{R}^d$ is an open set, then there exists a countable collection $\mathcal{Q}_0 \subseteq \mathcal{Q}$ of disjoint balls such that

$$\bigcup_{B \in \mathcal{Q}_0} B \subseteq U$$

and

$$\mu\left((A\cap U)\setminus\left(\bigcup_{B\in\mathcal{Q}_0} B\right)\right)=0.$$

Proof. i. We start by showing that for any $\varepsilon\in(0,1/N(d))$ there exists a family $\{B_i : i=1,\ldots,n_1\}\subseteq U$ of balls such that

$$\mu\left((A\cap U)\setminus\left(\bigcup_{i=1}^{n_1} B_i\right)\right)\le(1-\varepsilon)\mu(A\cap U). \tag{8.17}$$

Indeed, let $\mathcal{Q}^1=\{B\in\mathcal{Q} : \operatorname{diam}(B)\le 1,\ B\subseteq U\}$. Since U is open, $A\cap U$ coincides with the set of centers of balls in $\mathcal{Q}^1$. By Theorem 8.3.2, since the supremum of the set of diameters of balls in $\mathcal{Q}^1$ is finite, there exist families $\mathcal{Q}_1,\ldots,\mathcal{Q}_{N(d)}$, consisting of disjoint balls in $\mathcal{Q}^1$, such that

$$A\cap U\subseteq\bigcup_{k=1}^{N(d)}\bigcup_{B\in\mathcal{Q}_k} B.$$

Consequently,

$$\mu(A\cap U)\le\sum_{k=1}^{N(d)}\mu\left(A\cap U\cap\bigcup_{B\in\mathcal{Q}_k} B\right),$$

and so there exists k_0 such that

$$\frac{1}{N(d)}\mu(A\cap U)\le\mu\left(A\cap U\cap\bigcup_{B\in\mathcal{Q}_{k_0}} B\right).$$

It follows from Theorem 2.4.3*d* that there exists a subset $\{B_i : i=1,\ldots,n_1\}\subseteq\mathcal{Q}_{k_0}$ for which

$$\varepsilon\mu(A\cap U)\le\mu\left(A\cap U\cap\bigcup_{i=1}^{n_1} B_i\right). \tag{8.18}$$

On the other hand, the Carathéodory definition of measurability implies that

$$\mu(A\cap U)=\mu\left(A\cap U\cap\bigcup_{i=1}^{n_1} B_i\right)+\mu\left((A\cap U)\setminus\left(\bigcup_{i=1}^{n_1} B_i\right)\right),$$

which, combined with (8.18), yields (8.17):

$$\mu\left((A\cap U)\setminus\left(\bigcup_{i=1}^{n_1} B_i\right)\right)\le\mu(A\cap U)-\varepsilon\mu(A\cap U).$$

ii. Now that we have proved (8.17), let $U_2 = U \setminus (\bigcup_{i=1}^{n_1} B_i)$ and let $\mathcal{Q}^2 = \{B \in \mathcal{Q} : \text{diam }(B) \le 1,\ B \subseteq U_2\}$. Clearly, U_2 is open. From part i of the proof, it follows that there exists a sequence $\{B_i : i = n_1+1, \ldots, n_2\} \subseteq \mathcal{Q}^2$ of balls for which

$$\mu\left((A \cap U) \setminus \left(\bigcup_{i=1}^{n_2} B_i\right)\right) = \mu\left((A \cap U_2) \setminus \left(\bigcup_{i=n_1+1}^{n_2} B_i\right)\right)$$
$$\le (1-\varepsilon)\mu(A \cap U_2) \le (1-\varepsilon)^2 \mu(A \cap U).$$

We finish the proof by continuing this process and noting that $(1-\varepsilon)^j \to 0$ as $j \to \infty$. □

We note that Theorem 8.3.3 is true not only for a measure μ on a σ-algebra $\mathcal{A}$, but also for the outer measure μ^* associated with μ.

Corollary 8.3.4. *Let $(\mathbb{R}^d, \mathcal{A}, \mu)$ be a regular Borel measure space. Let μ^* be the outer measure associated with μ, let $\mathcal{Q}$ be a collection of closed balls in $\mathbb{R}^d$, and let A denote the set of centers of balls in $\mathcal{Q}$. Assume that $\mu^*(A) < \infty$ and that, for each $a \in A$,*

$$\inf_{\overline{B}(a,r)\in\mathcal{Q}} r = 0.$$

There exists a countable collection $\mathcal{Q}_0$ of disjoint balls in $\mathcal{Q}$ such that

$$\mu^*\left(A \setminus \bigcup_{B\in\mathcal{Q}_0} B\right) = 0.$$

8.4 Differentiation of measures

Let $f : \mathbb{R} \to \mathbb{R}$ be an increasing right-continuous function. In Chapter 3 we considered a set function μ_f defined by

$$\mu_f((a,b]) = f(b) - f(a),$$

which extends to a measure called the Lebesgue–Stieltjes measure associated with the function f. Using this notation, we see that the derivative of f at x can be defined as

$$f'(x) = \lim_{(x,x+h]\to\{x\}} \frac{\mu_f((x,x+h])}{m((x,x+h])}, \tag{8.19}$$

whenever the limit exists. Similarly, if $f \in L^1_{\text{loc}}(\mathbb{R})$, the Lebesgue differentiation theorem (Theorem 4.3.2) and FTC–I allow us to assert that

$$\lim_{(x,x+h]\to\{x\}} \frac{\sigma_f((x,x+h])}{m((x,x+h])} = f(x), \quad m\text{-a.e.}, \tag{8.20}$$

where $\sigma_f(A) = \int_A f\, dm$, $A \in \mathcal{M}(\mathbb{R})$. Thus, if $F(x) = \int_a^x f\, dm$, then $F' = f$ m-*a.e.* (FTC–I).

These observations motivate us to generalize the notion of derivative in this section. Our setting will be $\mathbb{R}^d$, but the differentiation of measures and integrals, such as in (8.19) and (8.20), is a well-established theory in abstract measure spaces going back to the 1930s; see [170] and [133].

We begin by defining a notion of convergence of a sequence of sets to a point in $\mathbb{R}^d$. This is meant to generalize the convergence "$(x, x+h] \to \{x\}$" in $\mathbb{R}$. Our choice is certainly not the simplest possible generalization. However, it has the advantage of allowing us to prove substantial results about the differentiation of measures in $\mathbb{R}^d$, as well as the most natural generalization of the Lebesgue differentiation theorem to $\mathbb{R}^d$.

Definition 8.4.1. Measure-metrizable convergence of sets to a point
Let $(\mathbb{R}^d, \mathcal{A}, \nu)$ be a Borel measure space. Let $x \in \mathbb{R}^d$. A sequence $\{A_j : j = 1, \dots\} \subseteq \mathcal{A}$ of sets *converges measure-metrizably* to x if the following conditions hold:

i. $\forall\, j = 1, \dots,\ \exists\, r_j$ such that $A_j \subseteq B(x, r_j)$,
ii. $\lim_{j\to\infty} r_j = 0$,
iii. $\exists\, \alpha \in (0, 1]$ such that $\forall\, j = 1, \dots,\ \nu(A_j) \geq \alpha\nu(B(x, r_j))$.

We note here that x need not belong to any of the sets A_j. The definition states only that the sequence $\{A_j : j = 1, \dots\}$ is comparable, in terms of the measure ν, with a sequence of balls converging to x.

It follows from the definition that the sequence $\{B(x, r_j) : j = 1, \dots\}$ converges measure-metrizably to $x \in \mathbb{R}^d$ if and only if $\lim_{j\to\infty} r_j = 0$.

Definition 8.4.2. Differentiation of measures
Let $(\mathbb{R}^d, \mathcal{A})$, $\mathcal{B}(\mathbb{R}^d) \subseteq \mathcal{A}$, be a measurable space. Let $\mu \in M_b(\mathbb{R}^d)$, let ν be a measure on $\mathbb{R}^d$, and let $x \in \mathbb{R}^d$. If there exists $z \in \mathbb{C}$ such that

$$\lim_{j\to\infty} \frac{\mu(A_j)}{\nu(A_j)} = z,$$

for each sequence $\{A_j : j = 1, \dots\}$ convergent measure-metrizably to x with respect to ν, then z is the *derivative of μ with respect to ν at x*. We denote this derivative by

$$D_\nu(\mu)(x).$$

The following restricted version of the notion of differentiation of measures is often considered. Let μ and ν be as in Definition 8.4.2. If the limit

$$\lim_{r\to 0} \frac{\mu(B(x, r))}{\nu(B(x, r))}$$

exists, it is the *symmetric derivative of μ with respect to* ν at $x \in \mathbb{R}^d$. We shall use this notion in the proof of the next lemma.

Lemma 8.4.3. *Let* $(\mathbb{R}^d, \mathcal{A})$, $\mathcal{B}(\mathbb{R}^d) \subseteq \mathcal{A}$, *be a measurable space. Let* μ *and* ν *be regular Borel measures. If* $\mu(A) = 0$, $A \in \mathcal{A}$, *then there exists a set* $B \subseteq A$, $B \in \mathcal{A}$, *such that*

$$\nu(A \setminus B) = 0$$

and

$$\forall\, x \in B, \quad D_\nu(\mu)(x) = 0.$$

Proof. i. Without loss of generality we can assume that $A \subseteq O$, where $O \subseteq \mathbb{R}^d$ is open and bounded. Indeed, let $\mathbb{R}^d = \bigcup_{n=1}^{\infty} O_n$, where O_n is open and bounded for $n = 1, \ldots,$ and let B_n be the set obtained by applying Lemma 8.4.3 for $A \cap O_n$ with open and bounded O_n, $n = 1, \ldots.$ Since each O_n is open, it follows from the construction in part *ii* that the sets B_i and $O_i \cap B_n$, $i \leq n$, coincide modulo a set of measure zero. Therefore, we set $B = B_1 \cup \bigcup_{n=2}^{\infty}(B_n \setminus O_{n-1})$.

ii. Define B' to be the set of all points $x \in A$ for which the symmetric derivative of μ with respect to ν exists and equals 0, i.e.,

$$\lim_{r \to 0} \frac{\mu(B(x,r))}{\nu(B(x,r))} = 0.$$

Let B'_j, $j = 1, \ldots,$ be the set of all points $x \in A \setminus B'$ for which

$$\overline{\lim_{r \to 0}} \frac{\mu(B(x,r))}{\nu(B(x,r))} > \frac{1}{j}.$$

Fix $j > 0$ and let $\varepsilon > 0$. Since μ is regular and $\mu(A) = 0$, there exists an open set $U \subseteq \mathbb{R}^d$ such that $A \subseteq U$ and $\mu(U) < \varepsilon$. By definition, each point $x \in B'_j$ can be considered as a center of an open ball $B(x, r_x) \subseteq U$ for which

$$\nu(B(x, r_x)) < j\mu(B(x, r_x)).$$

In fact, each $x \in B'_j$ is a center of infinitely many such balls.

Using the version of the Besicovitch covering theorem formulated in Corollary 8.3.4 (here we use the assumption that A is contained in a bounded open set to guarantee that $\nu^*(A) < \infty$), and writing $r_k = r_{x_k}$, we see that there exists a countable collection $\{B(x_k, r_k) : k = 1, \ldots\}$ of disjoint balls that satisfies the above properties and such that

$$\nu^*\left(B'_j \setminus \left(\bigcup_{k=1}^{\infty} B(x_k, r_k)\right)\right) = 0.$$

Here, ν^* denotes the outer measure associated with the measure ν. Thus,

$$\begin{aligned} \nu^*(B'_j) &\leq \sum_{k=1}^{\infty} \nu(B(x_k, r_k)) \leq \sum_{k=1}^{\infty} j\mu(B(x_k, r_k)) \\ &\leq j\mu\left(\bigcup_{k=1}^{\infty} B(x_k, r_k)\right) \leq j\mu(U) \leq j\varepsilon. \end{aligned} \tag{8.21}$$

Since (8.21) is true for each $\varepsilon > 0$, we have

$$\nu^*(B_j') = 0.$$

Thus, by the regularity of the measure ν there exists a Borel set B_j such that $B_j' \subseteq B_j$ and $\nu(B_j) = 0$; cf. Problems 2.4 and 2.5.

Let $B \in \mathcal{A}$ be defined as

$$B = A \setminus \bigcup_{j=1}^{\infty} B_j.$$

Then

$$\nu(A \setminus B) \leq \nu\left(\bigcup_{j=1}^{\infty} B_j\right) = 0,$$

and this completes the proof of the first assertion of the lemma.

iii. In order to prove the second assertion, we fix $x \in B$ and consider a sequence $\{A_j : j = 1, \dots\} \subseteq \mathbb{R}^d$ of Borel sets convergent measure-metrizably to x with respect to ν. This implies that there exist a sequence $\{B(x, r_j) : j = 1, \dots\}$ of Euclidean balls and a constant $\alpha \in (0, 1]$ such that

$$\forall\, j = 1, \dots, \quad \frac{\mu(A_j)}{\nu(A_j)} \leq \frac{1}{\alpha} \frac{\mu(B(x, r_j))}{\nu(B(x, r_j))}. \tag{8.22}$$

For (8.22) we used the hypothesis that $\mu \geq 0$. It is clear from the construction in part *ii* that $B \subseteq B'$. Thus, from the definition of the set B', the terms on the right-hand side of (8.22) converge to 0. Consequently,

$$\lim_{j \to \infty} \frac{\mu(A_j)}{\nu(A_j)} = 0$$

for each sequence $\{A_j\}$ convergent measure-metrizably to x with respect to ν; equivalently,

$$D_\nu(\mu)(x) = 0. \qquad \square$$

In Chapter 5 we commented on the equivalence of FTC and R–N. We develop this a bit more now. Indeed, an argument for such an equivalence could be made if the generalized notion of differentiation of measures yielded

$$\mu(A) = \int_A D_\nu(\mu)\, d\nu.$$

This is a consequence of the next theorem.

Theorem 8.4.4. FTC and R–N on $\mathbb{R}^d$

Let $(\mathbb{R}^d, \mathcal{A})$ be a Borel measurable space, let $\mu \in M_b(\mathbb{R}^d)$, and let ν be a regular Borel measure.

a. *$D_\nu(\mu)(x)$ exists ν-a.e. and*

$$D_\nu(\mu) \in L^1_\nu(\mathbb{R}^d).$$

b. *μ can be decomposed as*

$$\forall\, A \in \mathcal{A},\ \mu(A) = \mu_s(A) + \int_A D_\nu(\mu)\, d\nu,$$

where $\mu_s \perp \nu$ and $D_\nu(\mu_s) = 0$ ν-a.e.

Proof. Before we proceed with the proof we shall make two observations. First, due to the Lebesgue decomposition theorem (Theorem 5.2.6), it is enough to consider two cases: $\mu \perp \nu$ and $\mu \ll \nu$. Second, since $\mu = \mu_r + i\mu_i$, without loss of generality we can assume that μ is a bounded signed regular Borel measure.

a. We start with the case $\mu \perp \nu$. Let $\mu = \mu^+ - \mu^-$ be the Jordan decomposition of μ. It follows from Problem 5.34 that if $\mu \perp \nu$ then $\mu^+ \perp \nu$. Thus, if C_ν is the set on which ν is concentrated, then $\nu(C_\nu^\sim) = \mu^+(C_\nu) = 0$. Hence, we can use Lemma 8.4.3 to obtain that $D_\nu(\mu^+)(x) = 0$ for ν-*a.e.* $x \in C_\nu$, and, consequently, for ν-*a.e.* $x \in \mathbb{R}^d$. Similarly, we obtain that $D_\nu(\mu^-)(x) = 0$ for ν-*a.e.* $x \in \mathbb{R}^d$, and so $D_\nu(\mu)(x) = 0$ ν-*a.e.*

In the case $\mu \ll \nu$, we recall the observation before Theorem 5.3.1 that allows us to use R–N when μ is a signed measure. This implies that there exists a unique function $f \in L^1_\nu(\mathbb{R}^d)$ such that

$$\forall\, A \in \mathcal{A}, \quad \mu(A) = \int_A f\, d\nu.$$

Therefore, our assertion will follow if we can prove that

$$D_\nu(\mu) = f \quad \nu\text{-a.e.} \tag{8.23}$$

In order to establish (8.23), we define a countable family of sets parametrized by $r \in \mathbb{Q}$:

$$N_r = \{x \in \mathbb{R}^d : f(x) < r\}, \quad P_r = \{x \in \mathbb{R}^d : f(x) \geq r\}.$$

We also define associated measures λ_r, $r \in \mathbb{Q}$, defined as

$$\forall\, A \in \mathcal{A}, \quad \lambda_r(A) = \int_{A \cap P_r} (f(x) - r)\, d\nu(x).$$

Since $\lambda_r(N_r) = 0$, Lemma 8.4.3 implies that for each $r \in \mathbb{Q}$ there exists $N_r' \subseteq N_r$, $N_r \in \mathcal{A}$, such that $\nu(N_r \setminus N_r') = 0$ and

$$\forall\, x \in N_r', \quad D_\nu(\lambda_r)(x) = 0. \tag{8.24}$$

From the definition of f and λ_r we obtain

$$\forall\, A \in \mathcal{A}, \quad \mu(A) - r\nu(A) = \int_A (f(x) - r)\, d\nu(x) \leq \lambda_r(A). \tag{8.25}$$

Let

$$E = \bigcup_{r \in \mathbb{Q}} (N_r \setminus N_r').$$

Clearly, $E \in \mathcal{A}$ and $\nu(E) = 0$.

Take $x_0 \in \mathbb{R}^d \setminus E$, a rational number $r > f(x_0)$, and a sequence $\{A_j : j = 1, \dots\} \subseteq \mathcal{A}$ convergent measure-metrizably to x_0 with respect to ν. Since $x_0 \notin E$ and $x_0 \in N_r$, we must have $x_0 \in N_r'$. From (8.25) it follows that

$$\forall\, j = 1, \dots, \quad \frac{\mu(A_j)}{\nu(A_j)} \leq \frac{\lambda_r(A_j)}{\nu(A_j)} + r.$$

Thus, (8.24) and (8.25) imply that

$$\overline{\lim_{j \to \infty}} \frac{\mu(A_j)}{\nu(A_j)} \leq r.$$

Since this inequality is true for each $r > f(x_0)$, we conclude that

$$\overline{\lim_{j \to \infty}} \frac{\mu(A_j)}{\nu(A_j)} \leq f(x_0). \tag{8.26}$$

Analogously, we prove (taking $-\mu$ instead of μ and $-f$ instead of f) that for each sequence $\{A_j : j = 1, \dots\} \subseteq \mathcal{A}$, convergent measure-metrizably to ν-*a.e.* $x_0 \in \mathbb{R}^d$ with respect to ν, we have

$$\underline{\lim_{j \to \infty}} \frac{\mu(A_j)}{\nu(A_j)} \geq f(x_0). \tag{8.27}$$

Inequalities (8.26) and (8.27) imply (8.23), and thus part *a* of our theorem is established.

b. Part *b* follows immediately from the proof of part *a*. □

For expositions on these matters we refer to [79], [129], [405]. Here we have followed the presentation given in [405].

An interesting result in this area is due to DIETRICH KÖLZOW (1968), e.g., [79], pages 35–37, who proved that whenever a measure space admits R–N then the R–N derivative has the form given in Definition 8.4.2.

Corollary 8.4.5. *Let $(\mathbb{R}^d, \mathcal{A})$, $\mathcal{B}(\mathbb{R}^d) \subseteq \mathcal{A}$, be a measurable space, let $\mu \in M_b(\mathbb{R}^d)$, and let ν be a regular Borel measure. Then*

a. *$\mu \perp \nu$ if and only if $D_\nu(\mu) = 0$ ν-a.e.*

b. *$\mu \ll \nu$ if and only if*

$$\forall\, A \in \mathcal{A}, \quad \mu(A) = \int_A D_\nu(\mu)\, d\nu.$$

The following is a consequence of Theorem 8.4.4. Because of Definition 8.4.1 it is also a generalization of Theorem 8.2.4.

Theorem 8.4.6. Lebesgue differentiation theorem on $\mathbb{R}^d$
Let $(\mathbb{R}^d, \mathcal{A}, \mu)$ be a regular Borel measure space, and let $f \in L^1_\mu(\mathbb{R}^d)$. Then

$$\lim_{j\to\infty} \frac{1}{\mu(A_j)} \int_{A_j} f\, d\mu = f(x), \quad \mu\text{-a.e. } x \in \mathbb{R}^d, \tag{8.28}$$

for each sequence $\{A_j : j = 1, \dots\} \subseteq \mathcal{A}$ convergent measure-metrizably to x with respect to μ.

Proof. Define the complex measure ν by

$$\forall\, A \in \mathcal{A}, \quad \nu(A) = \int_A f\, d\mu.$$

By Theorem 5.3.3 and Theorem 5.3.5, ν is a complex regular Borel measure that is absolutely continuous with respect to μ. We already know from the proof of Theorem 8.4.4 that, in such cases, $D_\mu(\nu) = f$ μ-*a.e.* Thus, by definition of $D_\mu(\nu)$, we obtain (8.28). (Note that in order to state (8.28) in terms of "μ", we have switched the roles of "μ" and "ν" in applying Theorem 8.4.4.) □

In the case of $\mathbb{R}$ an important way to view FTC, as we have seen in Chapter 4, is Theorem 4.4.5. We now provide an analogue of this latter result in $\mathbb{R}^d$.

Theorem 8.4.7. Strong Lebesgue differentiation theorem on $\mathbb{R}^d$
Let $(\mathbb{R}^d, \mathcal{A}, \mu)$ be a regular Borel measure space, and let $f \in L^1_\mu(\mathbb{R}^d)$. Then

$$\lim_{j\to\infty} \frac{1}{\mu(A_j)} \int_{A_j} |f(x) - f(x_0)|\, d\mu(x) = 0, \quad \mu\text{-a.e. } x_0 \in \mathbb{R}^d, \tag{8.29}$$

for each sequence $\{A_j : j = 1, \dots\} \subseteq \mathcal{A}$ convergent measure-metrizably to x_0 with respect to μ.

Proof. Let $\{r_n : n = 1, \dots\} \subseteq \mathbb{C}$ be dense in $\mathbb{C}$. Define the following functions on $\mathbb{R}^d$:

$$\forall\, n = 1, \dots, \quad f_n = |f - r_n|.$$

These functions are not integrable with respect μ, but they have the property that for each compact set $K \subseteq \mathbb{R}^d$, $f_n \in L^1_\mu(K)$. Thus, the measures μ_n, defined by

$$\forall\, A \in \mathcal{A}, \quad \mu_n(A) = \int_A |f - r_n|\, d\mu,$$

are regular Borel measures, absolutely continuous with respect to μ. For each n, let L_n denote the set of points for which

$$D_\mu(\mu_n)(x) = |f(x) - r_n|. \tag{8.30}$$

In view of Theorem 8.4.4, $\mu(L_n^\sim) = 0$ for all $n = 1, \ldots$. Let $L = \bigcap_{n=1}^\infty L_n$. Clearly, $\mu(L^\sim) = 0$.

Fix $x_0 \in L$ and $\varepsilon > 0$. Choose r_n such that $|f(x_0) - r_n| < \varepsilon/2$. Therefore, for all $A \in \mathcal{A}$, $\mu(A) > 0$, we have

$$\begin{aligned} &\frac{1}{\mu(A)} \int_A |f - f(x_0)|\, d\mu \\ &\quad \le \frac{1}{\mu(A)} \int_A |f - r_n|\, d\mu + \frac{1}{\mu(A)} \int_A |f(x_0) - r_n|\, d\mu \\ &\quad < \frac{\mu_n(A)}{\mu(A)} + \frac{\varepsilon}{2}. \end{aligned} \tag{8.31}$$

If $\{A_j : j = 1, \ldots\} \subseteq \mathcal{A}$ converges measure-metrizably to x_0 with respect to μ, then, since (8.30) holds, (8.31) implies that, for j sufficiently large,

$$\frac{1}{\mu(A_j)} \int_{A_j} |f - f(x_0)|\, d\mu \le \varepsilon. \tag{8.32}$$

Since (8.32) is true for μ-*a.e.* $x_0 \in \mathbb{R}^d$ and for all $\varepsilon > 0$, (8.29) holds. □

Definition 8.4.8. Lebesgue set and set of differentiability
a. Let $(\mathbb{R}^d, \mathcal{A}, \mu)$ be a regular Borel measure space, and let f be μ-measurable. The set of points $x_0 \in \mathbb{R}^d$ for which

$$\lim_{r \to 0} \frac{1}{\mu(B(x_0, r))} \int_{B(x_0,r)} f\, d\mu = f(x_0)$$

is the *set of differentiability* of the function f.

b. Let $(\mathbb{R}^d, \mathcal{A}, \mu)$ be a regular Borel measure space, and let f be μ-measurable. The set of points $x_0 \in \mathbb{R}^d$ for which

$$\lim_{r \to 0} \frac{1}{\mu(B(x_0, r))} \int_{B(x_0,r)} |f - f(x_0)|\, d\mu = 0$$

is the *Lebesgue set* of the function f.

Since sequences of balls centered at x_0 with radii decreasing to 0 converge measure-metrizably to x_0, and in view of Theorem 8.4.7, we can make the following assertion: *If* $f \in L^1_\mu(\mathbb{R}^d)$, *then* μ-*a.e. points in* $\mathbb{R}^d$ *belong to the Lebesgue set of* f. It is not difficult to see that the same is true for locally integrable functions.

Further, *the Lebesgue set is a subset of the set of differentiability of f, whereas the converse statement is false.*

8.5 Bounded variation and the divergence theorem

In Section 7.5 we showed that distributional derivatives of complex-valued functions of one variable that are of bounded variation and right continuous can be identified with the space $M_b(\mathbb{R})$ of all complex regular Borel measures. Not only is this result an interesting fact for the reasons stated in Chapter 7, but it serves as a motivation for an extension of the definition of bounded variation from one to several dimensions. Our treatment is just an outline, and we refer to [521] for a detailed study of the properties of functions of bounded variation in $\mathbb{R}^d$. We repeat that the following definition is compatible with our formulation, Theorem 7.5.10, of the RRT in $\mathbb{R}$ in terms of the distributional derivative of a function of bounded variation.

Definition 8.5.1. Bounded variation
a. Let $X \subseteq \mathbb{R}^d$ be an open set and let $F \in L^1_{\text{loc}}(X)$ be a complex-valued function whose distributional partial derivatives $D_i(F)$, $i = 1, \ldots, d$, defined as

$$\forall\, i = 1, \ldots, d,\ \forall\, f \in C_c^\infty(X), \quad D_i(F)(f) = -\int_X F(x) \frac{\partial f}{\partial x_i}(x)\, dx,$$

are complex regular Borel measures on X. In this case we say that F is a *function of bounded variation on X*.

The space of functions of bounded variation on X is denoted by $BV(X)$. If $F \in BV(X)$, we shall write $\nabla(F)$ to denote the vector-valued complex measure

$$\nabla(F) = (D_1(F), \ldots, D_d(F)).$$

b. We say that a function $F \in L^1_{\text{loc}}(X)$ is a *function of locally bounded variation on X* if

$$\forall\, U \subseteq X,\ U \text{ open and } \overline{U} \text{ compact}, \quad F \in BV(U).$$

In this case we write $F \in BV_{\text{loc}}(X)$.

Remark. For any function $F \in BV(\mathbb{R}^d)$, we can consider the *total variation* $|D_i(F)|$ of each associated complex regular Borel measure $D_i(F)$, $i = 1, \ldots, d$. These total variations, $|D_i(F)|$, are bounded regular Borel measures on $\mathbb{R}^d$, with *total variation norms*, $\|D_i(F)\|_1$, defined as

$$\forall\, i = 1, \ldots, d, \quad \|D_i(F)\|_1 = |D_i(F)|(\mathbb{R}^d).$$

Then, for $F \in BV(\mathbb{R}^d)$, we can define the *total variation, $\|\nabla(F)\|_1$, of the vector-valued measure* $\nabla(F)$ so that

$$\|\nabla(F)\|_1 = \sup\left\{\int_{\mathbb{R}^d} F\cdot \operatorname{div}(V)\,dx : V \in \left(C_c^\infty(\mathbb{R}^d)\right)^d, \|V(x)\| \le 1, x\in\mathbb{R}^d\right\}$$
$$< \infty,$$

where div (V) is the *divergence* of the vector field $V = (V_1, \ldots, V_d)$, i.e.,

$$\operatorname{div}(V) = \sum_{i=1}^{d} \frac{\partial V_i}{\partial x_i};$$

see, e.g., [521], Section 5.1. This formulation of total variation is compatible with the functional-analytic equivalence to set-theoretic total variation used in the proof of Theorem 7.2.7 (RRT for complex measures).

It is interesting to observe that the notion of bounded variation from Definition 8.5.1 is equivalent to an older, classical definition of bounded variation as studied by LAMBERTO CESARI [96] and TONELLI [473], among others. To introduce this concept, we shall need the notion of approximate continuity. A measurable function $f : U \to \mathbb{C}$, $U \subseteq \mathbb{R}^d$, is *approximately continuous at* $x \in U$ if there exists $A \in \mathcal{M}(U)$ such that

$$\lim_{r\to 0} \frac{m^d(A\cap B(x,r))}{m^d(B(x,r))} = 1,$$

and f is continuous at x relative to A.

For $\phi : [a,b] \to \mathbb{C}$, define

$$essV(\phi,[a,b]) = \sup\left\{\sum_{j=1}^{n} |\phi(x_j) - \phi(x_{j-1})| : \ a \le x_0 \le \cdots \le x_n \le b\right\},$$

where $x_0, \ldots, x_n$ are points of approximate continuity of ϕ. Given $f : \mathbb{R}^d \to \mathbb{C}$ and $1 \le i \le d$, we write $\phi_i(x_i) = f(x_1, \ldots, x_d)$. Clearly, ϕ_i depends on $\tilde{x} = (x_1, \ldots, x_{i-1}, x_{i+1}, \ldots, x_d)$.

Theorem 8.5.2. Equivalence of definitions of bounded variation
Let $f \in L^1_{\mathrm{loc}}(\mathbb{R}^d)$. Then $f \in BV_{\mathrm{loc}}(\mathbb{R}^d)$ if and only if

$$\forall\, i = 1, \ldots, d,\ \forall\, Q \subseteq \mathbb{R}^{d-1},\ Q \text{ a rectangular cell, and } \forall\, a < b,$$
$$\int_Q essV(\phi_i,[a,b])\,d\tilde{x} < \infty.$$

The study of measures associated with functions of bounded variation plays an important role in geometric measure theory. In particular, they are a useful tool in the generalization of the fundamental theorem of calculus in the spirit of Section 8.1. We refer the interested reader to [163], [342], [159], [5], [303], [521].

Definition 8.5.3. Finite perimeter, reduced boundary, and exterior normal

a. A Borel set $A \subseteq \mathbb{R}^d$ has *finite perimeter in an open set* $U \subseteq \mathbb{R}^d$ if

$$\mathbb{1}_A \in BV(U).$$

A set that has a finite perimeter in $\mathbb{R}^d$ is called a *set of finite perimeter.*

b. A Borel set $A \subseteq \mathbb{R}^d$ has *locally finite perimeter* if

$$\forall\, U \subseteq \mathbb{R}^d,\ U \text{ open and bounded}, \quad \|\nabla(\mathbb{1}_A)\|_1(U) < \infty.$$

c. If $A \subseteq \mathbb{R}^d$ has locally finite perimeter, then the *reduced boundary of* A, $\partial^-(A) \subseteq \mathbb{R}^d$, is a set such that, for all $x \in \partial^-(A)$, the following two conditions are satisfied:

i. $\|\nabla(\mathbb{1}_A)\|_1(B(x,r)) > 0$,

ii. $\exists\, n^-(x,A) = -\lim_{r\to 0} \dfrac{\nabla(\mathbb{1}_A)(B(x,r))}{\|\nabla(\mathbb{1}_A)\|_1(B(x,r))} \in \mathbb{R}^d \quad \text{with } \|n^-(x,A)\| = 1,$

where $B(x,r)$ denotes the Euclidean ball of radius $r > 0$ centered at x.

The vector $n^-(x,A)$ is called the *exterior normal to* A.

The use of the minus sign in $\partial^-(A)$ indicates that the normal vector is directed "outward" from the set A, i.e., in the direction opposite to the gradient of $\mathbb{1}_A$.

Remark. Using the theory that we developed in Section 8.4, it becomes clear that the vector-valued function, defined by

$$\forall\, x \in \mathbb{R}^d, \quad n(x) = \lim_{r\to 0} \frac{\nabla(\mathbb{1}_A)(B(x,r))}{\|\nabla(\mathbb{1}_A)\|_1(B(x,r))},$$

whenever this limit exists, is the Radon–Nikodym derivative of $\nabla(\mathbb{1}_A)$ with respect to $\|\nabla(\mathbb{1}_A)\|_1$. Therefore, for all Borel sets $E \subseteq \mathbb{R}^d$, we have

$$\nabla(\mathbb{1}_A)(E) = \int_E n \, d\|\nabla(\mathbb{1}_A)\|_1.$$

Consequently,

$$\int_A \operatorname{div}(V)\, dx = \int_{\mathbb{R}^d} V \cdot n \, d\|\nabla(\mathbb{1}_A)\|_1,$$

for $V \in (C_c^\infty(\mathbb{R}^d))^d$, which implies that $\|n\| = 1$, $\|\nabla(\mathbb{1}_A)\|_1$-*a.e.* Since

$$\|\nabla(\mathbb{1}_A)\|(\mathbb{R}^d \setminus \partial^-(A)) = 0,$$

we see that $n^-(x,A)$ is a multidimensional analogue of the Radon–Nikodym derivative of $\nabla(\mathbb{1}_A)$ with respect to $\|\nabla(\mathbb{1}_A)\|$.

The next result is a preliminary version of a classical generalization of FTC. For its proof see [163], [521]. We write μ_{d-1} to denote $(d-1)$-dimensional *Hausdorff measure*, which will be formally defined in Section 9.3, but see Problem 2.47. For the moment, we think of it as a restriction of the d-dimensional Lebesgue measure m^d to sets of Hausdorff dimension $d-1$, which is also defined in Section 9.3. With so many undefined terms the following theorem, although precise, is really only an impressionistic FTC driven by a suggestive formula.

Theorem 8.5.4. Generalization of FTC to $\mathbb{R}^d$
Let $A \subseteq \mathbb{R}^d$ be a set of locally finite perimeter and let $f \in C_c^\infty(\mathbb{R}^d)$. Then, for each $x \in \mathbb{R}^d$ and for m-a.e. $r > 0$, we have

$$\forall\, i = 1, \ldots, d, \quad \int_{A \cap B(x,r)} D_i(f)\, dm^d$$
$$= -\int_{B(x,r)} f\, d(D_i(\mathbb{1}_A)) + \int_{A \cap S(x,r)} f(y) n_i^-(y, B(x,r))\, d\mu_{d-1}(y),$$

where $S(x,r)$ denotes the sphere of radius r centered at x, and $n_i^-(y, B(x,r))$ is the ith component of $n^-(y, B(x,r))$.

We shall now generalize the notion of a boundary of a Lebesgue measurable set and of a vector normal to the boundary. This new definition utilizes the notion of a vector being "outside" of a set in the following measure-theoretic sense.

Definition 8.5.5. Measure-theoretic normal and boundary
a. Let $A \in \mathcal{M}(\mathbb{R}^d)$. A unit vector n is a *measure-theoretic normal to A at x* if the following two conditions hold:

$$\lim_{r \to 0} \frac{1}{r^d} m^d \left(B(x,r) \cap \{y : (y-x) \cdot n < 0,\ y \notin A\}\right) = 0$$

and

$$\lim_{r \to 0} \frac{1}{r^d} m^d \left(B(x,r) \cap \{y : (y-x) \cdot n > 0,\ y \in A\}\right) = 0.$$

We write $n(x, A)$ to denote the measure-theoretic normal to A at $x \in \mathbb{R}^d$.
b. The *measure-theoretic boundary of $A \in \mathcal{M}(\mathbb{R}^d)$* is defined to be the set,

$$\partial(A) = \{x \in \mathbb{R}^d : n(x, A) \text{ exists}\}.$$

Definition 8.5.5 generalizes the notion of reduced boundary, since, for sets with locally finite perimeter, we have

$$\partial^-(A) \subseteq \partial(A);$$

see, e.g., Theorem 5.6.5 in [521].

We have introduced all the terms needed to state the following generalization of FTC, referred to classically as the *divergence theorem*.

Theorem 8.5.6. Gauss–Ostrogradsky theorem
Let $A \subseteq \mathbb{R}^d$ be a set with locally finite perimeter. Then

$$\forall\, V \in (C_c^1(\mathbb{R}^d))^d, \quad \int_A \operatorname{div}(V)\, dm^d = \int_{\partial(A)} n(x,A)\cdot V(x)\, d\mu_{d-1}(x).$$

It is not difficult to see that in the case $d = 1$ and $A = [a, b]$, Theorem 8.5.6 reduces to

$$\int_a^b f'(x)\, dx = f(b) - f(a),$$

see the Remark at the end of Section 4.5 and Example 8.1.1*b*.

8.6 Rearrangements and the maximal function theorem

One of the goals of this section is to prove the Hardy–Littlewood maximal theorem for $L^p(\mathbb{R}^d)$ (Theorem 8.6.14). Our presentation is due to the virtuosic HERZ [232], who emphasized Hardy–Littlewood rearrangement functions rather than so-called strong- or weak-type L^p-estimates. One of HERZ' points of view was that in most cases there are stronger, explicit inequalities involving rearrangements than exist for weak-type estimates. For us it is an opportunity to develop some of the technology of rearrangements, which gives a penetrating measure-theoretic insight and has many applications.

In Section 7.6.5 we defined the distribution function of a probability measure. We now modify this definition for the setting at hand, but the *concept* is the same even though one function is increasing and the other is decreasing.

Definition 8.6.1. Distribution function and decreasing rearrangements

a. Let $(X, \mathcal{A}, \mu)$ be a σ-finite measure space and let $f : X \to \mathbb{C}$ be μ-measurable. The *distribution function* $D_f : \mathbb{R}^+ \to \mathbb{R}^+$ of f is defined by

$$\forall\, s \in \mathbb{R}^+, \quad D_f(s) = \mu(\{x \in X : |f(x)| > s\}).$$

The *decreasing rearrangement* $f^* : \mathbb{R}^+ \to \mathbb{R}^+$ is defined by

$$\forall\, t \in R^+, \quad f^*(t) = \inf\{s \in \mathbb{R}^+ : D_f(s) \le t\}.$$

We use the convention $\inf \emptyset = \infty$, so that if $D_f(s) > t$ for all $s \in \mathbb{R}^+$ then $f^*(t) = \infty$.

b. Consider the Lebesgue measure space $(\mathbb{R}^+, \mathcal{M}(\mathbb{R}^+), m)$ and the setting of part *a*. Then f and f^* are *equimeasurable*, that is, $D_f = D_{f^*}$ on $\mathbb{R}^+$. Further,

$$D_{f^*}(s) = \sup\{t \in \mathbb{R}^+ : f^*(t) > s\}.$$

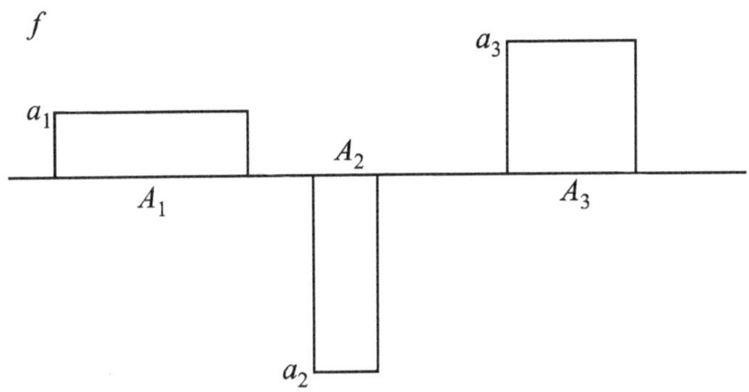

Fig. 8.1. A function f.

Example 8.6.2. Computation of D_f and f^*

a. Let $f = a_1 \mathbb{1}_{A_1} + a_2 \mathbb{1}_{A_2} + a_3 \mathbb{1}_{A_3}$ be defined on $(\mathbb{R}, \mathcal{M}(\mathbb{R}), m)$, where $A_1, A_2, A_3 \in \mathcal{M}(\mathbb{R})$ and $a_1 < a_3 < |a_2|$; see Figure 8.1. A direct application of the definitions of D_f and f^* gives the graphs of D_f and f^* illustrated in Figure 8.2.

b. Let $(X, \mathcal{A}, \mu)$ be a σ-finite measure space and let $f : X \to \mathbb{C}$ be μ-measurable. If D_f is a continuous and strictly decreasing surjection $D_f : [0, \infty) \to [0, \infty)$, then $f^* = D_f^{-1}$.

c. Let $f : \mathbb{R}^d \to \mathbb{C}$ be *radial*, i.e., $f(x) = f(\|x\|)$, $\|x\| = (x_1^2 + \cdots + x_d^2)^{1/2}$. If f takes values in $\mathbb{R}^+$ and $f(\|x\|)$ is strictly decreasing on $[0, \infty)$, then

$$D_f(s) = m^d(B(0, f^{-1}(s))),$$

where $B(0, f^{-1}(s)) \subseteq \mathbb{R}^d$ is the Euclidean ball of radius $f^{-1}(s)$ centered at the origin; see Example 2.3.11 for the volume of the unit ball $B_d = B(0,1) \subseteq \mathbb{R}^d$.

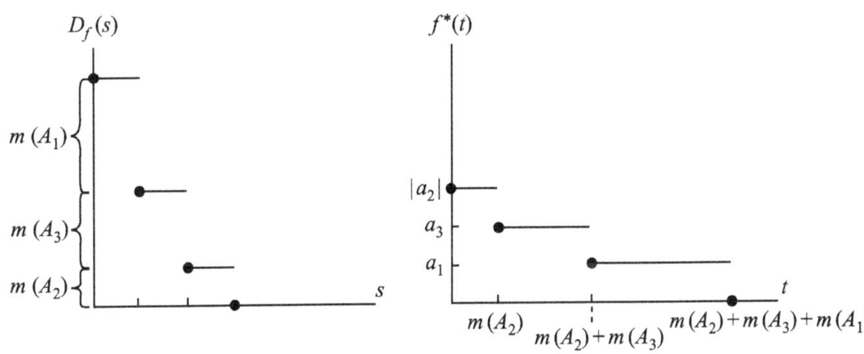

Fig. 8.2. D_f and f^*.

Theorem 8.6.3 collects some basic properties of D_f and f^*. We are omitting the proof, not because details are not required, but because verification follows from the definitions of D_f and f^*. Some basic references dealing with this result as well as with finer points about D_f and f^* and their role in analysis are [524], [450], [220], [50].

Theorem 8.6.3. Elementary properties of D_f and f^*
Let $(X, \mathcal{A}, \mu)$ be a σ-finite measure space, and assume that all the functions $X \to \mathbb{C}$ in the following statements are μ-measurable.

a. *D_f and f^* are nonnegative, decreasing, right-continuous functions on $\mathbb{R}^+$.*

b. *If $|g| \leq |f|$ μ-a.e., then $D_g \leq D_f$ and $g^* \leq f^*$ on $\mathbb{R}^+$.*

c. *If $c \in \mathbb{C} \setminus \{0\}$, then $D_{cf}(s) = D_f(s/|c|)$ and $(cf)^*(t) = |c|f^*(t)$, $s, t \in \mathbb{R}^+$.*

d. *$D_{f+g}(s_1+s_2) \leq D_f(s_1)+D_g(s_2)$ and $(f+g)^*(t_1+t_2) \leq f^*(t_1)+g^*(t_2)$, $s_j, t_j \in \mathbb{R}^+$.*

e. *If $\{|f_n|\}$ increases to f μ-a.e., then $\{D_{f_n}\}$ increases to D_f and $\{f_n^*\}$ increases to f^*.*

We sketch the proof of the following elementary and useful result for $0 < p < \infty$. Note that $\| \dots \|_p$, $0 < p < 1$, is not a norm as defined in Definition A.1.9.

Theorem 8.6.4. $\|f\|_p$ in terms of D_f and f^*
Let $(X, \mathcal{A}, \mu)$ be a σ-finite measure space, and let $f : X \to \mathbb{C}$ be μ-measurable. If $0 < p < \infty$, then

$$\|f\|_p = p \int_0^\infty s^{p-1} D_f(s)\, ds = \int_0^\infty (f^*)^p\, dm, \tag{8.33}$$

and, if $p = \infty$, then

$$\|f\|_\infty = \inf\{s \in \mathbb{R}^+ : D_f(s) = 0\} = f^*(0).$$

Proof. By the Levi–Lebesgue theorem (Theorem 3.3.6) and $\underline{\lim}$ calculations associated with the verification of Theorem 8.6.3*e*, it is sufficient to prove (8.33) for nonnegative simple functions $f = \sum_{j=1}^n a_j \mathbb{1}_{A_j}$, where $a_1 > a_2 > \cdots > a_n > 0$, the sequence $\{A_j\}$ is pairwise disjoint, and each $\mu(A_j) < \infty$. The result follows from the following calculations:

$$\|f\|_p^p = \sum_{j=1}^n (a_j^p - a_{j+1}^p)\mu_j = p \sum_{j=1}^n \mu_j \int_{a_{j+1}}^{a_j} s^{p-1}\, ds = p \int_0^\infty s^{p-1} D_f(s)\, ds$$

and

$$\|f\|_p^p = \sum_{j=1}^n a_j^p \mu(A_j) = \sum_{j=1}^n a_j^p (\mu_j - \mu_{j-1}) = \int_0^\infty (f^*)^p\, dm,$$

where $\mu_j = \sum_{k=1}^j \mu(A_k)$ and $\mu_0 = 0$. □

We could continue in the setting of relatively general measure spaces for some of the following results, but, since our goal here is analysis on $\mathbb{R}^d$, we shall deal with measure spaces $(\mathbb{R}^d, \mathcal{A}, \mu)$ for the remainder of this section.

Decreasing rearrangements D_f defined in Definition 8.6.1 are not additive. In fact, they are not even subadditive; cf. Theorem 8.6.3*d*. We can correct this problem in Theorem 8.6.7 by introducing the notion of *averaged decreasing rearrangement* f^{**} defined by

$$\forall\, t > 0, \quad f^{**}(t) = \frac{1}{t}\int_0^t f^*(u)\, du.$$

Remark. *a*. The averaged decreasing rearrangement has properties analogous to those stated in Theorem 8.6.3. Assume that all functions in this Remark are μ-measurable functions defined on the measure space $(\mathbb{R}^d, \mathcal{A}, \mu)$. The following properties are more or less clear:

i. $f^{**} \geq 0$ is either everywhere finite or everywhere infinite on $(0, \infty)$, since f^* is decreasing;

ii. $f^{**} = 0$ on $(0, \infty) \iff f = 0$ μ-*a.e.*;

iii. f^{**} is continuous on $(0, \infty)$.

The following properties follow as in Theorem 8.6.3:

i. $|g| \leq |f|$ μ-*a.e.* $\iff g^{**} \leq f^{**}$ on $(0, \infty)$;

ii. $(cf)^{**} = |c| f^{**}$ on $(0, \infty)$ for $c \in \mathbb{C}$;

iii. $\{|f_n|\}$ increases to f μ-*a.e.* $\implies \{f_n^{**}\}$ increases to f^{**} on $(0, \infty)$.

***b*.** Note that $f^* \leq f^{**}$ on $(0, \infty)$ since f^* is decreasing. In fact,

$$f^*(t) = f^*(t)\frac{1}{t}\int_0^t ds \leq \frac{1}{t}\int_0^t f^*(s)\, ds = f^{**}(t).$$

Further, f^{**} is decreasing on $(0, \infty)$ since f^* is decreasing. In fact, if $0 < t \leq s$, then

$$f^{**}(s) \leq \frac{1}{s}\int_0^s f^*\left(\frac{tu}{s}\right)\, du = f^{**}(t).$$

We shall use the following result to prove Theorem 8.6.7.

Theorem 8.6.5. f^{**} as a minimizer

Let $(\mathbb{R}^d, \mathcal{A}, \mu)$ be a measure space, and let f be a nonnegative μ-measurable function on $\mathbb{R}^d$ with σ-finite support. We define $\mathcal{R}(f)$ to be the set of all nonnegative functions ϕ on $(0, \infty)$ with the properties that

i. $t\phi(t)$ is concave, and

ii. $\forall\, A \in \mathcal{A}$, such that $0 < \mu(A) < \infty$,

$$\frac{1}{\mu(A)}\int_A f\, d\mu \leq \phi(\mu(A)).$$

Then f^{**} is a minimal element of $\mathcal{R}(f)$.

Proof. First, we note that, since f^* is decreasing, the function $tf^{**}(t)$ is concave. Indeed, since for any $t_1 < t_2$,

$$\int_{(t_1+t_2)/2}^{t_2} f^*(u)\, du \le \int_{t_1}^{(t_1+t_2)/2} f^*(u)\, du,$$

it follows that

$$\frac{1}{2}\left(\int_0^{t_1} f^*(u)\, du + \int_0^{t_2} f^*(u)\, du\right) \le \int_0^{(t_1+t_2)/2} f^*(u)\, du.$$

This inequality implies, in turn, the concavity of $tf^{**}(t) = \int_0^t f^*(u)\, du$.

Now consider a simple function $f = \sum_{j=1}^n a_j \mathbb{1}_{A_j}$, with $a_1 > \cdots > a_n > 0$, and with pairwise disjoint sets $A_j \in \mathcal{A}$, $\mu(A_j) < \infty$, $j = 1, \ldots, n$. Let $A^k = \bigcup_{j=1}^k A_j$ and let $\alpha_k = \mu(A^k)$. Moreover, let $\alpha_0 = 0$ and $\alpha_{n+1} = \infty$. If $t \in [\alpha_k, \alpha_{k+1})$ and $A \in \mathcal{A}$ satisfy $\mu(A) = t$, then

$$\int_A f\, d\mu \le \sum_{j=1}^{k} a_j(\alpha_j - \alpha_{j-1}) + a_{j+1}(t - \alpha_j) = tf^{**}(t),$$

and so $f^{**} \in \mathcal{R}(f)$. Also,

$$\int_0^{\alpha_k} f^*(u)\, du = \int_{A^k} f\, d\mu.$$

Therefore, for any $\phi \in \mathcal{R}(f)$, $\alpha_k \phi(\alpha_k) \ge \alpha_k f^{**}(\alpha_k)$. Since $t\phi(t)$ is concave and greater than or equal to $tf^{**}(t)$ at all points $t = \alpha_k$, we have $t\phi(t) \ge tf^{**}(t)$ for all $t \in (0, \infty)$; this follows by the linearity of $tf^{**}(t)$ on each interval (α_k, α_{k+1}). A consequence of this inequality is that if f is a simple function, then f is a minimal element of $\mathcal{R}(f)$.

To prove the assertion of the theorem for arbitrary functions f, we consider a sequence $\{f_n : n = 1, \ldots\}$ of simple functions such that $f_n \le f_{n+1}$ and $f = \lim_{n\to\infty} f_n$ μ-*a.e.* From Theorem 8.6.3*e* and from the Levi–Lebesgue theorem (Theorem 3.3.6) it follows that

$$\forall\, t > 0, \quad f^{**}(t) = \lim_{n\to\infty} f_n^{**}(t).$$

Thus, for any $A \in \mathcal{A}$, the result for simple functions combined with Theorem 3.3.6 implies that

$$\frac{1}{\mu(A)} \int_A f\, d\mu = \lim_{n\to\infty} \frac{1}{\mu(A)} \int_A f_n\, d\mu \le \lim_{n\to\infty} f_n^{**}(\mu(A)) = f^{**}(\mu(A)),$$

and so $f^{**} \in \mathcal{R}(f)$. Since $f_n \le f$, we have $\mathcal{R}(f) \subseteq \mathcal{R}(f_n)$. Thus, for any $\phi \in \mathcal{R}(f)$, we obtain $\phi \ge f_n^{**}$ for all $n = 1, \ldots,$ which, in turn, implies $\phi \ge f^{**}$. □

The following corollary of Theorem 8.6.5 is used in the proof of Theorem 8.6.10. Its proof depends on the fact that f^{**} is decreasing.

Corollary 8.6.6. *Let $(\mathbb{R}^d, \mathcal{A}, \mu)$ be a measure space and let f be a nonnegative μ-measurable function on $\mathbb{R}^d$ with σ-finite support. Then*

$$\forall\, t > 0, \quad \sup\left\{\frac{1}{\mu(A)}\int_A f\, d\mu : A \in \mathcal{A}, \mu(A) \geq t\right\} \leq f^{**}(t).$$

We are now ready to verify that the averaged decreasing rearrangements are subadditive.

Theorem 8.6.7. Subadditivity of f^{**}
Let $(\mathbb{R}^d, \mathcal{A}, \mu)$ be a measure space and let f, g be nonnegative μ-measurable functions on $\mathbb{R}^d$ with σ-finite supports. Then

$$\forall\, t > 0, \quad (f+g)^{**}(t) \leq f^{**}(t) + g^{**}(t).$$

Proof. Let $\phi = f^{**} + g^{**}$. The function $t\phi(t)$ is concave as a sum of two concave functions. Moreover, if $A \in \mathcal{A}$ satisfies $0 < \mu(A) < \infty$, then

$$\begin{aligned}\frac{1}{\mu(A)}\int_A (f+g)\, d\mu &= \frac{1}{\mu(A)}\int_A f\, d\mu + \frac{1}{\mu(A)}\int_A g\, d\mu \\ &\leq f^{**}(\mu(A)) + g^{**}(\mu(A)).\end{aligned}$$

Thus, $\phi \in \mathcal{R}(f+g)$, and we use Theorem 8.6.5 to conclude that

$$(f+g)^{**} \leq f^{**} + g^{**}.$$ □

Let $(X, \mathcal{A}, \mu)$ be a measure space, let $1 \leq p < \infty$, and let $f \in L^p_\mu(\mathbb{R}^d)$. Since f^* is decreasing, we have $f^* \leq f^{**}$. Thus, by Theorem 8.6.4, we obtain the following inequality:

$$\|f\|_p \leq \left(\int_0^\infty f^{**p}(t)\, dt\right)^{1/p}. \tag{8.34}$$

It is interesting that the opposite inequality holds for some constant strictly greater than 1.

Theorem 8.6.8. Hardy inequality
Let $(\mathbb{R}^d, \mathcal{A}, \mu)$ be a measure space, and let f be a nonnegative μ-measurable function on $\mathbb{R}^d$ with σ-finite support. Let $1 < p < \infty$. Then

$$\left(\int_0^\infty f^{**p}(t)\, dt\right)^{1/p} \leq \frac{p}{p-1}\|f\|_p \tag{8.35}$$

and

$$\sup_{t>0} f^{**}(t) \leq \|f\|_\infty. \tag{8.36}$$

Proof. i. We first prove (8.35). For this purpose, we consider the multiplicative group $(0,\infty)$ equipped with the Haar measure dt/t; see Appendix B.9. We can define convolution on this group by the formula

$$g * h(t) = \int_0^\infty g(ts^{-1})h(s)\,\frac{ds}{s} = \int_0^\infty g(s)h(s^{-1}t)\,\frac{ds}{s}.$$

Further, if $1 \le p < \infty$, then the multiplicative L^p-norm of g is

$$\|g\|_{p,\times} = \int_0^\infty |g(t)|^p\,\frac{dt}{t}.$$

Using the integral Minkowski inequality (Section 5.6.5) and the properties of Haar measure, we obtain

$$\begin{aligned}\|g*h\|_{p,\times} &= \left(\int_0^\infty \left|\int_0^\infty g(s)h(s^{-1}t)\,\frac{ds}{s}\right|^p \frac{dt}{t}\right)^{1/p}\\ &\le \int_0^\infty \left(\int_0^\infty |g(s)|^p|h(s^{-1}t)|^p\,\frac{dt}{t}\right)^{1/p}\frac{ds}{s}\\ &= \int_0^\infty |g(s)|\,\frac{ds}{s}\left(\int_0^\infty |h(t)|^p\,\frac{dt}{t}\right)^{1/p} = \|g\|_{1,\times}\|h\|_{p,\times}.\end{aligned} \tag{8.37}$$

Let $g(t) = t^{(1-p)/p}$ for $t \ge 1$ and let $g(t) = 0$ for $t < 1$, and let $h(t) = t^{1/p}f^*(t)$. Thus,

$$\begin{aligned}g*h(t) &= \int_0^t t^{(1-p)/p}s^{(p-1)/p}f^*(s)s^{1/p}\,\frac{ds}{s}\\ &= t^{(1-p)/p}\int_0^t f^*(s)\,ds = t^{1/p}f^{**}(t).\end{aligned} \tag{8.38}$$

We now have

$$\left(\int_0^\infty f^{**p}(t)\,dt\right)^{1/p} = \|g*h\|_{p,\times} \le \|g\|_{1,\times}\|h\|_{p,\times} = \frac{p}{p-1}\|f\|_p,$$

where the first equality is a consequence of (8.38), the inequality follows from (8.37), and the second equality is a consequence of Theorem 8.6.4.

ii. Inequality (8.36) follows immediately from the definition of f^{**}. □

We do not have an analogue of the Hardy inequality in case $p = 1$. However, we do note that $\|f\|_1 = \sup_{t>0} tf^{**}(t)$, and that there is a smaller norm, $\|\ldots\|_*$, which agrees with $\|\ldots\|_1$ on spaces of measure not greater than 1. In fact, we define

$$\|f\|_* = f^{**}(1) = \int_0^1 f^*(s)\,ds. \tag{8.39}$$

We define $L_*(\mathbb{R}^d)$ to be the space of equivalence classes of μ-measurable functions f on $\mathbb{R}^d$ with σ-finite support that satisfy the following two properties:

$$\|f\|_* < \infty$$

and

$$\lim_{t\to\infty} f^*(t) = 0.$$

Proposition 8.6.9. *The space $L_*(\mathbb{R}^d)$ is a Banach space with norm $\|\ldots\|_*$, and the set of simple functions with σ-finite support is a dense subset of $L_*(\mathbb{R}^d)$.*

We set $\|f\|_{p,\infty} = \sup_{t>0} t^{1/p} f^{**}(t)$, $1 \le p < \infty$. It is straightforward to verify

$$\|f\|_* \le \|f\|_{p,\infty}.$$

On the other hand, using the Hölder inequality, we obtain

$$\int_0^t f^*(s)\, ds \le \left(\int_0^t f^{*p}(s)\, ds\right)^{1/p} \left(\int_0^t 1\, ds\right)^{(p-1)/p} \le t^{(p-1)/p}\|f\|_p. \quad (8.40)$$

We can rewrite (8.40) as

$$\|f\|_{p,\infty} \le \|f\|_p. \quad (8.41)$$

In particular, $L^p_{m^d}(\mathbb{R}^d) \subseteq L_*(\mathbb{R}^d)$ for each $1 \le p < \infty$.

Theorem 8.6.10. Semimartingale maximal theorem
Let $(\mathbb{R}^d, \mathcal{A}, \mu)$ be a measure space and let f be a nonnegative μ-measurable function with σ-finite support. Let $\{\mathcal{A}_n : n \in \mathbb{Z}\}$ be an increasing family of subalgebras of $\mathcal{A}$, and let $\{f_n : n \in \mathbb{Z}\}$ be a sequence of nonnegative functions with σ-finite supports such that each f_n is μ-measurable with respect to $\mathcal{A}_n$. Assume that

$$\forall\, A \in \mathcal{A}_n \text{ and } \forall\, n \in \mathbb{Z}, \quad \int_A f_n\, d\mu \le \int_A f\, d\mu.$$

Define

$$\tilde{f}(x) = \sup_{n\in\mathbb{Z}} f_n(x). \quad (8.42)$$

Then

$$\forall\, t > 0, \quad \tilde{f}^*(t) \le f^{**}(t). \quad (8.43)$$

Proof. We can write

$$\tilde{f} = \lim_{n\to-\infty} \sup_{k>n} f_k.$$

In order to prove (8.43) we observe, by Theorem 8.6.3e and the fact that $\{\sup_{k>n} f_k\}$ increases to $\tilde{f}$, that it suffices to show that

$$\forall\, n \in \mathbb{Z} \text{ and } \forall\, t > 0, \quad \left(\sup_{k>n} f_k\right)^*(t) \le f^{**}(t). \tag{8.44}$$

To verify (8.44) we fix $n \in \mathbb{Z}$. Take $t > 0$ and consider $0 \le \lambda < (\sup_{k>n} f_k)^*(t)$. For each $k > n+1$ we define the set

$$A_{k,\lambda} = \{x : f_k(x) > \lambda \text{ and } f_l(x) \le \lambda \text{ for } n < l < k\}.$$

We also let $A_{n+1,\lambda} = \{x : f_{n+1}(x) > \lambda\}$. Then $A_{k,\lambda} \in \mathcal{A}_k$ and, for $k \ne l$, $A_{k,\lambda} \cap A_{l,\lambda} = \emptyset$. Let $A_\lambda = \bigcup_{k\in\mathbb{Z}} A_{k,\lambda}$. It is not difficult to see that

$$A_\lambda = \{x : \left(\sup_{k>n} f_k\right)(x) > \lambda\},$$

and we deduce that $\mu(A_\lambda) > t$. Thus, from Corollary 8.6.6, we have

$$f^{**}(t) \ge \frac{1}{\mu(A_\lambda)} \int_{A_\lambda} f\, d\mu. \tag{8.45}$$

On the other hand,

$$\int_{A_\lambda} f\, d\mu = \sum_{k\in\mathbb{Z}} \int_{A_{k,\lambda}} f\, d\mu \ge \sum_{k\in\mathbb{Z}} \int_{A_{k,\lambda}} f_k\, d\mu \ge \sum_{k\in\mathbb{Z}} \lambda\mu(A_{k,\lambda}) = \lambda\mu(A_\lambda),$$

or, equivalently,

$$\frac{1}{\mu(A_\lambda)} \int_{A_\lambda} f\, d\mu \ge \lambda. \tag{8.46}$$

Combining (8.45) and (8.46), we obtain

$$\forall\, 0 \le \lambda < \left(\sup_{k>n} f_k\right)^*(t), \quad f^{**}(t) \ge \lambda,$$

which yields (8.44). □

Definition 8.6.11. Dyadic cubes in $\mathbb{R}^d$
A *dyadic cube* in $\mathbb{R}^d$ of side 2^{-k}, $k \in \mathbb{Z}$, is a set of the form

$$\{x \in \mathbb{R}^d : 2^{-k} l_i < x_i \le 2^{-k}(l_i + 1), i = 1, \dots, d\}$$

for $l = (l_1, \dots, l_d) \in \mathbb{Z}^d$. The family of dyadic cubes of side 2^{-k} forms a partition of $\mathbb{R}^d$. See the Remark after Theorem 2.5.5.

Let $\mathcal{A}_k$ be the collection of all the unions (necessarily countable) of dyadic cubes of side 2^{-k}. For each $k \in \mathbb{Z}$, $\mathcal{A}_k$ is a σ-algebra. Moreover, $\mathcal{A}_k \subseteq \mathcal{A}_{k+1}$. For a nonnegative function $f \in L^1_{\text{loc}}(\mathbb{R}^d)$ and for $n \in \mathbb{Z}$ we define

$$f_n(x) = \frac{1}{m^d(Q)} \int_Q f\, dm^d, \tag{8.47}$$

where Q is the unique dyadic cube of side 2^{-n} that contains x. Such functions f_n satisfy the property,

$$\forall\, A \in \mathcal{A}_n,\ m^d(A) < \infty, \quad \int_A f_n\, dm^d = \int_A f\, dm^d.$$

Proposition 8.6.12. *Let $(\mathbb{R}^d, \mathcal{M}(\mathbb{R}^d), m^d)$ be the Lebesgue measure space, and let $f \in L^1_{\mathrm{loc}}(\mathbb{R}^d)$. Let the sequence $\{f_n : n \in \mathbb{Z}\}$ be defined by (8.47), and let $\tilde{f}$ be defined as in (8.42). Then*

$$\forall\, t > 0, \quad (M(f))^*(t) \le 4^d \tilde{f}^*(2^{-d}t).$$

Proof. Let $A_\lambda = \{x \in \mathbb{R}^d : M(f)(x) > \lambda\}$. For each $x \in A_\lambda$ there exists a cube $Q(x)$ (not necessarily dyadic) centered at x with the property that

$$\int_{Q(x)} f\, dm^d > \lambda m^d(Q(x)).$$

If τ is the side of $Q(x)$ then there are 2^d dyadic cubes of side s, $\tau < s \le 2\tau$, that cover $Q(x)$. Among these cubes there must exist at least one cube Q with the following properties:

$$\int_Q f\, dm^d \ge \frac{1}{2^d} \int_{Q(x)} f\, dm^d > \frac{1}{2^d}\lambda m^d(Q(x)) > \frac{1}{4^d}\lambda m^d(Q)$$

and

$$Q(x) \cap Q \neq \emptyset.$$

The second condition implies that $x \in Q^*$, where Q^* denotes a cube with the same center as Q and twice the side. If we let

$$B_\lambda = \bigcup \left\{ Q : Q \text{ is a dyadic cube and } \int_Q f\, dm^d \ge 4^{-d}\lambda m^d(Q) \right\},$$

then $A_\lambda \subseteq B^*_\lambda$, and so

$$m^d(A_\lambda) \le 2^d m^d(B_\lambda).$$

On the other hand, if $x \in B_\lambda$ then $\tilde{f}(x) > 4^{-d}\lambda$. Thus, for any $\lambda < (M(f))^*(t)$ we must have $m^d(A_\lambda) > t$, which implies $m^d(B_\lambda) > 2^{-d}t$. This, in turn, implies that

$$\tilde{f}^*(2^{-d}t) > 4^{-d}\lambda.$$

These latter assertions are elementary to verify, but do require writing out the definitions of decreasing rearrangement and distribution functions several times. □

Theorem 8.6.13. Maximal function theorem
Let $(\mathbb{R}^d, \mathcal{M}(\mathbb{R}^d), m^d)$ be the Lebesgue measure space, and let $f \in L^1_{\mathrm{loc}}(\mathbb{R}^d)$. Then

$$\forall\, t > 0, \quad (M(f))^*(t) \le 4^d f^{**}(2^{-d}t).$$

Proof. Using Proposition 8.6.12 and Theorem 8.6.10 for the family of algebras $\mathcal{A}_k$ of unions of dyadic cubes of side 2^k, $k \in \mathbb{Z}$, and functions f_k defined by (8.47), we obtain

$$(M(f))^*(t) \le 4^d \tilde{f}^*(2^{-d}t) \le 4^d f^{**}(2^{-d}t). \qquad \square$$

Theorem 8.6.14. Hardy–Littlewood maximal theorem for $L^p(\mathbb{R}^d)$
Let $(\mathbb{R}^d, \mathcal{M}(\mathbb{R}^d), m^d)$ be the Lebesgue measure space, let $1 < p \le \infty$, and let $f \in L^p(\mathbb{R}^d)$. Then $M(f) \in L^p(\mathbb{R}^d)$,

$$\forall\, 1 < p < \infty, \quad \|M(f)\|_p \le 8^d \frac{p}{p-1} \|f\|_p, \tag{8.48}$$

and

$$\|M(f)\|_\infty \le 4^d \|f\|_\infty.$$

Proof. For $1 < p < \infty$, we use Theorem 8.6.13 and Theorem 8.6.8 to write

$$\begin{aligned}\|M(f)\|_p &= \left(\int_0^\infty M(f)^{*p}(t)\, dt \right)^{1/p} \le 4^d \left(\int_0^\infty f^{**p}(2^{-d}t)\, dt \right)^{1/p} \\ &= 4^d 2^{d/p} \left(\int_0^\infty f^{**p}(t)\, dt \right)^{1/p} \le 4^d 2^{d/p} \frac{p}{p-1} \|f\|_p.\end{aligned}$$

For $p = \infty$, we use Theorem 8.6.13 to obtain

$$\begin{aligned}\|M(f)\|_\infty &= \sup_{t>0} (M(f))^*(t) \le 4^d \sup_{t>0} f^{**}(2^{-d}t) = 4^d \sup_{t>0} f^{**}(t) \\ &= 4^d \sup_{t>0} \frac{1}{t} \int_0^t f^*(s)\, ds \le 4^d \sup_{t>0} \sup_{t \ge s > 0} f^*(s) \\ &= 4^d \sup_{t>0} f^*(t) = 4^d \|f\|_\infty. \qquad \square\end{aligned}$$

As a consequence of Theorem 8.6.14 we have the following classical *Hardy inequality: If $f \in L^p_m((0,\infty))$, $1 < p < \infty$, and $F(X) = \frac{1}{x}\int_0^x f$, then*

$$\|f\|_{L^p((0,\infty))} \le \frac{p}{p-1} \|f\|_{L^p((0,\infty))};$$

and thus the averaging mapping $f \mapsto F$ defines a continuous linear function, $L : L^p((0,\infty)) \to L^p((0,\infty))$.

The maximal theorem stated in the form of Theorem 8.6.13 allows us to write the following converse result. For the details of its proof see [232].

Theorem 8.6.15. Converse maximal function theorem
Let $(\mathbb{R}^d, \mathcal{M}(\mathbb{R}^d), m^d)$ be the Lebesgue measure space, and let $f \in L^1_{\mathrm{loc}}(\mathbb{R}^d)$. Then

$$\forall\, t > 0, \quad \frac{1}{2^d + 4^d} f^{**}(t) \le (M(f))^*(t).$$

8.7 Change of variables and surface measure

In this section we shall omit or sketch the proofs of the more familiar change of variables formulas treated in advanced calculus or undergraduate analysis courses; see [7], [179], [235], [407], [426], [445], [459].

The most elementary change of variables theorem from the calculus is the following; see Problem 4.26 for this result in the context of $L^1_m([a,b])$.

Theorem 8.7.1. Change of variables on $[a,b]$
Let $f : [a,b] \to \mathbb{C}$ be continuous and let $\phi : [c,d] \to [a,b]$ be continuously differentiable. If $\phi(c) = a$ and $\phi(d) = b$ then

$$\int_a^b f(x)\,dx = \int_c^d f(\phi(t))\phi'(t)\,dt. \tag{8.49}$$

It is a natural question to ask about generalizations of Theorem 8.7.1 in the context of $\mathbb{R}^d$. First, we consider the special case that $\phi = A : \mathbb{R}^d \to \mathbb{R}^d$ is an invertible linear transformation. The set of invertible linear transformations on $\mathbb{R}^d$ is called the *general linear group.* It is identified with the set of $d \times d$ invertible matrices with real entries, it is a group under matrix multiplication, and it is denoted by $\mathrm{GL}(d,\mathbb{R})$. The determinant of $A \in \mathrm{GL}(d,\mathbb{R})$ is denoted by $\det(A)$.

Proposition 8.7.2. *Let $\phi = A \in \mathrm{GL}(d,\mathbb{R})$ and let $f \in L^1_{m^d}(\mathbb{R}^d)$. Then $f \circ \phi \in L^1_{m^d}(\mathbb{R}^d)$ and*

$$\int_{\mathbb{R}^d} f\,dm^d = |\det(A)| \int_{\mathbb{R}^d} f \circ \phi\,dm^d. \tag{8.50}$$

Proof. Let Φ denote the set of those $A \in \mathrm{GL}(d,\mathbb{R})$ for which (8.50) is satisfied for all $f \in L^1_{m^d}(\mathbb{R}^d)$. We start by noting that Φ is nonempty. Indeed, the identity matrix satisfies (8.50). Next, we observe that because of the properties of the determinant, Φ is closed under matrix multiplication and inversion. This implies that Φ is a subgroup of $\mathrm{GL}(d,\mathbb{R})$. Thus, to complete the proof it is enough to show that Φ contains the generators of $\mathrm{GL}(d,\mathbb{R})$: row additions, row multiplications, and row-exchange transformations.

The result for row additions is a consequence of Fubini's theorem. To show that (8.50) holds for row multiplications and row-exchange transformations, it is sufficient to prove the result for characteristic functions of arbitrary rectangles, and then to use this fact to prove it for all integrable functions. □

Proposition 8.7.2 can be extended to hold for smooth but nonlinear transformations. This result, Theorem 8.7.3, requires the following notation for smooth transformations $\phi : \mathbb{R}^d \to \mathbb{R}^d$:

$$D(\phi) = (\partial_i \phi_j)_{i,j=1}^d,$$

where $\phi = (\phi_1, \ldots, \phi_d)$ and each $\phi_j : \mathbb{R}^d \to \mathbb{R}$. $D(\phi)$ is the *Jacobian matrix* of ϕ.

Theorem 8.7.3. Change of variables on $\mathbb{R}^d$
Let $U \subseteq \mathbb{R}^d$ be open and let $\phi : U \to \mathbb{R}^d$ be an injective continuously differentiable transformation. Let $f \in L^1_{m^d}(\mathbb{R}^d)$. Then $f \circ \phi \cdot |\det D(\phi)| \in L^1_{m^d}(\mathbb{R}^d)$ and

$$\int_{\phi(U)} f \, dm^d = \int_U f \circ \phi \cdot |\det D(\phi)| \, dm^d. \tag{8.51}$$

The following result can be viewed as a special case of the Lebesgue differentiation theorem, and it gives insight into the Jacobian matrix.

Proposition 8.7.4. *Let $\phi : \mathbb{R}^d \to \mathbb{R}^d$ be continuously differentiable at some point $x_0 \in \mathbb{R}^d$. Then*

$$\lim_{r \to 0} \frac{m^d(\phi(B(x_0, r)))}{m^d(B(x_0, r))} = |\det D(\phi)(x_0)|.$$

Proof. We consider three cases. The first case is illustrative and it assumes that $D(\phi)$ is the identity matrix. In this case, we can show that for each $\varepsilon \in (0,1)$ there exists $r > 0$, sufficiently small, such that

$$B(\phi(x_0), (1-\varepsilon)r) \subseteq \phi(B(x_0, r)) \subseteq B(\phi(x_0), (1+\varepsilon)r).$$

This implies that

$$\frac{m^d(\phi(B(x_0, r)))}{m^d(B(x_0, r))} \to 1 = |\det(Id)|, \quad r \to 0.$$

For the next case we assume that $D(\phi)(x_0)$ is invertible and we consider the mapping $D(\phi)(x_0)^{-1} \circ \phi$, which has the property that $D(D(\phi)(x_0)^{-1} \circ \phi)$ is the identity matrix. Thus,

$$\begin{aligned} \frac{m^d(\phi(B(x_0, r)))}{m^d(B(x_0, r))} &= |\det(D(\phi)(x_0))| \frac{m^d((D(\phi)(x_0)^{-1} \circ \phi)(B(x_0, r)))}{m^d(B(x_0, r))} \\ &\to |\det(D(\phi)(x_0))|, \quad r \to 0. \end{aligned}$$

For the last case, we let $D(\phi)(x_0)$ be singular, and then change the coordinate system in $\mathbb{R}^d$ so that

$$\frac{m^d(\phi(B(x_0, r)))}{m^d(B(x_0, r))} \to 0, \quad r \to 0.$$

□

Next, we consider an extension of Proposition 8.7.2 to the case that the integrand is a product of two functions, only one of which is subject to a change of variables. This is the setting for the duality arising in distribution theory.

Theorem 8.7.5. Change of variables on $\mathbb{R}^d$ for two functions
Let $f \in L^1_{m^d}(\mathbb{R}^d)$ and let $g \in L^\infty_{m^d}(\mathbb{R}^d)$. Let $\phi = A \in \mathrm{GL}(d,\mathbb{R})$. Then $(\det(A))^{-1} = \det(A^{-1}) \neq 0$ and

$$(f \circ A)(g) = \int_{\mathbb{R}^d} (f \circ \phi)\, g\, dm^d = |\det(A^{-1})| \int_{\mathbb{R}^d} f\left(g \circ \phi^{-1}\right) dm^d.$$

Let $A \in \mathrm{GL}(d,\mathbb{R})$. Define the *adjoint mapping*

$$A^* : \mathbb{R}^d \to \mathbb{R}^d$$

by

$$\forall\, x \in \mathbb{R}^d \text{ and } \forall\, y \in \mathbb{R}^d, \quad \langle A(x), y\rangle = \langle x, A^*(y)\rangle,$$

where $\langle \ldots, \ldots \rangle$ denotes the Euclidean inner product on $\mathbb{R}^d \times \mathbb{R}^d$. Then $A^* \in \mathrm{GL}(d,\mathbb{R})$, $\det(A^*) = \overline{\det(A)}$ and

$$(A^*)^{-1} = (A^{-1})^*;$$

see [185], pages 337–358.

Because of Theorem 8.7.5, if $(T,g) \in \mathcal{D}'(\mathbb{R}^d) \times C_c^\infty(\mathbb{R}^d)$ or $(T,g) \in \mathcal{S}'(\mathbb{R}^d) \times \mathcal{S}(\mathbb{R}^d)$ and if $A \in \mathrm{GL}(d,\mathbb{R})$, then $T \circ A$ is defined by

$$(T \circ A)(g) = |\det(A^{-1})| T(g \circ A^{-1}). \tag{8.52}$$

The *Fourier transform* of $T \in \mathcal{S}'(\mathbb{R}^d)$ is well defined by the Parseval formula in the following way:

$$\forall\, f \in \mathcal{S}(\mathbb{R}^d), \quad \widehat{T}(\overline{\widehat{f}}) = T(\overline{f}).$$

Theorem 8.7.6. Change of variables and the Fourier transform
Let $A \in \mathrm{GL}(d,\mathbb{R})$.

a. *If $f \in L^1_{m^d}(\mathbb{R}^d)$, then $f \circ A \in L^1_{m^d}(\mathbb{R}^d)$ and*

$$\widehat{(f \circ A)} = |\det(A^{-1})| \left(\hat{f} \circ (A^{-1})^*\right).$$

b. *If $T \in \mathcal{S}'(\mathbb{R}^d)$, then $T \circ A \in \mathcal{S}'(\mathbb{R}^d)$ and*

$$\widehat{(T \circ A)} = |\det(A^{-1})| \left(\widehat{T} \circ (A^{-1})^*\right).$$

Proof. **a.** By Theorem 8.7.5 and the definition of A^*, we have

$$\begin{aligned}
\widehat{(f \circ A)}(\xi) &= \int_{\mathbb{R}^d} (f \circ A)(x) e^{-2\pi i \langle x, \xi\rangle}\, dm^d(x) \\
&= |\det(A^{-1})| \int_{\mathbb{R}^d} f(x) e^{-2\pi i \langle A^{-1}(x), \xi\rangle}\, dm^d(x) \\
&= |\det(A^{-1})| \int_{\mathbb{R}^d} f(x) e^{-2\pi i \langle x, (A^{-1})^*(\xi)\rangle}\, dm^d(x) \\
&= |\det(A^{-1})| (\hat{f} \circ (A^{-1})^*)(\xi).
\end{aligned}$$

b. If $f \in \mathcal{S}(\mathbb{R}^d)$ then

$$\begin{aligned}(\widehat{T \circ A})(\overline{\hat{f}}) &= (T \circ A)(\overline{f}) = |\det(A^{-1})| T(\overline{f} \circ A^{-1}) \\ &= |\det(A^{-1})| \widehat{T} \left(|\det(A)| \overline{\hat{f}} \circ A^* \right) = |\det((A^{-1})^*)| \left(\widehat{T} \circ (A^{-1})^* \right) (\overline{\hat{f}}) \\ &= |\det(A^{-1})| \left(\widehat{T} \circ (A^{-1})^* \right) (\overline{\hat{f}}),\end{aligned}$$

where the first equality is the definition of the Fourier transform of distributions, the second and fourth equalities are a consequence of (8.52), the third is from part *a*, and the last equality is a consequence of the properties of the adjoint mapping $A \to A^*$. □

We can now use the change of variables formula as a motivation to introduce the notion of measure on smooth parametric surfaces (manifolds) in $\mathbb{R}^d$. This results in the natural *surface measure* σ on these surfaces.

We call $M \subseteq \mathbb{R}^d$ an *m-dimensional smooth surface* if, for each $x \in M$, there exist an open set $V \subseteq M$ such that $x \in V$, an open set $U \subseteq \mathbb{R}^m$, and a bijective, open, and infinitely differentiable mapping $\phi : U \to V$ with injective derivative at each point of U. We say that ϕ is a *coordinate chart* on $V \subseteq M$, and V is a *coordinate patch* of M containing $x \in M$. The *metric tensor* of M on U is the matrix Φ of functions on U defined by

$$\Phi(j,k)(x) = D(\phi)(x)^* D(\phi)(x).$$

The metric tensor Φ is infinitely differentiable and positive definite.

Example 8.7.7. Spherical polar coordinates
a. We consider the following coordinate system in $\mathbb{R}^d$:

$$\begin{aligned}x_1 &= r\sin(\phi_{d-2}) \cdots \sin(\phi_2)\sin(\phi_1)\cos(\theta), \\ x_2 &= r\sin(\phi_{d-2}) \cdots \sin(\phi_2)\sin(\phi_1)\sin(\theta), \\ x_3 &= r\sin(\phi_{d-2}) \cdots \sin(\phi_2)\cos(\phi_1), \\ &\vdots \\ x_k &= r\sin(\phi_{d-2}) \cdots \sin(\phi_{k-1})\cos(\phi_{k-2}), \\ &\vdots \\ x_d &= r\cos(\phi_{d-2}),\end{aligned}$$

where $r \in (0,\infty)$, $\theta \in (0, 2\pi)$, and $\phi_i \in (0,\pi)$, $i = 1, \ldots, d-2$. This defines the *spherical polar coordinate system* on $\mathbb{R}^d \setminus E$, for some $E \subseteq \mathbb{R}^d$, for which $m^d(E) = 0$. Then, for $f \in L^1_{m^d}(\mathbb{R}^d)$,

$$\int_{\mathbb{R}^d} f dm^d = \int_0^{\pi} \cdots \int_0^{\pi}\int_0^{2\pi}\int_0^{\infty} f(x_1, \ldots, x_d) J(r, \theta, \phi_1, \ldots, \phi_{d-2}) dr\, d\theta\, d\phi_1 \cdots d\phi_{d-2}, \tag{8.53}$$

where $x_i = x_i(r, \theta, \phi_1, \ldots, \phi_{d-2})$, $i = 1, \ldots, d$, and

$$J(r, \theta, \phi_1, \ldots, \phi_{d-2}) = r^{d-1} \sin(\phi_1) \sin^2(\phi_2) \cdots \sin^{d-2}(\phi_{d-2}) \geq 0.$$

Here J is the Jacobian determinant $\det(D(\phi))$ defined above, and we do not have to worry about the absolute value in (8.51), since $J \geq 0$ in this case.

b. For fixed $r > 0$, let $\psi_r : \mathbb{R}^{d-1} \to \mathbb{R}^d$ be defined as

$$\psi_r(\theta, \phi_1, \ldots, \phi_{d-2}) = (x_1, \ldots, x_d).$$

Let $S_r^{d-1} \subseteq \mathbb{R}^d$ be the sphere of radius r centered at the origin. Then there exists a finite collection $\{V_i\} \subseteq \mathbb{R}^{d-1}$ of open sets such that $S_r^{d-1} \setminus \{(0, \ldots, 0, \pm r)\} \subseteq \bigcup \psi_r(V_i)$ and ψ_r is a coordinate chart on each V_i. Thus, $S_r^{d-1} \setminus \{(0, \ldots, 0, \pm r)\}$ is a $(d-1)$-dimensional manifold in $\mathbb{R}^d$.

Note that for any permutation of the coordinates $x_1, \ldots, x_d$, we define a different spherical polar coordinate system, which covers a different subset of S_r^{d-1}. As a consequence, we have that S_r^{d-1} is a $(d-1)$-dimensional smooth surface in $\mathbb{R}^d$.

Given this setting of a smooth surface M and a coordinate patch V, let $f : V \to \mathbb{C}$ be a Borel measurable function; see Appendix A.1. We define the *surface integral* $\int_V f \, d\sigma$ of f as

$$\int_V f \, d\sigma = \int_U f \circ \phi \sqrt{\det \Phi} \, dm^d.$$

This definition is independent of the choice of a coordinate chart, i.e., if $\psi : W \to V$ is another coordinate chart on V then

$$\int_U f \circ \phi \sqrt{\det \Phi} \, dm^d = \int_W f \circ \psi \sqrt{\det \Psi} \, dm^d.$$

Definition 8.7.8. Smooth partition of unity

Given $A \subseteq \mathbb{R}^d$ and an open covering $\mathcal{V} = \{\tilde{V}\}$ of A, we consider a collection $\mathcal{G}$ of infinitely differentiable functions g_i, $i \in I$, defined on some open set containing A, that satisfies the following properties:

i. $\forall\, x \in A$, and $\forall\, g \in \mathcal{G}$, $0 \leq g(x) \leq 1$,

ii. $\forall\, x \in A$, $\exists\, V$, open, $x \in V$, such that all but finitely many $g \in \mathcal{G}$ satisfy $g = 0$ on V,

iii. $\forall\, x \in A$, $\sum_{i \in I} g_i(x) = 1$,

iv. $\forall\, g \in \mathcal{G}$, $\exists\, \tilde{V} \in \mathcal{V}$ such that $g = 0$ outside of some closed set contained in $\tilde{V}$.

Such a collection $\mathcal{G}$ is a *smooth partition of unity for A subordinate to the covering $\mathcal{V}$*. A countable smooth partition of unity subordinate to any given open covering of $A \subseteq \mathbb{R}^d$ exists, e.g., [447].

We note that the set of coordinate patches V forms an open covering of a smooth surface M. Moreover, each V is of the form $V = \tilde{V} \cap M$, for some open set $\tilde{V} \subseteq \mathbb{R}^d$. Thus, there exists a countable smooth *partition of unity* $\{g_i : i = 1, \dots\}$ subordinate to the open covering $\{\tilde{V}_i\}$ of M in $\mathbb{R}^d$.

For this partition of unity and for a Borel measurable function $f : M \to \mathbb{R}^+$ we define the *surface integral* $\int_M f \, d\sigma$ of f to be

$$\int_M f \, d\sigma = \sum_{i=1}^{\infty} \int_{V_i} g_i f \, d\sigma,$$

where $\operatorname{supp} g_i \subseteq \tilde{V}_i$ and $V_i = \tilde{V}_i \cap M$. This definition is independent of the choice of partition of unity. In fact, if h_i is another partition of unity subordinate to the open covering $\{\tilde{V}_i\}$ of M, then, since f is nonnegative,

$$\begin{aligned}
\sum_{i=1}^{\infty} \int_{V_i} g_i f \, d\sigma &= \sum_{i=1}^{\infty} \int_{V_i} \sum_{j=1}^{\infty} \int_{V_j} h_j g_i f \, d\sigma \\
&= \sum_{j=1}^{\infty} \int_{V_j} \sum_{i=1}^{\infty} \int_{V_i} h_j g_i f \, d\sigma \\
&= \sum_{j=1}^{\infty} \int_{V_j} h_j f \, d\sigma.
\end{aligned}$$

For a complex-valued Borel measurable function $f : M \to \mathbb{C}$, we write f as a combination of nonnegative Borel measurable functions, i.e., $f = (\operatorname{Re} f)^+ - (\operatorname{Re} f)^- + i(\operatorname{Im} f)^+ - i(\operatorname{Im} f)^-$, and we let

$$\int_M f \, d\sigma = \int_M (\operatorname{Re} f)^+ \, d\sigma - \int_M (\operatorname{Re} f)^- \, d\sigma + i \int_M (\operatorname{Im} f)^+ \, d\sigma - i \int_M (\operatorname{Im} f)^- \, d\sigma.$$

We define the *surface measure* σ on a smooth surface M by letting

$$\sigma(A) = \int_M \mathbb{1}_A \, d\sigma,$$

for any Borel set $A \subseteq M$. This notion of surface measure, which is defined only on M, is formulated precisely in terms of Hausdorff measure on $\mathbb{R}^d$ in Section 9.3.

Proposition 8.7.9. *Let M be a smooth m-dimensional surface in $\mathbb{R}^d$, and let $A \subseteq M$ be closed in M and such that $\sigma(A) = 0$. Then, for any Borel measurable function $f : M \to \mathbb{C}$, we have*

$$\int_M f \, d\sigma = \int_{M \setminus A} f \, d\sigma.$$

Proof. We begin by writing $\int_M f\, d\sigma = \int_M \mathbb{1}_A f\, d\sigma + \int_M \mathbb{1}_{M\setminus A} f\, d\sigma$. Since A is closed in M, $M \setminus A$ is a smooth surface with coordinate charts and coordinate patches induced by those for M, and so $\int_M \mathbb{1}_{M\setminus A} f\, d\sigma = \int_{M\setminus A} f\, d\sigma$. On the other hand, since $\sigma(A) = 0$, we also have $\int_M \mathbb{1}_A f\, d\sigma = 0$. □

Proposition 8.7.9 is used in the proof of the following theorem; cf. (8.53).

Theorem 8.7.10. Integration in polar coordinates
Let $S^{d-1} \subseteq \mathbb{R}^d$ be the unit sphere in $\mathbb{R}^d$. Then

$$\forall\, f \in L^1_{m^d}(\mathbb{R}^d), \quad \int_{\mathbb{R}^d} f\, dm^d = \int_{S^{d-1}} \left(\int_0^\infty f(r\omega) r^{d-1}\, dr \right) d\sigma_{d-1}(\omega),$$

where σ_{d-1} is the surface measure on the smooth surface S^{d-1}.

Example 8.7.11. Surface measure of the sphere in $\mathbb{R}^d$
In Example 2.3.11 we computed the Lebesgue measure of the unit ball in $\mathbb{R}^d$ to be

$$m^d(B_d) = \frac{\pi^{d/2}}{(d/2)\Gamma(d/2)}, \quad d = 1, 2, \ldots.$$

Similarly, one can compute

$$\forall\, x \in \mathbb{R}^d \text{ and } \forall\, r > 0, \quad m^d(B(x,r)) = r^d \frac{\pi^{d/2}}{(d/2)\Gamma(d/2)}.$$

By Theorem 8.7.10, we can write

$$\begin{aligned} m^d(B(0,r)) &= \int_{\mathbb{R}^d} \mathbb{1}_{B(0,r)}\, dm^d = \int_{S^{d-1}} \left(\int_0^\infty \mathbb{1}_{B(0,r)}(r\omega) r^{d-1}\, dr \right) d\sigma_{d-1}(\omega) \\ &= \int_{S^{d-1}} \left(\int_0^r r^{d-1}\, dr \right) d\sigma_{d-1}(\omega) = \sigma_{d-1}(S^{d-1}) \frac{r^d}{d}. \end{aligned}$$

Thus,

$$\sigma_{d-1}(S^{d-1}) = \frac{2\pi^{d/2}}{\Gamma(d/2)}.$$

8.8 Potpourri and titillation

1. It is easier to see how Stokes' theorem is a generalization of FTC if we write it in differential geometric terms.

Let M be a compact, orientable, d-dimensional manifold with boundary ∂M, and let ω be a $(d-1)$-form on M. Then

$$\int_M d\omega = \int_{\partial M} \omega.$$

This result includes all classical versions of the Stokes' theorem. We refer the reader to [447] for the necessary definitions of integrals of forms on manifolds and for the proof of this and other related results.

2. In Section 4.7.5 we stated CARLESON's theorem: *If $f \in L^2_m(\mathbb{T})$ then its Fourier series converges m-a.e. to f.* The depth and intricacy of his proof, hinted at superficially by one of his lemmas also stated in Section 4.7.5, belies the straightforward plan of proof, which we shall now describe. For further insight into CARLESON's approach we refer to his comments upon receiving the 1984 Steele Prize; see [39], pages 175–176.

In analogy with the Hardy–Littlewood maximal function of Sections 8.2 and 8.6, CARLESON considered the maximal function M_{FS}, defined on $L^2_m(\mathbb{T})$, as follows:

$$M_{FS}(f)(x) = \sup_{N \geq 0} |S_N(f)(x)|,$$

where $S_N(f)(x) = \sum_{|n| \leq N} \hat{f}(n) e^{-2\pi i n x}$; see Appendix B.5. He then proved the fundamental theorem,

$$\exists\, C_2 > 0 \text{ such that } \forall\, f \in L^2_m(\mathbb{T}), \quad \|M_{FS}(f)\|_2 \leq C_2 \|f\|_2. \tag{8.54}$$

A beautifully written and complete proof of (8.54) is due to CHARLES J. MOZZOCHI [351].

Statement (8.54) yields the fact that $\lim_{N\to\infty} S_N(f) = f$ *m-a.e.* by the following argument. For a given $\varepsilon \in (0,1)$, we can choose a sequence $\{\varepsilon_k\} \subseteq \mathbb{R}^+$ and a sequence $\{p_k\} \subseteq L^2_m(\mathbb{T})$ of polynomials such that $0 < \varepsilon_k^2 < \varepsilon^k$ and $\|f - p_k\| < \varepsilon_k^2$ for each k, and $p_k \to f$ *m-a.e.* For this sequence $\{p_k\}$, and for each fixed k and $x \in \mathbb{T}$, we have

$$\begin{aligned} \overline{\lim_{N\to\infty}} |S_N(f)(x) - f(x)| &\leq \overline{\lim_{N\to\infty}} |S_N(f - p_k)(x)| + |f(x) - p_k(x)| \\ &\leq M_{FS}(f - p_k)(x) + |f(x) - p_k(x)|. \end{aligned} \tag{8.55}$$

Let $A_k = \{x \in \mathbb{T} : M_{FS}(f - p_k)(x) > \varepsilon_k\}$. Using (8.54), it is elementary to check that

$$m(A_k) \leq C_2^2 \varepsilon_k^2;$$

and, by definition of A_k, $M_{FS}(f - p_k)(x) \leq \varepsilon_k$ if $x \notin A_k$. Thus, $M_{FS}(f - p_k)(x) \to 0$ if $x \notin A_\varepsilon = \bigcup_{k=1}^{\infty} A_k$. Note that $m(A_\varepsilon) \leq C_2^2(\varepsilon/(1-\varepsilon))$ and $p_k(x) \to f(x)$ for $x \notin B$, where B has the property that $m(B) = 0$. Hence, taking the limsup of (8.55), we have $S_N(f)(x) \to f(x)$ for $x \notin A_\varepsilon \cup B$, where $m(A_\varepsilon \cup B) \leq C_2^2(\varepsilon/(1-\varepsilon))$.

Therefore, $S_N(f) \to f$ *m-a.e.* In fact, if $r > 0$, we set $\varepsilon = r/(C_2^2 + r) < 1$, and form ε_k, $\{p_k\}$, A_ε, B as above. Set $X_r = A_\varepsilon \cup B$, so that

$$\forall\, r > 0, \exists\, X_r \subseteq \mathbb{T} \text{ such that } m(X_r) < r \text{ and } S_N(f) \to f \text{ off of } X_r.$$

Thus, if $S_N(f) \to f$ on X, $m(X) > 0$, we obtain a contradiction by choosing $r < m(X)$.

3. The Hardy–Littlewood theorem (Theorem 8.6.14) and CARLESON's theorem (8.54) in Section 8.8.2 are special cases of inequalities,

$$\exists\, C > 0 \text{ such that } \forall\, f \in X, \quad \|L(f)\|_Y \le C\|f\|_X, \tag{8.56}$$

where X and Y are Banach spaces and $L : X \to Y$ is linear. Such inequalities are equivalent to the continuity of L (see Appendix A.8).

Plancherel's theorem (Theorem B.4.2), $\|\hat{f}\|_2 = \|f\|_2$, is a special case of (8.56). In the case $1 < p < 2$, WILLIAM BECKNER (1975) proved that

$$\forall\, f \in L^p_{m^d}(\mathbb{R}^d) \cap L^1_{m^d}(\mathbb{R}^d), \quad \|\hat{f}\|_q \le (p^{1/p}q^{-1/q})^{d/2}\|f\|_p. \tag{8.57}$$

Clearly, (8.57) is a special case of (8.56). There was an earlier important contribution by KONSTANTIN I. BABENKO (1961). Note that if g is a Gaussian of the form

$$g(x) = Ce^{-\langle x, A(x)\rangle + \langle B, x\rangle},$$

where $C \in \mathbb{C}$, $B \in \mathbb{C}^d$, and A is a symmetric, real, positive definite matrix, then g is an equalizer for (8.57). ELLIOT H. LIEB (1990) proved the converse. This is relevant since best constants are useful in uncertainty principle inequalities, which, in turn, can often be proved using Fourier transform norm inequalities of the form (8.57) or (8.59), e.g., [225], [181], and [45], Chapter 7.

The spaces X and Y can be quite general topological vector spaces, but they can also have a particular structure. We shall comment on specific weighted spaces. A *weight* w on $\mathbb{R}^d$ is a nonnegative Lebesgue measurable function on $\mathbb{R}^d$, and, if $1 \le p < \infty$, then

$$L^p_w(\mathbb{R}^d) = \{f \in L^1_{\text{loc}}(\mathbb{R}^d) : \|fw^{1/p}\|_p < \infty\}.$$

If $1 < p < \infty$, then w is an A_p-*weight* if there is $a_p > 0$ such that, for every cube $Q \subseteq \mathbb{R}^d$,

$$\left(\frac{1}{m^d(Q)} \int_Q w \, dm^d\right)^{1/p} \left(\int_Q w^{1-q} \, dm^d\right)^{1/q} \le a_p.$$

In this case, $w \in L^1_{\text{loc}}(\mathbb{R}^d)$. BENJAMIN MUCKENHAUPT (1972) proved the following fundamental theorem.

Theorem 8.8.1. Weighted norm inequality for *M*
Let w be a weight on $\mathbb{R}^d$. Then w is an A_2-weight if and only if

$$\begin{gathered}\exists\, C_2 > 0 \text{ such that } \forall\, f \in L^2_w(\mathbb{R}^d), \\ \int_{\mathbb{R}^d} M(f)(x)^2 w(x) \, dm^d(x) \le C_2^2 \int_{\mathbb{R}^d} |f(x)|^2 w(x) \, dm^d(x).\end{gathered} \tag{8.58}$$

This theorem is part of a monumental effort in twentieth-century classical harmonic analysis; see [190]. There are also inequalities of the form (8.58), where M is replaced by various singular integral operators. What is surprising, after dwelling on the problem, is that some Fourier-transform-weighted norm inequalities are also characterized by A_p-weights. In fact, one of the authors, along with HANS HEINIG and RAYMOND L. JOHNSON, proved the following result (1987).

Theorem 8.8.2. Fourier-transform-weighted norm inequality
Let w be an even weight on $\mathbb{R}$ that is nondecreasing on $(0, \infty)$. Then w is an A_2-weight if and only if

$$\begin{gathered} \exists\, C_2 > 0 \ \textit{such that} \ \forall\, f \in L^2_w(\mathbb{R}^d), \\ \int_{\widehat{\mathbb{R}}} |\hat{f}(\xi)|^2 w(1/\xi)\, d\xi \leq C_2^2 \int_{\mathbb{R}} |f(x)|^2 w(x)\, dx. \end{gathered} \tag{8.59}$$

The terms "$w(1/\xi)$" and "$w(x)$" reflect the role of the uncertainty principle in Fourier analysis. Once again, this theorem is part of a major effort associated with Fourier-transform-weighted norm inequalities; see [46] for a perspective of results in the area.

4. The Lebesgue differentiation theorem (Theorem 8.2.4) is true for converging cubes, as stated, with a similar proof for converging balls. Besides the generalization to measure-metrizable convergence of $\{A_j\}$ in Section 8.4, the result is true in more geometric terms for sequences $\{A_j\}$ of parallelepipeds $A_j = \{x = (x_1, \ldots, x_d) : a_i^j \leq x_i \leq b_j^j, i = 1, \ldots, d\}$, where $\lim_{j \to \infty}(b_i^j - a_i^j) = 0$ for each $i = 1, \ldots, d$, and for which

$$\exists\, c > 0 \ \text{such that} \ \forall j = 1, \ldots, \ \text{and} \ \forall\, i, k = 1, \ldots, d, \quad \frac{b_i^j - a_i^j}{b_k^j - a_k^j} \leq c. \tag{8.60}$$

Condition (8.60) asserts that the ratio of each two dimensions remains bounded as $j \to \infty$. This leads to several *questions*, and we address two of them in the following two paragraphs.

Recall from Example 4.4.7 that the arithmetic means $F_N * f$ of the partial sums of the Fourier series $S(f)$ of $f \in L^1_m(\mathbb{T})$ converge to f on the Lebesgue set $L(f)$. The notion of Lebesgue set arose in Theorem 4.4.5, which itself is a strong Lebesgue differentiation theorem on $\mathbb{R}$; cf. Theorem 8.4.7 and Definition 8.4.8. We *ask* in what way the pointwise convergence of $\{F_N * f\}$ generalizes for $f \in L^1_m(\mathbb{T}^d)$, $d > 1$. An answer is that the same result is valid for sequences of parallelepipeds satisfying (8.60). The proof is difficult and is due to MARCINKIEWICZ and ZYGMUND (1939); see [524], Chapter XVII, Section 3 for details as well as a reference by ZYGMUND to an earlier related approach by C. N. MOORE (1913).

Further, we *ask* to what extent the Lebesgue differentiation theorem is valid for sequences $\{A_j\}$ of parallelepipeds converging to $x \in \mathbb{R}^d$ when condition (8.60) is not imposed. SAKS (1934) proved that the set of $f \in L^1_{m^d}(\mathbb{R}^d)$, $d > 1$, for which

$$\exists\, x \in \mathbb{R}^d \text{ such that } \quad \lim_{j\to\infty} \frac{1}{m^d(A_j)} \int_{A_j} f \, dm^d = f(x) \tag{8.61}$$

is a set of first category in $L^1_{m^d}(\mathbb{R}^d)$. On the other hand, the limit in (8.61) is valid m^d-*a.e.* for each $f \in L^1_{m^d}(\mathbb{R}^d) \cap L^\infty_{m^d}(\mathbb{R}^d)$; see [412]. These results can be viewed as background for ZYGMUND's theorem (1934): *If* $L^p_{m^d}(\mathbb{T}^d)$, $d > 1$, *for all* $p > 1$, *then the limit (8.61) for the setting of* $\mathbb{T}^d$ *is valid* m^d*-a.e.* ZYGMUND's theorem was then proven for a larger space, denoted by $L(\log L)^{d-1}(\mathbb{T}^d)$, by JOHAN JESSEN, MARCINKIEWICZ, and ZYGMUND (1935); see [524], Chapter XVII, Section 3 for details.

5. In his testimonial to ZYGMUND and ALBERTO CALDERÓN [449], ELIAS M. STEIN notes that "It is ironic that complex methods with their great power and success in the one-dimensional theory actually stood in the way of progress to higher dimensions. ...The only way ..., as ZYGMUND foresaw, required a further development of 'real' methods." At this period, in mid-twentieth-century, CALDERÓN and ZYGMUND wrote their celebrated paper [86], "On the existence of certain singular integrals". In it they proved the so-called *Calderón–Zygmund decomposition* of $\mathbb{R}^d$ for nonnegative $f \in L^1_m(\mathbb{R}^d)$ and $\alpha > 0$: *There is a decomposition of* $\mathbb{R}^d$ *with the properties that* $\mathbb{R}^d = F \cup \Omega$, $F \cap \Omega = \emptyset$, $f \leq \alpha$ m^d*-a.e. on* F, *and* Ω *is a union of cubes* Q_k *with disjoint interiors such that*

$$\forall\, k = 1, \ldots, \quad \alpha < \frac{1}{m^d(Q_k)} \int_{Q_k} f \, dm^d \leq 2^d \alpha.$$

The proof depends on the Lebesgue differentiation theorem, and it is used to prove weak-type L^1 and L^p, $1 < p < \infty$, estimates for important singular integral operators such as the Hilbert transform when $d = 1$. STEIN waxes eloquent about [86], asserting that there is probably no paper in the second half of the twentieth-century "which had such widespread influence in analysis"; cf. [128] and the wonderful book review [168] by ROBERT A. FEFFERMAN.

6. Closely related to the semimartingale maximal theorem (Theorem 8.6.10) is EBERHARD HOPF's *maximal ergodic theorem* (1954), rooted in HARDY–LITTLEWOOD [219]. DOOB [144] proved the semimartingale theorem in his classic mid-century treatise on stochastic processes.

Even earlier (1939), WIENER [506] understood the relation between ergodic theory and differentiation theory. One might say he brought a brilliant synthesis to the table, beginning with GEORGE D. BIRKHOFF's profound ergodic theorem, showing the relevance and importance of covering lemmas of Vitali type, proving maximal function inequalities, and showing how to proceed from maximal function inequalities to ergodic measure-preserving analogues.

7. We defined radial functions on $\mathbb{R}^d$ in Example 8.6.2*c*. Similarly, we can define radial elements of $M_b(\mathbb{R}^d)$. In fact, let $\mathrm{SO}(d, \mathbb{R}) \subseteq \mathrm{GL}(d, \mathbb{R})$ be the subgroup of elements S for which $\det(S) = 1$ and whose transpose is S^{-1}. Then radial functions f are characterized by the property that $f(S(x)) = f(x)$ for all $S \in \mathrm{SO}(d, \mathbb{R})$. We say that $\mu \in M_b(\mathbb{R}^d)$ is *radial* if $S(\mu) = \mu$ for all $S \in \mathrm{SO}(d, \mathbb{R})$, where $S(\mu)$ is defined by $S(\mu)(f) = \mu(f \circ S)$, $f \in C_c(\mathbb{R}^d)$.

Further, in Problem 3.5, we defined the convolution $f * g \in L^1_m(\mathbb{R})$ for $f, g \in L^1_m(\mathbb{R})$; and this definition extends to $\mathbb{R}^d$ and locally compact groups. Also, in Section 7.6.3, equation (7.47), we gave a natural formal definition of the convolution of distributions, noting that, in this generality, convolution is not always well defined. It is not difficult to prove that if $\mu, \nu \in M_b(\mathbb{R}^d)$, then the *convolution*,

$$\forall\, f \in C_c(\mathbb{R}^d), \quad \mu * \nu(f) = \mu_x(\nu_y(f(x+y))),$$

is a well-defined element $\mu * \nu \in M_b(\mathbb{R}^d)$. This definition is a natural extension of L^1-convolution and a particular case of (7.47).

We know that the Radon–Nikodym theorem (Theorem 5.3.1) gives an embedding $L^1_{m^d}(\mathbb{R}^d) \subseteq M_b(\mathbb{R}^d)$. Under convolution, $L^1_{m^d}(\mathbb{R}^d)$, as well as the space of continuous measures in $M_b(\mathbb{R}^d)$, is a closed ideal of $M_b(\mathbb{R}^d)$. The space of all discrete measures is a closed subalgebra of $M_b(\mathbb{R}^d)$.

In problem 3.6, the goal was to prove STEINHAUS' result: If $X \in \mathcal{M}(\mathbb{R})$ and $m(X) > 0$, then $X - X$ is a neighborhood of 0; cf. STEINHAUS' comparable result in Problem 1.13 for the Cantor set C. A significant extension of this idea is the assertion *that if $\mu, \nu \in M_b(\mathbb{R}^d)$ are radial, then the convolution $\mu * \nu$ is an element of $L^1_{m^d}(\mathbb{R}^d)$.*

9 Self-Similar Sets and Fractals

9.1 Self-similarity and fractals

We began this book with the definition of the *ternary Cantor set* in Example 1.2.7. This set, the more general *perfect symmetric sets*, and the associated *Cantor functions* serve as a source of examples and counter-examples in various areas of analysis; recall Example 1.3.5, Example 1.3.17, Problem 1.12, Example 2.4.14, Problem 2.2, Example 4.2.4, Example 4.5.1, Problem 4.5, Problem 4.24, Problem 4.42, and Example 5.2.9. In this final chapter, we return to the Cantor set, and view it and all perfect symmetric sets in the broader context of self-similarity and fractals.

We start with two classical examples.

Example 9.1.1. Sierpiński carpet
The *Sierpiński carpet* (sometimes referred to as the *Sierpiński curve*), see [433], is an analogue of the Cantor set in $\mathbb{R}^2$. We start with $F_0 = [0,1] \times [0,1]$. In the next step we represent this square as a union of nine compact squares, each with sides of length $1/3$:

$$F_0 = \bigcup_{j=1}^{9} F_1^j.$$

Clearly, this union is *not* disjoint. We take F_1^1 to be the "middle" square, and we define

$$F_1 = F_0 \setminus \operatorname{int} F_1^1.$$

Next, we divide each of the squares F_1^j, $j = 2, \ldots, 9$, into the union of nine compact squares, each with sides of length $1/9$, and let F_2^j, $j = 2, \ldots, 9$, denote the "middle" squares of this decomposition. We define

$$F_2 = F_1 \setminus \left(\bigcup_{j=2}^{9} \operatorname{int} F_2^j \right).$$

We proceed inductively, denoting by F_n the set obtained from F_{n-1} by removing the interiors of the 8^{n-1} "middle" squares, each with sides of length

J.J. Benedetto and W. Czaja, *Integration and Modern Analysis*, Birkhäuser Advanced Texts, DOI 10.1007/978-0-8176-4656-1_9,

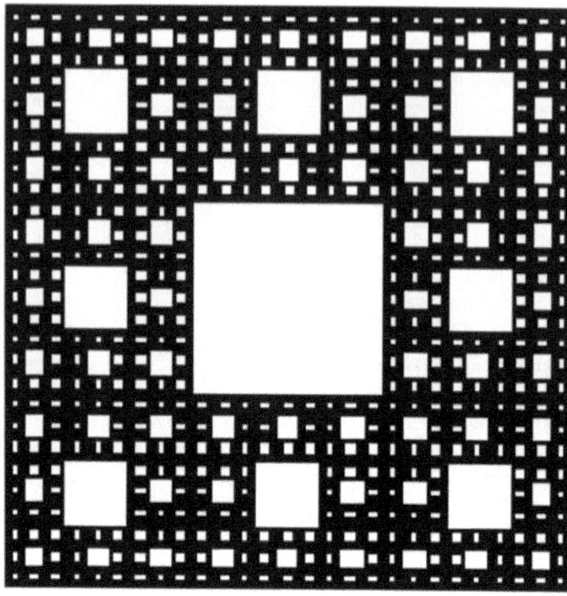

Fig. 9.1. Sierpiński carpet.

$1/3^n$; see Figure 9.1. The *Sierpiński carpet* is defined to be the set

$$S = \bigcap_{j=0}^{\infty} F_j.$$

S is a nonempty, compact, and connected set of Lebesgue measure $m^2(S) = 0$. It is not difficult to observe that S cuts the plane into infinitely many disjoint parts. An interesting feature of the *Sierpiński carpet* is that its boundary has infinite length. Indeed, the length of the boundary of F_n is

$$m(\partial F_n) = 4 + 4 \sum_{j=1}^{n} \frac{8^{n-1}}{3^n} \longrightarrow \infty, \quad n \to \infty.$$

An analogous set may be constructed with equilateral triangles replacing squares. The resulting object is called the *Sierpiński triangle*, although Benoît Mandelbrot coined a phrase that is commonly used today, viz., the *Sierpiński gasket*, since it reminded him of "the part that prevents leaks in motors"; see [337].

Example 9.1.2. Menger sponge
The *Menger sponge* in $\mathbb{R}^3$ is another example of a set constructed in analogy with the Cantor set. We start with the cube $I_0 = [0,1] \times [0,1] \times [0,1]$ and

divide it into 27 cubes, each of side length 1/3. We remove the seven cubes from it that are disjoint from the edges of I_0; and we define I_1 to be the closure of the remaining set in the Euclidean topology. Then I_1 may be represented as a union of 20 cubes, and to each one of them we apply the analogous procedure in order to obtain I_2; and we proceed inductively. The set

$$M = \bigcap_{j=0}^{\infty} I_j$$

is the *Menger sponge*; see Figure 9.2 for an approximation. Again, M is a nonempty, compact, and connected set of Lebesgue measure $m^3(S) = 0$; but it does not cut $\mathbb{R}^3$; i.e., $R^3 \setminus M$ is connected.

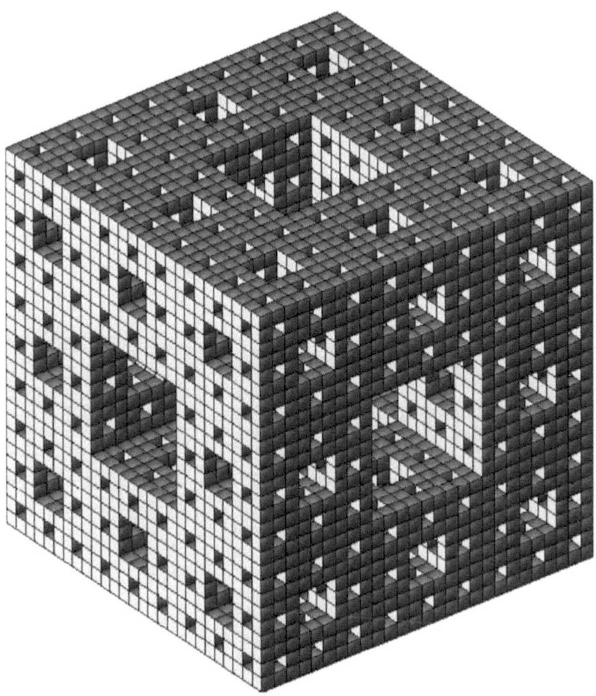

Fig. 9.2. Menger sponge.

Examples 9.1.1 and 9.1.2, as well as many totally disconnected sets, possess a certain feature not shared by generic open sets in $\mathbb{R}^d$: If we look through a magnifying glass at these objects, they always seem to look the same, no matter how closely we look. This serves as the motivation to introduce the notion of *self-similarity*.

Definition 9.1.3. Self-similarity and self-affinity
a. A transformation $S : \mathbb{R}^d \to \mathbb{R}^d$ of the form

$$S(x) = rO(x) + y,$$

where $r > 0$, $y \in \mathbb{R}^d$, and $O : \mathbb{R}^d \to \mathbb{R}^d$ is an orthogonal transformation of $\mathbb{R}^d$, is called a *similarity transformation* or *similarity*. The *orthogonal transformations* of $\mathbb{R}^d$ can be identified with the set of $d \times d$ matrices Q with real entries that satisfy

$$Q^*Q = QQ^* = Id,$$

where Q^* denotes the transpose of Q. We write $\mathrm{SO}(d, \mathbb{R}) = \{Q \in \mathrm{GL}(d, \mathbb{R}) : Q^*Q = QQ^* = Id \text{ and } \det Q = 1\}$; cf. Section 8.8.7.

If $\mathcal{S} = \{S_j : j = 1, \ldots, n\}$ is a family of similarities, we write

$$\forall\, A \subseteq \mathbb{R}^d, \quad \mathcal{S}(A) = \bigcup_{j=1}^{n} S_j(A).$$

A set $A \subseteq \mathbb{R}^d$ is *invariant under* $\mathcal{S}$ if

$$\mathcal{S}(A) = A.$$

b. A set $A \subseteq \mathbb{R}^d$ is *strictly self-similar* if there exists a family $\mathcal{S} = \{S_j : j = 1, \ldots, n\}$ of similarities, $S_j(x) = rO_j(x) + y_j$, $j = 1, \ldots, n$, $r < 1$, such that A is invariant under $\mathcal{S}$ and

$$\forall\, i \neq j, \quad S_i(A) \cap S_j(A) = \emptyset.$$

This notion of *strict self-similarity* in Definition 9.1.3 was introduced by MANDELBROT [337]. The problem with this approach is that it excludes the *Sierpiński carpet* and *gasket* and the *Menger sponge* from the family of self-similar sets. In order to extend the definition of self-similarity to include such examples we shall use the notions of *Hausdorff measure* and *Hausdorff dimension*, which we shall introduce in Section 9.3. Until then the reader should think of *self-similar sets* as sets with "negligible" overlaps $S_i(A) \cap S_j(A)$, that is, small sets when compared to the original set A.

Example 9.1.4. Self-similar sets
a. The *ternary Cantor set* C is invariant under the following family $\mathcal{S}_C = (S_1, S_2)$ of similarities:

$$S_1(x) = \frac{1}{3}x, \quad S_2(x) = \frac{1}{3}x + \frac{2}{3}.$$

Clearly, $S_1(C) \cap S_2(C) = \emptyset$, and the ternary Cantor set is strictly self-similar.

b. The family of similarities associated with the Sierpiński carpet S consists of the following transformations:

$$S_j(x) = \frac{1}{3}x + y_j, \quad j = 1, \ldots, 8,$$
$$y_1 = (0,0),\ y_2 = (1/3,0),\ y_3 = (2/3,0),\ y_4 = (2/3,1/3),$$
$$y_5 = (2/3,2/3),\ y_6 = (1/3,2/3),\ y_7 = (0,2/3),\ y_8 = (1/3,1/3).$$

It is not difficult to see that S is invariant under $\mathcal{S}_S = \{S_j : j = 1, \ldots, 8\}$.

The overlaps $S_i(S) \cap S_j(S)$, $i \neq j$, are line segments. Thus, S is *not* strictly self-similar. On the other hand, although it is not immediately clear why these overlaps are negligible with respect to S, it is intuitively obvious that the dimension of S is greater than 1 (which is the dimension of a line segment), and so S should be self-similar; see Example 9.4.10.

c. Let K_0 be the piecewise linear curve described as the graph of the following function:

$$k(x) = \begin{cases} 0, & \text{if } x \in [0,1], \\ \sqrt{3}x - \sqrt{3}, & \text{if } x \in [1,3/2], \\ -\sqrt{3}x + 2\sqrt{3}, & \text{if } x \in [3/2,2], \\ 0, & \text{if } x \in [2,3]. \end{cases}$$

Let $\mathcal{S}_K$ be the family of similarities consisting of the following transformations:

$$S_1(x) = \frac{1}{3}x,\ S_2(x) = \frac{1}{3}O_1(x) + (1,0),$$
$$S_3(x) = \frac{1}{3}O_2(x) + (3/2, \sqrt{3}/2),\ S_4(x) = \frac{1}{3}x + (2,0),$$

where O_1 is the rotation by $\pi/3$ and O_2 is the rotation by $-\pi/3$. We define the set K_1 to be the set K_0 with each line segment replaced by $S_j(K_0)$, $j = 1, \ldots, 4$, and we proceed inductively, defining the set K_n to be the set K_{n-1} with every line segment replaced by $S_j(K_{n-1})$, $j = 1, \ldots, 4^n$, respectively. Geometrically, plot k on $[0,3]$ and see Figure 9.3. The set

$$K = \lim_{n \to \infty} K_n$$

is the *triadic Koch curve*. Clearly, K is invariant under $\mathcal{S}_K$ and the overlaps $S_i(K) \cap S_i(K)$, $i \neq j$, are single points, and thus negligible. The *Koch curve* is sometimes called the *snowflake curve*, since it is a part of a boundary of a "snowflake" constructed by Helge von Koch (1870–1924).

d. Let $E \subseteq [0,1]$ be a perfect symmetric set and let C_E be the associated *Cantor function*, defined in Example 1.3.17. This function is sometimes referred to as the *devil's staircase*. Let D be the graph of C_E. Clearly, it is built from scaled copies of itself and from straight lines that are inserted between them. Thus, D fails to be self-similar because of these straight lines.

It may seem disappointing that although the graph of the Cantor function seems to possess some self-similar features it is not a self-similar set. There are

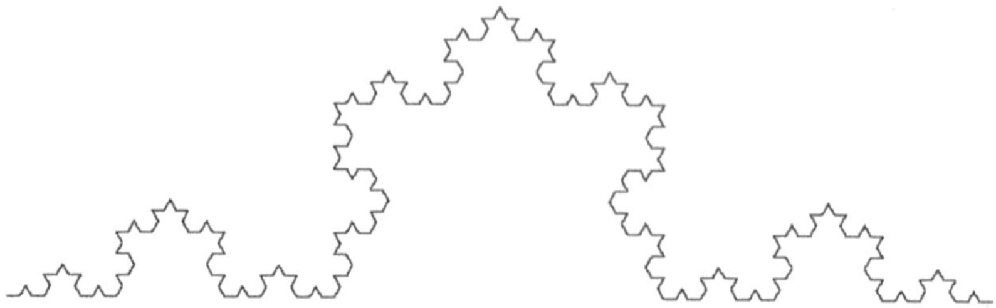

Fig. 9.3. Triadic Koch curve.

other analogous and well-known examples of sets that are not self-similar but that one would like to consider in this context. One family of such examples is the collection of *Koch islands.* The original *Koch island* is the set bounded by three joined copies of the *Koch curve* in the same fashion that three equal intervals build an equilateral triangle.

It was the study of appearances of scalings in various sciences that brought MANDELBROT [336] to formulate a notion of self-similar objects called *fractals.* His definition of a *fractal set* is that it is a set whose topological dimension is strictly less than its *Hausdorff dimension*; see Section 9.3. However, this definition is not ideal. MANDELBROT himself wrote,

> I continue to believe that one would do better without a definition. ... the immediate reason is that the present definition will be seen to exclude certain sets one would prefer to see included. More fundamentally, my definition involves D [Hausdorff dimension] and D_T [topological dimension], but it seems that the notion of fractal structure is more basic than either D or D_T. ... In other words, one should be able to define fractal structures as being invariant under some suitable collection of smooth transformations. But this task is unlikely to be an easy one.

Another, inequivalent, definition of a *fractal set* F is that its Hausdorff dimension is not an integer; see the wavelet fractal set in Figure 9.4 from [49].

The theory of fractals has been studied extensively from various points of view, and the references are numerous. Besides the work of MANDELBROT, we refer the interested reader to [83], [248], [250], [160], [161], [258], [153], [342], [460], and the references included therein, as well as the popular and authoritative tour de force [423] by MANFRED SCHROEDER.

9.2 Peano curve

Example 9.2.1. Peano curve

We shall construct a continuous function from the interval $[0,1]$ *onto* the square $[0,1] \times [0,1]$. Consider the partition of $[0,1]$ into nine intervals of

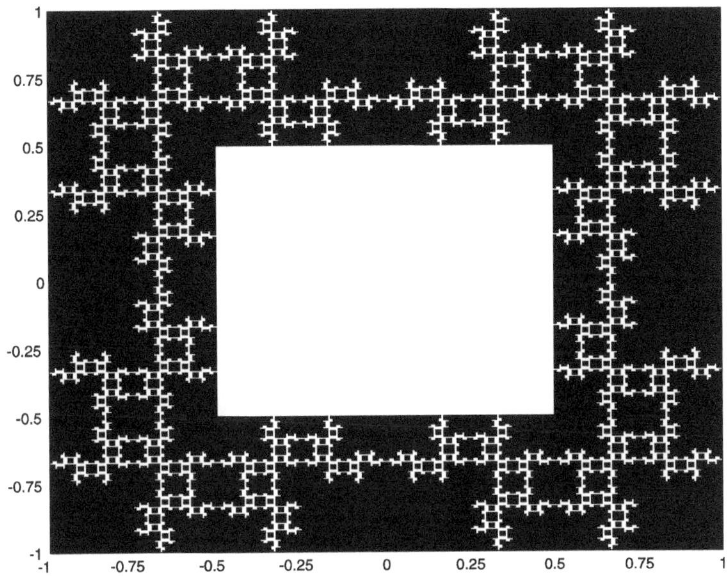

Fig. 9.4. Wavelet fractal set.

equal lengths $1/9$, and divide $[0,1] \times [0,1]$ into nine equal squares of side lengths $1/3$. We start by defining the function $p_1 : [0,1] \to [0,1] \times [0,1]$ as

$$p_1(0) = (0,0),\ p_1(1/9) = (1/3,1/3),\ p_1(2/9) = (0,2/3),\ p_1(1/3) = (1/3,1),$$
$$p_1(4/9) = (2/3,2/3),\ p_1(5/9) = (1,1/3),\ p_1(2/3) = (2/3,0),$$
$$p_1(7/9) = (1/3,1/3),\ p_1(8/9) = (2/3,2/3),\ p_1(1) = (1,1),$$

and we extend p_1 to $[0,1]$ linearly on the intervals $(j/9, (j+1)/9)$. The function p_1 has the following properties:

i. each interval in the partition of $[0,1]$ is transformed onto a diagonal of one of the squares in the partition of $[0,1] \times [0,1]$;

ii. each square in the partition of $[0,1] \times [0,1]$ has a diagonal that is an image of some interval in the partition of $[0,1]$.

Next, we proceed by induction. Our construction is similar to the inductive construction of the *snowflake curve* in Example 9.1.4*c*: we shall use scaled copies of the original graph of p_1 to replace the edges of consecutive approximations.

Suppose we have a partition of $[0,1]$ into 9^n equal intervals each of length 9^{-n}; and assume that $[0,1] \times [0,1]$ is divided into 9^n equal squares each of side length 3^{-n}. Assume that a function $p_n : [0,1] \to [0,1] \times [0,1]$ satisfies the following two properties:

i_n. each interval in the partition of $[0,1]$ into 9^n intervals is transformed onto a diagonal of one of the squares in the partition of $[0,1] \times [0,1]$;

Fig. 9.5. Peano curve.

ii_n. each square in the partition of $[0,1] \times [0,1]$ into 9^n equal squares has a diagonal that is an image of some interval in the partition of $[0,1]$.

Let $[a,b]$ be a generic interval in the partition of $[0,1]$, and assume that $p_n(a) = (x_1, x_2)$ and $p_n(b) = (y_1, y_2)$. Moreover, let q_j denote the composition of p_1 with a rotation by $\pi(j-1)/2$, $j = 1, \ldots, 4$, respectively. For $x \in [a,b]$ we define

$$p_{n+1}(x) = \begin{cases} (x_1, x_2) + 3^{-n} q_1(x), & \text{if } x_1 < y_1,\ x_2 < y_2, \\ (y_1, x_2) + 3^{-n} q_2(x), & \text{if } x_1 > y_1,\ x_2 < y_2, \\ (y_1, y_2) + 3^{-n} q_3(x), & \text{if } x_1 > y_1,\ x_2 > y_2, \\ (x_1, y_2) + 3^{-n} q_4(x), & \text{if } x_1 < y_1,\ x_2 > y_2. \end{cases}$$

In this way we have constructed a continuous function $p_{n+1} : [0,1] \to [0,1] \times [0,1]$ that satisfies the properties i_{n+1} and ii_{n+1}; see Figure 9.5. From our construction we deduce that

$$\forall\, x \in [0,1] \text{ and } \forall\, m \geq n, \quad \|p_m(x) - p_n(x)\| \leq 3^{-n}. \tag{9.1}$$

This implies that, for each $x \in [0,1]$, the sequence $\{p_n(x) : n = 1, \ldots\} \subseteq [0,1] \times [0,1]$ is a Cauchy sequence, and so it is convergent. Define

$$p(x) = \lim_{n \to \infty} p_n(x).$$

Statement (9.1) implies that

$$\forall\, x \in [0,1], \quad \|p(x) - p_n(x)\| \leq 3^{-n},$$

i.e., p_n converges uniformly to p on $[0,1]$. Thus, Proposition 1.3.15a implies that $p : [0,1] \to [0,1] \times [0,1]$ is continuous.

From the properties ii_n, $n = 1, \ldots$, we deduce that the image of p is dense in $[0,1] \times [0,1]$. Since a continuous image of a compact set is closed, we conclude that p is *onto*, i.e.,

$$p([0,1]) = [0,1] \times [0,1].$$

The function p is the *Peano transformation* or the *Peano curve*; see [368] for PEANO's original proof and see [7], whose proof, like ours, is based on SCHOENBERG's proof (1938).

For the purpose of the next theorem we introduce the notion of a *locally connected set* X, which is the set such that for each $x \in X$ and for each open neighborhood $U \subseteq X$ of x, there exists a connected set $Z \subseteq U$ such that $x \in \operatorname{int} Z$. Perhaps surprisingly, there exist connected sets that are *not* locally connected.

Proposition 9.2.2. *Let X be a compact, connected, locally connected metric space. Then,*

a. *X is path connected,*

b. *$\forall\, \varepsilon > 0$, $\exists\, \delta > 0$ such that $\forall\, x \neq y$ with $\rho(x,y) < \delta$, $\exists$ a continuous path q from x to y with* $\operatorname{diam}(q) < \varepsilon$.

The following theorem provides the classification of all metric spaces that are continuous images of $[0,1]$. It was obtained independently by MAZURKIEWICZ (1913) and by HAHN (1914).

Theorem 9.2.3. Hahn–Mazurkiewicz theorem
Let X be a nonempty metric space. Then X is a continuous image of $[0,1]$ if and only if X is a compact, connected, and locally connected space.

Proof. ($\Longrightarrow$) A continuous image of a compact and connected set is compact and connected. Moreover, using the fact that a compact metric space X is locally connected if and only if for each $\varepsilon > 0$ there exists a finite cover of X by compact connected sets of diameter not exceeding ε, we see that any continuous image of $[0,1]$ is also locally connected.

($\Longleftarrow$) From Corollary A.3.6 it follows that there exist a closed subset A of the Cantor set C and a surjective continuous function $\phi : A \to X$. We shall extend this function to a continuous function on the whole interval $[0,1]$ by a construction analogous to the construction of the Volterra example, Example 1.3.1. Let $a = \inf A$ and let $b = \sup A$. Since A is closed, the set $[a,b] \setminus A$ is an open set in $\mathbb{R}$, which implies that its components of connectivity are open intervals $(a_i, b_i), i = 1, \ldots$.

Assume that there are infinitely many intervals (a_i, b_i). Then $\lim_{i\to\infty} (b_i - a_i) = 0$. Thus, by the uniform continuity of ϕ, see Proposition 1.3.13, and from Proposition 9.2.2*b* it follows that there exists a sequence $\{i_n : n = 1, \ldots\}$ such that for each $i \geq i_n$ there is a continuous path from $\phi(a_i)$ to $\phi(b_i)$ with diameter less than $1/n$.

Without loss of generality we can assume that $i_n \leq i_{n+1}$, $n = 1, \ldots$. For $i_n \leq i < i_{n+1}$, let $\phi_i : [0,1] \to X$ be a continuous function defining such a path. For $i < i_1$, let $\phi_i : [0,1] \to X$ be any continuous path from $\phi(a_i)$ to $\phi(b_i)$. Define $f : [0,1] \to X$ to be

$$f(x) = \begin{cases} \phi(a), & \text{if } x \in [0,a], \\ \phi(x), & \text{if } x \in A, \\ \phi_i(x), & \text{if } x \in [a_i, b_i],\ i = 1, \ldots, \\ \phi(b), & \text{if } x \in [b,1]. \end{cases} \tag{9.2}$$

Thus, $f : [0,1] \to X$ is a continuous surjection.

When the number of components of connectivity is finite, we consider any family of continuous paths joining points $\phi(a_i)$ and $\phi(b_i)$, and we define the function f similarly to (9.2). □

The construction of the *Peano curve* may be viewed as a special example of a generalization of VON KOCH's construction of Example 9.1.4*c*. This generalized construction is generated by two piecewise linear shapes. One is the *initiator* and the other is the *generator*. The process starts by replacing each straight line of the initiator by an appropriately rescaled and rotated copy of the generator. At each next stage we repeat the procedure, that is, we replace each straight line of the previous stage with a copy of the generator, and we proceed ad infinitum. The *snowflake curve* is an example of this procedure, with the initiator equal to the generator. Other well-known examples include PÓLYA's and CESÀRO's *triangle sweeping curves*, e.g., [28], [98], which are modifications of PEANO's square-filling construction (Example 9.2.1).

The generalized VON KOCH construction yields many shapes of interest, besides plane-filling curves. Among the most important are various *fractals*, as defined in Section 9.1. We refer the interested reader to [337] for more examples and illustrations.

9.3 Hausdorff measure

Let $\mathcal{H}$ be a family of homeomorphisms of $\mathbb{R}^+$; e.g., Appendix A.1; cf. Problem 2.47.

Definition 9.3.1. Hausdorff outer measure

a. For each $h \in \mathcal{H}$ and $\varepsilon > 0$, define

$$\forall\, A \subseteq \mathbb{R}^d, \quad \mu_h^\varepsilon(A) = \inf\left\{\sum_{j=1}^{\infty} h(\text{diam}\,(A_j)) : A \subseteq \bigcup_{j=1}^{\infty} A_j, \text{diam}\,(A_j) < \varepsilon\right\},$$

and

$$\forall\, A \subseteq \mathbb{R}^d, \quad \mu_h^*(A) = \sup_{\varepsilon>0} \mu_h^\varepsilon(A).$$

b. If $h \in \mathcal{H}$ is an increasing function, then for all $\varepsilon_1 > \varepsilon_2$ we have

$$\forall\, A \subseteq \mathbb{R}^d, \quad \mu_h^{\varepsilon_1}(A) \leq \mu_h^{\varepsilon_2}(A).$$

In this case,

$$\forall\, A \subseteq \mathbb{R}^d, \quad \mu_h^*(A) = \lim_{\varepsilon\to 0} \mu_h^\varepsilon(A).$$

c. If $h(t) = t^p$, $p > 0$, we write $\mu_h^\varepsilon = \mu_p^\varepsilon$ and $\mu_h^* = \mu_p^*$. The set function $\mu_p^* : \mathcal{P}(\mathbb{R}^d) \to \mathbb{R}^+ \cup \{\infty\}$ is *Hausdorff outer measure* on $\mathbb{R}^d$, depending on p.

It is not difficult to verify that for $p = 0$ the associated Hausdorff outer measure μ_0^* is the counting measure c.

Remark. In the definition of μ_p^ε we have assumed that the covering sets A_j, $j = 1, \ldots,$ are arbitrary. However, we may without loss of generality consider either closed sets, since the closure of a set has the same diameter as the original set, or open sets, since the set $\{x : \text{dist}\,(x, A) < \delta\}$ is open and its diameter is diam $(A) + 2\delta$ for any $\delta > 0$ (dist (x, A) is defined as $\inf\{\|x - y\| : y \in A\}$).

Proposition 9.3.2. *Let $A \subseteq \mathbb{R}^d$, and let $p, q > 0$.*
a. *If $\mu_p^*(A) < \infty$, then $\mu_q^*(A) = 0$ for all $q > p$.*
b. *If $\mu_p^*(A) > 0$, then $\mu_q^*(A) = \infty$ for all $q < p$.*

Proof. We begin by noting that statements a and b are equivalent. Thus, it suffices to prove part a.

Fix $\varepsilon > 0$ and $q > p$. Let $A \subseteq \mathbb{R}^d$ and let $\{A_j : j = 1, \ldots\}$ be a sequence of sets with each diam $(A_j) < \varepsilon$, and such that $A \subseteq \bigcup A_j$. Then

$$\sum_{j=1}^{\infty} \text{diam}\,(A_j)^q \leq \varepsilon^{q-p} \sum_{j=1}^{\infty} \text{diam}\,(A_j)^p.$$

This implies

$$\mu_q^\varepsilon(A) \leq \varepsilon^{q-p} \mu_p^\varepsilon(A),$$

and, consequently, if $\mu_p^*(A) < \infty$, then

$$\mu_q^*(A) = 0.$$

□

Proposition 9.3.2 allows us to define a generalized notion of dimension, which has already been mentioned in Sections 8.5 and 9.1.

Definition 9.3.3. Hausdorff dimension
Let $A \subseteq \mathbb{R}^d$. The *Hausdorff dimension* of A is defined to be

$$\dim_H(A) = \sup\{p \geq 0 : \mu_p^*(A) > 0\} = \inf\{p \geq 0 : \mu_p^*(A) = 0\}.$$

Remark. In [223], HAUSDORFF (1868–1942) generalized CARATHÉODORY's definition of dimension to the noninteger case. In fact, he proved a fundamental property of Hausdorff dimension for irregular curves such as the Koch curve.

For perspective, note that the integral definition of the length L of the graph of a smooth function f defined on $[0,1]$ depends on the following idea. We approximate L by $L(\varepsilon) = \varepsilon N_\varepsilon$. Here, N_ε is the number of contiguous line segments of length ε needed to "travel" from $f(0)$ to $f(1)$, where each line segment intersects the graph at least at the endpoint of the segment. Because of the smoothness, we can prove that $L(\varepsilon) \to L$ as $\varepsilon \to 0$, and L is designated the *length* of the graph.

The convergence described in the previous paragraph does not occur in the case of the Koch curve, which is a graph in $\mathbb{R}^2$. In fact, in this case there is a critical exponent D $(= \dim_H(K)) > 1$ for which $\varepsilon^D N_\varepsilon$ stays finite, whereas $\lim_{\varepsilon\to 0} \varepsilon^C N_\varepsilon = \infty$ if $C < D$ and $\lim_{\varepsilon\to 0} \varepsilon^C N_\varepsilon = 0$ if $C > D$. It can be calculated that $\dim_H(K) = \log(4)/\log(3)$. This phenomenon is an essential feature of Hausdorff dimension.

Example 9.3.4. Examples of Hausdorff dimension
a. Let $Q = [0,1]^d$ be a unit cube in $\mathbb{R}^d$. We shall verify that $0 < \mu_d^*(Q) < \infty$.

On the one hand, we can cover Q by a union of N^d adjacent cubes of side length $1/N$, which implies

$$\mu_d^\varepsilon(Q) \leq N^d \left(\text{diam}\left([0,1/N]^d\right)\right)^d \leq \sqrt{d}^d < \infty.$$

Thus, $\mu_d^*(Q) < \infty$.

On the other hand, we first note that, for each set $A \subseteq \mathbb{R}^d$ for which $\text{diam}\,(A) < \infty$, we have $m^{d*}(A) \leq 2^d\,(\text{diam}\,(A))^d$. Thus,

$$\begin{aligned}\mu_d^\varepsilon(Q) &\geq 2^{-d}\inf\left\{\sum_{j=1}^{\infty} m^{d*}(A_j) : Q \subseteq \bigcup_{j=1}^{\infty} A_j, \text{diam}\,(A_j) < \varepsilon\right\}\\ &\geq 2^{-d} m^{d*}(Q) = 2^{-d},\end{aligned}$$

and so $\mu_d^*(Q) \geq 2^{-d} > 0$.

Combining these two statements, we obtain

$$\dim_H(Q) = d.$$

b. Let C be the ternary Cantor set, defined in Section 1.2. The collection of intervals $\{C_n^j : j = 1, \ldots, 2^n\}$, defined in Example 1.2.7c, forms a covering of C for each $n = 1, \ldots$. Thus,

$$\forall\, n \text{ such that } \frac{1}{3^n} \leq \varepsilon, \quad \mu_p^\varepsilon(C) \leq \sum_{j=1}^{2^n}\left(\frac{1}{3^n}\right)^p = \frac{2^n}{3^{np}},$$

and, consequently,

$$\mu_p^*(C) \leq \lim_{n\to\infty} \frac{2^n}{3^{np}}.$$

Therefore, in order for $\mu_p^*(C) < \infty$, we must assume that

$$p \geq \frac{\log(2)}{\log(3)} = \log_3(2). \tag{9.3}$$

On the other hand, for the lower estimate, we shall verify the inequality,

$$\mu_{\log_3(2)}^*(C) > 0. \tag{9.4}$$

In fact, it is enough to show that

$$\sum_{j=1}^{\infty} \text{diam } (A_j)^{\log_3(2)} \geq 1 \tag{9.5}$$

for any open covering $\{A_j : j = 1, \dots\}$ of C; cf. the Remark after Definition 9.3.1. Since C is compact, we assume that there are only finitely many sets A_j, $j = 1, \dots, N$, and that they are all open intervals. Next, let $n > 0$ be sufficiently large that each of the sets C_n^k, $k = 1, \dots, 2^n$, is contained in some A_j. For such n and for any A_j, we note that

$$\text{diam } (A_j)^{\log_3(2)} \geq \sum_{C_n^k \subseteq A_j} \text{diam } (C_n^h)^{\log_3(2)}.$$

This observation, in turn, implies (9.5), since

$$\sum_{j=1}^{N} \text{diam } (A_j)^{\log_3(2)} \geq \sum_{j=1}^{N} \sum_{C_n^k \subseteq A_j} \text{diam } (C_n^k)^{\log_3(2)} \geq \sum_{k=1}^{2^n} \text{diam } (C_n^k)^{\log_3(2)} = 1.$$

Combining (9.3) with (9.4) yields that

$$\dim_H(C) = \frac{\log(2)}{\log(3)} = \log_3(2).$$

We now verify that, for all $p \geq 0$, the *Hausdorff outer measure* μ_p^* is indeed an *outer measure* as defined in Section 2.4. Clearly, μ_p^* is nonnegative and monotone. Thus, we need to check only that it is subadditive, i.e.,

$$\mu_p^*\left(\bigcup_{n=1}^{\infty} A_n\right) \leq \sum_{n=1}^{\infty} \mu_p^*(A_n).$$

This, in turn, follows from the fact that the set functions μ_p^{ε}, $p \geq 0$, $\varepsilon > 0$, are subadditive, since

$$\inf \left\{ \sum_{j=1}^{\infty} \operatorname{diam}(B_j)^p : \bigcup_{n=1}^{\infty} A_n \subseteq \bigcup B_j, \ \operatorname{diam}(B_j) < \varepsilon \right\}$$

$$\leq \inf \left\{ \sum_{n=1}^{\infty} \sum_{j=1}^{\infty} \operatorname{diam}(B_{j,n})^p : A_n \subseteq \bigcup B_{j,n}, n = 1, \ldots, \operatorname{diam}(B_{j,n}) < \varepsilon \right\}.$$

We conclude, using the Carathéodory theory of constructing measures from outer measures, see Theorem 2.4.19, that, for each $p \geq 0$, there exist a σ-algebra $\mathcal{A}_p \subseteq \mathcal{P}(\mathbb{R}^d)$ and a nonnegative σ-additive set function μ_p on $\mathcal{A}_p$ that is the restriction of μ_p^* to $\mathcal{A}_p$, i.e., μ_p is a measure on $\mathcal{A}_p$, $p \geq 0$. We call μ_p the *Hausdorff measure* on $\mathbb{R}^d$, depending on p.

Some standard references on Hausdorff measure are [223], [248], [271], Chapter 2, [163], [400]. There is also CARLESON's book [92], which has an extensive list of references assembled by HANS WALLIN. For more recent references see, e.g., [160], [342].

Proposition 9.3.5. *Let* $A, B \subseteq \mathbb{R}^d$ *have the property that* $\operatorname{dist}(A, B) > 0$. *Then*

$$\forall\, p \geq 0, \quad \mu_p^*(A \cup B) = \mu_p^*(A) + \mu_p^*(B).$$

Theorem 9.3.6. Hausdorff measure
For each $p \geq 0$, $(\mathbb{R}^d, \mathcal{A}_p, \mu_p)$ *is a regular Borel measure space.*

Proof. In order to prove that $\mathcal{B}(\mathbb{R}^d) \subseteq \mathcal{A}_p$, it is enough to show that closed subsets of $\mathbb{R}^d$ are μ_p-measurable. This reduces to showing that

$$\forall\, p \geq 0, \forall\, A \subseteq \mathbb{R}^d,\ A \text{ closed}, \forall\, E \subseteq \mathbb{R}^d, \quad \mu_p^*(E) \geq \mu_p^*(E \cap A) + \mu_p^*(E \cap A^\sim).$$

We can assume without loss of generality that $\mu_p^*(E) < \infty$. Let $A^{1/n} = \{x \in \mathbb{R}^d : \operatorname{dist}(x, A) \leq 1/n\}$. By monotonicity of μ_p^* and from Proposition 9.3.5 we deduce that

$$\begin{aligned} \mu_p^*(E) &\geq \mu_p^*(E \cap A) + \mu_p^*(E \cap (A^{1/n})^\sim) \\ &\geq \mu_p^*(E \cap A) + \mu_p^*(E \cap A^\sim) - \mu_p^*(E \cap A^{1/n} \cap A^\sim). \end{aligned}$$

The sets $E \cap A^{1/n} \cap A^\sim$ are decreasing, their intersection is the empty set since A is closed, and we can write

$$E \cap A^{1/n} \cap A^\sim = \bigcup_{k=n}^{\infty} E \cap A^{1/k} \cap (A^{1/(k+1)})^\sim.$$

Thus, it suffices to show that $\sum \mu_p^*(E \cap A^{1/k} \cap (A^{1/(k+1)})^\sim) < \infty$. Indeed, let $B_k = E \cap A^{1/k} \cap (A^{1/(k+1)})^\sim$. Then, using Proposition 9.3.5, we obtain

$$\begin{aligned}\sum_{k=1}^{\infty} \mu_p^*(B_k) &= \lim_{n\to\infty} \left(\sum_{k=1}^{n} \mu_p^*(B_{2k}) + \sum_{k=1}^{n} \mu_p^*(B_{2k+1}) \right) \\ &= \lim_{n\to\infty} \left[\mu_p^* \left(\bigcup_{k=1}^{n} B_{2k} \right) + \mu_p^* \left(\bigcup_{k=1}^{n} B_{2k+1} \right) \right] \\ &\le \lim_{n\to\infty} 2\mu_p^*(E \cap A^1) < \infty.\end{aligned}$$

The regularity of μ_p follows from the Remark following Definition 9.3.1 and from the fact that compact sets in $\mathbb{R}^d$ are bounded. □

Example 9.3.7. Examples of Hausdorff measure

The set L of (transcendental) Liouville numbers was defined in Problem 1.25. It is not difficult to prove that not only is $m(L) = 0$ but also that $\mu_p(L) = 0$ for each $p > 0$, e.g., [362], page 9.

By comparison, if C is the ternary Cantor set, then

$$\mu_p(C) = \begin{cases} \infty, & \text{if } 0 < p < \log(2)/\log(3), \\ 1, & \text{if } p = \log(2)/\log(3), \\ 0, & \text{if } p > \log(2)/\log(3); \end{cases}$$

cf. Example 9.3.4*b* and [160], Section 1.5.

Remark. Hausdorff measure μ_p, $p \geq 0$, like Lebesgue measure, is translation- and rotation-invariant; and if $D_a(A) = \{ax \in \mathbb{R}^d : x \in A\}$, $a \in \mathbb{R}^+$, is the dilation of a set A by a factor $a > 0$, then

$$\mu_p(D_a(A)) = a^p \mu_p(A).$$

This observation, about translation invariance of both Hausdorff and Lebesgue measure, implies in particular that they are both *Haar measures* on $\mathbb{R}^d$; see Appendix B.9. Thus, the uniqueness of Haar measure implies that Hausdorff measure μ_d is a constant multiple of Lebesgue measure m^d on the Borel sets $\mathcal{B}(\mathbb{R}^d)$. In fact,

$$\forall\, A \in \mathcal{B}(\mathbb{R}^d), \quad \mu_d(A) = \frac{2^d \Gamma((d/2)+1)}{\pi^{d/2}} m^d(A).$$

Example 9.3.8. Hausdorff measure and surface measure

Hausdorff measure in $\mathbb{R}^d$ is a natural tool to define the surface measure for m-dimensional manifolds in $\mathbb{R}^d$. In particular, if M is a subset of an m-dimensional linear subspace of $\mathbb{R}^d$, then Hausdorff measure μ_m coincides with a constant multiple of the m-dimensional Lebesgue measure. We observe that, as opposed to the surface measures constructed in Section 8.7, Hausdorff measure μ_m is defined for all Borel subsets of $\mathbb{R}^d$, and not just for the subsets of a fixed manifold M.

We write

$$\forall\, A \in \mathcal{B}(\mathbb{R}^d) \text{ and } \forall\, 0 < m \le d, \quad \alpha_m(A) = \frac{\pi^{m/2}}{2^m \Gamma((m/2)+1)} \mu_m(A).$$

With this notation, we have, for every nonnegative Borel measurable function $f : \mathbb{R}^d \to \mathbb{R}$, that

$$\int_{\mathbb{R}^d} f\, dm^d = \int_{S^{d-1}} \left(\int_0^\infty f(r\omega) r^{d-1}\, dr \right) d\alpha_{d-1}(\omega);$$

cf. Theorem 8.7.10.

We close this section with a few results that analyze sets of measure zero for any sort of thickness they might have. A profound study in this area has been made by BOREL [67], Chapter 4, and there are still many answers (and questions) to be found; see the Remark after Problem 2.47.

Proposition 9.3.9. ***a.*** *Let $A \in \mathcal{B}(\mathbb{R}^d)$. Then, $\mu_h(A) = 0$ for all $h \in \mathcal{H}$ if and only if for every strictly decreasing sequence $\{a_n : n = 1, \ldots\} \subseteq \mathbb{R}^+$ tending to 0, there exists a sequence of intervals I_n such that*

$$A \subseteq \bigcup_{n=1}^{\infty} I_n \quad \textit{and} \quad m(I_n) \le a_n,\ n = 1, \ldots.$$

b. *Let $F \subseteq \mathbb{R}^d$ be concentrated in a neighborhood of a countable set H in the following sense:*

$$\forall\, U \text{ open}, \quad H \subseteq U \Longrightarrow F \subseteq U.$$

Let $\mathcal{K} = \{k : \mathbb{R}^+ \to \mathbb{R}^* : k(0) = 0, k(t) > 0$ for $t > 0$, and k is continuous and increasing $\}$. Then

$$\forall\, k \in \mathcal{K}, \quad \mu_k(F) = 0.$$

Let $E \subseteq [0,1]$ be a Borel set with $m(E) = 0$. As a strengthening of Proposition 9.3.9*a*, we have the following assertion: There exists a sequence $\{a_n : \sum a_n < \infty\} \subseteq \mathbb{R}^+$ with the property that

$$\forall\, \varepsilon > 0,\ \exists\, \{I_n = (c_n, d_n) : n = 1, \ldots\} \text{ such that}$$

$$E \subseteq \bigcup_{n=1}^{\infty} I_n \quad \text{and} \quad m(I_n) \le \varepsilon a_n,\ n = 1, \ldots.$$

Indeed, let $\{I_{j,k} : j,k = 1, \ldots\}$ be a covering of E by open intervals such that $\sum_k m(I_{j,k}) < (1/4)^j$. Enumerate $\{I_{j,k} : j,k = 1, \ldots\}$ as $\{J_n : n = 1, \ldots\}$, and let I_n be an interval with the same center as J_n such that $2^j m(J_n) = m(I_n)$, where J_n corresponds to some $I_{j,k}$. Let $a_n = m(I_n)$.

We say that the sequence $\{a_n : n = 1, \ldots\}$ is *associated* with the set E. Part *b* of the following result is a consequence of part *a*.

Proposition 9.3.10. ***a.*** *Let $E \subseteq [0,1]$ be a Borel set with Hausdorff dimension $\dim_H(E) > 0$. If*

$$\forall\, p > 0, \quad \sum_{n=1}^{\infty} a_n^p < \infty,$$

then $\{a_n : n = 1, \dots\}$ cannot be associated with E.

b. *There exists $\varepsilon > 0$ such that for any cover of the Cantor set C by open intervals $I_n, n = 1, \dots,$ there exists n_0 for which $m(I_{n_0}) > \varepsilon/2^{n_0}$.*

c. *Let $0 < r < \log_2(3) - 1$ and set $a_n = 1/n^{\log_2(3)-1}$, $n = 1, \dots$. Then the sequence $\{a_n : n = 1, \dots\}$ is associated with the Cantor set C.*

In light of part *a* it is natural to ask whether for any convergent series $\sum a_n$ of positive numbers there is a set $E \subseteq [0,1]$, $m(E) = 0$, such that $\{a_n : n = 1, \dots\}$ is *not* associated with E. The answer is that such a set E always exists, and the proof depends on a result of ARYEH DVORETZKY (1948) [400], pages 68 ff.; the details are due to E. BOARDMAN [63].

9.4 Hausdorff dimension and self-similar sets

Having defined the notions of Hausdorff measure and Hausdorff dimension in the previous section, we can now make a precise definition of *self-similar sets*. The approach we use is due to JOHN E. HUTCHINSON [250].

Definition 9.4.1. Self-similar sets
A set $A \subseteq \mathbb{R}^d$ is *self-similar* if

i. $\exists\, \mathcal{S} = \{S_j : j = 1, \dots, n\}$, a family of similarities $S_j(x) = rO_j(x) + y_j$, $j = 1, \dots, n$, $r < 1$, such that A is invariant under $\mathcal{S}$,

ii. $\mu_p(A) > 0$, for some $p > 0$, and

iii. $\forall\, i \neq j, \quad \mu_p(S_i(A) \cap S_j(A)) = 0$, where $p = \dim_H(A)$.

In case more precision is needed, we say that A is *self-similar with respect to $\mathcal{S}$*.

It turns out that there are not too many invariant sets for a given family of similarities, as the following result of HUTCHINSON shows.

Theorem 9.4.2. Existence and uniqueness of invariant sets
Let $\mathcal{S}$ be a family of similarities on $\mathbb{R}^d$, with $r > 0$. Then there exists a unique compact set $F_{\mathcal{S}} \subseteq \mathbb{R}^d$ invariant under $\mathcal{S}$.

Although for any $\mathcal{S}$ there is only one closed and bounded invariant set, there may be many sets that are not closed and are unbounded, e.g., $\mathbb{R}^d$.

If a closed nonempty set $A \subseteq \mathbb{R}^d$ satisfies the inclusion,

$$\bigcup_{j=1}^{n} S_j(A) \subseteq A,$$

then $F_{\mathcal{S}} \subseteq A$.

Definition 9.4.3. Invariant measures
Let $(\mathbb{R}^d, \mathcal{B}(\mathbb{R}^d), \mu)$ be a Borel measure space. The measure μ is *invariant* or *self-similar* under a given family of similarities $\mathcal{S} = \{S_j : j = 1, \ldots, n\}$ on $\mathbb{R}^d$ if

$$\forall\, A \in \mathcal{B}(\mathbb{R}^d), \quad \mu_{\mathcal{S}}(A) = \frac{1}{n}\sum_{j=1}^{n} \mu(S_j^{-1}(A)).$$

The analysis and harmonic analysis of these invariant or self-similar measures is an exciting recent development with many applications and with roots based in the study of some of the classical functions defined in Chapter 1; see [458], [257], [255], [256].

Theorem 9.4.4. Existence and uniqueness of invariant measures
a. *Let $\mathcal{S}$ be a family of similarities on $\mathbb{R}^d$ with $r < 1$. There exists a unique regular Borel measure $\mu_{\mathcal{S}}$ with compact support that is invariant under $\mathcal{S}$ and for which $\mu_{\mathcal{S}}(\mathbb{R}^d) = 1$.*

b. *The support of $\mu_{\mathcal{S}}$ is the unique compact invariant set $F_{\mathcal{S}}$.*

For the proof of Theorem 9.4.4 we again refer the reader to [250].

Proposition 9.4.5. *Let $\mathcal{S} = \{S_j : j = 1, \ldots, n\}$ be a family of similarities on $\mathbb{R}^d$ with $r < 1$, and let $F_{\mathcal{S}}$ be the compact set invariant under $\mathcal{S}$.*

a. $\mu_{\log_{1/r}(n)}(F_{\mathcal{S}}) < \infty$, *i.e.,* $\log_{1/r}(n) \geq \dim_H(F_{\mathcal{S}})$.

b. *If* $0 < \mu_p(F_{\mathcal{S}}) < \infty$, *then $F_{\mathcal{S}}$ is self-similar if and only if* $p = \log_{1/r}(n)$.

Proof. **a.** Clearly,

$$\forall\, m \in \mathbb{N}, \quad F_{\mathcal{S}} = \bigcup_{j_1,\ldots,j_m=1}^{n} S_{j_1} \circ \cdots \circ S_{j_m}(F_{\mathcal{S}}).$$

Moreover,

$$\sum_{j_1,\ldots,j_m=1}^{n} \operatorname{diam}\,(S_{j_1} \circ \cdots \circ S_{j_m}(F_{\mathcal{S}}))^{\log_{1/r}(n)} = \sum_{j_1,\ldots,j_m=1}^{n} r^{\log_{1/r}(n)} \operatorname{diam}\,(F_{\mathcal{S}})^{\log_{1/r}(n)}$$
$$= (\operatorname{diam}\,(F_{\mathcal{S}}))^{\log_{1/r}(n)}\,.$$

Since the diameters of $S_{j_1} \circ \cdots \circ S_{j_m}(F_{\mathcal{S}})$ are arbitrarily small as $m \to \infty$, a covering of $F_{\mathcal{S}}$ by $\{S_{j_1} \circ \cdots \circ S_{j_m}(F_{\mathcal{S}}) : j_1, \ldots, j_m = 1, \ldots, n\}$ yields $\mu_{\log_{1/r}(n)}(F_{\mathcal{S}}) < \infty$.

b. ($\Longrightarrow$) Let $F_{\mathcal{S}}$ be a self-similar set such that $0 < \mu_p(F_{\mathcal{S}}) < \infty$. In particular, $\mu_p(S_i(F_{\mathcal{S}}) \cap S_j(F_{\mathcal{S}})) = 0$ for $i \neq j$. Thus,

$$\mu_p(F_{\mathcal{S}}) = \sum_{j=1}^{n} \mu_p(S_j(F_{\mathcal{S}})) = \sum_{j=1}^{n} r^p \mu_p(F_{\mathcal{S}}) = n r^p \mu_p(F_{\mathcal{S}}),$$

which implies that $p = \log_{1/r}(n)$.

($\Longleftarrow$) Assume $0 < \mu_{\log_{1/r}(n)}(F_{\mathcal{S}}) < \infty$. Then

$$\begin{aligned}\mu_{\log_{1/r}(n)}(F_{\mathcal{S}}) &\leq \sum_{j=1}^{n} \mu_{\log_{1/r}(n)}(S_j(F_{\mathcal{S}})) = \sum_{j=1}^{n} r^{\log_{1/r}(n)} \mu_{\log_{1/r}(n)}(F_{\mathcal{S}}) \\ &= n r^{\log_{1/r}(n)} \mu_{\log_{1/r}(n)}(F_{\mathcal{S}}) = \mu_{\log_{1/r}(n)}(F_{\mathcal{S}}).\end{aligned}$$

Therefore, $\mu_{\log_{1/r}(n)}(F_{\mathcal{S}}) = \sum_{j=1}^{n} \mu_{\log_{1/r}(n)}(S_j(F_{\mathcal{S}}))$, and it follows from a modification of Problem 2.18 that $\mu_{\log_{1/r}(n)}(S_i(F_{\mathcal{S}}) \cap S_j(F_{\mathcal{S}})) = 0$ for $i \neq j$.

□

Proposition 9.4.5 provides a simplification in verifying whether an invariant set is in fact self-similar, since we no longer have to measure the overlaps. We can make this task even simpler if we impose an additional assumption of the family of similarities; see [250].

Definition 9.4.6. Open set condition
A family of similarities $\mathcal{S} = \{S_j : j = 1, \ldots, n\}$ on $\mathbb{R}^d$ satisfies the *open set condition* if there exists a nonempty, bounded, open set $U_{\mathcal{S}} \subseteq \mathbb{R}^d$ such that

i. $\bigcup_{j=1}^{n} S_j(U_{\mathcal{S}}) \subseteq U_{\mathcal{S}}$, and
ii. $\forall\, i \neq j, \quad S_i(U_{\mathcal{S}}) \cap S_j(U_{\mathcal{S}}) = \emptyset$. We call $U_{\mathcal{S}}$ a *separating set* for $\mathcal{S}$.

It is not difficult to verify that the invariant sets we have considered so far satisfy the open set condition. In fact, for the ternary Cantor set C we can take $U = (0,1)$; a separating set associated with the Sierpiński carpet is the open square $(0,1) \times (0,1)$; and a separating set for the Koch curve is the open triangle with vertices $(0,0)$, $(3,0)$, and $(\sqrt{3}/2, 3/2)$.

Lemma 9.4.7. *Let $0 < c_1 < c_2 < \infty$ and let $0 < r < \infty$. Let $\{U_j : j = 1, \ldots\}$ be a disjoint family of open subsets of $\mathbb{R}^d$. If each U_j contains a ball of radius rc_1 and is contained in a ball of radius rc_2, then, for each $x \in \mathbb{R}^d$, at most $(1+2c_2)^d c_1^d$ of the sets $\overline{U_j} \cap B(x,r)$ are nonempty.*

Proof. If $\overline{U_j} \cap B(x,r) \neq \emptyset$, then $\overline{U_j} \subseteq B(x, r(1+2c_2))$. If there are N such sets U_j, then, since each of them contains a ball of radius rc_1, there are N such disjoint balls contained in $B(x, r(1+2c_2))$. By comparing the volumes, we obtain the required inequality,

$$N(c_1)^d \leq (1+2c_2)^d.$$

□

Theorem 9.4.8. Hausdorff dimension and the open set condition
Let $\mathcal{S} = \{S_j : j = 1, \ldots, n\}$ be a family of similarities on $\mathbb{R}^d$ with $r < 1$ that satisfies the open set condition. Then

$$0 < \mu_{\log_{1/n}(r)}(F_{\mathcal{S}}) < \infty,$$

i.e., $\dim_H(F_{\mathcal{S}}) = \log_{1/r}(n)$.

Proof. i. From the invariance of $F_{\mathcal{S}}$ we have that $F_{\mathcal{S}} = \bigcup (S_{j_1} \circ \cdots \circ S_{j_m})(F_{\mathcal{S}})$. Let $\varepsilon_m = \text{diam}\,(S_{j_1} \circ \cdots \circ S_{j_m})(F_{\mathcal{S}}) = r^m \text{diam}\,(F_{\mathcal{S}})$. Thus, we have

$$\begin{aligned} \mu^{\varepsilon_m}_{\log_{1/r}(n)}(F_{\mathcal{S}}) &\leq \sum_{j_1,\ldots,j_m=1}^{n} \varepsilon_m^{\log_{1/r}(n)} \\ &= n^m r^{m \log_{1/r}(n)} \text{diam}\,(F_{\mathcal{S}}) = \text{diam}\,(F_{\mathcal{S}}). \end{aligned} \tag{9.6}$$

Because we have assumed $r < 1$, it follows that $\varepsilon_m \to 0$ as $m \to \infty$. Therefore, (9.6) implies that $\mu_{\log_{1/r}(n)}(F_{\mathcal{S}}) < \infty$.

ii. Let $c_1 > 0$ be the radius of a ball contained in $U_{\mathcal{S}}$, and let $c_2 < \infty$ be the radius of a ball containing $U_{\mathcal{S}}$. We shall show that for any $\varepsilon < 1$, $\mu_{\mathcal{S}}(B(x,\varepsilon)) \leq (1+2c_2)^d \varepsilon^{\log_{1/r}(n)}/(rc_1)^d$ for all $x \in \mathbb{R}^d$. Indeed, let m be chosen such that $r^m < \varepsilon \leq r^{m-1}$. Invariance of $\mu_{\mathcal{S}}$ implies that

$$\mu_{\mathcal{S}}(B(x,\varepsilon)) = \frac{1}{n^m} \sum_{j_1,\ldots,j_m=1}^{n} \mu_{\mathcal{S}}((S_{j_1} \circ \cdots \circ S_{j_m})^{-1}(B(x,\varepsilon))).$$

Theorem 9.4.4*b* implies that the supports of $\mu_{\mathcal{S}} \circ (S_{j_1} \circ \cdots \circ S_{j_m})^{-1}$ are contained in $\overline{S_{j_1} \circ \cdots \circ S_{j_m}(U_{\mathcal{S}})}$. These sets, in turn, are disjoint, and each one of them contains a ball of radius $r^m c_1$ and is contained in a ball of radius $r^m c_2$. In particular, these sets contain balls of radius $r\varepsilon c_1$ and they are contained in balls of radius εc_2. Thus, Lemma 9.4.7 implies that at most $(1+2c_2)^d (rc_1)^{-d}$ of the sets $\overline{S_{j_1} \circ \cdots \circ S_{j_m}(U_{\mathcal{S}})} \cap B(x,\varepsilon)$ are nonempty. Hence,

$$\begin{aligned} \mu_{\mathcal{S}}(B(x,\varepsilon)) &\leq \frac{1}{n^m} \frac{(1+2c_2)^d}{(rc_1)^d} = \frac{(1+2c_2)^d}{(rc_1)^d} r^{m \log_{1/r}(n)} \\ &\leq \frac{(1+2c_2)^d}{(rc_1)^d} \varepsilon^{\log_{1/r}(n)}. \end{aligned}$$

iii. In order to prove that $\mu_{\log_{1/r}(n)}(F_{\mathcal{S}}) > 0$, it is enough to show that, for any covering $\{A_j\}$ of $F_{\mathcal{S}}$, $\sum \text{diam}\,(A_j)^{\log_{1/r}(n)}$ is bounded away from 0. Since every set of diameter δ is contained in a closed ball of diameter δ, we may assume, without loss of generality, that each A_j is a ball. Thus, using the result of part *ii*, we obtain

$$\begin{aligned} \sum_{j=1}^{\infty} \text{diam}\,(A_j)^{\log_{1/r}(n)} &\geq \frac{(rc_1)^d}{(1+2c_2)^d} \frac{1}{2^{\log_{1/r}(n)}} \sum_{j=1}^{\infty} \mu_{\mathcal{S}}(A_j) \\ &\geq \frac{(rc_1)^d}{(1+2c_2)^d} \frac{1}{2^{\log_{1/r}(n)}} \mu_{\mathcal{S}}(F_{\mathcal{S}}) > 0. \end{aligned}$$ □

We can now use the observation from Proposition 9.4.5*b* to restate Theorem 9.4.8 as a result about self-similar sets. In particular, we do not need to measure the overlaps $S_i(A) \cap S_j(A)$.

Corollary 9.4.9. *Let $\mathcal{S} = \{S_j : j = 1, \ldots, n\}$ be a family of similarities on $\mathbb{R}^d$ with $r < 1$ that satisfies the open set condition. Then $F_{\mathcal{S}}$ is self-similar.*

Example 9.4.10. Self-similar sets revisited
a. The generalized Cantor set C_ξ is a special example of a perfect symmetric set, see Example 1.2.8, where $\xi_j = \xi$, $j = 1, \ldots,$ for some $\xi \in (0, 1/2)$. The set C_ξ is self-similar and its Hausdorff dimension is $\dim_H(C_\xi) = \log_{1/\xi}(2) = \log(2)/\log(1/\xi) < 1$. See Example 9.3.4*b* for the ternary Cantor set C, where $\xi = 1/3$.

b. Corollary 9.4.9 implies that the Sierpiński carpet S is a self-similar set with Hausdorff dimension

$$\dim_H(S) = \log_3(8) < 2.$$

The Menger sponge M is a self-similar set of Hausdorff dimension

$$\dim_H(M) = \log_3(20) < 3.$$

The Koch curve K is also a self-similar set with Hausdorff dimension

$$\dim_H(K) = \log_3(4) > 1.$$

Remark. Theorem 9.4.8 and Corollary 9.4.9 are stated for sets invariant under similarity transformations with fixed scaling $r < 1$. We can generalize these results to self-affine sets, that is, the sets invariant under families of similarities with different scaling factors, r_j, $j = 1, \ldots, n$. Then the role of the number $\log_{1/n}(r)$ is played by the unique number D for which

$$\sum_{j=1}^{n} r_j^D = 1;$$

see [250].

9.5 Lipschitz mappings and generalizations of fractals

The open set condition, see Definition 9.4.6, allows us to deduce additional properties of sets invariant under families of similarities. In order to state the result, we need to define an *n-dimensional C^1-manifold in $\mathbb{R}^d$* to be the set $M \subseteq \mathbb{R}^d$ such that for each $x \in M$ there exist an open set $U \subseteq \mathbb{R}^d$ containing x, an open set $V \subseteq \mathbb{R}^n$, and a differentiable homeomorphism $\phi : V \to M \cap U$ such that its inverse $\phi^{-1} : M \cap U \to V$ is also differentiable.

Theorem 9.5.1. Self-similar sets are unrectifiable
Let $\mathcal{S} = \{S_j : j = 1, \ldots, N\}$ be a family of similarities on $\mathbb{R}^d$ that satisfies the open set condition with sets U and V, where $U \subseteq V$. Let $F_{\mathcal{S}} \subseteq \mathbb{R}^d$ be

the compact set invariant under $\mathcal{S}$. Assume that every n-dimensional affine subspace $W \subseteq \mathbb{R}^d$ satisfies the following implication: If there exists $i \neq j$ such that $W \cap S_i(U) \neq \emptyset$ and $S_j(U) \neq \emptyset$, then

$$W \cap \left(V \setminus \left(\bigcup_{j=1}^{N} \overline{S_j(U)} \right) \right) \neq \emptyset.$$

Then, for every n-dimensional C^1-manifold M in $\mathbb{R}^d$, we have

$$\mu_n(M \cap F_{\mathcal{S}}) = 0.$$

Theorem 9.5.1 is due to HUTCHINSON [250], Theorem 5.4.(1). It shows that self-similar sets are really different from planes or smooth surfaces. Thus, we are led to the notions of *rectifiability* and *unrectifiability*, which essentially measure how much various sets differ from "straight" sets. It turns out that as a model for rectifiable sets one uses more general objects, i.e., images of Lipschitz mappings instead of differentiable manifolds. For a rationale, see [342], Chapter 15, whose definitions we follow.

Definition 9.5.2. Rectifiable and unrectifiable sets
a. A set $A \subseteq \mathbb{R}^d$ is *n-rectifiable* if there exist Lipschitz functions $f_j : \mathbb{R}^n \to \mathbb{R}^d$, $j = 1, \ldots,$ such that

$$\mu_n \left(A \setminus \left(\bigcup_{j=1}^{\infty} f_j(\mathbb{R}^n) \right) \right) = 0.$$

b. A set $B \subseteq \mathbb{R}^d$ is called *purely n-unrectifiable* if, for every n-rectifiable set $A \subseteq \mathbb{R}^d$,

$$\mu_n(A \cap B) = 0.$$

The notion of rectifiability was introduced by BESICOVITCH [53], [55], [56]; see also [163], [160], [161], [342] for extended expositions on these ideas.

Rectifiable sets play an important role in the study of singular integral operators, as in the Calderón–Zygmund theory [86]. A typical example of a singular integral operator is C_A, defined by

$$C_A(f)(z) = \lim_{\varepsilon \to 0} \frac{1}{2\pi i} \int_{\zeta \in A, |z-\zeta| > \varepsilon} \frac{1}{z - \zeta} f(\zeta)\, d\mu_1(\zeta),$$

where $A \subseteq \mathbb{C}$ is μ_1-measurable and we identify $\mathbb{C}$ with $\mathbb{R}^2$, and where $f \in L^2_{\mu_1}(A)$. We call C_A the *Cauchy transform* associated with the *Cauchy kernel* $\mathbb{1}_A \cdot \frac{1}{z-\zeta}$.

CALDERÓN [85] proved that if A is the graph of a Lipschitz function with small Lipschitz norm, then the operator $C_A : L^2_{\mu_1}(A) \to L^2_{\mu_1}(A)$ is bounded. This result was generalized to all Lipschitz graphs by RONALD R. COIFMAN,

Alan G. R. McIntosh, and Meyer [111]. For a general theory of singular integral operators see, e.g., [450] and [448].

The study of the relationship between geometric properties of sets and analytic properties of functions and operators defined on these sets is of great interest in analysis. A theory that generalizes the above example to that of the Cauchy transform on rectifiable curves was developed by Guy David, Peter W. Jones, and Stephen W. Semmes. We shall now briefly describe some of their results, referring the interested reader to beautiful accounts of this theory in [121], [122], [123], [342].

First, we note that instead of working with sets of Hausdorff dimension n, we shall need a more quantitative condition.

Definition 9.5.3. Regular sets
A set $A \subseteq \mathbb{R}^d$ is *regular* with Hausdorff dimension n if it is closed and if there exists a constant $C > 0$ such that

$$\forall\, x \in A \text{ and } \forall\, r > 0, \quad \frac{1}{C} r^n \leq \mu_n(A \cap B(x,r)) \leq C r^n.$$

The smallest such constant C is the *regularity constant* for A.

Definition 9.5.4. Odd kernels
a. Let $\mathcal{K}_n(\mathbb{R}^d)$ be the class of all smooth real-valued functions K on $\mathbb{R}^d \setminus \{0\}$ that satisfy the conditions:

$$\forall\, x \in \mathbb{R}^d \setminus \{0\}, \quad K(-x) = -K(x)$$

and

$$\forall\, j \in \mathbb{N},\ \exists\, C_j > 0 \text{ such that } \forall\, x \in \mathbb{R}^d \setminus \{0\}, \quad \|x\|^{n+j} \|\nabla^j(K)(x)\| \leq C_j.$$

b. We say that a μ_n-measurable set $A \subseteq \mathbb{R}^d$ is *good for all the kernels in* $\mathcal{K}_n(\mathbb{R}^d)$ if, for each $K \in \mathcal{K}_n(\mathbb{R}^d)$, the operators,

$$\forall\, \varepsilon > 0, \quad T_\varepsilon(f)(x) = \int_{y \in A, |y-x| > \varepsilon} K(x-y) f(y)\, d\mu_n(y),$$

are bounded as operators on $L^2_{\mu_n}(A)$, uniformly in $\varepsilon > 0$.

Before we introduce the main result of this section we need to define one more piece of the puzzle.

Definition 9.5.5. Carleson measures and Carleson sets
a. Let $A \subseteq \mathbb{R}^d$ be a Hausdorff n-dimensional regular set and let μ be a Borel measure on $A \times \mathbb{R}^+$. We say that μ is a *Carleson measure* on $A \times \mathbb{R}^+$ if there exists a constant $C > 0$ such that

$$\forall\, x \in A \text{ and } \forall\, r > 0, \quad \mu(B(x,r) \times (0,r)) \leq C r^n.$$

The smallest constant C such that the above inequality holds is called the *Carleson norm* of μ.

b. Let $A \subseteq \mathbb{R}^d$ be a Hausdorff n-dimensional regular set. We say that a Borel set $\Omega \subseteq A \times \mathbb{R}^+$ is a *Carleson set* if there exists a constant $C > 0$ such that

$$\forall\, x \in A \text{ and } \forall\, r > 0, \quad \iint_{(A\cap B(x,r))\times(0,r)} \mathbb{1}_\Omega(y,t)\, d\mu_n(y)\, \frac{dt}{t} \leq Cr^n.$$

In view of part *a*, this is equivalent to saying that the measure,

$$\mathbb{1}_\Omega(x,t)\, d\mu_n(x)\, \frac{dt}{t},$$

is a *Carleson measure*.

c. Carleson measures were introduced by CARLESON (1958), and played a role in his solution of the celebrated Corona problem; see [89], [90]. Carleson measure techniques have important applications in partial differential equations and Fourier multiplier problems, see [263] by JOHNSON, as well as with BMO and its ramifications, e.g., [190] and [202], Chapters 7 and 8.

Theorem 9.5.6. Characterizations of uniform rectifiability
Let $A \subseteq \mathbb{R}^d$ be a Hausdorff n-dimensional regular set. The following are equivalent:

a. *A is good for all kernels in $\mathcal{K}_n(\mathbb{R}^d)$.*

b. *The measure $\beta_1^2(x,t)\, d\mu_n(x)\, \frac{dt}{t}$ is a Carleson measure on $A \times \mathbb{R}^+$, where*

$$\beta_1(x,t) = \inf_P \left\{ \frac{1}{t^n} \int_{A\cap B(x,t)} \frac{1}{t} \operatorname{dist}(y,P)\, d\mu_n(y) \right\},$$

and the infimum is taken over all n-dimensional affine subspaces P of $\mathbb{R}^d$.

c. *There are constants $C, M > 0$ such that, for each $x \in A$ and for each $0 < r \leq \operatorname{diam}(A)$, there exist a set $E \subseteq A \cap B(x,r)$ and a function $f : E \to \mathbb{R}^n$ for which*

$$\forall x, y \in E, \quad M^{-1}\|x - y\| \leq \|f(x) - f(y)\| \leq M\|x - y\|$$

and

$$\mu_n(E) \geq Cr^n.$$

d. *For all compactly supported, infinitely differentiable, odd functions ψ, the measure defined by*

$$\sum_{j=-\infty}^{\infty} \left| \int_A 2^{-jn} \psi(2^{-j}(x-y))\, d\mu_n(y) \right|^2 d\mu_n(x)\, d\delta_{2^k}(t)$$

is a Carleson measure on $A \times \mathbb{R}^+$, where δ_z denotes the Dirac measure concentrated at z.

e. *For all $\varepsilon > 0$ the set*

$$\{(x,t) \in A \times \mathbb{R}^+ : \beta_b(x,t) > \varepsilon\}$$

is a Carleson set, where

$$\beta_b(x,t) = \inf_P \left\{ \sup_{y \in A \cap B(x,t)} \frac{1}{t} \text{dist}\,(y,P) + \sup_{z \in A \cap B(x,t)} \frac{1}{t} \text{dist}\,(z,A) \right\},$$

and the infimum is taken over all n-dimensional affine subspaces P of $\mathbb{R}^d$.

We say that a set $A \subseteq \mathbb{R}^d$ that is regular and satisfies any of the equivalent conditions *a–e* is *uniformly rectifiable*. These are some of the equivalent definitions of uniform rectifiability, and we refer the interested reader to [120], [121], [122], [430].

The fact that the notion of uniform rectifiability is much stronger than the notion of rectifiability is seen from part *c* of Theorem 9.5.6.

Related to the notion of uniform rectifiability is the notion of BPI, introduced by David and Semmes; see, e.g., [123]. This is yet another way of defining self-similarity. The acronym BPI stands for "big pieces of itself".

Definition 9.5.7. BPI sets

A set $X \subseteq \mathbb{R}^d$ is a *BPI set* if it is regular of Hausdorff dimension n and if the following condition holds: There exist constants $\theta, C > 0$ such that, for each pair of balls $B(x_1,r_1)$, $B(x_2,r_2) \subseteq \mathbb{R}^d$, where x_1, $x_2 \in X$ and r_1, $r_2 \leq$ diam (X), there exist a closed set $A \subseteq B(x_1,r_1) \cap X$ with $\mu_n(A) \geq \theta r_1^n$ and a function $f : A \to B(x_2,r_2) \cap X$ for which

$$\forall\, x,y \in A, \quad \frac{r_2}{Cr_1}\|x-y\| \leq \|f(x)-f(y)\| \leq \frac{Cr_2}{r_1}\|x-y\|.$$

The notion of BPI sets allows measure theory to enter into the study of self-similarity in a new and significant way. David and Semmes write, "By asking only that pairs of balls contain substantial subsets which look alike, rather than the whole replicas of each other, we have allowed a greater role for measure theory in this notion of self-similarity than is customary. We have increased the possibility for repetition, distortion, rupture, and clutter"; see [123].

The next definition introduces the possibility for identifying two BPI sets that have "similar" structures.

Definition 9.5.8. BPI equivalence

Let X, $Y \subseteq \mathbb{R}^d$ be two BPI sets with the same Hausdorff dimension n. We say that X and Y are *BPI equivalent* if there are constants $\delta, K > 0$ such that, for any $x \in X$, $0 < r \leq$ diam (X), $y \in Y$, and $0 < s \leq$ diam (Y), there exist a closed set $A \subseteq B(x,r) \cap X$ with $\mu_n(A) \geq \delta r^n$ and a function $f : A \to B(y,s) \cap Y$ for which

$$\forall\, x, y \in A, \quad \frac{s}{Kr}\|x-y\| \le \|f(x)-f(y)\| \le \frac{Ks}{r}\|x-y\|.$$

BPI equivalence is, in fact, an equivalence relation on the family of BPI sets.

Remark. Let $E \subseteq \mathbb{R}^d$. The function $f : E \to \mathbb{R}^d$ is *bilipschitz with respect to the Euclidean metric, with constant* M, if there exists a constant $M > 0$ such that the following condition holds:

$$\forall\, x,\, y \in E, \quad M^{-1}\|x-y\| \le \|f(x)-f(y)\| \le M\|x-y\|;$$

cf. the definition of Lipschitz functions in Problem 1.31 and in Problem 4.36. This is the condition that appears in Theorem 9.5.6*c*.

An analogous notion, which we have used in the definitions of BPI sets and of BPI equivalence, Definitions 9.5.7 and 9.5.8, respectively, uses the following condition:

$$\exists\, K, \theta > 0, \text{such that } \forall\, x, y \in E,\ M^{-1}\theta\|x-y\| \le \|f(x)-f(y)\| \le M\theta\|x-y\|.$$

Functions satisfying this condition are called *conformally bilipschitz* in [123]. Such a definition allows for better uniform comparison of sets with respect to families of sets of different sizes.

We close this section with several results relating BPI equivalence and uniform rectifiability.

Proposition 9.5.9. *Let* X, $Y \subseteq \mathbb{R}^d$ *be two BPI sets with the same Hausdorff dimension* n. *The sets* X *and* Y *are BPI equivalent if and only if there exist sets* $A \subseteq X$ *and* $B \subseteq Y$ *of positive measure, a constant* $K > 0$, *and a surjective function* $f : A \to B$ *such that*

$$\forall\, x, y \in A, \quad \frac{1}{K}\|x-y\| \le \|f(x)-f(y)\| \le K\|x-y\|.$$

An immediate consequence of Propostion 9.5.9 and of the definition of uniform rectifiability is the following result.

Proposition 9.5.10. *Let* $X \subseteq \mathbb{R}^d$ *be a BPI set that is uniformly rectifiable. Then* X *is BPI equivalent to* $\mathbb{R}^n$, *where* n *is the Hausdorff dimension of* X.

Proof. Indeed, according to Proposition 9.5.9, we need to find two sets, $A \subseteq X$ and $B \subseteq \mathbb{R}^n$, a constant $K > 0$, and a surjective function $f : A \to B$ such that

$$\forall\, x, y \in A, \quad \frac{1}{K}\|x-y\| \le \|f(x)-f(y)\| \le K\|x-y\|.$$

This, in turn, follows directly from part c of Theorem 9.5.6. □

The assumption in Proposition 9.5.10 about the set X being both uniformly rectifiable and BPI is redundant, as the next result shows; see [123], Proposition 3.3.

Proposition 9.5.11. *Uniformly rectifiable sets are BPI.*

Finally, we have the following characterization of uniformly rectifiable subsets of $\mathbb{R}^d$ in terms of the notion of BPI.

Theorem 9.5.12. BPI characterization of uniform rectifiability
Let $X \subseteq \mathbb{R}^d$ be a regular set of Hausdorff dimension n. Then X is uniformly rectifiable if and only if X is BPI and BPI equivalent to $\mathbb{R}^n$.

Proof. ($\Longrightarrow$) If X is uniformly rectifiable, then it follows from Proposition 9.5.11 that X is a BPI set. Thus, we may use Proposition 9.5.10 to conclude that X is BPI equivalent to $\mathbb{R}^n$, where n is the Hausdorff dimension of X.

($\Longleftarrow$) This follows immediately from the definition of uniform rectifiability (Theorem 9.5.6*c*) by an argument analogous to the one used in the proof of Proposition 9.5.10. □

9.6 Potpourri and titillation

1. Let $A(\mathbb{T})$ be the space of absolutely convergent Fourier series, $F = \hat{f}$, i.e., $f \in \ell^1(\mathbb{Z})$. As in Section 7.6.2, we define closed sets $E \subseteq \mathbb{T}$ of *spectral synthesis*, so-called S-sets, by the condition $A'(E) = A'_S(E)$. Clearly, $M_b(E) \subseteq A'_S(E) \subseteq A'(E)$, where $M_b(E) = \{\mu \in M_b(\mathbb{T}) : \text{supp}\ (\mu) \subseteq E\}$. We shall investigate the relations between these three spaces. To begin, recall the definition of RIEMANN's sets of uniqueness, or U-sets, E from Section 3.8. These will play a role in stating the relations between $M_b(E)$, $A'_S(E)$, and $A'(E)$, as will various number-theoretic topics such as Kronecker's theorem and Pisot–Vigayraghavan numbers.

LEOPOLD KRONECKER proved that *if $\{x_1, \ldots, x_n, \pi\} \subseteq \mathbb{R}$ is linearly independent over the rationals, $\{y_1, \ldots, y_n\} \subseteq \mathbb{R}$, and if $\varepsilon > 0$, then there is an integer m such that for each j,*

$$\left| e^{ix_j m} - e^{iy_j} \right| < \varepsilon;$$

see [221], [271], and [33], pages 239–241, where the proof depends on the Bohr compactification.

Remark. A *special* case of KRONECKER's theorem is the assertion that *if $x \in \mathbb{R} \setminus \mathbb{Q}$, then $\{nx \ (\text{mod} 1) : n \in \mathbb{N}\}$ is dense in $[0, 1]$.* Problem 3.29*b* gives a stronger statement of this assertion in terms of uniform distribution.

The problem of the distribution of $\{nx \ (\text{mod} 1) : n \in \mathbb{N}\}$ in $[0, 1]$ was studied in the theory of secular perturbations in astronomy going back to PTOLEMY and explicitly with JOSEPH-LOUIS LAGRANGE, who posed a basic problem about mean angular motion of trigonometric polynomials. The so-called Kronecker–Weyl theorem was proved by WEYL to treat LAGRANGE's problem. The theorem itself can be viewed as a generalization of WEYL's characterization of uniform distribution modulo 1 in terms of Riemann integrals that we stated in Problem 3.29. In fact, WEYL proved the Kronecker–Weyl theorem in his fundamental number-theoretic paper [500] on uniform distribution.

The Kronecker–Weyl theorem can be viewed as a result about replacing time averages by space averages. It is an early example of an "ergodic theorem", and many view it as the catalyst for the dynamic (sic) development of ergodic theory; cf. Section 8.8.6. An expert, readable, brilliant exposition on this material is due to SHLOMO Z. STERNBERG [453] (1969).

In his 1948 Josiah Willard Gibbs lecture, WEYL [501] (1950) has written a dazzling midcentury vision and retrospective integrating some of this material.

Because of KRONECKER's theorem we say that a closed set $E \subseteq \mathbb{T}$ is a *Kronecker set* if for each $\varepsilon > 0$ and each continuous function $f : E \to \mathbb{C}$, satisfying $|f| = 1$, there is an integer m for which

$$\sup_{x \in E} \left| f(x) - e^{imx} \right| < \varepsilon.$$

An *algebraic integer* is an algebraic number where the corresponding polynomial is monic; see Problem 1.24. A *Pisot–Vijayaraghavan (P–V) number* is a real algebraic integer $\alpha > 1$ with the property that all of the other roots of its minimal polynomial have modulus less than 1. As an example, note that the *golden ratio* $\alpha = (1 + \sqrt{5})/2$ is a P–V number since $|1 - \sqrt{5}|/2 < 1$.

Remark. ***a.*** Pisot–Vigayraghavan numbers come into the picture for the investigation of badly distributed sequences, as opposed to WEYL's uniform distribution; see Problem 3.29. KRONECKER's theorem can be proved using WEYL's results on uniform distribution.

b. Independently of each other, AXEL THUE (1912 in Norske Vid. Selsk. Skr.) and HARDY [214] (1919 in J. of Indian Math. Soc.) observed that

$$\alpha \text{ P–V} \implies \lim_{n \to \infty} \alpha^n \pmod 1 = 0, \tag{9.7}$$

and proved one of the key properties of P–V numbers: *If* $\alpha > 1$ *is an algebraic integer and* $\lim_n \alpha^n \pmod 1 = 0$ *then* α *is a P–V number.* For quadratic P–V numbers α_1 with conjugate α_2 we easily verify (9.7) as follows:

$$(\alpha_1^n + \alpha_2^n) = (\alpha_1^{n-1} + \alpha_2^{n-1})(\alpha_1 + \alpha_2) - \alpha_1\alpha_2(\alpha_1^{n-2} + \alpha_2^{n-2}),$$
$$\alpha_1 + \alpha_2 = 1, \quad \alpha_1\alpha_2 = -1;$$

and an induction argument shows that

$$\lim_{n\to\infty} \alpha_1^n \pmod 1 = 0.$$

c. The work of both THUE and HARDY does not seem to have been properly advertised until the early 1960s. P–V numbers have been studied extensively by CHARLES PISOT beginning with his thesis in 1938, and a good bibliography on the subject up to 1962 is in [373].

SALEM proved the following spectacular theorem: *Let $E = E_\xi \subseteq \mathbb{T}$ be the perfect symmetric set with $\xi_k = \xi \in (0, 1/2)$ for each k; then E is a U-set if and only if $1/\xi$ is a P–V number.* SALEM announced this result in 1943, and an error in the sufficiency conditions was found by members of the theory of functions seminar at the University of Moscow in 1945. In 1948 SALEM published some special cases in which the sufficiency is true, and in 1954 PIATETSKI–SHAPIRO proved that *if β is a P–V number of degree n and $\beta > 2^n$, then $E = E_\xi$ is a U-set where $\xi = 1/\beta$.* Finally, in 1955, SALEM and ZYGMUND, using PIATETSKI–SHAPIRO's method, proved the full generality of the originally stated result; see [413].

RAPHAËL SALEM (1898–1963) was the key figure in the revival of the Paris (Orsay) school of Fourier analysis. SALEM returned to Paris from MIT after World War II, and his lectures in 1948 on unsolved problems in Fourier series were the catalyst for great activity. SALEM's career is warmly sketched by ZYGMUND [414] from his birth in Thessaloniki, his banking profession (manager of the Banque de Paris et des Pays-Bas by 1938), to the days in Cambridge, and to Paris.

The following theorem summarizes several striking results dealing with notions we have just discussed; see [32], [33], [269], [347], [296], [297], [322], [234] for excursions through this field.

Theorem 9.6.1. Kronecker, U, and S sets
Let $E \subseteq \mathbb{T}$ be closed.

a. *(MALLIAVIN, 1962) If $M_b(E) = A'(E)$ then every closed subset of E is an S-set (easy), and if this latter condition holds then E is a U-set.*

b. *(VAROPOULOS, 1965) If E is a Kronecker set then $M_b(E) = A'(E)$.*

c. *(KÖRNER, 1972) There is E such that $M_b(E) = A'_S(E)$ and $A'_S(E) \neq A'(E)$.*

2. The notion of Hausdorff dimension provides a finer, intuitive notion of dimension than ordinary Euclidean dimension, and there are many examples in which this notion is relevant, e.g., [336], [337], [423]. It is often all but impossible to compute the Hausdorff dimension of a physical object. In order to address this issue, the method of a *multifractal formalism* has arisen along with the concept of the *spectrum of singularities* of a given function. For instance, there is the *structure function method* of U. FRISCH and G. PARISI

[186]. There are also other methods that involve the wavelet transform. The deepest results in this area are due to JAFFARD [257], [255], [256], [258]; and the outline below is also due to JAFFARD.

Let $f : \mathbb{R}^d \to \mathbb{R}$ and let $\alpha > 0$. Given $x_0 \in \mathbb{R}^d$, we write $f \in C^\alpha(x_0)$ if there are a polynomial P_{x_0} of degree less than α and a constant $C_{x_0} > 0$ such that

$$|f(x) - P_{x_0}(x - x_0)| \leq C_{x_0}\|x - x_0\|^\alpha \tag{9.8}$$

in a neighborhood of x_0. It can be shown that if $f \in C^\alpha(x_0)$, then $f \in C^\beta(x_0)$ for any $\beta < \alpha$. Thus, the following definition makes sense.

The *pointwise Hölder exponent*, $h_f(x_0)$, of $f : \mathbb{R}^d \to \mathbb{R}$ at x_0 is

$$h_f(x_0) = \sup\{\alpha > 0 : f \in C^\alpha(x_0)\},$$

where we say that $h_f(x_0) = 0$ if $f \notin C^\alpha(x_0)$ for any $\alpha > 0$. See the classical definition of *Hölder exponent* in Section 1.5.1. The *spectrum of singularities* of $f : \mathbb{R}^d \to \mathbb{R}$ is the function $d_f : [0, \infty] \to \{-\infty\} \cup [0, d]$ defined as

$$d_f(\alpha) = \dim_H (\{x : h_f(x) = \alpha\}),$$

where we adopt the convention that $\dim(\emptyset) = -\infty$.

A *multifractal function* $f : \mathbb{R}^d \to \mathbb{R}$ is a function whose spectrum of singularities has the property that $\{\alpha > 0 : d_f(\alpha) \in (0, d]\}$ contains an interval.

A modification of the Weierstrass everywhere continuous nowhere differentiable function defined in Section 1.5.1 is

$$K_\alpha(x) = \sum_{k=0}^{\infty} \frac{\sin(2^k \pi x)}{2^{k\alpha}}.$$

A direct calculation shows that

$$\forall\, x \in \mathbb{R}, \quad h_{K_\alpha}(x) = \alpha,$$

i.e., for any given $x \in \mathbb{R}$, $\beta < \alpha$, and $\alpha < \gamma$, we have $K_\alpha \in C^\beta(x)$ and $K_\alpha \notin C^\gamma(x)$.

The spectrum of singularities d_R of RIEMANN's function R, defined in Section 1.5.1, is

$$d_R(\alpha) = \begin{cases} 4\alpha - 2, & \text{if } \alpha \in [1/2, 3/4], \\ 0, & \text{if } \alpha = 3/2, \\ -\infty, & \text{elsewhere.} \end{cases}$$

In particular, R is a multifractal function.

We also mention the remarkable articles [82] and [148] on the history of R and on the role of self-similarity in analyzing its behavior, respectively.

Finally, RIEMANN's ruler function r, defined after Proposition 1.3.6, is also multifractal with spectrum of singularities

$$d_r(\alpha) = \begin{cases} 2\alpha, & \text{if } \alpha \in [0, 1/2], \\ -\infty, & \text{elsewhere.} \end{cases}$$

The Hölder exponent of r vanishes on the rationals.

3. Wavelets can be used in the analysis of Hölder exponents. Let $\alpha > 0$ and set $k = [\alpha]$. Choose a "wavelet" $\psi \in L^1_m(\mathbb{R})$ satisfying the following properties:

$$\exists\, C > 0 \text{ such that } \forall\, j = 0, \ldots, k+1, \quad \forall\, x \in \mathbb{R}, \quad |\psi^{(j)}(x)| \le C(1+|x|)^{-k-2};$$

$$\forall\, j = 0, \ldots, k, \quad \int_{\mathbb{R}} x^j \psi(x)\, dx = 0,$$

and

$$\int_0^\infty \frac{|\widehat{\psi}(\xi)|^2}{\xi}\, d\xi = 1 \quad \text{with } \hat{\psi}(\xi) = 0 \text{ if } \xi < 0.$$

If $f \in L^\infty_m(\mathbb{R})$, then the *continuous wavelet transform* $C_f(a, b)$ of f is defined as

$$\forall\, a \in (0, \infty) \text{ and } \forall\, b \in \mathbb{R}, \quad C_f(a, b) = \frac{1}{a} \int_{\mathbb{R}} f(x) \overline{\psi((x-b)/a)}\, dx.$$

In this setting, JAFFARD proved the following result.

Theorem 9.6.2. Wavelet characterization of $C^\alpha(x_0)$
a. *If $f \in C^\alpha(x_0)$, then*

$$\exists\, C > 0 \text{ such that } \forall\, (a, b) \in (0, \infty) \times \mathbb{R}, \quad |C_f(a, b)| \le C a^\alpha \left(1 + \frac{|b - x_0|}{a}\right)^\alpha.$$

b. *Conversely, if there exists $\beta < \alpha$ for which*

$$\exists\, C > 0 \text{ such that } \forall\, (a, b) \in (0, \infty) \times \mathbb{R}, \quad |C_f(a, b)| \le C a^\alpha \left(1 + \frac{|b - x_0|}{a}\right)^\beta,$$

then

$$|x - x_0| \le 1/2 \quad \Longrightarrow \quad |f(x) - f(x_0)| \le C|x - x_0|^\alpha.$$

This theorem can be used to give a characterization of Hölder exponents [8], and with it the authors were able to give a wavelet characterization of the so-called oscillating singularity exponent. We mention this since such exponents are the underlying idea behind so-called *chirps*, which have a long history in signal analysis, e.g., [39]. Examples of chirps are the functions $f_{\alpha,\beta}(x) = x^\alpha \sin(1/x^\beta)$, $x \ne 0$, that were analyzed in Problems 4.4 and 4.5.

To fix ideas, consider $f = f_{1,1}$. Clearly, $h_f(x) = \infty$ if $x \neq 0$, and it is not to difficult to calculate that $h_f(0) = 1$. The study of singular oscillating behavior in terms of $f_{\alpha,\beta}$, wavelets, and microlocal analysis was initiated by MEYER (1992); see [259].

4. There are thousands of measures on the mathematical landscape, custom-made to clarify and analyze and create. Some may seem to have more global impact that others, but who can really say as new ideas evolve? We give a short list of measures, exclusive of those already mentioned, to illustrate the diversity of this landscape.

• There are *Riesz product measures*, designed to construct singular measures with applications as far-reaching as ergodic theory and multifractal analysis; e.g., [391], [77], [285], [415]; see [42] for an overview, references dealing with both probability and engineering, and a surprising comparison with the Cantor measure.

• There is *Wiener measure* and Brownian motion, e.g., [510], [151], [305]; cf. [32] for one of its roles in harmonic analysis as well as [363], Chapters IX and X, and the extraordinary program of JEAN-PIERRE KAHANE [268] (1968) continuing to the present.

• There are the complex measures $E_{x,y}$ inherent in the definition of a resolution of the identity $E : \mathcal{A} \to \mathcal{L}(H)$ used to prove the spectral theorem for certain operators $A \in \mathcal{L}(H)$, e.g., [392], [465], [406]. Here $\mathcal{A}$ is a σ-algebra in a set X, $\mathcal{L}(H)$ is the space of bounded linear operators $A : H \to H$, H is a complex Hilbert space, and, for given $x, y \in H$, $E_{x,y}(A) = \langle E(A)(x), y \rangle$, where $A \in \mathcal{A}$. The setting can be generalized, and the result for Hilbert space, due to HILBERT in 1912, is a generalization of the spectral theorem for $\mathbb{C}^d$.

• *Positive operator-valued measures* (POVMs) are related to the measures $E_{x,y}$, and they play an influential role in quantum information processing and quantum measurement, e.g., [492], [73]. Tight frames, with all of their applications in signal processing, can be used to construct POVMs.

• The *power spectrum* of $f \in L^\infty(\mathbb{R})$ is a positive measure whose Fourier transform is the deterministic autocorrelation,

$$\lim_{T \to \infty} \frac{1}{2T} \int_{-T}^{T} f(t+u)\overline{f(u)}\, du.$$

(Some hypotheses are required regarding convergence.) The notion of power spectrum was central to some of WIENER's most profound work, e.g., Appendix B.12, [505], [33], and it provides a foundation for KOLMOGOROV's prediction theory, e.g., [292], [38]. In probabilistic formulation, this theory deals with the Wiener–Khinchin theorem and the Wold decomposition, and uses analytic tools such as the Szegő alternative, see [39], Section 3.7, and [152].

• The notion of *spectral measure* is not unrelated to KOLMOGOROV's prediction theory. Whereas one of KOLMOGOROV's prediction theorems characterizes the completeness condition,

$$L^2_\mu(\mathbb{R}) = \overline{\text{span}}\,\{e^{2\pi i x\xi} : \xi \leq 0\},$$

we define a nonnegative $\mu \in M_b(\mathbb{R})$ to be a *spectral measure* if there is a *spectral* set $\Lambda \subseteq \mathbb{R}$ defined by the property that $\{e^{2\pi i x\xi} : \xi \in \Lambda\}$ is an orthonormal basis for $L^2_\mu(\mathbb{R})$. We call $S \subseteq \mathbb{R}$ a *spectral set* if $\mu = m|_S$ is a spectral measure. The *Fuglede conjecture* states that $S \in \mathcal{M}(\mathbb{R})$ is a spectral set if and only if it tiles $\mathbb{R}$ by integer translates. TERRENCE TAO gave a counterexample to the conjecture when $\mathbb{R}$ is replaced by $\mathbb{R}^5$. There are now counterexamples for $d \geq 5$ as well as many positive results and related problems. See [309] for a recent contribution as well as a list of results by BÁLINT FARKAS, ALEX IOSEVICH, PALLE JORGENSEN, NETS KATZ, MIHAIL KOLOUNTZAKIS, JEFFREY LAGARIAS, IZABELLA ŁABA, MÁTÉ MATOLCSI, PETER MÓRA, STEEN PEDERSEN, SZILÁRD RÉVÉSZ, ROBERT STRICHARTZ, SÁNDOR SZABÓ, and YANG WANG.

A Functional Analysis

A.1 Definitions of spaces

This appendix lists results from functional analysis that are used in this book. There are many excellent texts and expositions including [138], [156], [242], [200], [311], [372], [392], [406], [465].

Definition A.1.1. Topological space
A *topological space* X is a pair $(X, \mathcal{T})$, where X is a nonempty set, $\mathcal{T} \subseteq \mathcal{P}(X)$, and $\mathcal{T}$ satisfies the following conditions:

i. $\emptyset \in \mathcal{T},\ X \in \mathcal{T}$,

ii. $\{U_\alpha : \alpha \in I,$ an index set$\} \subseteq \mathcal{T} \Longrightarrow \bigcup_{\alpha \in I} U_\alpha \in \mathcal{T}$,

iii. $\{U_j : j = 1, \ldots, n\} \subseteq \mathcal{T} \Longrightarrow \bigcap_{j=1}^{n} U_j \in \mathcal{T}$.

The elements of $\mathcal{T}$ are called *open sets* and $\mathcal{T}$ is a *topology* for the set X. The *interior* of $S \subseteq X$, denoted by int S, is the largest open set contained in S. The complement of an open set is a *closed set.* A set S in a topological space $(X, \mathcal{T})$ is a *neighborhood* of $x \in X$ if $x \in U \subseteq S$ for some $U \in \mathcal{T}$. We call $\mathcal{B} \subseteq \mathcal{T}$ a *basis* for the topological space $(X, \mathcal{T})$ if for each $x \in X$ and each neighborhood S of x, we have $x \in V \subseteq S$ for some $V \in \mathcal{B}$. We say that $\mathcal{B}_x$ is a *basis at* $x \in X$ if each element of $\mathcal{B}_x$ is a neighborhood of x, and, for every neighborhood S of x, we have $x \in B \subseteq S$ for some $B \in \mathcal{B}_x$.

Theorem A.1.2. Characterization of a basis
A family $\mathcal{B}$ is a basis for some topology $\mathcal{T}$ for $X = \bigcup\{B : B \in \mathcal{B}\}$ if and only if

$$\forall\, U,\ V \in \mathcal{B} \text{ and } \forall\, x \in U \cap V,\ \exists\, W \in \mathcal{B} \text{ such that } x \in W \subseteq U \cap V.$$

In this case, $\mathcal{T}$ is the family of all unions of members of $\mathcal{B}$.

We shall *assume* that all of our topological spaces X are *Hausdorff*, i.e., that they satisfy the following property:

$$\forall\, x, y \in X,\ x \neq y,\ \exists\, U_x, U_y \in \mathcal{T} \text{ such that } x \in U_x, y \in U_y, \text{ and } U_x \cap U_y = \emptyset.$$

Let $(X, \mathcal{T})$ be a topological space and let $Y \subseteq X$. Define $\mathcal{T}_Y = \{U \cap Y : U \in \mathcal{T}\}$. As such, $(Y, \mathcal{T}_Y)$ is a topological space, and $\mathcal{T}_Y$ is the *induced topology*

J.J. Benedetto and W. Czaja, *Integration and Modern Analysis*, Birkhäuser Advanced Texts,
DOI 10.1007/978-0-8176-4656-1_10,

on Y from $(X, \mathcal{T})$. Among other natural situations, the concept of induced topology allows us to discuss the Borel algebra $\mathcal{B}(Y)$ in terms of $\mathcal{B}(X)$, as well as Borel measurable functions on Y when we are given $\mathcal{B}(X)$; e.g., see Sections 2.4 and 8.7.

A set $Y \subseteq X$ is *dense* in X if for each $x \in X$ and each open set U containing x there is a point $y \in Y \cap U$. According to Definition 1.2.11*a*, $K \subseteq X$ is *compact* if every covering of K by open sets contains a finite subcovering; and $K \subseteq X$ is *relatively compact* if its *closure* (the smallest closed set containing it) is compact. A topological space X is *locally compact* if every point has at least one compact neighborhood, i.e., if

$$\forall\, x \in X,\ \exists\, K \subseteq X, \text{compact, and } \exists\, V \in \mathcal{T} \text{ such that } x \in V \subseteq K.$$

Recall that a function $f : X_1 \to X_2$ is a *bijection* if f is one-to-one (injective) and onto (surjective). Two topological spaces $(X_i, \mathcal{T}_i)$, $i = 1, 2$, are *homeomorphic* if there is a bijection $f : X_1 \to X_2$ such that

$$\forall\, U \in \mathcal{T}_1,\ f(U) \in \mathcal{T}_2 \quad \text{and} \quad \forall\, V \in \mathcal{T}_2,\ f^{-1}(V) \in \mathcal{T}_1.$$

In this case, f is a *homeomorphism*. These two conditions define the *continuity* of f^{-1} and f on X_2 and X_1, respectively; cf. the equivalent definition of continuity for metric spaces X and Y in Definition A.4.2. This latter definition emphasizes the local nature of continuity by defining continuity at a point.

Theorem A.1.3. Urysohn lemma
Let X be a locally compact Hausdorff space. If $K \subseteq X$ is compact and $U \subseteq X$ is an open set containing K, then there is a continuous function $f : X \to [0, 1]$ such that $f = 1$ on K and $f = 0$ on $U^\sim$.

A topological space X is *connected* if it cannot be represented as a disjoint union of two nonempty closed sets. A space X is *locally connected* if it has the following property at each $x \in X$: Every neighborhood of x contains a connected neighborhood of x. If X is a locally compact Hausdorff space and if for every two points $x, y \in X$ there exists a continuous function $p : [0, 1] \to X$ such that $p(0) = x$ and $p(1) = y$, we say that X is *path connected.*

Standard references for topological spaces include [279], [307].

Definition A.1.4. Metric space
a. A *metric space* X is a pair (X, ρ), where X is a nonempty set and $\rho : X \times X \to \mathbb{R}^+$ satisfies the following conditions:

i. $\forall\, x, y \in X, \quad \rho(x, y) \geq 0,$
ii. $\forall\, x, y \in X, \quad \rho(x, y) = \rho(y, x),$
iii. $\forall\, x, y, z \in X, \quad \rho(x, z) \leq \rho(x, y) + \rho(y, z)$ (triangle inequality),
iv. $\forall\, x, y \in X, \quad \rho(x, y) = 0 \Longleftrightarrow x = y.$

We call ρ a *metric.* If only the first three conditions are satisfied we say that ρ is a *pseudometric.*

b. The *open ball* $B(x,r)$, with center x and radius r, in a metric space X is

$$B(x,r) = \{y \in X : \rho(x,y) < r\}.$$

A metric space is a topological space and U is defined to be *open* if

$$\forall\, x \in U, \quad \exists\, B(x,r) \subseteq U.$$

Equivalently, we can define a basis $\mathcal{B}$ for the topology in a metric space X to be $\{B(x,r) : x \in X, r > 0\}$.

In particular, metric spaces are Hausdorff.

c. A sequence $\{x_n : n = 1, \dots\} \subseteq X$, where X is a metric space, is *Cauchy* if

$$\forall\, \varepsilon > 0,\ \exists\, N \text{ such that } \forall\, m,n > N, \quad \rho(x_m, x_n) < \varepsilon.$$

If X is a metric space in which every Cauchy sequence $\{x_n : n = 1, \dots\}$ converges to some element $x \in X$, i.e., $\rho(x_n, x) \to 0$, then X is *complete.*

d. Two metric spaces (X_i, ρ_i), $i = 1, 2$, are *isometric* if there is a bijection $f : X_1 \to X_2$ such that

$$\forall\, x,y \in X_1, \quad \rho_1(x,y) = \rho_2(f(x), f(y)).$$

In this case, f is an *isometry.*

e. Let (X, ρ) be a metric space. A subset $V \subseteq X$ is *closed* if, whenever $\{x_n : n = 1, \dots\} \subseteq V$ and $\rho(x_n, x) \to 0$ for some $x \in X$, we can conclude that $x \in V$. The *closure* $\overline{Y}$ of a subset $Y \subseteq X$ is the set of all elements $x \in X$ for which there is a sequence $\{x_n : n = 1, \dots\} \subseteq Y$ such that $\rho(x_n, x) \to 0$. The complement of a closed set V is open and vice versa.

f. The *diameter* of a subset Y of a metric space (X, ρ), denoted by diam (Y), is

$$\text{diam}\,(Y) = \sup\{\rho(x,y) : x, y \in Y\}.$$

Theorem A.1.5. Compact metric spaces
Let (X, ρ) be a metric space. A subset $K \subseteq X$ is compact if and only if every sequence has a convergent subsequence.

Remark. If a topological space X has a countable basis then X is said to satisfy the *second axiom of countability.* If X is a locally compact Hausdorff space, then the second axiom of countability is equivalent to the existence of a metric ρ on X and a sequence of compact sets F_n such that $X = \bigcup_{n=1}^{\infty} F_n$, $F_n \subseteq \text{int}\, F_{n+1}$; see [278], Theorem I.5.3.

Example A.1.6. Hilbert cube
An important example of a metric space is the *Hilbert cube* $[0,1]^{\aleph_0}$. It is defined to be the Cartesian product of countably infinitely many copies of $[0,1]$ equipped with the metric

$$\rho(x,y) = \left(\sum_{i=1}^{\infty}(\min(1/i, |x_i - y_i|))^2\right)^{1/2},$$

for $x = \{x_i : i = 1, \dots\}$ and $y = \{y_i : i = 1, \dots\}$.

The following result should be compared with our construction of the *Cantor function* in Example 1.2.7*d*.

Proposition A.1.7. *There exists a continuous function $f : C \to [0,1]^{\aleph_0}$ that maps the Cantor set C onto the Hilbert cube $[0,1]^{\aleph_0}$.*

An excellent reference for metric spaces is [198].

It is sometimes necessary to consider topological vector spaces in which the topology cannot be described by a metric, e.g., in the theory of distributions; see Chapter 7, Example A.6.5, [242], [428], or [39], Chapter 2. In such cases we would still like to have a notion of completeness, and this is accomplished through the theory of uniform spaces, e.g., [279].

Definition A.1.8. Uniform space
A *uniform structure* on a set X is a family $\mathcal{X}$ of subsets of $X \times X$ that satisfies the following conditions:
i. $\forall\, V \in \mathcal{X},\ \{(x,x) : x \in X\} \subseteq V$,
ii. $\forall\, V \in \mathcal{X},\ \{(y,x) : (x,y) \in V\} \in \mathcal{X}$,
iii. $\forall\, V \in \mathcal{X},\ \exists\, V' \in \mathcal{X}$ such that

$$\{(x,y) : \exists\, z \in X \text{ such that } (x,z),(z,y) \in V'\} \in \mathcal{X},$$

iv. $\forall\, V, V' \in \mathcal{X},\ V \cap V' \in \mathcal{X}$,
v. $\forall\, V \subseteq X \times X$, for which $\exists\, V' \in \mathcal{X}$ and $V' \subseteq V$, we have $V \in \mathcal{X}$.

A *uniform space* $(X, \mathcal{X})$ is a topological space $(X, \mathcal{T})$ with the topology $\mathcal{T}$ defined by sets of the form

$$U = \{x : x \in X \text{ and } \exists\, y \in A \subseteq X \text{ such that } (y,x) \in V \in \mathcal{X}\},$$

for all subsets $A \subseteq X$ and for all sets $V \in \mathcal{X}$.

A uniform structure $\mathcal{X}$ is *pseudometrizable* if its corresponding topology $\mathcal{T}$ has a countable basis.

If $\{\rho_i,\ i = 1, \dots\}$ is a family of pseudometrics, respectively, metrics, on a nonempty set X, consider the uniform structure $\mathcal{X}$ on X defined by the collection of sets

$$\{(x,y) \in X \times X : \rho_i(x,y) < \varepsilon\}, \quad \varepsilon > 0,\ i = 1, \dots.$$

If the topology $\mathcal{T}$ corresponding to this uniform structure $\mathcal{X}$ has a countable basis, then there exists a pseudometric, respectively, metric, ρ that induces the same topology as $(X, \mathcal{X})$.

Standard references for uniform spaces are [279], [70].

Definition A.1.9. Normed vector space and Banach space
Let X be a vector space over $\mathbb{F}$, $\mathbb{F} = \mathbb{R}$ or $\mathbb{F} = \mathbb{C}$. We call X a *normed vector space* if there is a function $\|...\| : X \to \mathbb{R}^+$ such that

i. $\forall\, x \in X, \quad \|x\| = 0 \Longleftrightarrow x = 0,$

ii. $\forall\, x, y \in X, \quad \|x + y\| \leq \|x\| + \|y\|$ (triangle inequality),

iii. $\forall\, a \in \mathbb{F},\ \forall\, x \in X, \quad \|ax\| = |a|\|x\|.$

The function $\|...\|$ is a *norm*. A normed vector space is a metric space with metric $\rho(x, y) = \|x - y\|$.

A complete normed vector space is a *Banach space*.

Let X be a normed vector space. We say that $\sum x_n$ *converges* to $x \in X$, for $x_n \in X$, $n = 1, \ldots,$ if

$$\lim_{N \to \infty} \left\| x - \sum_{n=1}^{N} x_n \right\| = 0.$$

Moreover, $\sum x_n$ is *absolutely convergent* if

$$\sum_{n=1}^{\infty} \|x_n\| < \infty.$$

We have the following characterization of Banach spaces.

Proposition A.1.10. *A normed vector space X is a Banach space if and only if every absolutely convergent series is convergent.*

Proof. ($\Longrightarrow$) Take $\{x_n : n = 1, \ldots\} \subseteq X$ for which $\sum \|x_n\| < \infty$, and choose $\varepsilon > 0$. If $\sum_{n=N}^{\infty} \|x_n\| < \varepsilon/2$, then, for each $n > m \geq N$,

$$\left\| \sum_{j=m}^{n} x_j \right\| < \varepsilon.$$

Thus, $\sum x_n$ converges to some $x \in X$ since X is complete.

($\Longleftarrow$) Let $\{x_n : n = 1, \ldots\} \subseteq X$ be a Cauchy sequence in X. Hence, for each k there is $n_k \in \mathbb{N}$ such that

$$\forall\, m, n \geq n_k, \quad \|x_m - x_n\| < \frac{1}{2^k};$$

we can also choose $n_{k+1} > n_k$. Set $y_k = x_{n_k} - x_{n_{k-1}}$ for $k = 1, \ldots,$ where $x_{n_0} = 0$. Therefore, $\sum y_k$ is absolutely convergent, so that by hypothesis and the fact that

$$\sum_{k=1}^{m} y_k = x_{n_m},$$

$\{x_{n_m} : m = 1, \ldots\}$ converges to some $x \in X$. It is easy to check that

$$\lim_{n \to \infty} \|x - x_n\| = 0.$$

□

Let X be a Banach space over $\mathbb{F} = \mathbb{R}$ or $\mathbb{F} = \mathbb{C}$. A subset $V \subseteq X$ is a *linear subspace* of X if V is a vector space over $\mathbb{F}$. If $V \subseteq X$ is a linear subspace then its closure $\overline{V}$ in X is also a linear subspace.

The *span* of a subset $Z \subseteq X$, designated span Z, is the set of all finite linear combinations $x = \sum c_n x_n$, where $c_n \in \mathbb{F}$ and $x_n \in Z$. (The notion of span Z can be defined in any vector space.) Clearly, span Z is a linear subspace of X; its closure is designated by $\overline{\text{span}}\ Z$.

Definition A.1.11. Hilbert space
Let $\mathbb{F} = \mathbb{R}$ or $\mathbb{F} = \mathbb{C}$. A *Hilbert space* H is a Banach space with a function $\langle ..., ... \rangle : H \times H \to \mathbb{F}$ that satisfies the following conditions:

i. $\forall\, x, y \in H, \quad \overline{\langle x, y \rangle} = \langle y, x \rangle,$
ii. $\forall\, x, y, z \in H, \quad \langle x + y, z \rangle = \langle x, z \rangle + \langle y, z \rangle,$
iii. $\forall\, a \in \mathbb{F}$ and $\forall\, x, y \in H, \quad \langle ax, y \rangle = a\langle x, y \rangle,$
iv. $\forall\, x \in H, \quad \|x\| = \sqrt{\langle x, x \rangle}.$

$\langle ..., ... \rangle$ is an *inner product.*

The following result is straightforward to verify, and it does not require completeness.

Proposition A.1.12. *Let H be a Hilbert space. Then*

$$\forall\, x, y \in H, \quad |\langle x, y \rangle| \leq \|x\|\|y\| \tag{A.1}$$

and

$$\forall\, x, y \in H, \quad \|x + y\|^2 + \|x - y\|^2 = 2(\|x\|^2 + \|y\|^2). \tag{A.2}$$

Remark. Inequality (A.1) is the *Schwarz inequality,* which in the case of $H = L^2_\mu(X)$ is the Hölder inequality; see Theorem 5.5.2*b* and Example A.2.3. Of course, this does not mean that there is a simple proof of the Hölder inequality by means of the elementary inequality (A.1). In fact, the Schwarz inequality *assumes* the existence of an inner product; and the Hölder inequality *shows* the existence of an inner product for H.

Equation (A.2) is the *parallelogram law*; see Example A.2.5.

Excellent references for Banach and Hilbert spaces are [19], [200], [392], [406], [465], [517].

A.2 Examples

1. ***a.*** Let X be a topological space and let $C(X)$ be the vector space of continuous functions $f : X \to \mathbb{C}$. We let $C_b(X)$ denote the vector space of functions $f \in C(X)$ such that

$$\|f\|_\infty = \sup_{x \in X} |f(x)| < \infty, \tag{A.3}$$

i.e., f is bounded on X.

b. Now let X be a locally compact Hausdorff space and let $C_0(X)$ be the vector space of functions $f \in C_b(X)$ such that

$$\forall\, \varepsilon > 0, \ \exists K_f \subseteq X, \ \text{compact, for which } \forall\, x \notin K_f, \quad |f(x)| < \varepsilon. \tag{A.4}$$

Intuitively, f "vanishes at infinity". We denote by $C_c(X)$ the vector space of functions $f \in C_0(X)$ such that f vanishes outside of some compact set $K_f \subseteq X$. Equation (A.3) defines a norm on $C_b(X)$, and, with this norm, $C_b(X)$ and $C_0(X)$ are Banach spaces; $C_0(X)$ is a closed subspace of $C_b(X)$. With this norm on $C_0(X)$, Urysohn's lemma gives

$$\overline{C_c(X)} = C_0(X).$$

In fact, for $f \in C_0(X)$, $\varepsilon > 0$, and K_f as in (A.4), choose $g \in C_c(X)$ with $0 \le g \le 1$ and $g = 1$ on K_f by Theorem A.1.3, set $h = fg \in C_c(X)$, and obtain $\|f - h\|_\infty < \varepsilon$.

If X is compact we write $C(X) = C_b(X)$.

c. Let $(X, \mathcal{A}, \mu)$ be a measure space that is also a topological space. Since the uniform limit of continuous functions is continuous, $C_b(X)$ can be regarded as a closed subspace of $L^\infty_\mu(X)$, defined in Definition 2.5.9. It is for this reason we use the notation $\|\ldots\|_\infty$ from Definition 2.5.9 in (A.3).

2. $L^p_\mu(X)$, $1 \le p < \infty$, with L^p-norm $\|\ldots\|_p$ defined in Definition 5.5.1, is a Banach space (Theorem 5.5.2). Further, the set of simple functions $\sum_{j=1}^n a_j \mathbb{1}_{A_j}$, $\mu(A_j) < \infty$, is dense in $L^p_\mu(X)$ (Theorem 5.5.3). In the case that X is a locally compact Hausdorff space and μ is a regular Borel measure, we noted that

$$\overline{C_c(X)} = L^p_\mu(X)$$

(Theorem 7.2.6).

3. For any measure space $(X, \mathcal{A}, \mu)$, $L^2_\mu(X)$ is a Hilbert space with inner product

$$\langle f, g\rangle = \int_X f(x)\overline{g(x)}\, d\mu(x).$$

The fact that the integral is defined follows from the Hölder inequality (Theorem 5.5.2*b*). The structurally important converse is this: *Let H be a nonzero Hilbert space; then there are a set X and a linear bijection,*

$$L : H \to \ell^2(X),$$

such that

$$\langle x, y\rangle = \sum_{t\in X} (L(x))(t)\overline{(L(y))(t)}.$$

The fundamental elementary results of Hilbert space theory are used to prove this fact and to determine card X uniquely in terms of the cardinality of *orthonormal sets*; see Definition A.12.1.

4. If the measure space $(X, \mathcal{A}, \mu)$ is also a compact Hausdorff space and if $\mathcal{A}$ contains the Borel algebra, then

$$C(X) \subseteq L_\mu^\infty(X) \subseteq \cdots \subseteq L_\mu^p(X) \subseteq L_\mu^r(X) \subseteq \cdots \subseteq L_\mu^1(X), \quad 1 \leq r \leq p.$$

In $(X, \mathcal{P}(X), c)$, where X is topologized with the metric $\rho(x,y) = 0$ if $x = y$ and $\rho(x,y) = 1$ if $x \neq y$, we have

$$\ell^1(X) \subseteq \cdots \subseteq \ell^p(X) \subseteq \ell^r(X) \subseteq \cdots \subseteq \ell^\infty(X) = C_b(X), \quad 1 \leq p \leq r.$$

In both cases we have the inequality $\|\ \|_p \geq \|\ \|_r$, so that the corresponding injection is continuous (continuous functions are defined in Definition A.4.2).

5. ***a.*** A Banach space is a Hilbert space if and only if the *parallelogram law*, $\|x+y\|^2 + \|x-y\|^2 = 2(\|x\|^2 + \|y\|^2)$, is valid; see Proposition A.1.12. Using this fact we see that there are Banach spaces that are not Hilbert spaces.

b. Next, we give a standard example of a nontrivial complete metric vector space that is not a Banach space. Let X be the space of C^∞-functions on $[0,1]$. Define the metric ρ by

$$\rho(f,g) = \sum_{k=0}^{\infty} \frac{\|f-g\|_{(k)}}{2^k(1 + \|f-g\|_{(k)})},$$

where

$$\|f\|_{(k)} = \sup_{0 \leq j \leq k} \|f^{(j)}\|_\infty.$$

As such, X is complete. If the complete metric space X is a normed vector space with norm $\|...\|$, it is possible to show that

$$\forall\, n = 1, \ldots, \ \exists\, C_n \text{ such that } \forall\, f \in X, \text{ for which } \|f\| \leq 1,$$
$$\|f^{(n)}\|_\infty \leq C_n.$$

It is then not difficult to find $f \in X$ such that

$$\forall\, n = 1, \ldots, \quad \|f^{(n)}\|_\infty > nC_n,$$

from which we obtain the desired contradiction to the hypothesis that X is normed.

6. If (X, ρ) is a metric space there is a complete metric space $(\widetilde{X}, \tilde{\rho})$ such that $X \subseteq \widetilde{X}$, $\tilde{\rho} = \rho$ on $X \times X$, and X is dense in $\widetilde{X}$. We may take $\widetilde{X}$ to be the set of equivalence classes of Cauchy sequences from X, where $\{x_n : n = 1, \ldots\}$ is said to be *equivalent* to $\{y_n : n = 1, \ldots\}$ if $\rho(x_n, y_n) \to 0$. Let $\{x_n\} \subseteq X$ be a Cauchy sequence, and let $\{\{x_n\}\}$ be the equivalence class of all Cauchy sequences $\{z_n\} \subseteq X$ equivalent to $\{x_n\}$, i.e., $\lim_{n\to\infty} \rho(x_n, z_n) = 0$. For two equivalence classes, $\{\{x_n\}\}$ and $\{\{y_n\}\}$, $\tilde{\rho}$ is defined by

$$\tilde{\rho}(\{\{x_n\}\}, \{\{y_n\}\}) = \lim_{n\to\infty} \rho(x'_n, y'_n),$$

where $\{x'_n : n = 1, \ldots\}$ and $\{y'_n : n = 1, \ldots\}$ are any representatives of the equivalence classes $\{\{x_n\}\}$ and $\{\{y_n\}\}$, respectively. We call $(\widetilde{X}, \tilde{\rho})$ the *completion* of (X, ρ). A relevant theorem using this concept is the following. *Define*

$$\forall\, f, g \in C([a,b]), \quad \rho(f,g) = R\int_a^b |f-g|;$$

then $\widetilde{C}([a,b]) = L^1_m([a,b])$ *and* $\tilde{\rho}(f,g) = \int_a^b |f-g|$; cf. Theorem 7.1.1 and the Remark at the end of Section 7.3.

An even more basic example of the completion of a metric space is the construction of real numbers from rational numbers mentioned in Chapter 1.

For an alternative way to describe the completion of a metric space (X, ρ), let $B(X)$ be the Banach space of bounded real functions on X with metric $\sigma(f,g) = \sup\{|f(x) - g(x)| : x \in X\}$. Fix $x_0 \in X$ and define the function $F : X \to B(X)$, $x \mapsto f_x$, where

$$f_x(y) = \rho(x,y) - \rho(x_0, y).$$

Then F is an isometry $X \to F(X) \subseteq B(X)$ and $\widetilde{X} = \overline{F(X)}$.

7. Let $p \geq 2$ be a prime number. Any $x \in \mathbb{Q} \setminus \{0\}$ has the unique factorization $x = p^r q$, where $r \in \mathbb{Z}$ and where the numerator and denominator of $q \in \mathbb{Q}$ are both relatively prime to p. The *p-adic norm* $\|x\|$ of x is $\|x\| = p^{-r}$ and we define $\|0\| = 0$. It is elementary to check that

$$\forall\, x, y \in \mathbb{Q}, \quad \|x+y\| \leq \max(\|x\|, \|y\|) \quad \text{and} \quad \|xy\| = \|x\|\,\|y\|. \tag{A.5}$$

The function $\rho_p(x,y) = \|x - y\|$ defines the *p-adic metric* on $\mathbb{Q}$, and the completion of $\mathbb{Q}$ with respect to ρ_p is the field $\mathbb{Q}_p$ of *p-adic numbers*. The completion $\mathbb{Z}_p$ of $\mathbb{Z}$ with respect to ρ_p is the ring of *p-adic integers*. Note the analogy with the construction of $\mathbb{R}$ from $\mathbb{Q}$, as the completion of $\mathbb{Q}$ with respect to the usual absolute value norm. For one entry into p-adic analysis, see [398], [382].

We point out that $\mathbb{Q}_p$ consists of all formal Laurent series in p with coefficients $0, 1, \ldots, p-1$, with addition and multiplication as usual for Laurent series, except with carrying of digits. For example, in $\mathbb{Q}_5$, we have

$$(3 + 2\cdot 5) + (4 + 3\cdot 5) = 2 + 1\cdot 5 + 1\cdot 5^2.$$

The field $\mathbb{Q}_p$ is a locally compact abelian group under addition, with topology induced by the p-adic norm; see Appendix B.9 for a definition of a locally compact group. However, an important distinction between $\mathbb{R}$ and $\mathbb{Q}_p$, driven by (A.5), is the fact that $\mathbb{Z}_p$ is a compact open subgroup of $\mathbb{Q}_p$. This property leads to fascinating analysis with far-reaching applications in subjects as diverse as number theory, quantum field theory, and wavelet theory. In this last area, see [40].

A.3 Separability

A topological space is *separable* if it contains a countable dense subset. It is not difficult to prove the following theorem.

Theorem A.3.1. Separability of some L^p-spaces, $p \in [1, \infty)$
Let $(X, \mathcal{M}(\mathbb{R}^d), m^d)$ be a measure space, where $X \subseteq \mathbb{R}^d$ and where m^d is Lebesgue measure on X. If $p \in [1, \infty)$, then $L^p_{m^d}(X)$ is separable.

Example A.3.2. $L^\infty_m([0,1])$ is not separable
We shall prove that $L^\infty_m([0,1])$ is not separable. Let $\{f_n : n = 1, \ldots\}$ be an arbitrary sequence in $L^\infty_m([0,1])$ and write

$$(0,1] = \bigcup_{n=1}^\infty \left(\frac{1}{2^n}, \frac{1}{2^{n-1}}\right] = \bigcup_{n=1}^\infty E_n.$$

If

$$\operatorname*{ess\,sup}_{x \in E_n} |f_n(x)| \leq \frac{1}{2},$$

define $g = 1$ on E_n. ("ess sup" was defined in Definition 2.5.9.) Otherwise set $g = 0$ on E_n. Consequently, $g \in L^\infty_m([0,1])$ and

$$\forall\, n = 1, \ldots, \quad \|f_n - g\|_\infty \geq \frac{1}{2}.$$

Thus, $\{f_n : n = 1, \ldots\}$ is not dense in $L^\infty_m([0,1])$. Since $\{f_n : n = 1, \ldots\}$ is arbitrary, $L^\infty_m([0,1]$ cannot be separable.

Example A.3.3. Nonseparability of some L^p-spaces, $p \in [1, \infty)$
The fact that a given space X is separable has no bearing on the separability of $L^p_\mu(X)$, $1 \leq p < \infty$. Take $([0,1], \mathcal{P}([0,1]), c)$, where c is counting measure. If $f \in L^1_c([0,1])$, then $f = 0$ outside of a countable set. Thus, if $\{f_n : n = 1, \ldots\} \subseteq L^1_c([0,1])$, then there is $y \in [0,1]$ such that $f_n(y) = 0$ for each n. Define $g = \mathbb{1}_{\{y\}} \in L^1_c([0,1])$ so that

$$\forall\, n = 1, \ldots, \quad \|f_n - g\|_1 \geq 1.$$

Theorem A.3.4. Sequential pointwise convergence of simple functions
Let $(X, \mathcal{A}, \mu)$ be a measure space, let Y be a separable complete metric space, and let $(Y, \mathcal{C}, \nu)$ be a measure space, where $\mathcal{B}(Y) \subseteq \mathcal{C}$. If $f : X \to Y$ is measurable, then there is a sequence $\{g_k : k = 1, \ldots\}$ of simple functions $X \to Y$ such that $\{g_k : k = 1, \ldots\}$ converges pointwise to f.

Historically, a separable complete metric space is *Polish*.

The next theorem states that the Hilbert cube is "universal" for separable metric spaces.

Theorem A.3.5. Urysohn theorem
Every separable metric space X is homeomorphic to a subset of the Hilbert cube $[0,1]^{\aleph_0}$.

Corollary A.3.6. *For every separable metric space X there exist a subset A of the Cantor set C and a continuous surjective function $f : A \to X$. If X is compact, then A can be chosen to be closed.*

Corollary A.3.6 is an extension of Proposition A.1.7 and Example 1.2.7*d*.

A.4 Moore–Smith and Arzelà–Ascoli theorems

Let (X, ρ) be a metric space. We say that $\{x_{m,n}\} \to x$, i.e., $\lim_{m,n\to\infty} x_{m,n} = x$, if

$$\forall\, \varepsilon > 0,\ \exists\, N \text{ such that } \forall\, m, n > N, \quad \rho(x_{m,n}, x) < \varepsilon.$$

The following result can be generalized to uniform spaces with essentially the same proof.

Theorem A.4.1. Moore–Smith theorem
Let $\{x_{m,n} : m, n = 1, \dots\}$ be a sequence in a complete metric space (X, ρ). Assume
i. $\exists \lim_{n\to\infty} x_{m,n} = y_m$ *uniformly in* m,
ii. $\forall\, n = 1, \dots, \exists \lim_{m\to\infty} x_{m,n} = z_n$.
Then $\lim_{n\to\infty} \lim_{m\to\infty} x_{m,n}$, $\lim_{m\to\infty} \lim_{n\to\infty} x_{m,n}$, *and* $\lim_{m,n\to\infty} x_{m,n}$ *all exist and are equal.*

Proof. Assumption i means that

$$\forall\, \varepsilon > 0,\ \exists\, K > 0 \text{ such that } \forall\, n > K \text{ and } \forall\, m, \quad \rho(y_m, x_{m,n}) < \varepsilon.$$

Using i and ii we show that $\{y_m : m = 1, \dots\}$ is Cauchy by computing

$$\rho(y_m, x_{m,n}) < \frac{\varepsilon}{4} \quad \text{and} \quad \rho(z_k, x_{p.k}) < \frac{\varepsilon}{4}.$$

Since X is complete, $y_m \to w \in X$; and it is easy to check that

$$\lim_{m,n\to\infty} x_{m,n} = w. \tag{A.6}$$

Thus,

$$\lim_{m\to\infty} \lim_{n\to\infty} x_{m,n} = \lim_{m,n\to\infty} x_{m,n} = w.$$

Finally, in order to prove that $\lim_n z_n = w$, take $\varepsilon > 0$ and write

$$\rho(z_n, w) \le \rho(z_n, x_{m,n}) + \rho(x_{m,n}, w).$$

Since (A.6) holds,

$$\exists\, N \in \mathbb{N} \text{ such that } \forall\, m, n > N, \quad \rho(x_{m,n}, w) < \varepsilon;$$

and so

$$\forall\, n > N, \quad \rho(z_n, w) \leq \overline{\lim_{m\to\infty}}\, \rho(z_n, x_{m,n}) + \varepsilon = \varepsilon. \qquad \square$$

Definition A.4.2. Continuity and equicontinuity
Let (X, ρ) and (Y, θ) be metric spaces. A function $F : X \to Y$ is *continuous at* $x \in X$ if

$$\forall\, \varepsilon > 0,\ \exists\, \delta > 0 \text{ such that } \rho(z, x) < \delta \Longrightarrow \theta(f(z), f(x)) < \varepsilon;$$

and f is *continuous on* X if it is continuous at each $x \in X$.

A sequence $\{f_n : n = 1, \dots\}$ of continuous functions is *equicontinuous at* $x \in X$ if

$$\forall\, \varepsilon > 0,\ \exists\, \delta > 0 \text{ such that } \forall\, n, \quad \rho(z, x) < \delta \Longrightarrow \theta(f_n(z), f_n(x)) < \varepsilon.$$

The sequence $\{f_n : n = 1, \dots\}$ is *equicontinuous on* X if it is equicontinuous at each $x \in X$.

The notion of equicontinuity was introduced by Giulio Ascoli in 1883 [13], and the following theorem was proved by Arzelà in 1895 and 1899 [11], [12]; cf. [237], [238]. Clearly, the theorem generalizes the Bolzano–Weierstrass property of $\mathbb{R}$.

Theorem A.4.3. Arzelà–Ascoli theorem
Let X be a separable metric space, let Y be a compact metric space, and let $\{f_n : n = 1, \dots\}$ be an equicontinuous sequence of functions $X \to Y$. Then there is a subsequence of $\{f_n : n = 1, \dots\}$ that converges pointwise to a continuous function.

Proof. Let $\{x_n : n = 1, \dots\} \subseteq X$ be dense. Since Y is compact,

$$\exists\, J_1 \subseteq \mathbb{N} \text{ such that } \{f_n(x_1) : n \in J_1\} \text{ is convergent.}$$

Pick $J_2 \subseteq J_1$ such that $\{f_n(x_2) : n \in J_2\}$ is convergent, and continue in this way. Consequently,

$$\forall\, j = 1, \dots,\ \exists\ \lim_{k\to\infty} f_{n_k}(x_j) = g(x_j),$$

where $n_k \in J_k$ and $\lim_{k\to\infty} n_k = \infty$. Let $z \in X \setminus \{x_n : n = 1, \dots\}$ with $x_{q_p} \to z$ as $p \to \infty$. Then

$$\lim_{p\to\infty} f_{n_k}(x_{q_p}) = f_{n_k}(z), \text{ uniformly in } k = 1, \dots,$$

by the equicontinuity hypothesis. Also,

$$\forall\, p = 1, \dots, \quad \lim_{k\to\infty} f_{n_k}(x_{q_p}) = g(x_{q_p}).$$

Consequently, by the Moore–Smith theorem,

$$\exists \lim_{k\to\infty} f_{n_k}(z) = g(z).$$

The continuity of g is straightforward to check. □

Obviously the result is still true if, instead of assuming that Y is compact, we assume that the range of each f_n is compact in Y. It is also easy to prove that the convergence of $\{f_{n_k} : k = 1, \dots\}$ is uniform on compact subsets of X.

Remark. The notion of equicontinuity of a sequence can be generalized to an equicontinuous set by replacing "$\forall\, n$" in Definition A.4.2 with "for all elements of the set". Specifically, if X is a compact set and $S \subseteq C(X)$, then Theorem A.4.3 can be formulated as follows: *If S is pointwise bounded and equicontinuous, then S is relatively compact in the sup norm topology on $C(X)$, and every sequence in S has a uniformly convergent subsequence.*

A.5 Uniformly continuous functions

Definition A.5.1. Uniform continuity
Let (X, ρ) and (Y, θ) be metric spaces. A function $f : X \to Y$ is *uniformly continuous* if

$$\forall\, \varepsilon > 0,\ \exists\, \delta > 0 \text{ such that } \rho(x, y) < \delta \Longrightarrow \theta(f(x), f(y)) < \varepsilon.$$

Remark. If X is a compact metric space and $f : X \to \mathbb{R}$ is continuous then f is uniformly continuous. The function $f(x) = \sin(1/x)$ is a bounded continuous mapping $(0, 1] \to [-1, 1]$, but it is not uniformly continuous. Observe that

$$f : [0, 1) \to [0, \infty),$$
$$x \mapsto \frac{x}{1 - x},$$

is bijective and bicontinuous, i.e., a homeomorphism, whereas the Cauchy sequence $\{1 - (1/n) : n = 1, \dots\}$ in $[0, 1)$ is transformed into the sequence $\{n - 1 : n = 1, \dots\}$, which is not Cauchy. In this case the range space is complete and $[0, 1)$ is not complete. Such a phenomenon leads us to distinguish between topological properties, dealing with homeomorphisms, and uniform properties, dealing with Cauchyness, uniform continuity, and completeness.

Generally there are no relations between these two categories except the following: *Let X be a metric space; X is compact if and only if it is complete and totally bounded* ((X, ρ) is *totally bounded* if

$$\forall\, \varepsilon > 0,\ \exists\, x_1, \ldots, x_n \in X \text{ such that } \quad X \subseteq \bigcup_{j=1}^{n} B(x_j, \varepsilon)).$$

The proof of this theorem can be obtained by means of a circular chain of implications in which Theorem A.1.5 is also proved; e.g., [198], pages 267–268. In any case, for perspective, recall from Definition 1.2.11 that the compact subsets of $\mathbb{R}^d$ are precisely the closed and bounded subsets of $\mathbb{R}^d$.

Theorem A.5.2. Unique uniformly continuous extensions
Let X be a metric space and let Y be a complete metric space. Assume that $Z \subseteq X$ and that $f : Z \to Y$ is a uniformly continuous function. Then f has a unique uniformly continuous extension to $\overline{Z}$.

Definition A.5.3. Absolute continuity
Let (X, ρ) and (Y, θ) be metric spaces and let $f : X \to Y$ be a continuous function. Then f is *absolutely continuous* if

$$\forall\, \varepsilon > 0,\ \exists\, \delta > 0 \text{ such that } \forall\, \{x_1, \ldots, x_n\} \subseteq X,$$
$$\sum_{j=1}^{n-1} \rho(x_j, x_{j+1}) < \delta \Longrightarrow \sum_{j=1}^{n-1} \theta(f(x_j), f(x_{j+1})) < \varepsilon.$$

Remark. Let (X, ρ) and (Y, θ) be metric spaces and let $f : X \to Y$ be absolutely continuous. If $\sigma : X \times X \to \mathbb{R}$ is defined by $\sigma(x, y) = \rho(x, y) + \theta(f(x), f(y))$, then (X, ρ) and (X, σ) have the same topologies and

$$f : (X, \sigma) \to (Y, \theta)$$

is absolutely continuous. For $X = Y = \mathbb{R}$, taken with the absolute value metric, this definition of absolute continuity characterizes the class of Lipschitz functions, which, in turn, is properly contained in the class of absolutely continuous functions on $\mathbb{R}$ as defined in Chapter 4.

Example A.5.4. Comparison between absolute and uniform continuity
Let (X, ρ) and (Y, θ) be metric spaces and let $f : X \to Y$ be a continuous function. We shall show that it is not generally possible to find metrics σ and τ on X and Y, respectively, such that $f : (X, \sigma) \to (Y, \tau)$ is absolutely and uniformly continuous. Take $f : (0, 1] \to [1, \infty)$, $f(x) = 1/x$, with the usual metrics. Assume that we can find σ, τ that yield both absolute and uniform continuity. Then, from Theorem A.5.2, f has a unique uniformly continuous extension $[0, 1] \to [1, \infty)$, and this is obviously false.

A.6 Baire category theorem

An excellent reference for the Baire category theorem is [362]. A metric space is *Baire* if every countable intersection of open dense sets is dense. Since $\mathbb{R}$ is a complete metric space, Theorem A.6.1 and Theorem A.6.2*b* yield the fact that $\mathbb{R}$ is not a set of first category; see (1.12).

Theorem A.6.1. Baire category theorem I
Every complete metric space X is Baire.

Proof. i. We give CANTOR's necessary conditions for the completeness of a metric space, as promised in Section 2.1. (The converse is true and easy.) Take $\{A_n : n = 1, \dots\} \subseteq X$, where each A_n is closed, nonempty, and $A_1 \supseteq A_2 \supseteq \cdots$. Assuming that $\lim_{n\to\infty} \sup\{\rho(y,z) : y, z \in A_n\} = 0$ we verify that $\bigcap A_n = \{x\} \subseteq X$ for some $x \in X$.

For all n, let $x_n \in A_n$. The sequence $\{x_n : n = 1, \dots\}$ is Cauchy, for if $m \geq n$, then

$$\rho(x_m, x_n) \leq \sup\{\rho(y,z) : y, z \in A_n\} = \text{diam}\,(A_n) \to 0, \qquad n \to \infty.$$

Here, for $A \subseteq \mathbb{R}^d$, we write

$$\text{diam}\,(A) = \sup\,\{|x - y| : x, y \in A\}.$$

By the completeness of X there is a point $x \in X$ such that $\rho(x_n, x) \to 0$. Now, for each n, $x_m \in A_n$ when m is sufficiently large. Consequently, $x \in \bigcap A_n$ since A_n is closed. If $y \in \bigcap A_n$, then $\rho(x,y) \leq \text{diam}\,(A_n)$ for each n, so that, by hypothesis, $\rho(x,y) = 0$. Thus, $x = y$.

ii. Let U_n be an open and dense subset of X. Thus, $A_n = U_n^{\sim}$ is *nowhere dense*, i.e., int $A_n = \emptyset$. Moreover, $(\bigcap U_n)^{\sim} = \bigcup A_n = A$, where each A_n is closed. We prove that if V is open, then

$$V \cap \left(\bigcap_{n=1}^{\infty} U_n\right) \neq \emptyset.$$

Choose an open set V_1 such that $\overline{V}_1 \subseteq V$ and diam $(\overline{V}_1) < 1$. Since V_1 is not a subset of A_1,

$$V_1 \cap U_1 \neq \emptyset,$$

and $V_1 \cap U_1$ is open.

Choose an open set V_2 such that $\overline{V}_2 \subseteq V_1 \cap U_1$ and diam $(\overline{V}_2) < 1/2$. Generally, then, we choose open sets V_n with $\overline{V}_n \subseteq V_{n-1} \cap U_{n-1}$ and diam $(\overline{V}_n) < 1/n$. The hypotheses of part *i* are satisfied for $\overline{V}_n$, and hence $\bigcap \overline{V}_n = \{x\}$. Therefore,

$$x \in \bigcap_{n=1}^{\infty} (V_n \cap U_n) \subseteq V_1 \cap \left(\bigcap_{n=1}^{\infty} U_n\right). \qquad \square$$

Let X be a metric space. A set $A \subseteq X$ is of *first category* if it is the countable union of nowhere dense sets, i.e., sets having empty interior. Any other subset of X is a set of *second category*. RENÉ BAIRE introduced these notions in 1899. Among other results, he proved that the countable intersection of open dense sets (in $\mathbb{R}$) is dense, and this is our definition of a Baire metric space. The following is straightforward to prove.

Theorem A.6.2. Baire category theorem II
The following are equivalent for a metric space X.

a. *X is Baire.*

b. *Every countable union of closed nowhere dense sets has empty interior.*

c. *Every nonempty open set is of second category.*

d. *If $\bigcup A_n$, A_n closed, contains an open set, then some A_j contains an open set.*

e. *The complement of every set of first category is dense in X.*

Example A.6.3. Sets of first and second category

a. First category sets are not necessarily nowhere dense. In fact, take $\mathbb{Q} \subseteq \mathbb{R}$, noting that $\mathbb{Q}$ is of first category and $\overline{\mathbb{Q}} = \mathbb{R}$.

b. Let $S \subseteq \mathbb{R}$. If $\{x - y : x, y \in S\}$ is a set of first category, then S is a set of first category; and so, if S is of second category, then $\{x - y : x, y \in S\}$ is of second category.

c. It is easy to construct a first-category set of Lebesgue measure 1 in $[0,1]$. Let E_n be a perfect symmetric set with $m(E_n) \geq 1 - (1/n)$. Then $E = \bigcup E_n$ does the trick; cf. Problem 2.9 and Problem 2.10.

d. Clearly, $[0,1]$ does not contain a countable dense $\mathcal{G}_\delta$, $D = \bigcap U_j$. In fact, if $D = \{d_j : j = 1, \dots\}$ were such a set, then $V_j = U_j \setminus (\bigcup_{n=1}^{j} d_n)$ would be open and dense, and $\bigcap V_j = \emptyset$. This contradicts Baire category theorem I.

Example A.6.4. Open coverings of accessible points
Let $E \subseteq [0,1]$ be any perfect symmetric set. As such, it is associated with a countable set A of accessible points. (A point $a \in A \subseteq \mathbb{R}$ is *accessible* if it is the endpoint of a contiguous open interval.) Note that if $\{U_\alpha\}$ is an open covering of A, it does not necessarily follow that $E \subseteq \bigcup U_\alpha$. For example, if $x \in E \setminus A$ consider $[0,x) \cup (x,1]$. For each $a_n \in A$ let $\{I_{m,n} : m = 1, \dots\}$ be a sequence of open intervals about a whose lengths tend to 0. Then $\{a_n : n = 1, \dots\} = \bigcap_{m=1}^{\infty} I_{m,n}$. Now let $V_m = \bigcup_{n=1}^{\infty} I_{m,n}$, so that $E \cap V_m$ is open and dense in E. Observe that $A \subseteq \bigcap_{m=1}^{\infty}(E \cap V_m)$, properly. To prove this note that $U_m = E \cap V_m \setminus \{a_1, \dots, a_m\}$ is open and dense in E, so that $\bigcap U_m$ is dense. On the other hand, $A \cap (\bigcap U_m) = \emptyset$ and $\bigcap (E \cap V_m) = A \cup (\bigcap U_m)$.

Example A.6.5. A complete nonmetric space
Let $C_c(\mathbb{R})$ be the vector space of continuous functions $f : \mathbb{R} \to \mathbb{C}$ that vanish outside of some compact set, depending on f. We define sequential convergence in $C_c(\mathbb{R})$ as follows:

$$f_n \to f \text{ in } C_c(\mathbb{R}),\ f_n, f \in C_c(\mathbb{R}), \text{if } \|f_n - f\|_\infty \to 0 \text{ and}$$
$$\exists\, r > 0 \text{ such that } \forall\, n,\ f_n = 0 \text{ on } [-r,r]^\sim.$$

We shall prove that, with this convergence, $C_c(\mathbb{R})$ cannot be a complete metric space $(C_c(\mathbb{R}), \rho)$. If such a metric ρ exists, then $C_c(\mathbb{R})$ is a Baire space. We shall show that $(C_c(\mathbb{R}), \rho)$ is of first category to obtain the contradiction.

First, note that

$$C_c(\mathbb{R}) = \bigcup_{n=1}^{\infty} C_{c,n}, \quad C_{c,n} = \{f \in C_c(\mathbb{R}) : f = 0 \text{ on } [-n,n]^{\sim}\}.$$

Clearly, $\overline{C_{c,n}} = C_{c,n}$, and it is sufficient to check that int $C_{c,n} = \emptyset$. Assume not, and let $V \subseteq C_{c,n}$ be an open neighborhood of 0 in $C_c(\mathbb{R})$. Choose $f_k \in C_{c,n+1} \setminus C_{c,n}$ such that $\rho(f_k, 0) \to 0$. Consequently, $f_k \in V \subseteq C_{c,n}$, and this contradicts the definition of f_k. There is, in fact, a (completely regular) topology on $C_c(\mathbb{R})$ whose uniform structure renders $C_c(\mathbb{R})$ complete and whose sequential convergence is that given above; cf. Section 7.3.

Example A.6.6. Everywhere continuous nowhere differentiable functions

In Chapter 1 we discussed everywhere continuous nowhere differentiable functions. The soft analysis proof of their existence uses Baire category theorem I. Take $C([0,1])$ with the $\|\ldots\|_\infty$ norm, so that $C([0,1])$ is complete with the metric $\rho(f,g) = \|f-g\|_\infty$. Define

$$F_n = \left\{ f \in C([0,1]) : \exists\, x \in [0,1] \text{ such that } \forall\, h > 0, \left| \frac{f(x+h)-f(x)}{h} \right| < n \right\}.$$

Each F_n is closed and nowhere dense, and so $C([0,1]) \neq \bigcup F_n$. Consequently, the set of continuous nowhere differentiable functions is dense in $C([0,1])$.

Example A.6.7. Sets A such that $0 < \mu(A \cap I) < \mu(I)$ for all I

The proof of Problem 2.45*b* is elementary. First, let $A_1 \subseteq [0,1]$ be a perfect symmetric set of measure $1/4$. Then let $A_2 = \bigcup_{j=1}^{\infty} A_{j,2}$, where the measure of A_2 is $1/8$ and where each $A_{j,2}$ is a perfect symmetric set of positive measure in the jth contiguous interval of A_1. Define all of the A_j in this way, and set $A = \bigcup_{j=1}^{\infty} A_j$. A generalization of this result is due to R. B. Kirk [286]: *Let the measure space $(X, \mathcal{A}, \mu)$ be a separable metric space with metric ρ, assume that μ is continuous, and suppose $\mathcal{B}(X) \subseteq \mathcal{A}$; then there is $A \in \mathcal{B}(X)$ such that for each open set I of positive measure we have*

$$0 < \mu(A \cap I) < \mu(I).$$

A.7 Uniform Boundedness Principle and Schur lemma

The Uniform Boundedness Principle, also known as the Banach–Steinhaus theorem, is one of fundamental results in functional analysis. The first result of this type was proved by Banach and Steinhaus in 1927 [24].

Theorem A.7.1. Banach–Steinhaus Uniform Boundedness Principle
Let (X, ρ) be a complete metric space and let $\mathcal{F}$ be a set of continuous functions $X \to \mathbb{C}$. Assume

$$\forall\, x \in X,\ \exists\, M_x > 0 \text{ such that } \forall\, f \in \mathcal{F}, \quad |f(x)| \leq M_x.$$

Then there are a nonempty open set $U \subseteq X$ and a constant M such that

$$\forall\, x \in U \text{ and } \forall\, f \in \mathcal{F}, \quad |f(x)| \leq M.$$

Proof. For each $f \in \mathcal{F}$ and $m \in \mathbb{N}$, define

$$A_{m,f} = \{x : |f(x)| \leq m\} \quad \text{and} \quad A_m = \bigcap_{f \in \mathcal{F}} A_{m,f}.$$

Since f is continuous, A_m is closed. We show that $X = \bigcup A_m$. In fact, if $x \in X$ choose $m = M_x$, so that $x \in A_m$.

Consequently, from Baire category theorems I and II, $U = \text{int}\, A_n \neq \emptyset$ for some n and we take $M = n$. □

See Theorem A.8.6 for a statement of the Uniform Boundedness Principle in terms of Banach spaces.

As noted after Definition 6.3.1, sequential weak convergence in $L^1_\mu(X)$ is actually sequential convergence for a certain topology (called the *weak* topology) on $L^1_\mu(X)$. We shall discuss the weak topology generally in Appendix A.9, but for now consider a special result for the case of $\ell^1(\mathbb{N})$. This result, Theorem A.7.3 (Schur's lemma), is studied in greater detail in Chapter 6. The proof we outline uses the Baire category theorems. To formulate Schur's lemma we need the following definition.

Definition A.7.2. The weak topology $\sigma(\ell^1(\mathbb{N}), \ell^\infty(\mathbb{N}))$
Let $F \subseteq \ell^\infty(\mathbb{N})$ be a finite set and let $x \in \ell^1(\mathbb{N})$. Define

$$U(F, x, \varepsilon) = \{y \in \ell^1(\mathbb{N}) : \forall\, x' \in F,\ |x'(y - x)| < \varepsilon\},$$

where, if $x' = \{x'_j\} \subseteq \mathbb{C}$, $y = \{y_j\} \subseteq \mathbb{C}$, and $x = \{x_j\} \subseteq \mathbb{C}$, then the operation of x' on $(y - x)$ is $x'(y - x) = \sum_{j=1}^\infty x'_j(y_j - x_j)$. The family $\{U(F, x, \varepsilon) : F \subseteq \ell^\infty(\mathbb{N}), x \in \ell^1(\mathbb{N}),\ \text{and } \varepsilon > 0\}$ is a basis for a topology $\sigma(\ell^1(\mathbb{N}), \ell^\infty(\mathbb{N}))$ on $\ell^1(\mathbb{N})$. This is the *weak topology* for $\ell^1(\mathbb{N})$.

If U is open for the $\|\ldots\|_1$ topology on $\ell^1(\mathbb{N})$, then $U \in \sigma(\ell^1(\mathbb{N}), \ell^\infty(\mathbb{N}))$. It is not difficult to verify that

$$S = \left\{ x \in \ell^1(\mathbb{N}) : \sum_{j=1}^\infty |x_j| = 1 \right\} \text{ is } \|\ldots\|_1 \text{ closed}$$

and

$$\text{the weak closure of } S \text{ is the } \|\ldots\|_1 \text{ closure of } B(0,1),$$

where the ball $B(0,1)$ is defined in terms of $\|\ldots\|_1$. Consequently, the topology $\sigma(\ell^1(\mathbb{N}), \ell^\infty(\mathbb{N}))$ is *strictly* weaker than the norm topology on $\ell^1(\mathbb{N})$. Note that sequential weak convergence for $\ell^1(\mathbb{N})$ is precisely the analogue for $\mathbb{N}$ of that defined for $[0,1]$ immediately after Theorem A.7.1.

Schur's lemma [424] tells us that *$\sigma(\ell^1(\mathbb{N}), \ell^\infty(\mathbb{N}))$ and $\|\ldots\|_1$ yield the same convergent sequences* in $\ell^1(\mathbb{N})$; see [19], pages 137–139; cf. [216], Section 3.2, and [465], pages 327–329, for selected results from [424].

Theorem A.7.3. Schur lemma
If $\{x^{(n)} : n = 1,\ldots\} \subseteq \ell^1(\mathbb{N})$ converges to 0 in $\sigma(\ell^1(\mathbb{N}), \ell^\infty(\mathbb{N}))$, then $\|x^{(n)}\|_1 \to 0$.

Proof. Let $Y = \{x' \in \ell^\infty(\mathbb{N}) : \sup |x'_j| \leq 1\}$ and define

$$\forall\, x', y' \in Y, \quad \rho(x', y') = \sum_{j=1}^{\infty} \frac{|x'_j - y'_j|}{2^j}.$$

Then (Y, ρ) is complete, and sets of the form

$$\forall\, x' \in Y, \quad S_{J,\delta} = \{y' : |x'_j - y'_j| < \delta,\ |j| \leq J\}$$

are a basis at x' for the topology of (Y, ρ). Next we define

$$\forall\, \varepsilon > 0 \text{ and } \forall\, m, \quad A_m = \left\{ y' \in Y : \forall\, n \geq m, \left| \sum_{j=1}^{\infty} y'_j x_j^{(n)} \right| \leq \varepsilon, \right\}.$$

It can be shown that A_m is closed in (Y, ρ) and $Y = \bigcup A_m$, so that by Baire category theorems I and II there is m such that $\operatorname{int} A_m \neq \emptyset$. From this point it is straightforward to prove that $\|x^{(n)}\|_1 \to 0$. □

A.8 Hahn–Banach theorem

Our presentation of the Hahn–Banach theorem (Theorem A.8.3) is standard. There are basically three distinct parts to the proof. The first and crucial step is Lemma A.8.2 which allows us to extend continuous linear functionals from a closed subspace Y to the closed subspace generated by Y and an element x (the setting here is necessarily with real vector spaces). Second, an axiom of choice argument is used to expand this finite procedure to extend maps in the infinite-dimensional case. Finally, an ingenious trick due to Henri F. Bohnenblust and Sobczyk yields the result for the complex case.

Remark. Let X and Y be vector spaces over $\mathbb{F} = \mathbb{R}$ or $\mathbb{F} = \mathbb{C}$, let $V \subseteq X$ be a linear subspace, and let $L : V \to Y$ be a linear function. Then there exists $K : X \to Y$ such that K is linear on X and $K = L$ on V. In infinite dimensions the proof requires the axiom of choice, usually in the form of Zorn's lemma; see Section 2.6.1. The general problem of extending continuous linear functions $L : Z \to Y$, $Z \subseteq X$, is usually intractable.

Our setting for Appendix A.8 through Appendix A.11 will be nonzero normed vector spaces, although many of the results are true for Hausdorff locally convex topological vector spaces. There are just a few cases in which we need this generality, e.g., in Chapter 7, so we have chosen to be efficient spacewise (sic), at least in this case, and not write out all the details. See BOURBAKI's *Espaces Vectoriels Topologiques* or [242] for classic and classical presentations.

Proposition A.8.1. *Let X and Y be normed vector spaces, and let $L : X \to Y$ be a linear function.*

a. *L is continuous either at every $x \in X$ or at no $x \in X$.*

b. *L is continuous on X if and only if*

$$\exists\, C > 0 \text{ such that } \forall\, x \in X, \quad \|L(x)\| \le C\|x\|.$$

Proof. Part *a* and the sufficient condition for continuity in part *b* are straightforward.

For the necessary condition in part *b*, assume that L is continuous at 0. Thus,

$$\exists\, \delta > 0 \text{ such that } \|x\| < \delta \Longrightarrow \|L(x)\| < 1.$$

If $x \neq 0$, let $x_\delta = (\delta x)/(2\|x\|)$, and so $\|L(x_\delta)\| < 1$. Hence,

$$\|L(x)\| = \frac{2}{\delta}\|x\|\|L(x_\delta)\| < \frac{2}{\delta}\|x\|,$$

and so we set $C = 2/\delta$. □

If X and Y are normed vector spaces, $Z \subseteq X$ is a linear subspace, and $L : Z \to Y$ is linear, we define

$$\|L\| = \sup\{\|L(x)\| : \|x\| \le 1,\ x \in Z\}. \tag{A.7}$$

Thus, $\|L\|$ is the smallest constant C such that $\|L(x)\| \le C\|x\|$ for all $x \in Z$. If $\|L\| < \infty$, then, because of Proposition A.8.1, we say that L is a *continuous* or *bounded linear function* $Z \to Y$. Clearly, *$\|L\|$ depends on the subspace Z*. This is important in what follows.

The space of continuous linear functions $X \to \mathbb{C}$ is denoted by X'. It is the *dual space* of X, and its elements are usually called *continuous* or *bounded linear functionals*.

Hilbert spaces H have the property that

$$H' = H. \tag{A.8}$$

Equation (A.8) is the Riesz representation theorem for the case of Hilbert spaces; see the last comment in Section 7.1.

Lemma A.8.2. *Let X be a real normed vector space, $Y \subseteq X$ a closed linear subspace, and Z the closed linear subspace of X generated by Y and some $x \in X \setminus Y$. If $L : Y \to \mathbb{R}$ is linear and continuous, then there is a continuous linear functional $K : Z \to \mathbb{R}$ such that $K = L$ on Y and $\|K\| = \|L\|$.*

Proof. If $x, y \in Y$, then

$$L(x) - L(y) \leq \|L\|\|x + z\| + \|L\|\|y + z\|;$$

and so

$$\sup_{u \in Y} (-\|L\|\|u + z\| - L(u)) = a \leq b = \inf_{u \in Y} (\|L\|\|u + z\| - L(u)) .$$

For fixed $c \in [a, b]$ we define $K(y + rz) = L(y) + rc$, where $r \in \mathbb{R}$, $y \in Y$, and $\{y + rz : r \in \mathbb{R}, y \in Y\} = Z$. □

Theorem A.8.3. Hahn–Banach theorem
a. *Let $Y \subseteq X$ be a linear subspace of the normed vector space X, and assume that $L : Y \to \mathbb{C}$ is linear and continuous. Then there is $K \in X'$ such that $K = L$ on Y and $\|K\| = \|L\|$.*

b. *If $Y \subseteq X$ is a closed linear subspace of the normed vector space X and $z \notin Y$, then there is $L \in X'$ such that $L(z) \neq 0$ and $L = 0$ on Y.*

Proof. ***a.****i.* We choose Y to be closed without any loss of generality. In fact, it is easy to extend L to $\overline{Y}$ by Theorem A.5.2.

ii. We now prove part *a* for the real case, assuming that Y is closed and that $Y \subseteq X$ properly.

Let $\mathcal{L}$ be the family of all continuous linear functions $K : Z \to \mathbb{R}$ such that $Y \subseteq Z$, $K = L$ on Y, and $\|K\| = \|L\|$. From Lemma A.8.2, $\mathcal{L}$ is nontrivial. We order $\mathcal{L}$ by setting $K \leq K_1$ if $Z \subseteq Z_1$ and $K_1 = K$ on Z. From Zorn's lemma (Section 2.6.1), i.e., the axiom of choice, there is a maximal element $K : Z \to \mathbb{R}$, and we easily check that $Z = X$.

iii. Let W be a complex vector space. If $K : W \to \mathbb{C}$ is real linear then K is complex linear if and only if $K(ix) = iK(x)$. Let $L : Y \to \mathbb{C}$ be complex linear, as in part *a*. Set $L_1 = \operatorname{Re} L$, $L_2 = \operatorname{Im} L$, and note that L is real linear. Thus, $L(iy) = iL(y)$ on Y, and, using this fact, we compute that

$$\forall\, y \in Y, \quad L_2(y) = -L_1(iy).$$

Because of part *ii* we can extend L_1 to K_1 on X, considered as a real vector space, such that $\|L_1\| = \|K_1\|$. Set $K(x) = K_1(x) - iK_1(ix)$ on X.

Similar computations show that K has the desired properties.

b. Part b is a consequence of part a. Indeed, define $L_z(y+rz) = r$, where $y+rz$, for $y \in Y$ and $r \in \mathbb{R}$, is a typical element of the closed linear subspace generated by Y and z. Note that

$$a = \inf_{y \in Y} \|z + y\| > 0$$

and

$$|L_z(y + rz)| \leq \frac{1}{a}\|y + rz\|.$$

Thus, we apply part a directly. □

Remark. The Hahn–Banach theorem allows us to assert that *if* $\{x_n : n = 1, \ldots\} \subseteq X$*, a Banach space, then* $\overline{\operatorname{span}}\ \{x_n\} = X$ *if and only if, whenever* $L(x_n) = 0$ *for all* n *for any given* $L \in X'$*, we can conclude that* $L = 0$.

By (A.8), the equivalent assertion for a Hilbert space H is that $\overline{\operatorname{span}}\ \{x_n\}$ $= H$ *if and only if, whenever* $\langle y, x_n \rangle = 0$ *for all* n *and any given* $y \in H$*, we can conclude that* $y = 0$.

Example A.8.4. L^p-duality
a. Let $1 \leq p < \infty$, let $1/p + 1/q = 1$, and let μ be a σ-finite measure on $\mathbb{R}$. Then $(L^p_\mu(\mathbb{R}))' = L^q_\mu(\mathbb{R})$, where $g : L^p_\mu \to \mathbb{C}$ is well defined by $g(f) = \int_\mathbb{R} f(t)\overline{g(t)}\, d\mu(t)$ for all $f \in L^p_\mu(\mathbb{R})$; see Theorem 5.5.5. In particular, the Hilbert space $H = L^2_\mu(\mathbb{R})$ has the property that $(L^2_\mu(\mathbb{R}))' = L^2_\mu(\mathbb{R})$.

b. Let $(\mathbb{R}, \mathcal{A}, \mu)$ be a measure space. Then, according to Theorem 5.5.7, $(L^\infty_\mu(\mathbb{R}))'$ is the space of complex-valued finitely additive bounded set functions on $\mathcal{A}$; see also Example A.11.3 as well as [150], part I, Chapter IV, Section 8.

Example A.8.5. Sufficiently many elements in X'
Let X be a normed vector space, and choose $x, y \in X$, $x \neq y$. By the Hahn–Banach theorem we see that there is $L \in X'$ such that $L(x) \neq L(y)$. In fact, let Y be the linear subspace generated by $x - y$, define $K(r(x-y)) = r\|x-y\|$, observe that $\|K\| = 1$, and use Theorem A.8.3.

The contrapositive equivalent assertion for $x \in X$ is the following:

$$\forall\, L \in X', \quad L(x) = 0 \Longrightarrow x = 0.$$

The following restatement of Theorem A.7.1 does not require the Hahn–Banach theorem, but it does use the terminology defined in this section.

Theorem A.8.6. Uniform Boundedness Principle for Banach spaces
Let X *be a Banach space, let* Y *be a normed vector space, and let* $\mathcal{L}$ *be a set of continuous linear functions* $X \to Y$*. Assume that*

$$\forall\, x \in X,\ \exists\, C_x > 0 \text{ such that } \forall\, L \in \mathcal{L}, \quad \|L(x)\| \leq C_x.$$

Then

$$\exists\, C > 0 \text{ such that } \forall\, L \in \mathcal{L}, \quad \|L\| \leq C. \tag{A.9}$$

Corollary A.8.7. *Let X be a Banach space, let Y be a normed vector space, and let $\{L_n : n = 1, \dots\}$ be a sequence of continuous linear functions $X \to Y$. Assume*

$$\forall\, x \in X,\ \exists\, L(x) \in Y \text{ such that } \quad \lim_{n\to\infty} \|L_n(x) - L(x)\| = 0.$$

Then $L : X \to Y$ is a continuous linear function.

Corollary A.8.8. *Let X be a Banach space and let $\{y_k : k = 1, \dots\} \subseteq X'$, $y \in X'$. The following are equivalent.*

a. $\sup_{k\geq 1} \|y_k\| < \infty$ *and $y_k \to y$ on a dense subset of X.*

b. *$y_k \to y$ uniformly on each compact subset of X.*

c. $\forall\, x \in X,\ y_k(x) \to y(x)$.

A typical application of Theorem A.8.6 is to the case $Y = \mathbb{C}$ and $\mathcal{L} \subseteq X'$.

Theorems A.7.1 and A.8.6, as well as Corollaries A.8.7 and A.8.8, can be formulated in somewhat different settings, e.g., [138], page 83, [406], pages 43–46. Corollary A.8.7 is a useful form of the Banach–Steinhaus theorem. BANACH and STEINHAUS' original assertion in 1927 is more general; see [19], pages 79–80.

Remark. Given the setting but not the assumption of Theorem A.8.6, the Uniform Boundedness Principle is the following dichotomous assertion: Either (A.9) holds or there is a nonempty set $Z \subseteq X$ for which $\overline{Z} = X$ and

$$\forall\, x \in Z, \quad \sup_{L\in\mathcal{L}} \|L(x)\| = \infty.$$

The set Z is also the intersection of a countable family of open sets.

Example A.8.9. Computation of $\|L\|$

a.i. Let X and Y be normed vector spaces, let $Z \subseteq X$ be a linear subspace, and let $L : Z \to Y$ be a nonzero linear function. The quantity $\|L\|$ defined by (A.7) can also be written as

$$\|L\| = \sup\{\|L(x)\| : \|x\| = 1, x \in Z\} \tag{A.10}$$

and

$$\|L\| = \sup\left\{\frac{\|L(x)\|}{\|x\|} : x \in Z \setminus \{0\}\right\}. \tag{A.11}$$

a.ii. We shall verify the assertions of part *a.i.* If $x \in Z \setminus \{0\}$, then

$$\frac{\|L(x)\|}{\|x\|} = \left\| L\left(\frac{x}{\|x\|}\right)\right\| \leq \sup\left\{\|L(z)\| : \|z\| = 1, z \in Z\right\};$$

and, if $x \in Z$, $\|x\| = 1$, then

$$\|L(x)\| = \frac{\|L(x)\|}{\|x\|} \leq \sup\left\{\frac{\|L(y)\|}{\|y\|} : y \in Z \setminus \{0\}\right\}.$$

Thus, the right sides of (A.10) and (A.11) are equal.

Next, label either of these suprema as $r > 0$, and let $\|L\|$ be defined by (A.7). For any $r > \varepsilon > 0$, we can choose $y \in Z \setminus \{0\}$ such that $(r - \varepsilon)\|y\| < \|L(y)\|$ by (A.11). Thus, $(r - \varepsilon) \leq \|L\|$, and so $r \leq \|L\|$.

We shall assume $r < \|L\|$ and obtain a contradiction. Since $\|L\| - r = p > 0$ we have $r < \|L\| - p/2$, so that

$$\forall\, x \in Z \setminus \{0\}, \qquad \frac{\|L(x)\|}{\|x\|} \leq r < \|L\| - \frac{p}{2},$$

i.e., $\|L(x)\| < (\|L\| - p/2)\|x\|$. This contradicts the definition of $\|L\|$ as the smallest constant C for which (A.7) holds.

b. The following situation frequently arises and the result is useful. *Let X be a Banach space over $\mathbb{C}$, let $Z \subseteq X$ be a dense linear subspace, and let $L : Z \to \mathbb{C}$ be a linear function for which*

$$r = \sup\left\{\frac{|L(x)|}{\|x\|} : x \in Z \setminus \{0\}\right\} < \infty.$$

Then $L \in X'$ and $\|L\| = r$. The proof is not difficult and first requires proving that L is a well-defined linear function $X \to \mathbb{C}$. The hypotheses to the claim can also be weakened.

c. Let $X = \mathbb{C}$ and let $Z = \mathbb{C} \setminus \{z \in \mathbb{C} : |z| = 1\}$. Let $L \in X' \setminus \{0\}$. Note that with $\|L\|$ defined by (A.7), we have

$$\|L\| = \sup_{y \in Z \setminus \{0\}} \frac{|L(y)|}{\|y\|} \neq \sup_{y \in Z, \|y\| = 1} |L(y)|$$

since $\{y \in Z : \|y\| = 1\} = \emptyset$.

Example A.8.10. Hilbert–Schmidt operators and Schur lemma
a.i. Let $(X, \mathcal{A}, \mu)$ be a σ-finite measure space, and let $\mu \times \mu$ be the corresponding product measure on $X \times X$. If $K \in L^2_{\mu \times \mu}(X \times X)$, then we define the operator L as

$$(L(f))(x) = \int_X K(x, y) f(y)\, d\mu(y).$$

It is not difficult to prove that $L \in \mathcal{L}(L^2_\mu(X))$, the space of continuous linear functions $L^2_\mu(X) \to L^2_\mu(X)$, where $L^2_\mu(X)$ is given the L^2-norm; see Appendix A.10. In fact, one makes the estimate,

$$\|L\|^2 \leq \int_X \int_X |K(x, y)|^2\, d\mu(x) d\mu(y),$$

using the definition of $\|L\|$. We call L a *Hilbert–Schmidt integral operator.*

a.ii. A natural generalization of the notion of Hilbert–Schmidt integral operators is that of *Hilbert-Schmidt operators* acting on a separable Hilbert space H. We say that $A \in \mathcal{L}(H)$, the space of continuous linear functions on H, is a *Hilbert–Schmidt operator* if there exists an ONB $\{e_n : n = 1, \dots\}$ for H such that

$$\sum_{n=1}^{\infty} \|A(e_n)\|^2 < \infty.$$

We define the *Hilbert–Schmidt norm* of A to be

$$\|A\|_{HS} = \left(\sum_{n=1}^{\infty} \|A(e_n)\|^2\right)^{1/2}.$$

b.i. We have seen versions of Schur's lemma in Theorem 6.2.1 and Theorem A.7.3. In this realm of ideas, SCHUR proved the following result. *If* $\{c_{m,n} : m, n \in \mathbb{Z}\}$ *is a double sequence of complex numbers with the properties,*

$$\sum_{n\in\mathbb{Z}} |c_{m,n}| \leq C_1, \quad \textit{independent of } m$$

and

$$\sum_{m\in\mathbb{Z}} |c_{m,n}| \leq C_2, \quad \textit{independent of } n,$$

then the linear operator L *defined by the matrix* $(c_{m,n})_{m,n\in\mathbb{Z}}$ *is an element of* $\mathcal{L}(\ell^2(\mathbb{Z}))$. In fact, *we have*

$$\|L\|^2 \leq C_1 C_2.$$

b.ii. Using Schur's lemma from part *b.i* and Fubini's theorem, we shall prove that $L \in \mathcal{L}(L^2_\mu(X))$, where the hypothesis on K from part a is replaced by the conditions,

$$\exists\, C_1 > 0 \text{ such that } \forall\, x \in X, \quad \int_X |K(x,y)|\, d\mu(y) \leq C_1$$

and

$$\exists\, C_2 > 0 \text{ such that } \forall\, y \in X, \quad \int_X |K(x,y)|\, d\mu(x) \leq C_2.$$

The proof is based on the following calculation for $f, g \in L^2_\mu(X)$:

$$\begin{aligned}
&\left(\int_X \int_X |K(x,y)||f(x)||g(y)|\, d\mu(x)d\mu(y)\right)^2 \\
&\leq \left(\int_X\int_X |K(x,y)||f(x)|^2\, d\mu(x)d\mu(y)\right)\left(\int_X\int_X |K(x,y)||g(y)|^2\, d\mu(y)d\mu(x)\right) \\
&= \left(\int_X \left[\int_X |K(x,y)|\, d\mu(y)\right] |f(x)|^2\, d\mu(x)\right) \\
&\quad \times \left(\int_X \left[\int_X |K(x,y)|\, d\mu(x)\right] |g(y)|^2\, d\mu(y)\right) \leq C_1 C_2 \|f\|_2^2 \|g\|_2^2.
\end{aligned}$$

The following result depends on the compactness criteria in terms of sets being totally bounded (Appendix A.5) as well as the Uniform Boundedness Principle.

Theorem A.8.11. Compact subsets of a Banach space
Let X be a Banach space. A set $Y \subseteq X$ is compact in X if and only if for every sequence of linear functionals $L_n : X \to \mathbb{C}$, for which

$$\forall\, x \in X, \quad L_n(x) \to 0, \tag{A.12}$$

we have

$$L_n \to 0 \text{ uniformly on } Y. \tag{A.13}$$

Proof. We prove the necessary conditions for compactness, which require X to be only a normed vector space. By the Uniform Boundedness Principle, (A.12) implies that

$$\exists\, M \text{ such that } \forall\, n \in \mathbb{N}, \quad \|L_n\| \le M,$$

and, in particular, each $L_n \in X'$.

Since Y is compact and hence totally bounded, we have

$$\forall\, \varepsilon > 0,\ \exists\, y_1, \ldots, y_m \in Y \text{ such that}$$
$$\forall\, y \in Y,\ \exists\, y_{j(y)} \in \{y_1, \ldots, y_m\} \text{ for which } \|y - y_{j(y)}\| < \frac{\varepsilon}{2M}.$$

By hypothesis,

$$\exists\, N = N(\varepsilon) \text{ such that } \forall\, n \ge N \text{ and } \forall\, k = 1, \ldots, m, \quad \|L_n(y_k)\| < \frac{\varepsilon}{2}.$$

Thus,

$$\forall\, n \ge N \text{ and } \forall\, y \in Y,$$
$$\|L_n(y)\| \le \|L_n(y_{j(y)})\| + \|L_n(y - y_{j(y)})\| < \frac{\varepsilon}{2} + M\frac{\varepsilon}{2M}.$$

This is the desired uniform convergence in Y. □

See [275], pages 300–301, for a proof of the sufficient conditions; cf. the Arzelà–Ascoli (Theorem A.4.3) and Kolmogorov compactness (Theorem 6.6.1) theorems.

A.9 The weak and weak* topologies

Let X be a normed vector space. Then X' is a Banach space normed by

$$\forall\, x' \in X', \quad \|x'\| = \sup\{|x'(x)| : \|x\| \le 1\}. \tag{A.14}$$

As such, X' is the *dual* of X. We then consider $(X')' = X''$, normed analogously, noting that X'' is a Banach space and that X can be embedded isometrically and algebraically isomorphically onto a linear subspace of X''. The mapping defining this isomorphism is given by

$$\forall\, x \in X, \quad x(x') = x'(x).$$

It should be pointed out that the proof that the natural mapping $X \to (X')^*$, $x \mapsto L_x$, defined by $L_x(x') = x'(x)$, is injective requires the Hahn–Banach theorem in the form of Example A.8.5. ($(X')^*$ is the space of linear functions (functionals) $X' \to \mathbb{C}$.)

The space X is *reflexive* if $X = X''$ under this canonical mapping.

Theorem A.9.1. The norm in terms of the dual space
Let X be a normed vector space, and let

$$B' = \{x' \in X' : \|x'\| \leq 1\}.$$

Then

$$\forall\, x \in X, \quad \|x\| = \sup\{|x'(x)| : x' \in B'\}.$$

In particular, for each fixed $x \in X$, the linear functional $L_x : X' \to \mathbb{C}$, $x' \mapsto x'(x)$, is continuous, so that $L_x \in X''$ and $\|L_x\| = \|x\|$.

Proof. Let $x \in X$. It is a consequence of the Hahn–Banach theorem (Theorem A.8.3) that

$$\exists\, y' \in B' \text{ such that } \quad y'(x) = \|x\|.$$

Also,

$$\forall\, x' \in B', \quad |x'(x)| \leq \|x\|\,\|x'\| \leq \|x\|.$$

Thus,

$$\|x\| = y'(x) = |y'(x)| \leq \sup\{|x'(x)| : x' \in B'\} \leq \|x\|,$$

and we have the result. □

Definition A.9.2. The weak and weak* topologies
Let X be a normed vector space.

The *weak topology* on X, denoted by $\sigma(X, X')$, has a basis at $0 \in X$ given by sets of the form

$$\left\{x \in X : \left|x_j'(x)\right| < \varepsilon,\ j = 1, \ldots, n\right\},$$

where $\varepsilon > 0$ and $\{x_1', \ldots, x_n'\}$ is an arbitrary finite subset of X'. Similarly, we define $\sigma(X', X'')$. See [465], pages 149, 151–154, 227–231, for a clear rationale and exposition of the weak topology.

The *weak* topology* on X', denoted by $\sigma(X', X)$, is defined analogously with corresponding sets

$$\{x' \in X' : |x'(x_j)| < \varepsilon,\ j = 1, \ldots, n\},$$

where $\varepsilon > 0$ and $x_j \in X$, $j = 1, \ldots, n$. Clearly, $\sigma(X', X)$ is generally weaker than $\sigma(X', X'')$, i.e., $\sigma(X', X) \subseteq \sigma(X', X'')$.

The following theorem is a consequence of the Hahn–Banach theorem, a finite-dimensional algebraic result, and the definitions of the weak and weak* topologies; see [399], pages 31–33, for a most efficient proof. It is also true, with analogous proof, for Hausdorff locally convex topological vector spaces (LCTVSs).

Theorem A.9.3. Weak and weak* dual spaces
Let X be a normed vector space with dual space X'.
a. *The dual space of X taken with the weak topology $\sigma(X, X')$ is X'.*
b. *The dual space of X' taken with the weak* topology $\sigma(X', X)$ is X.*

A subset K of a vector space X is *convex* if, for each $x, y \in K$ and $0 \leq r \leq 1$,

$$rx + (1 - r)y \in K.$$

An important application of Theorem A.8.3 is the following fact.

Theorem A.9.4. Equivalent norm and weak closures
Let X be a normed vector space and let $K \subseteq X$ be convex. Then K has the same norm and $\sigma(X, X')$ closure.

Banach proved the following result for the case of separable spaces in 1932 [19], Chapter VIII, Theorem 3. The general version was obtained by Leonidas Alaoglu in 1940 [3].

Theorem A.9.5. Banach–Alaoglu theorem
Let X be a normed vector space. Then B' is weak compact.*

Proof. For each $x \in X$ define

$$D_x = \{z \in \mathbb{C} : |z| \leq \|x\|\}.$$

Clearly, $B' \subseteq D = \prod_{x \in X} D_x$. Since *the product of compact spaces is compact* (this statement is equivalent to the axiom of choice and it is called the *Tychonov theorem*) and since it is easy to check that B' is closed in D, B' is a compact subset of D.

It is immediate from the definition that the induced product topology on B' is its weak* topology. (For the definition of the product topology see, e.g., [279].) □

In this regard we note the following fact.

Theorem A.9.6. Characterization of weak* compactness
Let X be a Banach space. Then $Y \subseteq X'$ is weak compact if and only if Y is weak* closed and norm bounded.*

Proof. The sufficient condition for weak* compactness follows from Theorem A.9.5.

For the necessary condition we must verify that weak* boundedness implies norm boundedness (since weak* compactness yields weak* boundedness). This follows from Theorem A.7.1 or Theorem A.8.6, noting that X is complete. □

Since weak* boundedness implies norm boundedness in a Banach space, we see that every *weak* convergent sequence is norm bounded.*

Remark. We require X to be complete in Theorem A.9.6. For a counterexample let X be the vector space of all finite sequences of complex numbers normed by $\|x\| = \sup\{|x_n|\}$, $x = \{x_n : n = 1, \ldots\}$. Set $x'_n(x) = n|x_n|$ and $Y = \{0\} \cup \{x'_n : n = 1, \ldots\} \subseteq X'$. Then $x'_n \to 0$ in $\sigma(X', X)$, whereas $\|x'_n\| = n$. The situation is corrected by the following result: *Let X be a normed vector space and let $Y \subseteq X'$ be weak* compact; Y is norm bounded if and only if the weak* closure of the smallest convex set containing Y is weak* compact.*

A useful result, e.g., [32], page 141, concerning weak* closures is the *Krein–Smulian theorem*: *Let X be a Banach space and let $K \in X'$ be convex; by definition, a net $\{x'_\alpha\} \subseteq X'$ converges to 0 in the Krein–Smulian topology if $x'_\alpha \to 0$ uniformly on compact sets of X ([279], page 65); then the Krein–Smulian and weak* closures of K are identical.*

Note, of course, that the finite subsets of X are compact.

Example A.9.7. A weak* closure of characteristic functions

Define Y to be the space of functions $f : [0,1] \to [0,1]$ having the form $f = \mathbb{1}_A$, where the subset $A \subseteq [0,1]$ is a finite disjoint union of intervals. The weak* closure of Y, as a subset of $L^\infty_m([0,1])$, is

$$\{f \in L^\infty_m([0,1]) : 0 \leq f \leq 1\}.$$

Theorem A.9.8. Sequential weak* compactness

Let X be a separable normed vector space. Then B' is sequentially compact in the $\sigma(X', X)$ topology; cf. Theorem A.9.5.

Proof. Let $\{x'_k : k = 1, \ldots\} \subseteq B'$ and let $\{x_n : n = 1, \ldots\}$ be a countable dense subset of X. By the expected diagonal argument there is a subsequence $\{x'_{k_j} : j = 1, \ldots\} \subseteq \{x'_k : k = 1, \ldots\}$ such that

$$\forall\, n \in \mathbb{N}, \quad \lim_{j \to \infty} x'_{k_j}(x_n) = x'(x_n).$$

For $x \in X$, let $x_{n_p} \to x$, so that

$$\lim_{p\to\infty} x'_{k_j}(x_p)$$

exists uniformly in j. We complete the proof by the Moore–Smith theorem. □

Note that a compact topological space is metrizable if and only if it has a countable basis. Thus, *if a normed vector space X is separable, then B' with the weak* topology is metrizable* (by definition of the weak* topology). However, X' is never metrizable in its weak* topology if X is infinite-dimensional.

If X is a normed vector space, then, as noted at the beginning of this section, X' is a Banach space with the norm defined by (A.14).

Convergence criteria compatible with the weak topology require nets. However, by definition, a sequence $\{x_n : n = 1, \ldots\} \subseteq X$ converges to $0 \in X$ if for every weak neighborhood U of 0, there is N_U such that $x_n \in U$ for all $n \geq N_U$. Thus, by the definition of U, not only do we have

$$\|x_n\| \to 0 \Longrightarrow x_n \to 0 \text{ in } \sigma(X, X'),$$

but we also have the following satisfying result; cf. Section 6.3.

Theorem A.9.9. Sequential weak convergence
Let X be a normed vector space and $\{x_n : n = 1, \ldots\}$ a sequence in X. Then $x_n \to 0$ in the weak topology $\sigma(X, X')$ if and only if

$$\forall\, x' \in X', \quad \lim_{n\to\infty} x'(x_n) = 0.$$

An immediate corollary of the Uniform Boundedness Principle (Theorem A.8.6) for the Banach space X' is the following result.

Theorem A.9.10. Boundedness of weakly convergent sequences
Let X be a normed vector space and assume $x_n \to x$ in $\sigma(X, X')$. Then $\{\|x_n\| : n = 1, \ldots\}$ is bounded.

Using Theorem A.9.8 we can prove the following "converse" to Theorem A.9.10.

Theorem A.9.11. Weak convergence of norm bounded sequences
Let X be a reflexive Banach space and let $\{x_n : n = 1, \ldots\}$ be a norm bounded sequence in X. Then there is a subsequence that converges to some $x \in X$ in the $\sigma(X, X')$ topology; cf. Theorem 6.5.5.

If "subsequence" is replaced by "subnet" in Theorem A.9.11 the result is immediate from the Banach–Alaoglu theorem. It is interesting to compare this result with Theorem 6.3.2 noting that $L^1_m([0,1])$ is not reflexive. Because of Theorem A.9.11 it is easy to check that reflexive Banach spaces are sequentially weakly complete.

By Theorem A.9.3 and since Theorem A.8.3*b* is valid for Hausdorff locally convex topological vector spaces, we obtain the following result.

Theorem A.9.12. Hahn–Banach in the weak* setting
Let X be a normed vector space and let $Y \subseteq X'$ be a $\sigma(X', X)$ closed linear subspace. If $y' \in X' \setminus Y$, then there is $x \in X$ such that $y'(x) \neq 0$ and

$$\forall\, x' \in Y, \quad x'(x) = 0.$$

A.10 Linear maps

If X and Y are Banach spaces, $\mathcal{L}(X, Y)$ denotes the space of continuous linear functions $X \to Y$. If $X = Y$ we write $\mathcal{L}(X)$.

Proposition A.10.1. *If X and Y are Banach spaces over $\mathbb{F} = \mathbb{R}$ or $\mathbb{F} = \mathbb{C}$ (X a normed vector space is sufficient), then $\mathcal{L}(X, Y)$ is a Banach space over $\mathbb{F}$, where $L = c_1 L_1 + c_2 L_2$ is defined by $L(x) = c_1 L_1(x) + c_2 L_2(x)$, $x \in X$ and $c_1, c_2 \in \mathbb{F}$, and where $\|L\|$ is defined by (A.7).*

By Theorem A.5.2 we have the following result.

Proposition A.10.2. a. *Let X and Y be Banach spaces and let $L \in \mathcal{L}(X, Y)$. Then L is uniformly continuous.*

b. *Let X and Y be Banach spaces and let $Z \subseteq X$ be a linear subspace of X. If $L \in \mathcal{L}(Z, Y)$, then L has a unique continuous linear extension to $\overline{Z}$.*

Parts a and b of the following result are the *Banach open mapping theorem* and *Banach closed graph theorem*, respectively.

Theorem A.10.3. Banach open mapping and closed graph theorems
a. *Let $L \in \mathcal{L}(X, Y)$ be bijective. Then $L^{-1} \in \mathcal{L}(X, Y)$.*

b. *Let X and Y be Banach spaces and let $L : X \to Y$ be linear. Assume that*

$$\|x_n - x\| \to 0 \quad \text{and} \quad \|L(x_n) - y\| \to 0 \tag{A.15}$$

imply $y = L(x)$. Then $L \in \mathcal{L}(X, Y)$.

The proof of part a depends on the Baire category theorem. Part b is clear from part a by applying part a to the setting

$$\begin{aligned} X \times L(X) &\to X, \\ (x, L(x)) &\mapsto x, \end{aligned}$$

where the norm on $X \times L(X)$ is given by $\|(x, L(x))\| = \|x\| + \|L(x)\|$. Criterion (A.15) is used to check that $X \times L(X)$ is complete.

Remark. Let $L \in \mathcal{L}(X, Y)$ be injective. If L is not surjective, then $L(X)$ is of first category in Y. To verify this fact, we assume that $L(X)$ is of second category in Y. Then it is not difficult to verify that

$$\forall\, r > 0, \exists\, R > 0 \text{ such that } \overline{B(0,R)} \subseteq L(\overline{B(0,r)}). \tag{A.16}$$

Take any $y \in Y$. For any $r > 0$, choose R as in (A.16). There is N_R such that

$$\forall\, n > N_R, \quad \frac{1}{n} y \in \overline{B(0,R)} \subseteq L(\overline{B(0,r)}).$$

Thus, for any such n, there is $x \in \overline{B(0,r)}$ for which $L(nx) = y$. We conclude that $L(X) = Y$.

Example A.10.4. What the Banach closed graph theorem asserts
The Banach closed graph theorem does not say that if $X \times L(X)$ is closed in $X \times Y$ then L is continuous. It asserts the continuity of L if each $(x, y) \in \overline{X \times L(X)} \subseteq X \times Y$ can be approximated by $\{(x_n, L(x_n)) : n = 1, \dots\}$, for some sequence $\{x_n\} \subseteq X$.

Assume that $\{x_\alpha\} \subseteq X$ and $\{y_\alpha\} \subseteq Y$ are Hamel bases with $\|x_\alpha\| \leq 1$, $\sup \|y_\alpha\| = \infty$. Taking card X = card Y we define $L(x_\alpha) = y_\alpha$, and extend L linearly to all of X. Then L is a linear surjection and $X \times L(X) = X \times Y$, but $X \times L(X)$ does not satisfy (A.15). Clearly, L is not continuous.

Example A.10.5. Discontinuous identity mappings on $\ell^\infty(\mathbb{N})$
We shall put two norms on $\ell^\infty(\mathbb{N})$ such that $\ell^\infty(\mathbb{N})$ is a Banach space for each norm but such that neither identity mapping $\ell^\infty(\mathbb{N}) \to \ell^\infty(\mathbb{N})$ is continuous. Choose $\|\dots\|_\infty$ for the first norm. To define the second norm first observe that

$$\text{card } \ell^1(\mathbb{N}) = \text{card } \ell^\infty(\mathbb{N}). \tag{A.17}$$

To prove (A.17) consider the injection

$$\ell^\infty(\mathbb{N}) \to \ell^1(\mathbb{N}),$$
$$\{x_n : n = 1, \dots\} \mapsto \{x_n/2^n : n = 1, \dots\}.$$

Thus, card $\ell^\infty(\mathbb{N}) \leq$ card $\ell^1(\mathbb{N})$. On the other hand, card $\ell^1(\mathbb{N}) \leq$ card $\ell^\infty(\mathbb{N})$ since $\ell^1(\mathbb{N}) \subseteq \ell^\infty(\mathbb{N})$. Equation (A.17) follows from the Schröder–Bernstein theorem, e.g., Problem 1.6. Consequently, if H_p is a Hamel basis for $\ell^p(\mathbb{N})$, then card H_∞ = card H_1, and so we choose any bijection $b : H_\infty \to H_1$. We extend b by linearity to a bijection $L : \ell^\infty(\mathbb{N}) \to \ell^1(\mathbb{N})$. By Theorem A.10.3*a*, the nonseparability of $\ell^\infty(\mathbb{N})$, and the separability of $\ell^1(\mathbb{N})$, we see that $L \notin \mathcal{L}(\ell^\infty(\mathbb{N}), \ell^1(\mathbb{N}))$.

The second norm on $\ell^\infty(\mathbb{N})$ is then defined by

$$\|x\| = \|L(x)\|_1.$$

It is easy to check that $\ell^\infty(\mathbb{N})$ with this norm is complete.

The following was given by Carleson with regard to an interpolation problem [88].

Theorem A.10.6. Carleson open mapping theorem
Let X and Y be Banach spaces with norms $\|\dots\|_X$ and $\|\dots\|_Y$, respectively. Assume $Y \subseteq X$ and $\|\dots\|_Y \geq \|\dots\|_X$ on Y. If

$$\exists\, M > 0 \text{ and } \exists\, \{x_n : n = 1, \dots\} \subseteq Y \text{ such that,}$$
$$\forall\, x' \in X', \quad \|x'\| \leq M \sup_{n \in \mathbb{N}} |x'(x_n)|,$$

then $X = Y$ and $\|\dots\|_Y \leq M\|\dots\|_X$.

Let X and Y be Banach spaces. If $L \in \mathcal{L}(X, Y)$, then the *adjoint*, L', of L is the element of $\mathcal{L}(Y', X')$ defined by

$$\forall\, x \in X \text{ and } \forall\, y' \in Y', \quad (L'(y'))(x) = y'(L(x)).$$

We call L' an *open mapping* if $L'(U) \subseteq X'$ is open for every open set $U \subseteq Y'$, i.e., if

$$\exists\, C > 0 \text{ such that } \forall\, y' \in Y', \quad \|y'\|_{Y'} \leq C\|L'(y')\|_{X'}.$$

Theorem A.10.7. Surjectivity consequences of the Banach open mapping theorem
Let X and Y be Banach spaces and assume that $L \in \mathcal{L}(X, Y)$ is injective and $\overline{L(X)} = Y$. The following are equivalent:
- **a.** *$L(X) = Y$,*
- **b.** *L' is an open mapping,*
- **c.** *$L'(Y') = X'$.*

Part *a* of the following result is true when X and Y are normed vector spaces. As in Theorem A.10.7, it depends on the Banach open mapping theorem.

Theorem A.10.8. Injectivity and surjectivity duality
Let X and Y be Banach spaces and let $L \in \mathcal{L}(X, Y)$.
- **a.** *$L'(Y') = X' \iff L^{-1}$ exists and $L^{-1} \in \mathcal{L}(L(X), X)$.*
- **b.** *$L(X) = Y \iff (L')^{-1}$ exists and $(L')^{-1} \in \mathcal{L}(L'(Y'), Y')$. Further, if L^{-1} exists then $L^{-1} \in \mathcal{L}(L(X), X)$.*

A.11 Embeddings of dual spaces

Let $B_1 \subseteq B_2$, where B_1 and B_2 are normed vector spaces, and let Id : $B_1 \to B_2$textitId be the identity mapping with adjoint $Id' : B_2' \to B_1'$ acting between the dual Banach spaces. By definition,

$$\forall x \in B_1 \text{ and } \forall y \in B_2', \quad (Id'(y))(x) = y(x),$$

i.e., $Id'(y) = y$ on $B_1 \subseteq B_2$.

Assume that Id, and hence Id', are continuous. Note that *if* $y \in B_2'$, *then* $y|_{B_1}$, *the restriction of* y *to* B_1, *is an element of* B_1'. To see this first note that since $B_1 \subseteq B_2$ and $y \in B_2'$, then $y|_{B_1}$ is linear on B_1. The mapping $y|_{B_1}$ is also continuous on B_1 because of the continuity of Id. In fact, since y is continuous on B_1 with the induced topology from B_2, then it is continuous on B_1 with its given norm convergence because this latter topology is stronger (finer) than the B_2 criterion. (*Continuity of a function for a given topology on its domain implies continuity for any stronger topology on that domain.*)

Definition A.11.1. Embedding of dual spaces
We say that B_2' is *embedded* in B_1', in which case we write $B_2' \subseteq B_1'$, if Id' is a continuous injection. This means that whenever $Id'(y) = 0 \in B_1'$, then $y = 0$, i.e., $y(x) = 0$ for all $x \in B_2$.

With the above assumptions, we further assume that $\overline{B_1} = B_2$. Let $y \in B_2'$ have the property that $Id'(y) = 0 \in B_1'$. Suppose $x \in B_2$ and $\lim_{n\to\infty} \|x_n - x\|_{B_2} = 0$, where $\{x_n\} \subseteq B_1$. Then $\lim_{n\to\infty} y(x_n) = y(x)$, and $y(x_n) = (Id'(y))(x_n) = 0$. Thus, $y(x) = 0$, and so $y \in B_2'$ is the 0-element. Hence, Id' is a continuous injection. Moreover, Id' is also the identity function, i.e., for all $y \in B_2'$, $Id'(y) = y$ on a dense linear subspace of B_2.

We can summarize what has been said by the following embedding theorem.

Theorem A.11.2. Embedding theorem
Let B_1 *and* B_2 *be normed vector spaces. If* $B_1 \subseteq B_2$ *in the sense that the identity mapping* $Id : B_1 \to B_2$ *is continuous, and if* $\overline{B_1} = B_2$, *then* $B_2' \subseteq B_1'$.

Example A.11.3. $C_b(\mathbb{R}), M_b(\mathbb{R})$, and duality
a. Let $C_b(\mathbb{R})$ be the Banach space of continuous bounded functions on $\mathbb{R}$ taken with the L^∞-norm $\|...\|_\infty$; and let $C_0(\mathbb{R})$ be the closed linear subspace of $C_b(\mathbb{R})$ whose elements f satisfy the condition that $\lim_{|t|\to\infty} f(t) = 0$. Recall from Theorem 7.2.7 (RRT) that $(C_0(\mathbb{R}))' = M_b(\mathbb{R})$.

Then $C_b(\mathbb{R})$ is a closed linear subspace of $L^\infty(\mathbb{R})$, and so it is natural to describe the relation between $(C_b(\mathbb{R}))'$ and the dual space of $L^\infty_\mu(\mathbb{R})$; see Theorem 5.5.7. To this end, let $\mathcal{A}$ be the algebra generated by the closed subsets of $\mathbb{R}$. Then let $FR(X) \subseteq F(X)$, defined before Theorem 5.5.7, be the set of those elements $\nu \in F(X)$ for which $|\nu|$ is regular in the sense of Definition 2.5.12. It is not difficult to prove that $(C_b(\mathbb{R}))' = FR(X)$; see [150], Part I, Chapter IV, Section 6.

b. Since the continuous identity mapping $Id : C_0(\mathbb{R}) \to C_b(\mathbb{R})$ is not dense, we can *not* conclude that $(C_b(\mathbb{R}))' \subseteq M_b(\mathbb{R})$, as is apparent from the characterizations of $(C_0(\mathbb{R}))'$ and $(C_b(\mathbb{R}))'$; cf. Theorem A.11.2.

c. The characterizations of $(C_0(\mathbb{R}))'$ and $(C_b(\mathbb{R}))'$ do imply, however, that

$$M_b(\mathbb{R}) \subseteq (C_b(\mathbb{R}))'. \tag{A.18}$$

In this regard, if $\mu \in (C_0(\mathbb{R}))'$, then μ extends to an element $\mu_e \in (C_b(\mathbb{R}))'$ by the Hahn–Banach theorem. Of course, there is no a priori guarantee of a unique extension.

On the other hand, and without invoking the characterization of $(C_b(\mathbb{R}))'$, we can see the validity of (A.18) in the following way.

Let $\mu \in M_b(\mathbb{R})$. If $f \in C_b(\mathbb{R})$ we can choose $\{f_n\} \subseteq C_0(\mathbb{R})$ for which $\lim_{n\to\infty} f_n = f$ pointwise on $\mathbb{R}$ and $\sup_n \|f_n\|_\infty = \|f\|_\infty < \infty$. Then we apply LDC for $L^1_{|\mu|}(\mathbb{R})$, which allows us to assert that $f \in L^1_{|\mu|}(\mathbb{R})$ and $\lim_{n\to\infty} \|f_n - f\|_1 = 0$. The integral $\mu(f)$ is well defined, i.e., it is independent of the sequence $\{f_n\} \subseteq C_0(\mathbb{R})$. Further, $\mu : C_b(\mathbb{R}) \to \mathbb{C}$ is linear. To prove the continuity of μ on $C_b(\mathbb{R})$, let $f \in C_b(\mathbb{R})$, let $\varepsilon > 0$, and choose $\{f_n\}$ as above. Then,

$$\exists N > 0 \text{ such that } \forall n \geq N, \quad |\mu(f - f_n)| < \varepsilon;$$

and so, for such n,

$$|\mu(f)| \leq \varepsilon + |\mu(f_n)| \leq \varepsilon + \|\mu\|_1 \|f\|_\infty.$$

This is true for all $\varepsilon > 0$, and so $\mu \in (C_b(\mathbb{R}))'$. We designate μ so defined on $C_b(\mathbb{R})$ by μ^*.

The inclusion (A.18) is accomplished by the mapping $\mu \mapsto \mu^*$. The fact that many extensions μ_e of μ exist does not contradict (A.18). In fact, $\nu_e = \mu_e - \mu^* \in (C_b(\mathbb{R}))'$ vanishes on $C_0(\mathbb{R})$; and if ν_e is not identically 0 on $C_b(\mathbb{R})$, then μ_e is not countably additive on $\mathcal{B}(\mathbb{R})$ and so it does not correspond to an element of $M_b(\mathbb{R})$.

A.12 Hilbert spaces

Definition A.12.1. Orthonormal set and orthonormal basis (ONB)

a. Let H be a Hilbert space. Elements $x, y \in H$ are *orthogonal* if $\langle x, y\rangle = 0$; and this property is denoted by $x \perp y$. An element $x \in H$ is *orthogonal to the set* $S \subseteq H$, denoted by $x \perp S$, if $\langle x, y\rangle = 0$ for all $y \in S$. A set $S \subseteq H$ is an *orthogonal set* if $x \perp y$ for all $x, y \in S$ for which $x \neq y$. A set $S \subseteq H$ is an *orthonormal set* if it is orthogonal and if $\|x\| = 1$ for each $x \in S$.

b. A countable orthonormal set $S = \{x_n : n = 1, \ldots\}$ is an *orthonormal basis* (ONB) for H if

$$\forall\, x \in H,\ \exists\, \{c_n : n = 1, \ldots\} \subseteq \mathbb{C} \text{ such that } \quad x = \sum_{n=1}^{\infty} c_n x_n \ \text{ in } \ H.$$

Proposition A.12.2. *Let $S = \{x_\alpha\}$ be an orthonormal set in a separable Hilbert space H. Then S is a countable set.*

Proof. By separability, let $D = \{y_n : n = 1, \ldots\}$ be a countable dense subset of H. Since S is orthonormal, we can assert that

$$\forall\, \alpha, \beta \text{ for which } \alpha \neq \beta, \quad \|x_\alpha - x_\beta\| = \sqrt{2}. \tag{A.19}$$

Using the density, we have

$$\forall\, \alpha,\ \exists\, n = n(\alpha) \in \mathbb{N}, \text{ such that } \|x_\alpha - y_{n(\alpha)}\| < \frac{\sqrt{2}}{2}. \tag{A.20}$$

We are forced into choosing a different $n(\alpha)$ for each α, for, otherwise, if $y_{n(\alpha)}$ corresponds to both x_α and x_β in the sense of (A.20), then

$$\|x_\alpha - x_\beta\| \leq \|x_\alpha - y_{n(\alpha)}\| + \|y_{n(\alpha)} - x_\beta\| < \sqrt{2},$$

and this contradicts (A.19). Thus, (A.20) gives rise to an injective mapping $S \to D$, and hence S is countable. □

Example A.12.3. Hilbert spaces and ONBs

a. $H = L^2(\mathbb{T}_{2\Omega})$ is a Hilbert space with inner product defined by $\langle F, G\rangle = \int_{\mathbb{T}_{2\Omega}} F(x)\overline{G(x)}\, dx$, where $\mathbb{T}_{2\Omega} = \mathbb{R}/(2\Omega\mathbb{Z})$, F and G are 2Ω-periodic on $\mathbb{R}$, and $\int_{\mathbb{T}_{2\Omega}} F(x)\, dx$ is defined as the Lebesgue integral $\frac{1}{2\Omega}\int_{-\Omega}^{\Omega} F(x)\, dx$. The sequence $\{e^{-\pi i n x/\Omega} : n \in \mathbb{Z}\}$ is an ONB for $L^2(\mathbb{T}_{2\Omega})$, e.g., Proposition B.8.1.

b. $H = \ell^2(\mathbb{Z}^d)$ is defined to be the vector space of all sequences $f : \mathbb{Z}^d \to \mathbb{C}$ with the property that

$$\|f\|_{\ell^2(\mathbb{Z}^d)} = \left(\sum_{n\in\mathbb{Z}^d} |f[n]|^2\right)^{1/2} < \infty.$$

With this norm, $\ell^2(\mathbb{Z}^d)$ is a Hilbert space, and its inner product is given by

$$\forall\, f, g \in \ell^2(\mathbb{Z}^d), \quad \langle f, g\rangle = \sum_{n\in\mathbb{Z}^d} f[n]\overline{g[n]}.$$

Let $u_n \in \ell^2(\mathbb{Z}^d)$ be defined by $u_n[m] = \delta(m, n)$, for $m, n \in \mathbb{Z}^d$, where

$$\delta(m,n) = \begin{cases} 1, \text{ if } & m = n, \\ 0, \text{ if } & m \neq n. \end{cases}$$

It is easy to check that the sequence $\{u_n\}$ is an ONB for $\ell^2(\mathbb{Z}^d)$.

c. $H = L^2(\mathbb{R})$ is a Hilbert space with inner product defined by $\langle f, g\rangle = \int_{\mathbb{R}} f(t)\overline{g(t)}\, dt$. The *Hermite functions*, $h_n(x) = e^{-\pi x^2} H_n(2\sqrt{\pi}x)$, $n = 0, \ldots,$ where

$$\forall\, n = 0, \ldots, \quad H_n(x) = (-1)^n e^{x^2/2} \frac{d^n}{dx^n} e^{-x^2/2},$$

are an ONB for $L^2(\mathbb{R})$; see, e.g., [505], [35], and Remark 2.4.11 in [39]. The concept of a multiresolution analysis in wavelet theory leads to the construction of many other ONBs for $L^2(\mathbb{R})$, e.g., [118], [348]; cf. [475].

d. $H = PW_\Omega = \{f \in L^2(\mathbb{R}) : \text{supp}\, \hat{f} \subseteq [-\Omega, \Omega]\}$ is a closed linear subspace of $L^2(\mathbb{R})$, and is the so-called *Paley–Wiener space* of Ω-band-limited functions. We denote by $\hat{f}$ the Fourier transform of f; see Appendix B. Furthermore, PW_Ω is a Hilbert space with inner product induced from $L^2(\mathbb{R})$. The sequence $\{(\frac{1}{\sqrt{2\Omega}})\tau_{n/(2\Omega)}(d_{2\pi\Omega}) : n \in \mathbb{Z}\}$ is an ONB for PW_Ω, where $\tau_x(f)(y) = f(y-x)$ and where

$$d_{2\pi\Omega}(t) = \frac{\sin(2\pi\Omega t)}{\pi t}.$$

In light of our mention of multiresolution in part *c*, we note that PW_Ω can be considered as part of a multiresolution analysis of $L^2(\mathbb{R})$ for the so-called Shannon wavelet system, e.g., [118], [348], [112], [113]; cf. [475]

The following is an immediate, useful consequence of the Schwarz inequality.

Proposition A.12.4. Continuity of the inner product
Let H be a Hilbert space. The inner product is continuous on $H \times H$, i.e., if $\{x_n : n = 1, \dots\} \subseteq H$ converges to $x \in H$ and $\{y_n : n = 1, \dots\} \subseteq H$ converges to $y \in H$, then

$$\lim_{n\to\infty} \langle x_n, y_n \rangle = \langle x, y \rangle.$$

Theorem A.12.5. Consequences of orthonormality
Let H be a Hilbert space and let $\{x_n : n = 1, \dots\}$ be an orthonormal sequence.

a. *Bessel inequality. The mapping*

$$\begin{aligned} L : H &\longrightarrow \ell^2(\mathbb{N}), \\ y &\longmapsto \{\langle y, x_n \rangle\}, \end{aligned} \tag{A.21}$$

is well defined, linear, and continuous; in fact

$$\forall\, y \in H, \quad \sum_{n=1}^{\infty} |\langle y, x_n \rangle|^2 \leq \|y\|^2.$$

b. *For each $y \in H$, $\sum \langle y, x_n \rangle x_n$ converges in H.*

c. *$\sum c_n x_n$ converges in H if and only if $c = \{c_n : n = 1, \dots\} \in \ell^2(\mathbb{N})$.*

d. *If $y = \sum c_n x_n$ converges in H, then each $c_n = \langle y, x_n \rangle$.*

Proof. i. Let $y \in H$, let $F \subseteq \mathbb{N}$ be finite, and suppose $\{c_n : n \in F\} \subseteq \mathbb{C}$. Using orthonormality, two direct calculations yield

$$\left\| \sum_{n\in F} c_n x_n \right\|^2 = \sum_{n\in F} |c_n|^2 \tag{A.22}$$

and

$$0 \le \left\| y - \sum_{n \in F} \langle y, x_n \rangle x_n \right\|^2 = \|y\|^2 - \sum_{n \in F} |\langle y, x_n \rangle|^2. \qquad \text{(A.23)}$$

ii. The Bessel inequality (part a) is immediate from (A.23). In particular, $\{\langle y, x_n \rangle\} \in \ell^2(\mathbb{N})$. Since H is complete, to prove part b we need only show that

$$\{s_N\} = \left\{ \sum_{n=1}^{N} \langle y, x_n \rangle x_n \right\}, \quad N > 0,$$

is a Cauchy sequence in H. This is a consequence of (A.22) and the fact that $\{\langle y, x_n \rangle\} \in \ell^2(\mathbb{N})$, which, in turn, was a consequence of part a. Part c also follows from (A.22).

iii. To prove part d, we use the orthonormality and the continuity of inner products (Proposition A.12.4) to compute

$$\langle y, x_n \rangle = \lim_{N \to \infty} \left\langle \sum_{m=1}^{N} c_m x_m, x_n \right\rangle = c_n. \qquad \square$$

The following result is also elementary to verify. One efficient route is to prove the implications a implies b implies c implies d implies e implies a.

Theorem A.12.6. Parseval formula and ONB
Let H be a Hilbert space and let $\{x_n : n = 1, \dots\}$ be an orthonormal sequence. The following are equivalent.

a. *$\{x_n\}$ is an ONB for H.*

b. *Parseval formula:*

$$\forall\, x, y \in H, \quad \langle x, y \rangle = \sum_{n=1}^{\infty} \langle x, x_n \rangle \overline{\langle y, x_n \rangle}.$$

c. *The mapping L of (A.21) (in Theorem A.12.5a) is a linear surjective isometry, and, in fact,*

$$\forall\, y \in H, \quad \|y\| = \left(\sum_{n=1}^{\infty} |\langle y, x_n \rangle|^2 \right)^{1/2}.$$

d. $\overline{\text{span}}\ \{x_n\} = H$.

e. *If $\langle y, x_n \rangle = 0$ for each $n \in \mathbb{N}$, then $y = 0$.*

Because of Example A.12.3a, the coefficients $\langle y, x_n \rangle$ for an ONB $\{x_n\} \subseteq H$ are called the *Fourier coefficients* of $y \in H$; cf. Definition B.5.1.

Theorem A.12.7. Hilbert space Fourier series
Let H be a Hilbert space and let $\{x_n : n = 1, \ldots\}$ be an ONB for H. Then

$$\forall\, y \in H, \quad y = \sum_{n=1}^{\infty} \langle y, x_n \rangle x_n \quad \text{in } H.$$

Proof. If $y \in H$, then $\sum \langle y, x_n \rangle x_n = x$ in H for some $x \in H$ by Theorem A.12.5*b*. Hence,

$$\forall\, n \in \mathbb{Z}^d, \quad \langle x, x_n \rangle = \langle y, x_n \rangle,$$

by Theorem A.12.5*d*. The result follows by Theorem A.12.6, using the equivalence either of parts *a* and *c* or of parts *a* and *e*. □

Remark. *a.* By the definition of an ONB, *if H contains an ONB, then H is separable*. The converse is also true: *If H is a separable Hilbert space, then H contains an* ONB. The proof of the converse has four elementary steps. First, if $S = \{x_n : n = 1, \ldots\}$ is a countable dense subset of H, then $\overline{\text{span}}\ \{x_n\} = H$. Next, we choose a linearly independent subset $\{y_n\}$ of $\{x_n\}$, that also has the property that $\overline{\text{span}}\ \{y_n\} = H$. This can be accomplished both constructively and iteratively by throwing out those x_n that are linear combinations of finite sets $\{x_j : j \in F \text{ and } j \neq n\}$. Third, the *Gram–Schmidt orthogonalization* procedure, e.g., [200], pages 21–22, constructs $\{u_n\}$ in terms of $\{y_n\}$ with the properties that $\{u_n\}$ is orthonormal and $\overline{\text{span}}\ \{u_n\} = H$. Finally, we invoke Theorem A.12.6 to complete the proof.

b. Let X be a Banach space over $\mathbb{C}$. A sequence $\{x_n : n = 1, \ldots\} \subseteq X$ is a *Schauder basis* for X if each $x \in X$ has a unique representation $x = \sum_{n=1}^{\infty} c_n(x) x_n$, where each $c_n(x) \in \mathbb{C}$ and where the series converges in X in the ordinary sense that

$$\lim_{N \to \infty} \sum_{n=1}^{N} c_n(x) x_n = x.$$

If $\{x_n\}$ is a Schauder basis for X and if we consider $\{c_n(x)\}$ as a sequence of mappings $c_n : X \to \mathbb{C}$, $x \mapsto c_n(x)$, then each $c_n \in X'$, e.g., [443], page 20. The situation in part *a* leads to the question (the *basis problem*), posed by BANACH in 1932 [19], whether every separable Banach space contains a Schauder basis. Using Walsh functions and lacunary Fourier series, PER ENFLO proved in 1973 that there are separable Banach spaces having no Schauder basis [158]; see [419], especially Sections 1.1 and 5.6.

Definition A.12.8. Direct sum and orthogonal complement
a. Let B be a Banach space, and let $X, Y \subseteq B$ be linear subspaces of B for which $X \cap Y = \{0\}$ and $X + Y = \{x + y : x \in X, y \in Y\} = B$. We denote this situation by

$$B = X \oplus Y,$$

and B is the *direct sum* of X and Y.

b. Suppose $B = X \oplus Y$. Let $z \in B$ and assume $z = x_1 + y_1 = x_2 + y_2$, where $x_j \in X$ and $y_j \in Y$. Then $x_1 - x_2 = y_2 - y_1$. Thus, $x_1 - x_2,\, y_2 - y_1 \in X \cap Y$, and so $x_1 = x_2$ and $y_1 = y_2$. Therefore, *if $B = X \oplus Y$, then each $z \in B$ has a unique representation $z = x + y$ for some $x \in X$ and $y \in Y$.*

c. Let H be a Hilbert space, and let $X \subseteq H$ be a subset of H. The *orthogonal complement* $X^{\perp}$ of X is the set $\{y \in H : \forall x \in X, x \perp y\}$.

d. Let X be a closed linear subspace of H. It is not difficult to prove that *for each $z \in H$ there are unique elements $x \in X$ and $y \in X^{\perp}$ such that $z = x + y$.* The proof requires the following two results.

i. For each $z \in H$ there is a unique element $x \in X$ such that

$$\|z - x\| = \inf\{\|z - w\| : w \in X\}.$$

ii. If $z \in H$ and $x \in X$, then $\langle z - y, w\rangle = 0$ for all $w \in X$ if and only if

$$\|z - x\| = \inf\{\|z - u\| : u \in X\}.$$

e. From part *d*, we see that *if $X \neq \{0\}$ is a closed linear subspace of H, then $X^{\perp}$ is a closed linear subspace of H,*

$$H = X \oplus X^{\perp},$$

and $(X^{\perp})^{\perp} = X$. We refer to $X \oplus X^{\perp}$ as an *orthogonal complement direct sum.*

A.13 Operators on Hilbert spaces

In the case of Hilbert spaces H_1 and H_2 over $\mathbb{F} = \mathbb{R}$ or $\mathbb{F} = \mathbb{C}$, we write

$$\forall\, x \in H_1 \text{ and } \forall\, y \in H_2, \quad \langle L(x), y\rangle_{H_2} = \langle x, L'(y)\rangle_{H_1} \tag{A.24}$$

to define the *adjoint* L' of $L \in \mathcal{L}(H_1, H_2)$. The adjoint was defined for Banach spaces after Theorem A.10.6. In (A.24) we have used the fact that Hilbert spaces H have the property that $H' = H$ of (A.8); cf. Example A.8.4*b*.

We shall now make use of the orthogonal complement (Definition A.12.8) and of the range and kernel of an operator $L \in \mathcal{L}(H_1, H_2)$. The *range* of L, also called the *image* of L, is defined as $\mathcal{R}(L) = \{Lx : x \in H_1\} \subseteq H_2$; and the *kernel* of L, also called the *null space* of L, is defined as the closed linear subspace $\ker L = \{x \in H_1 : Lx = 0 \in H_2\}$.

Theorem A.13.1. Kernel and range properties for Hilbert space operators
Let H_1 and H_2 be Hilbert spaces over $\mathbb{F} = \mathbb{R}$ or $\mathbb{F} = \mathbb{C}$, and let $L \in \mathcal{L}(H_1, H_2)$.

a. *$L'(H_2) = H_1$ if and only if L^{-1} exists and $L^{-1} \in \mathcal{L}(L(H_1), H_1)$.*

b. *$L(H_1) = H_2$ if and only if $(L')^{-1}$ exists and $(L')^{-1} \in \mathcal{L}(L'(H_2), H_2)$. Further, if L^{-1} exists, then it is in $\mathcal{L}(L(H_1), H_1)$.*

c. *If L^{-1} exists and $L^{-1} \in \mathcal{L}(H_2, H_1)$, then $(L')^{-1} \in \mathcal{L}(H_1, H_2)$ and*

$$(L')^{-1} = (L^{-1})'.$$

d. $\ker L = (\mathcal{R}(L'))^{\perp}$, $\ker L' = (\mathcal{R}(L))^{\perp}$, $\overline{\mathcal{R}(L)} = (\ker L')^{\perp}$, *and* $\overline{\mathcal{R}(L')} = (\ker L)^{\perp}$.

Let H be a Hilbert space over $\mathbb{C}$ and let $X \neq \{0\}$ be a closed linear subspace of H. Then $P : H \to H$ is the *orthogonal projection onto* X if

$$\forall\, x \in X \text{ and } \forall\, y \in X^{\perp}, \quad P(x+y) = x.$$

The orthogonal projection onto X is an element of $\mathcal{L}(H)$, and, in fact, $\|P\| = 1$. We say that $L \in \mathcal{L}(H)$ is *self-adjoint* or *Hermitian* if $L' = L$. The orthogonal projections of H are the building blocks for the theory of self-adjoint operators in the sense that every $L \in \mathcal{L}(H)$ is the limit in norm of a sequence of linear combinations of orthogonal projections; see [200].

An elementary calculation yields the following result.

Proposition A.13.2. *Let H be a Hilbert space over $\mathbb{C}$. Then $P \in \mathcal{L}(H)$ is an orthogonal projection onto some closed linear subspace of H if and only if P is self-adjoint and $P^2 = P$.*

Let $X \subseteq H \setminus \{0\}$ be a closed linear subspace of the Hilbert space H over $\mathbb{C}$, and let $L \in \mathcal{L}(H)$. We call X an *L-invariant subspace* if $L(X) \subseteq X$. There is the following relationship between invariant subspaces and orthogonal projections.

Theorem A.13.3. Invariant subspaces and orthogonal projections
Let $X \subseteq H \setminus \{0\}$ be a closed linear subspace of the Hilbert space H over $\mathbb{C}$, and let $L \in \mathcal{L}(H)$. Then X is L-invariant if and only if $LP = PLP$, where P is the orthogonal projection onto X.

Remark. The *invariant subspace problem* is to determine whether for any given Hilbert space H over $\mathbb{C}$, every $L \in \mathcal{L}(H)$ has a nontrivial L-invariant subspace. There is a spectacular positive solution due to VICTOR LOMONOSOV, which is valid even for Banach spaces for the case of *compact* operators. (Compact operators $L \in \mathcal{L}(H)$ are those for which any sequence $\{x_n : n = 1, \ldots\} \subseteq H$ of unit norm elements has the property that $\{L(x_n) : n = 1, \ldots\}$ has a convergent subsequence in H.) There has been progress since LOMONOSOV, but the general problem is open; see [369].

Remark. It is elementary to check that $L \in \mathcal{L}(H)$ *is self-adjoint if and only if $\langle L(x), x\rangle \in \mathbb{R}$ for all $x \in H$.* One direction is immediate: $L' = L$ implies $\langle L(x), x\rangle = \langle x, L(x)\rangle = \overline{\langle L(x), x\rangle}$. Conversely, $\langle L(z), z\rangle \in \mathbb{R}$ implies $\langle L(x+cy), x+cy\rangle = \langle x+cy, L(x+cy)\rangle$ for all $x, y \in H$ and $c \in \mathbb{C}$; and using the hypothesis again on this equality we can calculate that $\operatorname{Im}\langle cL(y), x\rangle = \operatorname{Im}\langle cy, L(x)\rangle$, which in turn gives $L' = L$ by considering $c = 1$ and $c = i$.

We shall say that $L \in \mathcal{L}(H)$ is *positive* if $\langle L(x), x\rangle \geq 0$ for all $x \in H$. By the above observation we see that *if H is a complex Hilbert space and $L \in \mathcal{L}(H)$ is positive, then L self-adjoint.*

Let H_1, H_2 be Hilbert spaces over $\mathbb{C}$. We call $U \in \mathcal{L}(H_1, H_2)$ an *isometry* if $\|U(x)\|_{H_2} = \|x\|_{H_1}$ for all $x \in H_1$. If $H = H_1 = H_2$ and $U \in \mathcal{L}(H)$ is a surjective isometry, then U is a *unitary operator.*

Proposition A.13.4. *Let H_1 and H_2 be Hilbert spaces over $\mathbb{C}$, and let $U \in \mathcal{L}(H_1, H_2)$.*

a. *The following are equivalent:*

i. U is an isometry;

ii. $U'U$ is the identity mapping Id on H_1;

iii. $\forall\, x, y \in H_1,\ \langle U(x), U(y)\rangle_{H_2} = \langle x, y\rangle_{H_1}$.

Further, if U is a surjective isometry then $U'U$ is the identity mapping Id on H_2.

b. *Let $H = H_1 = H_2$. Then $U \in \mathcal{L}(H)$ is unitary if and only if U^{-1} exists on H and $U^{-1} = U'$. Thus, unitary operators U are characterized by the property that*

$$UU' = Id = U'U.$$

Example A.13.5. Unitary operators

The Fourier transform mapping $\mathcal{F}$ on $L^2_{m^d}(\mathbb{R}^d)$, the DFT mapping $\mathcal{F}_N$ on $L^2_c(\mathbb{Z}_N)$, and the Hilbert transform mapping $\mathcal{H}$ on $L^2(\mathbb{R})$ are all unitary operators; see Appendix B.

A.14 Potpourri and titillation

1. At the beginning of the Preface we referred to this book as a paean to twentieth-century real analysis. This development of real variables, measure theory, and integration theory was one of several interleaving intellectual threads through the century. One such journey is the theory of frames.

With Theorem A.12.6 as a backdrop we make the following definition, which, at first blush, may seem to be an effete generalization of an ONB. Let H be a separable Hilbert space. A sequence $\{x_n : n = 1, \dots\} \subseteq H$ is a *frame* for H if there are $A, B > 0$ such that

$$\forall\, x \in H, \quad A\|x\|^2 \leq \sum_{n=1}^{\infty} |\langle x, x_n\rangle|^2 \leq B\|x\|^2.$$

The constants A and B are *frame bounds*, and a frame is tight if $A = B$. A frame is an *exact frame* if it is no longer a frame whenever any of its elements is removed. The following is the basic decomposition theorem for frames.

Theorem A.14.1. Frame decomposition
Let $\{x_n : n = 1, \dots\} \subseteq H$ *be a frame for* H*, and define the mapping*

$$S : H \to H,$$
$$x \mapsto \sum_{n=1}^{\infty} \langle x, x_n \rangle x_n.$$

Then S *is a continuous bijection onto* H*, and*

$$\forall\, x \in H, \quad x = \sum_{n=1}^{\infty} \langle x, S^{-1}(x_n) \rangle x_n = \sum_{n=1}^{\infty} \langle x, x_n \rangle S^{-1}(x_n). \tag{A.25}$$

In Theorem A.14.1, the assertion that S is a continuous bijection onto H implies that $S^{-1} : H \to H$ is continuous by Theorem A.10.3. Generally, we refer to any continuous bijection $L : H \to H$ as a *topological isomorphism*; see [45], Chapters 3 and 7.

Expositions of the theory of frames are found in [518], [118], [45], and [103]. The theory was explicitly formulated by RICHARD J. DUFFIN (1909–1996) and ALBERT C. SCHAEFFER (1952) [147]. What is truly remarkable is the genuine applicability of the theory of frames in addressing sampling problems, erasure problems associated with the Internet, quantization problems arising in audio, image-processing problems, and a host of other problems, e.g., see [99], [100], [320], [466]. A reason for this applicability is the effectiveness of frames in providing numerically stable, robust, and generally "inexpensive" decompositions; and this reason is due to the fact that frames are generally not ONBs or even Schauder bases, even though there are representations such as (A.25).

In order to describe some of the early developments of frames, we first expand on the definition of a Schauder basis in the setting of a separable Hilbert space H. A Schauder basis $\{x_n : n = 1, \dots\}$ for H is an *unconditional basis* for H if

$$\exists\, C > 0 \text{ such that } \forall\, F \subseteq \mathbb{N}, \text{ where card } F < \infty,$$
$$\text{and } \forall\, b_n, c_n \in \mathbb{C}, \text{ where } n \in F \text{ and } |b_n| < |c_n|,$$
$$\left\| \sum_{n \in F} b_n x_n \right\| \leq C \left\| \sum_{n \in F} c_n x_n \right\|.$$

An unconditional basis is a *bounded unconditional basis* for H if

$$\exists\, A, B > 0 \text{ such that } \forall\, n \in \mathbb{N}, \quad A \leq \|x_n\| \leq B.$$

Finally, a Schauder basis $\{x_n : n = 1, \dots\}$ for H is a *Riesz basis* for H if there is a topological isomorphism on H mapping $\{x_n : n = 1, \dots\}$ onto an ONB for H.

In 1936 KÖTHE [301] proved that bounded unconditional bases are exact frames, and the converse is straightforward. Also, the category of Riesz bases is precisely that of exact frames. Thus, the following three notions are equivalent: Riesz bases, exact frames, and bounded unconditional bases. Besides the article by DUFFIN and SCHAEFFER, BARI's characterization of Riesz bases (1951) [26] is fundamental in this realm of ideas. From our point of view, her work has all the more impact because it was motivated in part by her early research, with others in the Russian school, in analyzing RIEMANN's sets of uniqueness for trigonometric series; see Section 3.8.4.

From a functional-analytic point of view, in 1921 VITALI [487] proved that if $\{x_n : n = 1, \dots\}$ is a tight frame with $A = B = 1$ and with $\|x_n\| = 1$ for all n, then $\{x_n : n = 1, \dots\}$ is an ONB. Actually, VITALI's result is stronger for the setting $H = L^2([a, b])$ in which he dealt.

Frames have also been studied in terms of the celebrated Naimark dilation theorem (1943), a special case of which asserts that any frame can be obtained by "compression" from a basis. The rank-1 case of the Naimark theorem is the previous assertion for tight frames. The finite-decomposition rank-1 case of the Naimark theorem antedates MARK A. NAIMARK's paper, and it is due to HUGO HADWIGER (1940) and GASTON JULIA (1942). This is particularly interesting in light of modern applications of finite unit-norm tight frames in communications theory. In this context, we mention CHANDLER DAVIS' use of Walsh functions to give explicit constructions of dilations [125]. DAVIS [126] also provides an in-depth perspective of the results referred to in this paragraph.

Other applications of the Naimark theorem in the context of frames include feasibility issues for von Neumann measurements in quantum signal processing.

The general theory of frames was inspired by the study of nonharmonic Fourier series and Fourier frames. Just as we define Fourier series in Example 3.3.4 and Appendix B, we define *nonharmonic Fourier series* to be of the form $\sum_{\lambda \in \Lambda} c_\lambda e_\lambda$, where $\Lambda \subseteq \mathbb{R}$ is countable and $e_\lambda = e^{-2\pi i x \lambda}$. Typically, we investigate the elements of $L^2([-R, R])$ that can be represented in L^2-norm by such series in a manner analogous to Theorem A.14.1. As such, Fourier frames can be thought of as going back to DINI (1880) and his book on Fourier series [140], pages 190 ff. There he gives Fourier expansions in terms of the set $\{e_\lambda : \lambda \in \Lambda\}$ of harmonics, where each λ is a solution of the equation

$$x \cos(\pi x) + a \sin(\pi x) = 0. \tag{A.26}$$

Equation (A.26) was chosen because of a problem in mathematical physics from RIEMANN and HEINRICH WEBER's classical treatise [387], pages 158–167. DINI returned to this topic in 1917, just before his death, with a significant generalization including Fourier frames that are not ONBs [141].

The inequalities defining a Fourier frame $\{e_\lambda : \lambda \in \Lambda\}$ for $L^2([-R, R])$ (of which our definition of a frame is a natural generalization) were explicitly written by PALEY and WIENER [363], page 115, inequalities (30.56).

The book by PALEY and WIENER (1934), and to a lesser extent a stability theorem by BIRKHOFF (1917), had tremendous influence on mid-twentieth-century harmonic analysis. Although nonharmonic Fourier series expansions were developed, the major effort in the study of Fourier systems emanating from [363] addressed completeness problems of sequences $\{e_\lambda : \lambda \in \Lambda\} \subseteq L^2([-R, R])$, i.e., on determining when the closed linear span of $\{e_\lambda : \lambda \in \Lambda\}$ is all of $L^2([-R, R])$. This culminated in the profound work of BEURLING and MALLIAVIN in 1962 and 1966 [59], [60], [295]; see [43], Chapter 1, for a technical overview.

A landmark on the road to the results of BEURLING and MALLIAVIN is the article by DUFFIN and SHAEFFER. In retrospect, their paper was underappreciated when it appeared in 1952. The authors defined Fourier frames as well as the general notion of a frame for a Hilbert space H. They emphasized that frames $\{x_n : n = 1, \dots\} \subseteq H$ provide discrete representations $x = \sum_{n=1}^{\infty} c_n x_n$ in norm, as opposed to the previous emphasis on completeness. These discrete representations for Fourier frames provide a natural setting for nonuniform sampling, e.g., [43], Chapter 1, [229], [515], [169], [222]. DUFFIN and SHAEFFER understood that the Paley–Wiener theory for Fourier systems is equivalent to the theory of exact Fourier frames. (We noted above that PALEY and WIENER used precisely the inequalities defining Fourier frames.) DUFFIN and SHAEFFER also knew that generally they were dealing with overcomplete systems, a useful feature in noise-reduction problems and the other applications we have mentioned.

The next step on this path created by DUFFIN and SHAEFFER is the article by INGRID DAUBECHIES, ALEX GROSSMANN, and MEYER [119]. From the point of view of the affine and Heisenberg groups, and inspired by DUFFIN and SHAEFFER, this article establishes the basic theory of wavelet and Gabor frames. About 1990, DUFFIN expressed satisfactory surprise to one of the authors that the theory of frames had risen like a phoenix almost 40 years after its creation.

B Fourier Analysis

B.1 Fourier transforms

Throughout the text of this book there are many examples and problems dealing with the notion of Fourier series and Fourier transforms; see, e.g., Examples 3.3.4, 3.6.6, 4.4.7, 4.5.7, and Problems 3.14, 3.28, 4.5, 4.17, 4.44, 5.20. In this appendix we give a brief outline of the basic elementary theory of classical Fourier analysis. There are many excellent texts and expositions including [39], [66], [276], [298], [450], [524], [155], [270], [469], [27].

Definition B.1.1. Fourier transform on $L^1_m(\mathbb{R})$
a. The *Fourier transform* of $f \in L^1_m(\mathbb{R})$ is the function F defined as

$$F(\xi) = \int_{\mathbb{R}} f(x) e^{-2\pi i x \xi}\, dx, \quad \xi \in \widehat{\mathbb{R}} = \mathbb{R}. \tag{B.1}$$

Notationally, we write the pairing between the functions f and F in the following way:

$$\hat{f} = F.$$

The space of Fourier transforms of $L^1_m(\mathbb{R})$ functions is denoted by $A(\widehat{\mathbb{R}})$, i.e.,

$$A(\widehat{\mathbb{R}}) = \{F : \widehat{\mathbb{R}} \to \mathbb{C} : \exists\, f \in L^1_m(\mathbb{R}) \text{ such that } \hat{f} = F\}.$$

b. Let $f \in L^1_m(\mathbb{R})$. The *Fourier transform inversion formula* is

$$f(x) = \int_{\widehat{\mathbb{R}}} F(\xi) e^{2\pi i x \xi}\, d\xi.$$

We use the notation $\check{F} = f$ to denote this inversion. See [39], pages 2–3, for a formal intuitive derivation of the Fourier transform inversion formula by means of a form of the uncertainty principle. The *Jordan pointwise inversion formula* gives an explicit theorem. Also, see Theorem B.3.7.

Theorem B.1.2. Jordan inversion formula
Let $f \in L^1_m(\mathbb{R})$. Assume that $f \in BV([x_0 - \varepsilon, x_0 + \varepsilon])$, for some $x_0 \in \mathbb{R}$ and $\varepsilon > 0$. Then

$$\frac{f(x_0+) + f(x_0-)}{2} = \lim_{M \to \infty} \int_{-M}^{M} \hat{f}(\xi) e^{2\pi i x \xi}\, d\xi.$$

J.J. Benedetto and W. Czaja, *Integration and Modern Analysis*, Birkhäuser Advanced Texts, DOI 10.1007/978-0-8176-4656-1_11,

Example B.1.3. $f \in L^1_m(\mathbb{R})$ does not imply $\hat{f} \in L^1_m(\widehat{\mathbb{R}})$
Let

$$f(x) = H(x)e^{-2\pi i r x},$$

where $r > 0$ and H is the *Heaviside function*, i.e., $H = \mathbb{1}_{[0,\infty)}$. Then

$$\hat{f}(\xi) = \frac{1}{2\pi(r + i\xi)} \notin L^1_m(\widehat{\mathbb{R}}).$$

Theorem B.1.4. Algebraic properties of Fourier transforms
a. *Let $f_1, f_2 \in L^1_m(\mathbb{R})$, and assume $c_1, c_2 \in \mathbb{C}$. Then*

$$\forall\, \xi \in \widehat{\mathbb{R}}, \quad (c_1 f_1 + c_2 f_2)\hat{}(\xi) = c_1 \hat{f}_1(\xi) + c_2 \hat{f}_2(\xi).$$

b. *Let $f \in L^1_m(\mathbb{R})$ and assume $F = \hat{f} \in L^1_m(\widehat{\mathbb{R}})$. Then*

$$\forall\, x \in \mathbb{R}, \quad \widehat{F}(x) = f(-x).$$

c. *Let $f \in L^1_m(\mathbb{R})$. Then*

$$\forall\, \xi \in \widehat{\mathbb{R}}, \quad \widehat{\overline{f}}(\xi) = \overline{\hat{f}(\xi)}.$$

For a fixed $\gamma \in \mathbb{R}$ we set

$$e_\gamma(x) = e^{2\pi i x \gamma}.$$

For a fixed $t \in \mathbb{R}$ and for a given function $f : \mathbb{R} \to \mathbb{C}$, the *translation operator* τ_t is defined as

$$\tau_t(f)(x) = f(x - t),$$

and, for a fixed $\lambda \in \mathbb{R} \setminus \{0\}$ and for a given function $f : \mathbb{R} \to \mathbb{C}$, the *dilation operator* is defined by the dilation formula,

$$f_\lambda(x) = \lambda f(\lambda x).$$

Example B.1.5. Dilation and the Poisson function
If $f(x) = e^{-2\pi r|x|}$, $r > 0$, then

$$\hat{f}(\xi) = \frac{1}{r} P_{1/r}(\xi) = \frac{1}{r} \frac{1}{\pi(1 + \xi^2/r^2)} \in L^1_m(\widehat{\mathbb{R}}),$$

where $P(\xi) = 1/(\pi(1 + \xi^2))$ is the *Poisson function.*

Proposition B.1.6. *Let $f \in L^1_m(\mathbb{R})$, let $t \in \mathbb{R}$, let $\gamma \in \widehat{\mathbb{R}}$, and let $\lambda \in \widehat{\mathbb{R}} \setminus \{0\}$. Then*

i. $(e_\gamma f)\hat{}(\xi) = \tau_\gamma(\hat{f})(\xi)$,
ii. $(\tau_t(f))\hat{}(\xi) = e_{-t}(\xi)\hat{f}(\xi)$,
iii. $(f_\lambda)\hat{}(\xi) = (\lambda/|\lambda|)\hat{f}(\xi/\lambda)$.

In the following result we shall assume the existence of a pair (f, F) of functions such that

$$F = \hat{f} \quad \text{and} \quad f = \check{F}. \tag{B.2}$$

We shall not be concerned with correct mathematical hypotheses for asserting the existence of the Fourier transform or of the inversion formula.

Proposition B.1.7. *Assume that there exists a pair (f, F) of functions that satisfies (B.2).*

a. *f is real if and only if*

$$\overline{F(\xi)} = F(-\xi).$$

In this case,

$$F(\xi) = \int_{\mathbb{R}} f(x)\cos(2\pi x\xi)\, dx - i\int_{\mathbb{R}} f(x)\sin(2\pi x\xi)\, dx$$

and

$$f(x) = 2\mathrm{Re}\int_0^\infty F(\xi)e^{2\pi i x\xi}\, d\xi.$$

b. *f is real and even if and only if F is real and even. In this case,*

$$F(\xi) = 2\int_0^\infty f(x)\cos(2\pi x\xi)\, dx$$

and

$$f(x) = 2\int_0^\infty F(\xi)\cos(2\pi x\xi)\, d\xi.$$

c. *f is real and odd if and only if F is odd and imaginary. In this case,*

$$F(\xi) = -2i\int_0^\infty f(x)\sin(2\pi x\xi)\, dx$$

and

$$f(x) = 2i\int_0^\infty F(\xi)\sin(2\pi x\xi)\, d\xi.$$

Example B.1.8. The Gaussian

Let $f(x) = e^{-\pi r x^2}$, $r > 0$. We could calculate $\hat{f}$ by means of contour integrals, but we choose a real, and by now classical, approach [172], page 476. By definition of $\hat{f}$, which is real and even, we have

$$(\hat{f})'(\xi) = -2\pi i\int_{\mathbb{R}} te^{-\pi r x^2}e^{-2\pi i x\xi}\, dx. \tag{B.3}$$

Noting that

$$\frac{d}{dx}(e^{-\pi r x^2}) = -2\pi r x\, e^{-\pi r x^2},$$

we rewrite (B.3) as

$$\begin{aligned}(\hat{f})'(\xi) &= -2\pi i \int \frac{-1}{2\pi r}(e^{-\pi r x^2})' e^{-2\pi i x\xi}\, dx \\ &= \frac{i}{r}\left[e^{-\pi r x^2} e^{-2\pi i x \xi}\Big|_{-\infty}^{\infty} - \int_{\mathbb{R}} e^{-\pi r x^2}(-2\pi i \xi) e^{-2\pi i x\xi}\, dx\right] \\ &= \frac{-2\pi\xi}{r}\hat{f}(\xi).\end{aligned}$$

Thus, $\hat{f}$ is a solution of the differential equation,

$$F'(\xi) = -\frac{2\pi\xi}{r}F(\xi), \tag{B.4}$$

and (B.4) is solved by elementary means with solution

$$F(\xi) = Ce^{-\pi\xi^2/r}.$$

Taking $\xi = 0$ and using the definition of the Fourier transform, we see that

$$C = \int_{\mathbb{R}} e^{-\pi r x^2}\, dx.$$

In order to calculate C we first evaluate $a = \int_0^\infty e^{-u^2}\, du$:

$$\begin{aligned} a^2 &= \int_0^\infty e^{-s^2}\, ds \int_0^\infty e^{-t^2}\, dt \\ &= \int_0^\infty \int_0^\infty e^{-(s^2+t^2)}\, dsdt = \int_0^{\pi/2}\int_0^\infty e^{-r^2}\, rdrd\theta \\ &= \frac{\pi}{4}\int_0^\infty e^{-u}\, du = \frac{\pi}{4}. \end{aligned}$$

Thus, $a = \frac{\sqrt{\pi}}{2}$, and so

$$\int_{\mathbb{R}} e^{-u^2}\, du = \sqrt{\pi}.$$

Consequently,

$$C = \int_{\mathbb{R}} e^{-\pi r x^2}\, dt = \frac{1}{\sqrt{\pi r}}\int_{\mathbb{R}} e^{-u^2}\, du = \frac{1}{\sqrt{r}}.$$

Therefore, we have shown that

$$\hat{f}(\xi) = \frac{1}{\sqrt{r}}e^{-\pi\xi^2/r}.$$

We write

$$G(x) = \frac{1}{\sqrt{\pi}} e^{-x^2},$$

so that if $\lambda > 0$, then

$$(G_\lambda)\hat{}(\xi) = e^{-(\pi\xi/\lambda)^2}.$$

In particular,

$$\frac{1}{\sqrt{r}} (G_{\sqrt{\pi r}})\hat{} = G_{\sqrt{\pi/r}}$$

and hence $(G_{\sqrt{\pi}})\hat{} = G_{\sqrt{\pi}}$. We refer to G as the *Gauss function* or *Gaussian*, and note that $\int_{\mathbb{R}} G(x)\,dx = 1$.

B.2 Analytic properties of Fourier transforms

Theorem B.2.1. Boundedness and continuity of Fourier transforms
Assume $f \in L^1_m(\mathbb{R})$.

a. $\forall\, \xi \in \widehat{\mathbb{R}},\ |\hat{f}(\xi)| \leq \|f\|_1$.

b. $\forall\, \varepsilon > 0,\ \exists\, \delta > 0$ *such that* $\forall\, \xi$ *and* $\forall\, \zeta$, *for which* $|\zeta| < \delta$, *we have* $|\hat{f}(\xi + \zeta) - \hat{f}(\xi)| < \varepsilon$, *i.e.,* $\hat{f}$ *is uniformly continuous.*

Proof. **a.** Part *a* is immediate from the definition of Fourier transform.

b. First, we note that

$$|\hat{f}(\xi + \zeta) - \hat{f}(\xi)| \leq \int_{\mathbb{R}} |f(x)| \left|e^{-2\pi i x\zeta} - 1\right|\, dx.$$

Let $g_\zeta(x) = |f(x)| \left|e^{-2\pi i x\zeta} - 1\right|$. Since $\lim_{\zeta \to 0} g_\zeta(x) = 0$ for all $x \in \mathbb{R}$, and since $|g_\zeta(x)| \leq 2|f(x)|$, we can use LDC to obtain

$$\lim_{\zeta \to 0} \int_{\mathbb{R}} g_\zeta(x)\, dx = 0.$$

This limit holds independently of ξ. Consequently, we have

$$\forall \varepsilon > 0,\ \exists \zeta_0 > 0 \text{ such that } \forall\, \zeta \in (-\zeta_0, \zeta_0) \text{ and } \forall\, \xi \in \widehat{\mathbb{R}},\quad |\hat{f}(\xi + \zeta) - \hat{f}(\xi)| < \varepsilon.$$

This is the desired uniform continuity. □

The next result has essentially the same proof as the *Riemann–Lebesgue lemma for* $L^1_m(\mathbb{T})$; see Theorem 3.6.4.

Theorem B.2.2. Riemann–Lebesgue lemma for $L^1_m(\mathbb{R})$
Assume $f \in L^1_m(\mathbb{R})$. *Then*

$$\lim_{|\xi| \to \infty} \hat{f}(\xi) = 0.$$

Theorem B.2.3. Differentiation of Fourier transforms
a. *Assume that $f^{(n)}$, $n \geq 1$, exists everywhere and that*

$$f(\pm\infty) = \cdots = f^{(n-1)}(\pm\infty) = 0.$$

Then

$$(f^{(n)})\hat{}(\xi) = (2\pi i\xi)^n \hat{f}(\xi).$$

b. *Assume that $x^n f(x) \in L^1_m(\mathbb{R})$, for some $n \geq 1$. Then $x^k f(x) \in L^1_m(\mathbb{R})$, $k = 1, \ldots, n-1$, $(\hat{f})', \ldots, (\hat{f})^{(n)}$ exist everywhere, and*

$$\forall\, k = 0, \ldots, n, \quad ((-2\pi i\cdot)^k f(\cdot))\hat{}(\xi) = \hat{f}^{(k)}(\xi).$$

Proof. **a.** In the statement of the theorem, $f(\pm\infty) = 0$ denotes the facts that $\lim_{x\to\pm\infty} f(x) = 0$. Using integration by parts (Theorem 4.6.3) we compute

$$\int_{-S}^{T} f^{(n)}(x)e^{-2\pi i x\xi}\,dx = f^{(n-1)}(x)e^{-2\pi i x\xi}\Big|_{-S}^{T} + 2\pi i\xi \int_{-S}^{T} f^{(n-1)}(x)e^{-2\pi i x\xi}\,dx.$$

Iterating this procedure we obtain that

$$\int_{-S}^{T} f^{(n)}(x)e^{-2\pi i x\xi}\,dx = \sum_{j=0}^{n-1}(2\pi i\xi)^j\, f^{(n-(j+1))}(x)e^{-2\pi i x\xi}\Big|_{-S}^{T} + (2\pi i\xi)^n \int_{-S}^{T} f(x)e^{-2\pi i x\xi}\,dx.$$

Letting $S, T \to \infty$, the right-hand side converges to $(2\pi i\xi)^n\hat{f}(\xi)$ and the result is proved.

b. Without loss of generality we assume that $n = 1$ and we fix $\xi \in \widehat{\mathbb{R}}$. Then

$$\frac{\hat{f}(\xi+\zeta) - \hat{f}(\xi)}{\zeta} = \int_{\mathbb{R}} f(x)e^{-2\pi i x\xi}\left(\frac{e^{-2\pi i x\zeta} - 1}{\zeta}\right)\,dx.$$

If we denote the integrand on the right-hand side by $f(x,\zeta)$, then we have

$$|f(x,\zeta)| \leq 4\pi|xf(x)|,$$

which follows from the inequality

$$\left|\frac{e^{-2\pi i x\zeta} - 1}{\zeta}\right| \leq 4\pi|x|.$$

Moreover, we have that

$$\lim_{\zeta\to 0} f(x,\zeta) = -2\pi i x f(x)e^{-2\pi i x\xi}.$$

Thus, we can use LDC to assert that

$$\lim_{\zeta\to 0}\frac{\hat{f}(\xi+\zeta)-\hat{f}(\xi)}{\zeta}=\int_{\mathbb{R}}(-2\pi i x)f(x)e^{-2\pi i x\xi}\,dx. \qquad \square$$

Example B.2.4. The role of absolute continuity

a. We recall from Section 4.6 that the facts that f is differentiable m-*a.e.* and that $f, f' \in L^1_m(\mathbb{R})$ do not imply that f' is absolutely continuous. Therefore, it is *not* necessarily true that

$$\forall\, \xi \in \widehat{\mathbb{R}}, \quad (f')\hat{}\,(\xi) = 2\pi i\xi\hat{f}(\xi). \tag{B.5}$$

Consider the Cantor function C_C associated with the usual Cantor set C; see Example 1.3.17. Let

$$f(x) = C_C(x+1)\mathbb{1}_{[-1,0]} + (1 - C_C(x))\mathbb{1}_{[0,1]}.$$

Clearly, f is a continuous compactly supported function of bounded variation on $\mathbb{R}$ for which $f' = 0$ m-*a.e.* In particular, $f, f' \in L^1_m(\mathbb{R})$ and (B.5) fails; cf. Theorem 4.6.8.

b. In obtaining the formula

$$\int_{\mathbb{R}} f^{(n)}(x)e^{-2\pi i x\xi}\,dx = (2\pi i\xi)^n\hat{f}(\xi), \tag{B.6}$$

we use a subtle fact that everywhere differentiability of $f^{(n-1)}$ allows us to deduce that it is absolutely continuous; see Theorem 4.6.7.

Equation (B.6) is also valid, without the aforementioned subtlety, if the hypotheses, that $f^{(n-1)}$ is everywhere differentiable and $f^{(n)} \in L^1_m(\mathbb{R})$, are replaced by the hypothesis that $f^{(n)}$ is piecewise continuous.

c. The assumption in Theorem B.2.3 that $f(\pm\infty) = \cdots = f^{(n-1)}(\pm\infty) = 0$ is not necessary. For simplicity, let $n = 1$, and assume that $f, f' \in L^1_m(\mathbb{R})$ and that f is absolutely continuous. For fixed $a \in \mathbb{R}$ and $c \in \mathbb{C}$, set $F(x) = c + \int_a^x f'(t)\,dt$. By FTC-I, F is absolutely continuous and $F' = f'$ m-*a.e.* Since f is absolutely continuous, we have $f = F + C$ on $[a, \infty)$, for some $C \in \mathbb{C}$, and consequently,

$$\forall\, x \in [a,\infty), \quad f(x) = F(a) + C + \int_a^x f'(t)\,dt.$$

Therefore, $f(a) = F(a) + C$ and

$$f(x) - f(a) = \int_a^x f'(t)\,dt.$$

This observation, combined with the fact that $f' \in L^1_m(\mathbb{R})$, implies that $\lim_{x\to\pm\infty} f(x)$ exist. Moreover, since $f \in L^1_m(\mathbb{R})$, $\lim_{x\to\pm\infty} f(x) = 0$.

Example B.2.5. $\boldsymbol{C_0(\widehat{\mathbb{R}}) \setminus A(\widehat{\mathbb{R}}) \neq \emptyset}$

Theorems B.2.1 and B.2.2 allow us to conclude that $A(\widehat{\mathbb{R}}) \subseteq C_0(\widehat{\mathbb{R}})$. It is not difficult to see that the inclusion is proper. Indeed, let F be defined as

$$F(\xi) = \begin{cases} \frac{1}{\log(\xi)}, & \text{if } \xi > e, \\ \frac{\xi}{e}, & \text{if } 0 \leq \xi \leq e, \end{cases}$$

on $[0, \infty)$ and as $-F(-\xi)$ on $(-\infty, 0]$. Then $F \in C_0(\widehat{\mathbb{R}})$. The fact that $F \notin A(\widehat{\mathbb{R}})$ depends on the divergence of

$$\int_e^{\infty} \frac{1}{\xi \log(\xi)}\, d\xi;$$

cf. [39], [201].

This function F is not an isolated example. In fact, $A(\widehat{\mathbb{R}})$ is a set of first category in $C_0(\widehat{\mathbb{R}})$; see the Remark after Theorem A.10.3. Even more, a Baire category argument can also be used to show the existence of $F \in C_c(\widehat{\mathbb{R}})$ for which $F \notin A(\widehat{\mathbb{R}})$. Explicit examples of such functions are more difficult to construct, but it is possible to do. For example, define the *butterfly function*,

$$B(\xi) = \begin{cases} \frac{1}{n} \sin(2\pi 4^n \xi), & \text{if } \frac{1}{2^{n+1}} < |\xi| \leq \frac{1}{2^n}, \\ 0, & \text{if } \xi = 0 \text{ or } |\xi| > \frac{1}{2}; \end{cases}$$

see [230].

B.3 Approximate identities

In Problem 3.5 we defined the convolution $f * g$ of $f, g \in L^1_m(\mathbb{R})$ to be

$$f * g(x) = \int_{\mathbb{R}} f(t) g(x-t)\, dt = \int_{\mathbb{R}} f(x-t) g(t)\, dt.$$

Proposition B.3.1. *Let $f, g \in L^1_m(\mathbb{R})$. Then $f * g \in L^1_m(\mathbb{R})$ and*

$$(f * g)\hat{}(\xi) = \hat{f}(\xi)\hat{g}(\xi).$$

Proof. The assertion that $f * g \in L^1_m(\mathbb{R})$ is part of Problem 3.5. Thus, we can use the Fubini–Tonelli theorems (Theorem 3.7.5 and Theorem 3.7.8) to compute

$$\begin{aligned} (f * g)\hat{}(\xi) &= \int_{\mathbb{R}} \int_{\mathbb{R}} f(x-t) g(t) e^{-2\pi i x \xi}\, dt\, dx \\ &= \int_{\mathbb{R}} \int_{\mathbb{R}} f(x-t) g(t) e^{-2\pi i (x-t)\xi} e^{-2\pi i t \xi}\, dt\, dx \end{aligned}$$

$$= \int_{\mathbb{R}} \left(\int_{\mathbb{R}} f(x-t) e^{-2\pi i (x-t)\xi} \, dx \right) g(t) e^{-2\pi i t \xi} \, dt$$

$$= \int_{\mathbb{R}} \hat{f}(\xi) g(t) e^{-2\pi i t \xi} \, dt = \hat{f}(\xi)\hat{g}(\xi). \qquad \square$$

This innocent proposition is actually a raison d'être for transform methods generally and for the Fourier transform in particular, see [39].

The following notion is critical in approximating $\delta \in M_b(\mathbb{R})$, and for providing examples in applications including signal processing and spectral estimation.

Definition B.3.2. Approximate identity
An *approximate identity* is a family $\{K_{(\lambda)} : \lambda > 0\} \subseteq L^1_m(\mathbb{R})$ of functions with the following properties:

i. $\forall \lambda > 0, \quad \int_{\mathbb{R}} K_{(\lambda)}(x)\, dx = 1;$
ii. $\exists\, M > 0$ such that $\forall\, \lambda > 0, \quad \|K_{(\lambda)}\|_1 \leq M,$
iii. $\forall\, \delta > 0, \quad \lim_{\lambda \to \infty} \int_{|x| \geq \delta} |K_{(\lambda)}(x)|\, dx = 0;$

cf. Problem 3.28*c*.

Proposition B.3.3. *Let $K \in L^1_m(\mathbb{R})$ have the property that $\int_{\mathbb{R}} K(x)\, dx = 1$. Then the family $\{K_\lambda : K_\lambda(x) = \lambda K(\lambda x),\ \lambda > 0\} \subseteq L^1_m(\mathbb{R})$ of dilations of K is an approximate identity.*

Example B.3.4. Approximate identities
a. The *Fejér function* W is defined as

$$W(x) = \frac{1}{2\pi} \left(\frac{\sin(x/2)}{x/2} \right)^2;$$

cf. Problem 3.28. The *Fejér function* W is nonnegative and $\int_{\mathbb{R}} W(x)\, dx = 1$. Thus, the *Fejér kernel* $\{W_\lambda : \lambda > 0\} \subseteq L^1_m(\mathbb{R})$, defined as the family of dilations of W, is an approximate identity. FEJÉR's name at birth was WEISZ, whence W. (In Hungarian, "white" is "fehér".)

b. The *Dirichlet function* D is the function

$$D(x) = \frac{\sin(x)}{\pi x}.$$

Although $\int_{\mathbb{R}} D(t)\, dt = 1$, we have $D \notin L^1_m(\mathbb{R})$, and so the *Dirichlet kernel* $\{D_\lambda : \lambda > 0\}$ is not an approximate identity.

c. The *Poisson function*, defined in Example B.1.5, is

$$P(x) = \frac{1}{\pi(1+x^2)},$$

and it satisfies $\int_{\mathbb{R}} P(x)\, dx = 1$. Thus, the *Poisson kernel* $\{P_\lambda : \lambda > 0\} \subseteq L^1_m(\mathbb{R})$ is an approximate identity.

d. The *Gauss function* defined in Example B.1.8 is

$$G(x) = \frac{1}{\sqrt{\pi}} e^{-x^2}.$$

We see that G is positive and $\int_{\mathbb{R}} G(x)\, dx = 1$; and, thus, the *Gauss kernel* $\{G_\lambda : \lambda > 0\} \subseteq L^1_m(\mathbb{R})$ is an approximate identity.

Theorem B.3.5. Approximation and uniqueness
Let $f \in L^1_m(\mathbb{R})$.

a. *If $\{K_{(\lambda)} : \lambda > 0\} \subseteq L^1_m(\mathbb{R})$ is an approximate identity, then*

$$\lim_{\lambda\to\infty} \|f - f * K_{(\lambda)}\|_1 = 0.$$

b. *We have*

$$\lim_{\lambda\to\infty} \int_{\mathbb{R}} \left| f(x) - \int_{-\lambda/2\pi}^{\lambda/2\pi} \left(1 - \frac{2\pi|\xi|}{\lambda}\right) \hat{f}(\xi) e^{2\pi i x\xi}\, d\xi \right| dx = 0.$$

c. *If $\hat{f} = 0$ on $\widehat{\mathbb{R}}$, then f is the 0 function.*

Proof. **a.** We use the fact that $\int_{\mathbb{R}} K_{(\lambda)}(x)\, dx = 1$ to compute

$$\begin{aligned}\|f - f * K_{(\lambda)}\|_1 &= \int_{\mathbb{R}} \left| \int_{\mathbb{R}} K_{(\lambda)}(t) f(x)\, dt - \int_{\mathbb{R}} K_{(\lambda)}(t) f(x-t)\, dt \right| dx \\ &\leq \int_{\mathbb{R}} |K_{(\lambda)}(t)| \left(\int_{\mathbb{R}} |f(x) - f(x-t)|\, dx \right) dt.\end{aligned}$$

Let $\varepsilon > 0$. Using the result from Problem 3.14*b*, there is $\delta > 0$ with the property that

$$\forall\, |t| < \delta, \quad \int_{\mathbb{R}} |f(x) - f(x-t)|\, dx < \frac{\varepsilon}{M},$$

where $\|K_{(\lambda)}\|_1 \leq M$. Therefore, we have the estimate

$$\begin{aligned}\|f - f * K_{(\lambda)}\|_1 &\leq 2\|f\|_1 \int_{|t|\geq\delta} |K_{(\lambda)}(t)|\, dt + \frac{\varepsilon}{M} \int_{|t|\leq\delta} |K_{(\lambda)}(t)|\, dt \\ &\leq 2\|f\|_1 \int_{|t|\geq\delta} |K_{(\lambda)}(t)|\, dt + \varepsilon.\end{aligned}$$

Consequently, by the definition of approximate identity, we have

$$\overline{\lim_{\lambda\to\infty}} \|f - f * K_{(\lambda)}\|_1 \leq \varepsilon.$$

Since ε was arbitrary, the proof of part *a* is complete.

b. It is not difficult to calculate that the Fejér kernel satisfies

$$W_\lambda(x) = \int_{-\lambda/2\pi}^{\lambda/2\pi} \left(1 - \frac{2\pi|\xi|}{\lambda}\right) e^{2\pi i x\xi}\, d\xi;$$

see, e.g., [39]. Then, by the definition of convolution and an application of the Fubini–Tonelli theorems, we compute

$$f * W_\lambda(x) = \int_{-\lambda/2\pi}^{\lambda/2\pi} \left(1 - \frac{2\pi|\xi|}{\lambda}\right) \hat{f}(\xi) e^{2\pi i x\xi}\, d\xi.$$

Since $\{W_\lambda\}$ is an approximate identity, part *b* follows from part *a*.

c. Part *c* is immediate from part *b*. □

Proposition B.3.6. *Let $f \in L^\infty_m(\mathbb{R})$ be continuous on $\mathbb{R}$. If $\{K_{(\lambda)} : \lambda > 0\} \subseteq L^1_m(\mathbb{R})$ is an approximate identity, then*

$$\forall\, x \in \mathbb{R}, \quad \lim_{\lambda\to\infty} f * K_{(\lambda)}(x) = f(x).$$

If $f \in L^1_m(\mathbb{R})$ and $\hat{f} \in L^1_m(\widehat{\mathbb{R}})$, we can use Theorem B.3.5 to obtain the following pointwise inversion theorem. What we explicitly mean in its statement is that if $f \in L^1_m(\mathbb{R})$ and $\hat{f} \in L^1_m(\widehat{\mathbb{R}})$, then the formula in (B.7) is true *m-a.e.*, and that if f is continuous then (B.7) is true for all $x \in \mathbb{R}$. Compare the proof of Theorem B.3.7 with that in [39], pages 38–39.

Theorem B.3.7. Inversion formula for $L^1_m(\mathbb{R}) \cap A(\mathbb{R})$
Let $f \in L^1_m(\mathbb{R}) \cap A(\mathbb{R})$. Then

$$\forall\, x \in \mathbb{R}, \quad f(x) = \int_{\widehat{\mathbb{R}}} \hat{f}(\xi) e^{2\pi i x\xi}\, d\xi. \tag{B.7}$$

Proof. The statement of the theorem follows from two observations. First, if $\{K_{(\lambda)} : \lambda > 0\} \subseteq L^1_m(\mathbb{R})$ is an approximate identity, then there exists a subsequence $\{\lambda_n : n = 1, \dots\}$ such that

$$\lim_{n\to\infty} f * K_{(\lambda_n)} = f \quad m\text{-}a.e.$$

This fact is a consequence of Theorem B.3.5*a*. Second, assume that $\hat{f} \in L^1_m(\widehat{\mathbb{R}})$, that $(K_{(\lambda)})\hat{} \in L^1_m(\widehat{\mathbb{R}})$, and that

$$\forall\, x \in \mathbb{R}, \quad K_{(\lambda)}(x) = \int_{\widehat{\mathbb{R}}} (K_{(\lambda)})\hat{}(\xi) e^{2\pi i x\xi}\, d\xi.$$

Then

$$\lim_{\lambda\to\infty} \left\| \int_{\widehat{\mathbb{R}}} \hat{f}(\xi) e^{2\pi i x\xi}\, d\xi - f * K_{(\lambda)}(x) \right\|_\infty = 0.$$ □

Remark. It follows from the first observation in the proof of Theorem B.3.7 that if $f \in L^1_m(\mathbb{R})$, then there exists a sequence $\{\lambda_n : n = 1, \ldots\} \subseteq (0, \infty)$ such that

$$\lim_{\lambda_n \to \infty} \int_{-\lambda_n/2\pi}^{\lambda_n/2\pi} \left(1 - \frac{2\pi|\xi|}{\lambda_n}\right) \hat{f}(\xi) e^{2\pi i x \xi}\, d\xi = f(x) \quad m\text{-a.e.}$$

It turns out that λ_n can be replaced by λ in the above, and that the convergence m-a.e. can be enlarged to include all x in the *Lebesgue set* for f; see, e.g., [201], as well as Example 4.4.7, Definition 8.4.8, and Section 8.8.4.

B.4 The $L^2_m(\mathbb{R})$ theory of Fourier transforms

We have defined the Fourier transform for functions $f \in L^1_m(\mathbb{R})$. Now our goal is to extend this transform to the space $L^2_m(\mathbb{R})$. Clearly, $L^2_m(\mathbb{R}) \not\subseteq L^1_m(\mathbb{R})$, and so we cannot use the formula (B.1), since the function under the integral sign may not be integrable.

Lemma B.4.1. *Let $f \in C_c(\mathbb{R})$. Then $\hat{f} \in L^2_m(\widehat{\mathbb{R}})$ and*

$$\|f\|_2 = \|\hat{f}\|_2.$$

Proof. Let $\tilde{f}(t) = \overline{f(-t)}$. Clearly, $\hat{f} \in A(\widehat{\mathbb{R}})$, since $C_c(\mathbb{R}) \subseteq L^1_m(\mathbb{R})$. Define $g = f * \tilde{f}$. Thus, g is continuous, $g \in L^1_m(\mathbb{R}) \cap L^\infty_m(\mathbb{R})$, and

$$g(0) = \|f\|_2^2.$$

Moreover, using Fubini's theorem and the translation invariance of Lebesgue measure, we have

$$\forall \xi \in \widehat{\mathbb{R}}, \quad \hat{g}(\xi) = |\hat{f}(\xi)|^2.$$

By Proposition B.3.6 and since g is continuous, we deduce that

$$g(0) = \lim_{\lambda \to \infty} \int_{-\lambda/2\pi}^{\lambda/2\pi} \left(1 - \frac{2\pi|\xi|}{\lambda}\right) |\hat{f}(\xi)|^2\, d\xi.$$

Finally, the Levi–Lebesgue theorem allows us to assert that $\hat{f} \in L^2_m(\widehat{\mathbb{R}})$ and that

$$\|\hat{f}\|_2^2 = g(0) = \|f\|_2^2.$$ □

Theorem B.4.2. Plancherel theorem
There is a unique linear bijection $\mathcal{F} : L^2_m(\mathbb{R}) \to L^2_m(\widehat{\mathbb{R}})$ with the following properties:

a. $\forall\, f \in L^1_m(\mathbb{R}) \cap L^2_m(\mathbb{R})$ *and* $\forall\, \xi \in \widehat{\mathbb{R}}$, $\hat{f}(\xi) = \mathcal{F}(f)(\xi)$;
b. $\forall\, f \in L^2_m(\mathbb{R})$, $\|f\|_2 = \|\mathcal{F}(f)\|_2$.

Proof. i. We first define the action of the operator $\mathcal{F}$ on $C_c(\mathbb{R})$ by

$$\mathcal{F}(f) = \hat{f}.$$

It follows from Lemma B.4.1 that, for $f \in C_c(\mathbb{R})$, $\mathcal{F}(f) \in L^2_m(\widehat{\mathbb{R}})$.

ii. Next, we shall prove that $\mathcal{F}(C_c(\mathbb{R})) \subseteq A(\widehat{\mathbb{R}}) \cap L^2_m(\widehat{\mathbb{R}})$ is a dense subspace of $L^2_m(\widehat{\mathbb{R}})$. Indeed, let $g \in L^2_m(\widehat{\mathbb{R}})$ and suppose that

$$\forall\, f \in C_c(\mathbb{R}), \quad \int_{\mathbb{R}} \hat{f}(\xi)\overline{g(\xi)}\, d\xi = 0. \tag{B.8}$$

If $f \in C_c(\mathbb{R})$, then the function $\tau_u(f)(x) = f(x-u)$ is also an element of $C_c(\mathbb{R})$, and so (B.8) implies that

$$\forall\, f \in C_c(\mathbb{R}) \text{ and } \forall\, u \in \mathbb{R}, \quad \int_{\mathbb{R}} \hat{f}(\xi)\overline{g(\xi)}e^{-2\pi i u\xi}\, d\xi = 0. \tag{B.9}$$

By Hölder's inequality, $\hat{f}\overline{g} \in L^1_m(\widehat{\mathbb{R}})$, and so (B.9) allows us to invoke the uniqueness theorem (Theorem B.3.5*c*) to conclude that $\hat{f}\overline{g} = 0$ *m-a.e.* for each $f \in C_c(\mathbb{R})$.

Note that

$$\forall\, f \in C_c(\mathbb{R}) \text{ and } \forall\, \xi \in \widehat{\mathbb{R}}, \quad e^{2\pi i x\xi} f(x) \in C_c(\mathbb{R}).$$

Thus, $\mathcal{F}(C_c(\mathbb{R}))$ is translation-invariant, i.e.,

$$\forall\, f \in C_c(\mathbb{R}) \text{ and } \forall\, \xi \in \widehat{\mathbb{R}}, \quad \tau_u(\hat{f}) \in \mathcal{F}(C_c(\mathbb{R})).$$

From this we conclude that, for each $\xi_0 \in \widehat{\mathbb{R}}$, there is $f \in C_c(\mathbb{R})$ for which $|\hat{f}| > 0$ on some interval centered about ξ_0. To verify this claim, suppose there is ξ_0 such that for each $f \in C_c(\mathbb{R})$ and for each interval I centered at ξ_0, $\hat{f}$ has a zero in I. Consequently, $\hat{f}(\xi_0) = 0$ for each $f \in C_c(\mathbb{R})$. By the translation invariance of $\mathcal{F}(C_c(\mathbb{R}))$, $\tau_\eta(\hat{f}) \in \mathcal{F}(C_c(\mathbb{R}))$ for each $\eta \in \widehat{\mathbb{R}}$, and so

$$\forall\, f \in C_c(\mathbb{R}) \text{ and } \forall\, \eta \in \widehat{\mathbb{R}}, \quad \tau_\eta(\hat{f})(\xi_0) = 0,$$

i.e., $\hat{f} = 0$ on $\widehat{\mathbb{R}}$ for each $f \in C_c(\mathbb{R})$. This contradicts Theorem B.3.5*c*, and the claim is proved.

Therefore, if we assume (B.8) we can conclude that $g = 0$ *m-a.e.* Consequently, by the Hahn–Banach theorem (Theorem A.8.3) and by the fact that $L^2_m(\widehat{\mathbb{R}})$ is its own dual, we have that $\mathcal{F}(C_c(\mathbb{R}))$ is dense in $L^2_m(\widehat{\mathbb{R}})$.

iii. We have shown that $\mathcal{F}$ is a continuous linear injection $C_c(\mathbb{R}) \to L^2_m(\widehat{\mathbb{R}})$ when $C_c(\mathbb{R})$ is endowed with the $L^2_m(\mathbb{R})$ norm, and so $\mathcal{F}$ has a unique linear injective extension to $L^2_m(\mathbb{R})$. Also, $\mathcal{F}(C_c(\mathbb{R}))$ is closed and dense in $L^2_m(\widehat{\mathbb{R}})$ by Lemma B.4.1 and by part *ii*. Thus, $\mathcal{F}$ is also surjective.

Property *a* now follows since $C_c(\mathbb{R})$ is dense in $L^1_m(\mathbb{R})$, when equipped with the $L^1_m(\mathbb{R})$ norm; and property *b* is an immediate consequence of the continuity of $\mathcal{F}$. □

Notationally, because of Plancherel's theorem, we refer to $\mathcal{F}(f)$ as the *Fourier transform of* $f \in L^2_m(\mathbb{R})$. We often write

$$\hat{f} = \mathcal{F}(f).$$

Theorem B.4.3. Parseval formula
Let $f, g \in L^2_m(\mathbb{R})$. *Then the following formulas hold:*

$$\int_{\mathbb{R}} f(x)\overline{g(x)}\, dx = \int_{\widehat{\mathbb{R}}} \hat{f}(\xi)\overline{\hat{g}(\xi)}\, d\xi \tag{B.10}$$

and

$$\int_{\mathbb{R}} f(x)g(x)\, dx = \int_{\widehat{\mathbb{R}}} \hat{f}(\xi)\hat{g}(-\xi)\, d\xi. \tag{B.11}$$

Proof. Equation (B.10) is a consequence of Theorem B.4.2*b* and the fact that

$$4f\overline{g} = |f+g|^2 - |f-g|^2 + i|f+ig|^2 - i|f-ig|^2.$$

Equation (B.11) can be proved similarly. □

We shall refer to Theorems B.4.2 and B.4.3 as the *Parseval–Plancherel theorem*. Marc-Antoine Parseval was a French engineer who gave a formal verification of the Fourier series version of Theorem B.4.2*b* in 1799; his publication is dated 1805.

Example B.4.4. An idempotent problem in $L^1_m(\mathbb{R})$ and $L^2_m(\mathbb{R})$
Consider the equation

$$f = f * f. \tag{B.12}$$

a. If we ask whether (B.12) has a solution $f \in L^1_m(\mathbb{R}) \setminus \{0\}$, the answer is no. Indeed, if there were such an f, then we would have $\hat{f} = (\hat{f})^2$ so that $\hat{f}$ would take only the values 0 and 1. If $\hat{f} = 0$ on $\widehat{\mathbb{R}}$, then $f = 0$ by Theorem B.3.5*c*. If $\hat{f} = 1$ on $\widehat{\mathbb{R}}$, then $f \notin L^1_m(\mathbb{R})$ since $A(\widehat{\mathbb{R}}) \subseteq C_0(\widehat{\mathbb{R}})$. If $\hat{f}$ takes both 0 and 1 values we contradict the continuity of $\hat{f}$.

b. If we ask whether (B.12) has a solution $f \in L^2_m(\mathbb{R}) \setminus \{0\}$, the answer is yes. In fact, let $\hat{f} = \mathbb{1}_A$, where $m(A) < \infty$. We are using here the Parseval–Plancherel theorem to assert the existence of $f \in L^2_m(\mathbb{R})$ for which $\hat{f} = \mathbb{1}_A$.

See the related discussion in [39], Example 1.10.6 and Remark 3.10.13.

Example B.4.5. Hilbert transform
a. *Formally,* the *Hilbert transform* $\mathcal{H}(f)$ of $f : \mathbb{R} \to \mathbb{C}$ is the convolution

$$\mathcal{H}(f)(t) = \lim_{\varepsilon \to 0} \frac{1}{\pi} \int_{|t-x| \geq \varepsilon} \frac{f(x)}{t-x}\, dx.$$

The Hilbert transform opens the door to a large and profound area of harmonic analysis associated with the theory, relevance, and importance of *singular integrals*, e.g., [448], [190]; cf. [357] for a magnificent introduction.

b. As mentioned in Example A.13.5, $\mathcal{H} \in \mathcal{L}(L^2_m(\mathbb{R}))$, and $\mathcal{H}$ is a unitary operator on $L^2_m(\mathbb{R})$. Further, $\mathcal{H} \circ \mathcal{H} = -Id$ on $L^2_m(\mathbb{R})$, and

$$\mathcal{H} = \mathcal{F}^{-1}\sigma(\mathcal{H})\mathcal{F},$$

where $\sigma(\mathcal{H})(\gamma) = -i\mathrm{sgn}\,(\gamma)$.

c. Let $f : \mathbb{R} \to \mathbb{C}$ satisfy $\mathrm{supp}\,(f) \subseteq [0, \infty)$, and define the unilateral *Laplace transform* of f as $\mathcal{L}(f)(t) = \int_0^\infty f(x)e^{-tx}\,dx$. A formal calculation, which is valid under mild hypotheses, shows that

$$\forall\, t > 0, \quad \mathcal{L}(\mathcal{L}(f))(t) = -\pi\mathcal{H}(f)(-t).$$

See [39], Problem 2.57, for a role of $\mathcal{H}$ in signal processing as related to the Paley–Wiener logarithmic integral theorem [363].

B.5 Fourier series

In Section 3.3 we defined the Fourier series of a function $f \in L^1_m(\mathbb{T})$, where $\mathbb{T} = \mathbb{R}/\mathbb{Z}$. We now elaborate.

Definition B.5.1. Fourier series

a. Let $f \in L^1_m([-\Omega, \Omega))$. The *Fourier series* of f is the series

$$\forall\, x \in [-\Omega, \Omega), \quad S(f)(x) = \sum_{n \in \mathbb{Z}} c_n e^{-\pi i n x/\Omega}, \tag{B.13}$$

where the coefficients are defined as

$$c_n = \frac{1}{2\Omega}\int_{-\Omega}^{\Omega} f(x)e^{\pi i n x/\Omega}\,dx.$$

The c_n are the *Fourier coefficients* of f.

b. If the sequence $c = \{c_n : n \in \mathbb{Z}\}$ satisfies $\sum_{n \in \mathbb{Z}} |c_n| < \infty$, then the right-hand side of (B.13) is well defined, and we say that it is the *Fourier transform* $\hat{c}$ of $\{c_n : n \in \mathbb{Z}\}$.

In Definition B.5.1, instead of thinking about functions defined on a finite interval, we can think of functions f that are 2Ω-periodic, Lebesgue measurable functions that are integrable on all compact subsets of $\mathbb{R}$, i.e., locally integrable functions. The set of all locally integrable functions is denoted by $L^1_{\mathrm{loc}}(\mathbb{R})$, e.g., Section 5.5. If $\Omega > 0$ and $f \in L^1_{loc}(\mathbb{R})$ is 2Ω-periodic, then we write $f \in L^1(\mathbb{T}_{2\Omega})$. We identify $\mathbb{T}_{2\Omega}$ with $\mathbb{R}/2\Omega\mathbb{Z}$. The L^1*-norm* of $f \in L^1(\mathbb{T}_{2\Omega})$ is

$$\|f\|_{L^1(\mathbb{T}_{2\Omega})} = \frac{1}{2\Omega}\int_{-\Omega}^{\Omega} |f(x)|\,dx.$$

If f is a 2Ω-periodic, Lebesgue measurable function, and $f^2 \in L^1(\mathbb{T}_{2\Omega})$, then we write $f \in L^2(\mathbb{T}_{2\Omega})$. The L^2-*norm* of $f \in L^2(\mathbb{T}_{2\Omega})$ is

$$\|f\|_{L^2(\mathbb{T}_{2\Omega})} = \left(\frac{1}{2\Omega} \int_{-\Omega}^{\Omega} |f(x)|^2 \, dx \right)^{1/2}.$$

By Hölder's inequality, $L^2(\mathbb{T}_{2\Omega}) \subseteq L^1(\mathbb{T}_{2\Omega})$, and we have

$$\forall\, f \in L^2(\mathbb{T}_{2\Omega}), \quad \|f\|_{L^1(\mathbb{T}_{2\Omega})} \le \|f\|_{L^2(\mathbb{T}_{2\Omega})}.$$

Definition B.5.2. Fourier transform on $L^1(\mathbb{T}_{2\Omega})$
If $f \in L^1(\mathbb{T}_{2\Omega})$, its *Fourier transform* is the sequence $\hat{f} = \{\hat{f}(n) : n \in \mathbb{Z}\}$, where

$$\forall\, n \in \mathbb{Z}, \quad \hat{f}(n) = \frac{1}{2\Omega} \int_{-\Omega}^{\Omega} f(x) e^{-\pi i n x/\Omega} \, dx.$$

In Appendix B.1, we defined the Fourier transform of $f \in L^1_m(\mathbb{R})$ to be a function on $\widehat{\mathbb{R}} = \mathbb{R}$. Here we have two dual settings. First, for a sequence $c = \{c_n : n \in \mathbb{Z}\}$ such that $\sum |c_n| < \infty$, the Fourier transform of c is defined on $\mathbb{T}_{2\Omega}$. Second, for $f \in L^1(\mathbb{T}_{2\Omega})$, the Fourier transform of f is defined on $\mathbb{Z}$. Mathematically, $\mathbb{R}$ and $\widehat{\mathbb{R}}$ are *locally compact abelian groups* (LCAGs) that are *dual*, in a technical sense, to each other; see Appendix B.9 and Appendix B.10. Similarly, the discrete LCAG $\mathbb{Z}$ is the dual group of the compact LCAG $\mathbb{T}_{2\Omega}$, and vice versa, e.g., [404], [155], [33].

We shall use the Riemann–Lebesgue lemma, Theorem 3.6.4, to verify DIRICHLET's fundamental theorem, which provides sufficient conditions on a function $f \in L^1(\mathbb{T}_{2\Omega})$ so that $S(f)(x_0) = f(x_0)$ for a given point x_0. The following ingenious proof is due to PAUL CHERNOFF [101]; cf. [323] and the classical proof as found in [524]. The Dirichlet theorem for Fourier series naturally preceded the analogous inversion theorem for Fourier transforms, as formulated in Theorem B.1.2 and Section B.3.

Theorem B.5.3. Dirichlet theorem
If $f \in L^1(\mathbb{T}_{2\Omega})$ and f is differentiable at x_0, then $S(f)(x_0) = f(x_0)$ in the sense that

$$\lim_{M,N\to\infty} \sum_{n=-M}^{N} c_n e^{-\pi i n x_0/\Omega} = f(x_0),$$

where $c = \{c_n : n \in \mathbb{Z}\}$ is the sequence of Fourier coefficients of f.

Proof. i. Without loss of generality, assume $x_0 = 0$ and $f(x_0) = 0$. In fact, if $f(x_0) \neq 0$, then consider the function $f - f(x_0)$ instead of f, which is also an element of $L^1(\mathbb{T}_{2\Omega})$, and then translate this function to the origin.

ii. Since $f(0) = 0$ and $f'(0)$ exists, we can verify that

$$g(x) = \frac{f(x)}{e^{-\pi i x/\Omega} - 1}$$

is bounded in some interval centered at the origin. To see this note that

$$g(x) = \frac{f(x)}{x} \frac{1}{\sum_{j=1}^{\infty}(-\pi i/\Omega)^j (1/j!) x^j},$$

and hence $g(x)$ is close to $-\Omega f'(0)/(\pi i)$ in a neighborhood of the origin.

The boundedness near the origin, coupled with integrability of f on $\mathbb{T}_{2\Omega}$, yields the integrability of g on $\mathbb{T}_{2\Omega}$. Therefore, since $f(x) = g(x)(e^{-\pi i x/\Omega} - 1)$, we compute $c_n = d_{n+1} - d_n$, where $d = \{d_n : n \in \mathbb{Z}\}$ is the sequence of Fourier coefficients of g. Thus, the partial sum $\sum_{n=-M}^{N} c_n e^{-\pi i n x_0/\Omega}$ is the telescoping series

$$\sum_{n=-M}^{N} (d_{n+1} - d_n) = d_{N+1} - d_{-M}.$$

Consequently, we can apply the Riemann–Lebesgue lemma to the sequence of Fourier coefficients of g to obtain

$$\lim_{M,N \to \infty} \sum_{n=-M}^{N} c_n e^{-\pi i n x/\Omega} = 0. \qquad \square$$

With regard to Theorem B.5.3, we can further assert that *if* $f \in BV_{\mathrm{loc}}(\mathbb{R})$, f *is* 2Ω*-periodic, and* f *is continuous on a closed subinterval* $I \subseteq \mathbb{T}_{2\Omega}$, *then*

$$\sum_{n=-N}^{N} \hat{f}(n) e^{-\pi i n x/\Omega}$$

converges uniformly to f *on* I; cf. [524], Volume I, pages 57–58. The *Dirichlet theorem* and this version of it for intervals of continuity are often referred to as the *Dirichlet–Jordan test.*

B.6 The $L^1(\mathbb{T}_{2\Omega})$ theory of Fourier series

Definition B.6.1. $A(\mathbb{T}_{2\Omega})$ and $A(\mathbb{Z})$

a. If $c = \{c_n : n \in \mathbb{Z}\}$ is a sequence such that $\sum |c_n| < \infty$ and if $\Omega > 0$, then $\hat{c}$ is an *absolutely convergent Fourier series*, and the space of such series is denoted by $A(\mathbb{T}_{2\Omega})$. By definition, the norm of $\hat{c} \in A(\mathbb{T}_{2\Omega})$ is

$$\|\hat{c}\|_{A(\mathbb{T}_{2\Omega})} = \|c\|_1 = \sum_{n \in \mathbb{Z}} |c_n|.$$

We have the proper inclusions

$$A(\mathbb{T}_{2\Omega}) \subseteq C(\mathbb{T}_{2\Omega}) \subseteq L^\infty(\mathbb{T}_{2\Omega}) \subseteq L^2(\mathbb{T}_{2\Omega}) \subseteq L^1(\mathbb{T}_{2\Omega}).$$

b. Let $A(\mathbb{Z})$ be the space of all sequences $c = \{c_n : n \in \mathbb{Z}\}$ such that $\hat{c} \in L^1(\mathbb{T}_{2\Omega})$. The space $\ell^2(\mathbb{Z})$ of square-summable sequences (defined in Section 5.5) is a subset of $A(\mathbb{Z})$.

Example B.6.2. Trigonometric series

a. The Riemann–Lebesgue lemma asserts that if $f \in L^1(\mathbb{T}_{2\Omega})$, then

$$\lim_{n\to\pm\infty} c_n = 0,$$

where $c = \{c_n : n \in \mathbb{Z}\}$ is the sequence of Fourier coefficients of f. On the other hand, suppose we are give a trigonometric series $\sum c_n e^{-2\pi i n x}$ for which $\lim_{n\to\pm\infty} c_n = 0$. Then it is not necessarily true that this series is the Fourier series of some function $f \in L^1(\mathbb{T}_{2\Omega})$. Indeed, the trigonometric series,

$$\sum_{n=2}^{\infty} \frac{\sin(\pi n x/\Omega)}{\log(n)},$$

converges pointwise for each $x \in \mathbb{R}$, but it is *not* the Fourier series of an element $f \in L^1(\mathbb{T}_{2\Omega})$. This is an analogue of Example B.2.5.

b. Let $\Omega > 0$. Then the series,

$$\sum_{n=3}^{\infty} \frac{\sin(\pi n x/\Omega)}{n\log(n)},$$

converges uniformly on $\mathbb{R}$ to a function $f \in C(\mathbb{T}_{2\Omega}) \setminus A(\mathbb{T}_{2\Omega})$.

Remark. If $f \in L^1(\mathbb{T}_{2\Omega})$ and if we let

$$S_N(f)(x) = \sum_{n=-N}^{N} c_n e^{-\pi i n x/\Omega},$$

where $c = \{c_n : n \in \mathbb{Z}\}$ is the sequence of Fourier coefficients of f, then it is known that the desirable statement,

$$\lim_{N\to\infty} \|S_N(f) - f\|_{L^1(\mathbb{T}_{2\Omega})} = 0, \tag{B.14}$$

is *not* true for all $f \in L^1(\mathbb{T}_{2\Omega})$. On the other hand, a sequence $\{f_n : n = 1, \ldots\} \subseteq L^1(\mathbb{T}_{2\Omega})$ converges to $f \in L^1(\mathbb{T}_{2\Omega})$ *weakly*, i.e.,

$$\forall\, g \in L^\infty(\mathbb{T}_{2\Omega}), \quad \lim_{n\to\infty} \frac{1}{2\Omega} \int_{-\Omega}^{\Omega} (f_n(x) - f(x))g(x)\, dx = 0,$$

cf. Definition 6.3.1, if and only if

$$\lim_{n\to\infty} \int_A (f_n(x) - f(x))\, dx = 0 \tag{B.15}$$

for every Lebesgue measurable set $A \subseteq \mathbb{T}_{2\Omega}$. If we have weak convergence, or, equivalently, (B.15), then it follows from Theorem 6.5.1 that (B.14) is true for $S_n(f) = f_n$ if $\{f_n : n \in \mathbb{N}\}$ converges to f in measure.

Definition B.6.3. Convolution
a. Let $f, g \in L^1(\mathbb{T}_{2\Omega})$. The *convolution* of f and g, denoted by $f * g$, is

$$f * g(x) = \frac{1}{2\Omega} \int_{-\Omega}^{\Omega} f(x-y)g(y)\, dy = \frac{1}{2\Omega} \int_{-\Omega}^{\Omega} f(y)g(x-y)\, dy;$$

cf. Problem 3.28. As with $L^1_m(\mathbb{R})$, see Problem 3.5, it is not difficult to prove that $f * g \in L^1(\mathbb{T}_{2\Omega})$ and

$$\forall\, f, g \in L^1(\mathbb{T}_{2\Omega}), \quad \|f * g\|_{L^1(\mathbb{T}_{2\Omega})} \leq \|f\|_{L^1(\mathbb{T}_{2\Omega})} \|g\|_{L^1(\mathbb{T}_{2\Omega})}.$$

b. With the operations of addition and convolution, $L^1(\mathbb{T}_{2\Omega})$ is a *commutative algebra*, i.e., it is a complex vector space under addition, and convolution is distributive with respect to addition, as well as associative and commutative; $L^1(\mathbb{T}_{2\Omega})$ is a commutative Banach algebra when normed by $\|...\|_{L^1(\mathbb{T}_{2\Omega})}$.

Proposition B.6.4. *Let $f, g \in L^1(\mathbb{T}_{2\Omega})$, with corresponding sequences $c = \{c_n : n \in \mathbb{Z}\}$, $d = \{d_n : n \in \mathbb{Z}\} \in A(\mathbb{Z})$ of Fourier coefficients. Then the sequence $c \cdot d = \{c_n d_n : n \in \mathbb{Z}\} \in A(\mathbb{Z})$ is the sequence of Fourier coefficients of $f * g \in L^1(\mathbb{T}_{2\Omega})$, i.e.,*

$$\forall\, n \in \mathbb{Z}, \quad c_n d_n = \frac{1}{2\Omega} \int_{-\Omega}^{\Omega} f * g(x) e^{\pi i n x/\Omega}\, dx.$$

Definition B.6.5. Approximate identity
An *approximate identity* is a family $\{K_N : N = 1, \ldots\} \subseteq L^1(\mathbb{T}_{2\Omega})$ of functions with the following properties:

i. $\forall\, N = 1, \ldots,\ \frac{1}{2\Omega} \int_{-\Omega}^{\Omega} K_N(x)\, dx = 1$,
ii. $\exists\, M > 0$ such that $\forall\, N = 1, \ldots,\ \|K_N\|_{L^1(\mathbb{T}_{2\Omega})} \leq M$,
iii. $\forall\, \delta \in (0, \Omega],\ \lim_{N\to\infty} \frac{1}{2\Omega} \int_{\delta \leq |x| \leq \Omega} |K_N(x)|\, dx = 0$.

The following theorem is the analogue for Fourier series of Theorem B.3.5 and Proposition B.3.6; see Problem 3.28.

Theorem B.6.6. Approximation and uniqueness
a. *Let $f \in C(\mathbb{T}_{2\Omega})$, and let $\{K_N : N = 1, \ldots\} \subseteq L^1(\mathbb{T}_{2\Omega})$ be an approximate identity. Then*

$$\lim_{N\to\infty} \|f - f * K_N\|_{L^\infty(\mathbb{T}_{2\Omega})} = 0.$$

b. *Let $f \in L^1(\mathbb{T}_{2\Omega})$, and let $\{K_N : N = 1, \ldots\} \subseteq L^1(\mathbb{T}_{2\Omega})$ be an approximate identity. Then*

$$\lim_{N\to\infty} \|f - f * K_N\|_{L^1(\mathbb{T}_{2\Omega})} = 0;$$

cf. Theorem B.3.5a.

c. *Let $f \in L^1(\mathbb{T}_{2\Omega})$, and let $c = \{c_n : n \in \mathbb{Z}\}$ be its sequence of Fourier coefficients. If $c_n = 0$ for each $n \in \mathbb{Z}$, then $f = 0$ m-a.e.*

B.7 The Stone–Weierstrass theorem

In 1904, FEJÉR proved the following fundamental approximation theorem using the fact that the Fejér kernel $\{W_N : N \geq 0\}$, defined in Problem 3.28, is an approximate identity, without explicitly defining the general concept of an approximate identity.

Theorem B.7.1. Fejér theorem
a. *Let $f \in C(\mathbb{T}_{2\Omega})$. Then*

$$\lim_{N\to\infty} \|f - f * W_N\|_{L^\infty(\mathbb{T}_{2\Omega})} = 0. \tag{B.16}$$

b. *Let $f \in L^p_m(\mathbb{T}_{2\Omega})$, $1 \leq p < \infty$. Then*

$$\lim_{N\to\infty} \|f - f * W_N\|_{L^p(\mathbb{T}_{2\Omega})} = 0.$$

Theorem B.7.1 and the Weierstrass approximation theorem are conceptually closely related. We note that WEIERSTRASS' original proof (1885) also used a convolution approximate identity argument [81], pages 269–273.

Theorem B.7.2. Weierstrass approximation theorem
Let $f : [\alpha, \beta] \to \mathbb{C}$ be a continuous function. There is a sequence $\{Q_N : N = 1, \dots\}$ of polynomials for which

$$\lim_{N\to\infty} \|f - Q_N\|_{L^\infty_m([\alpha,\beta])} = 0. \tag{B.17}$$

Equation (B.17) can be derived from (B.16) in the following way. By translation we can take $f \in C([-\Omega, \Omega])$. Next choose c such that $g(-\Omega) = g(\Omega)$, where $g(x) = f(x) - c$ for $x \in [-\Omega, \Omega]$. In fact, let

$$c = \frac{f(\Omega) - f(-\Omega)}{2\Omega}.$$

Apply (B.16) to g considered as an element of $C(\mathbb{T}_{2\Omega})$. Finally, uniformly approximate the trigonometric polynomials $g * W_N$ on $[-\Omega, \Omega]$ by polynomial approximants of their Taylor series expansions.

A monumental journey in effective abstraction was undertaken by MARSHALL H. STONE in 1937 [454], and has resulted in the Stone–Weierstrass theorem.

STONE's own works [455], [456] are a readable paradigm of the creative inquiry required to formulate fundamental abstract ideas resulting from and embedded in classical results.

In order to state a useful version of STONE's theorem, let X be a locally compact Hausdorff space, and note that the complex Banach space $C_0(X)$ (defined in Section A.2) is also in *algebra*. This means that $C_0(X)$ is not only a vector space, but that it is closed under pointwise multiplication, and that the commutative, associative, and distributive laws hold. A subset $S \subseteq C_0(X)$ is *separating* if

$$\forall\, x_1, x_2 \in X,\ \exists\, f \in S \text{ such that } \quad f(x_1) \neq f(x_2).$$

Theorem B.7.3. Stone–Weierstrass theorem
Let X be a locally compact Hausdorff space, and let $S \subseteq C_0(X)$ be a separating subalgebra with the following properties:
i. $\forall\, f \in S,\ \overline{f} \in S,$
ii. $\forall\, x \in X,\ \exists f \in S$ *such that* $f(x) \neq 0$.
Then S is dense in $C_0(X)$.

If X is compact we refer to [329], pages 11–12, for a brief proof based on [455] by STONE, and to [235], pages 90–99, for a more complete treatment.

B.8 The $L^2(\mathbb{T}_{2\Omega})$ theory of Fourier series

Recall that according to Definition A.12.1 an *orthonormal basis* for $L^2(\mathbb{T}_{2\Omega})$ is an orthonormal sequence $\{e_n : n \in \mathbb{Z}\} \subseteq L^2(\mathbb{T}_{2\Omega})$ such that

$$\forall\, f \in L^2(\mathbb{T}_{2\Omega}),\ \exists\, \{c_n : n \in \mathbb{Z}\} \subseteq \mathbb{C} \text{ such that } \quad f = \sum_{n \in \mathbb{Z}} c_n e_n \ \text{ in } \ L^2(\mathbb{T}_{2\Omega}).$$

In fact, $L^2(\mathbb{T}_{2\Omega})$ is a Hilbert space with inner product defined in Example A.12.3*a*, where the following result was also stated.

Proposition B.8.1. *The sequence $\{e_n(x) = e^{-\pi i n x/\Omega} : n \in \mathbb{Z}\} \subseteq L^2(\mathbb{T}_{2\Omega})$ is an orthonormal basis for $L^2(\mathbb{T}_{2\Omega})$.*

Theorem B.8.2 is a special case of Theorem A.12.6, once it is proved that $L^2(\mathbb{T}_{2\Omega})$ is a Hilbert space. The following proof is self-contained.

Theorem B.8.2. Parseval formula
Let $f, g \in L^2(\mathbb{T}_{2\Omega})$ with corresponding sequences $c = \{c_n : n \in \mathbb{Z}\}$, $d = \{d_n : n \in \mathbb{Z}\}$ of Fourier coefficients of f and g. Then $c, d \in \ell^2(\mathbb{Z})$ and

$$\frac{1}{2\Omega} \int_{\Omega}^{\Omega} f(x)\overline{g(x)}\, dx = \sum_{n \in \mathbb{Z}} c_n \overline{d_n};$$

and, in particular,

$$\frac{1}{2\Omega} \int_{\Omega}^{\Omega} |f(x)|^2\, dx = \sum_{n \in \mathbb{Z}} |c_n|^2.$$

Proof. We first observe that for any $f \in L^2(\mathbb{T}_{2\Omega})$ and for any $N \in \mathbb{N}$,

$$0 \leq \|f - S_N(f)\|^2_{L^2(\mathbb{T}_{2\Omega})} = \|f\|^2_{L^2(\mathbb{T}_{2\Omega})} - \sum_{|n| \leq N} |c_n|^2,$$

which implies

$$\sum_{|n| \leq N} |c_n|^2 \leq \|f\|^2_{L^2(\mathbb{T}_{2\Omega})}. \tag{B.18}$$

Further, if $N > M$, then

$$\|S_N(f) - S_M(f)\|^2_{L^2(\mathbb{T}_{2\Omega})} = \sum_{M<|n|\le N} |c_n|^2,$$

and so, by (B.18), $\{S_N(f) : N \in \mathbb{N}\}$ is a Cauchy sequence in $L^2(\mathbb{T}_{2\Omega})$. Thus, $\sum c_n e^{-\pi inx/\Omega}$ converges to some $h \in L^2(\mathbb{T}_{2\Omega})$ since $L^2(\mathbb{T}_{2\Omega})$ is complete. Now, for any $f \in L^2(\mathbb{T}_{2\Omega})$ and corresponding h we have, by Proposition B.8.1, that

$$\begin{aligned}&\frac{1}{2\Omega}\int_{-\Omega}^{\Omega}(f(x) - h(x))e^{\pi inx/\Omega}\,dx\\ &\quad = c_n - \lim_{N\to\infty}\sum_{|m|\le N} c_m \frac{1}{2\Omega}\int_{-\Omega}^{\Omega} e^{-\pi i(m-n)x/\Omega}\,dx = 0.\end{aligned}$$

Therefore, $f = h$ *m-a.e.* in $\mathbb{T}_{2\Omega}$, or, in other words,

$$\lim_{N\to\infty}\|f - S_N(f)\|_{L^2(\mathbb{T}_{2\Omega})} = 0.$$

Using this fact, we obtain

$$\begin{aligned}\frac{1}{2\Omega}\int_{\Omega}^{\Omega} f(x)\overline{g(x)}\,dx &= \lim_{N\to\infty}\frac{1}{2\Omega}\int_{\Omega}^{\Omega} S_N(f)(x)\overline{S_N(g)(x)}\,dx\\ &= \lim_{N\to\infty}\sum_{|m|,|n|\le N} c_m\overline{d_n}\frac{1}{2\Omega}\int_{\Omega}^{\Omega} e^{-\pi i(m-n)x/\Omega}\,dx\\ &= \sum_{n\in\mathbb{Z}} c_n\overline{d_n},\end{aligned}$$

where the last equality again follows from Proposition B.8.1. □

The inequality (B.18) is called the *Bessel inequality*; cf. Theorem A.12.5*a*. It implies that if $f \in L^2(\mathbb{T}_{2\Omega})$, then the sequence $c = \{c_n : n \in \mathbb{Z}\}$ of Fourier coefficients of f is square summable, i.e.,

$$\sum_{n\in\mathbb{Z}} |c_n|^2 < \infty.$$

F. RIESZ' formulation of the *Riesz–Fischer theorem* completes the picture as follows; see Section 5.6.2.

Theorem B.8.3. $l^2(\mathbb{Z}) \longrightarrow L^2(\mathbb{T}_{2\Omega})$
There is a unique linear bijection $\mathcal{F} : l^2(\mathbb{Z}) \longrightarrow L^2(\mathbb{T}_{2\Omega})$ with the following properties:

a. $\forall\, c \in l^2(\mathbb{Z})$, $\|c\|_{l^2(\mathbb{Z})} = \|\mathcal{F}(c)\|_{L^2(\mathbb{T}_{2\Omega})}$,
b. $\forall\, f \in L^2(\mathbb{T}_{2\Omega})$, $\mathcal{F}^{-1}(f)$ *is the sequence of Fourier coefficients of* f.

B.9 Haar measure

An additive group G with a locally compact Hausdorff topology is a *locally compact group* if the function $G \times G \to G$, $(x, y) \mapsto x - y$, is continuous. For significant classical treatises on locally compact groups, that were begun in the 1930s, see [375] and [497] by LEV S. PONTRYAGIN and ANDRÉ WEIL, respectively.

A complex vector space X that is also a topological space is a *topological vector space* if the functions $\mathbb{C} \times X \to X$, $(c, x) \mapsto cx$, and $X \times X \to X$, $(x, y) \mapsto x + y$, are continuous. Let X be a Hausdorff topological vector space. Then $K \subseteq X$ is *absorbing* if

$$\forall\, x \in X, \exists\, \varepsilon > 0 \quad \text{such that} \quad 0 < |c| \leq \varepsilon \implies cx \in K.$$

Moreover, $K \subseteq X$ is *balanced* if

$$|c| \leq 1 \implies cK \subseteq K.$$

In both cases, $c \in \mathbb{C}$.

A fundamental result in harmonic analysis is that *if G is a locally compact group then there is a Borel measure m_G on G such that*

$$\forall\, B \in \mathcal{B}(G) \text{ and } \forall\, x \in G, \quad m_G(B) = m_G(B + x),$$

where $B + x = \{y + x : y \in B\}$. In this case m_G is a *right Haar measure* on G, and, when $B + x$ is replaced by $x + B$, m_G is a *left Haar measure* on G. Thus, the crucial feature of translation invariance for Lebesgue measure on the line extends to locally compact groups. If every right Haar measure is a left Haar measure on a locally compact group G, and vice versa, then G is *unimodular*. Every compact group and every locally compact abelian group (LCAG) is unimodular.

We shall prove the existence of a Haar measure on G *compact* and *abelian* using the Markov–Kakutani fixed-point theorem, which we shall also prove. It is easy to show that there is only one such m_G that ensures that $m_G(G) = 1$ for G compact; see Theorem B.9.3.

Theorem B.9.1. Markov–Kakutani fixed-point theorem
Let X be a Hausdorff topological vector space, take a compact and convex set $K \subseteq X$, and let $\{T_\alpha\}$ be a family of continuous linear maps $T_\alpha : X \to X$ that satisfies

$$\forall\, \alpha, \quad T_\alpha(K) \subseteq K$$

and

$$\forall\, \alpha, \beta \quad T_\alpha \circ T_\beta = T_\beta \circ T_\alpha. \tag{B.19}$$

Then there is $k \in K$ such that

$$\forall\, \alpha, \quad T_\alpha(k) = k.$$

Proof. Let

$$T_\alpha^{(n)} = \frac{Id + T_\alpha + \cdots + T_\alpha^{n-1}}{n},$$

where $Id : X \to X$ is the identity map and T_α^j is the composition $T_\alpha \circ \cdots \circ T_\alpha$, j times. Clearly, each $T_\alpha^{(n)} : X \to X$ is continuous and linear. We write $T = \{T_\alpha^n\}$ and let $\tilde{T}$ be the set of finite products, under composition, of elements from T. Note that for each $u \in \tilde{T}$, $u(K) \subseteq K$. This follows by the convexity of K. Thus, $u \circ v(K) \subseteq u(K)$ for $u, v \in \tilde{T}$. Further, because of (B.19), $u \circ v = v \circ u$ for $u, v \in \tilde{T}$; and hence $v \circ u(K) \subseteq v(K)$ implies $u \circ v(K) \subseteq v(K)$.

Let $\tilde{K} = \bigcap\{u(K) : u \in \tilde{T}\}$ and note that

$$\forall\, u, v \in \tilde{T}, \quad u(K) \cap v(K) \neq \emptyset;$$

in fact,

$$u \circ v(K) = u \circ v(K) \cap u \circ v(K) \subseteq u(K) \cap u \circ v(K) \subseteq u(K) \cap v(K).$$

Since K is compact, $u(K)$ is compact for each $u \in \tilde{T}$. Consequently, because $u(K) \cap v(K) \neq \emptyset$ for $u, v \in \tilde{T}$, since each $u(K) \subseteq K$, and because K is compact, there is $k \in K$ such that for all $u \in \tilde{T}$, $k \in u(K)$. Therefore, $k \in \tilde{K}$.

In particular, for each α and n, $k \in T_\alpha^{(n)}(K)$, so that there is $t \in K$ (depending on α and n) such that $T_\alpha^{(n)}(t) = k$. Thus,

$$T_\alpha(k) = \frac{T_\alpha(t) + \cdots + T_\alpha^n(t)}{n},$$

and hence

$$T_\alpha(k) - k = -\frac{t}{n} + \frac{T_\alpha^n(t)}{n} \tag{B.20}$$

since $T_\alpha^{(n)}(t) = k$.

Note that the function $K \times K \to X$, $(s, t) \mapsto s - t$, takes $K \times K$ into a compact set E.

Because of (B.20) and the fact that $T_\alpha^n(t) \in K$ we have

$$T_\alpha(k) - k = \frac{T_\alpha^n(t) - t}{n} \in \frac{1}{n}E. \tag{B.21}$$

We shall show that

$$\bigcap_{n=1}^{\infty} \left(\frac{1}{n}E\right) = \{0\}, \tag{B.22}$$

so that since (B.21) is true for each n, $T_\alpha(k) - k = 0$, and we are done.

Let $V \subseteq U$ be a balanced set for which $0 \in \text{int } V$. By the definition of a topological vector space there is a balanced and absorbing open set $W \subseteq X$ containing 0 such that $W + W \subseteq V$. $\{x + W : x \in X\}$ is an open cover of E, so that by the compactness there are points x_i, $i = 1, \ldots, m$, such that

$$E \subseteq \bigcup_{i=1}^{m} (x_i + W).$$

Since W is absorbing, we have $rx_i \in W$ for $i = 1, \ldots, m$ and for some $r \in (0,1]$. Therefore,

$$r(E \cap (x_i + W)) \subseteq W + rW, \quad i = 1, \ldots, m. \tag{B.23}$$

Because the right-hand side of (B.23) is independent of i,

$$rE = \bigcup_{i=1}^{m} r(E \cap (x_i + W)) \subseteq W + rW.$$

Thus, $E \subseteq (1/r)V$, noting that $W + rW \subseteq W + W \subseteq V$; and so if $1/n < r$, then

$$\frac{1}{n}E \subseteq \frac{1}{nr}V \subseteq \frac{r}{r}V = V.$$

Consequently, $\bigcap_{n=1}^{\infty}(1/n)E \subseteq V$, so that since X is Hausdorff and V is arbitrary we have (B.22). □

Theorem B.9.2. Existence of Haar measure
Let G be a compact abelian group. Then there is a Haar measure m_G on G.

Proof. Let $M_1(G) = \{\mu \in M_b(G) : \|\mu\|_1 \leq 1\}$. By the Banach–Alaoglu theorem, $M_1(G)$ is weak* compact in $M_b(G)$. Let $M_1^+(G) = \{\mu \in M_1(G) : \mu(1) = 1\}$.

Note that μ is positive if $\mu \in M_1^+(G)$; to prove this we assume the opposite and obtain a contradiction using the fact that $\|\mu\|_1 = \mu(1)$; e.g., [69], page 101.

If $M_b(G)$ is taken with the weak* topology, then the map $M_b(G) \to \mathbb{C}$, $\mu \mapsto \mu(1)$, is continuous. Hence, $\{\mu \in M_b(G) : \mu(1) = 1\}$ is weak* closed. Thus, $M_1^+(G)$ is weak* compact. It is easy to check that $M_1^+(G)$ is convex. For $x \in G$ and $\mu \in M_b(G)$ we define the translation $\tau_x(\mu)$ as

$$\tau_x(\mu)(f) = \int f(y - x)\, d\mu(y),$$

where $f \in C(G)$. Then for each $x \in G$ we define the map $T_x : M_b(G) \to M_b(G)$, $\mu \mapsto \tau_x(\mu)$. Note that T_x is continuous with the weak* topology on both domain and range, linear, and

$$\forall\, x, y \in G, \quad T_x \circ T_y = T_{x+y} = T_y \circ T_x,$$

since G is abelian.

It is also elementary to check that, for each $x \in G$,

$$T_x(M_1^+(G)) \subseteq M_1^+(G).$$

Therefore, by Theorem B.9.1, there is $m_G \in M_1^+(G)$ such that $\tau_x(m_G) = m_G$, for all $x \in G$, the required translation invariance. Further, $\|m_G\|_1 = 1$, $m_G(1) = 1$, and m_G is positive. □

Remark. The question of existence of Haar measures goes back to SOPHUS LIE. Haar (1933) [209] proved the existence of translation-invariant measures on separable compact groups. As a matter of fact, HAAR credits ADOLF HURWITZ for a remark in [249] that is essential for proving the existence of a Haar measure on a Lie group; see [209]. Existence of a Haar measure on a general locally compact group was first proved by WEIL [497] and, later the same year, by HENRI CARTAN [94].

Besides the existence, it is natural to ask about the uniqueness of Haar measure on locally compact groups. This question was first answered by VON NEUMANN for compact groups [490]. VON NEUMANN later extended his own result to second countable locally compact groups [491] (employing a different technique). Here we prove the uniqueness of Haar measure in the simple context of an LCAG. We follow the proof of [404]; see also [72], [328], [354], [416], and [497]. For an short proof in the nonabelian case, which uses a notion of an approximate identity, we refer the reader to [262].

Theorem B.9.3. Uniqueness of Haar measure
Let G be an LCAG. Let m_G^1 and m_G^2 be two Haar measures on G. Then there exists $C > 0$ such that $m_G^1 = C m_G^2$.

Proof. Let $g_1 \in C_c^+(G)$ be chosen such that $\int_G g_1 \, dm_G^1 = 1$, and let $C = \int_G g_1(-x) \, dm_G^2(x)$. Then, for all $g_2 \in C_c^+(G)$, we have

$$\begin{aligned}
\int_G g_2 \, dm_G^2 &= \int_G g_1(x_1) \, dm_G^1(x_1) \int_G g_2(x_2) \, dm_G^2(x_2) \\
&= \int_G \left(\int_G g_2(x_2) \, dm_G^2(x_2) \right) g_1(x_1) \, dm_G^1(x_1) \\
&= \int_G \left(\int_G g_2(x_1 + x_2) \, dm_G^2(x_2) \right) g_1(x_1) \, dm_G^1(x_1) \\
&= \int_G \int_G g_1(x_1) g_2(x_1 + x_2) \, dm_G^2(x_2) \, dm_G^1(x_1) \\
&= \int_G \int_G g_1(x_1) g_2(x_1 + x_2) \, dm_G^1(x_1) \, dm_G^2(x_2) \\
&= \int_G \int_G g_1(y_1 - y_2) g_2(y_1) \, dm_G^1(y_1) \, dm_G^2(y_2) \\
&= \int_G \int_G g_1(y_1 - y_2) g_2(y_1) \, dm_G^2(y_2) \, dm_G^1(y_1) \\
&= \int_G \left(\int_G g_1(y_1 - y_2) \, dm_G^2(y_2) \right) g_2(y_1) \, dm_G^1(y_1) \\
&= \left(\int_G g_1(-y_2) \, dm_G^2(y_2) \right) \int_G g_2(y_1) \, dm_G^1(y_1) = C \int_G g_2 \, dm_G^1.
\end{aligned}$$

□

Example B.9.4. Examples of Haar measures
a. Let $G = \mathbb{R}^d$ be considered as an additive group. With topology defined by the Euclidean norm, G is an LCAG. Then the Lebesgue measure m^d is a Haar measure on G.

b. Let $G = \mathbb{R}\setminus\{0\}$ be the multiplicative group of real numbers taken with the topology induced from the Euclidean norm topology from $\mathbb{R}$. Then G is an LCAG. Further, $m_G(A) = \int_A (1/|x|)\, dm(x)$ defines a Haar measure on G.

c. Let $G = \mathbb{C}$ be the additive group of complex numbers with the usual topology of the complex plane. Then G is an LCAG, and $m_G(A) = \int_A dx\, dy = \int_A dm^2(z)$, $z = x + iy$, is a Haar measure. This is the product measure on $\mathbb{R} \times \mathbb{R}$, which is isomorphic to $\mathbb{C}$.

d. Let $G = \mathbb{C}\setminus\{0\}$, $0 \in \mathbb{C}$, be the multiplicative group of complex numbers with the induced topology from $\mathbb{C}$. Then G is an LCAG, and $m_G(A) = \int_A (1/|z|^2)\, dm^2(z)$ defines a Haar measure on G.

The group $\mathbb{T} = \mathbb{R}/\mathbb{Z}$, identified with $\{z \in \mathbb{C} : |z| = 1\}$, is a compact subgroup of G. Note that $m_G(\mathbb{T}) = 0$. However, there is a natural locally compact topology on $\mathbb{T}$ and corresponding Haar measure $m_\mathbb{T}$ when $\mathbb{T}$ is identified with the additive group $[0, 1)$ with addition defined mod 1. In this case, $\mathbb{T}$ and $m_\mathbb{T}$ can be identified with $([0, 1), \mathcal{M}, m)$. We mention this because of the discussion of the *dual group* in Appendix B.10.

e. Let $G = (\mathbb{R}^+ \setminus \{0\}) \times \mathbb{R}$ be a group with group action defined by $(a, b) \cdot (a', b') = (aa', ab' + b)$. Note that, in general, $(aa', ab' + b) \neq (a'a, a'b + b')$. G is a nonabelian locally compact group, and it is called the *affine group*. In this case, a left Haar measure is defined by $m_G^L(A) = \int_A (1/a^2)\; dm(a)\; dm(b)$; and a right Haar measure on G is defined by $m_G^R(A) = \int_A (1/a)\; dm(a)\; dm(b)$. They are distinct.

f. Let G be a discrete group. Then, $m_G = \sum_{g \in G} \delta_g$.

g. The general linear group $G = \mathrm{GL}(d, \mathbb{R})$ with matrix multiplication is a nonabelian locally compact group, whose topology is induced from the product topology on the d^2-direct product $\mathbb{R} \times \cdots \times \mathbb{R}$. The group G played a role in Section 8.7. In this case, $m_G(A) = \int_A |\det(X)|^{-d} dm^{d^2}(X)$ defines both a left and a right Haar measure on G.

It is well known that *every locally compact topological group is complete with its right uniform structure.* We shall verify this general statement for the metrizable setting.

Theorem B.9.5. Completeness of locally compact groups
Let G be a metrizable locally compact group with metric $\rho : G \times G \to [0, \infty)$. Then G is complete.

Proof. Since G is a metrizable topological space, it is completely regular and has a countable open basis at the origin $0 \in G$. Therefore, since G is a T_0 topological group, the metric can be taken to be right-translation-invariant, i.e.,

$$\forall\, x, y, z \in G, \quad \rho(x, y) = \rho(x + z, y + z);$$

see [233], pages 68–70, for a proof of translation invariance using a countable basis at the origin.

To prove that G is complete, let $\{x_n : n = 1, \dots\} \subseteq G$ be Cauchy and find $x_0 \in G$ such that

$$\forall\, \varepsilon > 0,\ \exists\, N_\varepsilon > 0 \quad \text{such that} \quad \forall\, n \geq N_\varepsilon, \quad \rho(x_n, x_0) < \varepsilon. \tag{B.24}$$

By local compactness let C be a compact neighborhood of the origin and choose $r > 0$ such that $B(0,r) = \{x \in G : \rho(x, 0) < r\}$. Since $\{x_n : n = 1, \dots\}$ is Cauchy,

$$\exists\, N_r \text{ such that } \forall\, n \geq N_r, \quad \rho(x_n, x_{N_r}) = \rho(x_n - x_{N_r}, 0) < r,$$

and so $\{y_n = x_n - x_{N_r} : n \geq N_r\} \subseteq B(0,r) \subseteq C$. Thus, there are a subsequence $\{x_{m_n} : n = 1, \dots\}$ and $y_r \in G$ such that $\lim_{n\to\infty} \rho(y_{m_n}, y_r) = 0$. We set $x_0 = y_r + x_{N_r}$, and hence

$$0 = \lim_{n\to\infty} \rho(x_{m_n} - x_{N_r}, x_0 - x_{N_r}) = \lim_{n\to\infty} \rho(x_{m_n}, x_0). \tag{B.25}$$

We now verify (B.24). By (B.25),

$$\exists\, N_1 \text{ such that } \forall\, m_n > N_1, \quad \rho(x_{m_n}, x_0) < \frac{\varepsilon}{2}. \tag{B.26}$$

Also,

$$\exists\, N_2 \text{ such that } \forall\, m, n > N_2, \quad \rho(x_n, x_m) < \frac{\varepsilon}{2} \tag{B.27}$$

since $\{x_n : n = 1, \dots\}$ is Cauchy; and hence if $N_\varepsilon = \max(N_1, N_2)$, then (B.27) implies that

$$\forall\, n, m_k > N_\varepsilon, \quad \rho(x_n, x_{m_k}) < \frac{\varepsilon}{2}. \tag{B.28}$$

Therefore, by (B.26) and (B.28), if $n > N_\varepsilon$, then

$$\rho(x_n, x_0) \leq \rho(x_n, x_{m_k}) + \rho(x_{m_k}, x_0) < \varepsilon,$$

where we have chosen $x_{m_k} > N_\varepsilon$. □

Section B.9 is the beginning of the story of abstract harmonic analysis. RUDIN's [404] and EDWIN HEWITT and KENNETH A. ROSS' [233] treatises are a superb next step.

B.10 Dual groups and the Fourier analysis of measures

Let G be an LCAG. The collection Γ of all continuous group homomorphisms $\gamma : G \to \mathbb{T}$ is the *dual group* of G. With pointwise multiplication, Γ is an abelian group. The fact that Γ is a group follows from the definition of $\gamma_1 + \gamma_2$ by the rule $(\gamma_1 + \gamma_2)(x) = \gamma_1(x)\gamma_2(x)$, $x \in G$. Then if Γ is equipped with the "weak* topology" $\sigma(\Gamma, G)$, it becomes an LCAG. In fact, we have not defined the weak* topology on groups, so, to be precise, let us define a basis $\mathcal{B}$ for such a topology. For each compact $K \subseteq G$ and $r > 0$, let $N(K, r) = \{\gamma \in \Gamma : \forall\, x \in K,\ |1 - \gamma(x)| < r\}$. Set $\mathcal{B} = \{N(K, r) + \gamma : \gamma \in \Gamma,\ K \subseteq G$ is compact, and $r > 0\}$.

Given a Haar measure m_G on G, we define the *Fourier transform* of $f \in L^1_{m_G}(\mathcal{G})$ as

$$\forall\, \gamma \in \Gamma, \quad \mathcal{F}(f)(\gamma) = \int_G f(x)\overline{\gamma(x)}\, dm_G(x). \tag{B.29}$$

Then $\mathcal{F} : L^1_{m_G}(G) \to C_0(\Gamma)$ is a homomorphism, where multiplication is convolution in $L^1_{m_G}(G)$ and pointwise multiplication of functions in $C_0(\Gamma)$. This notion of Fourier transform can be extended to bounded Radon measures on G by means of the formula,

$$\forall\, \mu \in M_b(G), \quad \mathcal{F}(\mu)(\gamma) = \int_G \overline{\gamma(x)}\, d\mu(x). \tag{B.30}$$

Thus $\mathcal{F}(\mu)$ is a bounded uniformly continuous function on Γ.

Example B.10.1. Homomorphisms and transforms

a. If $G = \mathbb{R}$, then $L^1_m R) \subseteq M_b(\mathbb{R}) \subseteq \mathcal{S}'(\mathbb{R})$ with analogous inclusions for $G = \mathbb{R}^d$; see Definition 7.5.7. In this case, the dual group Γ can be identified with $\mathbb{R}$, which we denoted by $\widehat{\mathbb{R}}$ in Definition B.1.1. The identification can be made by proving that if $\gamma \in \Gamma$ is a continuous homomorphism $G \to \mathbb{T}$, then there is a real number $\xi = \xi_\gamma \in \widehat{\mathbb{R}}$ such that $\gamma(x) = e^{2\pi i x \xi}$. Clearly, if γ is of this form, then $\gamma \in \Gamma$. Thus, (B.29) is the same definition of Fourier transform as given in Definition B.1.1.

b. Let $\mathcal{A}$ be an algebra, such as $L^1_m(\mathbb{R})$ with multiplication defined by convolution. If $\mathcal{A}$ is a topological algebra such as the Banach algebra $L^1_m(\mathbb{R})$ with topology defined by the norm $\|...\|_1$, then we can consider the subspace $\mathcal{H} \subseteq \mathcal{A}'$ of continuous linear homomorphisms $\mathcal{A} \to \mathbb{C}$. This setting gives rise to a *transform* $\mathcal{T}$ for which there is an all-important *exchange formula*, $\mathcal{T}(f * g) = \mathcal{T}(f)\mathcal{T}(g)$, where $f, g \in \mathcal{A}$.

For example, if $\mathcal{A} = L^1_m(\mathbb{R})$ as in part *a*, then $\mathcal{H}$ can be defined by $\{e^{2\pi i x \xi} : \xi = \xi_\gamma$ and $\gamma \in \Gamma = \widehat{\mathbb{R}}\}$ and $\mathcal{T}$ is the usual Fourier transform $\mathcal{F}$ as given in Definition B.1.1.

This is the basis of the *Gelfand theory*, and this particular point of view is expanded in [33].

c. Another example of part *b* is the case of $\mathcal{E}'(\mathbb{R}) \subseteq \mathcal{S}'(\mathbb{R})$, the topological algebra of distributions having compact support with multiplication defined by convolution. When properly topologized, the dual of $\mathcal{E}'(\mathbb{R})$ is $C^\infty(\mathbb{R})$ [428], and $\mathcal{H}$ is defined by $\{e^{sx} : s \in \mathbb{C}\}$. In this case, the transform $\mathcal{T}$ is the classical bilateral *Laplace transform.*

d. Let $\mathbb{Z}_N$ be the set of integers $0, 1, \ldots, N-1$ under addition modulo N. Then $\mathbb{Z}_N$ is a commutative group with this definition of addition, and it is a compact abelian group when it is taken with the discrete topology, i.e., every element is defined to be an open set. Then, the dual group $\widehat{\mathbb{Z}_N}$ is $\mathbb{Z}_N$. Further, the Fourier transform defined by (B.29) is the classical *discrete Fourier transform* (DFT) $\mathcal{F}_N$ defined for $f \in L^1_c(\mathbb{Z}_N)$ by

$$\forall\, n = 0, 1, \ldots, N-1, \quad \mathcal{F}_N(f)[n] = \frac{1}{\sqrt{N}} \sum_{m=0}^{N-1} f[m] e^{-2\pi i mn/N}.$$

In Example A.13.5, we noted that $\mathcal{F}_N : L^2_c(\mathbb{Z}_N) \to L^2_c(\mathbb{Z}_N)$ is a unitary operator. Of course, in the setting of $\mathbb{Z}_N$, $L^1_c(\mathbb{Z}_N) = L^2_c(\mathbb{Z}_N) = \mathbb{C}^N$.

For a useful introduction to the DFT and its fast algorithm, the fast Fourier transform (FFT), see [39], Sections 3.8–3.10. For comprehensive treatments, see [74], [467], and [495].

Remark. One of the landmarks of SCHWARTZ' theory of distributions is that the Fourier transform can be defined in a meaningful and productive way. In fact, if $T \in \mathcal{S}'(\mathbb{R})$, then the Fourier transform $\widehat{T}$ of T is defined by the Parseval duality formula,

$$\forall\, f \in \mathcal{S}(\mathbb{R}), \quad \widehat{T}(\overline{\hat{f}}) = T(\overline{f}); \tag{B.31}$$

cf. Theorem 8.7.6. In this case, $\widehat{T} \in \mathcal{S}'(\mathbb{R})$. If $T = \mu \in M_b(\mathbb{R})$, then $\hat{\mu}$ defined by (B.31) is the same as $\mathcal{F}(\mu)$ defined by (B.30).

It should be pointed out that the theory of distributions and corresponding harmonic analysis is highly developed on locally compact groups.

Definition B.10.2. Positive definite functions
a. Let G be an LCAG with the dual group Γ. $R : \Gamma \to \mathbb{C}$ is *positive definite* if

$$\forall\, c_1, \ldots, c_n \in \mathbb{C} \text{ and } \forall\, \gamma_1, \ldots, \gamma_n \in \Gamma, \quad \sum_{j,k=1}^{n} c_j \overline{c_k} R(\gamma_j - \gamma_k) \geq 0.$$

b. Let $F \in L^2_{m_\Gamma}(\Gamma)$, and define the involution $\tilde{F}(\gamma) = \overline{F(-\gamma)}$. Then a straightforward calculation shows that $R = F * \tilde{F}$ is a continuous positive definite function.

c. Another elementary calculation shows that *if $\mu \in M_b(G)$ is positive, then $\mathcal{F}(\mu)$ is a continuous positive definite function.* The Bochner theorem, which we now state, is the converse. It was first published by GUSTAV HERGLOTZ (1911) for $\Gamma = \mathbb{Z}$. BOCHNER (1933) proved it for $\Gamma = \widehat{\mathbb{R}}$ and WEIL published the proof for any LCAG. For positive definite distributions and their extensions, see [189].

Theorem B.10.3. Bochner theorem
Let R be a continuous positive definite function on Γ. Then there exists a unique positive bounded Radon measure $\mu \in M_b(G)$ such that $R = \mathcal{F}(\mu)$ on Γ.

There are several accessible conceptually different proofs, e.g., [143], [276], [404]; cf. [428].

Remark. Because of our decomposition theorems for measures, the Bochner theorem allows us to assert that $F : \Gamma \to \mathbb{C}$ is of the form $\mathcal{F}(\mu)$, for some $\mu \in M_b(G)$, if and only if F is a finite linear combination of positive definite functions. This does not give a useful, implementable, *intrinsic* characterization of $\mathcal{F}(M_b(G))$. For perspective, such a characterization does exist for $\mathcal{F}(L^2_{m_G}(G))$ because of Plancherel's theorem, which is valid for LCAGs. This characterization for $\mathcal{F}(L^2_{m_G}(G))$ is that $F \in \mathcal{F}(L^2_{m_G}(G))$ *if and only if* $F \in L^2_{m_\Gamma}(\Gamma)$.

The problem for such an intrinsic characterization of $\mathcal{F}(M_b(G))$ or even $\mathcal{F}(L^1_m(\mathbb{R}))$ is *unsolved*. We know that $\mathcal{F}(L^1_m(\mathbb{R})) \subseteq C_0(\widehat{\mathbb{R}})$, but we do not have implementable conditions on $F \in C_0(\widehat{\mathbb{R}})$ as a function on $\widehat{\mathbb{R}}$ that are necessary and sufficient to assert that $F \in \mathcal{F}(L^1_m(\mathbb{R}))$.

Example B.10.4. Lévy continuity theorem
a. In Section 6.6.5 we defined a sequence $\{\mu_n : n = 1, \dots\} \subseteq M_1^+(\mathbb{R}^d)$ to be *tight* if

$$\forall\, \varepsilon > 0,\ \exists\, K_\varepsilon \subseteq \mathbb{R}^d, \text{ compact, such that } \forall\, n = 1, \dots, \quad \mu_n(K_\varepsilon) > 1 - \varepsilon.$$

(Besides being defined in Section 6.6.5, the space $M_1^+(\mathbb{R}^d)$ was also defined, in a slightly different way, in Theorem B.9.2.) We then stated the Prohorov theorem, which describes the relation between convergence in the sense of Bernoulli and tight sequences. These ideas have far-reaching consequences for the convergence of probability distributions, for example, in dealing with probability measures on certain infinite-dimensional spaces. See [335] for the "down-to-earth" theory on $\mathbb{R}^d$. An important result in this area is the Lévy continuity theorem proved by PAUL LÉVY in 1925, e.g., [61].

b. A central part of the Lévy continuity theorem, which is also appropriate for this section, is the following result: *Given $\{\mu_n, \mu : n = 1, \dots\} \subseteq M_1^+(\mathbb{R}^d)$, then $\{\mu_n : n = 1, \dots\}$ converges to μ in the sense of Bernoulli if and only if*

$$\forall\, \gamma \in \widehat{\mathbb{R}}^d, \quad \lim_{n\to\infty} \hat{\mu}_n(\gamma) = \hat{\mu}(\gamma). \tag{B.32}$$

For each fixed $\gamma \in \widehat{\mathbb{R}}^d$, $e^{-2\pi i x\cdot\gamma} \in C_b(\mathbb{R}^d)$ as a function of x, and so (B.32) is a consequence of Bernoulli convergence. The proof that (B.32) implies convergence in the sense of Bernoulli is more substantial, and we give an outline.

c. Assume (B.32). We know that $\hat{\mu}$ is continuous on $\widehat{\mathbb{R}}^d$, although for this proof we need it to be continuous only at $0 \in \widehat{\mathbb{R}}^d$. By (B.32), $\hat{\mu}(0) = \lim_{n\to\infty} \hat{\mu}_n(0) = 1$. Let $\varepsilon > 0$. By the continuity,

$$\exists\, \delta > 0 \text{ such that } \forall\, \xi \in Q_\delta, \quad |1 - \hat{\mu}(\xi)| < \frac{\varepsilon}{2}, \tag{B.33}$$

where $Q_\delta = \{\xi \in \widehat{\mathbb{R}}^d : \forall\, j = 1, \ldots, d,\ |\xi_j| \leq \delta\}$. Let $A = \{\xi \in \widehat{\mathbb{R}}^d : \forall\, j = 1, \ldots, d,\ |x_j| \leq \frac{1}{\pi\delta}\}$. Thus, if $x = (x_1, \ldots, x_d) \in A^\sim$, then some $|x_k| > 1/(\pi\delta)$, and so

$$\left| \prod_{j=1}^{d} \frac{\sin(2\pi x_j \delta)}{2\pi x_j \delta} \right| < \frac{1}{2}.$$

Further,

$$\forall\, x \in \mathbb{R}^d, \quad \frac{1}{m^d(Q_\delta)} \int_{Q_\delta} e^{-2\pi i x\cdot\xi}\, d\xi = \prod_{j=1}^{d} \frac{\sin(2\pi x_j \delta)}{2\pi x_j \delta}.$$

Therefore, for each $n \geq 1$,

$$\begin{aligned}\frac{1}{m^d(Q_\delta)} \int_{Q_\delta} (1 - \hat{\mu}(\xi))\, d\xi &= \int_{\mathbb{R}^d} \left(1 - \frac{1}{m^d(Q_\delta)} \int_{Q_\delta} e^{-2\pi i x\cdot\xi}\, d\xi\right) d\mu_n(x) \\ &\geq \int_{A^\sim} \left(1 - \frac{1}{m^d(Q_\delta)} \int_{Q_\delta} e^{-2\pi i x\cdot\xi}\, d\xi\right) d\mu_n(x) \\ &> \frac{1}{2}\mu_n(A^\sim).\end{aligned} \tag{B.34}$$

Because of (B.32) and the fact that $\|\hat{\mu}_n\|_\infty \leq \|\mu\|_1 = 1$, we can use (B.33), (B.34), and LDC to assert that

$$\begin{aligned}\overline{\lim_{n\to\infty}}\, \mu_n(A^\sim) &\leq \overline{\lim_{n\to\infty}}\, \frac{2}{m^d(Q_\delta)} \int_{Q_\delta} (1 - \hat{\mu}_n(\xi))\, d\xi \\ &= \frac{2}{m^d(Q_\delta)} \int_{Q_\delta} (1 - \hat{\mu}(\xi))\, d\xi < \varepsilon.\end{aligned}$$

Consequently, we choose N such that $\mu_n(A^\sim) < \varepsilon$ for all $n > N$. Next, choose compact $K_\varepsilon \supseteq A$ for which $\mu_n(K_\varepsilon) \geq 1 - \varepsilon$ for $1 \leq n \leq N$. Thus, $\mu_n(K_\varepsilon) \geq 1 - \varepsilon$ for all $n \geq 1$, and $\{\mu_n : n = 1, \ldots\}$ is tight.

Thus, by part *b* of the Prohorov theorem (Theorem 6.6.5), there are $\nu \in M_1^+(\mathbb{R}^d)$ and a subsequence $\{\mu_{n_k} : k = 1, \ldots\} \subseteq \{\mu_n : n = 1, \ldots\}$ for which $\{\mu_{n_k} : k = 1, \ldots\}$ converges to ν in the sense of Bernoulli. Now, by the Fourier uniqueness theorem for measures and another application (perhaps "corollary" is more accurate) of the Prohorov theorem, it is elementary to prove that $\{\mu_n : n = 1, \ldots\}$ converges to ν in the sense of Bernoulli and that $\nu = \mu$. This completes the proof.

B.11 Radial Fourier transforms

To outline Fourier analysis on $\mathbb{R}$ (or $\mathbb{Z}$ or $\mathbb{T}_{2\Omega}$) and then to define an essentially qualitative theory on LCAGs, as we have done, may be said to have missed a very big point. There remains the quantitative Fourier analysis of $\mathbb{R}^d$, with unresolved geometric intricacies in topics such as spectral synthesis and with analytic mysteries beyond extensions of the Paley–Littlewood theory and the singular integral operators that generalize the Hilbert transform. Some difficulties are implied in Sections 8.8.2–8.8.5, but, notwithstanding twentieth-century accomplishments, e.g., [202], much is yet to be fathomed; see [167].

On the other hand, the radial theory, with its one degree of freedom, is highly developed. A function $f : \mathbb{R}^d \to \mathbb{C}$ is *radial* if $f(x) = g(\|x\|)$, for some $g : [0, \infty) \to \mathbb{C}$, where $\|x\| = (x_1^2 + \cdots + x_d^2)^{1/2}$. An equivalent definition was given in Example 8.6.2*c*. Thus, f is radial if and only if $f(S(x)) = f(x)$ for all $S \in \mathrm{SO}(d, \mathbb{R})$. We record some useful facts in this area. To whet one's appetite we give two useful examples, compute the Fourier transform of a radial function, and state a fascinating theorem.

Definition B.11.1. Fourier transform on $L^1_{m^d}(\mathbb{R}^d)$
The *Fourier transform* of $f \in L^1_{m^d}(\mathbb{R}^d)$ is the function F defined as

$$\forall\, \xi \in \widehat{\mathbb{R}}^d = \mathbb{R}^d, \quad F(\xi) = \int_{\mathbb{R}^d} f(x) e^{-2\pi i x\cdot\xi}\, dx.$$

Also, the *dilation* f_λ, $\lambda > 0$, is $f_\lambda(x) = \lambda^d f(\lambda x)$. In particular, $\int_{\mathbb{R}^d} f_\lambda(x)\, dx = \int_{\mathbb{R}^d} f(x)\, dx$.

Example B.11.2. The Gaussian in $\mathbb{R}^d$
Let $f(x) = e^{-\pi r\|x\|^2}$, $r > 0$, $x \in \mathbb{R}^d$. By Example B.1.8 and Fubini's theorem, the Fourier transform F of f is

$$F(\xi) = r^{-d/2} e^{-\pi\|\xi\|^2/r}.$$

Setting $G^d(x) = G(x_1)\cdots G(x_d)$, we have $\int_{\mathbb{R}^d} G^d(x)\, dx = 1$ and

$$(G^d_\lambda)\hat{\ }(\xi) = e^{-(\pi\|\xi\|/\lambda)^2}.$$

Example B.11.3. The Poisson function
We shall define and compute the Fourier transform of the natural generalization to $\mathbb{R}^d$ of the Poisson function on $\mathbb{R}$, as defined in Example B.1.5 and Example B.3.4.

a. We begin on $\mathbb{R}$, using either Theorem B.1.2 or Theorem B.3.7, to compute

$$e^{-2\pi r|\gamma|} = \frac{2r}{\pi}\int_0^\infty \frac{\cos(2\pi t\gamma)}{r^2 + t^2}\, dt. \tag{B.35}$$

If $r = 1$, we use (B.35) to obtain

$$e^{-2\pi|\gamma|} = \frac{2}{\pi}\int_0^\infty \cos(2\pi t\gamma)\left(\int_0^\infty e^{-u}e^{-ut^2}\,du\right)dt$$
$$= \frac{2}{\pi}\int_0^\infty e^{-u}\left(\frac{1}{2}\int_{-\infty}^\infty e^{-ut^2}e^{2\pi it\gamma}\,dt\right)du.$$

We apply Example B.1.8 to the right side, so that

$$e^{-2\pi|\gamma|} = \frac{1}{\sqrt{\pi}}\int_0^\infty \frac{e^{-u}}{\sqrt{u}}e^{-(\pi\gamma)^2/u}\,du.$$

Thus, for $r > 0$ and $\gamma \in \mathbb{R}$,

$$e^{-2\pi r|\gamma|} = \frac{1}{\sqrt{\pi}}\int_0^\infty \frac{e^{-u}}{\sqrt{u}}e^{-(\pi r\gamma)^2/u}\,du. \tag{B.36}$$

b. Next, we recall the definition of the *gamma function* Γ:

$$\forall\, u > 0, \quad \Gamma(u) = \int_0^\infty x^{u-1}e^{-x}\,dx.$$

Using the Laplace transform $\mathcal{L}$ in Example B.4.5*c*, and for fixed $u > -1$, we have

$$\forall\, s \text{ such that } \operatorname{Re}(s) > 0, \quad \mathcal{L}(x^u)(s) = \frac{1}{s^{u+1}}\Gamma(u+1),$$

as a function of s. It is easy to see that if $u = m \in \mathbb{N}$ then $\mathcal{L}(x^m)(s) = m!/s^{m+1}$, and so $\Gamma(m+1) = m!$. Further, $\Gamma(u+1) = u\Gamma(u)$ and $\Gamma(1/2) = \sqrt{\pi}$.

c. Now consider $e^{-2\pi r\|\xi\|}$, $\xi \in \widehat{\mathbb{R}}^d$, and take its inverse Fourier transform using (B.36) and Example B.11.2:

$$\int_{\widehat{\mathbb{R}}^d} e^{-2\pi r\|\xi\|}e^{2\pi ix\cdot\xi}\,d\xi = \frac{1}{\sqrt{\pi}}\int_0^\infty \frac{e^{-u}}{\sqrt{u}}\left(\int_{\widehat{\mathbb{R}}^d} e^{-(\pi r\|\xi\|)^2/u}e^{2\pi ix\cdot\xi}\,d\xi\right)du$$
$$= \frac{1}{r^d\pi^{(d+1)/2}}\int_0^\infty e^{-u(1+\|x\|^2/r^2)}u^{(d-1)/2}\,du$$
$$= \frac{r}{\pi^{(d+1)/2}}\frac{1}{(r^2+\|x\|^2)^{(d+1)/2}}\int_0^\infty e^{-v}v^{(d-1)/2}\,dv.$$

Therefore,

$$\int_{\widehat{\mathbb{R}}^d} e^{-2\pi r\|\xi\|}e^{2\pi ix\cdot\xi}\,d\xi = \frac{r\Gamma((d+1)/2)}{\pi^{(d+1)/2}}\frac{1}{(r^2+\|x\|^2)^{(d+1)/2}}. \tag{B.37}$$

The right side of (B.37) is the *Poisson function* on $\mathbb{R}^d$ and $e^{-2\pi r\|\xi\|}$ is its Fourier transform on $\widehat{\mathbb{R}}^d$.

Example B.11.4. Bessel functions on $\mathbb{R}$
We shall use *Bessel functions of the first kind,*

$$J_\nu(x) = \frac{(x/2)^\nu}{\sqrt{\pi}\Gamma(\nu + 1/2)} \int_0^\pi e^{-ix\cos(\theta)} \sin^{2\nu}(\theta)\, d\theta,$$

for the case $\nu = (d-2)/2$. The functions J_ν arose as solutions of the differential equation,

$$z^2 \frac{d^2 w}{dz^2} + z\frac{dw}{dz} + (z^2 - \nu^2)w = 0;$$

see [360].

Theorem B.11.5. Fourier transform of radial functions
Let $f \in L^1_{m^d}(\mathbb{R}^d)$ be radial. Then its Fourier transform F is radial. Further,

$$F(\|\xi\|) = \frac{2\pi}{\|\xi\|^{(d-2)/2}} \int_0^\infty r^{d/2} J_{(d-2)/2}(2\pi r\|\xi\|)\phi(r)\, dr, \tag{B.38}$$

where $r = \|x\|$ and $\phi(r) = f(\|x\|)$.

The proof that F is radial comes down to checking that $F(\xi)$ is invariant with respect to rotations about the origin. The formula (B.38) results from Theorem 8.7.10, Example 8.7.11, and Example B.11.4, e.g., [426], pages 222–227.

Example B.11.6. Radial Fourier transforms for $d = 1, 2, 3$
Given the notation of Theorem B.11.5. If $d = 1$, then

$$F(|\xi|) = 2\int_0^\infty \cos(2\pi r|\xi|)\phi(r)\, dr$$

as predicted from Proposition B.1.7. If $d = 2$, then

$$F(\|\xi\|) = 2\pi \int_0^\infty r J_0(2\pi r\|\xi\|)\phi(r)\, dr.$$

If $d = 3$, then

$$F(\|\xi\|) = \frac{2}{\|\xi\|}\int_0^\infty r\sin(2\pi r\|\xi\|)\phi(r)\, dr.$$

We close this section with SCHOENBERG's surprising result (1938); see [420], [143], pages 201–206.

Theorem B.11.7. Schoenberg theorem
Let $F : [0,\infty) \to \mathbb{C}$ be continuous, and, for all $d \geq 1$, define $R(\xi) = F(\|\xi\|)$, $\xi \in \widehat{\mathbb{R}}^d$. Then R is positive definite on $\widehat{\mathbb{R}}^d$ for all $d \geq 1$ if and only if

$$\exists\, \mu \in M_b([0,\infty)),\ \mu \geq 0,\ \text{such that } \forall\, \gamma \geq 0, \quad F(\gamma) = \int_0^\infty e^{-x\gamma^2}\, d\mu(x),$$

noting that this integral is the Laplace transform $\mathcal{L}(\mu)(\gamma^2)$.

B.12 Wiener's Generalized Harmonic Analysis (GHA)

In this section not only do we have a chance to advertise WIENER's beautiful theory of Generalized Harmonic Analysis (GHA), but the theory itself combines many of the ideas we have introduced including the Bochner theorem and distribution theory. It is particularly exciting because of the continuing applicability of these notions, e.g., power spectra and spectrograms, in physics and signal processing. See the penultimate bullet in Section 9.6.4.

In 1930, WIENER [511], Volume II, pages 183–324, proved an analogue of the Parseval–Plancherel formula, $\|f\|_2 = \|\hat{f}\|_2$, for functions that are *not* elements of $L^2(\mathbb{R})$. We refer to his formula as the *Wiener–Plancherel formula*, e.g., (B.40). It became a beacon in his perception and formulation of the statistical theory of communication, e.g., [508], [317]. WIENER [511] even chose to have the formula appear on the cover of his autobiography, *I Am a Mathematician.* (This is a twentieth-century analogue of ARCHIMEDES' tombstone, which had a carving of a sphere inscribed in a cylinder to commemorate his "1:2:3" theorem; see Section 3.9.1 for details concerning the mathematical results, CICERO's role, and a recent update.)

Besides the use of GHA as an explanation of the polychromatic nature of sunlight, WIENER discussed the background for GHA in [511], Volume II, pages 183–324; and this background has been explained scientifically and historically in a virtuoso display of scholarship by MASANI, e.g., MASANI's remarkable commentaries in [511], Volume II, pages 333–379, as well as [341]. Two precursors, whose work Wiener studied and who should be mentioned vis-à-vis GHA, were Sir ARTHUR SCHUSTER and Sir GEOFFREY I. TAYLOR. SCHUSTER pointed out analogies between the harmonic analysis of light and the statistical analysis of hidden periods associated with meteorological and astronomical data. TAYLOR conducted experiments in fluid mechanics dealing with the *onset to turbulence*, and formulated a special case of correlation. A third scientist, whose work (1914) vis-à-vis GHA was not known to WIENER, was EINSTEIN. EINSTEIN writes, "Suppose the quantity y (for example, the number of sun spots) is determined empirically as a function of time, for a very large interval, T. How can one represent the statistical behavior of y?" In his heuristic answer to this question he came close to the notions of autocorrelation and power spectrum, e.g., B.12.5; cf. [341], pages 112–113, EINSTEIN's paper [157], and commentaries by MASANI [340] and A. M. YAGLOM [516].

The Fourier analysis of $L^1(\mathbb{R})$ or $L^2(\mathbb{R})$ and the theory of Fourier series were inadequate tools to analyze the issues confronting SCHUSTER, TAYLOR, and EINSTEIN. On the other hand, GHA became a successful device to gain some insight into their problems, as well as other problems in which the data and/or noises cannot be modeled by the Fourier transform decay, finite energy, or periodicity inherent in the above classical theories, e.g., [9], Chapter II, [29], [386].

The material in this section outlines GHA and is due to WIENER [511], Volume II, pages 183–324 and pages 519–619, and [505]; cf. [33] Chapter 2, [52]. The higher-dimensional theory, with its geometrical ramifications, is found in [41], [37]; cf. [6].

Definition B.12.1. Bounded quadratic means
The space $BQM(\mathbb{R})$ of functions having *bounded quadratic means* is the set of all functions $f \in L^2_{\mathrm{loc}}(\mathbb{R})$ for which

$$\sup_{T>0} \frac{1}{2T} \int_{-T}^{T} |f(t)|^2 \, dt < \infty.$$

The *Wiener space* $W(\mathbb{R})$ is the set of all functions $f \in L^2_{\mathrm{loc}}(\mathbb{R})$ for which

$$\int \frac{|f(t)|^2}{1+t^2} \, dt < \infty.$$

Theorem B.12.2. Inclusions for GHA
The following inclusions hold:

$$L^\infty_m(\mathbb{R}) \subseteq BQM(\mathbb{R}) \subseteq W(\mathbb{R}) \subseteq \mathcal{S}'(\mathbb{R}).$$

Moreover, all the inclusions are proper.

Definition B.12.3. The Wiener s-function
The *Wiener s-function* associated with $f \in BQM(\mathbb{R})$ is defined as the sum $s = s_1 + s_2$, where

$$s_1(\xi) = \int_{-1}^{1} f(t) \frac{e^{-2\pi i t\xi} - 1}{-2\pi i t} \, dt$$

and

$$s_2(\xi) = \int_{|t|\geq 1} f(t) \frac{e^{-2\pi i t\xi}}{-2\pi i t} \, dt.$$

Since $f \in L^1[-1,1]$, we have $s_1 \in C(\mathbb{R})$ and $|s_1(\xi)| \leq 2|\xi| \|f\|_{L^1[-1,1]}$. Since $f \in BQM(\mathbb{R})$, Theorem B.12.2 and the Parseval–Plancherel theorem allow us to conclude that $s_2 \in L^2_m(\widehat{\mathbb{R}})$. In particular, $s \in L^2_{\mathrm{loc}}(\widehat{\mathbb{R}}) \cap \mathcal{S}'(\widehat{\mathbb{R}})$.

Theorem B.12.4. The derivative of the Wiener s-function
Let $f \in BQM(\mathbb{R})$. Then $f \in \mathcal{S}'(\mathbb{R})$ and

$$s' = \hat{f},$$

where $s \in L^2_{\mathrm{loc}}(\widehat{\mathbb{R}}) \cap \mathcal{S}'(\widehat{\mathbb{R}})$ is the Wiener s-function associated with f.

Definition B.12.5. Deterministic autocorrelation
The *deterministic autocorrelation* R of a function $f : \mathbb{R} \to \mathbb{C}$ is formally defined as

$$R(t) = \lim_{T\to\infty} \frac{1}{2T} \int_{-T}^{T} f(u+t)\overline{f(u)}\, du.$$

To fix ideas, suppose R exists for each $t \in \mathbb{R}$. It is easy to prove that R is positive definite, and so $R = \widehat{S}$ for some positive bounded Radon measure $S \in M_b(\widehat{\mathbb{R}})$ by Theorem B.10.3. There is also a notion of *stochastic autocorrelation* arising in the study of stationary stochastic processes; see [293], [292], [144], [61]. Deterministic and stochastic autocorrelation are the same if the process has the property of being *correlation ergodic*; see [366]. This notion is not unrelated to the ergodic theorem discussed in Section 8.8.6. The measure S is called the *power spectrum* of f, and, in applications, letters such as S are used instead of μ.

The *Wiener–Plancherel formula* is equation (B.40) in the following result.

Theorem B.12.6. Wiener–Plancherel formula
Let $f \in BQM(\mathbb{R})$, and suppose its deterministic autocorrelation $R = \widehat{S}$ exists for each $t \in \mathbb{R}$.

a. *Then*

$$\forall\, t \in \mathbb{R}, \quad R(t) = \lim_{\varepsilon\to 0} \frac{2}{\varepsilon} \int |\Delta_\varepsilon s(\xi)|^2 e^{-2\pi i t\xi}\, d\xi, \tag{B.39}$$

where $\Delta_\varepsilon s(\xi) = \frac{1}{2}(s(\xi+\varepsilon) - s(\xi-\varepsilon))$.

b. *In particular,*

$$\lim_{T\to\infty} \frac{1}{2T} \int_{-T}^{T} |f(t)|^2\, dt = \lim_{\varepsilon\to 0} \frac{2}{\varepsilon} \int |\Delta_\varepsilon s(\xi)|^2\, d\xi. \tag{B.40}$$

Example B.12.7. Related formulas and spectral estimation
a. Because of (B.39) and assuming the setup of Theorem B.12.6, the following formulas are true under the proper hypotheses, e.g., [33], page 90, [36], page 847:

$$\lim_{\varepsilon\to 0} \frac{2}{\varepsilon} |\Delta_\varepsilon s(\xi)|^2 = S \tag{B.41}$$

and

$$\begin{aligned} \int |\hat{k}(\xi)|^2\, dS(\xi) &= \lim_{T\to\infty} \frac{1}{2T} \int_{-T}^{T} |k * f(t)|^2\, dt \\ &= \lim_{\varepsilon\to 0} \frac{2}{\varepsilon} \int |\hat{k}(\xi)\Delta_\varepsilon s(\xi)|^2\, d\xi. \end{aligned} \tag{B.42}$$

b. Formally, (B.42) is (B.40) for the case $k = \delta$. For $k \in C_c(\mathbb{R})$ the first equality of (B.42) is not difficult, e.g., [36], pages 847–848. The second equality, or, equivalently, Theorem B.12.6, requires the *Wiener Tauberian theorem* (Theorem B.12.9).

c. The following diagram illustrates the action and "levels" of the functions and measure in Theorem B.12.6 for a given signal f.

$$\begin{array}{ccc} f & \longleftrightarrow \hat{f} = s' & s \\ \downarrow & & \downarrow \\ R = \widehat{S} \longleftrightarrow & S & \{\tfrac{2}{\varepsilon}|\Delta_\varepsilon s|^2|\} \end{array}$$

d. Since S is the "power" spectrum, (B.40) and (B.41) allow us to assert that

$$\lim_{T\to\infty} \frac{1}{2T} \int_{-T}^{T} |f(t)|^2\, dt$$

is a measure of the total power of f; cf. WIENER's comparison of energy and power in [508], pages 39–40 and 42. In light of the spectral estimation problem, see, e.g., [39], Definition 2.8.6, the middle term of (B.42) is a measure of the power in a frequency band $[\alpha, \beta]$ if $\hat{k} = \mathbb{1}_{[\alpha,\beta]}$ in the first term of (B.42); cf. [36], Theorem 5.2.

Remark. The Parseval–Plancherel formula, $\|f\|_2 = \|\hat{f}\|_2$, allowed us to define the Fourier transform of a square-integrable function (Theorem B.4.2), and, at certain levels of abstraction, it is considered to characterize what is meant by a harmonic analysis of f. On the other hand, for most applications in $\mathbb{R}$, the formula assumes the workaday role of an effective tool used to obtain quantitative results. It is this latter role that was envisaged for the Wiener–Plancherel formula in dealing with the non-square-integrable case. After all, distribution theory gives the proper definition of the Fourier transform of tempered distributions. The real issue is to obtain quantitative results for problems where a harmonic analysis of a non-square-integrable function is desired. As mentioned above, a host of such problems comes under the heading of a harmonic (spectral) analysis of signals containing non-square-integrable noise and/or random components, whether it be speech recognition, image processing, geophysical modeling, or turbulence in fluid mechanics. Such problems can be attacked by BEURLING's profound theory of spectral synthesis, e.g., [33], as well as by the extensive multifaceted theory of time series, e.g., [378]. BEURLING's spectral synthesis does not deal with energy and power considerations, i.e., quadratic criteria; and the theory of time series relies on a stochastic point of view. The Wiener–Plancherel formula deals with these problems deterministically, and hence with potential for real implementation.

Example B.12.8. Elementary power spectra

a. The value of an autocorrelation R is that it can be measured in many cases in which the underlying signal f cannot be quantified. This is the basis of the Michelson interferometer. Also, the discrete part of the power spectrum S characterizes periodicities in f, e.g., [507], Chapter X. This can be illustrated by taking $f(t) = \sum_{k=1}^{n} r_k e^{-2\pi i t \lambda_k}$, $r_k \in \mathbb{C}$, $\lambda_k \in \widehat{\mathbb{R}}$. The

L^2-autocorrelation is not defined, but the deterministic autocorrelation is $R(t) = \sum_{k=1}^{n} |r_k|^2 e^{-2\pi i t \lambda_k}$ (by direct calculation); and hence the power spectrum is

$$S = \sum_{k=1}^{n} |r_k|^2 \, \delta_{\lambda_k}.$$

b. If $f : \mathbb{R} \to \mathbb{C}$ has the property that $\lim_{|t|\to\pm\infty} f(t) = 0$, then $S = 0$. It is elementary to construct examples f for which $S = 0$ whereas $\overline{\lim}_{|t|\to\pm\infty} |f(t)| > 0$; cf. [505], pages 151–154, [29], pages 99–100, [33], pages 84 and 87, [35], Section IV.

As mentioned in Example B.12.7, the following result is required to prove the Wiener–Plancherel formula. It was stated in the Remark after Problem 5.8.

Theorem B.12.9. Wiener Tauberian theorem
Let $f \in L^1_m(\mathbb{R})$ have a nonvanishing Fourier transform and let $g \in L^\infty_m(\mathbb{R})$. If

$$\lim_{t\to\infty} f * g(t) = r \int_{-\infty}^{\infty} f(u)\, du, \tag{B.43}$$

then

$$\forall h \in L^1_m(\mathbb{R}), \quad \lim_{t\to\infty} h * g(t) = r \int_{-\infty}^{\infty} h(u)\, du. \tag{B.44}$$

Remark. *a.* Theorem B.12.9 has the format of *classical Tauberian theorems*: A boundedness (or related) condition and "summability" by a certain method yield "summability" by other methods. In Theorem B.12.9, the boundedness or "Tauberian" condition is the hypothesis that $g \in L^\infty_m(\mathbb{R})$. The given summability is (B.43), where f represents a so-called summability method. The conclusion (B.44) of the theorem is summability for a whole class of summability methods, viz., for all $h \in L^1_m(\mathbb{R})$. A classical and masterful treatment of summability methods is due to Hardy [216].

If G is the Gaussian defined in Example B.1.8, then $\widehat{G}$ never vanishes. Thus, in this case, if $g \in L^\infty_m(\mathbb{R})$ has the property that

$$\lim_{t\to\infty} G * g(t) = r,$$

then

$$\forall\, \lambda, \quad \lim_{t\to\infty} W_\lambda * g(t) = r,$$

where $\{W_\lambda\}$ is the Fejér kernel.

The particular functions used by Wiener to prove his Wiener Tauberian formulas are found in [505], [33], pages 91–92.

b. *Modern* Tauberian theorems have an algebraic and/or functional-analytic flavor to them. For example, the Wiener Tauberian theorem is a special case of the fact that *if $\hat{f} \in A(\mathbb{R})$, $T \in A'(\mathbb{R})$, and $T\hat{f} = 0$, then $\hat{f} = 0$*

on supp T. In fact, the generalizations of Theorem B.12.9 are much more far-reaching than this. The book [33] gives an extensive treatment of both classical and modern Tauberian theory, as well as the history of the subject, and applications to spectral synthesis and analytic number theory; see also, e.g., [246].

Because of the importance of translation-invariant systems and the theory of multipliers, we define the *closed translation-invariant subspace* V_f *generated by* $f \in X$, where X is $L^1_m(\mathbb{R})$ or $L^2_m(\mathbb{R})$, to be the closure in X of the linear span of translations of f by $t \in \mathbb{R}$, i.e.,

$$V_f = \overline{\text{span}}\{\tau_t(f) : t \in \mathbb{R}\}. \tag{B.45}$$

Theorem B.12.10. Zero sets and dense subspaces

a. *If* $f \in L^1_m(\mathbb{R})$ *and* $\hat{f}$ *never vanishes, then* $V_f = L^1_m(\mathbb{R})$.

b. *If* $f \in L^2_m(\mathbb{R})$ *and* $|\hat{f}| > 0$ *a.e., then* $V_f = L^2_m(\mathbb{R})$.

Proof. Part *a* is the Wiener Tauberian theorem, and we refer to [505], [33], pages 25–26, 49–50, 94–95, and Section 2.3 for proofs.

The proof of part *b* is much simpler than that of part *a*, and so we shall give it. Suppose $V_f \neq L^2_m(\mathbb{R})$. Then there is $h \in L^2_m(\mathbb{R}) \setminus \{0\}$ such that

$$\forall\, t \in \mathbb{R}, \qquad \int (\tau_t(f))(u)\overline{h(u)}\, du = 0. \tag{B.46}$$

Equation (B.46) is a consequence of the Hahn–Banach theorem and the fact that $L^2_m(\mathbb{R})' = L^2_m(\mathbb{R})$. By the Parseval–Plancherel theorem,

$$\forall\, t \in \mathbb{R}, \qquad \int \hat{f}(\xi)\overline{\hat{h}(\xi)}e^{-2\pi i t\xi}\, d\xi = 0.$$

By the Hölder inequality $\hat{f}\overline{\hat{h}} \in L^1_m(\widehat{\mathbb{R}})$, and so, by the L^1-uniqueness theorem (Theorem B.3.5*c*), $\hat{f}\overline{\hat{h}} = 0$ a.e. Since $|\hat{f}| > 0$ a.e., we conclude that $\hat{h} = 0$ a.e., and this contradicts the hypothesis on h. Thus, $V_f = L^2(\mathbb{R})$. □

Subspaces such as V_f in (B.45) play an important role in Gabor and wavelet decompositions in the case that the set of translates $\tau_t(f)$ is reduced to $\{\tau_r(f) : r \in D\}$, where D is a discrete subset of $\mathbb{R}$, e.g., [348], [118], [45], [310].

Remark. ***a.*** Let $M_b^+(\widehat{\mathbb{R}})$ be the space of positive bounded Radon measures on $\widehat{\mathbb{R}}$. In GHA, a function f is analyzed for its frequency information by computing its autocorrelation R and its power spectrum $S = R^\vee \in M_b^+(\widehat{\mathbb{R}})$. Mathematically, this is a mapping between a class of functions f and a class of measures $S \in M_b^+(\widehat{\mathbb{R}})$. A natural question to ask is the following: For any $\mu \in M_b^+(\widehat{\mathbb{R}})$, does there exist f whose autocorrelation R exists, and for which $\check{R} = \mu$?

b. The question of part a is answered affirmatively in the case of weakly stationary stochastic processes (WSSPs) by the Wiener–Khinchin theorem: *A necessary and sufficient condition for R to be the stochastic autocorrelation of some* WSSP X *is that there exist $S \in M_b^+(\widehat{\mathbb{R}})$ for which $\widehat{S} = R$.* In one direction, if R is the stochastic autocorrelation of a WSSP X, then $S = R^{\vee} \in M_b^+(\widehat{\mathbb{R}})$ by Theorem B.10.3. The question in part a deals with the opposite direction, and the positive answer is not difficult to prove, e.g., [378], pages 221–222, [152], pages 62–63 and 72–73. ALEKSANDR Y. KHINCHIN's proof dates from 1934, and there were further probabilistic contributions by HERMAN O. A. WOLD (1938), CRAMÉR (1940), and KOLMOGOROV [292]; cf. [38].

c. The deterministic and *constructive* affirmative answer to the question in part a is the *Wiener–Wintner theorem* (1939) [511]. JEAN BASS and JEAN-PAUL BERTRANDIAS made significant contributions to this result, e.g., [29], and the multidimensional version is found in [36], [282].

Theorem B.12.11. Wiener–Wintner theorem
Let $\mu \in M_b^+(\widehat{\mathbb{R}})$. There is a constructible function $f \in L^{\infty}_{\text{loc}}(\mathbb{R})$ such that its deterministic autocorrelation R exists for all $t \in \mathbb{R}$, and $\widehat{R} = \mu$.

B.13 Epilogue

This appendix serves as a handmaiden to the book, but the material is really a preface to harmonic analysis as one of the goddesses of mathematics. There are magnificent, profound edifices from classical Fourier series to representation theory, from nonharmonic Fourier series to sampling, wavelets, and time-frequency analysis, from Fourier methods in classical partial differential equations to pseudodifferential operators, from the computation of Gauss sums to the role of Fourier analysis at all levels of analytic number theory, and from fast Fourier transforms to an ever expanding litany of genuine applications. We have referenced introductory texts and groundbreaking treatises, and encyclopedic works of scholarship.

References

1. A. Abian, *An example of a nonmeasurable set*, Boll. U.M.I. **1** (1968), 366–368.
2. ———, *The outer and inner measures of a nonmeasurable set*, Boll. U.M.I. **3** (1970), 555–558.
3. L. Alaoglu, *Weak topologies of normed linear spaces*, Annals of Math. **41** (1940), 252–267.
4. E. Albrecht and M. Neumann, *Automatische Stetigkeitseigenschaften einiger Klassen linearer Operatoren*, Math. Ann. **240** (1979), 251–280.
5. L. Ambrosio, N. Fusco, and D. Pallara, *Functions of Bounded Variation and Free Discontinuity Problems*, Oxford University Press, London, 2000.
6. K. Anzai, S. Koizumi, and K. Matsuoka, *On the Wiener formula of functions of two variables*, Tokyo Math. J. **3** (1980), 249–270.
7. T. M. Apostol, *Mathematical Analysis*, Addison–Wesley, Reading, MA, 1974 (1957).
8. A. Arneodo, E. Bacry, S. Jaffard, and J. F. Muzy, *Singularity spectrum of multifractal functions involving oscillating singularities*, J. Fourier Anal. Appl. **4** (1998), 159–174.
9. J. Arsac, *Fourier Transforms and the Theory of Distributions*, Prentice Hall, Englewood Cliffs, NJ, 1966.
10. C. Arzelà, *Sulla integrazione per serie*, Atti Accad. Lincei Roma **1** (1885), 532–537.
11. ———, *Sulle funzioni di linee*, Bologna Accad. Sci. Mem. **5** (1895), 225–244.
12. ———, *Sulle serie di funzioni*, Bologna Accad. Sci. Mem. **8** (1899), 131–186.
13. G. Ascoli, *Le curve limiti di una varietà data di curve*, Mem. Accad. dei Lincei **18** (1883), 521–586.
14. R. B. Ash and C. A. Doléans-Dade, *Probability and Measure Theory*, Academic Press, San Diego, CA, 2000.
15. E. Asplund and L. Bungart, *A First Course in Integration*, Holt, Rinehart, and Winston, New York, 1966.
16. D. Austin, *A geometric proof of the Lebesgue differentiation theorem*, Proc. Amer. Math. Soc. **16** (1965), 220–222.
17. S. Banach, *Sur une classe de fonctions d'ensemble*, Fund. Math. **6** (1924), 170–188.
18. ———, *Sur les lignes rectifiables et les surfaces dont l'aire est finie*, Fund. Math. **7** (1925), 225–236.
19. ———, *Théorie de Opérations Linéares*, (corrected and expanded reprint of original), Chelsea Publishing Company, New York, 1932.
20. ———, *Sur les suites d'ensembles excluant l'existence d'une mesure*, Coll. Math. **1** (1947), 103–108.

21. ———, *Oeuvres*, PWN, Warszawa, 1967.
22. S. Banach and K. Kuratowski, *Sur une généralisation du problème de la mesure*, Fund. Math. **14** (1929), 127–131.
23. S. Banach and S. Saks, *Sur la convergence forte dans les espaces* L^p, Studia Math. **2** (1930), 51–57.
24. S. Banach and H. Steinhaus, *Sur le principle de la condensation de singularités*, Fund. Math. **9** (1927), 50–61.
25. S. Banach and A. Tarski, *Sur la décomposition des ensembles de points en parties respectivement congruentes*, Fund. Math. **6** (1924), 244–277.
26. N. K. Bari, *Biorthogonal systems and bases in Hilbert space (Russian)*, Moskov. Gos. Univ. Učenye Zapiski Matematika **148** (1951), 69–107.
27. ———, *A Treatise on Trigonometric Series. Volumes I, II*, The Macmillan Co., New York, 1964 (1957).
28. P. Barry, R. Goldman, and T. DeRose, *B-splines, Pólya curves, and duality*, J. Approx. Theory **65** (1991), 3–21.
29. J. Bass, *Fonctions de Corrélation, Fonctions Pseudo-Aléatoires, et Applications*, Masson, Paris, 1984.
30. E. F. Beckenbach and R. Bellman, *Inequalities*, Springer-Verlag, New York, 1961.
31. E. M. Beesley, A. P. Morse, and D. C. Pfaff, *Lipschitzian points*, Amer. Math. Monthly **79** (1962), 603–608.
32. J. Benedetto, *Harmonic Analysis on Totally Disconnected Sets*, Lecture Notes in Mathematics, 202, Springer-Verlag, New York, 1971.
33. J. J. Benedetto, *Spectral Synthesis*, Academic Press, New York, 1975.
34. ———, *Fourier analysis of Riemann distributions and explicit formulas*, Math. Ann. **252** (1980), 141–164.
35. ———, *Harmonic analysis and spectral estimation*, J. Math. Anal. Appl. **91** **(2)** (1983), 444–509.
36. ———, *A multidimensional Wiener–Wintner theorem and spectral estimation*, Trans. Amer. Math. Soc. **327** (1991), 833–852.
37. ———, *The spherical Wiener–Plancherel formula and spectral estimation*, SIAM J. Math. Anal. **22** (1991), 1110–1130.
38. ———, *Stationary frames and spectral estimation*, in Probabilistic and Stochastic Methods in Analysis, ed. J.S. Byrnes et al., Kluwer Academic Publishers, Dordecht (1992), 117–161.
39. ———, *Harmonic Analysis and Applications*, CRC Press, Boca Raton, FL, 1997.
40. J. J. Benedetto and R. L. Benedetto, *A wavelet theory for local fields and related groups*, J. Geom. Analysis **14** (2004), 423–456.
41. J. J. Benedetto, G. Benke, and W. R. Evans, *An n-dimensional Wiener–Plancherel formula*, Adv. in Appl. Math. **10** (1989), 457–487.
42. J. J. Benedetto, E. Bernstein, and I. Konstantinidis, *Multiscale Riesz products and their support properties*, Acta Appl. Math. **88** (2005), 201–227.
43. J. J. Benedetto and P. J. S. G. Ferreira (eds.), *Modern Sampling Theory: Mathematics and Applications*, Birkhäuser, Boston, MA, 2001.
44. J. J. Benedetto and M. Fickus, *Finite normalized tight frames. frames*, Adv. Comput. Math. **18** (2003), 357–385.
45. J. J. Benedetto and M. W. Frazier (eds.), *Wavelets: Mathematics and Applications*, CRC Press, Boca Raton, FL, 1994.

46. J. J. Benedetto and H. P. Heinig, *Weighted Fourier inequalities: new proofs and generalizations*, J. Fourier Anal. Appl. **9** (2003), 1–37.
47. J. J. Benedetto and A. Kebo, *The role of frame force in quantum detection*, J. Fourier Anal. Appl. **14** (2008), 443–474.
48. J. J. Benedetto and M. Leon, *The construction of single wavelets in d-dimensions*, J. Geom. Analysis **11** (2001), 1–15.
49. J. J. Benedetto and S. Sumetkijakan, *A fractal set constructed from a class of wavelet sets*, Contemporary Math. **313** (2002), 19–35.
50. C. Bennett and R. Sharpley, *Interpolation of Operators*, Academic Press, New York, 1988.
51. H. Bergström, *Weak Convergence of Measures*, Academic Press, New York, 1982.
52. J. P. Bertrandias, J. Couot, J. Dhombres, M. Mendès France, Pham Phu Hien, and Kh. Vo-Khac, *Espaces de Marcinkiewicz, Corrélations, Mesures, Systèmes Dynamiques*, Masson, Paris, 1984.
53. A. S. Besicovitch, *On the fundamental geometric properties of linearly measurable plane sets of points*, Math. Ann. **98** (1928), 422–464.
54. ———, *Concentrated and rarified sets of points*, Acta Math. **62** (1934), 289–300.
55. ———, *On the fundamental geometric properties of linearly measurable plane sets of points II*, Math. Ann. **115** (1938), 296–329.
56. ———, *On the fundamental geometric properties of linearly measurable plane sets of points III*, Math. Ann. **116** (1939), 349–357.
57. ———, *A general form of the covering principle and relative differentiation of additive functions I*, Proc. Cambridge Philos. Soc. **41** (1945), 103–110.
58. ———, *A general form of the covering principle and relative differentiation of additive functions II*, Proc. Cambridge Philos. Soc. **42** (1946), 1–10.
59. A. Beurling and P. Malliavin, *On Fourier transforms of measures with compact support*, Acta Math. **107** (1962), 291–309.
60. ———, *On the closure of characters and the zeros of entire functions*, Acta Math. **118** (1967), 79–93.
61. P. Billingsley, *Probability and Measure*, John Wiley and Sons, New York, 1986.
62. W. W. Bledsoe, *An inequality about complex numbers*, Amer. Math. Monthly **77** (1970), 180–182.
63. E. Boardman, *Universal covering series*, Amer. Math. Monthly **79** (1972), 780–781.
64. R. P. Boas, *A Primer of Real Functions*, Carus Monographs, Mathematical Association of America, New York, 1960.
65. S. Bochner, *Additive set functions on groups*, Annals of Math. **40** (1939), 769–799.
66. S. Bochner and K. Chandrasekharan, *Fourier Transforms*, Princeton University Press, Princeton, N.J., 1949.
67. É. Borel, *Leçons sur les Fonctions Monogènes Uniformes d'une Variable Complexe*, Gauthier-Villars, Paris, 1917.
68. M. W. Botsko, *An elementary proof of Lebesgue's differentiation theorem*, Amer. Math. Monthly **110** (2003), 834–838.
69. N. Bourbaki, *Intégration (Chap. 1–4), 2nd edition*, Hermann, Paris, 1965.
70. ———, *General Topology, Part I*, Addison-Wesley, Reading, MA, 1966 (1953).
71. ———, *General Topology, Part II*, Addison-Wesley, Reading, MA, 1966 (1955).

72. ———, *Intégration (Chap. 7–8), 2nd edition*, Hermann, Paris, 1967.
73. H. E. Brandt, *Quantum measurement with a positive operator-valued measure*, J. Opt. B Quantum Semiclass. Opt., Wigner Centennial **5** (2003), S266–S279.
74. W. L. Briggs and V. E. Henson, *The DFT*, SIAM, Philadelphia, PA, 1995.
75. J. K. Brooks, *On the Vitali–Hahn–Saks and Nikodym theorems*, Proc. Nat. Acad. Sci. **64** (1969), 468–471.
76. J. K. Brooks and J. Mikusiński, *On some theorems in functional analysis*, Bull. Acad. Polon. Sci. **18** (1970), 151–155.
77. G. Brown, *Riesz products and generalized characters*, Proc. Math. London Soc. **30** (1975), 209–238.
78. G. Brown, I. Glicksberg, and E. Hewitt, *Indicator functions with large Fourier transforms*, Pacific J. Math. **105** (1983), 11–20.
79. A. M. Bruckner, *Differentiation of Integrals*, Amer. Math. Monthly **78 no. 9, Part II** (1971).
80. A. M. Bruckner and J. L. Leonard, *Derivatives*, Papers in Analysis, Amer. Math. Monthly, Part II **73** (1963), 24–56.
81. R. B. Burckel, *An Introduction to Classical Complex Analysis*, Birkhäuser, Boston, MA, 1979.
82. P. L. Butzer and E. L. Stark, *Riemann's example of a continuous nondifferentiable function in the light of two letters (1865) of Christoffel to Prym*, Bull. Société Mathématique Belgique **38** (1986), 45–73.
83. C. Cabrelli and U. Molter, *Harmonic Analysis and Fractal Geometry: A Dynamic Interplay*, Birkhäuser, Boston, MA, to appear in 2009.
84. C. A. Cabrelli, K. E. Hare, and U. M. Molter, *Sums of Cantor sets*, Ergodic Theory Dynam. Systems **17** (1997), 1299–1313.
85. A. P. Calderón, *Cauchy integrals on Lipschitz curves and related operators*, Proc. Nat. Acad. Sci. USA **74** (1977), 1324–1327.
86. A. P. Calderón and A. Zygmund, *On the existence of certain singular integrals*, Acta Math. **88** (1952), 85–139.
87. C. Carathéodory, *Algebraic Theory of Measure and Integration*, Chelsea Publishing Company, New York, 1963 (1927).
88. L. Carleson, *Representations of continuous functions*, Math. Zeit. **66** (1957), 447–451.
89. ———, *An interpolation problem for bounded analytic functions*, Amer. J. Math. **80** (1958), 921–930.
90. ———, *Interpolation of bounded analytic functions and the corona problem*, Annals of Math. **76** (1962), 547–559.
91. ———, *On convergence and growth of partial sums of Fourier series*, Acta Math. **116** (1966), 135–157.
92. ———, *Selected Problems on Exceptional Sets*, D. Van Nostrand Co., Princeton, NJ, 1967.
93. ———, *Some analytic problems related to statistical mechanics*, Euclidean Harmonic Analysis, Lecture Notes in Mathematics, 779, Springer-Verlag, 1980, pp. 5–45.
94. H. Cartan, *Sur la mesure de Haar*, Comptes Rendus Acad. Sci. Paris **211** (1940), 759–762.
95. ———, *Differential Forms*, Houghton Mifflin Co., Boston, MA, 1970.
96. L. Cesari, *Sulle funzioni a variazione limitata*, Ann. Scuola Norm. Sup. Pisa **5** (1936), 299–313.

97. ———, *Variation, multiplicity, and semicontinuity*, Amer. Math. Monthly **65** (1958), 317–331.
98. E. Cesàro, *Fonctions continues sans dérivée*, Archiv der Math. und Phys. **10** (1906), 57–63.
99. A. Chebira and J. Kovačević, *Life beyond bases: the advent of frames (Part I)*, IEEE Signal Processing Mag. **24** (2007), 86–104.
100. ———, *Life beyond bases: the advent of frames (Part II)*, IEEE Signal Processing Mag. **24** (2007), 115–125.
101. P. Chernoff, *Pointwise convergence of Fourier series*, Amer. Math. Monthly **87** (1980), 399–400.
102. G. Choquet, *Un ensemble paradoxal en théorie de la mesure*, Bull. Sci. Math. **94** (1970), 247–250.
103. O. Christensen, *An Introduction to Frames and Riesz Bases*, Birkhäuser, Boston, MA, 2003.
104. K. Ciesielski, *How good is Lebesgue measure?*, Math. Intelligencer **11** (1989), 54–58.
105. ———, *On some details of Stefan Banach's life*, Opuscula Math. **13** (1993), 71–74.
106. J. A. Clarkson, *Uniformly convex spaces*, Trans. Amer. Math. Soc. **40** (1936), 396–414.
107. L. W. Cohen, *A new proof of Lusin's theorem*, Fund. Math. **9** (1927), 122–123.
108. ———, *Measure and integration in the manner of Borel*, Scripta Math. **29** (1973), 417–435.
109. ———, *The fundamental theorem of the calculus and Borel measure*, 1975.
110. L. W. Cohen and G. Ehrlich, *The Structure of the Real Number System*, Huntington, New York, 1977 (1963).
111. R. R. Coifman, A. McIntosh, and Y. Meyer, *L'intégrale de Cauchy définit un opérateur borné sur L^2 pour les courbes lipschitziennes*, Annals of Math. **116** (1982), 361–388.
112. R. R. Coifman, Y. Meyer, and M. V. Wickerhauser, *Wavelet analysis and signal processing*, Wavelets and Their Applications, Jones and Bartlett, Boston, MA, 1992, pp. 153–178.
113. ———, *Signal processing and compression with wavelet packets*, Progress in Wavelet Analysis and Applications, Proceedings of the 3rd International Conference on Wavelets and Applications (Toulouse, France, June 8–13, 1992), Editions Frontières, Gif-sur-Yvette, 1993, pp. 77–93.
114. J. H. Conway and N. J. A. Sloane, *Sphere Packings, Lattices and Groups*, Springer-Verlag, New York, 1999.
115. Á. Császár and D. Petz, *A panorama of the Hungarian real and functional analysis in the 20th century*, in A Panorama of Hungarian Mathematics in the Twentieth Century, ed. J. Horváth, Bolyai Society Mathematical Studies, No. 14, Springer-Verlag, New York (2006), 211–244.
116. W. Czaja and A. M. Powell, *Recent developments in the Balian–Low theorem*, Harmonic Analysis and Applications, ed. C. Heil, Birkhäuser, Boston, MA, 2006, pp. 79–100.
117. R. B. Darst, *Perfect null sets in compact Hausdorff spaces*, Proc. Amer. Math. Soc. **16** (1965), 845.
118. I. Daubechies, *Ten Lectures on Wavelets, CBMS 61*, SIAM, Philadelphia, PA, 1992.

119. I. Daubechies, A. Grossmann, and Y. Meyer, *Painless nonorthogonal expansions*, J. Math. Phys. **21** (1986), 1271–1283.
120. G. David, *Wavelets and Singular Integrals on Curves and Surfaces*, Lecture Notes in Mathematics, 1465, Springer-Verlag, New York, 1991.
121. G. David and S. Semmes, *Singular integrals and rectifiable sets in R^n: au-delà des graphes lipschitziens*, Astérisque **193** (1991).
122. ———, *Analysis of and on Uniformly Rectifiable Sets*, Amer. Math. Soc. Surveys and Monographs 38, 1993.
123. ———, *Fractured Fractals and Broken Dreams*, Clarendon Press, Oxford, 1997.
124. R. O. Davies and A. O. Ostaszewski, *Denumerable compact metric squares admit isometry-invariant finitely additive measures*, Mathematika **26** (1979), 184–186.
125. C. Davis, *Geometric approach to a dilation theorem*, Linear Algebra and Appl. **18** (1977), 33–43.
126. ———, *Some dilation and representation theorems*, Proceedings of the Second International Symposium in West Africa on Functional Analysis and Its Applications, Kumasi (1979), 159–182.
127. K. Dąbrowski and E. Hensz-Chądzyńska, *Józef Marcinkiewicz (1910–1940). in commemoration of the 60th anniversary of his death*, Fourier Analysis and Related Topics (Będlewo, 2000), Banach Center Publ. **56** (2002), 31–35.
128. M. de Guzmán, *Differentiation of Integrals in R^n*, Lecture Notes in Mathematics, 481, Springer-Verlag, New York, 1975.
129. M. de Guzmán and B. Rubio, *Remarks on the Lebesgue differentiation theorem, the Vitali lemma and the Lebesgue–Radon–Nikodym theorem*, Amer. Math. Monthly **79** (1972), 341–348.
130. C. de la Vallée Poussin, *Intégrales de Lebesgue, 2nd edition*, Éditions Jacques Gabay, Sceaux, 1995 (1934; 1916 1st edition).
131. E. M. deJager, *Applications of Distributions in Mathematical Physics*, Mathematical Centre, Amsterdam, 1964.
132. A. Denjoy, *Leçons sur le Calcul des Coefficients d'une Série Trigonométrique*, Gauthier-Villars, Paris, 1941.
133. R. dePossel, *Sur la dérivation abstraite des fonctions d'ensembles*, J. Math. Pures Appl. **15** (1936), 391–409.
134. S. Diaz, *Weak compactness in $L^1(\mu, X)$*, Proc. Amer. Math. Soc. **124** (1996), 2685–2693.
135. E. DiBenedetto, *Real Analysis*, Birkhäuser, Boston, MA, 2002.
136. J. Dieudonné, *Sur la convergence des suites de mesures de Radon*, An. da Acad. Bras. Ci. **23** (1951), 21–38.
137. ———, *Sur les espaces de Köthe*, J. d'Analyse **1** (1951), 81–115.
138. J. Dieudonné, *History of Functional Analysis*, North-Holland, Amsterdam, 1981.
139. E. J. Dijksterhuis, *Archimedes*, Princeton University Press, Princeton, NJ, 1987 (1938, Chapters I–V).
140. U. Dini, *Serie di Fourier e Altre Rappresentazioni Analitiche delle Funzioni di una Variabile Reale*, Tip. T. Nistri e C., Pisa, 1880.
141. ———, *Sugli sviluppi in serie $\frac{1}{2}a_0 + \sum_1^\infty (a_n \cos \lambda_n x + b_n sen \lambda_n x)$ dove le λ_n sono radici dell'equazione trascendente $f(z) \cos \pi z + f_1(z) sen \pi z = 0$*, Opere Volume II, Edizioni Cremonese, Roma (1954), 183–207.

142. J. Dixmier, *L'Intégrale de Lebesgue*, Centre de Documentation Universitaire, Paris, 1964.
143. W. L. Donoghue Jr., *Distributions and Fourier Transforms*, Academic Press, New York, 1969.
144. J. L. Doob, *Stochastic Processes*, John Wiley and Sons, New York, 1953.
145. R. E. Dressler and R. B. Kirk, *Non-measurable sets of reals whose measurable subsets are countable*, Israel J. Math. **11** (1972), 265–270.
146. L. Dubins, *An elementary proof of Bochner's finitely additive Radon–Nikodym theorem*, Amer. Math. Monthly **76** (1969), 520–523.
147. R. J. Duffin and A. C. Schaeffer, *A class of nonharmonic Fourier series*, Trans. Amer. Math. Soc. **72** (1952), 341–366.
148. J. J. Duistermaat, *Self-similarity of "Riemann's nondifferentiable function"*, Nieuw Arch. Wisk. **9** (1991), 303–337.
149. N. Dunford and B. J. Pettis, *Linear operations on summable functions*, Trans. Amer. Math. Soc. **47** (1940), 323–392.
150. N. Dunford and J. T. Schwartz, *Linear Operators*, John Wiley and Sons, New York, 1957.
151. R. Durrett, *Brownian Motion and Martingales in Analysis*, Wadsworth Advanced Books, Belmont, CA, 1984.
152. H. Dym and H. P. McKean, *Gaussian Processes, Function Theory, and the Inverse Spectral Problem*, Academic Press, New York, 1976.
153. G. A. Edgar, *Measure, Topology, and Fractal Geometry*, Springer-Verlag, New York, 1990.
154. H. M. Edwards, *Riemann's Zeta Function*, Academic Press, New York, 1974.
155. R. E. Edwards, *Fourier Series. Volumes 1 and 2*, Holt, Rinehart, and Winston, New York, 1967.
156. ———, *Functional Analysis, Theory and Applications*, Dover Publications, New York, 1995 (1965).
157. A. Einstein, *Méthode pour la détermination de valeurs statistiques d'observations concernant des grandeurs soumises à des fluctuations irrégulières*, Archive des Sciences Physiques et Naturelles **37** (1914), 254–255.
158. P. Enflo, *A counterexample to the approximation problem in Banach spaces*, Acta Math. **130** (1973), 309–317.
159. L. C. Evans and R. F. Gariepy, *Measure Theory and Fine Properties of Functions*, CRC Press, Boca Raton, FL, 1992.
160. K. J. Falconer, *The Geometry of Fractal Sets*, Cambridge University Press, Cambridge, 1985.
161. ———, *Fractal Geometry: Mathematical Foundations and Applications*, John Wiley and Sons, New York, 1990.
162. P. Fatou, *Séries trigonométriques et séries de Taylor*, Acta Math. **30** (1906), 335–400.
163. H. Federer, *Geometric Measure Theory*, Springer-Verlag, New York, 1969.
164. C. Fefferman, *A Radon–Nikodym theorem for finitely additive set functions*, Pacific J. Math. **23** (1967), 35–45.
165. ———, *Characterization of bounded mean oscillation*, Bull. Amer. Math. Soc. **77** (1971), 587–588.
166. ———, *Pointwise convergence of Fourier series*, Annals of Math. **98** (1973), 551–571.

167. ———, *Recent progress in classical Fourier analysis*, Proceedings of the International Congress of Mathematicians (Vancouver, B.C., 1974) **1** (1975), 95–118.
168. R. Fefferman, *Review of Differentiation of Integrals in R^n by Miguel de Guzmán*, Bull. Amer. Math. Soc. **83** (1977), 196–200.
169. H. Feichtinger and K. Gröchenig, *Theory and practice of irregular sampling*, Wavelets: Mathematics and Applications, ed. J. J. Bendetto and M. W. Frazier, CRC Press, Boca Raton, FL, 1994, pp. 305–363.
170. W. Feller, *Bemerkungen zur Masstheorie in abstrakten Räumen*, Bull. Int. Acad. Yougoslav. **28** (1934), 30–45.
171. ———, *An Introduction to Probability Theory and Its Applications, Volume 1*, John Wiley and Sons, New York, 1950.
172. ———, *An Introduction to Probability Theory and Its Applications, Volume 2*, John Wiley and Sons, New York, 1966.
173. R. Feynman, *Surely You're Joking, Mr. Feynman*, Bantam Books, New York, 1986.
174. G. M. Fichtenholz, *Note sur les fonctions absolument continues*, Acad. Roy. Belgique Bull. Sci. **8** (1922), 430–443.
175. G. M. Fichtenholz and L. V. Kantorovich, *Sur les opérations linéaires dans l'espace des fonctions bornées*, Studia Math. **5** (1934), 69–98.
176. E. Fischer, *Applications d'un théorème sur la convergence en moyenne*, Comptes Rendus Acad. Sci. Paris **144** (1907), 1148–1151.
177. ———, *Sur la convergence en moyenne*, Comptes Rendus Acad. Sci. Paris **144** (1907), 1022–1024.
178. P. M. Fitzpatrick, *Advanced Calculus, 2nd edition*, Thomson Brooks/Cole, Belmont, CA, 2006.
179. W. H. Fleming, *Functions of Several Variables*, Addison-Wesley, Reading, MA, 1965.
180. G. B. Folland, *Real Analysis, 2nd edition*, John Wiley and Sons, New York, 1999.
181. G. B. Folland and A. Sitaram, *The uncertainty principle: a mathematical survey*, J. Fourier Anal. Appl. **3** (1997), 207–238.
182. R. Franck, *Sur une propriété des fonctions additives d'ensembles*, Fund. Math. **5** (1924), 252–261.
183. M. W. Frazier, B. Jawerth, and G. Weiss, *Littlewood–Paley Theory and the Study of Function Spaces*, CBMS 79, Amer. Math. Soc., Providence, RI, 1991.
184. M. Fréchet, *Sur les ensembles de fonctions et les opérations linéaires*, Comptes Rendus Acad. Sci. Paris **144** (1907), 1414–1416.
185. S. H. Friedberg, A. J. Insel, and L. E. Spence, *Linear Algebra, 3rd edition*, Prentice Hall, Upper Saddle River, NJ, 1997.
186. U. Frisch and G. Parisi, *Fully developed turbulence and intermittency*, Proceedings of the International School of Physics, Enrico Fermi, North-Holland (1985), 84–88.
187. G. Fubini, *Sugli integrali multipli*, Rend. Accad. Lincei Roma **16** (1907), 608–614.
188. S. A. Gaal, *Point Set Topology*, Academic Press, New York–London, 1964.
189. J.-P. Gabardo, *Extensions of Positive-Definite Distributions and Maximum Entropy*, Memoirs Amer. Math. Soc., 102, 1993.
190. J. Garcia-Cuerva and J. L. Rubio de Francia, *Weighted Norm Inequalities and Related Topics*, North-Holland, Amsterdam, 1985.

191. I. M. Gelfand and G. E. Shilov, *Generalized Functions*, Academic Press, New York–London, 1964, 1967, 1968.
192. J. Gerver, *The differentiability of the Riemann function at certain rational multiples of π*, Proc. Nat. Acad. Sci. **62** (1969), 668–670.
193. ———, *The differentiability of the Riemann function at certain rational multiples of π*, Amer. J. Math. **92** (1970), 33–55.
194. ———, *More on the differentiability of the Riemann function*, Amer. J. Math. **93** (1970), 33–41.
195. J. Gillis, *Note on a property of measurable sets*, J. London Math. Soc. **11** (1936), 139–141.
196. ———, *Some combinatorial properties of measurable sets*, Quart. J. Math. **7** (1936), 191–198.
197. C. C. Gillispie (ed.), *Dictionary of Scientific Biography*, Scribner, New York, 1970–1989.
198. A. M. Gleason, *Fundamentals of Abstract Analysis*, Addison-Wesley, Reading, MA, 1966.
199. C. Goffman, *On functions with summable derivative*, Amer. Math. Monthly **78** (1971), 874–875.
200. I. Gohberg and S. Goldberg, *Basic Operator Theory*, Birkhäuser, Boston, MA, 1981.
201. R. R. Goldberg, *Fourier Transforms*, Cambridge University Press, Cambridge, 1961.
202. L. Grafakos, *Classical and Modern Fourier Analysis*, Prentice Hall, Upper Saddle River, NJ, 2004.
203. C. C. Graham, *A uniform boundedness principle for compact sets and the decay of Fourier transforms*, Math. Zeit. **196** (1987), 331–342.
204. G. Green, *Mathematical Papers*, ed. N. M. Ferrers, Chelsea Publishing Company, New York, 1970 (1871).
205. L. Greengard and V. Rokhlin, *A fast algorithm for particle simulations*, J. Comput. Phys. **73** (1987), 325–348.
206. A. Grothendieck, *Sur les applications linéaires faiblement compactes d'espaces du type $C(K)$*, Canadian J. Math. **2** (1953), 129–173.
207. N. A. Gumerov and R. Duraiswami, *An Introduction to Fast Multipole Methods*, Birkhäuser, Boston, MA, 2009.
208. B.L Gurevich and G. E. Shilov, *Integral, Measure and Derivative: A Unified Approach*, Dover Publications, New York, 1977 (1966).
209. A. Haar, *Der Massbegriff in der Theorie der kontinuierlichen Gruppen*, Annals of Math. **34** (1933), 147–169.
210. J. W. Hagood and B. S. Thomson, *Recovering a function from a Dini derivative*, Amer. Math. Monthly **113** (2006), 34–46.
211. H. Hahn, *Über den Integralbegriff*, Anz. Acad. Wiss. Wien **66** (1929), 19–23.
212. P. R. Halmos, *Measure Theory*, D. Van Nostrand Co., New York, 1950.
213. ———, *Naive Set Theory*, D. Van Nostrand Co., Princeton, NJ, 1960.
214. G. H. Hardy, *A problem of Diophantine approximation*, J. Indian Math. Soc. **11** (1919), 162–166.
215. ———, *Pure Mathematics*, Cambridge University Press, Cambridge, 1963 (1908).
216. ———, *Divergent Series*, Oxford University Press, London, 1973 (1949).
217. G. H. Hardy and J. E. Littlewood, *Some problems in Diophantine approximation. The fractional part of $n^k\theta$*, Acta Math. **37** (1914), 155–191.

218. ———, *Some problems in Diophantine approximation. The trigonometrical series associated with the elliptic ϑ-functions*, Acta Math. **37** (1914), 193–239.
219. ———, *A maximal theorem with function-theoretic applications*, Acta Math. **54** (1930), 81–116.
220. G. H. Hardy, J. E. Littlewood, and G. Pólya, *Inequalities, 2nd edition*, Cambridge University Press, Cambridge, 1988 (1934, 1952).
221. G. H. Hardy and E. M. Wright, *Theory of Numbers*, Oxford University Press, London, 1965 (1938).
222. M. Harrison, *Frames and Irregular Sampling from a Computational Perspective*, Ph.D. thesis, University of Maryland, College Park, MD, 1998.
223. F. Hausdorff, *Dimension und äusseres Mass*, Math. Ann. **79** (1919), 157–179.
224. ———, *Set Theory*, Chelsea Publishing Company, New York, 1962.
225. V. Havin and B. Jöricke, *The Uncertainty Principle in Harmonic Analysis*, Springer-Verlag, New York, 1994.
226. T. Hawkins, *Lebesgue's Theory of Integration*, The University of Wisconsin Press, Madison, WI, 1970.
227. C. Heil (ed.), *Harmonic Analysis and Applications*, Birkhäuser, Boston, MA, 2006.
228. C. Heil and D. F. Walnut (eds.), *Fundamental Papers in Wavelet Theory*, Princeton University Press, Princeton, NJ, 2006.
229. W. Heller, *Frames of Exponentials and Applications*, Ph.D. thesis, University of Maryland, College Park, MD, 1991.
230. D. Herrero, *From my chest of examples of Fourier transforms*, Rev. Un. Mat. Argentina **32** (1985), 41–47.
231. H. Herrlich, *Axiom of Choice*, Springer-Verlag, Berlin, 2006.
232. C. Herz, *The Hardy–Littlewood Maximal Theorem*, Symp. on Harmonic Analysis, Warwick, 1968.
233. E. Hewitt and K. A. Ross, *Abstract Harmonic Analysis. Volume I*, Academic Press, New York, 1963.
234. ———, *Abstract Harmonic Analysis. Volume II*, Springer-Verlag, New York, 1970.
235. E. Hewitt and K. Stromberg, *Real and Abstract Analysis*, Springer-Verlag, Berlin–Heidelberg–New York, 1965.
236. T. H. Hildebrandt, *On bounded linear functional operations*, Trans. Amer. Math. Soc. **36** (1934), 868–875.
237. E. W. Hobson, *On modes of convergence of an infinite series of functions of a real variable*, Lond. M. S. Proc. (2) **1** (1904), 373–387.
238. ———, *The Theory of Functions of a Real Variable and the Theory of Fourier's Series*, Cambridge University Press, Cambridge, 1907.
239. K. Hoffman, *Banach Spaces of Analytic Functions*, Prentice Hall, Englewood Cliffs, NJ, 1962.
240. J. A. Hogan and J. D. Lakey, *Time-Frequency and Time-Scale Methods*, Birkhäuser, Boston, MA, 2005.
241. L. Hörmander, *The Analysis of Linear Partial Differential Operators*, Springer-Verlag, New York, 1983 (1963).
242. J. Horváth, *Topological Vector Spaces*, Addison-Wesley, Reading, MA, 1966.
243. ———, *A remark on the representation theorems of Frederick Riesz and Laurent Schwartz*, in Essays Dedicated to Laurent Schwartz on the Occasion of his 65th Birthday, ed. Leopoldo Nachbin (1981), 403–416.

244. ———, *On the Riesz–Fischer theorem*, Studia Sci. Math. Hungar. **41** (2004), 467–478.
245. T. Hrycak and V. Rokhlin, *An improved fast multipole algorithm for potential fields*, SIAM J. Sci. Comput. **19** (1998), 1804–1826.
246. A. Hulanicki and F. Ricci, *A Tauberian theorem and tangential convergence for bounded harmonic functions on balls in C^n*, Invent. Math. **62** (1980), 325–331.
247. A. Hulanicki and C. Ryll-Nardzewski, *Invariant extensions of the Haar measure*, Colloq. Math. **42** (1979), 223–227.
248. W. Hurewicz and H. Wallman, *Dimension Theory*, Princeton University Press, Princeton, NJ, 1948.
249. A. Hurwitz, *Über die Entwickelungscoefficienten der lemniskatischen Functionen*, Gött. Nachr. (1897), 273–276.
250. J. E. Hutchinson, *Fractals and self similarity*, Indiana Univ. Math. J. **30** (1981), 713–747.
251. A. Ivić, *The Riemann Zeta Function*, John Wiley and Sons, New York, 1985.
252. S. Jackson and D. R. Mauldin, *On the σ-class generated by open balls*, Math. Proc. Cambridge Philos. Soc. **127** (1999), 99–108.
253. S. Jaffard, *Local behavior of Riemann's function*, Contemporary Math. **189** (1995), 287–307.
254. ———, *The spectrum of singularities of Riemann's function*, Rev. Mat. Iberoamericana **12** (1996), 441–460.
255. ———, *Multifractal formalism for functions. I. General results*, SIAM J. Math. Anal. **28** (1997), 944–970.
256. ———, *Multifractal formalism for functions. II. Self-similar functions*, SIAM J. Math. Anal. **28** (1997), 971–998.
257. ———, *Old friends revisited: the multifractal nature of some classical functions*, J. Fourier Anal. Appl. **3** (1997), 1–22.
258. ———, *Multifractal Analysis of Functions*, Birkhäuser, Boston, MA, to appear.
259. S. Jaffard and Y. Meyer, *Wavelet Methods for Pointwise Regularity and Local Oscillations of Functions*, Memoirs Amer. Math. Soc., 123, 1996.
260. R. L. Jeffery, *The Theory of Functions of a Real Variable*, University of Toronto Press, Toronto, 1951.
261. ———, *Trigonometric Series*, University of Toronto Press, Toronto, 1956.
262. D. L. Johnson, *A short proof of the uniqueness of Haar measure*, Proc. Amer. Math. Soc. **55** (1976), 250–251.
263. R. Johnson, *Application of Carleson measures to partial differential equations and Fourier multiplier problems*, Harmonic Analysis (Cortona, 1982), Lecture Notes in Mathematics, 992, Springer-Verlag, New York, 1983, pp. 16–72.
264. F. Jones, *Lebesgue Integration on Euclidean Space*, Jones and Bartlett, Boston, MA, 1993.
265. D. Joyner, *Distribution Theorems of L-Functions*, Longman Scientific and Technical, Essex, England, 1986.
266. M. Kac, *Statistical Independence in Probability, Analysis and Number Theory*, Carus Monographs, Mathematical Association of America, New York, 1959.
267. J. P. Kahane, *Lacunary Taylor and Fourier series*, Bull. Amer. Math. Soc. **70** (1964), 199–213.
268. J.-P. Kahane, *Some Random Series of Functions*, D. C. Heath and Raytheon Education Co., Lexington, MA, 1968.

269. ———, *Séries de Fourier Absolument Convergentes*, Springer-Verlag, New York, 1970.
270. ———, *Review of Trigonometric Series, by Antoni Zygmund, 3rd edition, with foreword by Robert Fefferman*, Bull. Amer. Math. Soc. **41** (2004), 377–390.
271. J.-P. Kahane and R. Salem, *Ensembles Parfaits et Séries Trigonométriques*, Hermann, Paris, 1963.
272. S. Kakutani and K. Kodaira, *A non-separable translation invariant extension of the Lebesgue measure space*, Annals of Math. **52** (1950), 574–579.
273. S. Kakutani and J. C. Oxtoby, *Construction of a non-separable invariant extension of the Lebesgue measure space*, Annals of Math. **52** (1950), 580–590.
274. R. Kałuża, *The Life of Stefan Banach*, Birkhäuser, Boston, MA, 1996.
275. L. V. Kantorovich and G. P. Akilov, *Functional Analysis in Normed Spaces*, Macmillan Co., New York, 1964.
276. Y. Katznelson, *An Introduction to Harmonic Analysis*, Dover Publications, New York, 1976 (1968).
277. R. Kaufman and N. Rickert, *An inequality concerning measures*, Bull. Amer. Math. Soc. **72** (1966), 672–676.
278. A. S. Kechris, *Classical Descriptive Set Theory*, Springer-Verlag, New York, 1995.
279. J. Kelley, *General Topology*, D. Van Nostrand Co., New York, 1955.
280. ———, *Decomposition and representation theorems in measure theory*, Math. Ann. **163** (1966), 89–94.
281. S. Kempisty, *Fonctions d'Intervalle Non Additives*, Hermann, Paris, 1939.
282. R. B. Kerby, *The Correlation Function and the Wiener–Wintner Theorem in Higher Dimensions*, Ph.D. thesis, University of Maryland, College Park, MD, 1990.
283. H. Kestelman, *Modern Theories of Integration, 2nd edition*, Dover Publications, New York, 1960 (1937).
284. A. Khintchine, *Asymptotische Gesetze der Wahrscheinlichkeitsrechnung*, Chelsea Publishing Company, New York, 1948.
285. S. J. Kilmer and S. Saeki, *On Riesz product measures: mutual absolute continuity and singularity*, Ann. Inst. Fourier, Grenoble **38** (1988), 63–93.
286. R. B. Kirk, *Sets which split families of measurable sets*, Amer. Math. Monthly **79** (1972), 884–886.
287. A. W. Knapp, *Basic Real Analysis*, Birkhäuser, Boston, MA, 2005, along with a companion volume *Advanced Real Analysis*.
288. J. F. Koksma and L. Kuipers, *Asymptotic Distribution Modulo 1*, NUFFIC International Summer Session in Science, P. Noordhoff, the Netherlands, 1962.
289. A. N. Kolmogorov, *Une contribution à l'étude de la convergence des séries de Fourier*, Fund. Math. **5** (1924), 96–97.
290. ———, *Über Kompaktheit der Funktionenmengen bei der Konvergenz im Mittel*, Nachr. Ges. Wies. Göttingen **1, No. 37** (1931), 60–63.
291. ———, *Zur Normierbarkeit eines allgemeinem topologischen linearen Raumes*, Studia. Math. **5** (1934), 29–33.
292. ———, *Stationary sequences in Hilbert spaces*, Bull. Moscow State U., Math. **2 (6)** (1941), 40 pages.
293. ———, *Foundations of the Theory of Probability*, Chelsea Publishing Company, New York, 1956 (1931).

294. J. Komlós, *A generalization of a problem of Steinhaus*, Acta Math. Acad. Sci. Hungar. **18** (1967), 217–229.

295. P. Koosis, *Leçons sur le théorème de Beurling et Malliavin*, Université de Montréal, Les Publications CRM, Montreal, QC, 1996.

296. T. W. Körner, *A pseudofunction on a Helson set I*, Astérisque **5** (1973), 3–224.

297. ———, *A pseudofunction on a Helson set II*, Astérisque **5** (1973), 231–239.

298. ———, *Fourier Analysis, 2nd edition*, Cambridge University Press, Cambridge, 1989.

299. ———, *Decreasing rearranged Fourier series*, J. Fourier Anal. Appl. **5** (1999), 1–19.

300. A. Kosiński, *A proof of the Auerbach–Banach–Mazur–Ulam theorem on convex bodies*, Coll. Math. **4** (1957), 216–218.

301. G. Köthe, *Das Trägheitsgesetz der quadratischen Formen im Hilbertschen Raum*, Math. Zeit. **41** (1936), 137–152.

302. S. G. Krantz, *Real Analysis and Foundations*, CRC Press, Boca Raton, FL, 1991.

303. S. G. Krantz and H. R. Parks, *The Geometry of Domains in Space*, Birkhäuser, Boston, MA, 1999.

304. L. Kuipers and H. Niederreiter, *Uniform Distribution of Sequences*, John Wiley and Sons, New York, 1974.

305. H.-H. Kuo, *White Noise Distribution Theory*, CRC Press, Boca Raton, FL, 1996.

306. K. Kuratowski, *Une méthode d'élimination des nombres transfinis des raisonnements mathématiques*, Fund. Math. **3** (1922), 76–108.

307. ———, *Topology, Volumes I and II*, Academic Press and PWN, New York and Warszawa, 1966 and 1968.

308. K. Kuratowski and A. Mostowski, *Set Theory*, PWN, Warszawa, 1968.

309. I. Łaba and Y. Wang, *Some properties of spectral measures*, Appl. Comput. Harmon. Anal. **20** (2006), 149–157.

310. J. Lawrence, G. E. Pfander, and D. Walnut, *Linear independence of Gabor systems in finite dimensional vector spaces*, J. Fourier Anal. Appl. **11** (2005), 715–726.

311. P. Lax, *Functional Analysis*, John Wiley and Sons, New York, 2002.

312. H. Lebesgue, *Intégrale, longueur, aire*, Annali di Mat. **7** (1902), 231–358.

313. ———, *Leçons sur l'Intégration et la Recherche des Fonctions Primitives, 1st and 2nd editions*, Gauthier-Villars, Paris, 1904 and 1928.

314. ———, *Leçons sur les Séries Trigonométriques*, Gauthier-Villars, Paris, 1906.

315. ———, *Sur l'intégration des fonctions discontinues*, Ann. École Norm. Sup. **27** (1910), 361–450.

316. ———, *Measure and the Integral*, Holden-Day, San Francisco, CA, 1966.

317. Y. Lee, *Statistical Theory of Communication*, John Wiley and Sons, New York, 1960.

318. G. Letta, *Une démonstration élémentaire du théorème de Lebesgue sur la derivation des fonctions croissantes*, L'Enseignement Math. **16** (1970), 177–184.

319. P. Lévy, *Processus Stochastiques et Mouvement Brownien*, Gauthier-Villars, Paris, 1948.

320. S. Li and H. Ogawa, *Pseudo-duals of frames with applications*, Appl. Comp. Harm. Anal. **11** (2001), 289–304.

321. E. H. Lieb and M. Loss, *Analysis*, Amer. Math. Soc. Graduate Studies in Mathematics 14, Providence, RI, 1997.
322. L.-Å. Lindahl and F. Poulsen (eds.), *Thin Sets in Harmonic Analysis*, Marcel Dekker, New York, 1971.
323. G. Lion, *A simple proof of the Dirichlet–Jordan convergence test*, Amer. Math. Monthly **93** (1986), 281–282.
324. J. Lipiński, *On zeros of a continuous nowhere differentiable function*, Amer. Math. Monthly **73** (1966), 166–168.
325. ———, *On derivatives of singular functions*, Bull. l'Acad. Pol. Sci. **20** (1972), 625–628.
326. L. Liusternik and V. Sobolev, *Elements of Functional Analyis*, F. Ungar Publishing Co., New York, 1961.
327. L. Loomis and S. Sternberg, *Advanced Calculus*, Addison-Wesley, Reading, MA, 1968.
328. L. H. Loomis, *An Introduction to Abstract Harmonic Analysis*, D. Van Nostrand Co., Princeton, NJ, 1953.
329. G. G. Lorentz, *Approximation of Functions, 2nd edition*, Chelsea Publishing Company, New York, 1986 (1966).
330. S. M. Lozinskii, *On the indicatrix of Banach (Russian)*, Doklady Akad. Nauk SSSR (N.S.) **60** (1948), 765–767.
331. ———, *On the indicatrix of Banach (Russian)*, Vestnik Leningrad. Univ. **13** (1958), 70–87.
332. N. Lusin, *Sur les propriétés des fonctions mesurables*, Comptes Rendus Acad. Sci. Paris **154** (1912), 1688–1690.
333. ———, *Sur la notion de l'intégral*, Ann. di Mat. **26** (1917), 77–129.
334. W. A. J. Luxemburg, *Arzelà's dominated convergence theorem for the Riemann integral*, Amer. Math. Monthly **78** (1971), 970–979.
335. P. Malliavin, *Integration and Probability*, Springer-Verlag, New York, 1995 (1982).
336. B. B. Mandelbrot, *Fractals: Form, Chance, and Dimension*, W. H. Freeman and Co., San Francisco, CA, 1977.
337. ———, *The Fractal Geometry of Nature*, W. H. Freeman and Co., New York, 1982.
338. P. Masani, *The prediction theory of multivariate stochastic processes, III*, Acta Math. **104** (1960), 141–162.
339. ———, *An outline of the theory of stationary measures over R^q*, On covering conditions and convergence, Proceedings of the Conference on Measure Theory (Oberwolfach, 1979), Lecture Notes in Mathematics, 794, Springer-Verlag, New York, 1980, pp. 295–309.
340. ———, *Einstein's contribution to generalized harmonic analysis and his intellectual kinship with Norbert Wiener*, Jahrbuch Überblicke Mathematik 1986 **19** (1986), 191–209.
341. ———, *Norbert Wiener*, Birkhäuser, Boston, MA, 1990.
342. P. Mattila, *Geometry of Sets and Measures in Euclidean Spaces. Fractals and Rectifiability*, Cambridge University Press, Cambridge, 1995.
343. R. D. Mauldin, *The set of continuous nowhere differentiable functions*, Pacific J. Math. **80** (1979), 199–205.
344. ———, *The Scottish Book. Mathematics from the Scottish Café*, Birkhäuser, Boston, MA, 1981.

345. S. Mazurkiewicz, *Sur les suites de fonctions continues*, Fund. Math **18** (1932), 114–117.
346. ———, *Über die Menge die differenzierbaren Funktionen*, Fund. Math. **27** (1936), 244–249.
347. Y. Meyer, *Algebraic Numbers and Harmonic Analysis*, American Elsevier Publishing Co., New York, 1972.
348. ———, *Ondelettes et Opérateurs*, Hermann, Paris, 1990.
349. A. P. Morse, *A theory of covering and differentiation*, Trans. Amer. Math. Soc. **55** (1944), 205–235.
350. Y. Moschovakis, *Descriptive Set Theory: A Foundational Approach*, North-Holland, Amsterdam, 1980.
351. C. J. Mozzochi, *On the Pointwise Convergence of Fourier Series*, Lecture Notes in Mathematics, 199, Springer-Verlag, Berlin–New York, 1971.
352. J. Mycielski, *Remarks on invariant measures in metric spaces*, Coll. Math. **32** (1974), 105–112.
353. ———, *A conjecture of Ulam on the invariance of measure in Hilbert's cube*, Studia Math. **60** (1977), 1–10.
354. L. Nachbin, *The Haar Integral*, D. Van Nostrand Co., Princeton, NJ, 1965.
355. I. P. Natanson, *Theory of Functions of a Real Variable, Volume I*, F. Ungar Publishing Co., New York, 1955.
356. ———, *Theory of Functions of a Real Variable, Volume II*, F. Ungar Publishing Co., New York, 1960.
357. U. Neri, *Singular Integrals,*, Lecture Notes in Mathematics, 200, Springer-Verlag, New York, 1971.
358. D.J. Newman, J.H. Neuwirth, and J. Ginsberg, *Approximation by* $\{f(kx)\}$, J. Functional Analysis **5** (1970), 194–203.
359. G. Ólafsson and S. Zheng, *Function spaces associated with Schrödinger operators: the Pöschel–Teller potential*, J. Fourier Anal. Appl. **12** (2006), 653–674.
360. F. W. J. Olver, *Asymptotics and Special Functions*, Academic Press, New York, 1974.
361. W. F. Osgood, *Non-uniform convergence and the integration of series term by term*, Amer. J. Math. **19** (1897), 155–190.
362. J. C. Oxtoby, *Measure and Category*, Springer-Verlag, Berlin–Heidelberg–New York, 1970.
363. R. E. A. C. Paley and N. Wiener, *Fourier Transforms in the Complex Domain*, Amer. Math. Soc. Colloquium Publications, Volume 19, Providence, RI, 1934.
364. J. Palis, *Homoclinic orbits, hyperbolic dynamics and dimension of Cantor sets*, Contemp. Math. **58** (1987), 203–216.
365. A. Papoulis, *Signal Analysis*, McGraw-Hill, New York, 1977.
366. ———, *Probability, Random Variables, and Stochastic Processes, 2nd edition*, McGraw-Hill, New York, 1984.
367. J. Pawlikowski, *The Hahn–Banach theorem implies the Banach–Tarski paradox*, Fund. Math. **138** (1991), 21–22.
368. G. Peano, *Sur une courbe, qui remplit une aire plane*, Math. Ann. **36** (1890), 157–160.
369. C. Pearcy and A. L. Shields, *A Survey of the Lomonosov Technique in the Theory of Invariant Subspaces, Topics in Operator Theory, Mathematical Surveys, No. 13*, Amer. Math. Soc., Providence, RI, 1974.
370. I. N. Pesin, *Classical and Modern Integration Theories*, Academic Press, London–New York, 1970.

371. J. Pfanzagl, *Convergent sequences of regular measures*, Manuscripta Math. **4** (1971), 91–98.
372. A. Pietsch, *History of Banach Spaces and Linear Operators*, Birkhäuser, Boston, MA, 2007.
373. C. Pisot, *Eine merkwürdige Klasse ganzer algebraischen Zahlen*, J. Reine Angew. Math. **209** (1962), 82–83.
374. G. Pólya and G. Szegő, *Aufgaben und Lehrsätze aus der Analysis, Volumes I and II*, Springer-Verlag, Berlin–Heidelberg–New York, 1964 (1924).
375. L. S. Pontryagin, *Topological Groups, 2nd edition*, Gordon and Breach, New York, 1966 (English translation).
376. P. Porcelli, *Linear Spaces of Analytic Functions*, Rand McNally, Chicago, 1966.
377. E. E. Posey and J. E. Vaughan, *Functions with a proper local maximum in each interval*, Amer. Math. Monthly **90** (1983), 281–282.
378. M. B. Priestley, *Spectral Analysis and Time Series, Volumes I and II*, Academic Press, New York, 1981.
379. T. Radó, *Length and Area*, Amer. Math. Soc. Colloquium Publications, Volume 30, 1948.
380. J. Rainwater, *Weak convergence of bounded sequences*, Proc. Amer. Math. Soc. **14** (1963), 999.
381. A. Rajchman, *Sur l'unicité du developpment trigonométrique*, Fund. Math. **3, 4** (1922, 1923), 207–302, 366–367.
382. D. Ramakrishnan and R. J. Valenza, *Fourier Analysis on Number Fields*, Springer-Verlag, New York, 1999.
383. K. P. S. Bhaskara Rao and M. Bhaskara Rao, *Theory of Charges: A Study of Finitely Additive Measures*, Academic Press, New York, 1983.
384. H. Reiter and J. D. Stegeman, *Classical Harmonic Analysis and Locally Compact Groups. 2nd edition*, The Clarendon Press, Oxford University Press, New York, 2000 (1968 by Reiter).
385. F. Ricci and G. Weiss, *A characterization of* $H^1(\Sigma_{n-1})$, Harmonic Analysis in Euclidean Spaces (Proc. Sympos. Pure Math., Williams Coll., Williamstown, MA, 1978), Amer. Math. Soc., Providence, RI, 1979, pp. 289–294.
386. S. O. Rice, *Mathematical analysis of random noise*, Noise and Stochastic Processes, Dover Publications, New York, 1954, originally published in Bell System Technical J., Volumes 23 and 24.
387. B. Riemann, *Partielle Differentialgleichungen, 2nd edition*, Braunschweig, 1876.
388. ———, *Über die Darstellbarkeit einer Function durch eine trigonometrische Reihe (1854), in Collected Works of Bernhard Riemann, ed. H. Weber, new introduction by H. Lewy*, Dover Publications, New York, 1953. Published by R. Dedekind in Göttingen Abh. 13 (1867). See French translation, Sur la possibilité de répresenter une fonction par une série trigonométrique, in Bull. de Sciences (1873), 20–96.
389. F. Riesz, *Sur les opérations fonctionnelles linéaires*, Comptes Rendus Acad. Sci. Paris **149** (1909), 974–977.
390. ———, *Untersuchungen über Systeme integrierbarer Funktionen*, Math. Ann. **69** (1910), 449–497.
391. ———, *Über die Fourierkoeffizienten einer stetigen Funktion von beschränkter Schwankung*, Math. Zeit. **2** (1918), 312–315.

392. F. Riesz and B. Sz.-Nagy, *Functional Analysis*, F. Ungar Publishing Co., New York, 1955.
393. M. Riesz, *Sur les systèmes orthogonaux de fonctions*, Comptes Rendus Acad. Sci. Paris **144** (1907), 615–619.
394. ———, *Sur les systèmes orthogonaux de fonctions et l'équation de Fredholm*, Comptes Rendus Acad. Sci. Paris **144** (1907), 734–736.
395. ———, *Sur une espèce de géométrie analytique des systèmes de fonctions sommables*, Comptes Rendus Acad. Sci. Paris **144** (1907), 1409–1411.
396. ———, *Über orthogonale Funktionensysteme*, Gött. Nachr. (1907), 116–122.
397. ———, *Sur les ensembles compactes de fonctions sommables*, Acta Sci. Math. Szeged **6** (1933), 136–142.
398. A. M. Robert, *A Course in p-adic Analysis*, Springer-Verlag, New York, 2000.
399. A. P. Robertson and W. J. Roberston, *Topological Vector Spaces*, Cambridge University Press, Cambridge, 1964.
400. C. A. Rogers, *Hausdorff Measures*, Cambridge University Press, Cambridge, 1970.
401. ———, *A linear Borel set whose difference set is not a Borel set*, Bull. Lond. Math. Soc. **2** (1970), 41–42.
402. H. L. Royden and P. M. Fitzpatrick, *Real Analysis*, Prentice Hall, New York, 2009.
403. W. Rudin, *Continuous functions on compact sets without perfect subsets*, Proc. Amer. Math. Soc. **8** (1957), 39–42.
404. ———, *Fourier Analysis on Groups*, John Wiley and Sons, New York, 1962.
405. ———, *Real and Complex Analysis*, McGraw-Hill, New York, 1966.
406. ———, *Functional Analysis*, McGraw-Hill, New York, 1973.
407. ———, *Principles of Mathematical Analysis*, McGraw-Hill, New York, 1976 (1953).
408. ———, *Lebesgue's first theorem*, Mathematical Analysis and Applications, Adv. in Math. Suppl. Stud. (1981), 741–747.
409. M. Saadoune, *A new extension of Komlós' theorem in infinite dimensions. Application: weak compactness in* L^1_X, Portugal. Math. **55** (1998), 113–128.
410. S. Saks, *Addition to the note: on some functionals*, Trans. Amer. Math. Soc. **35** (1933), 965–970.
411. ———, *On some functionals*, Trans. Amer. Math. Soc. **35** (1933), 549–556.
412. ———, *Theory of the Integral*, Dover Publication, New York, 1937 (1933).
413. R. Salem, *Algebraic Numbers and Fourier Analysis*, D. C. Heath and Co., Boston, MA, 1963.
414. ———, *Oeuvres Mathématiques*, Hermann, Paris, 1967.
415. S. Saliani, *Exceptional sets and wavelet packets orthonormal bases*, J. Fourier Anal. Appl. **5** (1999), 421–430.
416. P. J. Sally, *Harmonic analysis and group representations*, MAA Studies in Harmonic Analysis, ed. J. M. Ash **13** (1976), 224–256.
417. A. Sannami, *An example of a regular Cantor set whose difference set is a Cantor set with positive measure*, Hokkaido Math. J. **21** (1992), 7–24.
418. H. H. Schaefer, *Weak convergence of measures*, Math. Ann. **193** (1971), 57–64.
419. F. Schipp, P. Simon, and W. R. Wade, *Walsh Series*, Adam Hilger, New York, 1990.
420. I. J. Schoenberg, *Metric spaces and completely monotone functions*, Annals of Math. **39** (1938), 811–841.

421. K. Schrader, *A pointwise convergence theorem for series of continuous functions*, Trans. Amer. Math. Soc. **159** (1971), 155–163.
422. ———, *A generalization of the Helly selection theorem*, Bull. Amer. Math. Soc. **78** (1972), 415–419.
423. M. Schroeder, *Fractals, Chaos, Power Laws*, W. H. Freeman and Company, New York, 1991.
424. I. Schur, *Über lineare Transformationen in der Theorie der unendlichen Reihen*, J. Reine Angew. Math. **151** (1921), 79–111.
425. L. Schwartz, *Sur une propriété de synthèse spectrale dans les groupes noncompacts*, Comptes Rendus Acad. Sci. Paris **227** (1948), 424–426.
426. ———, *Méthodes Mathématiques pour les Sciences Physiques*, Hermann, Paris, 1961, with the assistance of D. Huet.
427. ———, *Some applications of the theory of distributions, Volume 1*, Lectures on Modern Mathematics, ed. T. L. Saaty, John Wiley and Sons, New York, 1963.
428. ———, *Théorie des Distributions*, Hermann, Paris, 1966 (1957).
429. ———, *A Mathematician Grappling with His Century*, Birkhäuser, Boston, MA, 2001.
430. S. Semmes, *Finding structure in sets with little smoothness*, Proceedings of the International Congress of Mathematicians, Zürich, 1994 (1995), 875–885.
431. J. Serrin and D. E. Varberg, *A general chain rule for derivatives and the change of variables formula for the Lebesgue integral*, Amer. Math. Monthly **76** (1969), 514–520.
432. H. Shapiro, *Uncertainty principles for bases in* $L^2(R)$, preprint (unpublished) (1991).
433. W. Sierpiński, *Sur une courbe Cantorienne qui contient une image biunivoque et continue de toute courbe donnée*, Comptes Rendus Acad. Sci. Paris **162** (1916), 629.
434. ———, *Sur la question de la mesurabilité de la base de M. Hamel*, Fund. Math. **1** (1920), 105–111.
435. ———, *Sur une problème concernant les ensembles mesurables superficiellement*, Fund. Math. **1** (1920), 112–115.
436. ———, *Démonstration de quelques théorèmes fondamentaux sur les fonctions mesurables*, Fund. Math. **3** (1922), 314–321.
437. ———, *Démonstration d'un théorème sur les fonctions additives d'ensembles*, Fund. Math. **5** (1924), 262–264.
438. ———, *Sur une problème conduisant à un ensemble non mesurable*, Fund. Math. **10** (1927), 177–179.
439. ———, *Remarque sur les suites infinies de fonctions*, Fund. Math. **18** (1932), 110–113.
440. ———, *Hypothèse du Continu*, Monografie Matematyczne, Volume IV, Warszawa, 1934.
441. ———, *On the Congruence of Sets and Their Equivalence by Finite Decomposition*, in Congruence of Sets and Other Monographs, Chelsea Publishing Company, New York, 1967 (1953).
442. S. M. Simmons, *A converse Steinhaus theorem for locally compact groups*, Proc. Amer. Math. Soc. **49** (1975), 383–386.
443. I. Singer, *Bases in Banach Spaces. Volume I*, Springer-Verlag, New York–Berlin, 1970.

444. A. N. Singh, *The Theory and Construction of Non-differentiable Functions, in Squaring the Circle and Other Monographs*, Chelsea Publishing Company, New York, 1969 (1935).
445. K. T. Smith, *Primer of Modern Analysis*, Springer-Verlag, New York–Berlin, 1983 (1971).
446. R. M. Solovay, *A model of set-theory in which every set of reals is Lebesgue measurable*, Annals of Math. **92** (1970), 1–56.
447. M. Spivak, *Calculus on Manifolds*, W. A. Benjamin, New York-Amsterdam, 1965.
448. E. M. Stein, *Singular Integrals and Differentiability Properties of Functions*, Princeton University Press, Princeton, NJ, 1970.
449. ———, *Singular Integrals: The Roles of Calderón and Zygmund*, Notices Amer. Math. Soc. **45** (1999), 1130–1140.
450. E. M. Stein and G. Weiss, *Introduction to Fourier Analysis on Euclidean Spaces*, Princeton University Press, Princeton, NJ, 1971.
451. S. K. Stein, *Archimedes. What did he do besides cry Eureka?* Mathematical Association of America, Washington, DC, 1999.
452. H. Steinhaus, *Stefan Banach's life*, Studia Math. (Ser. Specjalna) **1** (1963), 7–15.
453. S. Sternberg, *Celestial Mechanics, Part I*, W. A. Benjamin, New York, 1969.
454. M. H. Stone, *Applications of the theory of Boolean rings to general topology*, Trans. Amer. Math. Soc. **41** (1937), 375–481.
455. ———, *The generalized Weierstrass approximation theorem*, Math. Magazine **21** (1948), 237–254.
456. ———, *A generalized Weierstrass approximation theorem*, MAA Studies in Modern Analysis **1** (1962), 30–87.
457. R. S. Strichartz, *A Guide to Distribution Theory*, CRC Press, Boca Raton, FL, 1994.
458. ———, *Self-similarity in harmonic analysis*, J. Fourier Anal. Appl. **1** (1994), 1–37.
459. ———, *The Way of Analysis*, Jones and Bartlett, Boston, MA, 1995.
460. ———, *Analysis on Fractals*, Notices Amer. Math. Soc. **46 (10)** (1999), 1199–1208.
461. K. Stromberg, *Topics in Harmonic Analysis*, Kansas State University, 1988.
462. J. M. Sullivan, *Sphere packings give an explicit bound for the Besicovitch covering theorem*, J. Geom. Anal. **4** (1994), 219–231.
463. P. Suppes, *Axiomatic Set Theory*, D. Van Nostrand Co., Princeton, NJ, 1960.
464. M. Taibleson, *Fourier coefficients of functions of bounded variation*, Proc. Amer. Math. Soc. **18** (1967), 766.
465. A. E. Taylor, *Introduction to Functional Analysis*, John Wiley and Sons, New York, 1958.
466. A. Teolis, *Computational Signal Processing with Wavelets*, Birkhäuser, Boston, MA, 1998.
467. A. Terras, *Fourier Analysis on Finite Groups and Applications*, Cambridge University Press, Cambridge, 1999.
468. T. M. Thompson, *From Error-correcting Codes Through Sphere Packings to Simple Groups*, Mathematical Association of America, Washington, DC, 1983.
469. E. C. Titchmarsh, *Introduction to the Theory of Fourier Integrals, 2nd edition*, Oxford University Press, London, 1948 (1937).

470. ———, *The Theory of the Riemann Zeta Function*, Oxford University Press, London, 1951.
471. ———, *Theory of Functions, 2nd edition*, Oxford University Press, London, 1960.
472. L. Tonelli, *Sull'integrazione per parti*, Atti Accad. Naz. Lincei (5) **18** (1909), 246–253.
473. ———, *Sulla quadratura delle superficie*, Atti Reale Accad. Naz. Lincei (6) **3** (1926), 633–638.
474. A. Tonolo, *Vitali's obituary*, Rendiconti Padua U. Sem. Mat. **3** (1932), 67–81.
475. O. Treiber, *Affine Data Representations and Filter Banks*, Ph.D. thesis, University of Maryland, College Park, MD, 1999.
476. A. N. Tulaikov, *Zur Kompaktheit im Raum* L_p *für* $p = 1$, Nachr. Ges. Wiss. Göttingen 1 (1933), no. 39, 167–170.
477. S. Ulam, *Zur Mass theorie in der allgemeinen Mengenlehre*, Fund. Math. **16** (1930), 140–150.
478. ———, *Problems in Modern Mathematics*, John Wiley and Sons, New York, 1964.
479. ———, *Adventures of a Mathematician*, Charles Scribner's Sons, New York, 1976.
480. D. E. Varberg, *On absolutely continuous functions*, Amer. Math. Monthly **72** (1965), 831–841.
481. N. Varopoulos, *Algèbres tensorielles et analyse harmonique*, Sem. d'Analyse Harmonique, Orsay, 1967–1968.
482. G. Vitali, *Sulle funzioni integrali*, Atti. Acc. Sci. Torino **40** (1904–05), 1021–1034.
483. ———, *Sul problema della misura dei gruppi di punti di una retta*, Gamberini e Parmeggiani, Bologna, 1905.
484. ———, *Una proprietà delle funzioni misurabili*, Rend. Ist. Lombardo **38** (1905), 599–603.
485. ———, *Sui gruppi di punti e sulle funzioni di variabili reali*, Atti. Acc. Sci. Torino **43** (1907), 229–246.
486. ———, *Sull' integrazione per serie*, Rend. Circ. Matem. Palermo **23** (1907), 137–155.
487. ———, *Sulla condizione di chiusura di un sistema di funzioni ortogonali*, Atti Reale Accad. Naz. Lincei **30** (1921), 498–501.
488. ———, *Sulle funzioni continue*, Fund. Math. **8** (1926), 175–188.
489. V. Volterra, *Sui principii del calcolo integrale*, G. Battaglini **19** (1891), 333–372.
490. J. von Neumann, *Zum Haarschen Mass in topologischen Gruppen*, Compos. Math. **1** (1934), 106–114.
491. ———, *The uniqueness of Haar's measure*, Mat. Sbornik **1** (1936), 721–734.
492. ———, *Mathematical Foundations of Quantum Mechanics*, Princeton University Press, Princeton, NJ, 1949.
493. H. von Weizsäcker, *Can one drop* L^1*-boundedness in Komlós's subsequence theorem?* Amer. Math. Monthly **111** (2004), 900–903.
494. S. Wagon, *The Banach-Tarski Paradox*, Cambridge University Press, Cambridge, 1985.
495. J. S. Walker, *Fast Fourier Transforms, 2nd edition*, CRC Press, Boca Raton, FL, 1996.

496. C. R. Warner, A class of spectral sets, *Proc. Amer. Math. Soc.* **57** (1976), 99–102.
497. A. Weil, *L' Intégration dans les Groupes Topologiques et ses Applications, 2nd edition*, Hermann, Paris, 1953 (1940, 1st edition).
498. B. B. Wells, *Weak compactness of measures*, Proc. Amer. Math. Soc. **20** (1969), 124–130.
499. O. Wesler, *An infinite packing theorem for the sphere*, Proc. Amer. Math. Soc. **11** (1960), 324–326.
500. H. Weyl, *Über die Gleichverteilung von Zahlen mod. Eins*, Math. Annalen **77** (1916), 313–352.
501. ———, *Ramifications, old and new, of the eigenvalue problem*, Bull. Amer. Math. Soc. **56** (1950), 115–139.
502. R. Wheeden and A. Zygmund, *Measure and Integral: An Introduction to Real Analysis*, Marcel Dekker, New York, 1977.
503. D. V. Widder, *The Laplace Transform*, Princeton University Press, Princeton, NJ, 1946.
504. ———, *Advanced Calculus*, Dover Publications, New York, 1989 (1947).
505. N. Wiener, *The Fourier Integral and Certain of Its Applications*, Cambridge University Press, Cambridge, 1933.
506. ———, *The ergodic theorem*, Duke Math. J. **5** (1939), 1–18.
507. ———, *Cybernetics*, MIT Press, Cambridge, MA, 1948.
508. ———, *Time Series*, MIT Press, Cambridge, MA, 1949.
509. ———, *I Am a Mathematician*, MIT Press, Cambridge, MA, 1956.
510. ———, *Nonlinear Problems in Random Theory*, MIT Press, Cambridge, MA, 1958.
511. ———, *Collected Works*, MIT Press, Cambridge, MA, 1981 (1976, 1979).
512. J. H. Williamson, *Lebesgue Integration*, Holt, Rinehart, and Winston, New York, 1962.
513. W. H. Woodin, *The continuum hypothesis, I*, Notices Amer. Math. Soc. **48** (2001), 567–576.
514. ———, *The continuum hypothesis, II*, Notices Amer. Math. Soc. **48** (2001), 681–690.
515. H.-C. Wu, *Multidimensional Irregular Sampling in Terms of Frames*, Ph.D. thesis, University of Maryland, College Park, MD, 1998.
516. A. Yaglom, *Einstein's 1914 paper on the theory of irregularly fluctuating sums of observations*, Problemy Peredachi Informatsii **21** (1985), 101–107, English translation in IEEE ASSP Magazine, October 1987.
517. K. Yosida, *Functional Analysis*, Springer-Verlag, Berlin–New York, 1980.
518. R. M. Young, *An Introduction to Nonharmonic Fourier Series*, Academic Press, New York, 1980.
519. A. C. Zaanen, *Some examples in weak sequential convergence*, Amer. Math. Monthly **69** (1962), 85–93.
520. M. A. Zaretsky, *On one theorem on absolutely continuous functions (Russian)*, Doklady Akad. Nauk (1925), 88–90.
521. W. P. Ziemer, *Weakly Differentiable Functions. Sobolev Spaces and Functions of Bounded Variation*, Springer-Verlag, New York, 1989.
522. G. Zimmermann, *Normalized tight frames in finite dimensions*, Recent Progress in Multivariate Approximation, eds. K. J. W. Haussmann and M. Reimer, Birkhäuser, Boston, MA, 2001, pp. 249–252.
523. M. Zorn, *A remark on a method in transfinite algebra*, Bull. Amer. Math. Soc. **41** (1935), 667–670.

524. A. Zygmund, *Trigonometric Series*, Cambridge University Press, London, 1959 (1935, 1st edition).
525. ———, *On certain lemmas of Marcinkiewicz and Carleson*, J. of Approximation Theory **2** (1969), 249–257.

Subject Index

Index of Names

Page numbers are not listed after some of the names. These names are associated with some of the theorems, examples, etc., used in the book, e.g., the Heaviside function.

Index of Notation

Kolibri 360 GmbH
Ferdinand-Jühlke-Straße 7
99095 Erfurt, Deutschland
produktsicherheit@kolibri360.de